MW00845097

Cyclic 3',5'-Nucleotides: Mechanisms of Action

Edited by

Hinrich Cramer
Neurologische Universitätsklinik mit Abteilung für Neurophysiologie, Freiburg, Federal Republic of Germany

Joachim Schultz
Pharmazeutisches Institut der Universität, Tübingen, Federal Republic of Germany

A Wiley–Interscience Publication

JOHN WILEY & SONS

London · New York · Sydney · Toronto

Library of Congress Cataloging in Publication Data:
Main entry under title:

Cyclic nucleotides.

'A Wiley–Interscience publication.'
Bibliography: p.
Includes index.
1. Cyclic nucleotides. I. Cramer, Hinrich.
II. Schultz, J.
QP625.N89C9 574.8′732 76-45361

ISBN 0 471 99456 1

Typeset by Preface Ltd
Salisbury, Wilts. and printed in Great Britain
by Unwin Brothers Ltd.,
The Gresham Press, Old Woking, Surrey

Foreword

It's a great pleasure to write a foreword to this important volume on cyclic nucleotides. The state of the art is such in the late 1970s that there is an urgent need for critical reviews and overview presentations of what has become an enormously complex field. Judging by the titles and organization of this book, I am encouraged in the belief that younger investigators or those with more seasoning but still new to cyclic nucleotides will benefit greatly by having it available. Further, cyclic nucleotides have been implicated in so many systems that it is essential for those of us who have been in the field for a while to be able to step back from the bench and read authoritative accounts of what is going on in other subareas.

It seems to one who first started hearing about cyclic AMP in the late 1950s (when it was still known as 'X' and only to a very few people) that the development of research on cyclic nucleotides has been divided into three temporal stages.

The first period covered roughly fifteen years, during which Sutherland and Rall, Berthet and other associates carried out their truly remarkable experiments that were consumated by the discovery of cyclic AMP in 1957. From the point of view of this observer, this era — which might be extended until 1962 with the publication of four major papers on adenylate cyclase by Sutherland and his coworkers — remains as a landmark of and even a benchmark for excellence in research. Indeed, the discoveries of that group (along with the discovery of protein phosphokinases by E. G. Krebs and his coworkers) provided all of the principles upon which everything else in the area has been built.

Sutherland and Rall, with impressive reasoning and almost prescience, forecasted the second phase in cyclic nucleotide research, which lasted roughly from 1962 to 1970. This phase was one of exploitation of the earlier discoveries of Sutherland and Rall in an expansion of the role of cyclic AMP from the simple one in the control of glycogenolysis in liver and muscle to that of being a rather generalized intermediate in the actions of hormones in a wide variety of tissues.

Working in this expansionist phase was not always easy, for the assays required were very laborious. Further, the enthusiasm shared by the few laboratories involved in cyclic AMP research was not necessarily shared by those who had interest in the actions of 'their' hormones or drugs on 'their' systems. I recall hearing Earl express amusement (and a little wonder) at a noted pharmacologist who asked why a bunch of biochemists were 'messing around with the heart'. Indeed, at a meeting in 1967, the notion that decreased cyclic AMP levels might underlie the actions of α-adrenergic agonists was termed 'a dangerous hypothesis'.

By 1970, technology had caught up with theory to a greater degree than was the case in the second phase. The radioimmunoassay and the protein binding assay for cyclic AMP were reported, and these important tools made it possible for a tremendous growth of the field. Further, dibutyryl cyclic AMP and other derivatives became commercially available during the late 1960s. What followed these events was an almost incredible increase in activity. Cyclic AMP was measured, always with gusto, often with expertise, sometimes without, in a whole variety of tissues and systems. Further, gramme upon gramme or even kilogramme of dibutyryl cyclic AMP bathed tissues all over the world. While such experiments are not always without complications (including outright toxicology), they are always exciting.

The third period, which continues now, has been marked by surges of interest in a variety of subareas of cyclic nucleotide research. Cyclic GMP and the Yin–Yang hypothesis caused a great stir, followed by a steady increase in the number of publications on the subject and also an increase in our as-yet imperfect knowledge of the role of this nucleotide. Further, the relationships between calcium, cyclic AMP, cyclic GMP and a whole variety of cell functions are being enthusiastically studied in laboratories all over the world. This area, which is rife with controversy and which is extremely complex, is almost incomprehensible unless approached first via a well-reasoned review. Likewise, there have been hundreds of reports supporting or refuting the hypothesis that cyclic nucleotides are involved in the control of cell growth in normal cells and that defects in cyclic nucleotide metabolism underlie the loss of control in transformed cells.

These few examples serve to highlight the dilemma confronting anyone approaching cyclic nucleotide research at this time in history. There are so many actions, so many systems, so many contradictory data and so much detail that a volume like this one, outlining the state of the art in a variety of systems, seems absolutely essential.

Earl Sutherland felt — and this is by no means a sentiment unique to him — that there are far too few truly critical reviews or syntheses available to the scientific community. As a matter of fact, he felt that a critical review was probably worth more than several excellent research papers in terms of the well-being of science and the advancement of knowledge. One can only hope that this volume will serve such a role.

Reginald W. Butcher

Preface

Our knowledge of the existence of cyclic phosphates of purine and pyrimidine nucleosides is of fairly recent origin. The starting point of cyclic nucleotide research was the brilliant series of investigations by which Sutherland and his associates in the late fifties explained that the glycogenolytic action of adrenaline proceeds through the activation of a membrane bound enzyme which produces adenosine 3′,5′-monophosphate. Cyclic AMP, in turn, activates the initial step in a series of protein phosphorylation reactions. Since then, this field of biological research has rapidly expanded to produce a wealth of knowledge about the metabolism and behaviour of cyclic AMP and, to a hitherto lesser degree, of cylic GMP in living cells. Several unifying hypotheses about the roles of cyclic nucleotides have been proposed, e.g. that a common denominator of cyclic nucleotide action is the activation of protein kinases, or the dualism hypothesis of cyclic AMP and cyclic GMP behaviour and action. This has not decreased but rather intensified the challenge to study both the metabolic regulation and the mechanisms of action of cyclic nucleotides on various levels in normal, pathologic and pharmacologically treated cells.

This book is intended to survey some of the important areas of cyclic nucleotide research and the impact of cyclic nucleotides in ortho- and patho-physiologic regulation in a number of organs and states. Basic chemical and biochemical aspects of cyclic nucleotide formation and degradation, enzymic regulation, interactions of cyclic nucleotides with protein kinases, and cyclic nucleotide analogues as research tools are treated in the first part of the book. Then, cyclic nucleotides are considered as possible regulators and messengers in living systems from cultured cells to various organs including endocrine, exocrine and secretory systems, and functional entities such as circulation, degradation mechanisms in lysosomes and immune responses in mammals. In a third part, the impact of cyclic nucleotide research in the understanding of disease processes is discussed by specialists in several fields of human pathology. Finally, the most intriguing problem of the role of cyclic AMP and cyclic GMP in the functions of the nervous system, is treated from physiologic, biochemical and pharmacologic standpoints, showing that the unravelling of cyclic nucleotide roles in neurotransmission may well become a key point in the explanation of some nervous diseases.

The editors would like to thank all the authors who have contributed to this volume for presenting not only their own work but also reviewing the state of the art in their respective fields. We hope that our intention to present the highlights of cyclic nucleotide research to a larger readership including biochemists, pharmacologists, biologists and physicians has thus been met.

Hinrich Cramer
Joachim Schultz

Contributors

M. Samir Amer	*Mead Johnson Research Center, Evansville, Indiana 47712, USA*
Larry D. Barnes	*Department of Biochemistry, University of Texas, San Antonio, Texas 78284, USA*
Michael J. Berridge	*Unit of Invertebrate Chemistry and Physiology, Department of Zoology, Downing Street, Cambridge, England*
Jay E. Birnbaum	*Metabolic Disease Therapy Research Section, Lederle Laboratories, Division of American Cyanamid Company, Pearl River, New York 10965, USA*
Lutz Birnbaumer	*Department of Cell Biology, Baylor College of Medicine, Houston, Texas 77025, USA*
Jean-Marie Boeynaems	*Institut de Recherche Interdisciplinaire, School of Medicine, University of Brussels, and Biology Department, Euratom, B-1000 Brussels, Belgium*
John W. Daly	*National Institute of Arthritis, Metabolism and Digestive Diseases, National Institutes of Health, Bethesda, Maryland 20014, USA*
Thomas P. Dousa	*Nephrology Research Laboratories, Departments of Physiology and Biophysics and Medicine, Mayo Clinic and Foundation, Rochester, Minnesota 55901, USA*
Elizabeth A. Duell	*University of Michigan Medical School, Departments of Dermatology and Biological Chemistry, Ann Arbor, Michigan 48109, USA*
Jacques E. Dumont	*Institut de Recherche Interdisciplinaire, School of Medicine, University of Brussels, and Biology Department, Euratom, B-1000 Brussels, Belgium*
John N. Fain	*Section of Physiological Chemistry, Division of Biology and Medicine, Brown University, Providence, Rhode Island 02912, USA*

Louis J. Ignarro *Department of Pharmacology, Tulane University School of Medicine, New Orleans, Louisiana 70112, USA*

Toshio Kaneko *The First Department of Medicine, Faculty of Medicine, University of Tokyo, Japan*

James J. Keirns *Metabolic Disease Therapy Research Station, Lederle Laboratories, Division of American Cyanamid Company, Pearl River, New York 10965, USA*

Ernst-Georg Krause *Division of Cellular and Molecular Cardiology, Central Institute of Heart and Circulatory Regulation Research, Academy of Sciences of the DDR, 1115 Berlin-Buch, German Democratic Republic*

Denis C. Lehotay *Division of Endocrinology and Metabolism, Department of Medicine, University of Miami, School of Medicine, Miami, Florida 33152, USA*

Gerald S. Levey *Division of Endocrinology and Metabolism, Department of Medicine, University of Miami, School of Medicine, Miami, Florida 33152, USA*

Hilda K. Lo *Division of Endocrinology and Metabolism, Department of Medicine, University of Miami, School of Medicine, Miami, Florida 33152, USA*

Jon P. Miller *Nucleic Acid Research Institute, ICN Pharmaceuticals, Inc., 2727 Campus Drive, Irvine, California 92715, USA*

William Montague *Biochemistry Department, University of Leicester, Leicester, England*

John B. Moore, Jr *Metabolic Disease Therapy Research Section, Lederle Laboratories, Division of American Cyanimid Company, Pearl River, New York 10965, USA*
Present address: Department of Biochemistry and Biophysics, Texas A and M University, College Station, Texas 77843, USA

Charles W. Parker *Division of Allergy and Immunology, Department of Medicine, Washington University, Medical School, St Louis, MO 63110, USA*

Hans D. Peters *Institut für Pharmakologie, Abteilung II, Medizinische Hochschule Hannover, D-3000 Hannover 61, German Federal Republic*

Paul Rapp *Department of Applied Mathematics and Theoretical Physics, Silver Street, Cambridge, England*

Hans-Jürgen Ruoff *Department of Pharmacology, University of Tübingen, 7400 Tübingen, German Federal Republic*

Peter S. Schönhöfer *Institut für Pharmakologie, Abteilung II, Medizinische Hochschule Hannover, D-3000 Hannover 61, German Federal Republic*

Karl-Friedrich Sewing *Department of Pharmacology, University of Tübingen, 7400 Tübingen, German Federal Republic*

George R. Siggins *Arthur Vining Davis Center for Behavioral Neurobiology, The Salk Institute, P.O. Box 1809, San Diego, California 92112, USA*

Phil Skolnick *National Institute of Arthritis, Metabolism and Digestive Diseases, National Institutes of Health, Bethesda, Maryland 20014, USA*

Fridolin Sulser *Vanderbilt University School of Medicine, and Tennessee Neuropsychiatric Institute, Nashville, Tennessee 37272, USA*

Stéphane Swillens *Institut de Recherche Interdisciplinaire, School of Medicine, University of Brussels and Biology Department, Euratom, B-1000 Brussels, Belgium*

Edward L. Tolman *Metabolic Disease Therapy Research Section, Lederle Laboratories, Division of American Cyanamid Company, Pearl River, New York 10965, USA*

Jerry Vetulani *Vanderbilt University School of Medicine, and Tennessee Neuropsychiatric Institute, Nashville, Tennessee 37271, USA*
Present address: Institute of Pharmacology, Polish Academy of Sciences, Krakow, Poland

John J. Voorhees *University of Michigan Medical School, Departments of Dermatology and Biological Chemistry, Ann Arbor, Michigan 48109, USA*

Jerry H. Wang *Department of Biochemistry, Faculty of Medicine, University of Mannitoba, Winnipeg, Canada, R3E OW3*

Albert Wollenberger *Division of Cellular and Molecular Cardiology, Central Institute of Heart and Circulatory Regulation Research, Academy of Sciences of the DDR, 1115 Berlin-Buch, German Democratic Republic*

Contents

CHAPTER 1

Activation of Adenylate Cyclase: Requirement for Phospholipids

D. C. Lehotay, H. K. Lo and G. S. Levey

I INTRODUCTION

The isolation of adenosine 3′,5′-monophosphate (cAMP) by Rall and Sutherland (1958) marked the first step in the development of a comprehensive theory of hormone action. cAMP is produced by the enzyme adenylate cyclase (E.C. 4.6.1.1) from ATP. Mg^{2+} is required for the reaction in which pyrophosphate is formed in equimolar amounts with the cyclic nucleotide. The initial isolation of the cyclic nucleotide was followed by a great deal of research which revealed that an extraordinary number of hormonal effects is mediated by cAMP (Rall and Sutherland, 1962; Murad *et al.*, 1962; Klainer *et al.*, 1962; Butcher and Sutherland, 1962). Sutherland *et al.*, (1965) developed the two-messenger theory of hormone action: This theory proposes that after a hormone is released from its storage organ, it travels via the blood stream to its target tissue and there binds to a highly specific surface receptor. Binding of the hormone to its receptor site is followed by activation of adenylate cyclase with subsequent increases in the intracellular concentration of cAMP, the second messenger. cAMP then triggers a variety of intracellular reactions characteristic of the particular cell and hormone.

Adenylate cyclase is a highly regulated, complex enzyme system. The problem of whether recognition of hormones and catalytic activity reside within the same molecule or in separate but interacting components of the enzyme system has interested many workers. It has been postulated that there are separate receptor, or regulatory, and catalytic components of the enzyme (Robison *et al.*, 1967; Rodbell *et al.*, 1971a). A third component also exists which couples the events occurring at the receptor subunit to those taking place at the catalytic site. Since the hormones which stimulate the enzyme are extracellular, the receptor portion of the enzyme has to be on the outside of the plasma membrane. The events affected by cAMP, on the other hand, occur within the cell, so the catalytic component must discharge its product intracellularly. The adenylate cyclase complex must, therefore, span the plasma membrane. Since the enzyme is embedded in the predominantly lipid matrix of a membrane, the role of lipids may be of great importance in the functioning of the hormone-responsive adenylate cyclase.

The purpose of this chapter is to review briefly the important characteristics of this critical enzyme, and to discuss the role of phospholipids in the coupling of the receptor to the catalytic site of adenylate cyclase.

II GENERAL PROPERTIES OF ADENYLATE CYCLASE

Adenylate cyclase is among the most widely distributed enzymes. It has been shown to be present in every animal organ, except dog erythrocytes, examined to date (Sutherland *et al.*, 1962). While the response of the enzyme in broken-cell preparations from higher animals corresponds closely to the hormonal responsiveness of that particular tissue in the intact organism (Robison *et al.*, 1971), there are important differences between the enzyme at higher and lower phylogenetic levels. In most mammals adenylate cyclase is stimulated by hormones but in unicellular organisms it responds to the changing nutritional state of the organism (Pastan and Perlman, 1970); vertebrate hormones have no effect on these organisms.

Studies of the subcellular distribution of adenylate cyclase clearly show that in mammalian cells the enzyme is located in the plasma membrane. Little, if any, activity can be found in the nucleus, microsomes or mitochondria (Rosen and Rosen, 1969; Wolff and Jones, 1971), although definite enzymic activity has been detected in cardiac sarcoplasmic reticulum (Entman *et al.*, 1969; Katz *et al.*, 1974). A small number of observations has also been made which indicate that purified nuclei from some tissues may possess adenylate cyclase activity (Liano *et al.*, 1971; Soifer and Hechter, 1971). The biological significance of these data remains to be explored.

In the reaction catalysed by adenylate cyclase one molecule of ATP is converted to a molecule of cAMP with the release of pyrophosphate. Mg^{2+} is required, and the substrate for the reaction is probably a Mg^{2+}–ATP complex. It has been shown by Greengard *et al.* (1969a) that the reaction is readily reversible under certain conditions. They calculated that the equilibrium constant of the reaction is 0.065 M at pH 7.3 and 25°C, and that the free energy of hydrolysis of the phosphodiester bond of cAMP is equal to –11.9 kcal/mole (49.8 J/mole). Using solubilized enzyme from *Brevibacterium liquefaciens*, Kurashina *et al.* (1974) have demonstrated that the production of cAMP is an endergonic reaction. The standard Gibbs' free energy change ($\Delta G°$) varies as a function of pH and is +0.8 – 2.1 kcal/mole between pH 7.7 and 6.2 at 25°C in the presence of 5 mM $MgSO_4$. By measuring K_{eq} as a function of temperature, the enthalpy change of the reaction was found to be +5.0 kcal/mole. The inability of some workers to observe that the reaction is reversible may be due to the use of unfavourable reaction conditions, such as low concentrations of cAMP and pyrophosphate, low enzymic concentrations and short incubation times. The reversal of the reaction with the formation of ATP is unlikely under physiological situations where conditions favour the synthesis of cAMP (Hayaishi *et al.*, 1971).

The kinetics of the adenylate cyclase reaction have been thoroughly investigated. The enzyme requires Mg^{2+} which is bound at two sites. One of these is at the catalytic site, and the other is at a different location which is altered by the action of hormones and F^-, and which allosterically enhances the reaction of the catalytic site with its substrate, Mg^{2+}–ATP (Birnbaumer *et al.*, 1969). The dependence of activity on hormone concentration in fat cells was found to be sigmoid, while F^- concentration curves are hyperbolic. It has been suggested by Drummond *et al.* (1971) that hormones and F^- act by increasing the V_{max} of the reaction without affecting K_m for the substrate. The data of Rosen and Rosen (1969), however, seems to indicate that it is the K_m which is altered during activation of adenylate cyclase, and not the V_{max}. The K_m of the enzyme for ATP has been found in a variety of preparations to be around $1–5 \times 10^{-4}$M (Birnbaumer *et al.*, 1969; Bär and Hechter, 1969c; Drummond and Duncan, 1970). As the substrate for adenylate cyclase is Mg^{2+}–ATP rather than ATP itself and the two compounds are capable of forming complexes, it is not surprising that the ratio of Mg^{2+}/ATP exerts a profound effect on the production of cAMP. Maximal activity is obtained with a Mg^{2+}/ATP ratio close to two (Wolff and Jones, 1971), while ATP in excess of Mg^{2+} is usually inhibitory (Birnbaumer *et al.*, 1969; Severson *et al.*, 1972; Perkins and Moore, 1971).

It has been suggested (DeHaën, 1974) that hormones and F^- decrease the sensitivity of the enzyme toward inhibition by free ATP, thereby favouring the formation of the enzyme Mg^{2+}–ATP complex. Adenylate cyclase may be considered an allosteric enzyme on which hormones act as heterotropic activators. Enzyme activity is optimal between pH 7.0–8.0 (Perkins 1973).

Many ions have been shown to modulate adenylate cyclase activity. Na^+ and K^+ have both stimulatory and inhibitory effects on a renal vasopressin-sensitive adenylate cyclase (Dousa, 1972). Concentrations below 150 mM cause stimulation, while those higher than 200 mM levels are inhibitory. Mixtures of solutes simulating antidiuresis *in vivo* inhibit vasopressin-stimulated activity; mixtures simulating water diuresis permit the normal response. Li^+ stimulates basal cAMP production in fat cells (Birnbaumer *et al.*, 1969), but inhibits the activity of vasopressin and TSH-stimulated adenylate cyclase (Dousa and Hechter, 1970; Wolff *et al.*, 1970) and basal activity of the enzyme from rat renal cortex (Marcus and Aurbach, 1971).

When considering divalent cations, Mg^{2+}, Mn^{2+} and Co^{2+} activate the enzyme while Ca^{2+}, Zn^{2+} and Cu^{2+} are inhibitory (Perkins 1973). Mn^{2+} can replace Mg^{2+} in activating the enzyme (Birnbaumer *et al.*, 1969; Drummond *et al.*, 1971), and since Mn^{2+} has a greater affinity for free ATP, the V_{max} of the reaction is higher with Mn^{2+}–ATP than with Mg^{2+}–ATP as the substrate (Perkins and Moore, 1971).

Ca^{2+} plays an important role in the action of adenylate cyclase. While it inhibits activity in most tissues (Robison *et al.*, 1971) ACTH-stimulated AMP production exhibits a strong dependence on the presence of small amounts of Ca^{2+} (Bar and Hechter, 1969b). It was proposed that ACTH activates the enzyme by displacing bound Ca^{2+} from a site near the enzyme (Rubin *et al.*, 1972).

The mechanism of F^- stimulation is poorly understood. No effect is observed in intact cells, although F^- can easily enter the cell. It is probable that homogenization of cells alters the enzyme structure and makes it susceptible to F^- stimulation. While it has been demonstrated that both hormones and F^- stimulate the same enzyme, they do so by different means. The two effects are noncompetitive (Birnbaumer *et al.*, 1971), and it is the maximum velocity of the enzyme reaction (V_{max}) which is markedly increased by F^- leaving the K_m unaltered (Drummond *et al.*, 1971). The effect of F^- is related to the concentration of Mg^{2+} which forms complexes with F^-, at high concentrations of the anion the decrease of fat-cell ghosts adenylate cyclase activity can be reversed by increasing the Mg^{2+} concentration (Birnbaumer *et al.*, 1969). These workers have postulated that F^- may stimulate adenylate cyclase by reducing the apparent K_{dissoc} for Mg^{2+} at the proposed second binding site on the enzyme. However, this suggestion has not been confirmed in all systems tested (Wolff and Jones, 1971; Drummond and Duncan, 1970; Drummond *et al.*, 1971; Severson *et al.*, 1972).

Rodbell *et al.* (1971c), in a study of the glucagon-sensitive adenylate cyclase in rat liver plasma membranes, noted that GTP and GDP decrease the binding of [^{125}I-]glucagon to the plasma membrane receptor sites. The initial rates of both basal and glucagon-stimulated adenylate cyclase activities are enhanced by GTP, while stimulation by F^- is inhibited. Krishna *et al.* (1972) observed similar activation of adenylate cyclase by concentrations of GTP as low as 1 μM in platelet membranes.

It has been suggested that there may be an enzyme system in fat-cell plasma membranes capable of transferring the terminal phosphate of GTP to the catalytic component of the enzyme resulting in an inhibited form of adenylate cyclase (Harwood *et al.*, 1973). Rodbell *et al.* (1974) observed that cAMP production in the presence of 0.1 mM glucagon exhibits a lag of four minutes. This lag is abolished in the presence of 0.1 mM GTP. It was also noted that the K_m for glucagon is different in the presence and absence of GTP.

Using 5′-guanylyl imidodiphosphate, an analogue of GTP with a terminal phosphate resistant to enzymic cleavage, a purine nucleotide-binding site has been detected on the enzyme from turkey erythrocytes (Bilezikian and Aurbach, 1974; Spiegel and Aurbach, 1974). This enzyme has a higher apparent affinity for GTP than for ATP and may be involved in the regulation of the catalytic function of the adenylate cyclase complex. GTP enhances the catecholamine-stimulated activation of adenylate cyclase, and it has been suggested that the receptor site specific for the ethanolamine function of the catecholamine molecule is also the site influenced by the nucleotides (Bilezikian and Aurbach 1974).

In a recent study, Rodbell *et al.* (1975) have suggested that hormones, GTP and $HATP^{3-}$ (protonated ATP not complexed with Mg^{2+}) are the ligands which regulate adenylate cyclase activity. In addition to the catalytic and receptor sites, they have suggested that a third regulatory binding site specific for GTP exists, which they called the nucleotide-regulatory site. The binding of hormone, GTP, or H-ATP^{3-} to these sites causes the enzyme to undergo conformational transitions. Experiments with 5′-guanylyl imidodiphosphate suggested that the crucial step in the activation of adenylate cyclase is the binding of guanine nucleotides at the nucleotide-regulatory site which transforms the inactive enzyme to an active state with a high V_{max} and low K_i. This active form is very susceptible to inhibition by the protonated substrate. Hormones relieve this inhibition by inducing another form of the enzyme with a high K_i. At the normal *in vivo* concentrations of GTP and protonated substrate, the enzyme is thus maintained in its inhibited state, but is primed for conversion by a hormone to the high Ki state.

III ROLE OF PHOSPHOLIPIDS IN HORMONE ACTIVATION

Phospholipids are essential components of many membrane-bound enzymes. The ‘fluid mosaic model’ of membrane structure proposed by Singer and Nicholson (1972) provides a useful framework for the consideration of the relationship of phospholipids to various membrane-bound enzymes including adenylate cyclase. Phospholipids are thought to be arranged in a discontinuous bilayer with the ionic and polar heads in contact with the aqueous phase, whereas the nonpolar, fatty-acid chains are oriented inward away from the aqueous phase. Globular proteins and enzymes, such as adenylate cyclase, float within this lipid matrix and may partially protrude on either or both sides of the membrane. The ionic and polar groups of the proteins are in contact with water, while the nonpolar amino acid residues are directed toward the lipid phase. A small portion of the phospholipid is thought to interact with the protein specifically. This specific interaction may involve both the polar and nonpolar portions of the phospholipid, providing the conformation appropriate to enable the membrane-bound enzymes to interact with substrates, activators and inhibitors. The lipids may also confer certain properties such as stability on the enzyme. The structural and biochemical functions of lipids in the cell membrane and their diverse roles in the activity of membrane-bound enzymes have recently been reviewed by Coleman (1973).

Phospholipids have been demonstrated to play key roles in the hormone-induced activation of adenylate cyclase. Sutherland *et al.* (1962) found that solubilization of the adenylate cyclase from brain, heart, skeletal muscle and liver abolishes or diminishes the hormone responsiveness of the enzyme. These workers have suggested that adenylate cyclase may in fact be a lipoprotein.

The purpose of the remainder of this chapter is to review current investigations pertaining to phospholipids and their role in the hormone-induced activation of adenylate cyclase. For the most part, the discussion will focus on the solubilized cardiac adenylate cyclase derived from cat heart. Hormonal activation of soluble adenylate cyclases from liver, thyroid and kidney will also be reviewed.

A Solubilized Myocardial Adenylate Cyclase

The non-ionic detergent, Lubrol–PX, an ethylene oxide condensate of dodecanol, has been utilized in a simple, one-step method for solubilizing the adenylate cyclase from cat myocardium (Levey, 1970; 1971a). Most of the adenylate cyclase in homogenates of heart sediments at 12,000 *g* with only trace amounts remaining in the supernatant. Homogenization in the presence of Lubrol–PX reverses this distribution and almost all of the enzyme is found in the supernatant even after centrifugation at 250,000 *g* for two hours. The solubilized adenylate cyclase has a molecular weight as determined by gel filtration, of approximately 100,000 to 200,000 a figure in close agreement to that of 160,000 found for the solubilized canine heart enzyme (Lefkowitz *et al.*, 1973). This molecular size is appropriate for the localization of the enzyme in the plasma membrane about 7.5–9.0 nm thick.

The membrane-bound cardiac adenylate cyclase is activated by catecholamines (Murad *et al.*, 1962), glucagon (Murad and Vaughan, 1969; Levey and Epstein, 1969a), histamine (Klein and Levey, 1971a), thyroid hormone (Levey and Epstein, 1969b) and prostaglandins (Klein and Levey, 1971b). Solubilization abolishes the responsiveness to all these hormones (except prostaglandins) but the F^- activation of the enzyme, which occurs by direct stimulation of the catalytic site (Birnbaumer *et al.*, 1971; Drummond *et al.*, 1971), is not affected. Removal of the detergent from the solubilized enzyme is necessary to prevent its possible interference with the binding of the hormone to the receptor site when investigating the specific role of phospholipids in the hormone-activation step. Adenylate cyclase is adsorbed by DEAE–cellulose (Sutherland *et al.*, 1962) and it can be separated from Lubrol–PX by sequential elution with neutral buffers of increasing ionic strength. The enzyme remains in a soluble state after elution, as defined by previously established criteria, and continues not to respond to hormonal stimulation with the exception of prostaglandins.

B Monophosphatidylinositol and the Catecholamine Activation

The membrane-bound adenylate cyclase is activated by noradrenaline, while the solubilized enzyme is nonresponsive. The addition of highly purified bovine monophosphatidylinositol restores noradrenaline responsiveness to the solubilized enzyme, and also results in a marked increase in the sensitivity of the enzyme to the hormone. Half-maximal activation occurs at a concentration of noradrenaline of 1×10^{-7} M, which is about a 100 times less than that required by the membrane-bound enzyme. This hormonal sensitivity approaches that observed for intact cells; for example the sensitivity of contractile response in isolated papillary muscles is similar (Buccino *et al.*, 1967). The sensitivity of the particulate membrane-bound preparations to noradrenaline also increases upon addition of phosphatidylinositol to an extent similar to that which is observed with detergent-free soluble systems (Levey, 1973). Data suggest that the considerably lower *in vitro* sensitivity of broken-cell adenylate cyclase systems to hormones, compared with intact tissue, may be due to the homogenization procedure which in some way alters the relationship of the enzyme to monophosphatidylinositol. Other acidic phospholipids, such as phosphatidylserine and phosphatidylethanolamine, do not restore responsiveness to the catecholamines.

In the presence of monophosphatidylinositol, the soluble enzyme also behaves in a manner characteristic of the β-adrenergic receptor system found in the membrane-bound adenylate cyclase preparation in intact hearts. The order of potency of the catecholamines for the activation of the solubilized adenylate cyclase is: isoprenaline $>$ adrenaline $>$ noradrenaline. This order is the same for their potency in increasing heart rate and contractility (Ahlquist, 1948) and activating membrane-bound adenylate cyclase (Murad *et al.*, 1962). In addition, the noradrenaline activation is totally abolished by the β-adrenergic blocker propranolol providing further evidence for β-receptor specificity (Levey, 1971b, 1973).

C Phosphatidylserine and Activation by Glucagon and Histamine

Although phosphatidylserine is ineffective in restoring catecholamine responsiveness to soluble adenylate cyclase it does restore responsiveness to glucagon and histamine. The concentration–response curves for these hormones utilizing the solubilized adenylate cyclase, in the presence of phosphatidylserine, are similar to those obtained with the membrane-bound enzyme (Levey and Epstein, 1969a; Klein and Levey, 1971a). Phosphatidylethanolamine and monophosphatidylinositol are ineffective in restoring responsiveness to these hormones. The antihistamine, diphenhydramine, abolishes the histamine activation, but not that produced by glucagon (Levey and Klein, 1972). These findings need further clarification, however, since McNeill and Muschek (1972) have shown that diphenhydramine is a non-competitive inhibitor of cardiac adenylate cyclase thus suggesting that cardiac tissue is not an H_1 receptor. Also, burimamide, a H_2-receptor blocker, competitively inhibits the histamine activation of adenylate cyclase (McNeill and Verma, 1974).

The concentrations of phosphatidylserine utilized to restore the activation of adenylate cyclase by glucagon are markedly higher than those present in heart membranes *in vivo* (Ansell *et al.*, 1973). The minimal effective concentration of phosphatidylserine is about 64 μg/ml. While that of monophosphatidylinositol (0.8 μg/ml) is closer to the concentrations found *in vivo.* The reason and significance of the concentration discrepancy are unexplained. A number of other lipids including phosphatidylcholine, lysolecithin and cardiolipin has also been tested; these are without effect (Levey, 1971a).

D Prostaglandins

The mechanism of action of these fatty acids is unclear. In many tissues, however, it appears to involve an interaction with the membrane adenylate cyclase. Kuehl and Humes (1972) have described a lipoprotein-receptor site for prostaglandins in fat cells and speculated that prostaglandins interact with either the coupler or the catalytic site of the enzyme. Prostaglandins may also act as allosteric effectors of adenylate cyclase (Johnson and Ramwell, 1973) with, presumably, the polar groups of the prostaglandins bringing about certain phase changes in membrane components resulting in enzyme activation. It is of considerable interest that the critical phospholipids in the action of prostaglandins appear to be phosphatidylserine and phosphatidylinositol. Johnson and Ramwell (1973) also noted that there must be additional requirements for prostaglandin action, such as proteins with functional sulfhydryl groups, as described by Kuehl and Humes (1972).

Klein and Levey (1971a) have demonstrated that prostaglandins E_1, E_2 and A_1 activate the membrane-bound cardiac adenylate cyclase over an extraordinarily sensitive concentration range ($1 \times 10^{-11} - 1 \times 10^{-9}$ M. In contrast, prostaglandins $F_{1\alpha}$ and $F_{2\alpha}$ are inactive. However, all prostaglandins stimulate the solubilized adenylate cyclase in the absence of added phospholipids over a concentration range similar to that observed for the membrane-bound enzyme. This non-specific activation of the soluble adenylate cyclase by prostaglandins suggests that solubilization alters the configuration of the enzyme compared with the membrane-bound enzyme, thus exposing a diverse group of sites and allowing a broader spectrum of prostaglandins to activate the enzyme.

The fact that prostaglandins do not require added phospholipids for activity implies that the unsaturated fatty acids in the various acidic phospholipids may be the critical factors in the activation process. For example, differences in fatty acid composition are responsible for certain differences in the physical properties of phospholipids with similar polar groupings (Coleman 1973). It has been observed that reactivation of lipid-dependent enzymes is more effective with lipids which have a high proportion of unsaturated fatty acids (Jurtshuk *et al.*, 1961; Jones and Wakil, 1967; Sartorelli *et al.*, 1969). Although the fatty acid composition of the phospholipids utilized for studies of the solubilized cardiac adenylate cyclase have not been characterized systematically, it has been shown that monophosphatidylinositol contains significant amounts of arachidonic acid, a fatty acid closely related to prostaglandins (Levey, 1973).

A number of reports have clearly demonstrated hormone-mediated increases in membrane phospholipid synthesis. Scott *et al.* (1968) demonstrated that TSH specifically increases the synthesis of phosphatidylinositol in isolated thyroid cells and in thyroid slices as early as two minutes after the initiation of the reaction; stimulation is apparently independent of cAMP. Gaut and Huggins (1966) showed that catecholamines increase the synthesis of monophosphatidylinositol in cardiac slices in agreement with the data on the solubilized preparation (Levey, 1973). Hokin (1969) has demonstrated that a diverse group of hormones are able to stimulate the synthesis of monophosphatidylinositol. On the basis of these data the effect of glucagon on the metabolism of phospholipids in cat heart slices was examined (Lo and Levey, 1976). It was shown that glucagon increases the incorporation of ^{32}P into phosphatidylserine in heart muscle slices as early as ten minutes after the initiation of phospholipid synthesis. The increase above control is approximately two-fold at ten minutes and six-fold after thirty minutes. Glucagon does not stimulate the incorporation of ^{32}P into total lipids, phosphatidylcholine, phosphatidylinositol or sphingomyelin, but significantly increases ^{32}P incorporation into phosphatidylethanolamine at ten minutes and thirty minutes, increases are three and four-fold respectively.

In an attempt to obtain additional specificity for the above findings heart muscle slices were also incubated with [^{14}C]serine in the presence and absence of glucagon. It would be expected that [^{14}C]serine should be incorporated into phosphatidylserine as the result of an exchange of nitrogenous bases with phosphatidylethanolamine thus providing greater specificity than the studies with ^{32}P incorporation. Glucagon (2×10^{-5} M) markedly stimulates the synthesis of phosphatidylserine as early as three minutes after the initiation of the incubation. The increase is approximately twelve-fold compared with the control at three minutes and double at ten minutes. In contrast, glucagon does not stimulate the incorporation of [^{14}C]serine into any other phospholipid at three minutes. A small incorporation was seen at ten minutes in the case of phosphatidylethanolamine and the fraction containing cardiolipinphosphatidic acid. In the latter case the increase is two-fold. Glucagon does not stimulate the incorporation of [^{14}C]serine into total lipid, phosphatidylcholine, lysophosphatidylcholine, phosphatidylinositol or sphingomyelin. Noradrenaline, a catecholamine which increases cAMP levels, heart rate and contractility, does not increase the incorporation of [^{14}C]serine into any fraction.

Since most of the actions of glucagon appear to be mediated by cAMP, the effects of dibutyryl-cAMP (Bt_2-cAMP) on the incorporation of [^{14}C]serine into phosphatidylserine have also been examined. Bt_2cAMP (1×10^{-4} M), in contrast to glucagon (2×10^{-5} M), does not increase the incorporation of [^{14}C]serine into phosphatidylserine over any time period examined.

At this point it is only possible to speculate as to the significance of these findings in relation to glucagon activation of adenylate cyclase. As noted earlier acidic phospholipids appear to serve a critical role in hormonal activation of adenylate cyclase. The data demonstrating that glucagon produces a rapid, cAMP-independent, specific increase in phosphatidylserine provide further support for the direct role of phospholipids in the activation process. The general hypothesis is that following hormone binding, a rapid increase in acidic phospholipid synthesis occurs which results in activation of the adenylate cyclase probably by an alteration in the configuration of the catalytic site. Inactivation may occur with further metabolism of the phospholipid. Whether or not the alteration in phospholipid synthesis occurs rapidly enough *in vivo* is not known but there have been reports of increased phosphatidylinositol synthesis in platelets within two seconds of the addition of hormone (Lloyd *et al.*, 1973).

E Phospholipids and Adenylate Cyclase from Liver, Thyroid and Kidney

The relationship of membrane phospholipids to the hormone-responsive adenylate cyclases in liver, thyroid and kidney has also been intensively investigated. Highly purified phospholipases A and C cause a loss of both glucagon responsiveness (Birnbaumer *et al.*, 1971; Rubalcava and Rodbell, 1973) and a decrease in the binding of [^{125}I]glucagon to the membranes (Rodbell *et al.*, 1971b). Pohl *et al.* (1971), using the same system, further observed that glucagon responsiveness of adenylate cyclase and the binding of [^{125}I]glucagon to its receptor can be partially restored by the addition of aqueous suspensions of membrane lipids to the liver membranes which have previously been treated with digitonin or phospholipase A. Similar effects have been observed when pure phospholipids were added to the membranes, with phosphatidylserine being more potent than phosphatidylethanolamine or phosphatidylcholine.

Hormonal responsiveness of liver adenylate cyclase has also been studied by Rethy *et al.* (1972), who found that mild lipid extraction and treatment of the membranes with phospholipase A or C abolishes adrenaline and glucagon responsiveness. Phosphatidylserine partially restores the glucagon sensitivity and almost totally restores the adrenaline activation. Phosphatidylinositol has no effect on hormonal responsiveness but significantly increases basal hepatic adenylate cyclase activity.

In investigations similar to those noted for liver, Yamashita and Field (1973) found that phospholipase A abolishes the responsiveness of thyroid plasma-membrane adenylate cyclase to thyroid-stimulating hormone. Phosphatidylcholine restores the hormonal sensitivity to a significant degree whereas phosphatidylserine is somewhat less effective. Phosphatidylcholine also partially restores the hormonal-responsiveness of the adenylate cyclase in thyroid plasma membrane treated with Lubrol–PX. The findings of Yamashita and Field (1973) demonstrating the important role of phosphatidylcholine in the hormone-sensitive thyroid adenylate cyclase are consistent with the observations of Macchia *et al.* (1970). They showed that lecithinase C, an enzyme with a high specificity for phosphatidylcholine, abolishes TSH-mediated increases in thyroid glucagon oxidation, ^{32}Pi incorporation into phospholipids. I accumulation and cAMP formation. It would appear, therefore, that in thyroid cell membranes phosphatidylcholine is the critical phospholipid required for the hormonal responsiveness of adenylate cyclase.

Finally, Neer, (1973) has succeeded in solubilizing renal medullary adenylate cyclase using Lubrol–PX. Solubilization abolishes the hormonal responsiveness of the enzyme. In contrast to cardiac tissue, removal of the detergent by DEAE–cellulose chromatography restores the responsiveness to vasopressin without the addition of exogenous phospholipids. About 12% of the material eluted from the DEAE–cellulose is phospholipid, and Neer speculated that this endogenous phospholipid alone may be sufficient to restore vasopressin responsiveness to the detergent-free adenylate cyclase.

Neer (1974) also found that soluble renal adenylate cyclase binds small amounts of detergent (about 0.2 mg/mg of protein) and that no greater than 5% of the enzyme surface is involved in hydrophobic interactions with the various components of the renal cell membrane. It would appear, therefore, that the adenylate cyclase is attached to the inner surface of the plasma membrane and not embedded deeply in the lipid matrix. In this scheme the hormone receptor on the exterior surface of the cell would have a large hydrophobic portion which spans the lipid bilayer and interconnects with the adenylate cyclase.

F Site and Possible Mechanism of Action of Phospholipids

Considerations of the potential sites and mechanism of action of phospholipids in relation to the activation of membrane adenylate cyclase by hormones have been enormously aided by two early models of the membrane-associated enzyme.

The first was proposed by Robison *et al.* (1967), who visualized the enzyme as consisting of two sites, a regulatory site and a catalytic site with the former associated with the latter so that the interaction of a hormone with the regulatory site results in an increase in catalytic activity. Rodbell *et al.*, (1971a) has more recently modified this model and proposed a three-component model of the hormone-responsive membrane adenylate cyclase which takes into full account the important role of membrane phospholipids. The regulatory site, or receptor, which is located on the exterior of the cell membrane, discriminates between hormones and provides the specific binding site for the hormone. The catalytic site, which contains the binding sites for ATP and Mg^{2+}, and which generates cAMP, is located on the interior of the cell membrane. Connecting the regulatory sites to the catalytic site is an intermediate coupler which transmits the message of hormonal binding to the catalytic site resulting in activation of the enzyme. The membrane phospholipids are thought to act mainly at the coupler site. The catalytic site appears to be independent of phospholipids since solubilized adenylate cyclases and the phospholipase-treated enzyme retain responsiveness to F^- under mild treatment conditions. Similarly, under these same mild conditions of solubilization, hormonal binding remains intact (Rubalcava and Rodbell, 1973; Lefkowitz and Levey, 1972; Levey, 1973). The mechanism of phospholipid action has not been thoroughly clarified and a distinct protein-coupler subunit has not been demonstrated. Therefore, phospholipids may themselves provide the connection between regulatory and catalytic sites. Alternatively, they may play a structural role by maintaining the adenylate cyclase in such a configuration in the membrane that it retains hormone-binding ability and responsiveness.

In any event, the models of Robison *et al.* (1967), and Rodbell *et al.* (1971a) and the data reviewed in the discussion of phospholipids clearly indicate the important role played by phospholipids in hormone-sensitive membrane adenylate cyclase systems. Elucidation of the precise mechanism and site of action of phospholipids requires both the isolation and purification of the respective subunits, and the reassociation of the components with the necessary phospholipids.

ACKNOWLEDGEMENT

This investigation was supported in part by grants 1R01 HL13715-05 and T01 AM05710-04 from the National Institutes of Health. Dr. Gerald S. Levey is an Investigator of the Howard Hughes Medical Institute.

CHAPTER 2

Regulatory Events at the Level of Membrane-Bound Adenylate Cyclase

L. Birnbaumer

I INTRODUCTION

Most of the actions of peptide and protein hormones, as well as of many of the biogenic amines, are the result of a primary interaction with a membrane-bound receptor which then initiates a train of events leading to activation of a nearby adenylate cyclase molecule and to formation of the second messenger (cAMP) (Robison *et al.*, 1971). The enzymic system is complex and the mechanism by which the occupied hormonal receptor stimulates its activity is still unknown. Since, with one exception, solubilization of the systems has resulted in loss of hormonal responsiveness, studies of the system have extensively utilized indirect approaches. Such approaches have had to take into account that (*i*) effects seen may be due to alteration of the phospholipid matrix surrounding adenylate cyclase and not due to a direct action on the enzyme, (*ii*) the system may be a polymeric one composed of several receptor and catalytic units, and (*iii*) kinetic modelling, under both pre- and post-steady state conditions, may lend itself to several interpretations which are seldom mutually exclusive. Progress, however, is being made in the understanding of adenylate cyclase systems in spite of all of these difficulties.

II PROPERTIES OF ADENYLATE CYCLASES

A General

In all nucleated cells so far studied, adenylate cyclases are enzymes associated with the plasma membranes of cells (Sutherland and Rall, 1958; Sutherland *et al.*, 1962; Rall and Sutherland, 1962; Murad *et al.*, 1962; Klainer *et al.*, 1962; Sutherland and Rall, 1960; Davoren and Sutherland, 1963). Indeed, adenylate cyclase appears to be a good plasma-membrane marker enzyme. There have been claims that adenylate cyclase may also exist in other subcellular organelles, such as adrenal microsomes, and the nuclei and sarcoplasmic reticula of liver and prostate. However, in none of these instances has convincing evidence been provided to indicate that the low activities detected are not due to plasma membrane contamination.

Several attempts have been made to solubilize adenylate cyclase while at the same time preserving hormonal action but these have met with only limited success. Murad *et al.* (1962) described the solubilization of bovine cerebral cortical adenylate cyclase with Triton X-100. However, during the solubilization hormonal response is lost, and the preparation is no longer soluble once the detergent is removed. Pastan *et al.* (1970) prepared a non-sedimentable adenylate cyclase from mouse adrenal tumours by dispersing membranes in a medium containing phospholipids and F^-. Although a response to ACTH remains after dialysis against F^--free media, no further work on this system has been done.

Lubrol-PX has proved to be useful in the preparation of a soluble heart adenylate cyclase (Levey, 1970). Removal of the bulk of Lubrol-PX by chromatography on DEAE-cellulose does not influence the loss of sedimentability (the solubility criterion used).

Except for studies by Levey (1971a,b; Lehotay *et al.*, this volume) who restored hormonal response with phospholipids, most other solubilization attempts have resulted in irreversible loss of hormonal responsiveness. Basal adenylate cyclase activity and F^- stimulation of adenylate cyclase, on the other hand, are more resistant to disruption of the membrane lipid than is hormonal stimulation. Thus, treatment of liver membranes with phospholipases results in loss of glucagon stimulation long before the F^- responsiveness is affected (Birnbaumer *et al.* 1971). Puchwein *et al.* (1974), in studies to test the effects of the polyene antibiotic filipin on catecholamine-stimulated adenylate cyclase in pigeon erythrocyte membranes showed a similar preferential lability of the receptor-mediated stimulation. The reasons for these differential effects are not understood at present, but are surely related to the facts that the natural habitat of adenylate cyclases is the phospholipid matrix of membranes and that the hormone binding to specific receptors on the external surface affects the conformation of the catalytic site of an enzyme located at the opposite side of the membrane.

Ryan and Storm (1974) have reported solubilization of liver membranes with Triton X-305 with good preservation of both glucagon and adrenaline responsiveness. Triton X-305 is a detergent with a higher hydrophobic-lipophilic balance number and therefore is more hydrophobic than Triton X-100. This finding — solubilization with preservation of hormonal action — should lead in the not too distant future to the determination of all the molecular characteristics of the system.

B Molecular Size and Thermodynamic Characteristics of the Reaction

Hormonally insensitive adenylate cyclase appears to be an asymmetric molecule. Neer (1974), working with canine renal adenylate cyclase solubilized with Triton X-100 or Lubrol-PX, found two forms of the enzyme with molecular weights of 158,000 and 38,000. A naturally soluble, hormonally insensitive adenylate cyclase from *Brevibacterium liquefaciens* has been crystallized (Takai *et al.*, 1974). The enzyme differs from mammalian adenylate cyclases in the fact that it does not respond to F^- but is stimulated by pyruvate. It can be resolved in sodium dodecyl sulphate into two subunits of 46,000 daltons, and was found to catalyse not only the formation of cAMP and pyrophosphate from ATP, but also the reverse reaction. Together with determinations of the enthalpies for the formation of AMP from cAMP and of AMP from ATP (Kurashina *et al.*, 1974; Greengard *et al.*, 1969b; Hayaishi *et al.*, 1971), it has been established that the hydrolysis of the cyclic phospho-ester link is associated with a free energy change of approximately 12 kcal/mole (50 J/mole). Thus cAMP could potentially be a high-energy donor similar to ATP.

C Effects of Cations

Cations appear to play a variable role or have varying effects on hormonal stimulation of adenylate cyclase. Ca^{2+} seems to be essential for the adenylate cyclase-activating effect of ACTH, both in the adrenals and in fat cell membranes (Birnbaumer *et al.*, 1970), as well as for activity in solubilized brain adenylate cyclase (Johnson and Sutherland, 1973). On the other hand, the action of oxytocin on toad bladder adenylate cyclase is totally inhibited at 1 mM $CaCl_2$, under conditions in which no effect on basal activity has been detected (Bockaert *et al.*, 1972). While selective effects of Ca^{2+} of this nature occur at sub-millimolar concentrations of the cation and restitution of ACTH sensitivity in EGTA-treated membranes requires only micromolar concentrations of Ca^{2+} (Bär and Hechter, 1969b), higher concentrations of Ca^{2+} are invariably found to inhibit both hormone-stimulated and basal activities.

Not only Ca^{2+} has variable effects on adenylate cyclase but also the effect of ionic strength on hormonal stimulation varies from system to system. While 100 mM NaCl has little or no effect on hormonal stimulation of either fat or liver adenylate cyclase (Birnbaumer *et al.*, 1969; Pohl *et al.*, 1971b), it has a profound inhibitory action on luteinizing hormone stimulation of rabbit corpus luteum adenylate cyclase, resulting in loss of responsiveness as well as a loss of affinity for the hormone (Birnbaumer *et al.*, 1976).

It is likely that except for the effect of low concentrations of Ca^{2+} and of nucleotides, other effects of ions seen in the adenylate cyclase systems are more a reflection of the susceptibility of the phospholipid matrix of the membrane to alteration, than an induction of a specific regulatory effect on the hormone-sensitive system.

In contrast to the variable effects of Ca^{2+} and ionic strength, Li^+ appears to affect adenylate cyclase systems by consistently inhibiting hormonal stimulation. This was shown first by Wolff *et al.* (1970) who found that Li^+

inhibits the action of thyroid-stimulating hormone in thyroid membranes. Dousa (1974) reported the inhibitory effects of Li^+ on vasopressin stimulation of renal adenylate cyclases. Interestingly, Li^+ also has both antithyroid and anti-antidiuretic actions and it may be that these pharmacological effects of the ion are due to its action on the adenylate cyclase systems in various organs.

D Effect of Fluoride Ion

Almost all of the adenylate cyclase systems described so far in eucaryotic cells have been found to be stimulated by F^-. The effect has been shown to be a direct one of the anion on the system and not secondary to its capacity to inhibit ATPase. Little is known about the mode of action of F^-. Its effect, first described by Rall and Sutherland (1962), requires the presence of Mg^{2+} (or Mn^{2+}) and is poorly reversible (Schramm and Naim, 1970). In contrast to hormones, F^- activation is much less dependent on the lipid environment surrounding the cyclase system since all of these enzymes but brain adenylate cyclase (Johnson and Sutherland, 1973) preserve the original response to F^- after solubilization with detergent (Levey, 1970a; Neer, 1973; Neer, 1974).

Many laboratories have studied the characteristics of F^- activation of adenylate cyclases in parallel with those of hormonal stimulation, with the hope of learning, either by analogy or by difference, the mechanism by which hormones exert their action. It has been found that in addition to its greater stability, the process by which F^- stimulates the system has very different properties from those involved in hormonal stimulation. Thus, Birnbaumer *et al.* (1969) showed that in fat cells the concentration of F^- giving half-maximal stimulation varies with temperature, while that of the hormone does not. Also, the magnitude of the F^- response is strongly dependent on temperature, but that of hormones is much less so; F^- concentration–effect curves are hyperbolic, those for hormones are not. In addition, Birnbaumer *et al.* (1970) showed that alteration of lipids of fat-cell membranes results in rapid and complex losses of hormonal responses, while F^- responses either remain unchanged or increase. This latter differential property has also been observed in liver membranes (Birnbaumer *et al.*, 1971). In addition, the response to glucagon is selectively increased 100% by EDTA, requires much higher concentrations of inorganic pyrophosphate to be inhibited and is sensitive to inhibition by 10 mM $MnCl_2$. In spite of all these differences, F^- activation of adenylate cyclase does not seem to proceed independently of hormonal activation as shown more recently by Harwood and Rodbell (1973).

The interaction between F^- and hormonal stimulation has also been explored recently (Vaughan, 1976). Working at 30° C with a particulate fat-cell adenylate cyclase preparation these workers reported a number of findings. Enzymic activity in the presence of isoprenaline is greater than that obtained in the presence of a maximally activating F^- concentration (5.3 mM), and a combination of both agents results — in agreement with the report of Harwood and Rodbell (1973) — in activities that equal those of the 'less' active anion alone. The stimulatory activity of 5.3 mM F^- can be totally inhibited by 1.5 mM pyrophosphate, but inhibition of isoprenaline stimulation by F^- cannot, indicating that pyrophosphate inhibits stimulatory but not inhibitory effects of the anion. F^--inhibited isoprenaline-stimulated activity, obtained in the presence of pyrophosphate, can be blocked by propranolol indicating that these activities are due to isoprenaline and not to a reversal of pyrophosphate inhibition. Also inhibition of isoprenaline-stimulated activity by F^- is only partial. Direct incubation of fat-cell particles with about 0.2 mM F^- results in no significant activation of adenylate cyclase. However preincubation with 5.3 mM F^- followed by dilution to 0.2 mM during the determination of adenylate cyclase results, as previously reported by Perkins and Moore (1971), in activities which are greater than those seen by direct incubation of particles with 5.3 mM F^-. Addition of 5.3 mM F^- during enzyme assay to the particles activated by preincubation results in inhibition of the activity. Levels produced are similar to those seen by direct assay with 5.3 mM F^-. These findings indicate 'normal' F^--stimulated activity is the result of two distinct actions: (*i*) stimulation of adenylate cyclase (sensitive to pyrophosphate and possibly independent of hormonal stimulation), and (*ii*) inhibition of adenylate cyclase, regardless of the stimulant used — hormone, F^-, cholera toxin (Manganiello and Vaughan, 1976a,b) — thus interfering with hormonal stimulation. Manganiello and Vaughan also showed that the inhibitory action of F^- is rapid in onset and inhibited by Mn^{2+}, and that the otherwise irreversible activation by the anion can be reversed by pyrophosphate. Stimulatory and inhibitory effects of F^- are illustrated in Figure 2.1.

For several years, it was thought that F^- stimulates adenylate cyclase systems maximally. However, by improving the incubation media used to test for

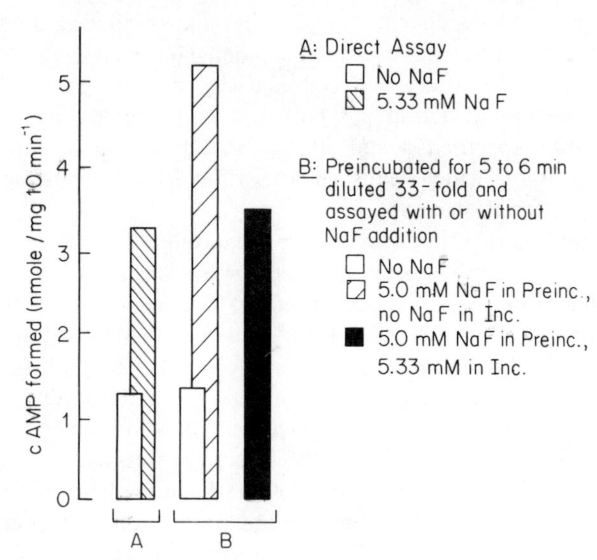

Figure 2.1 Stimulatory and inhibitory effects of F^- adapted from two experiments by Manganiello and Vaughan (personal communication)

hormonal responses, particularly by addition of critical nucleotides, it has frequently been possible to achieve hormone-stimulated activities which are higher than those obtained in the presence of the anion (Birnbaumer *et al.*, 1971). Furthermore, Rodbell and collaborators (Rodbell *et al.*, 1974; Rodbell *et al.*, 1975; Londos *et al.*, 1974; Rodbell, 1975), as well as other laboratories (Spiegel and Aurbach, 1974; Lefkowitz, 1975), have shown that 5′-guanylyl imidodiphosphate (GMP-P(NH)P), the imidodiphosphate analogue of GTP, is capable under appropriate conditions of increasing adenylate cyclase activity to levels that are far in excess of those achieved using F^-. This clearly demonstrates that F^- activity should not be considered as an expression of the catalytic capacity of the enzyme as previously assumed.

The mechanism by which F^- interacts with the adenylate cyclase remains to be established. Constantopoulos and Najjar (1973) have suggested that enzyme activation occurs via dephosphorylation of a phospho-enzyme, and earlier Schramm and Naim (1970) suggested that stimulation might be accompanied by dissociation of an inhibitory subunit from the holo-enzyme resulting in the 'release' of an active catalytic subunit.

E Stimulation by Hormones

1 General

Many hormones activate adenylate cyclase in a variety of tissues (Robison *et al.*, 1971). By doing so they increase intracellular cAMP levels which in turn alter cellular metabolism in such a way that the known hormonal response ensues. This seemingly nonspecific effect is made specific by the existence of hormone receptors which discriminate between the circulating hormones and assure by their presence and coupling to adenylate cyclase that only those hormones intended to affect the cell actually do so. Thus, the key to understanding hormonal action lies both in the discriminating process that assures hormonal specificity, and in the mechanism and nature of the signal–effect coupling which leads to activation of adenylate cyclase and, possibly, of other functions.

The discriminating function of receptors is exerted at the outer surface of the plasma membrane. This has been suggested by the results of experiments by Sato and collaborators (Schimmer *et al.*, 1968), who showed that ACTH covalently linked to cellulose is capable of stimulating cAMP-mediated steroidogenesis in intact cultured adrenal cells. Also, Rodbell *et al.* (1970) showed that trypsin treatment of intact adipocytes results in reduction in the ability of its adenylate cyclase system to respond to glucagon (100% loss), secretin (70%) and ACTH (30%) without affecting the ability either to express its basal activity or to respond to F^-. The trypsin sensitivity of the hormonal responses not only indicates localization of receptors, but also that hormone receptors are probably proteins.

Receptors are extremely specific for the topochemical characteristics of the hormones as shown in fat cells by the presence of five distinct types of receptors, all coupled to a single adenylate cyclase (Birnbaumer and Rodbell, 1969). Moreover, the receptors discriminate between specific local characteristics of the hormone molecule, as indicated by stereospecificity for (–)-isomers of catecholamines and β-adrenergic blockers (Kaumann and Birnbaumer, 1974), and by recognition of subtle differences, such as elimination or substitution of a single amino acid residue, in the structure of peptide hormones (Rodbell *et al.*, 1971b). The finding that elimination of the first amino acid of glucagon (histidine) results in reduction of affinity for the liver glucagon receptor by as much as fifteen-fold (Rodbell *et al.*, 1971b; Lin *et al.*, 1975a) indicates that selected areas of the peptide are responsible for binding characteristics of the peptide–receptor interaction.

2 Some Kinetic Properties of Hormone–Receptor Interactions

The interaction of hormones with their receptors which results in activation of adenylate cyclase is totally and rapidly reversible. This was originally shown in the glucagon-stimulated adenylate cyclase system of rat liver plasma membranes (Birnbaumer *et al.*, 1972). Since the activation of adenylate cyclases by hormones is rapidly reversible this implies not only that the hormone–receptor interaction is reversible, but also that the activated state of the enzyme has a half life that is as short as or shorter than that of the hormone–receptor complex. In both the hepatic glucagon-sensitive system and the myocardial catecholamine-sensitive system, de-activation of the active state by addition of competitive blockers does not result in permanent inactivation, but is overcome by the addition of excess native hormone. This lends strength to the assumption that all of the membrane components involved in hormonal stimulation of adenylate cyclase exist in several mutually interconvertible states of activity. The interconversion of states is probably under the influence of the hormonal receptor.

Based on the results of Birnbaumer *et al.* (1972) , as well as on the fact that in all systems but one the concentration–effect curves for hormonal action on adenylate cyclase activity appear to follow Michaelis–Menten kinetics with Hill coefficients close to one, it is assumed that hormone–receptor interactions follow the law of mass action. Further assumptions are that the reaction is totally reversible and that activation of the enzyme is in some way proportional to the formation of receptor–hormone complex. The one exception refers to the fact that stimulation of porcine renal adenylate cyclase by lysine-vasopressin has been found by Bockaert *et al.* (1973) to occur over several orders of magnitude with a Hill coefficient of approximately 0.4. Although this does not invalidate these basic assumptions, it does point towards the need to investigate the existence of co-operative effects in the binding of hormone to receptors — that is whether there are significant receptor–receptor interactions — and non-linear relationships between binding and enzyme activation. Evidence for the existence of both of these phenomena has been obtained. Co-operative effects between hormone-specific binding sites have been documented by Birnbaumer and Pohl (1973) who studied the binding of glucagon and deshistidine-glucagon to liver membranes and by DeMeyts *et al.* (1973) who studied binding of insulin to liver membranes, as well as by Limbird *et al.* (1975) who examined the binding of a β-adrenergic blocker to frog erythrocyte membranes. On the other hand, a non-linear relationship between binding and activation, has been suggested by Bockaert *et al.* (1973) who not only found a low Hill coefficient for activation of renal adenylate cyclase by lysine-vasopressin, but also showed that the binding of [^{3}H] lysine-vasopressin to porcine renal membranes proceeds with a Hill coefficient close to 1.0. Provided that binding of labelled material to membranes, demonstrated to be displaceable and specific for lysine-vasopressin, is representative of receptor binding, then this suggests the existence of non-linear coupling between receptor occupation and adenylate cyclase activation.

III COUPLING

A Definitions and Theoretical Considerations

Coupling is the unknown factor which intervenes between hormone–receptor interaction and adenylate cyclase activation. Elucidation of coupling is, therefore, synonymous with elucidation of the mechanism by which hormones stimulate adenylate cyclase. Since no secure information is available about the quaternary structure of the system, discussion of coupling and the mechanisms involved are by necessity speculative.

If one assumes that a hormone-sensitive adenylate cyclase system is a one receptor–one catalytic unit complex, that is it is a monomeric system composed of two subunits, then coupling must be linear, for the system is either under the

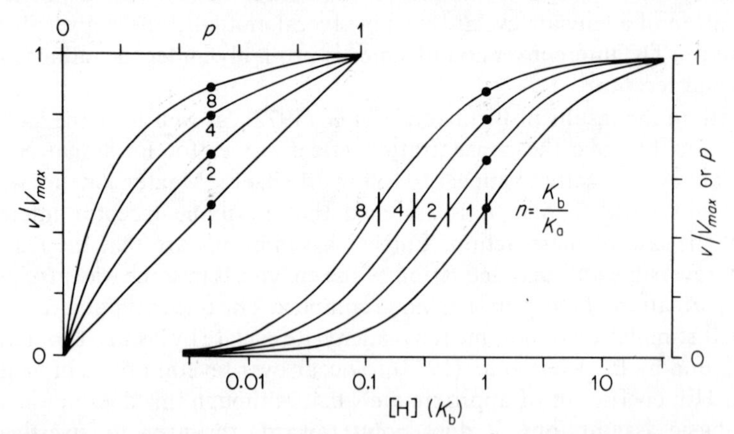

Figure 2.2 Examples of non-linear coupling. At a ratio of K_b (binding affinity) to K_a (activitation affinity) of 1 dose–response relationships coincide ($n = 1$) and coupling—relation between fractional occupancy (p) and fractional activation ($v/V_{\max}$) is linear. Hormonal concentrations (right panel) are expressed for convenience in K_b units. Both binding and activation are assumed to follow Michaelian behaviour without co-operative effects, thus, the slopes of Hill plots for binding and activation are equal to unity

influence of hormone or it is not, regardless of the mechanism by which activation occurs. However, if the system is polymeric such that several receptors are associated with each catalytic unit, then the coupling of the hormone–receptor interaction can either be linear or non-linear. Linear and non-linear coupling functions stemming from the comparison between hormonal binding to receptor and activation of enzyme activity by the hormone are shown in Figure 2.2. The example chosen depicts positive non-linear coupling — situations in which half-maximal activation is obtained with less than 50% receptor occupancy — but it is easy to see that negative coupling might also exist. In fact, most of the so-called homotropic positive co-operative enzyme systems, exhibiting substrate-concentration curves which are sigmoid as opposed to hyperbolic, are systems where coupling between substrate binding and enzyme activity is of the negative type. Suggestions to date have been that signal–effect coupling in adenylate cyclase systems can be of either type. From the above discussion, it becomes obvious that under ideal conditions, studies of coupling have to deal with both hormonal binding to receptor and hormonal activation of adenylate cyclase. Two important variables, which need to be explored in order to describe coupling, are the dependence of activation on hormonal concentration (steady-state analysis) and dependence of activation on time at submaximally stimulating hormonal concentrations (pre-steady-state analysis).

B Glucagon Binding and Action in Rat Hepatic Membranes

Activation of adenylate cyclase in hepatic plasma membranes by half maximally stimulating concentrations of glucagon is rapid. Under assay conditions which include a high concentration of ATP (3.2 mM), 1 mM EDTA and 5 mM $MgCl_2$, a steady state of activation is established within ten seconds of adding glucagon. The same result has been obtained with [^{125}I]glucagon (Birnbaumer and Pohl, 1973). De-activation of the enzyme, induced by addition of a competitive inhibitor or by dilution of the incubation medium (Birnbaumer *et al.*, 1972), has also been found to be rapid. Using unlabelled hormone or [^{125}I]glucagon, de-activation of half-maximally stimulated enzyme, by addition of deshistidine-glucagon consistently occurs within two minutes. Quantitative studies on the variation of the activation rate with time and hormonal concentration have not been carried out because activation is too fast to allow the determination of the rates under the assay conditions used in these experiments. Binding of labelled glucagon in this system has been found to be surprisingly slow both in terms of association with and dissociation from receptor-containing membranes. Thus, approximately fifteen minutes are needed to attain equilibrium of binding at hormonal concentrations which stimulate adenylate cyclase half-maximally within ten seconds, and not more than 15% and 50% of the bound glucagon is dissociated two and twenty minutes respectively, after addition of excess deshistidine-glucagon even though, de-activation of the enzyme occurs within two minutes. Rapid de-activation by deshistidine-glucagon does not appear to be due to a 'toxic' action, which would be effective after occupation of only a small fraction of the receptors, as indicated by the following findings: (*i*) blockade by deshistidine-glucagon is rapidly reversed by excess native glucagon; and (*ii*) the rates of dissociation of deshistidine-glucagon from specific binding sites are of the same order of magnitude as those found for glucagon (Rubalcava and Rodbell, 1973). This latter finding also rules out the possibility that the blockade is reversed due to significantly higher rates of dissociation for the analogue as compared to those for native glucagon.

Glucagon binding to hepatic membranes exhibits the characteristics that it is specific for glucagon and is displaced only by the competitive blocker deshistidine-glucagon (Rodbell *et al.*, 1971c) and that it occurs over a concentration range which is only two or three-fold higher than that over which activation of adenylate cyclase occurs. Also, it requires the presence of micromolar concentrations of GTP or millimolar concentrations of ATP for the dissociation to take place (Rodbell *et al.*, 1971d). Both nucleotides have subsequently been found to play important regulatory roles in hormonal stimulation not only of the glucagon-sensitive system in hepatic plasma membranes but also in many other adenylate cyclase systems (Rodbell *et al.*, 1971a; Birnbaumer *et al.*, 1972; Birnbaumer, 1973; Londos *et al.*, 1974). Bound glucagon has been found to be biologically active (Rodbell *et al.*, 1971c). Although these binding properties are consistent with the idea that glucagon binding represents receptor binding, it is clear from the studies of the rates of activation and de-activation of adenylate cyclase that only a small fraction — not more than 10–20% — of the glucagon-specific binding sites in hepatic membranes is related to adenylate cyclase activation. The role of the bulk of these binding sites is not known. It is possible that these sites are not related to receptor function — that is adenylate cyclase activation — thus providing no information about the binding properties of hormone receptors, since non-receptor binding masks receptor binding. Alternatively, theoretical considerations, prohibiting the existence of an effector system reacting faster in *both* directions than the binding reaction driving it, are incorrect.

Elucidation of the mechanism of activation of liver adenylate cyclase by glucagon and determination of the type of coupling which intervenes between receptor and adenylate cyclase will require further studies on both binding and activation. Data currently available indicate, if anything, that little of the binding observed in the absence of nucleotides is related to the cyclase. It will be interesting to see whether reduced binding, such as seen in the presence of nucleotides, — see Rodbell *et al.* (1974) — can be correlated with adenylate cyclase activation.

C Binding and Adenylate Cyclase Activation in Vasopressin-Sensitive Systems

Bockaert *et al.* (1973) have presented kinetic evidence which suggests that hormonal binding and hormonal activation of adenylate cyclase may correlate on a one-to-one basis under a given set of circumstances. They demonstrated that the time course of adenylate cyclase activation in porcine renal membranes by lysine-vasopressin (1×10^{-8} M) is identical to the time course of binding obtained under the same conditions. Both processes take about four minutes to achieve a steady state, and follow an exponential function in doing so, as predicted for a system with either linear or positive non-linear coupling. Since dose–response relationships for binding have been found to follow Michaelis–Menten kinetics and those for adenylate cyclase activation show negative co-operativity (slope of Hill plots is less than unity) and are shifted to the left of the binding curve, it seems that coupling in this system may be non-linear of the positive type. However, an assessment independent of binding has not been made, and adenylate cyclase activation by 2×10^{-9} M lysine-vasopressin also takes about four minutes to reach a steady state, in spite of the fact that the reduction of hormonal concentration would have predicted a two to four-fold increase in the time needed to achieve a steady state of activation. Since this time does not vary with even a five-fold variation of hormone concentration, it must be concluded that receptor-dependent activation follows complex kinetics and that 'correlation' seen at 1×10^{-8} M lysine-vasopressin cannot be interpreted until the activation process is explored further.

D Catecholamine-sensitive systems

Although much work has been carried out on catecholamine-sensitive adenylate cyclase systems (Robison *et al.*, 1971) not until recently has exploration of coupling been attempted in these systems. The reasons for this are: (*i*) the concentrations of catecholamines required to activate adenylate cyclase systems are of the order of 10^{-7} M and no 'slow' time courses of enzyme activation have been detected at these concentrations; (*ii*) membranes which contain catecholamine-sensitive adenylate cyclases and, therefore, catecholamine-specific receptor sites, also contain a multitude of non-receptor sites which make it impossible to explore directly the characteristics of the receptor–agonist interaction. Excess non-receptor binding sites (identified as sites which bind catecholamines in an almost irreversible manner without discriminating for stereo-isomers and exhibiting low affinity for highly active β-receptor blockers) were found in a variety of tissues, including liver, heart, fat, glial cells, avian erythrocytes, skeletal muscle and cultured myocardial cells (Tomasi *et al.*, 1970; Dunnick and Marinetti, 1971; Lefkowitz and Haber, 1971; Lefkowitz *et al.*, 1972; Lefkowitz *et al.*, 1973a; Jarett *et al.*, 1974; Koretz and Marinetti, 1974; Lesko and Marinetti, 1975; Maguire *et al.*, 1974; Bilezikian and Aurbach, 1973; Schramm *et al.*, 1972; Lefkowitz *et al.*, 1973b; Lefkowtiz *et al.*, 1974a; Aprille *et al.*, 1974). The nature of these non-receptor sites is not clear. Cuatrecasas *et al.* (1974) proposed that these sites are related to membrane-bound catechol-*O*-methyl transferase, and Maguire *et al.* (1974) and Wolfe *et al.* (1974) proposed that they are related to oxidative metabolism of catecholamines. It seems likely, as pointed out by Lefkowitz *et al.* (1976), that these sites are heterogeneous and include not only sites which metabolize catecholamines but also sites concerned with uptake, receptors other than β-adrenergic receptors and still other sites whose role in the disposition or action of catecholamines is at present, unknown. Binding studies carried out in the

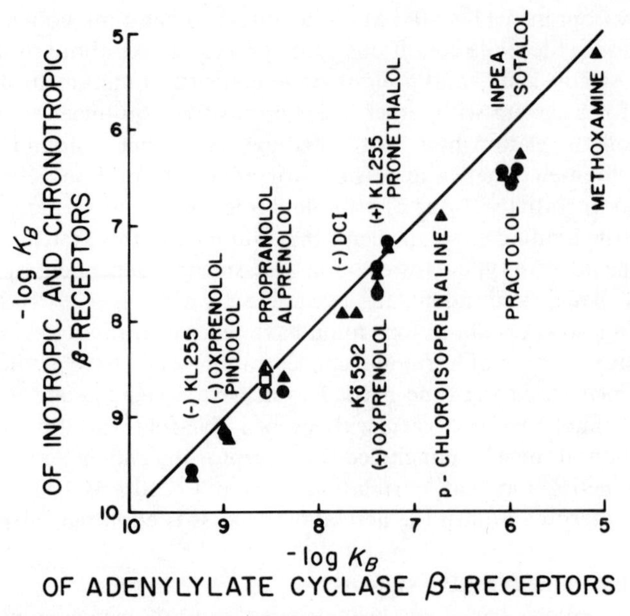

Figure 2.3 Correlation between affinity constants (K_b) of β-receptors for blockers as determined by their effects on inotropism and chronotropism and the same constants as determined by their effects antagonizing isoprenaline stimulation of adenylate cyclase in isolated myocardial membrane particles. Affinities for inotropic and chronotropic β-receptors are from Kaumann and Birnbaumer (1973); affinities for adenylate cyclase β-receptors are from Kaumann and Birnbaumer (1974)

presence of substances which either obliterate or compete with non-receptor binding, such as EDTA (Lacombe and Hanoune, 1974) or pyrocatechol (Pairault and Laudat, 1974), may allow identification of the characteristics of receptor–catecholamine interactions under adenylate cyclase assay conditions; such studies are still in progress.

In contrast to peptide and protein hormones, for which there are few competitive inhibitors, there is a large selection of catecholamine analogues available which interact competitively with β-adrenergic receptors, not only in intact cell systems, but also in subcellular particles containing β-receptors coupled to adenylate cyclase. Binding affinities of β-receptors for antagonists are similar in intact cell systems and membrane particles (Kaumann and Birnbaumer, 1974; Figure 2.3). This property was recognized by several laboratories (Atlas *et al.*, 1974; Aurbach *et al.*, 1974; Lefkowitz *et al.*, 1974b) and three probes, all β-receptor blockers, have been proposed and used to identify and quantify β-receptors in subcellular fractions. Thus, binding sites with characteristics of β-receptors were detected in erythrocyte membranes of turkeys and frogs using either [^{3}H]propanolol, [^{3}H]dihydro-alprenolol (DHA) or a radioiodinated hydroxybenzyl derivative of pindolol (IHYP). Detection of specific sites with labelled propranolol is associated with high 'backgrounds' — non-specific binding as well as non-receptor binding (Atlas *et al.*, 1974) — and is not always successful (Potter, 1967; Vatner and Lefkowitz, 1973) possibly because [^{3}H]propranolol is available only at relatively low specific radioactivities. Background values appear to be low with the more highly labelled DHA (20 Ci/mmole) and even lower with [^{125}I] IHYP (approximately 2000 Ci/mmole). Neither [^{3}H]DHA (Lefkowitz *et al.*, 1974b) nor [^{125}I]IHYP (Aurbach *et al.*, 1974) have yet been rigorously tested in mammalian systems, but it is predicted that these labelled antagonists will be as useful in detection of mammalian β-receptors as in avian and amphibian systems.

Probes of this type will also be useful in determining coupling of β-receptors to adenylate cyclase. Thus, the combined measurement of receptor binding by antagonist as a function of time in the presence of varying concentrations of agonist and measurement of receptor activity as seen through adenylate cyclase activity under the same conditions, will provide time courses relating receptor occupancy to activity. Indeed, Brown *et al.* (1976) have already determined the steady-state relationship between activation of turkey erythrocyte adenylate cyclase by agonist in the presence of 10^{-5} M 5′-guanylylimido-diphosphate (GMP-P(NH)P) and inhibition of [^{125}I]IHYP binding, and have found it to be linear.

Working with a mammalian system, it has recently been shown that dichloroisoprenaline is a partial agonist in kitten atria stimulating inotropism with an apparent affinity constant of $10^{-7.9}$, but it exhibits little if any agonistic effects on kitten myocardial adenylate cyclase (Kaumann and Birnbaumer, 1973; Kaumann and Birnbaumer, 1974). In this case the isoprenaline derivative behaves as a competitive inhibitor with an apparent affinity constant of $10^{-7.8}$ — not significantly different from the value of $10^{-7.9}$ found in intact organ studies. These findings suggest that most of the non-linearity seen in intact cell systems is likely to be due to effects occurring after adenylate cyclase stimulation; this must be confirmed by determination of receptor–cyclase coupling. It is clear that studies which elucidate coupling by using the techniques developed by Aurbach *et al.* (1974) and by Lefkowitz *et al.* (1974b) will soon be forthcoming.

E Regulation of the Coupling Process by Nucleotides and Nucleosides

1 General

In recent years, it has become apparent that the hormonal activation of adenylate cyclase systems is regulated by nucleotides and nucleosides. This is based on the observation that using as substrate either low ATP or AMP-P(NH)P the hepatic plasma-membrane adenylate cyclase exhibits an almost total requirement for GTP (or ATP) to show activation by glucagon (Rodbell *et al.*, 1971a). Similar findings have also been reported for other adenylate cyclase sytems activated by different hormones and has led to the conclusion that a nucleotide-dependent step is involved in hormonal activation (Rodbell *et al.*, 1971e).

The type of modulation effected by nucleotides varies with the system studied. Both positive and negative effects on hormonal stimulation have been observed with either nucleoside triphosphates or diphosphates. Moreover, in several systems, these nucleoside tri- and diphosphates can be shown to have opposite effects. Also, it has been found that the responsiveness of a bovine renal adenylate cyclase to vasopressin is enhanced by GDP and inhibited by GTP while that of a feline heart adenylate cyclase to catecholamines is enhanced by GTP and inhibited by GDP (Birnbaumer and Yang, 1974). Similar opposing effects have been found in the glucagon-responsive adenylate cyclase system of rat hepatic membranes (Salomon *et al.*, 1975). Not all adenylate cyclase systems are modulated bidirectionally and not all are regulated by guanyl nucleotides. Thus, a prostaglandin-responsive adenylate cyclase in bovine renal membranes is modulated positively by GTP, but appears to be unresponsive to GDP (Birnbaumer and Yang, 1974) and a prostaglandin-responsive adenylate cyclase in corpus luteum of rabbit is modulated positively by GDP, being unresponsive to GTP. Not only nucleotides, but also nucleosides appear to regulate adenylate cyclase. This was demonstrated in the vasopressin-sensitive system of bovine renal medulla, where adenosine exerts an important effect on hormonal stimulation. A summary of some of the effects of nucleotides on hormonal stimulation is presented in Table 2.1.

Nucleotides and nucleosides appear to exert their action by affecting not only the degree of stimulation by the hormone but also other basal activity and, possibly, the receptor–hormone interaction. Experimental evidence which is pertinent to various current views of the mechanism by which nucleotides affect adenylate cyclase systems and regulate hormonal stimulation will be reviewed.

2 *A Model for Guanyl Nucleotide Action*

a Principal Findings. Rodbell and collaborators have extensively explored the action of GTP and its analogue GMP-P(NH)P on the hepatic glucagon-sensitive adenylate cyclase system. Interpretation of their results is complex and necessitates a description of some details of assay conditions under which data were gathered. Initial studies were carried out in incubation media containing EDTA. When EDTA, ATP (3.2 mM) and $MgCl_2$ (5.0 mM) are present, activation of the enzyme by submaximally stimulating concentrations of glucagon ($2-4 \times 10^{-9}$ M) is too fast to allow for detection of a significant lag period between hormonal addition and reaching the steady state of activation.

Later studies (Rodbell *et al.*, 1974) were carried out in the absence of EDTA presumably resulting in the absence of added nucleotides in 'irreversible' binding of glucagon and in reduced maximal stimulation of the enzyme by saturating concentrations of glucagon. Under these conditions and using either a low concentration of ATP (0.1 mM) or AMP-P(NH)P (0.1 mM), the system is

Table 2.1 Examples of positive and negative modulation by nucleotides of hormonal stimulation of adenylate cyclase systems

Nucleotide	Hormone	System	Reference
Enhancement			
GTP	glucagon	rat liver	1
	glucagon	hamster β-cell	2
	glucagon	rat adipose tissue	3
	prostaglandins	human platelets	4
	prostaglandins	bovine thyroid gland	5
	prostaglandins	bovine renal medulla	6
	catecholamines	rat liver	7
	catecholamines	rat adipose tissue	3
	catecholamines	kitten myocardium	8
	ACTH	rat adipose tissue	3
	ACTH	rat adrenal cortex	9
	oxytocin	frog bladder epithelium	10
ITP	TSH	bovine thyroid	5
	TSH	horse thyroid	11
ATP	LH	corpus luteum of pregnant rabbit	6
	LH	pig graafian follicles	12
	arg-vasopressin	bovine renal medulla	13
	lys-vasopressin	bovine renal medulla	13
	oxytocin	bovine renal medulla	13
GDP	prostaglandins	corpus luteum of pregnant rabbit	6
	arg-vasopressin	bovine renal medulla	13
	lys-vasopressin	bovine renal medulla	13
	oxytocin	bovine renal medulla	13
Inhibition			
GTP	arg-vasopressin	bovine renal medulla	13
	lys-vasopressin	bovine renal medulla	13
	oxytocin	bovine renal medulla	13
GDP	catecholamines	kitten myocardium	6
	catecholamines	rat adipose tissue	3
	glucagon	rat liver	14
	ACTH	rat adrenal cortex	9
ADP	LH	rabbit corpus luteum	6

References: 1, Rodbell *et al.* (1971a); 2, Goldfine *et al.* (1972); 3, Harwood *et al.* (1973); 4, Krishna *et al.* (1972); 5, Wolff and Cook (1973); 6, Birnbaumer and Yang (1974); 7, Leray *et al.* (1972); 8, Kaumann and Birnbaumer (1974); 9, Londos and Rodbell (1975); 10, Bockaert *et al.* (1972); 11, Pochet *et al.* (1974); 12, Birnbaumer *et al.* (1976); 13, Birnbaumer *et al.* (1974); 14, Salomon *et al.* (1975).

no longer activated rapidly by submaximal concentrations of glucagon. In contrast, like the vasopressin-sensitive systems (Bockaert *et al.*, 1973; Nakahara and Birnbaumer, 1974), there are new significant lag periods between hormone addition and attainment of activation. These periods become shorter with increasing hormonal concentrations, but can not be reduced below about ten to twenty seconds. Addition of GTP or ATP under these conditions elicits three effects: (*i*) it accelerates activation of the enzyme by submaximal concentrations of glucagon; (*ii*) it eliminates the lag seen at saturating concentrations of glucagon; and (*iii*) while *not* affecting stimulation by high concentrations of glucagon — contrary to what would be observed in the presence of EDTA (see Rodbell *et al.*, 1971a), — it reduces the concentration of glucagon required for half-maximal stimulation five to eight-fold.

In the absence of EDTA and at low ATP or AMP–P(NH)P concentrations, the glucagon-sensitive system of rat liver resembles the vasopressin-sensitive bovine renal system described in detail in Chapter 14. In both cases the enzyme seems to react slowly to ligand binding as indicated by the presence of a lag between addition of high concentrations of glucagon and the onset of activation and it exhibits transient states of activity. Also, GTP can accelerate isomerization of one (initial) state into the other (final) state in both systems. These findings do not indicate whether the increased rates of activation by low concentrations of glucagon seen in the presence of GTP are due to increased rates of binding or whether they are due to an alteration of the coupling process. In addition, these experiments do not indicate if the increased apparent affinity of the system for glucagon seen in the presence of GTP is due to increased binding affinity, that is increased receptor occupancy.

In more recent studies, Rodbell and his colleagues (Salomon *et al.*, 1975; Lin *et al.*, 1975b; Rendell *et al.*, 1975) have described the effects of the GTP analogue GMP-P(NH)P on transient kinetics of the liver system. Assay conditions included 1 mM dithiotreitrol (DTT) and did not contain EDTA. As in the case of glucagon, GMP-P(NH)P activates basal adenylate cyclase activity slowly, with a lag period which decreases to a limiting value with increasing concentrations of the analogue. Furthermore, just as GTP increases the rate of activation by glucagon, glucagon increases the rate of activation by GMP–P(NH)P and, at sufficiently high concentrations, eliminates the lag period. Thus, assuming that inclusion of DTT does not introduce significant changes in behavior, the system appears to behave symmetrically with respect to glucagon and GMP–P(NH)P activation. This indicates (Rodbell *et al.*, 1974; Salomon *et al.*, 1975) that both ligands may be necessary for optimal activation acting either in an interdependent or in a concerted manner. The kinetic effect of the analogue differs however from that of GTP in two main aspects. While the analogue causes about a three-fold stimulation of basal activity, GTP only stimulates by between 30 and 50%. Also while the analogue's activation of the system proceeds after a discrete lag, activation by GTP does not, in the presence of DTT, show any such lag. The finding that the analogue activates adenylate cyclase in the absence of hormone has been interpreted by Rodbell and colleagues (1974) as indicating that occupation of the glucagon receptor is not a necessary step in activation of adenylate cyclase. This led Rodbell *et al.* (1975) to consolidate the above findings in a 'Three-State Model' for the steady-state kinetics of catalysis and regulation of adenylate cyclase by nucleotides.

In this model the enzyme can exist unaffected by nucleotides (basal or E state) or affected by nucleotides but under 'normal' assay conditions having low activity (transient or E′ state) or in equilibrium with the second and highly active under normal assay conditions (final or E″ state). According to this model, activation by GMP–P(NH)P proceeds with a lag due to the slow isomerization of E′ to E″. Also, hormones, such as glucagon in the case of the liver adenylate cyclase system, stimulate enzymic activity by accelerating the isomerization process between E′ and E″, and by shifting the equilibrium between these two states towards the highly active nucleotide-dominated E″ form. The model accounts for all of the basic features, such as the existence of lag in activation by high concentrations of GMP–P(NH)P, obliteration of this lag by addition of hormone, increased steady-state activity in the presence of nucleotide with the hormone compared with nucleotide alone, and, by assuming that isomerization in the presence of GTP is rapid, why no lag is seen when GTP or ATP is used.

b Three-State Model and Regulation of Adenylate Cyclase Activities by Magnesium. By studying enzymic activities in the absence and presence of GMP–P(NH)P, under varying pH and $MgCl_2$ concentrations, evidence was also presented that (*i*) the active substrate for the catalytic site is the Mg–ATP complex; (*ii*) the V_{max} of E is low and those of both E′ and E″ are high; and (*iii*) the protonated form of the substrate, $ATPH^{3-}$ is a competitive inhibitor with relatively low affinity for the basal state (E) of the enzyme, very high affinity for the transient state (E′) (thus accounting for the low activity of this state in spite of high enzymic capacity) and very low affinity for the final state (E″). Furthermore, based on the finding that the concentration of GMP–P(NH)P needed to activate the system half maximally is increased by increasing concentrations of $MgCl_2$, Rodbell *et al.* (1974) concluded that the active guanyl nucleotides interact with the system in the free rather than the Mg^{2}-chelated form.

3 Effects of Nucleotides on Catecholamine-Stimulated Adenylate Cyclase

All catecholamine-sensitive adenylate cyclase systems have been found to enhance their responsiveness to these amines when GTP is included in the incubation medium. This was first shown by Hanoune *et al.* (1975) in hepatic membranes, where addition of GTP results in responses which are comparable to those of glucagon (Leray *et al.*, 1973). Indeed, addition of GTP restores, in membranes, patterns of responsiveness to adrenaline and glucagon predicted from intact organ studies. Using cat cardiac adenylate cyclase with AMP–P(NH)P as substrate (thus avoiding addition of a nucleoside triphosphate-regenerating system), it has been found that GTP enhances and GDP inhibits the enzyme.

GMP–P(NH)P also affects the catecholamine-sensitive systems, altering not only the degree of responsiveness in human fat cells as reported by Cooper *et al.* (1975) but also the apparent sensitivity to stimulatory agents in turkey erythrocytes (Spiegel and Aurbach, 1974; Brown *et al.* 1976). In the latter system, it was demonstrated that sensitivity of the adenylate cyclase system to catecholamines in the cell-free membrane is ten times lower than sensitivity of the intact cell when tested in the absence of nucleotide additive, but is 'restored' to normal by the addition of GMP–P(NH)P *in vitro*.

In view of the fact that many of the findings relating to possible mechanisms of GMP-P(NH)P were made with catecholamine-sensitive systems, other features of catecholamine-stimulation in relation to the mechanism of action of this nucleotide will be described.

IV EFFECTS OF 5′-GUANYLYL IMIDODIPHOSPHATE

In a review of adenylate cyclase systems, including those of the thyroid, adrenal and fat cells, and frog erythrocytes, Londos *et al.* (1974) described the activation by GMP–P(NH)P of basal activity of all of these systems, and suggested that this synthetic nucleotide may stimulate all adenylate cyclase systems of eukaryotic cells. Subsequently, many other systems have been found to be activated by GMP-P(NH)P in the absence as well as the presence of hormones. Included are the catecholamine-sensitive systems of dog (Lefkowitz, 1974) and cat heart, turkey erythrocytes (Spiegel and Aurbach, 1974), pigeon erythrocytes (Pfeuffer and Helmreich, 1975), pig graafian follicles (Bockaert *et al.*, 1976), rabbit corpus luteum and a variety of cell lines.

Activation of adenylate cyclase by GMP-P(NH)P has two important characteristics which differ from those seen with GTP. One is that GMP-P(NH)P affects enzyme activity slowly in the absence of hormone and rapidly in its presence. In other words, hormones accelerate activation by GMP-P(NH)P. Since lag periods, which vary considerably in length depending on the system studied do not diminish with further increases of GMP-P(NH)P concentration (Salomon *et al.*, 1975), low initial activities are thought to be due to a form of the enzyme which has low activity and is in a relatively unstable conformation, isomerizing slowly to a more stable conformation with high activity. An alternative explanation would be that the lags seen after GMP-P(NH)P addition are due to hysteresis rather than to rapid formation of an unstable state.

Evidence supporting rapid formation of such a state was provided by Londos *et al.* (1974) and by Rodbell (1975), who found that in plasma membranes of rat fat cells, GMP-P(NH)P leads to an initial *decrease* of basal activity, followed in time by the customary marked increase in activity. Inhibition of basal activity was also observed by Cooper *et al.* (1975) in human fat cells membranes. Inhibition does not only occur using the synthetic nucleotide, as a transient inhibition of basal activity was also observed by Harwood *et al.* (1973) using GTP. The transient state of activity of the enzyme may exist for as long as ten or twenty minutes. It may, therefore, seem in some systems tested in ten-minute assays that GMP-P(NH)P is not stimulatory, but rather that it is without effect on basal activity or is inhibitory. In contrast to

GMP-P(NH)P, GTP appears to lead to states of activity which always have final catalytic capacities lower than those obtained with GMP-P(NH)P and which isomerize much more rapidly, so that lags in the presence of GTP are seldom observed.

As in the case of glucagon in the hepatic membrane adenylate cyclase, catecholamines increase the rate at which adenylate cyclase is stimulated by GMP-P(NH)P (Londos and Rodbell, 1975; Schramm and Rodbell, 1975). The effect of GMP-(NH)P on isoprenaline stimulation depends, however, upon the tissue studied, the assay conditions and the time necessary for basal adenylate cyclase to become activated by GMP-P(NH)P. The extent of isoprenaline stimulation in frog erythrocyte adenylate cyclase is reduced by the analogue (Lefkowitz and Carol, 1975; Schramm and Rodbell, 1975), whereas in turkey erythrocytes it is relatively unchanged although absolute activities are increased (Spiegel and Aurbach, 1974). In contrast, GMP-P(NH)P so dramatically sensitizes human fat cell adenylate cyclase to catecholamines that its presence is essential for the demonstration of catecholamine-stimulated activity (Cooper *et al.*, 1975).

Another, and perhaps more important, difference between the effects of GTP and GMP-P(NH)P is that the action of the analogue appears to be irreversible. Schramm and Rodbell (1975) reported that the combined interaction of GMP-P(NH)P and isoprenaline with the frog erythrocyte adenylate cyclase system results in formation of an active state, resistent to washing and propanolol treatment, and stable to solubilization by Lubrol–PX. Similar findings were also reported by Pfeuffer and Helmreich (1975) when studying the characteristics of the nucleotide activation of pigeon erythrocyte adenylate cyclase. The latter authors also studied a variety of other analogues for their effects on adenylate cyclase and found that GMP-P(CH_2)P and the thiotriphosphate analogue (GTPS) are also effective in stimulating adenylate cyclase both in the absence and presence of isoprenaline. Similarly, Hanoune *et al.* (1975) reported stimulatory effects of a variety of analogues on noradrenaline-stimulated liver adenylate cyclase, indicating that there does not seem to be a strict requirement for a fixed conformation of the triphosphate moiety of the guanyl nucleotide.

The finding that GMP-P(NH)P not only activates adenylate cyclases markedly but also stabilizes the activated states is of obvious significance in isolation experiments. Thus, experiments by Pfeuffer and Helmreich (1975), Lefkowitz and Carol (1975), and Schramm and Rodbell (1975) indicate that one way to purify active adenylate cyclase may be to pretreat membranes with a combination of stimulating hormone and GMP-P(NH)P, then to solubilize the system and finally to subject it to classical enzymic techniques. It might be hoped that re-introduction of purified material into phospholipid micelles (Racker, 1973), as has been done with purified acetylcholine receptors and purified Na^+/K^+-dependent ATPase (Hilden and Hokin, 1975), may restore some of the original conformation. Studies on the mechanism by which GMP-P(NH)P leads to activation, may give the answer to the subsequent question of how to reverse the process and restore hormonal sensitivity.

A fruitful approach to gain understanding of the mode of action of GMP-P(NH)P, as well as other guanyl nucleotides, is by studying their competitive interaction using adenylate cyclase as a read-out system. Using inhibition by GTP and GDP of GMP-P(NH)P-induced activation of adenylate cyclase, Lefkowitz (1975) found that although myocardial adenylate cyclase was activated to a much greater extent by GMP-P(NH)P than GTP, the affinity of the regulatory site is about ten-fold greater for GTP. GDP interacts with slightly less affinity than GMP-P(NH)P. The findings that GTP and GDP interact competitively with GMP-P(NH)P strongly suggest that all three nucleotides bind to a common regulatory site of the system, regardless of whether their final effect is activatory or inhibitory, or whether it is reversible or irreversible. These investigations are important not only because competitive interactions between GTP and GDP cannot be studied directly due to the presence of membrane nucleotidases, but also because they serve as a useful guide in isolation studies.

V ALTERNATIVE MODELS FOR ACTION OF GUANYL NUCLEOTIDES

Three models have been formally presented. One, contained in the Three-State Model of Rodbell *et al.* (1974b), has been discussed above. Pfeuffer and Helmreich (1975) and Cuatrecasas *et al.* (1975) proposed alternative possibilities.

Activation by GMP-P(NH)P may be associated with the dissociation of a regulatory subunit. Evidence for this was obtained by studying effects of various guanyl nucleotides on the catecholamine-sensitive adenylate cyclase system of pigeon erythrocyte membranes (Pfeuffer and Helmreich, 1975). It was found that chromatography of an isoprenaline and GMP-P(NH)P-stimulated, Lubrol–PX-solubilized enzyme on a Sephadex 4B column results in quantitative preservation of enzymic activity and in separation of a major GMP-P(NH)P binding peak from the enzyme. Moreover, treatment of solubilized, GMP-P(NH)P-sensitive adenylate cyclase with Sephadex to which GTPS had been covalently bound, results in loss of a large proportion of GMP-P(NH)P-binding sites and loss of the preparation's capacity to respond to GMP-P(NH)P without impairment of its response to F^-. While conclusive experiments are required to demonstrate restoration of GMP-P(NH)P sensitivity by re-addition of the separated GMP-P(NH)P-binding component, these studies suggest that guanyl nucleotides may activate by bringing about dissociation of an inhibitory subunit.

Cuatrecasas *et al.* (1975) reported on the activation of adipose-tissue adenylate cyclase by GTP and GMP-P(NH)P. They also found activation by GMP-P(NH)P to be persistent and proposed that activation by guanyl nucleotides is the result of the formation of a covalent enzyme–PP or enzyme–P(NH)P complex. Based on kinetic data they suggested that normal stimulation of the enzyme by hormones, in the presence of GTP, is the result of an increased rate of formation of a highly active and highly unstable enzyme–PP complex. In the presence of GMP-P(NH)P activation of the enzyme would be both slow and irreversible because of low rates of formation of active and highly stable enzyme-P(NH)P complex, and because it is impossible for this complex to decay releasing the enzyme and P(NH)P. In their model, hormonal stimulation of the enzyme in the presence of GMP-P(NH)P would be exclusively due to increased rates of formation of enzyme–P(NH)P complex and observed only early in the incubation.

The model proposed by Cuatrecasas *et al.* (1975) is attractive because like the Three-State-Model of Rodbell *et al.* (1975), it does account for experimental findings. Both models assume three basic states: a basal state, unaffected by nucleotides; a second state affected by nucleotides but not yet active (E′ in the model of Rodbell and reversible 'Michaelis' complex in the model of Cuatrecasas); and a third state, fully active (E″ or enzyme–PP complex). Also, it is interesting, that neither model invokes the formation of a complex between the receptor and the enzyme to account for hormonal stimulation. Instead it is assumed that the action of receptor is to modify the rate at which a final active state of the enzyme is formed. Both models have in common the concept that nucleotides play an intrinsic and obligatory role, and that it is by regulating the action of the nucleotide that hormones stimulate adenylate cyclase.

Neither of the models discussed above account for the transient kinetics observed in the vasopressin-sensitive adenylate cyclase of renal medullary membranes and the LH-sensitive system of porcine graafian follicle membranes. Clearly further investigation is necessary if a unified model is to be developed.

VI MODE OF ACTION OF HORMONES

None of the regulatory models already considered provides an explanation of how hormones exert their action in molecular terms. In other words, if occupation of hormonal receptor leads to an increased rate of E″ or enzyme–PP formation, how does it do this?

Cuatrecasas *et al.* (1975), in recognizing the complexity of plasma membranes and the fluid nature of the lipid matrix in which hormonal receptors and adenylate cyclases are embedded, proposed that hormonal activation is the result of a two-step process with the formation of an hormone–receptor complex, and then the search for and coupling to the catalytic unit(s) of adenylate cyclase. While attractive, there seems to be little, if any, experimental evidence for or against this model.

Constantopoulos and Najjar (1973) reported that treatment of dog platelet membranes previously activated with prostaglandin and then washed free of prostaglandin, with ATP plus protein kinase-containing cytoplasm results in a reduction in adenylate cyclase activity. The inhibition could be prevented by re-addition of PGE_1. Based on this observation, and on a similar finding in the case of F^- activation of rabbit polymorphonuclear granulocytic adenylate cyclase, Constantopoulos and Najjar (1973) proposed that activation of this enzyme is the result of a dephosphorylation reaction. They also suggested that inactive or basal adenylate cyclase activity is given by a phospho-form of the enzyme. Phosphorylation or dephosphorylation of membrane components has been proposed as a key element in hormonal action not only in the two systems mentioned, but also in the action of antidiuretic on toad bladder (DeLorenzo and Greengard, 1973), and of insulin in adipose tissue (Chang *et al.*, 1974). It is not clear, however, whether the phosphorylation reactions involved in these other systems are related to the primary action of the hormone, or related to the expression of final effects of these hormones. Direct evidence beyond that of Constantopoulos and Najjar (1973) for a general role of a phosphorylation–dephosphorylation cycle as an obligatory feature in hormonal stimulation of adenylate cyclases has not yet been provided.

Another model of adenylate cyclase activation suggests that cyclase systems are 'restrained'. The restraining elements would be subunits of the system, such as free hormonal receptors (Levey *et al.*, 1974; Klein *et al.*, 1973), a nucleotide-binding protein (Pfeuffer and Helmreich, 1975), a fluoride-binding component (Schramm and Naim, 1970) or the phospholipid matrix of the membrane itself. Schmidt *et al.* (1970) reported that the appearance of F^- response in the brain of the new-born rat is associated with a decrease in the 'basal' activity, suggesting that a constraining force exists. More recent studies on the appearance of responsiveness of adenylate cyclase to LH in developing rat ovaries, and in maturing rat and rabbit graafian follicles, (Hunzicker-Dunn and Birnbaumer, 1976) also showed that the appearance of hormonal responsiveness is associated with a decrease in the basal activity. Since appearance of hormonal responsiveness in some of these tissues also coincides with the appearance of hormone-specific binding, (Channing and Kammerman, 1973; Richards and Midgley, 1976) it may be argued that coupling of the hormonal receptor to adenylate cyclase results in restriction of catalytic activity. It is therefore possible to visualize stimulation by the hormone, at least in part, as the result of a release of inhibition via uncoupling of the receptor from the catalytic unit, similar to the mode of activation of protein kinase by cAMP (Brostrom *et al.*, 1970). However, it should be mentioned that simple 'release' of catalytic unit, thus restoring the situation which existed prior to the postulated coupling event, cannot account for all of the hormonal stimulation. Although the appearance of hormonal responsiveness is associated with marked reduction (50 – 70%) of basal activity, stimulation by hormone, or F^- is associated with an increase in absolute activity which exceeds by at least 100%, and frequently by as much as 200 and 300%, the activity seen prior to development of hormonal responsiveness. This indicates that if activation is due to a release of the catalytic unit, the conformation it acquires is different from the one it had before coupling had first occurred.

Levey *et al.* (1974) working with cat cardiac adenylate cyclase and studying activation by glucagon and effects of solubilization with Lubrol–PX, have provided some evidence in favour of dissociation being involved in hormonal activation. They reported that adenylate cyclase activity and glucagon-specific binding comigrate after chromatography of the solubilized system on Sephadex G–100, exhibiting an apparent molecular weight of between 100,000 and 200,000. However, the activities do not comigrate if the solubilized material is exposed to glucagon for one hour at 37°C prior to chromatography. Under these conditions, a complex of about 28,000 daltons is separated containing all of the bound glucagon and suggesting that the interaction of glucagon with the solubilized macromolecular complex results in dissociation of a much smaller glucagon-binding component. Levey *et al.* (1974) also reported a similar finding using materials solubilized from hepatic membranes, suggesting that this phenomenon may be a general one. Indeed, if it were possible to establish unequivocally that the glucagon-binding component separated by Sephadex chromatography is the receptor responsible for adenylate cyclase activation, then these findings would constitute good evidence in favour of activation being associated with subunit dissociation.

Finally, there still is the more 'classical' view of hormonal and nucleotide regulation of adenylate cyclase in which the system is composed of regulatory and catalytic subunits with allosteric receptor sites for hormones, regulatory nucleotides and, possibly, divalent cations and F^-. Based on a structure such as this, Robison *et al.* (1967) originally proposed activation by hormones to be the result of the interaction of hormone with the receptor subunit. Also, it is easy to account for the more recently discovered positive and negative modulation by nucleotides and for regulatory roles of a variety of ions by allosteric interactions. The key difference from the models of Rodbell *et al.* (1975) and Cuatrecasas *et al.* (1975) is that activity seen in the presence of nucleotide and hormone is the result of a tertiary complex and that nucleotides do not necessarily play an intrinsic obligatory role in hormonal stimulation.

VII CONCLUSION

From the above discussion it should be apparent that considerable information has been accumulated in recent years concerning the various regulatory features of adenylate cyclase systems. Probing the kinetics of adenylate cyclase has given an insight into possible modes of action of hormones and has led to the discovery of regulation by nucleotides. However, in view of the fact that every time a novel approach is used, totally new and unexpected findings emerge it must also be stressed that one or more fundamental aspects of the system are yet to be discovered. Very possibly the answers to the questions will have to come not only from a multi-disciplinary approach, that takes into consideration structure and function of plasma membranes, eukaryotic genetics and enzyme kinetics, but also from classical biochemical work, which investigates structural aspects by purification of the components which constitute these complex systems. These are formidable tasks that are definitely worth the efforts involved, for there are very few, if any, cellular functions which are not directly or indirectly influenced by an adenylate cyclase system.

ACKNOWLEDGMENT

The author is supported in part by grants from the U.S. Public Health Service (HD-06513, HD-09581 and HD-07495) and by the Clayton Foundation.

CHAPTER 3

Calcium-Regulated Protein Modulator in Cyclic Nucleotide Systems

J. H. Wang

I INTRODUCTION

Many physiological processes are regulated by both Ca^{2+} and cyclic nucleotides, and these regulatory agents in turn are involved in the regulation of each other's cellular metabolism. General aspects of this subject have been reviewed (Rasmussen *et al.*, 1972; Rubin, 1974; Rasmussen *et al.*, 1975; Berridge, 1975a,b) and are treated in detail elsewhere in this volume (see Berridge and Rapp, Chapter 5). Although specific modes of Ca^{2+}–cyclic nucleotides interaction vary among different cell types or different regulatory events even within a cell type, certain basic and universal mechanisms are likely to operate in a variety of cells. Indeed, general models have been postulated for the integration of the actions and metabolism of Ca^{2+} and cyclic nucleotides (Rasmussen *et al.*, 1975; Berridge, 1975a,b).

The present chapter deals with a specific protein factor which appears to participate in the Ca^{2+}–mediated regulation of cyclic nucleotide metabolism. This protein has been demonstrated in almost all mammalian tissues and animal species so far examined (Smoake *et al.*, 1974; Kakiuchi *et al.*, 1975; Waisman *et al.*, 1975). It was discovered by Cheung (1970) as a protein activator of bovine brain cyclic nucleotide phosphodiesterase, and independently by Kakiuchi and Yamazaki (1970a) as a modulatory factor which enhances the Ca^{2+} sensitivity of a Ca^{2+}-activated cyclic nucleotide phosphodiesterase from rat brain. Later, Kakiuchi *et al.* (1973), and Teo and Wang (1973) showed that Ca^{2+} and the protein factor are mutually dependent in the activation of the cyclic nucleotide phosphodiesterase. Using a homogenous preparation of this protein from bovine heart, Teo and Wang (1973) demonstrated that this factor is a Ca^{2+}-binding protein. Recently, Brostrom *et al.* (1975) observed that this same Ca^{2+}-binding protein also participates in the Ca^{2+}-mediated adenylate cyclase activation.

The protein factor has been referred to by different investigators as protein activator, modulatory protein, Ca^{2+}-binding protein, Ca^{2+}-dependent regulator and Ca^{2+}-dependent protein activator. In terms of its biological function, this protein appears to modulate the activity of specific enzymes and its modulatory action is regulated by Ca^{2+}. Therefore, it seems appropriate to designate this protein as a Ca^{2+}-regulated protein modulator. Some aspects of this protein modulator have been reviewed recently (Cheung *et al.*, 1975; Kakiuchi *et al.*, 1975; Wang *et al.*, 1975).

II GENERAL CHARACTERISTICS

A Assay and Detection

The protein modulator is usually assayed on the basis of its activation of the Ca^{2+}-activated cyclic nucleotide phosphodiesterase. For this assay, an enzyme preparation free of endogenous protein modulator and other forms of cyclic nucleotide phosphodiesterase is required. The enzyme can be separated from the protein modulator and other cyclic nucleotide phosphodiesterases by column chromatography using DEAE–cellulose or Sephadex G–200 or by polyacrylamide gel electrophoresis (Cheung, 1970; 1971; Wang *et al.*, 1972; Kakiuchi *et al.*, 1973; Uzunov and Weiss, 1972). The separation of the enzyme from its modulator may be greatly improved by the inclusion of EGTA in elution buffers (Teshima and Kakiuchi, 1974; Lin *et al.*, 1975; Ho *et al.*, 1976). Figure 3.1 presents an elution profile of bovine brain cyclic nucleotide phosphodiesterase from a DEAE–cellulose column showing the dependence of the enzyme activity on the modulator (Cheung, 1971).

Several specific assay methods and different definitions of the activity unit may be found in the literature on the protein modulator (Teo *et al.*, 1973; Lin *et al.*, 1974d; Kakiuchi *et al.*, 1975a; Wolff and Brostrom, 1974; Weiss *et al.*, 1974). Any of these described methods can give satisfactory results, but interconversions of activity units are often difficult. In the author's laboratory, the assay of the protein modulator involves the construction of a modulator dose–response curve using constant amounts of standard enzyme. The unknown sample is then used to activate the standard enzyme under the conditions of limiting modulator activity. One unit of modulator activity is defined as the amount which produces half maximal activation of the enzyme (Teo *et al.*, 1973).

Any of the assay methods may be used for the demonstration of the existence of the protein modulator in tissue extracts. Since the protein modulators from

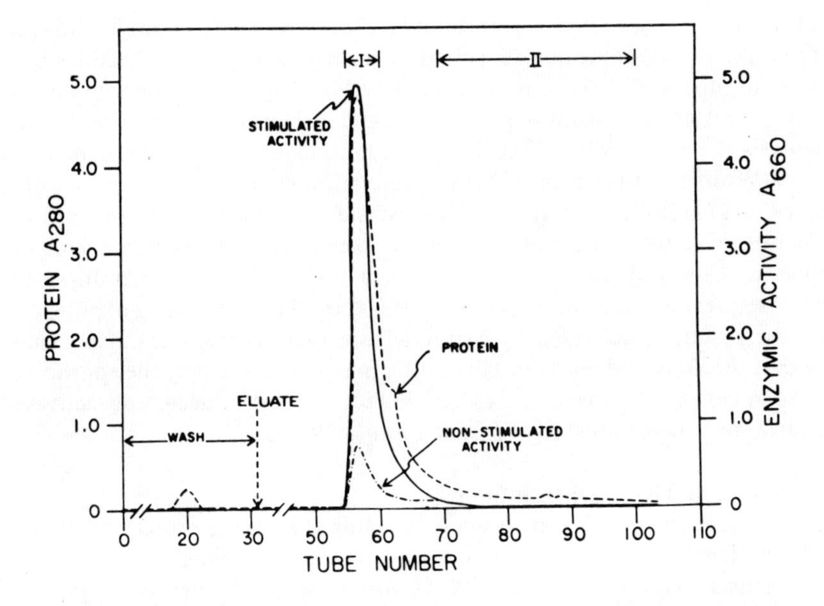

Figure 3.1 Demonstration of the activation of DEAE-cellulose phosphodiesterase chromatographed by the protein modulator. The nonstimulated and stimulated activities are obtained in the absence and presence of added protein modulator, respectively. (From Cheung, 1971; reproduced by permission of American Society of Biological Chemists, Inc.)

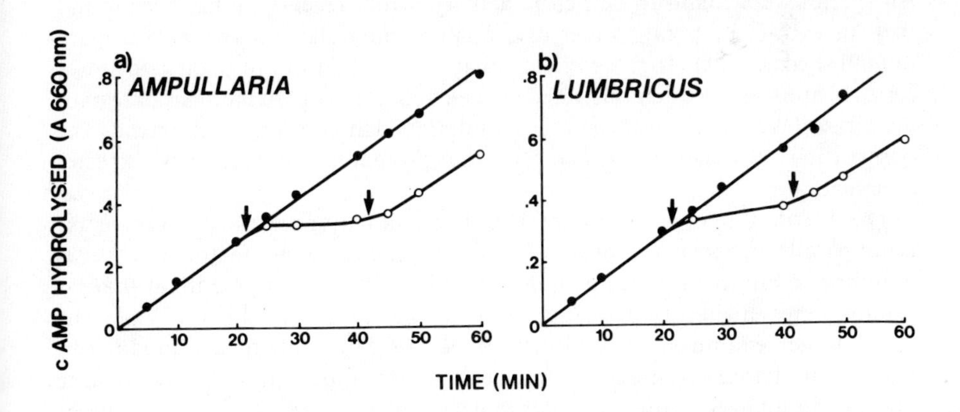

Figure 3.2 Ca^{2+} dependence and reversibility of phosphodiesterase activation by extracts of invertebrates. Arrows indicate the addition of EGTA at 22 min and of Ca^{2+} at 42 min during the course of enzyme reactions. (From Waisman *et al.*, 1975; reproduced by permission of Academic Press)

all sources have been shown to withstand brief boiling, such a procedure may be used to inactivate endogenous phosphodiesterase (Cheung, 1970). In addition to the protein modulator, the Ca^{2+}-activated cyclic nucleotide phosphodiesterase may be activated by phospholipids (Wolff and Brostrom, 1976) or by limited proteolysis (Cheung, 1970, 1971). The phospholipid activation is not dependent on Ca^{2+} (Wolff and Brostrom, 1976) and activation by proteolysis is irreversible (Cheung, 1970, 1971). Thus, to demonstrate the existence of the protein modulator in a tissue extract, the Ca^{2+}-dependency and reversibility of the enzyme activation should be established. For example, Figure 3.2 shows the effect of sequential addition of EGTA and Ca^{2+} on the time courses of phosphodiesterase reactions activated by extracts from two invertebrates. The results show that the activated reaction can be effectively suppressed by the addition of EGTA. When Ca^{2+} ions in excess of EGTA are introduced, the activated reaction rate is re-established (Waisman *et al.*, 1975).

B Species and Tissue Distributions

Since its discovery in bovine brain by Cheung (1970), the protein modulator has been demonstrated in all mammalian tissues examined, which include cardiac muscle (Goren and Rosen, 1971), adrenal gland, kidney, epididymal fat pad, blood, bone marrow, testis (Smoake *et al.*, 1974), liver, thymus (Kakiuchi *et al.*, 1975) and uterus (Smoake *et al.*, 1974; Kroeger *et al.*, 1976). Similarly, the Ca^{2+}-activated phosphodiesterase, which was discovered by Kakiuchi and Yamazaki (1970b) in rat brain, has been demonstrated in many tissues (Kakiuchi *et al.*, 1975; Lagarde and Colobert, 1972; Teo and Wang, 1973; Kroeger *et al.*, 1976). In addition, the protein modulator activity has been detected in many lower forms of animals. Waisman *et al.* (1975) have examined a dozen invertebrates belonging to eight different phyla for modulator activity. All species were found to be rich in activity. More recently, it has been found that an extract of potato tuber also contains modulator activity (Waisman, unpublished results). It is significant that the modulators in plant and lower forms of animals were demonstrated by their activation of bovine heart phosphodiesterase (Waisman *et al.*, 1975); a finding which confirms and extends the observation of Cheung (1971) that the protein modulator lacks species specificity.

The distribution of activities of the protein modulator and phosphodiesterase is not parallel in various animal tissues. For example, the modulator activity was found to be higher in rat testis than in rat brain, but the enzyme in rat brain is twenty times higher than in rat testis (Smoake *et al.*, 1974). While the invertebrates examined by Waisman *et al.* (1975) are all rich in modulator, phosphodiesterase activities are very low. Furthermore, invertebrate phosphodiesterases show little dependence on Ca^{2+} for activity. A similar situation exists in mammalian adrenal medulla (Egrie and Siegel, 1975).

C Ontogenetic Development

Smoake *et al.* (1974) have examined the ontogenetic development of the protein modulator and phosphodiesterase in rat brain, liver, thymus and testis from eight days before birth to maturity. The development was found to be characteristic of a particular tissue and in none of these tissues are the developments of the two activities parallel. Similarly, Strada *et al.* (1974) have made comparative studies of the multiple forms of phosphodiesterase and the protein modulator in cerebral cortex, cerebellum and brain stem of both newborn and adult rats. Again, the changes in modulator-activated enzyme and the modulator activities are not in parallel.

III PURIFICATION AND STRUCTURE

A Purification

The Ca^{2+}-regulated protein modulator has been purified to near homogeneity from bovine heart (Teo *et al.*, 1973), bovine brain (Lin *et al.*, 1974), porcine brain (Teshima and Kakiuchi, 1974), the electroplax of electric eels (Childers and Siegel, 1975) and earthworm. Most of the purification methods are modified from those of Teo *et al.* (1973) used for the bovine heart protein modulator. The method entails ammonium sulphate fractionation, isoelectric precipitation, heat treatment, and column chromatography on DEAE-cellulose and Sephadax G-100. An additional step of preparative disc-gel electrophoresis was found to be necessary for the purification of bovine brain modulator (Lin *et al.*, 1974). Wolff and Siegel (1972) purified a Ca^{2+}-binding phosphoprotein from porcine brain and this protein was later found to be the protein modulator (Wolff and Brostrom, 1974). The same protein has also been purified from bovine adrenal medulla and testis (Brooks and Siegel, 1973a,b). Vanaman *et al.* (1975) obtained from bovine brain a homogeneous preparation of a Ca^{2+}-binding protein which was found to be identical to the protein modulator (Stevens *et al.*, 1976; Watterson *et al.*, 1976).

B Physical Properties

Table 3.1 summarizes the physical parameters of a few purified Ca^{2+}-regulated protein modulators. It can be seen that the modulators from different sources have similar physical parameters. Molecular weights determined by hydrodynamic methods range from 15,000 to 19,000; differences in molecular weights observed for modulators from different sources are probably due to experimental variations. Recently, Watterson *et al.* (1976) have shown that determination of molecular weights by the sedimentation equilibrium method in low ionic strength solution results in low values for the bovine brain modulator. The marked difference in molecular weights determined by the ultracentrifugal and gel filtration methods probably arise because of the anomalous behaviour of the protein modulator on a gel-filtration column.

The ultraviolet absorption spectra of the protein modulators are not typical of globular proteins. Instead of having an absorption maximum at 280 nm, their spectra exhibit considerable vibrational structure in the region of 250–280 nm and absorption peaks occur at about 253, 259, 265, 268 and 276 nm (Wang *et al.*, 1975a; Stevens *et al.*, 1976; Watterson *et al.*, 1976; Liu and Cheung, 1976). This

Table 3.1 Physical properties of the protein modulators

parameter	bovine heart	bovine brain	earthworm
molecular weight:			
ultracentrifugation	17,000–19,000	15,000–18,000	17,000–19,000
gel filtration	27,000	31,000	–
sodium dodecyl sulphate gel electrophoresis	18,500	15,000	18,000
sedimentation constant (A20,w)	2.0	1.85	1.95
diffusion constant (D20,w) (cm^2/sec)	9×10^{-7}	1.09×10^{-6}	9×10^{-7}
frictional ratio	1.3	1.2	1.3
isolectric point (pI)	4.1	4.3	N.D.
absorbance index (f_{276}) 1% protein	2.0	1.8	N.D.
references	Teo *et al.* (1973a) Stevens *et al.* (1976)	Lin *et al.* (1974d) Watterson *et al.* (1976)	Waisman and Wang (1976)

unique optical property results from their amino-acid compositions (Table 3.2); no tryptophan is present and the phenylalanine/tyrosine ratio is high. Futhermore, the proteins have unusually low absorption indexes — about 2 at 275 nm for a 1% solution — (Watterson *et al.*, 1976; Stevens *et al.*, 1976).

Table 3.2 Amino-acid compositions of the protein modulator

	bovine brain	bovine heart	porcine brain	electroplax	earthworm
		(Residue/18,000 g)			
Lys	8	9	8	8	9
His	1	1	1	1–2	1
Arg	7	6	6	5	5
Asp	24	25	22	20	27
Thr	14	12	11	11	11
Ser	5	3	5	5	5
Glu	30	30	28	25	34
Pro	2	2	4	3	4
Gly	13	12	11	10	12
Ala	13	12	11	10	11
Cys	0	0	N.D.	N.D.	N.D.
Val	8	9	7	6	7
Met	11	9	5	4	7
Ile	9	8	8	9	8
Leu	11	10	10	13	8
Tyr	2	2	2	1	1–2
Phe	10	9	8	8	7
Try	0	0	N.D.	N.D.	N.D.
References	Lin *et al.* (1974d)	Wang *et al.* (1975a)	Wolff and Seigel (1972)	Childers and Seigel (1975)	Waisman and Wang (1976)

N.D. = not determined.

C Chemical Structure

Table 3.2 summarizes amino-acid compositions for protein modulators isolated from bovine brain and heart, porcine brain and earthworm. Compositions of modulators from different sources show a large degree of similarity and those for the two modulators from bovine heart and brain are virtually identical. Recently, Stevens *et al.*, (1976) have shown that peptide maps for tryptic hydrolysates of modulators from bovine heart and brain are indistinguishable, suggesting that these two modulators are identical proteins. In contrast, comparison of the peptide maps of bovine heart and earthworm modulators clearly establishes that these two proteins are structurally dissimilar.

Recently, Vanaman and coworkers (1975) have shown that both bovine heart and brain modulators contain an unusual amino-acid residue that co-chromatographs with lysine in the normal amino-acid analysis of protein acid hydrolysates, but can be separated from lysine when the analysis is carried out using the physiological fluid system. Thus, values for lysine in the case of the bovine modulators given in Table 3.2 may be inaccurate by one residue. The identity of this unusual amino-acid residue has not been established. It would be interesting to know if all Ca^{2+}-regulated protein modulators contain this unusual amino acid.

Determinations of the amino terminal of bovine brain protein modulator have led to conflicting observations. Lin *et al.* (1974d) using the dansyl chloride procedure have identified valine as the amino terminal. Watterson *et al.* (1976), however, have failed to detect any N-terminal amino acid using a sequencer and this result led them to suggest that the protein modulator has a blocked N-terminal. Stevens *et al.* (1976) have suggested that bovine heart protein modulators also have a blocked N-terminal.

It is generally thought that the Ca^{2+}-regulated mammalian protein modulators are simple proteins (Lin *et al.*, 1974d). Earlier claims that the bovine heart protein modulator is a glycoprotein (Teo *et al.*, 1973) and that porcine brain modulator is a phosphoprotein (Wolff and Seigel, 1972) may have been due to contaminants in the analysed samples (Wang *et al.*, 1975a; Wolff, personal communication). Further studies on the primary structure of the bovine brain protein modulator are now in progress and partial sequences of some tryptic peptides have been reported (Watterson *et al.*, 1976).

IV ACTIVATION OF CYCLIC NUCLEOTIDE PHOSPHODIESTERASE

A Calcium-Activated Cyclic Nucleotide Phosphodiesterase

Kakiuchi and Yamazaki (1970a) discovered the existence in rat brain of a Ca^{2+}-activated cyclic nucleotide phosphodiesterase which, it was subsequently shown, could be separated from a Ca^{2+}-independent form of the enzyme by gel filtration (Kakiuchi *et al.*, 1971). The Ca^{2+}-activation of the enzyme could be enhanced by the addition of a heat-stable endogenous protein factor (Kakiuchi and Yamazaki, 1970a) which, they postulated, might be identical to the protein activator, or modulator, discovered by Cheung (1970). This postulate has been substantiated by results from several laboratories (Kakiuchi *et al.*, 1973; Teo and Wang, 1973; Wolff and Brostrom, 1974; Lin *et al.*, 1974), thus indicating that Ca^{2+} and the modulator activate the same form of the phosphodiesterase.

All mammalian tissues contain multiple forms of cyclic nucleotide phosphodiesterase having different catalytic and molecular properties (Appleman *et al.*, 1973). Normally, a tissue contains at least three separable cyclic nucleotide phosphodiesterases (Appleman and Terasaki, 1975). Uzunov and Weiss (1972) separated rat cerebellum cyclic nucleotide phosphodiesterase into six fractions by polyacrylamide gel electrophoresis. However, only one of these multiple forms of the enzyme is markedly activated by Ca^{2+} and the protein modulator. The Ca^{2+}-activated cyclic nucleotide phosphodiesterases from different tissues are similar in catalytic and molecular properties and may be considered to be basically the same enzyme.

The enzyme is one of the soluble (cytosol) phosphodiesterases (Kakiuchi *et al.*, 1975; Appleman and Terasaki, 1975; Uzunov and Weiss, 1972; Ho *et al.*, 1976; Lindl *et al.*, 1976). Separation of the Ca^{2+}-activated phosphodiesterase from its endogenous protein modulator and Ca^{2+}-independent forms of cyclic nucleotide phosphodiesterase may be readily achieved (see Section A). However, attempts to purify extensively this enzyme have often resulted in the loss of its responses to Ca^{2+} and the modulator (Goren and Rosen, 1972; Cheung, 1975). Recently the enzyme has been purified from bovine heart to a very high specific activity; cAMP hydrolysis rate is 120 μmole/mg protein min^{-1}. Ca^{2+} activates this enzyme preparation five-fold. However, the enzyme preparation is not pure as judged by analytical polyacrylamide gel electrophoresis (Ho *et al.*, 1976). Since the Ca^{2+}-activated cyclic nucleotide phosphodiesterase has not been purified to homogeneity in its 'native state', little is known about its physical and chemical structure. From gel filtration studies, it has been suggested that the free enzyme has a molecular weight of 150,000 – 170,000 and the enzyme–modulator complex a molecular weight of 200,000 – 230,000 (Teshima and Kakiuchi, 1974; Lin *et al.*, 1975; Ho *et al.*, 1976).

B Characteristics of the Enzyme Activation

Kakiuchi *et al.* (1973), and Teo and Wang (1973) independently demonstrated that the enzyme activation depends on the simultaneous presence of Ca^{2+} and the protein modulator. This observation has been confirmed by many other investigators (Lin *et al.*, 1974; Wolff and Brostrom, 1974; Wickson *et al.*, 1975). As an example, the dual requirements of the cationic and protein activators in the activation of bovine heart cyclic nucleotide phosphodiesterase are shown in Figure 3.3. The apparent K_a for Ca^{2+} (concentration of Ca^{2+} required for half-maximal enzyme activation) depends on the level of the protein modulator present in the reaction mixture. At saturating amounts of the protein modulator, an apparent K_a of 2.3 μM has been found for the enzyme and the

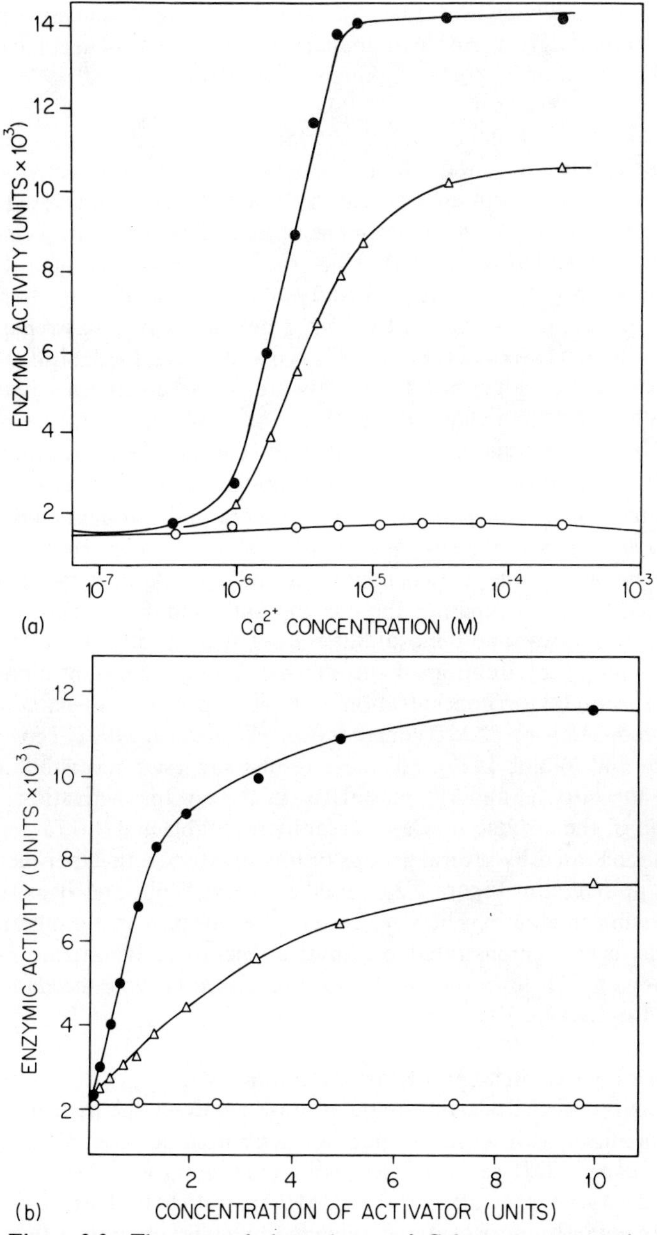

Figure 3.3 The mutual dependence of Ca^{2+} and the protein modulator in enzyme activation. (a) Ca^{2+} activation of the enzyme in the presence of 0(O). 1.3(Δ) and 13(●) units of the protein modulator. (b) Protein-modulator activation carried out in the presence of 0(O), 4(Δ) and 100 μM Ca^{2+}(●). (From Teo and Wang, 1973a; reproduced by permission of American Society of Biological Chemists, Inc.)

modulator from bovine heart (Teo and Wang, 1973). Other investigators using the protein modulators from mammalian brains and earthworm have obtained similar K_a values of 2 – 8 μM (Kakiuchi *et al.*, 1973; Lin *et al.*, 1974; Wolff and Brostrom, 1974; Wickson *et al.*, 1975).

Early studies by Butcher and Sutherland (1962) indicate that mammalian cyclic nucleotide phosphodiesterase depends on Mg^{2+} for activity and this cationic requirement can also be met by Mn^{2+}. The Ca^{2+}-activated phosphodiesterase still depends on the presence of millimolar concentrations of Mg^{2+} (Kakiuchi and Yamazaki, 1970a). Thus, the enzyme has two types of cationic requirements: Mg^{2+} for its activity and Ca^{2+} for its activation. A number of other metal ions, for example, Mn^{2+}, Sr^{2+}, Ba^{2+} and Co^{2+}, may replace Ca^{2+} as the activating ion (Kakiuchi *et al.*, 1972; Teo and Wang, 1973a; Lin *et al.*, 1974). The apparent K_a values for these ions are, however, at least one order of magnitude larger than that for Ca^{2+} (Teo and Wang, 1973a). Among these cations, Mn^{2+} is unique in that it can fulfil the metal requirements for both enzyme activity and enzyme activation (Kakiuchi *et al.*, 1972; Lin *et al.*, 1975).

The activation of phosphodiesterase by Ca^{2+} and the protein modulator may be characterized as a reversibly interacting system. Cheung (1971) showed that the enzyme activation is dependent on the modulator concentration but independent of preincubating the enzyme and modulator. Teo *et al.* (1973) observed that activation of phosphodiesterase by the protein modulator may be decreased simply by dilution of the enzyme. Using a constant amount of the enzyme, the modulator concentration required to provide half-maximal enzyme activation depends on cAMP concentration (Wang *et al.*, 1972; Teo *et al.*, 1973; Brostrom and Wolff, 1976). All these results suggest a reversible interaction between the enzyme and the modulator in the enzyme activation. Similarly, activation of the enzyme by Ca^{2+} is readily reversible and this reversibility has been demonstrated by several groups of investigators with experiments similar to those presented in Figure 3.2 (Lin *et al.*, 1974; Wolff and Brostrom, 1974). The reversible interactions between the enzyme, the protein modulator and Ca^{2+} have also been demonstrated by physical methods. Investigations of these interactions which form the basis for a postulated enzyme mechanism will be discussed in Section VI.

C Substrate Specificity and Kinetic Parameters

Brain cyclic nucleotide phosphodiesterase catalyses the hydrolysis of cyclic 3′,5′-nucleotides with a purine but not a pyrimidine base (Cheung, 1970b). Kakiuchi *et al.* (1973) were the first to point out that the Ca^{2+}-activated phosphodiesterase has a higher affinity for cGMP than cAMP. This observation has been subsequently confirmed and extended by many other investigators (Lin *et al.*, 1974d; Wang *et al.*, 1975a; Appleman and Terasaki, 1975; Wickson *et al.*, 1975; Ho *et al.*, 1976; Brostrom and Wolff, 1976). In several of these studies detailed kinetic characterizations have been carried out. Several lines of evidence support the notion that both these cyclic nucleotides are hydrolysed by the same Ca^{2+}-activated enzyme. Firstly, various separation techniques, including column chromatography, iso-electrofocusing and gel electrophoresis, have failed to resolve the two activities (Brostrom and Wolff, 1976; Ho *et al.*, 1976). Secondly, thermal inactivation of the enzyme has resulted in a parallel loss of the two activities (Brostrom and Wolff, 1976), and thirdly, each nucleotide serves as competitive inhibitor for the enzymic hydrolysis of the other nucleotide with the K_i value similar to its K_m (Ho *et al.*, 1976).

Kinetic constants of cAMP and cGMP hydrolysed by Ca^{2+}-activated cyclic nucleotide phosphodiesterases from several different tissues are listed in Table 3.3, and, when available, those for the Ca^{2+}-independent form of the cytosol enzymes are also listed for comparison. Results obtained using enzymes prepared by different methods from diverse sources all indicate that the Ca^{2+}-activated cyclic nucleotide phosphodiesterase has a much higher affinity toward cGMP than cAMP (Kakiuchi *et al.*, 1973; Brostrom and Wolff, 1974, 1976; Appleman and Terasaki, 1975; Ho *et al.*, 1976). These observations have led to the suggestion that the Ca^{2+}-activated enzyme is primarily a cGMP enzyme in the cells. In comparison with the Ca^{2+}-independent cytosol enzyme, the Ca^{2+}-

Table 3.3 Kinetic constants of phosphodiesterases

		K_m (m*M*)		$V(\%)$[a]		
source	enzyme form	cAMP	cGMP	cAMP	cGMP	references
bovine heart	Ca^{2+}-activated					
	basal	1.2–1.7	0.21–0.33	20	30	Ho *et al.* (1976)
	activated	0.16–0.27	0.007–0.011	100	30	
	Ca^{2+}-independent	0.053–0.061	0.030–0.048	100	85	
porcine brain	Ca^{2+}-activated					Brostrom and Wolff (1976)
	basal	4	0.2	100	33	
	activated	0.18	0.008	100	33	
rat brain	Ca^{2+}-activated					Kakiuchi *et al.* (1973)
	basal	0.004,0.09	0.004,0.05	14	9	
	activated	0.004,0.02	0.002	100	38	
rat liver	Ca^{2+}-activated	N.D.	0.005			
	Ca^{2+}-independent	0.04	0.03			
rat heart	Ca^{2+}-activated	N.D.	0.006			Appleman and Terasaki (1975)
	Ca^{2+}-independent	0.06	0.02			
rabbit heart	Ca^{2+}-activated	N.D.	0.010			
	Ca^{2+}-independent	0.03	0.02			

[a]Values are given as per cent of V_{max} values for cAMP of the independent forms of the enzyme or the activated enzyme at its activated state.
N.D. = not determined.

activated enzyme shows higher cGMP affinity and poorer cAMP affinity. Although good agreements in K_m values determined by different investigators are generally observed, significantly different K_m values are seen in some instances (see Table 3.3). It is not clear whether the differences are real or artifacts of certain experimental conditions. The possible existence of other factors which can alter kinetic properties of the enzyme should not be overlooked.

Conflicting observations on the kinetic mechanisms of the enzyme activation have been reported. Table 3.3 shows that the activation of the bovine heart phosphodiesterase toward cAMP hydrolysis involves increases in both substrate affinity (K_m) and maximal velocity (V_{max}) whereas the activation toward cGMP hydrolysis results in the change in K_m only. In contrast, the activation of porcine brain phosphodiesterase by increasing concentrations of the protein modulator results in the decrease in K_m values only, irrespective of the cyclic nucleotide substrate used (Brostrom and Wolff, 1976). However, Wickson *et al.* (1975) using cGMP as the substrate have shown that the enzyme activation results from the increase in V_{max} without any change in K_m, and Weiss *et al.* (1974) have demonstrated that activation of rat cerebral enzyme by the protein modulator results in increased V_{max} of cAMP hydrolysis with no change in K_m. The reason for these conflicting observations is not at all apparent at present. Presumably, factors which may influence the enzyme activation, such as assay conditions and enzyme preparations, have yet to be characterized and delineated.

D Other Effectors

Weiss and coworkers (1974) have examined the effect of several known inhibitors of cyclic nucleotide phosphodiesterase on the multiple forms of the rat cerebral enzyme and observed differential inhibitory effects. They have shown that trifluoperazine is a very potent inhibitor for the modulator-activated phosphodiesterase with K_i of about 10 μM. It is, however, a weak inhibitor for other forms of cyclic nucleotide phosphodiesterase, including the basal activity of the modulator-activatible form. Thus, at 25 μM trifluoperazine, the activated activity is 80% inhibited whereas the basal activity is not significantly affected. Since the inhibition of the enzyme by low concentrations of trifluoperazine can be overcome by increasing concentrations of the modulator, these investigators suggest that the drug is competitively blocking the modulator activation. They further suggest that the inhibitory effect of trifluoperazine may be related to its neuroleptic action since trifluoperazine sulphoxide and promethazine are much weaker inhibitors. In contrast to trifluoperazine, theophylline and papaverine inhibit the activated and basal activities of the enzyme equally. The studies of pharmacological agents on the multiple forms of rat brain phosphodiesterase have recently been reviewed (Weiss, 1975; Weiss and Greenberg, 1975).

Since the activation of the modulator-dependent form of cyclic nucleotide phosphodiesterase depends on the simultaneous presence of free Ca^{2+} and Mg^{2+}

(see Section IVB), any agents showing preferential chelation of Ca^{2+} rather than Mg^{2+} would be expected to inhibit the activated enzyme activity. Agents in this category include EGTA, citrate and ATP (Kakiuchi and Yamazaki, 1970a,b; Teshima, *et al.*, 1973). Earlier observations of the inhibition of brain cyclic nucleotide phosphodiesterase by high concentrations of Mg^{2+} may be due to the Mg^{2+} competing with the activation effect of Ca^{2+} (Cheung, 1967; Kakiuchi, 1972; Ho *et al.*, 1976).

Imidazole and ammonium sulphate, which are both known to activate mammalian cyclic nucleotide phosphodiesterase (Butcher and Sutherland, 1962; Nair, 1966), have been shown to potentiate specifically the activated state of the modulator-dependent enzyme (Ho *et al.*, 1976). These effects result from the increase in maximal levels of the enzyme activation by the modulator protein. The activation of brain phosphodiesterase by naturally occuring lipids has been observed by Bublitz (1973). Recently, Wolff and Brostrom (1976) have shown that the lipids may replace the protein modulator in the activation of the Ca^{2+}-activated phosphodiesterase. However, the enzyme activation is independent of Ca^{2+} and results from the increase in V_{max}. The most effective lipids in the enzyme activation are phosphatidylinositol and lysophosphatidylcholine.

V ACTIVATION OF BRAIN ADENYLATE CYCLASE

Millimolar concentrations of Ca^{2+} generally exert an inhibitory effect on the activity of adenylate cyclase (Perkins, 1973). However, Bradham and coworkers (Bradham *et al.*, 1970, Bradham 1972) have demonstrated that increasing concentrations of EGTA, in the presence of a molar excess of Mg^{2+}, result in a 70% inhibition of the adenylate cyclase of bovine cerebral cortex. This inhibition may be reversed by Ca^{2+} or Mn^{2+}. Johnson and Sutherland (1973), using a Lubrol–PX-solubilized adenylate cyclase preparation from rat cerebellum, demonstrated a 90% inhibition of the enzyme by EGTA and the reversal of the inhibition of Ca^{2+}, Sr^{2+} and Mn^{2+}.

Brostrom *et al.* (1975) showed that Ca^{2+}-activation of the Lubrol–PX-solubilized porcine brain adenylate cyclase is mediated by an endogenous factor which can be removed from the enzyme by passing it through an anion-exchange column. The chromatographed enzyme which possesses very low activity can be activated in the presence of Ca^{2+} by the homogeneous porcine brain protein modulator. The endogenous factor from adenylate cyclase can modulate Ca^{2+}-activation of cyclic nucleotide phosphodiesterase. Based on these results, Brostrom *et al.* (1975) suggested that the Ca^{2+}-regulated protein modulator participates in the regulation of both Ca^{2} -activated adenylate cyclase and phosphodiesterase. This suggestion is substantiated by the observations that the endogenous factors of adenylate cyclase and of the protein modulator have identical electrophoretic mobilities on polyacrylamide gel and their two activities, namely the activation of cyclase and of phosphodiesterase,

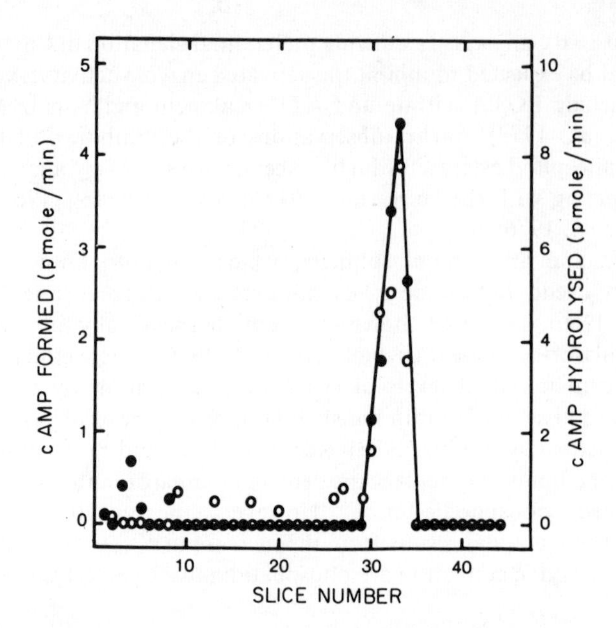

Figure 3.4 Identification of the endogenous factor of adenylate cyclase as the protein modulator. The endogenous factor and the protein modulator show identical mobility upon gel electrophoresis. The cyclase (O) and phosphodiesterase (●) activating activities of the electrophorized factor are located at same position on the gel. (From Brostrom *et al.*, 1975; courtesy of the authors and Proceedings of National Academy of Science)

correspond exactly on the gel. Results from these experiments are reproduced in Figure 3.4. Cheung and his associates (1975) obtained similar results using both bovine and rat brain adenylate cyclase solubilized by Lubrol–PX.

The activation of the brain adenylate cyclase by Ca^{2+} and the protein modulator is similar to the activation of phosphodiesterase in many respects. For example, the activation of adenylate cyclase depends on the simultaneous presence of both Ca^{2+} and the protein modulator. The effect of the protein modulator is specific; six other globular proteins do not affect the enzyme activity (Cheung *et al.*, 1975). The enzyme shows dual cationic requirements: Mg^{2+} for activity and Ca^{2+} for activation. Again, Mn^{2+} can fulfil both requirements (Johnson and Sutherland, 1973). Although the range of Ca^{2+} concentration required for the activation of adenylate cyclase has not been established, it is likely that micromolar concentrations of Ca^{2+} are essential.

Although solubilized brain adenylate cyclases were used in the identification of protein modulator as the Ca^{2+}-mediator of the enzyme, the fact that a particulate preparation of the enzyme is also inhibited by EGTA and activated by Ca^{2+} or Sr^{2+} (Bradham, 1972) strongly suggests that similar mechanisms operate for the membrane-bound form of the enzyme (see also Gnegy *et al.*, 1976). Of particular interest is the incomplete inhibition of particulate cyclase by

EGTA thus suggesting the existence of Ca^{2+}-dependent and Ca^{2+}-independent forms of adenylate cyclase. The Ca^{2+}-activated adenylate cyclase may not be restricted to mammalian brain; its existence has also been postulated in other cells (Bär and Hechter, 1969c; Birnbaumer and Rodbell, 1969; Franks *et al.*, 1974).

The presence of a factor which activates adenylate cyclase has been reported by other investigators (Kaufman *et al.*, 1972; Anderson and Pastan, 1975). Characterizations of these factors and their possible relationship to the Ca^{2+}-regulated protein modulator have not been carried out.

VI MECHANISM OF ACTION

A Protein Modulator – Calcium Interactions

As has been noted previously (Wang *et al.*, 1975a), the mutual dependence of the protein modulator and Ca^{2+} in the enzyme activations may be accounted for by at least two different mechanisms, one of which is that the activation results from the binding of both Ca^{2+} and the protein modulator to the enzyme. Alternatively, the protein modulator may have to bind Ca^{2+} to become effective. A simple experiment to distinguish between these alternatives involves an examination of the possible binding of Ca^{2+} to the enzyme and the protein modulator. Using a homogeneous preparation of bovine heart protein modulator, Teo and Wang (1973a) showed that the protein modulator binds Ca^{2+} strongly. A Scatchard plot for the equilibrium binding of Ca^{2+} to the protein modulator indicates the existence of two sets of binding sites: a high-affinity site with K_d of 2.9 μM and two or three low-affinity sites, K of approximately 12 μM. A similar observation has been reported by Lin *et al.* (1974d) for the bovine brain protein modulator. In contrast, these investigators found three high-affinity sites and one low-affinity site per molecule of the brain modulator, with K_d values of about 3.5 and 18 μM, respectively. That the Ca^{2+}-binding to the protein modulator is related to the enzyme activation is supported by the similarity between values for the apparent K_ds of Ca^{2+} and K_as of activation; 3 to 18 μM as compared to 2 to 8 μM. Also, these bindings may be inhibited by other activating cations (Teo and Wang, 1973a; Lin *et al.*, 1974d).

In contrast to the protein modulator, a partially purified phosphodiesterase preparation from bovine heart does not show significant Ca^{2+}-binding (Wang *et al.*, 1975a). The study of possible binding of Ca^{2+} to the enzyme should, however, be extended to cover different conditions and using homogeneous enzyme preparations.

The observation of the Ca^{2+}-binding to the protein modulator has led several investigators to suggest that the Ca^{2+}–modulator complex is the active form of this protein (Teo and Wang, 1973a; Teshima *et al.*, 1974a,b; Lin *et al.*, 1974d; Brostrom and Wolff, 1974). This suggestion is supported by the finding that the protein modulator upon Ca^{2+} binding undergoes changes in several of its physical properties. For example, the fluorescence emission intensity of bovine heart modulator exhibits a 30% enhancement at 315 nm in the presence of Ca^{2+}

(Wang *et al.*, 1975d; Teo, 1974). Also, the optical rotatory dispersion curve of the bovine brain protein modulator is changed by addition of Ca^{2+}, which indicates an increase in helical structure from 39 to 57% (Liu and Cheung, 1976). In addition to changes in optical properties, the protein modulator may be protected by Ca^{2+} against inactivation, particularly proteolytic degradation (Ho *et al.*, 1975a; Liu and Cheung, 1976). Some of these Ca^{2+}-induced changes have been used as conformational probes for the Ca^{2+}–protein modulator interaction; interaction constants derived are in good agreement with the K_d values obtained by equilibrium binding studies (Ho *et al.*, 1975; Teo, 1974).

B Protein Modulator–Enzyme Interactions

In addition to its dependency on Ca^{2+} for the activation of phosphodiesterase and cyclase, the protein modulator also depends on Ca^{2+} for its association with these enzymes. Kakiuchi *et al.* (1975) demonstrated this Ca^{2+}-dependence of the modulator–phosphodiesterase interaction on a gelfiltration column. Their results showed that the protein modulator and phosphodiesterase are well separated on a Sephadex G–200 column in the presence of EGTA, but are eluted together in the presence of Ca^{2+}. Similar observations have been reported by Lin *et al.* (1975) for the bovine brain phosphodiesterase and modulator. The dependency on Ca^{2+} for the interaction between phosphodiesterase and the protein modulator has also been demonstrated on a DEAE–cellulose column (Wang *et al.*, 1975d). The interaction of the solubilized adenylate cyclase and the protein modulator, which is similar to that of cyclic nucleotide phosphodiesterase, also depends on the presence of free Ca^{2+}, as has been demonstrated by Lynch *et al.* (1976) using the gel filtration technique.

The association of the enzymes and the protein modulator may be accompanied by changes in the enzyme conformation. Using partially purified phosphodiesterase from bovine heart, it has been shown that the enzyme is relatively stable in its free state at 55°C, but is rapidly inactivated when both Ca^{2+} and the protein modulator are present (Wang *et al.*, 1975d). The thermal stability has been used as a conformational probe to determine interaction constants for Ca^{2+} and the modulator and the values obtained are in good agreement with those for kinetic and binding constants. Other investigators (Kakiuchi *et al.*, 1975; Brostrom and Wolff, 1976; Wolff and Brostrom, 1976; Liu and Cheung, 1976) have also observed the destabilization of the Ca^{2+}-activated phosphodiesterase from mammalian brain by Ca^{2+} and the protein modulator.

Certain phospholipids also activate Ca^{2+}-activated phosphodiesterase (Wolff and Brostrom, 1976). Interestingly, these activators also decrease thermal stability of porcine brain phosphodiesterase and the extent of this enzyme destabilization is identical to that produced by the protein modulator and Ca^{2+} (Wolff and Brostrom, 1976). The result supports the view that the conformational change of phosphodiesterase, as manifested by the change in enzyme stability, is associated with the enzyme activation.

More recently, it has been found that the stability of the phosphodiesterase from bovine heart depends on the enzyme purity. With a highly purified enzyme preparation, Ca^{2+} and the protein modulator stabilize, rather than destabilize, the enzyme against thermal inactivation (Ho *et al.*, unpublished results). The observation, however, does not negate the suggestion that the interaction between the enzyme and the protein modulator results in conformational changes in the enzyme. It indicates that the manifestation of such conformational changes in terms of the enzyme stability depends on other factors.

C Molecular Mechanisms of Modulator Action

A mechanism for the activation of cyclic nucleotide phosphodiesterase by Ca^{2+} and the protein modulator has been proposed (Wang *et al.*, 1975d). In view of the similarity in the phosphodiesterase and adenylate cyclase activations, the same mechanisms may also be applied to the latter enzyme. It is postulated that both protein modulator and the target enzymes exist in interconvertible, inactive and active conformations. In the absence of sufficient amounts of free Ca^{2+}, the proteins are present separately and in the inactive states. When the Ca^{2+} concentration is increased to above 1 μM, the protein modulator associates with Ca^{2+} and is then converted to an active conformation. The activated modulator then complexes with the target enzyme to convert the enzyme from its less active state to a highly active state. This postulated mechanism outlines the stepwise events leading from the elevation of the Ca^{2+} concentration in the cells to the activation of the cyclic nucleotide enzymes. Reactions involved may be depicted as in Scheme 3.1

$$\underset{\text{(Inactive)}}{Ca^{2+} + \text{Modulator}} \rightleftharpoons \text{Modulator–}Ca^{2+} \rightleftharpoons \underset{\text{(Activated)}}{\text{Modulator*–}Ca^{2+}} \qquad \text{Stage 1}$$

$$\text{Modulator*–}Ca^{2+} + \underset{\text{(Inactive)}}{E} \rightleftharpoons \overset{E}{\overset{|}{\text{Modulator*–}Ca^{2+}}} \rightleftharpoons \underset{\text{(Activated)}}{\overset{E^{*}}{\overset{|}{\text{Modulator*–}Ca^{2+}}}} \qquad \text{Stage 2}$$

(Scheme 3.1)

Similar models have been suggested by other workers (Teo and Wang, 1973; Teshima and Kakiuchi, 1974; Lin *et al.*, 1974d; Wolff and Brostrom, 1974).

The reactions depicted in Scheme 3.1 indicate the sequence of events rather than show the exact chemical and physical reactions. Even in the case of phosphodiesterase, the stoichiometry of the interaction between the enzyme (E), Ca^{2+} and the modulator is not established. It has been suggested that one phosphodiesterase molecule binds two molecules of the modulator (Teshima and Kakiuchi, 1974). More exact measurements of these interactions are, however, necessary to define more completely the mechanism of reaction. It should be stressed that these studies require homogeneous enzyme preparations.

One question which, hopefully, a pure sample of phosphodiesterase may also help to answer is whether the enzyme activation is a result of relieving the enzyme inhibition by an inhibitory factor or an inhibitory subunit. In view of the proteolytic activation of the enzyme (Cheung, 1970b; Miki and Yoshida, 1972) and the loss of enzyme activation during purification (Cheung, 1975), the presence of an inhibitory subunit in the enzyme is quite possible. There are many examples of regulatory enzymes having such a subunit; degradation of the subunit by partial proteolysis results in enzyme activation (Graves *et al.*, 1974; Huang and Huang, 1975). In fact, the activation of the purified cyclic nucleotide phosphodiesterase from frog rod outer segments by treatment with trypsin has been shown to involve the degradation of one of the enzyme subunits (Miki *et al.*, 1975).

D Biological Mechanisms

The hypothesis of Rasmussen *et al.* (1972) that Ca^{2+}, in addition to cAMP, is a universal second messenger is gaining acceptance. General models have been proposed to integrate the functions and metabolisms of Ca^{2+} and cyclic nucleotides in cell activations (Rasmussen *et al.*, 1975; Berridge, 1975b). In these models, Ca^{2+} is suggested to be the major regulator in most cellular processes and cyclic nucleotides participate by modulating Ca^{2+} concentration. On the other hand, Ca^{2+} is involved in the regulation of cyclic nucleotide metabolism. Thus, the mutual regulation of Ca^{2+} and cyclic nucleotides are integral parts of the general models. Berridge in this volume (see Chapter 5) discusses the interactions between Ca^{2+} and cyclic nucleotides in detail.

The Ca^{2+}-regulated protein modulator plays important roles in the control of cellular levels of cyclic nucleotides. It may stimulate the hydrolysis of both cAMP and cGMP, and the synthesis of cAMP. As has been discussed in Section IVC, the Ca^{2+}-activated cyclic nucleotide phosphodiesterase has a much higher affinity for cGMP than cAMP. Kakiuchi *et al.* (1973, 1975b) have analysed several tissue preparations, such as cerebrum, cerebellum, heart, kidney and liver, and have shown that the Ca^{2+}-activated phosphodiesterase represents the predominant cGMP-degradation activity in all the tissues. In contrast, the enzyme contributes little to the degradation of cAMP in some tissues, for example, the liver and kidney. This is because in these tissues the enzyme has the lowest affinity for cAMP of the cyclic nucleotide phosphodiesterases. However, in cerebrum where this form of phosphodiesterase predominates, cAMP is hydrolysed mainly by this enzyme (Kakiuchi *et al.*, 1975a; Weiss *et al.*, 1974). Thus, depending on the tissue and the cell type, the Ca^{2+}-regulated protein modulator may control the degradation of cGMP, or both cGMP and cAMP.

Although the activation of adenylate cyclase by the Ca^{2+}-regulated protein modulator has been demonstrated only for the mammalian brain enzymes (Brostrom *et al.*, 1975; Cheung, *et al.*, 1975), the same mechanisms of the enzyme regulation may exist in other tissues (see Section V). In addition to the tissue distribution of the Ca^{2+}-activated adenylate cyclase, other properties of this enzyme have to be examined to define more fully its biological significance. Thus, the possible existence of Ca^{2+}-dependent and the Ca^{2+}-independent forms of brain adenylate cyclase and the relationship between Ca^{2+} activation and neurohumoral stimulation of adenylate cyclase should certainly be studied. The incomplete inhibition of particulate brain adenylate cyclase by EGTA observed by Bradham *et al.* (1970) suggests that Ca^{2+}-activated and Ca^{2+}-independent cyclases exist. Since the elevation of cAMP in guinea-pig brain slices in response to adenosine or putative neurotransmitters is not inhibited by 1 mM EGTA (Schultz, 1975a,b; Schultz and Kleefeld, 1975), it seems that the Ca^{2+}-activated adenylate cyclase may not be active under certain conditions.

Two biological mechanisms have been proposed to integrate the modulator-regulated reactions. Brostrom *et al.* (1975) have postulated that in response to a Ca^{2+} flux, the protein modulator stimulates simultaneously cAMP synthesis and cGMP degradation. This is based on the assumption that the Ca^{2+}-activated cyclic nucleotide phosphodiesterase is primarily a cGMP-hydrolysing enzyme. In the framework of the Yin-Yang hypothesis (Goldberg *et al.*, 1975), this modulator action represents a concerted regulation to raise the cellular cAMP/cGMP ratio. An alternative mechanism involving a sequential activation of cAMP synthesis and degradation by the protein modulator has been proposed (Cheung *et al.*, 1975; Gnegy *et al.*, 1976). Thus, a Ca^{2+} flux through the plasma membrane results in the activation of adenylate cyclase with a subsequent activation of phosphodiesterase as Ca^{2+} reaches the cytosol. Possibly, the two proposed mechanisms occur in different tissues or operate for different biological processes. In tissues where the Ca^{2+}-activated phosphodiesterase contributes little to the activity of cAMP hydrolysis (Kakiuchi *et al.*, 1975b), the mechanism of sequential activation of cAMP synthesis and degradation cannot operate. However, for any of these tissues, the existence of a protein modulator-regulated adenylate cyclase has yet to be demonstrated.

Data in support of any of these mechanisms operating in the cells are lacking or, at best only fragmentary. Schultz (1975a,b), and Schultz and Kleefeld (1975) have presented evidence for a Ca^{2+}-supported cAMP degradation in guinea-pig brain slices stimulated by adenosine or putative neurotransmitters. Under different conditions, Ca^{2+}-dependent stimulation of cAMP levels has been observed for brain and other tissues (Shimizu *et al.*, 1970c; Lefkowitz *et al.*, 1970). Schultz and Hardman (1975) have presented evidence for the Ca^{2+}-activated cGMP synthesis in smooth muscle, but not for a Ca^{2+}-activated cGMP hydrolysis. At least part of the difficulty in assessing the modulator-regulated reactions in the cells is that these reactions, if they do occur, constitute only a small portion of the very complex network of Ca^{2+}–cyclic nucleotide interaction. Many of the individual reactions in this network are still unknown or poorly understood.

There have been speculations that the protein modulator may have other functions. These are based on the observed non-parallel distribution of the protein modulator and phosphodiesterase, often with a great excess of the former (Waisman *et al.*, 1975; Egrie and Siegel, 1975). The non-parallel ontogenetic development of the two proteins (Smoake *et al.*, 1974; Strada *et al.*, 1974) and the apparent independent genetic control of the synthesis of the two proteins (Lynch *et al.*, 1975) also suggest this possibility. However, experimental evidence in support of the speculations has not been forthcoming.

VII COMPARISON WITH OTHER CALCIUM-MODULATED PROTEINS

Kretsinger (1975a, 1976) in his recent reviews has made a conceptual distinction between the Ca^{2+}-modulated intracellular proteins and Ca^{2+}-activated extracellular proteins. This distinction seems useful since interactions of Ca^{2+} with these two classes of proteins appear different both in biological and molecular mechanisms. The intracellular proteins have Ca^{2+} dissociation constants in the range of 10^{-7} to 10^{-5} M, and their activities are, therefore, regulated by the fluctuation in cellular Ca^{2+} concentrations. The extracellular enzymes have Ca^{2+} dissociation constants of about 10^{-3} M, and are activated when secreted into the extracellular fluid.

Comparison of several Ca^{2+}–modulated proteins with known primary sequences suggests that they are homologous protein. These include parvalbumin (a low molecular-weight Ca^{2+}-binding protein found in the muscles of all classes of invertebrates) (Pechere *et al.*, 1975) and tropinin C, (the Ca^{2+}-binding subunit of the muscle regulatory protein troponin and the light chains of myosin) (Collin *et al.*, 1973; Collin, 1974; Tufty and Kretsinger, 1975). Due to the striking similarity between the physical and chemical properties of rabbit skeletal muscle troponin C and bovine heart protein modulators, it has been suggested that the protein modulator is also homologous to these Ca^{2+}–modulated proteins (Wang *et al.*, 1975a; Stevens *et al.*, 1976; Watterson *et al.*, 1976). The recently determined partial sequence of the bovine protein modulator supports this suggestion (Watterson *et al.*, 1976). Kretsinger (1975a, 1976) has predicted that all the intracellular Ca^{2+}–modulated proteins and enzymes are evolved from a common ancestor protein. It seems that among these evolutionarily related proteins the protein modulator may be one of the earliest Ca^{2+}-binding proteins to emerge with a defined function. The protein has been detected in very primitive animals, such as echinoderms and porifers, (Waisman *et al.*, 1975) and even in a higher plant.

ACKNOWLEDGEMENTS

The author is grateful to Drs. Brostrom, Cheung, Kretsinger, Vanaman and Wolff for providing preprints of their work prior to publication.

CHAPTER 4

cAMP–Protein Kinase Interaction

S. Swillens and J. E. Dumont

I INTRODUCTION

Significant evidence has been gathered in support of the hypothesis that the action of cAMP in a wide variety of hormone-sensitive cells is expressed through the interaction with an ubiquitous receptor protein and through activation of a protein phosphorylating enzyme. The second messenger concept has been developed during investigations of the hormonal regulation of glycogen breakdown. It has been shown that this system consists of a 'cascade' involving two phosphorylation steps. Glycogen is broken down into glucose molecules by glycogen phosphorylase which is active when phosphorylated by phosphorylase kinase. The finding that this system is stimulated by cAMP was the key to the discovery of an intermediate protein which acts both as a receptor for the cyclic nucleotide and as an enzyme able to phosphorylate the phosphorylase kinase. As this protein exhibits some activity for other substrates than phosphorylase kinase, a generic name was chosen — 'cAMP-dependent protein kinase' or, more concisely, 'protein kinase'. It was suggested that any metabolic effect of an hormone-induced increase in cAMP level is mediated by this unique effector (Kuo and Greengard, 1969). The mode of action of the cAMP-dependent protein kinase is the transfer of the γ-phosphate group of ATP to a serine or threonine residue of a specific protein substrate in such a way that the physiological function of this substrate is altered.

In this chapter some current concepts of the control of protein kinases by cyclic nucleotides are summarized. For further details the reader is referred to the more extensive reviews by Krebs (1972), Walsh and Ashby (1973), Langan (1973) and Rubin and Rosen (1975).

II OCCURRENCE OF PROTEIN KINASE

cAMP-dependent protein kinases are distributed in various tissues and organs of animals. Until now no evidence has existed for another mode of action of the second messenger. The distribution of protein kinases in the subcellular fractions is also well documented. Either the cAMP-binding activity or the enzymic activity of protein kinase or both have been measured in the subcellular fractions of a number of different tissues (Langan, 1973). Generally, the majority (90%) of the protein kinase activity is found in the cytosol of, for example, skeletal muscle, liver, diaphragm, heart, uterus and mammary glands. On the contrary, 70% of the activity is present in the cell membrane in the case of human erythrocytes. In brain and anterior pituitary, half the total activity is found in the cytosol, but after treatment with detergents a large amount of latent activity associated with the particulate fraction is released.

Arthropods, in particular lobster tail muscle, contain cGMP-dependent protein kinases (Kuo and Greengard, 1970c). Several mammalian tissues contain cGMP-dependent protein kinase which may be also activated by cAMP as in the case of the rat cerebellum (Hofmann and Sold, 1972). Interestingly, many cAMP-dependent protein kinases may be activated by cGMP as well as by cCMP, cUMP and cIMP but the affinity of the protein kinase for these cyclic nucleotides is two to three orders of magnitude lower than the affinity for cAMP. Thus, the physiological role of such an activation is doubtful.

III MODE OF ACTION OF cAMP

Findings in cardiac muscle (Brostrom *et al.*, 1970) suggest that protein kinase is composed of two types of subunits the physiological roles of which are quite different. The regulatory subunit (R) is able to bind either cAMP or the catalytic subunit (C). The binding of the cyclic nucleotide to the regulatory subunit induces the release of the catalytic subunit C which is fully active as enzyme and able to phosphorylate a substrate. In the absence of cAMP, the two types of subunits are tightly bound together and the enzymic activity of the catalytic subunit is restrained to a large extent. A characteristic of this system is the necessary dissociation of the regulatory–catalytic complex to reveal the potential activity of the catalytic subunit. It is this dissociation which is under the control of cAMP. Such control is fundamentally different from the classical regulation of an enzymic activity by binding of some allosteric effector as an activator or inhibitor.

Essentially two cell types, bovine heart and the rabbit skeletal muscle, have been used to purify the protein kinase to homogeneity (Rosen *et al.*, 1973; Beavo *et al.*, 1974).

The regulatory subunit of protein kinase from bovine heart muscle has been described as a dimeric asymmetric protein (molecular weight $\simeq$ 40,000) capable of binding two monomeric globular catalytic subunits (molecular weight $\simeq$

40,000) (Rosen *et al.*, 1973). The protein kinase from rabbit skeletal muscle shares the same characteristics. The molecular weight of the catalytic subunit is identical and the regulatory subunit (molecular weight $\simeq$ 96,000) is able to bind two molecules of the catalytic subunit as well as two molecules of cAMP (Beavo *et al.*, 1974). Thus the stoichiometric description of the interaction of protein kinase with cAMP should be

$$2\,\text{cAMP} + R_2 - C_2 \rightleftharpoons R_2 - \text{cAMP}_2 + 2\text{C}. \quad (1)$$

Spontaneous dissociation of the complex R_2C_2 apparently cannot occur under physiological condition. The basal activity of protein kinase in the absence of cAMP is not inhibited by increasing the amount of free regulatory subunit (Walsh and Ashby, 1973) and thus is not due to free catalytic subunit which should be present if spontaneous dissociation occurs. It can be concluded that a conformational change of R_2 induced by the binding of cAMP is needed to release C. The converse situation also appears to be true, that is the release of cAMP from R_2 requires the binding of C. As cAMP degradation does not occur when phosphodiesterase is added to a mixture containing the complex R_2 – cAMP_2 but no catalytic subunit (Brostrom *et al.*, 1971a; O'Dea *et al.*, 1971), it is clear that no significant spontaneous dissociation of cAMP from the regulatory subunit occurs.

The intracellular concentration of protein kinases in different tissues seems to be very close to the basal level of cAMP (10^{-7} – 10^{-6} M) (Beavo *et al.*, 1974; Swillens *et al.*, 1976). If it is supposed that protein kinase activation parallels the occupancy level of the regulatory subunit by cAMP, it is clear that half-maximal activation level is obtained when half the receptor sites are occupied by cAMP. Thus the activation constant K_a, that is the total concentration of cAMP needed to obtain the half-maximal activity, is higher than half the total concentration of receptor sites. This fact does not depend on the model used (Swillens *et al.*, 1974). Thus it becomes clear that K_a depends on the concentration of cAMP-binding sites, and the relatively low K_a value of approximately 10^{-8} M obtained under *in vitro* conditions has no physiological significance because of the dilution of the protein kinase in the assay. This theoretical prediction has been confirmed experimentally (Beavo *et al.*, 1974). As a result of this, compartmentation of cAMP which has been suggested because of the discrepancy between the relatively low *in vitro* value of K_a and the high level of cAMP (Exton *et al.*, 1971) is not necessary to explain the effective physiological activation of protein kinase by cAMP.

The mode of action of cGMP on its specific protein kinase seems to follow a similar scheme as that for cAMP-dependent protein kinase, with the induced dissociation of the catalytic subunit due to the binding of cGMP. Interestingly, it has been shown that the regulatory subunit of cAMP-dependent protein kinase from mammalian brain can inhibit the activity of the catalytic subunit of cGMP-dependent protein kinase from lobster tail muscle. Such a cross specificity has led to a holoenzyme the activity of which is preferentially expressed under the control of cAMP. The specificity of the regulatory subunit is thus not lost after such a manipulation.

IV OTHER REGULATORY MECHANISMS

Depending on the type of protein kinases, different mechanisms besides the cyclic nucleotide action may regulate the protein kinase activity. Although their physiological significance remains obscure, the interest of such complementary controls has prompted some laboratories to examine these other regulatory mechanisms.

A Interaction with the Modulator

The discovery of an inhibitor of the phosphorylase kinase system in muscle (Posner *et al.*, 1965) led to the first demonstration of the cAMP-dependent protein kinase by Walsh (1968). Studies of this inhibitor have shown that it is a small (molecular weight 26,000), heat stable, trypsin-sensitive, non-proteolytic protein (Walsh *et al.*, 1971). Moreover, the inhibitor, which is widely distributed in animal tissues, causes a five-fold increase in the affinity of cAMP for protein kinase.

The action of the heat-stable inhibitor on protein kinase has been demonstrated to occur through the reversible binding of the free catalytic subunit of the kinase to the inhibitor (Walsh and Ashby, 1973). An advantage of the presence of this inhibitor is the possibility of storing protein kinase activity which may then be released by some unknown regulatory mechanism independent of the cAMP system.

A protein similar to the inhibitor has been found in lobster tail muscle (Donnelly *et al.*, 1973). This protein is referred to as a protein kinase modulator as it either stimulates or inhibits the protein kinase activity depending on the substrate to be phosphorylated, the requirement of the protein kinase for cAMP or cGMP and the source of the modulator. It has been suggested that such a modulator may enhance the activity of cGMP-dependent protein kinase and inhibit the activity of cAMP-dependent protein kinase in tissues where both the cGMP content and cGMP-dependent protein kinase activity are lower than cAMP content and cAMP-dependent protein kinase activity.

It has become evident that such a modulator is as ubiquitous as the protein kinase. Its physiological role remains obscure but it is of great interest as changes in inhibitory activity have been observed in rabbit heart during starvation (Walsh and Ashby, 1973) and in brown adipose tissue from rat during development (Skala *et al.*, 1974). Similar changes have also been found in the pancreas and epididymal fat cells of alloxan-induced diabetic rats and are in opposite directions for cAMP and for cGMP-dependent protein kinases (Kuo, 1975c).

B Interaction by Autophosphorylation

It has been reported that protein kinase from bovine heart muscle can catalyse the transfer of ^{32}P from [γ-^{32}P]ATP to two seryl residues in its regulatory subunit (Rosen *et al.*, 1973). This phenomenon was subsequently observed in various tissues such as porcine cardiac muscle and bovine brain (Maeno *et al.*, 1974). In the presence of cAMP, the phosphorylated regulatory subunit releases catalytic subunits much more easily suggesting that the autophosphorylation of protein kinase could be of physiological significance (Rosen and Erlichman, 1975). The phosphorylation depends on the protein kinase concentration, which suggests an intermolecular mechanism and does not affect the binding capacity for cAMP. The K_m for ATP is equal to 0.4 μM which is somewhat less than the K_m for ATP in the case of exogenous substrates. The phosphorylation is reversible, the dephosphorylation occurring under the action of a phosphatase or of the catalytic subunit itself which transfers the phosphate from the protein to ADP thereby forming ATP. The dephosphorylation is dependent on the addition of cAMP.

Such an endogenous 'phosphorylation–dephosphorylation' system has been demonstrated in the soluble and particulate fractions of a number of vertebrate tissues (Malkinson *et al.*, 1975). It was shown that autophosphorylation concerns only one of two different isozymes of bovine heart protein kinase (Corbin *et al.*, 1975).

C Interaction with MgATP Complex

It has been reported that the dissociation of cAMP from the regulatory subunit of protein kinase in the presence of the catalytic subunit is enhanced by the addition of the complex MgATP (Brostrom *et al.*, 1971a). Protein kinase from skeletal muscle has been shown to bind MgATP on both regulatory and catalytic subunits. The binding site of the regulatory subunit is different from the cAMP-binding site (Haddox *et al.*, 1972). Although the MgATP complex decreases the affinity of the cAMP-binding, the binding capacity is unaffected (Beavo *et al.*, 1974). Moreover, the complex seems to be essential for the recombination of regulatory and catalytic subunits. Thus, the following scheme has been proposed for protein kinase from skeletal muscle (Beavo *et al.*, 1974)

$$2\ \mathrm{cAMP} + \mathrm{R_2(MgATP)_2C_2} \rightleftharpoons \mathrm{R_2\text{–}cAMP_2} + 2(\mathrm{MgATP}) + 2\mathrm{C}. \quad (2)$$

In protein kinase from bovine heart, however, MgATP has an opposite effect on the binding affinity of cAMP (Hofmann *et al.*, 1975). It seems that physiological control by MgATP should be without regulatory significance as the intracellular concentration of MgATP is high enough to saturate the specific binding sites of the protein kinase. However, this mechanism further increases the K_a of the enzyme in the cell.

In summary, the complex MgATP exhibits a triple role with regard to the protein kinase: (*i*) it is a phosphate donor for the phosphorylation of substrates catalysed by active catalytic subunit; (*ii*) it is necessary for the autophosphorylation of the protein kinase; (*iii*) it stabilizes the binding of the regulatory and catalytic subunits possibly by the formation of a MgATP bridge between the subunits.

V PHYSIOLOGICAL SUBSTRATES

Although a number of proteins has been demonstrated to serve as protein kinase substrates in *in vitro* experiments, little is known about the physiological substrates. Krebs has proposed a set of criteria (reported by Langan, 1973) which must be satisfied to determine if a particular effect of cAMP is mediated by protein phosphorylation. (*i*) The cell type involved must contain a cAMP-dependent protein kinase. (*ii*) A protein substrate must exist which bears a functional relationship to the cAMP-mediated process. (*iii*) Phosphorylation of the substrate must alter its function *in vitro*. (*iv*) The protein substrate must be modified *in vivo* in response to cAMP. (*v*) A phosphoprotein phosphatase must exist to reverse the process.

Until now three systems fulfil, with sufficient evidence, the proposed criteria. First of all, the glycogen breakdown has, in fact, demonstrated the existence of the protein kinase as an obligatory mediator of cAMP, so that the physiological substrate — phosphorylase kinase — was discovered before the exact mechanism of activation was known. The situation is somewhat different for the two other substrates which are the liver and muscle glycogen synthetase, and the adipose tissue lipase. The phosphorylation of these two substrates leads to a decrease in glycogen synthesis and an increase in adipose tissue lypolysis, respectively.

Evidence has been obtained that the specificity of protein kinases for their physiological substrates is due to the three-dimensional configuration surrounding the serine residue of the substrate to be phosphorylated. The apparent non-specificity of protein kinase *in vitro* may result from a distortion of the substrate so that other serine or threonine residues are unmasked and thus available for the catalytic subunit. On the other hand, the extreme specificity of protein kinase for the substrate *in vivo* is supported by the findings that, for instance, two of the 200 seryl residues and none of 150 threonyl residues in the phosphorylase kinase and only one residue in lysine-rich histone and in glycogen synthetase are phosphorylated (Cohen *et al.*, 1975).

VI CONCLUSION

The regulatory function of cAMP in hormone-sensitive cells takes place through the activation of the phosphorylating enzyme — protein kinase. Protein kinase activity may be modulated by several mechanisms such as the interaction of cAMP and of a specific modulator with protein kinase and the autophosphorylation of protein kinase.

The distinguishing feature of the mode of action of cAMP is the induced dissociation of the protein kinase subsequent to cAMP binding. Because of the multiplicity of the receptor sites on the dimeric regulatory subunit, a possible interaction between these sites could lead to some co-operative effect. The action of the heat stable inhibitor allows a more efficient control of the protein kinase activity by cAMP and could be a regulator of this activity independent of the cAMP system. In order to define the repressive effect of the inhibitor, it will be necessary to know how high the protein kinase activity should be to express the physiological effects of a hormonal stimulus. The phosphorylation of the regulatory subunit by the protein kinase may lead to a much more efficient control of the activity by cAMP. Moreover, the dephosphorylation of the regulatory subunit could be under separate control so that it could directly effect the protein kinase activity.

It has to be emphasized that some protein kinases do not exhibit any dependence on cyclic nucleotides. Indeed, most of protein phosphorylation in tissues such as the thyroid is not regulated by cAMP (Lamy and Dumont, 1974). Such kinases must however be distinguished from the isolated catalytic subunit of cyclic nucleotide-dependent protein kinasess. The true independent protein kinases are not inhibited by either of the regulatory subunits nor by the heat-stable inhibitor and they may use GTP as phosphate donor for phosphorylation of substrates.

ACKNOWLEDGEMENT

This work was realized thanks to Contract of the Caisse Générale d'Epargne et de Retraite (Fonds Cancer) and to Contract of the Ministère de la Politique Scientifique within the framework of the Association Euratom — University of Brussels — University of Pisa (Actions Concertées). S. Swillens is a Fellow of the Institut pour la Recherche Scientifique dans l'Industrie et l'Agriculture (IRSIA).

The authors wish to thank Mrs. C. Rocmans for the preparation of the manuscript.

CHAPTER 5

Cyclic Nucleotides, Calcium and Cellular Control Mechanisms

M. J. Berridge and P. Rapp

I INTRODUCTION

The response of a cell to external stimuli is mediated by various intracellular second messengers. These internal signals act directly on the effector systems to bring about a change in cellular activity. The role of cAMP as a second messenger is covered in detail elsewhere in this volume. In this contribution the importance of calcium as a second messenger will be considered and special attention will be paid to the relationship between Ca^{2+} and the cyclic nucleotides.

It is commonly observed that cell activation is often associated with dramatic changes in the concentrations of both cAMP and Ca^{2+}. Control of muscle is an example in which Ca^{2+} clearly has a second messenger role in that it acts directly on troponin to initiate contraction. cAMP also has a precise second messenger function in muscle in that it stimulates a protein kinase which is responsible for activating phosphorylase (Cohen, 1974). In salivary glands, 5-hydroxytryptamine both stimulates adenylate cyclase and increases the uptake of Ca^{2+} into the cell through the plasma membrane (Prince *et al.*, 1972).

Often there has been confusion and controversy as to the relative importance of these different internal signals. It may be suggested that these are unproductive questions since Ca^{2+} and cyclic nucleotide concentrations are often interdependent; skeletal muscle is one of the few exceptions. In this paper the view that the object of study is not a specific messenger but rather the overall response to hormonal stimulation is adopted. This response may be effected by changes in the concentration of several messengers simultaneously.

II INTERACTION OF CYCLIC NUCLEOTIDES AND CALCIUM

In addition to the direct effects of Ca^{2+} and cAMP on key cellular events, it is possible to recognize more subtle actions of these internal signals. By means of positive and negative feedback loops they moderate each other's activity. Although the full details of these feedback loops have yet to be established, it is already evident that the interactions are extremely important in regulating the activity of many cells (Berridge, 1975a; Rasmussen *et al.*, 1972). Despite the fact that details are often lacking, it is possible after considering several examples to identify two control patterns which frequently appear (Figure 5.1). These feedback loops are usually employed to arrive at a new stable equilibrium in response to a hormone or some other stimulant. However, as will be discussed in a later section, in some systems, the resulting equilibrium may be unstable thus leading to sustained oscillations in Ca^{2+} and cAMP.

In loop A, cAMP causes an increase in cytosol Ca^{2+} by the release of Ca^{2+} from intracellular stores or by increasing entry through the plasma membrane. Thus, cAMP incrementally drives cytosol Ca^{2+}. Free Ca^{2+} forces a decrease in cAMP levels by either inhibiting adenylate cyclase and/or by stimulating phosphodiesterase. In loop B the opposite situation occurs; cAMP decreases Ca^{2+} levels either by increased deposition into intracellular stores or by extrusion from the cell across the plasma membrane. Thus, Ca^{2+} incrementally drives cAMP. In the overall structure the two loops are the same; an incremental process is followed by a decremental process. The experimental evidence for these control patterns will be considered briefly.

Indirect support that in loop A cAMP causes an increase in cytoplasmic Ca^{2+}, comes from experiments in which cAMP stimulation causes an increase in Ca^{2+} efflux as described in slime moulds (Chi and Francis, 1971), rat liver (Friedmann, 1972), β-cells (Brisson and Malaisse, 1973), and insect and mammalian salivary glands (Prince *et al.*, 1972; Nielsen and Petersen, 1972). More directly, Borle (1974) has shown that physiological concentrations of cAMP stimulate the release of Ca^{2+} from mitochondria of liver, kidney and heart. Howell and Montague (1975) have observed similar effects in particulate preparations of rat pancreatic islets. Another important site of action appears to be at the plasma membrane where cAMP can enhance the entry of Ca^{2+}. cAMP can prolong the influx of Ca^{2+} which occurs during the plateau phase of the cardiac action potential (Reuter, 1974b).

Ca^{2+} causes a decrease in cAMP concentration in loop A. Indirect experiments on intact cells have shown a depression in cAMP concentration associated with an increase in the intracellular Ca^{2+} level without reference to the site of Ca^{2+} action. This has been seen in toad bladder (Argy *et al.*, 1967), rat parotid (Butcher, 1975), rat renal tubules (Rasmussen and Nagata, 1970) and rat thyroid (van Sande *et al.*, 1975). In many preparations of adenylate cyclase, Ca^{2+} has been shown to be an inhibitor, for example, mouse adrenal tumours (Taunton *et al.*, 1969), frog and tadpole erythrocytes (Rosen and Rosen, 1969), turkey erythrocyte ghosts (Steer and Levitzki, 1975) and rat cerebellum

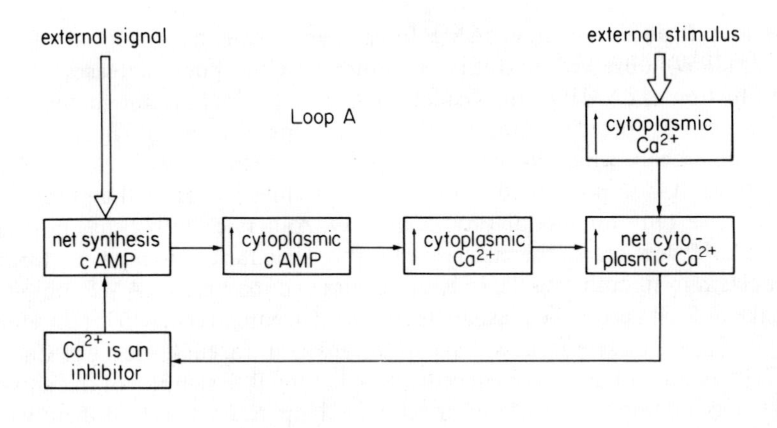

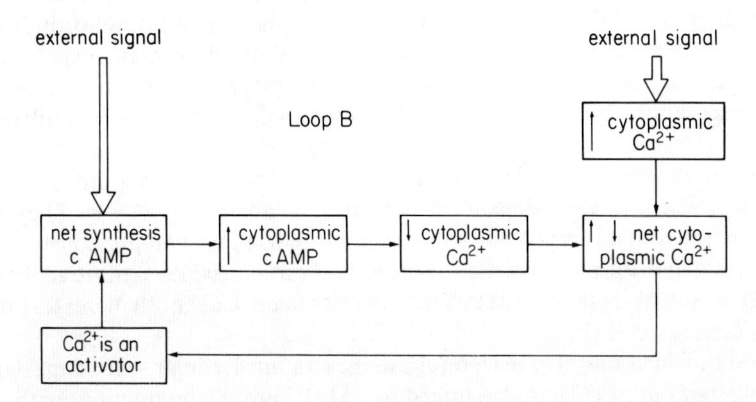

Figure 5.1 A summary of cAMP–Ca^{2+} feedback interactions arranged in the form of feedback loops. The experimental evidence for these different interactions is provided in the text. The concentrations of cAMP and Ca^{2+} can be influenced not only by external signals, but also by feedback loops within the cell. In loop A, an increase in the intracellular level of cAMP will lead to an increase in the level of Ca^{2+}in, for example, β-cells, liver, and insect and mammalian salivary glands. In loop B, an increase in cAMP will lower the level of Ca^{2+} and will thus oppose the action of external signals which act to increase cytoplasmic Ca^{2+}—such antagonistic effects are found in lymphocytes, mast cells and smooth muscle.

(Johnson and Sutherland, 1973). Ca^{2+} may also cause a decrease in cAMP by stimulating phosphodiesterase. This has been observed in rat cerebral cortex (Kakiuchi *et al.*, 1973), porcine cerebral cortex (Brostrom and Wolff, 1974) and bovine heart (Teo and Wong, 1973a). However, Kakiuchi *et al.* (1973) have shown that the resulting activated phosphodiesterase preferentially hydrolyses cGMP rather than cAMP. As will be pointed out in the discussion of loop B, this particular activation system is probably not significant in loop A.

In loop B, an increment in cAMP forces a decrement in cytosol Ca^{2+}. Entman *et al.* (1969) observed that the sarcotubular Ca^{2+} pool increases after the introduction of cAMP to the system. Tada *et al.* (1975) have shown that cAMP stimulates a protein kinase, which phosphorylates a protein called phospholamban, which in turn activates Ca^{2+} transport into the sarcoplasmic reticulum. It is important to note that cAMP does not drive the pump, but is responsible only for modulating its activity. Andersson (1972) has postulated that cAMP accelerates the mechanism responsible for removing Ca^{2+} from the cytoplasm in smooth muscle. In lymphocytes and mast cells, cAMP inhibits the uptake of Ca^{2+} across the plasma membrane (Freedman *et al.*, 1975; Foreman *et al.*, 1975). All these effects will act to lower the intracellular level of Ca^{2+}.

Ca^{2+} increases the cAMP concentration in loop B. Examples of this situation are not as common as the inhibitory effect in loop A. Part of the difficulty lies in experimental procedures. Sufficiently high concentrations of Ca^{2+} will cause a depression in cAMP either directly at the adenylate cyclase level or indirectly through a general inhibition of overall cell metabolism. In many experimental situations it is impossible, owing to membrane vesiculation during homogenization, to determine exactly how much Ca^{2+} bathes the adenylate cyclase. However, examples of Ca^{2+} activation of adenylate cyclase have been found. Brostrom *et al.* (1975) have isolated a small protein from porcine cerebral cortex which is a Ca^{2+}-dependent activator of adenylate cyclase. They have demonstrated that this protein is the same as the one they had previously shown to activate in the presence of Ca^{2+} a phosphodiesterase which is more active with cGMP as substrate than cAMP. Thus, in this system Ca^{2+} both increases cAMP and decreases cGMP.

cGMP, which has also been implicated as an intracellular second messenger (Goldberg *et al.*, 1973), is also linked to cAMP through interactions with Ca^{2+} (Berridge, 1975a). There is evidence to support the idea that guanylate cyclase is regulated by the intracellular level of Ca^{2+} (Schultz *et al.*, 1973; Schultz and Hardman, 1975). The ability of the divalent ionophore A-23187 to induce a Ca^{2+}-dependent increase in the intracellular level of cGMP in thyroid (van Sande *et al.*, 1975) and parotid (Butcher, 1975) lends support to this view.

The systems in Figure 5.1 have been referred to as control patterns. This phrase suggests a form of metabolic control frequently present, but not rigidly imposed under all circumstances. Indeed, while there are cell types in which components of either loop A or loop B appear to be the dominant factors, it is also true that some systems present the characteristics of either loop A or loop B depending on the stage in their life cycle and also depending on environmental factors, such as the degree of hormonal stimulation. It is possible for a cell to modulate between the dynamic elements of loop A and loop B both at the cAMP and Ca^{2+}-regulatory ends of the loop. For example, there is usually more than one intracellular reservoir containing bound Ca^{2+} (for example, microsomes and mitochondria). It has often been found that cAMP can have opposite effects on the Ca^{2+}-binding behaviour of different reservoirs. Andersson *et al.* (1975) have shown in rabbit colon that cAMP causes a decrease in mitochondrial Ca^{2+} but an increase in microsomal levels of the cation.

A consequence of coupling Ca^{2+} and cAMP concentrations with feedback loops (as in Figure 5.1) is that it might be possible for these systems to oscillate. This would result in sustained, uniform variations in concentrations with time. The existence of these oscillations may explain a variety of rhythmic phenomena, such as the cardiac pacemaker (Noble, 1975), myogenic activity in smooth muscle (Bolton, 1971; Ohba *et al.*, 1975) and fluctuations in membrane potential found in β-cells (Dean *et al.*, 1975) and insect salivary glands (Berridge *et al.*, 1975b). Measuring oscillations of cAMP or cGMP directly is extremely difficult and at present there is little direct evidence for such oscillations except in experiments on the frog heart to be described later. However, oscillations in membrane potential can be routinely measured and it is proposed that such oscillations are driven by periodic variations in the level of Ca^{2+} and cAMP (the potential is said to be entrained by cAMP and Ca^{2+}). This hypothesis could satisfactorily explain potential oscillations only if it is possible for Ca^{2+} and cAMP to bring about changes in the ionic permeability of membranes. This is indeed the case. Ca^{2+} is known to produce marked changes in K^+ (Romero and Whittam, 1971; Meech, 1974) and Cl^- permeability (Berridge *er al.*, 1975a). Similarly, cAMP has been implicated in the regulation of membrane permeability in toad bladder through its ability to stimulate the dephosphorylation of membrane proteins (Walton *et al.*, 1975). As described earlier, cAMP is capable of altering Ca^{2+} movement across the plasma membrane (Reuter, 1974b; Freedman *et al.*, 1975). It is not difficult to see, therefore, how oscillations in cAMP and Ca^{2+} could result in periodic changes in membrane potential. We do not wish to overstress the importance of second messenger oscillations. As the following specific cases show most cells probably use control systems, loop A and loop B, to maintain nearly constant concentrations of Ca^{2+} and cAMP. Oscillatory behaviour is exceptional. The maintenance of homeostasis is the principal object of most metabolic regulatory processes.

III THE ROLE OF CALCIUM AND CYCLIC NUCLEOTIDES IN CELL ACTIVATION

Ca^{2+} seems to play a central role in regulating the activity of many cells. Although cAMP has an essential role to play in certain cells, there is increasing evidence that many effects of cAMP are mediated indirectly through effects on Ca^{2+} (Berridge, 1975a). In some cells, for example, β-cells, insect salivary gland and liver, cAMP appears to cause an increase in the Ca^{2+} signal, as in loop A, whereas in others, for example, smooth muscle, mast cells and lymphocytes, cAMP acts to reduce the level of Ca^{2+}, as in loop B. The following examples have been selected to illustrate not only the central role of Ca^{2+} in cell activation but also the co-operation and interactions which exist between second messengers.

A Contraction

1 Skeletal Muscle

The contraction of skeletal muscle is triggered by the release of Ca^{2+} from the sarcoplasmic reticulum. The interaction of Ca^{2+} with troponin causes a displacement of tropomyosin, thus enabling myosin to interact with actin. In addition to stimulating contraction, Ca^{2+} can also stimulate metabolism through its ability to activate the enzyme phosphorylase kinase (Cohen, 1974). This ability of Ca^{2+} to increase glycogen metabolism means that there is close coupling between contraction and the supply of energy. Glycogen metabolism can also be stimulated by adrenaline using cAMP as an intermediary (Figure 5.2). cAMP acts by stimulating the conversion of phosphorylase kinase *b* to the active phosphorylase kinase *a*.

In skeletal muscle, therefore, Ca^{2+} stimulates contraction whereas metabolism can be activated by either cAMP or Ca^{2+}. In this system, however, there does not appear to be any direct interactions between cAMP and Ca^{2+}. Although they can both activate metabolism, they do so independently of each other (Figure 5.2).

2 Cardiac Muscle

The control mechanisms underlying the rhythmical beating of the heart are extremely complex. Not only is it necessary to understand excitation–contraction coupling during each heart beat, but also the subtle inotropic and chronotropic effects of adrenaline or acetylcholine should be considered. Some of the complexity can be avoided if the different regions of the heart are assessed separately.

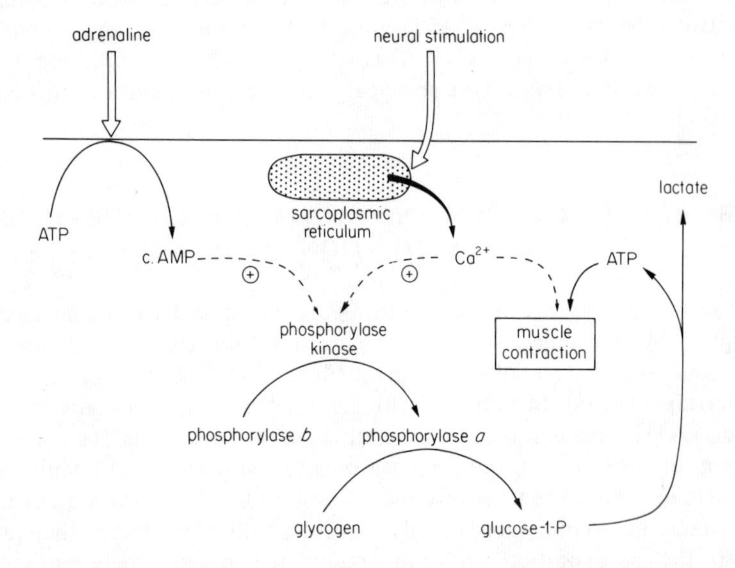

Figure 5.2 The control of contraction and glycogen metabolism in skeletal muscle

The contractile cells are driven by action potentials originating in the sino–atrial (S–A) node. This cardiac action potential brings about an increase in the intracellular Ca^{2+} level which then triggers contraction. The precise source of this activator Ca^{2+} is still being debated. Some of the Ca^{2+} appears to enter from the extracellular medium during the plateau phase of the action potential (Reuter, 1973), but the bulk of the cation is probably released from intracellular reservoirs which are most likely to be in the sarcoplasmic reticulum. However, it is still argued that the mitochondria may be a source of Ca^{2+}. The heart relaxes when this Ca^{2+} signal is removed by being pumped back into the reservoirs and by being removed from the cell in exchange for Na^+ (Reuter, 1974).

This regular ebb and flow of Ca^{2+} during each heartbeat can be modulated by adrenaline which can increase both the force of contraction and the rate of relaxation. These effects of adrenaline are mediated by cAMP through effects on Ca^{2+}. Firstly, cAMP prolongs the plateau phase of the action potential, thus increasing the amount of Ca^{2+} which enters the cells during excitation (Reuter, 1974). Secondly, cAMP stimulates the Ca^{2+}-pump on the sarcoplasmic reticulum—through effects on phospholamban as described earlier (Tada *et al.*, 1975)—thus accelerating the rate of relaxation. cAMP thus exerts a profound effect on Ca^{2+} metabolism during both contraction and relaxation. The feedback interactions between cAMP and Ca^{2+} may also exist in the S–A node where they could play an important role in generating the pacemaker potential.

The rhythmical trains of action potentials which drive the contractile cells originate in the S–A node. These action potentials are transmitted to the ventricles by way of the Purkinje cells. During the last few years there has been considerable progress in our understanding of how these spontaneous electrical signals are generated (Noble, 1975). Pacemaker activity in the S–A node seems to depend on the decay of a specific K^+ current (I_{K_i}) which is responsible for repolarizing the membrane at the end of the action potential (Noble, 1975). Once the membrane has been repolarized, this potassium current is slowly switched off, thus causing the membrane to depolarize gradually (the pacemaker potential). When the pacemaker potential reaches a critical level another action potential is induced and the whole process repeats itself.

The description of pacemaker activity relies solely on changes in membrane permeability being entrained to each action potential. However, McDonald and Sachs (1975) have found that the pacemaker activity of embryonic heart cells persists even when action potentials are blocked by tetrodotoxin. It is conceivable, therefore, that pacemaker activity may also depend on some internal metabolic oscillator resembling the feedback loops described earlier (Figure 5.1; see also Rapp and Berridge, 1976). In this model we propose that there is an interplay between internal oscillations in cAMP and Ca^{2+} and the changes in membrane permeability which are responsible for pacemaker activity. The exact nature of the feedback loop responsible for these second messenger oscillations is still not clear but, on the basis of the studies on the cardiac cells described earlier, it is likely to resemble loop B.

Some evidence for the existence of such oscillations has come from freeze-clamp studies on frog heart, which have revealed the existence of oscillations in both cAMP and cGMP during the course of a single heart beat (Wollenberger *et al.*, 1973; Brooker, 1975). If such oscillations in the levels of cAMP (and hence in Ca^{2+} through feedback interactions similar to those described earlier) exist in the S–A node, they could play an important role in driving pacemaker activity.

In order to relate oscillations in cAMP and intracellular Ca^{2+} to pacemaker activity, it is necessary to consider how this internal oscillator is linked to the membrane. There are at least two possibilities. Firstly, there is evidence that the I_{K_i} responsible for repolarization is switched on by Ca^{2+} (Isenberg, 1975). Similar effects of Ca^{2+} on K^+ conductance have been described in red blood cells (Romero and Whittam, 1971) and *Helix aspersa* nerve cells (Meech, 1974). Therefore, any fluctuation in the intracellular level of Ca^{2+} will be reflected in a corresponding fluctuation in K^+ permeability and hence membrane potential. Secondly, oscillations in cAMP may also cause changes in potential because this nucleotide is thought to enhance Ca^{2+} entry across the plasma membrane of heart cells (Reuter, 1974). These effects of cAMP and Ca^{2+} on membrane permeability may be particularly important in mediating the positive chronotropic effect of adrenaline. The increase in pacemaker activity seems to depend not only on the I_{K_i} being switched off faster but also on an increase in Ca^{2+} permeability (Noble, 1975). This acceleration of the membrane events responsible for the pacemaker potential may depend on the ability of adrenaline to increase the level of cAMP, thus leading to an acceleration of the internal cAMP–Ca^{2+} oscillator. It is interesting to note that Brooker (1975) has found that adrenaline can increase the amplitude of the cAMP oscillations in frog heart.

3 *Smooth Muscle*

The contraction of smooth muscle, like that of cardiac and skeletal muscle, is triggered by an increase in the intracellular level of Ca^{2+}. The source of this Ca^{2+} varies, some smooth muscles, for example, *Taenia coli* are very dependent on extracellular Ca^{2+} whereas various vascular smooth muscles seem to use intracellular Ca^{2+} (Berridge, 1975a). Such variations in the dependence of different muscles on external Ca^{2+} are well correlated with the relative volumes of the sarcoplasmic reticulum (Devine *et al.*, 1972). Most smooth muscles are excitable and contraction is triggered by an action potential. In *Taenia coli* the charge carrier is Ca^{2+} and during each action potential enough Ca^{2+} enters the cell to induce a small, but significant, contraction (Bülbring and Tomita, 1970; Tomita, 1970). These action potentials may arise either through nervous activity or as the result of an endogenous myogenic rhythm which has been shown to be present in many smooth muscles (Bolton, 1971; Tomita and Watanabe, 1973; Connor *et al.*, 1974; Ohba *et al.*, 1975) and which is reminiscent of the cardiac pacemaker potential described earlier. The mechanism of this myogenic activity remains a matter of speculation (Prosser, 1974).

When intestinal smooth muscles are voltage-clamped, there are rhythmical inward currents which have the same frequency as the slow potential waves

(Connor *et al.*, 1974; Ohba *et al.*, 1975). These slow waves are also relatively independent of variations in membrane potential, but seem to be very sensitive to changes in metabolic activity. This suggests that they may be driven by some internal oscillator similar to that postulated for the heart. As in the heart, interactions between cAMP and Ca^{2+} have been described (Andersson, 1972; Bär, 1974). The ability of β-adrenergic agents to relax various smooth muscles is thought to depend on the ability of cAMP to stimulate the removal of Ca^{2+} (Andersson *et al.*, 1975).

B Secretion

1 Insulin-Secreting β-Cells

The release of insulin by exocytosis from β-cells is triggered by an increase in the intracellular level of Ca^{2+} above a minimum threshold level (Matthews, 1975). This Ca^{2+} passes through the plasma membrane when the permeability is increased by the metabolic products of glucose (Dean *et al.*, 1975). As the level of glucose in the blood rises there is a progressive depolarization of the β-cell membrane which, at a critical level, suddenly displays Ca^{2+}-dependent action potentials (Dean and Matthews, 1970; Pace and Price, 1974; Matthews and Sakamoto 1975). As mentioned earlier in the case of heart and smooth muscle, these action potentials are caused by phasic increases in Ca^{2+} permeability. The Ca^{2+} which floods into the cell is then responsible for stimulating the release of insulin granules. In the presence of Ca^{2+}, the ionophore A-23187 is able to induce a sustained release of insulin (Wollheim *et al.*, 1975). The ability of glucose to depolarize the β-cell membrane is extremely complicated and seems to necessitate glucose entering the cells and being metabolized via the glycolytic pathway (Dean *et al.*, 1975). Under certain conditions, the membrane potential of β-cells oscillates and often shows spikes at the crests of the waves (Dean *et al.*, 1975; Matthews and Sakamoto, 1975). These oscillations are very reminiscent of the slow waves found in smooth muscle and it is suggested that they may be driven by an underlying oscillation of cAMP and Ca^{2+}.

There is an important feedback interaction operating between this cyclic nucleotide and Ca^{2+} which could form the basis of a feedback loop similar to that of loop A. There is considerable evidence to show that agents which increase the intracellular level of cAMP, such as glucagon or phosphodiesterase inhibitors, are capable of stimulating the release of insulin (Montague and Howell, 1975). An increase in the intracellular level of cAMP by itself is not sufficient to stimulate secretion. However, cAMP is capable of potentiating the effect of glucose (Montague and Cook, 1971; Cooper *et al.*, 1973). cAMP seems to act by stimulating the release of Ca^{2+} from an intracellular pool (Brisson and Malaisse, 1973). Evidently, the amount of the cation which can be released from intracellular stores does not exceed the secretion threshold. Howell and Montague (1975) have also observed that cAMP can inhibit the accumulation of Ca^{2+} in subcellular fractions. It thus appears that cAMP can modulate the intracellular level of Ca^{2+} in β-cells by causing a release of Ca^{2+} from internal reservoirs.

2 Insect salivary gland

Both cAMP and Ca^{2+} seem to play an important role in the regulation of fluid secretion by the salivary glands of the adult blowfly (Berridge, 1975a,b; Berridge and Prince, 1972). During the action of 5-hydroxytryptamine (5-HT) there appears to be an increase in the intracellular level of both cAMP and Ca^{2+}. The latter acts by simultaneously increasing the Cl^- permeability of both the apical and basal membranes (Berridge *et al.*, 1975a,b). There is also some indication that Ca^{2+} exerts a negative feedback effect on adenylate cyclase (Prince *et al.*, 1972). The precise mode of action of cAMP has not been established.

Electrophysiological experiments suggest that cAMP stimulates an electrogenic K^+ pump thought to be located on the apical membrane (Berridge and Prince, 1972; Prince and Berridge, 1973). However, it is not clear whether this is a direct effect of cAMP on the pump (as shown in Figure 5.3), or whether the system depends on the apparent ability of cAMP to release Ca^{2+} from intracellular reservoirs (Prince *et al.*, 1972). There is indirect evidence, therefore, for the feedback interactions depicted in Loop A (Figure 5.1). Thus, it is not surprising to find that under certain conditions the transepithelial potential oscillates (Berridge and Prince, 1972; Prince and Berridge, 1973). If these membrane oscillations reflect an underlying oscillation in the intracellular levels of Ca^{2+} then there should be corresponding changes in Cl^- permeability. One way of detecting such changes is to measure electrophysiologically trans-

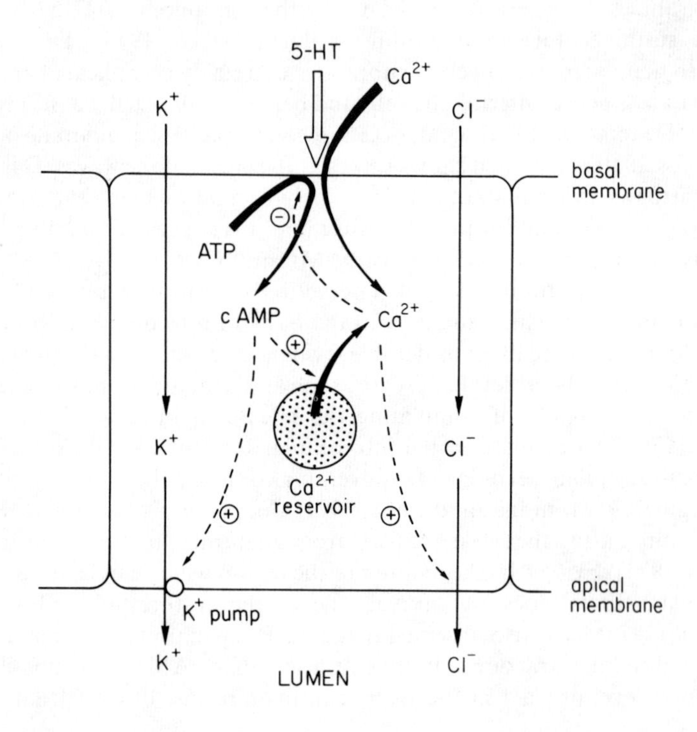

Figure 5.3 Control of fluid secretion in the blowfly salivary gland

epithelial resistance which has been found to vary in phase with the potential oscillations (Berridge, *et al.*, 1975a).

3 *Mast Cells*

The release of histamine from mast cells and leucocytes is triggered by the interaction of an antigen with a membrane-bound antibody. The result of this antigen–antibody interaction seems to be an increase in the membrane permeability to Ca^{2+}. The cation then floods into the cell triggering the release of histamine (Douglas, 1975; Foreman and Mongar, 1975). Other agents, such as the cationic compound 48/80 (Hogberg and Uvnäs, 1960), α-adrenergic and cholinergic agents (Kaliner *et al.*, 1972), and the ionophore A-23187 (Foreman *et al.*, 1973; Cochrane and Douglas, 1974) can also stimulate the release of histamine. These substances probably mimic the action of the antigen by increasing the Ca^{2+} permeability of the plasma membrane. In contrast to its stimulatory effect on secretion in β-cells and insect salivary gland, cAMP can markedly inhibit the Ca^{2+}-dependent release of histamine (Lichtenstein *et al.*, 1970; Loeffler *et al.*, 1971). The ability of β-adrenergic agents and prostaglandins to increase the intracellular level of cAMP correlates well with their ability to inhibit antigen-induced release of histamine (Bourne *et al.*, 1972; Orange *et al.*, 1971). Orange *et al.* (1971) have suggested that cAMP may act by decreasing the intracellular level of Ca^{2+}. Since Bt_2-cAMP was found to markedly inhibit the normal antigen-induced uptake of Ca^{2+} (Foreman *et al.*, 1975), it has been proposed that the action of cAMP may be exerted at the cell membrane where it closes a Ca^{2+} 'gate'.

C Cell Division

Lymphocytes

The ability of a variety of mitogens to stimulate cell division in lymphocytes appears to depend on an early rise in the intracellular level of Ca^{2+} (Berridge, 1975b). During the action of mitogens, such as phytohaemagglutinin and concanavalin A, there is an increase in the uptake of Ca^{2+} (Freedman *et al.*, 1975). When a wide range of compounds was tested, a close correlation was shown between the ability of agents to stimulate both the uptake of Ca^{2+} and cell division (Parker, 1974a). The divalent ionophore A-23187 is also capable of stimulating the synthesis of DNA (Maino *et al.*, 1974). The increase in Ca^{2+}, which is thought to occur during the action of mitogens, may account for the observed increases in cGMP levels (Hadden *et al.*, 1972). Ca^{2+} and cell division may be linked through cGMP which may initiate early events in the cell cycle by phosphorylating nuclear acidic proteins (Johnson and Hadden, 1975). There also appears to be an important interaction operating beween cAMP and Ca^{2+}. Elevation of the intracellular level of cAMP by direct application of the nucleotide or by treatment with prostaglandin E_1, cholera toxin or theophylline can prevent lectin-induced transformation of lymphocytes (DeRubertis *et al.*, 1974; Weinstein *et al.*, 1974). This ability to inhibit cell division seems to depend on the ability of cAMP to inhibit the influx of Ca^{2+} as described earlier (Freedman *et al.*, 1975).

IV CONCLUSIONS

In this contribution the properties of Ca^{2+}–cyclic nucleotide interactions have been outlined. Depending on the system being considered, Ca^{2+} increases or decreases cAMP and cGMP levels by activating or inhibiting adenylate cyclase and guanylate cyclase, respectively, or phosphodiesterase. cAMP can affect the cytoplasmic concentration of Ca^{2+} either by altering the plasma membrane permeability or by releasing the cation from intracellular reservoirs. From the examples considered, it is possible to construct two forms of Ca^{2+}–cAMP control loops which are consistent with the often indirect evidence available. In the past, studies have tended to concentrate on individual second messengers but the above discussions should underline the importance of integrated studies which consider not a single component of the control loop in isolation but rather the dynamical structure of the overall system.

The control loops presented here only begin to approach the level of complexity encountered in actual biological cases. These limitations are recognized by the authors. However, imperfect as our understanding is, these models have made it possible to produce a unifying picture of Ca^{2+}–cyclic nucleotide interactions. It has been shown that such interactions are probably central to the organization of three of the most basic physiological processes: contraction, secretion and cell division.

In addition, a mechanism has been postulated which allows an explanation of rhythmic behaviour in several different biological systems. Even though the proposed models are almost certainly incorrect in detail, they are probably accurate in terms of a broad outline; namely that coupled oscillations in Ca^{2+} and cyclic nucleotides constitute the driving oscillator of these observed rhythms.

CHAPTER 6

Cyclic Nucleotide Analogues

J. P. Miller

I INTRODUCTION

Analogues of cyclic nucleotides offer the possibility of mimicking the physiological actions of the naturally occuring cyclic nucleotides. A number of reviews have appeared which discuss these possibilities (Smith, 1966; Amer and McKinney, 1973, 1974, 1975). The ideal analogue for mimicking the biological action of a natural cyclic nucleotide would be one which (*i*) is able to penetrate cell membranes easily, (*ii*) is resistent to enzymic degradation by cyclic nucleotide phosphodiesterase, (*iii*) is equal to or better than the natural parent cyclic nucleotide in producing its biological effect, and (*iv*) is tissue or enzyme-specific. Another possibility is to develop analogues which are antagonistic to the physiological actions of a natural cyclic nucleotide. These analogues can also be used as prodrugs or drug-delivery forms of nucleosides or nucleoside 5′-phosphates.

A Some Comments on Dibutyryl-cAMP

Analogues of cAMP were first synthesized in the hopes of finding derivatives which could mimic the effects of cAMP in various biological systems. By far, the most extensively utilized analogue of cAMP is N^6, 2′-*O*-dibutyryl-cAMP (Bt_2-cAMP). This analogue was first reported by Posternak *et al.* (1962) and in a subsequent report (Henion *et al.*, 1967) was found to be among the more biologically active of a group of N^6-, 2′-*O*-acyl-, N^6-acyl- and 2′-*O*-acyl-cAMP derivatives. In many cases Bt_2-cAMP has proved to be more effective *in vivo* than cAMP, 2′-*O*-butyryl-cAMP or N^6-butyryl-cAMP, while only cAMP and

N^6-butyryl-cAMP are active *in vitro*. It has been postulated that the greater activity of Bt_2-cAMP *in vivo* is due to its greater cell permeability due to the acyl substitutions (Robison *et al.*, 1971). This conclusion has not always been verified by other investigators (Ryan and Durick, 1972). On the other hand, it is generally agreed that when used *in vivo*, N^6-butyryl-cAMP is the biologically active metabolite of Bt_2-cAMP inside the cell (Kaukel and Hilz, 1972; Kaukel *et al.*, 1972b; Miller *et al.*, 1973c; Nath and Rebhun, 1973).

This article will not consider Bt_2-cAMP separately. Rather, it will be considered within a chemical class of compounds: N^6,2′-*O*-diacyl-cAMP derivatives. In addition, when the data is available, the biological effects of Bt_2-cAMP will be compared to those of the other derivatives described.

II INTERACTION OF CYCLIC NUCLEOTIDE ANALOGUES WITH ISOLATED ENZYMES

The most systematic and complete data on analogues of cyclic nucleotides are on their interaction with isolated enzymes, most notably cyclic nucleotide-dependent protein kinases and cyclic nucleotide phosphodiesterases. In order to examine in a systematic, yet general manner, the effects of chemical substitutions and/or modifications of cyclic nucleotides on the biological activity of the resulting derivatives, the modifications in each of the four rings that make up the purine cyclic phosphate will be examined separately, and then the di- and tri-ring substitutions will be discussed.

The relative potency of the analogues as activators of cyclic nucleotide-dependent protein kinases can be conveniently compared by examining the K'_a values for the analogues. (The K'_a is the ratio of the activation constant (K_a) for the parent cyclic nucleotide, for example, cAMP or cGMP, to the K_a value for the analogue.) K'_a (cA) refers to values of K'_a where cAMP is used as the reference compound and K'_a (cG) to values refering to cGMP. Certain valid criticisms of K_a values determined using a single enzyme concentration have been pointed out (Soderling and Park, 1974; Swillens *et al.*, 1974). Some of the discrepancies between K'_a values reported by different investigators for the same analogues may be explained by the fact that the concentration of the protein kinase affects the equilibrium of the binding reaction. This shortcoming in the nature of the data should always be kept in mind. The results obtained in the author's laboratories have shown repeatedly that, when concentrations of protein kinase are varied, the K'_a values are affected considerably less than are the K_a values from which the corresponding K'_a values are derived. The K'_a values presented in Tables 6.1–6.4 are, in most cases, a compilation of values reported by different investigators using enzymic preparations from a variety of sources. In most cases the values are similar for a single analogue as an activator of two or more different protein kinases.

The ability of the analogues to serve as substrates for cyclic nucleotide phosphodiesterases has been compared by calculating the relative rate of hydrolysis. This is the ratio of the rate of hydrolysis for the analogue to the rate for the normal cyclic nucleotide. Again, the results are an accumulation of the results of many investigators using enzymic preparations from different sources. In this case, too, the results reported by different authors using a variety of enzymes with the same analogue are frequently consistant.

These hydrolysis data in most cases were determined at a cyclic nucleotide concentration of 0.5 - 5.0 mM and, therefore, represent hydrolysis by the 'high K_m' phosphodiesterases. This fact is important when one considers that the protein kinases are activated by 1 — 100 nM cyclic nucleotide. Analogues of cyclic nucleotides most probably would not normally reach a concentration of 0.5 — 5.0 mM in the cell. In addition, at concentrations of cyclic nucleotide of less than 1 μM the 'high K_m' phosphodiesterases would be expected to have little or no effect on the hydrolysis of the analogues. Only the 'low K_m' enzymes would be expected to affect significantly the intracellular concentration of analogues in the concentration range at which they can maximally stimulate protein kinase. An important limitation to the phosphodiesterase data, therefore, is that they have not been determined using the class of enzymes which most probably would hydrolyse the analogues inside the cell.

A Activation of Cyclic Nucleotide-Dependent Protein Kinases by Cyclic Nucleotide Analogues

The largest number of synthetic derivatives of cyclic nucleotides is derived from the modification of the purine ring. Table 6.1 summarizes the effects of some compounds with single modifications of the pyrimidine moiety of the purine ring on the ability of the resulting derivatives to stimulate cAMP-dependent protein kinases.

Certain N^1-substitutions of cAMP result in retention of activity, while other N^1-substitutions significantly diminish the activity (e.g., 1-Me- and 1-alkoxy-derivatives). It has been suggested that the relatively greater basicity of the 1-alkoxy-derivatives compared with cAMP is the cause of the loss of activity by these compounds (Meyer *et al.*, 1973b). In addition, the steric and electronic similarity of 1-alkoxy-cAMP and cAMP may account for the activity of the former (Meyer *et al.*, 1973b; Uno *et al.*, 1976a). That larger 1-alkyl derivatives (e.g. 1-$CH_3CH=CHCH_2$- and 1-$PhCH_2$-cAMP) demonstrate greater activity than 1-Me-cAMP may be accounted for by visualizing an interaction between the larger substitutions and the protein kinase which facilitates binding. The 1-Me-substitution of cGMP is also detrimental to its activity, while 1-NH_2-cGMP retains relatively more kinase-stimulating capacity.

Among the large number of 2-substituted cAMP derivatives reported, the 2-halo-, 2-*n*-alkyl- and 2-alkylthio-derivatives are the more active. The 2-aryl-, 2-amino-, 2-alkylamino-, 2-keto- and 2-thio- derivatives are significantly less active. Using multiple regression analysis, the K'_a (cA) data for 14 of these cAMP derivatives have been correlated with the chemical and physical

Table 6.1 Examples of activation of cAMP-dependent protein kinases by cyclic nucleotide analogues with *one* substitution in the *pyrimidine* ring

analogue	K_a'(cA) tissue	references
1 or 2-substituted-cAMP derivatives		
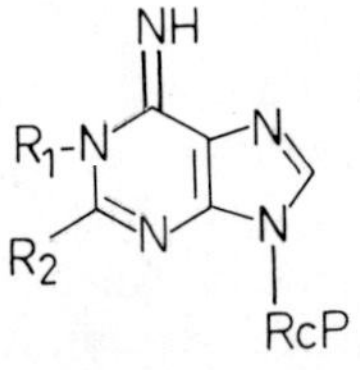 [structure: R_1-N, =NH, R_2, N-RcP]		
R_1 = $-CH_2CH=CHCH_3$, $-CH_2Ph$	0.2–0.8 bovine brain 0.008–0.051 bovine brain	Miller *et al.*, unpublished Meyer *et al.*, 1973b.
R_2 = *n*-alkyl: -Me, $-CF_3$, -Et, -*n*-octyl, -*n*-decyl, -CH=CHPh, $-CH_2CH_2Ph$.	0.16–1.6 bovine brain	Meyer *et al.*, 1975a
R_2 = $-NH_2$, $-NMe_2$, $-NHCH_2Ph$, $-NH(CH_2)_2NH_2$	0.007–0.12 bovine brain, heart; rabbit skeletal muscle	Meyer *et al.*, 1975a Panitz, 1974 Dills *et al.*, 1976
N^6*-substituted-cAMP derivatives*		
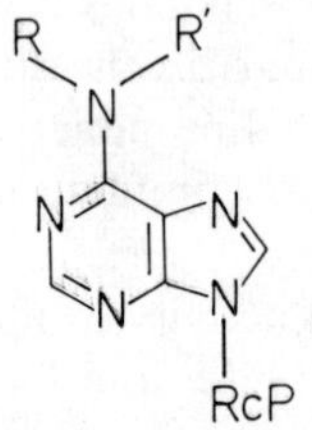 [structure: R, R' on N; N-RcP]		
N^6-monoalkyl, R = –H; R' = –Me, –Et, $-CH_2CH=CH_2$, –*n*-Bu, –*n*-Pentyl, $-CH_2CH_2CH=C(CH_3)_2$, $-CH_2CH=C(Cl)CH_3$, $-(CH_2)_2NH_2$, $-(CH_2)_4NH_2$, $-(CH_2)_6NH_2$, $-(CH_2)_9NH_2$, $-(CH_2)_{12}NH_2$, $-CH_2$–(2-furyl), $-CH_2Ph$, $-CH_2Ph$-*o*-Cl, $-CH_2Ph$-*o*-Me, $-CH_2Ph$-*p*-Me, -adamantyl	0.5–2.0 bovine brain, heart; human erythrocytes; procine brain; rabbit muscle	DuPlooy *et al.*, 1971; Meyer *et al.*, 1972; Kuo and Greengard, 1974; Kuo *et al.*, 1974a; Panitz, 1974; Hong *et al.*, 1975; Miller *et al.*, 1975; Severin *et al.*, 1975; Dills *et al.*, 1976
N^6-alkoxy, R = –H; R' = –OH, –OMe, –OEt, $-OCH_2Ph$	0.044–1.0 bovine brain; rat liver	Meyer *et al.*, 1973b; Miller *et al.*, 1975a
N^6-alkylcarbamoyl, R = –H; R' = –C(O)NHMe, –C(O)NH-*n*-Pr, –C(O)NH-*t*-Bu, –C(O)NH-*i*-Bu, –C(O)NH-*n*-$(CH_2)_7-CH_3$, $-C(O)NHCH_2CH=CH_2$, $-C(O)NHCH_2COOH$, –C(O)NHCH(COOH)CHOHCH$_3$	0.72–1.8 bovine brain, heart	Boswell *et al.*, 1973; Hong *et al.*, 1975

analogue	K_a'(cA) tissue	references
N^6-arylcarbamoyl, R = –H R′ = –C(O)NHPh, –C(O)NHPh-*o*-Cl	1.2–4.9 bovine brain, heart	Boswell *et al.*, 1973; Hong *et al.*, 1975
6-substituted-cNMP derivatives 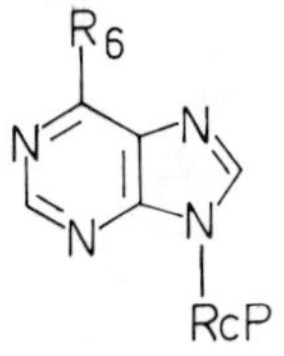		
R_6 = =O(cIMP), –OMe, =S, –SMe, –SEt, –SCH_2Ph, –Cl	0.38–2.0	DuPlooy *et al.*, 1971; Meyer *et al.*, 1972; Kuo and Greengard 1974; Kuo *et al.*, 1974a; Miller *et al.*, 1975

properties of various 2-substituents (Meyer *et al.*, 1975a). It is concluded that a ‘linear hydrophobic slot’ exists next to the binding locale of the 2-position in bovine brain cAMP-dependent protein kinase. In addition, the data clearly show that an electron-withdrawing substituent at C-2 of the purine ring gives greater K_a' (cA) values.

2-Substituted derivatives of cAMP are very poor cGMP-dependent protein kinase activators. 5-Amino-1-β-D-ribofuranosylimidazole-4-carboxamidine cyclic-3′,5′-monophosphate (2-nor-cAMP) and 6-amino-1-β-D-ribo-furanosylimidazole-4-carboxamide cyclic-3′,5′-monophosphate (2-nor-cGMP or 2-nor-cIMP) can be viewed as cAMP and cGMP/cIMP, respectively, with the atoms of the 2-position missing. Neither of these compounds is an effective stimulator of its respective kinase (Moyer *et al.*, 1973b; Kuo and Greengard, 1974, Kuo *et al.*, 1974a). In addition, N^2-Bt-cGMP demonstrates very low K_a' (cG) values.

A very large number of N^6-substituted derivatives of cAMP have been synthesized. As a group, N^6-mono-and dialkyl-, acyl-, aryl-, alkoxy-, alkoxycarbonyl- and substituted-carbamoyl derivatives are all good kinase activators. All 6-substituted cyclic nucleotide derivatives so far reported are also good activators (Table 6.1). These compounds are also able to activate a cGMP-dependent protein kinase, as are N^6-Et and N^6-Et_2-cAMP. In contrast, N^6-$PhCH_2$- and Ph(O)-C-cAMP demonstrate quite small K_a'(cG) values (Kuo and Greengard, 1974, Kuo *et al.*, 1974a).

A number of derivatives has been described which contain two or three substitutions in the pyrimidine moiety of the purine cyclic phosphates (Table 6.2). The known, 1,2-disubstituted-cAMP derivatives and substituted 2-nor-cAMP derivatives are inactive with both classes of kinases as are 2,6-disubstituted derivatives of cNMP (Jastorff and Freist, 1974; Miller *et al.*,

Table 6.2 Examples of activation of cyclic nucleotide-dependent protein kinases by cyclic nucleotide analogues with *two* or *three* substitutions in the *pyrimidine* ring

analogue	K_a'(cA) [K_a'(cG)]	tissue	references
1,2-disubstituted cAMP derivatives			
1,N^6-disubstituted-cAMP derivatives			
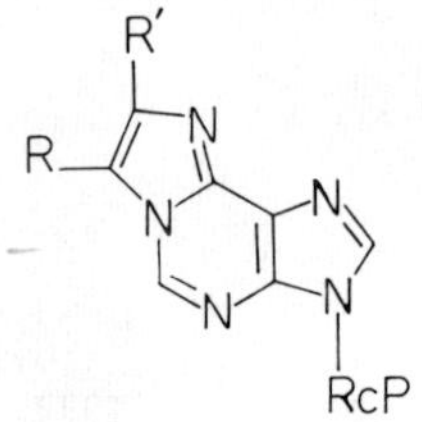			
R = H; R′ = –H, –*n*-Pr, –Ph R = Ph; R′ = –H	0.10–2.6	bovine brain, heart; rabbit muscle	Secrist *et al.*, 1972; Meyer *et al.*, 1973b; Jones *et al.*, 1973; Panitz, 1974
2,N^6-disubstituted-cAMP derivatives			
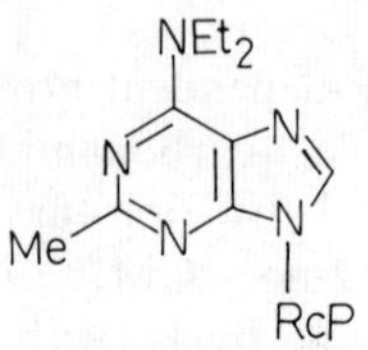	0.90	bovine brain	Miller *et al.*, unpublished results
2,6-disubstituted cNMP derivatives			
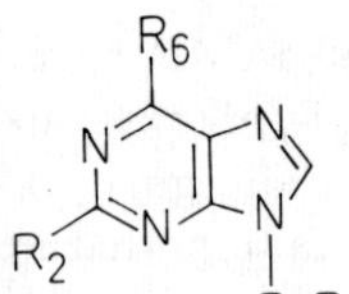			
R_2 = =O; R_6 = =O(cXMP) R_2 = –NH_2; R_6 = –H, =O (cGMP), = S, –Cl, –MNe_2	0.002–0.090 [0.010–0.18]	bovine brain, heart; rabbit muscle; rat brain lobster tail muscle	DuPlooy *et al.*, 1971; Miller *et al.*, 1973a; Fikus *et al.*, 1974; Kuo and Greengard, 1974; Kuo *et al.*, 1974; Scheit, 1974; Meyer *et al.*, 1975b; Miyamoto, 1975.

analogue	K_a'(cA) [K_a'(cG)]	tissue	references
1,2,N⁶-trisubstituted cAMP derivatives			
R_2 = –Me, –*n*-Bu, –SEt, –Ph	0.05–0.17	bovine brain	Uno *et al.*, 1976b

1973b, Uno *et al.*, 1976b). Some examples of this latter group show moderate K_a' (cG) values. 1,N^2-Etheno-cGMP derivatives are most interesting because they have apparently lost their kinase specificity and are able to produce significant activation of both cAMP and cGMP-dependent kinases. In comparison, the 1, N^6-etheno-cAMP derivatives, while known to be active as cAMP-dependent protein kinase activators, have apparently never been examined as possible cGMP-dependent protein kinase activators. With the exception of 1-O-cIMP, 1,6-disubstituted cNMP derivatives have moderate K_a' (cA) values but are quite inactive as cGMP analogues. The only 2,N^6-disubstituted cAMP derivative to be considered has activity equal to that of cAMP, while the 1,2,N^6-derivatives (2-substituted-1,-N^6-etheno-cAMP derivatives) have only moderate activities at best.

The results allow some tentative conclusions to be drawn. Single substitutions in the 1-, 2-, 6- or N^6-positions of cAMP or cNMP are generally accepted by the protein kinase structure. When a 1- and a 2-substitution, a 2- and a 6-substitution, or a 2- and a 1,N^6-etheno-substitution are present in the molecule, the resulting analogues are less active than either of the singly-modified parent compounds. The activity of 1, N^6-, 1,6-and 2,N^6-disubstituted derivatives falls within the same range as the singly-modified parent compounds.

The most easily substituted position in the imidazole portion of cAMP or cGMP is the 8-position. The K_a' values for the large number of 8-substituted derivatives of cAMP and cGMP which have been synthesized are in the range of 0.038 – 1.9 (Bauer *et al.*, 1971; Du Plooy *et al.*, 1971; Muneyama *et al.*, 1971; Miller *et al.*, 1973a; Kuo and Greengard, 1970b; Kuo *et al.*, 1974a; Panitz, 1974; Severin *et al.*, 1975; Dills *et al.*, 1976; Christensen *et al.*, 1975; Drummond and Powel, 1970; Jastorff and Bär, 1973; Gillen and Nagyrary, 1976; Eckstein *et al.*, 1974). The 8-keto-, thio-, halo-, alkylthio- and arylthio-derivatives are the most active, while the 8-alkyl- and alkoxy-(phenoxy)-derivatives are somewhat less active. Except for a few short chain examples, the 8-alkylamino-derivatives are only poor activators of the kinases. The same pattern of activity is seen with the

Table 6.3 Activation of cAMP-dependent protein kinases by cyclic nucleotide analogues with *two* substitutions in the *purine* ring

analogue	K_a'(cA)	tissue	references
1,8-disubstituted-cAMP derivatives			
NH_2 R_1-N N N N R_8 RcP			
R_1 = –O; R_8 = –Br, R_1 = –Me; R_8 = –SCH_2Ph	0.15–1.3	bovine brain	Uno *et al.*, 1976a
2,8-disubstituted-cAMP derivatives			
NH_2 N N N N R_2 R_8 RcP			
R_2 = –Me; R_8 = –Br, –SCH_2Ph, –SPh-*p*-Cl, –N_3 R_2 = –*n*-Bu; R_8 = –Br, –SCH_2Ph, –SPh-*p*-Cl	0.45–3.4	bovine brain	Uno *et al.*, 1976a
N^6,8-disubstituted-cAMP derivatives			
R R′ N N N N N R_8 RcP			
R_8 = –SMe; R = –H; R′ = –*n*-Bu, –CH_2Ph R_8 = –SCH_2Ph; R = –H; R′ = –Me, –Et, –*n*-Pr, –*n*-Bu, –*n*-Pentyl, –*i*-Pentyl, –*n*-Hexyl, –CH_2Ph, –C(O)NHPh R = R′; R′ = Me, –Et, –*n*-Pr R = Me; R′ = –*n*-Bu NRR′ = morpholino piperidino R_8 = –SPh-*p*-Cl; R = –H; R′ = –*n*-Bu, –*i*-Bu, –Ph, NRR′ = morpholino R_8 = –Br; R = –H; R′ = –*n*-Bu, –CH_2Ph	1.1–16		Boswell *et al.*, 1973; Miller *et al.*, 1976c
R_8 = –$NHCH_2$Ph; R = –H; R′ = –$(CH_2)_2NH_2$	0.003	rabbit muscle	Rieke, unpublished results

analogue	K_a'(cA)	tissue	references
6,8-disubstituted-cNMP derivatives			
(structure: purine ring with R_6 at position 6, R_8 at position 8, RcP at N-9)			
R_6 = =O; R_8 = –OH, –Br, –SPh-*p*-Cl R_6 = –Cl; R_8 = =OH, $-NH_2$, $-NEt_2$ R_6 = =S; R_8 = $-NH_2$ R_6 = –SMe; R_8 = –SMe	0.1–1.7	bovine brain, heart; rabbit muscle	DuPlooy *et al.*, 1971; Miller *et al.*, 1973a, 1976c; Kuo and Greengard, 1974; Kuo *et al.*, 1974a.

cAMP-dependent and the cGMP-dependent protein kinases. 8-Substituted derivatives of cAMP and cGMP retain their specificities for their respective cAMP-dependent kinases (Miller *et al.*, 1973a).

Table 6.3 presents some data on disubstituted derivatives of cyclic nucleotides which contain one substitution each in the pyrimidine and imidazole rings. Depending on the nature of substitutions, examples of the 1,8- and 2,8-disubstituted-cAMP derivatives exist having activities which are greater than, equal to, or less than cAMP. As a group, the N^6,8- and 6,8-disubstituted analogues of cAMP and cNMP are either intermediate in activity to the two parent compounds or equal in activity to the more active of the two parent compounds. Those 6,8-disubstituted derivatives containing a 6-keto group (8-substituted-cIMP derivatives) demonstrate significant activity with both the cAMP and cGMP-dependent kinases.

The 1,N^6,8-trisubstituted-cAMP derivatives have relative activities in the same range as the 8-substituted and 1,N^6-disubstituted parent compounds.

The activity of the 'aza' and 'deaza' analogues (Table 6.4) indicates that the electronic nature of *N*-3 is important for the interaction of cAMP with the cAMP-dependent protein kinase. In fact, Miller *et al.* (1976) have shown that the charge on the lone electron pair at the 3-position is proportional to the log [K_a' (cA)] with a correlation coefficient of 0.98. The necessity for the *N*-3 to interact with the enzyme would suggest an *anti* conformation, while the superior activity of cAMP derivatives with bulky 8-substitutents would suggest a *syn* conformation. Since both of these classes of compounds are efficient kinase activators, a transition between *syn* and *anti* conformation could explain the data.

7-Substitution of 7-deaza-cAMP is detrimental to the superior activity of the parent analogue while addition of a 1,N^6-etheno-substitution to the relatively

Table 6.4 Activation of cAMP-dependent protein kinases by cyclic nucleotide derivatives with a *modified purine* ring

analogue	K_a'(cA) [K_a'(cG)]	tissue	references
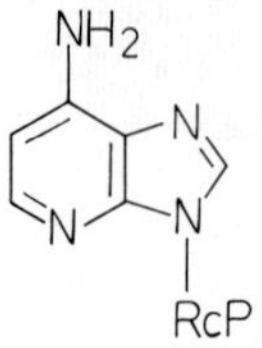 1-deaza-cAMP; 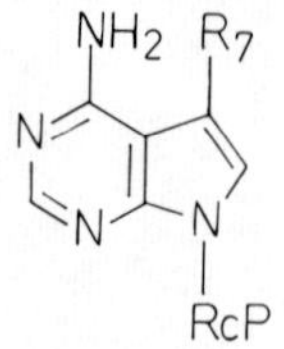 7-deaza-cAMP R_7 = –H, –CN, –C(O)NH_2	0.19–1.9	bovine brain, heart; rat liver	Drummond and Powell, 1970; Kuo and Greengard, 1970a; Miller *et al.*, 1973b, 1976d.
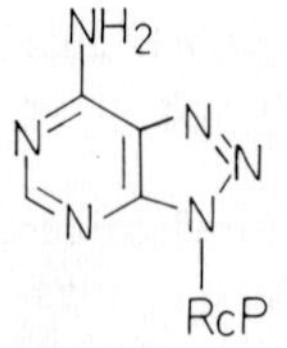 8-aza-cAMP	0.60–0.85	bovine brain, heart; rat liver	Miller *et al.*, 1976d
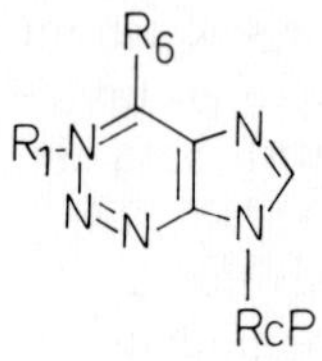 2-aza-cAMP R_1 = unsubstituted R_6 = –NH_2, =O; R_1 = =O; R_6 = –OH; 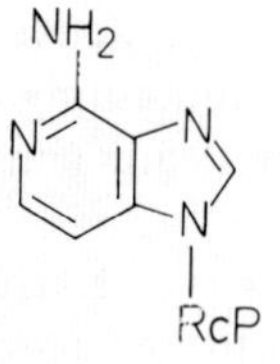 3-deaza-cAMP	0.038–0.11	bovine brain, heart; rat liver	Miller *et al.*, 1976d
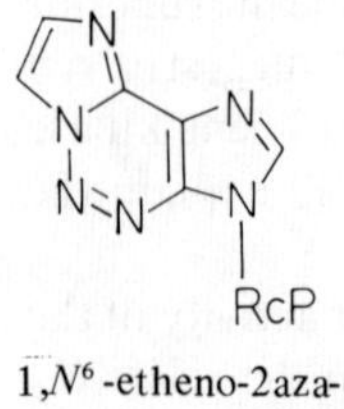 1,N^6-etheno-2aza-cAMP	0.68 [0.11]	bovine brain lobster tail muscle	Miller *et al.*, unpublished results

inactive 2-aza-cAMP increases the activity of the parent analogue. It is also interesting to note that 1,N^6-etheno-2-aza-cAMP is active with both the cAMP and cGMP-dependent kinases.

The K'_a values for cyclic nucleotide derivatives containing substitutions or modifications of the ribose ring show that, without exception, any changes in the 2′-position essentially eliminate the activity of the resulting analogues (Du Plooy *et al.*, 1971; Jastorff and Freist, 1974; Kuo *et al.*, 1975a; Miller *et al.*, 1973c; Miller *et al.*, 1976a). Severin *et al.* (1975) have reported that 2′-*O*-chloroacetyl-cAMP can form a covalent bond with the kinase. They have suggested that this may be the result of the alkylation of a nucleophilic group on the kinase close to the region where the 2′-OH group interacts with the enzyme. The only known modification of the ribose which is tolerated is that at the 4′-position.

The cyclic phosphate ring, like the ribose ring, is extremely sensitive to modification. All 3′-modifications result in inactive derivatives. Of a series of 5′-endocyclic derivatives, only 5′-deoxy-5′-thio-cAMP is active (Shuman *et al.*, 1973). The difference between this compound and all the other inactive 5′-endocyclic-cAMP derivatives is the lack of any additional atoms attached to the atom in the 5′-position (Drummond and Powell, 1970; Jastorff and Bär, 1973; Puni *et al.*, 1975). This suggests that the lack of activity of all other 5′-modified analogues is due to a steric effect of these additional atoms.

Substitutions of or on the phosphate itself are also deleterious to activity, with the sole exception of the phosphorothionate derivative of cAMP. All the inactive derivatives (-P-*O*-alkyl, -P-NMe_2 and cyclic sulphate) have in common the loss of the negative charge on this ring. This substantiates the importance of such a charge in the cyclic phosphate ring (Meyer *et al.*, 1975b; Severin *et al.*, 1975).

Derivatives containing two modifications in the cyclic phosphate ring show that the detrimental effect of one substitution cannot be overcome by the addition of an otherwise tolerated substitution (Eckstein *et al.*, 1974; Drummond and Powell, 1970; Gillen and Nagyvara, 1976; Jastorff and Bär, 1973; Kuo and Greengard, 1970; Miller *et al.*, unpublished results; Panitz *et al.*, 1975; Severin *et al.*, 1975; Shuman *et al.*, 1973).

Data from disubstituted derivatives with substitutions in both the purine and ribose moieties indicate that the general inactivity of 2′-derivatives of cyclic nucleotides with the protein kinases is not overcome by the addition of an otherwise tolerated substitution in the purine ring, even though the parent purine-substituted compound is equal to or even better than cAMP (or GMP) in activating protein kinase (Boswell *et al.*, 1973; Drummond and Powell, 1970; Jastorff and Freist, 1974; Khwaia *et al.*, 1975; Kuo and Greengard, 1974; Kuo *et al.*, 1974a; Kuo, 1975c; Miller *et al.*, 1973c, 1976a). Similar conclusions can be drawn for derivatives containing modifications in both the purine and cyclic phosphate moieties. In addition, in the case of 5′-deoxy-5′-thio-cIMP, both parent compounds are active, but the disubstituted derivative is inactive (Jones *et al.*, 1973; Meyer *et al.*, 1973b; Miller *et al.*, unpublished results; Murthy and Ellis, unpublished results; Shuman *et al.*, 1973).

B Hydrolysis of Cyclic Nucleotide Analogues by Cyclic Nucleotide Phosphodiesterases

In order that a derivative of a cyclic nucleotide may mimic the physiological action of the unmodified parent compound, it is most important that the phosphate ring remains intact. One of the aims of the medicinal chemist is to synthesize analogues which retain their biological potency, but which are resistant to enzymic degradation.

Substitution of the 1- and 2-positions of cyclic nucleotides and the 6-position in cNMP is the least effective of the purine-ring modifications; of these three modifications, the 2-substituted-cAMP derivatives are the least resistant to enzymic hydrolysis. However the results obtained for N^2-substituted cGMP derivatives do not allow any generalization (Michel *et al.*, 1974; Miller *et al.*, 1976).

Certain substitutions of the N^6-position confer almost complete resistance to hydrolysis by phosphodiesterase (N^6-acyl, carbamoyl- and alkoxycarbonyl derivatives), while other substitutions allow the resulting analogues (N^6-alkyl- and alkoxy derivatives) to be hydrolysed by the enzyme (Michal *et al.*, 1970; 1974; Meyer *et al.*, 1972; 1973b; Boswell *et al.*, 1973; Miller *et al.*, 1973c, 1975a; Hong *et al.*, 1975; Dills *et al.*, 1976). The available data also indicate that 6-substituted derivatives of cNMP are, in general, better substrates for the phosphodiesterases than N^6-substituted derivatives of cAMP. The relative rates of hydrolysis of the 1,2-, 1,N^6-, 1,6- and 2,6-disubstituted and of the 1,2,N^6- and 1,2,6-trisubstituted derivatives of cAMP or cNMP lend support to the general conclusion that substitution of the pyrimidine ring is not seriously detrimental to the interaction of cyclic nucleotides and the phosphodiesterase. Consistent with the former conclusion, 1,6-disubstituted and 1,2,6-trisubstituted derivatives of cNMP are, in general, better substrates than the 1,N^6-disubstituted and 1,2,N^6-trisubstituted derivatives (Meyer *et al.*, 1973b; Michal *et al.*, 1974; Secrist *et al.*, 1972; Uno *et al.*, 1976b).

Substitution at the 8-position confers the greatest resistance to hydrolysis (Chu *et al.*, 1974, 1975; Christensen *et al.*, 1975; Dills *et al.*, 1976; Michal *et al.*, 1970, 1974; Miller *et al.*, 1973a, 1975a; Muneyama *et al.*, 1971, 1974). Most 8-substituted analogues of cAMP, cGMP and cIMP which have been examined are completely resistant to phosphodiesterase within the limits of detection of the assays used. The notable exceptions to this generalization are the 8-amino- and 8-methyl-derivatives of cAMP and cGMP. The detrimental effect of 8-substitution on the ability of the resulting analogues to be hydrolysed by phosphodiesterase is not overcome by additional substitutions in the pyrimidine ring. This is demonstrated by the resistance to hydrolysis demonstrated by 2,8-disubstituted-cAMP derivatives and 2,6,8-trisubstituted-cNMP derivatives.

The results obtained with 'aza' and 'deaza' analogues show that substitution of a ring nitrogen with carbon, or *vice versa*, results in analogues which are able to be hydrolysed enzymically (Blecher *et al.*, 1971; Drummond and Powell, 1970; Miller *et al.*, 1973b, 1976d; Tsou *et al.*, 1974). These results along with the observation of general resistance to hydrolysis of N^6- and 8-substituted analogues suggest that the electronic nature of the purine ring is significantly less important for interaction with the phosphodiesterase than is the *lack* of a bulky group at the 8-position or a carbonyl group at the 6-position. The purine ring is most definitely necessary for interaction with the enzyme, since pyrimidine cyclic phosphates are relatively resistant to hydrolysis by most cyclic nucleotide phosphodiesterases (Drummond *et al.*, 1963; Michal, 1970, 1974; Long *et al.*, 1972). A possible exception to this generalization is cUMP. This cyclic nucleotide has been reported to be hydrolysed at significant rates by various enzyme preparations. An explanation for this may be the presence of a cUMP-specific phosphodiesterase in the enzymic preparation (Klotz and Stock, 1971).

Modifications to the ribose ring (2′ and 4′ positions) of cyclic nucleotides (Table 6.5), for the most part, have little effect on the rates of hydrolysis of the resulting derivatives. The only exception is the inactivity, as would be expected, of the L-diasteriomer of cAMP.

Certain modifications to the cyclic phosphate ring are tolerated by the phosphodiesterase (Table 6.5); 3′-deoxy-3′-amino-cAMP and 5′-deoxy-5′-thio-cAMP can be enzymically hydrolysed. All other reported modifications to this ring result in derivatives which are resistant to hydrolysis.

Analogues containing substitutions or modifications in both the ribose ring and the pyrimidine portion of the purine ring (for example N^2,2′-*O*-,2,2′-*O*-, N^6,2′-*O*- and 6,2′-*O*-disubstituted derivatives) are hydrolysed at only slightly slower rates than the corresponding N^2-, 2-, N^6- and 6-derivatives, respectively, (Boswill *et al.*, 1973; Drummond and Powell, 1970; Michal *et al.*, 1970, 1974; Miller *et al.*, 1976a). This suggests that substitution at the purine ring more or less determines how the analogue interacts with phosphodiesterase, and that the ribose modification has little or no additional effect on the ability of the resulting derivatives to be hydrolysed. Consistant with this latter conclusion are the results for 2′,8-disubstituted derivatives (Khwaja *et al.*, 1975; Miller *et al.*, 1976a); the tolerated 2′-modification is not able to countermand the detrimental effect of the 8-substitution. The data on the 4′,N^6-disubstituted and the 4′,N^6,2′-*O*, N^6,8,2′-*O*- and 6,8,2′-*O*-trisubstituted derivatives further substantiate this generalization (Anisuzzanam *et al.*, 1973; Boswell *et al.*, 1973). In the case of 5′,6- and P,N^6-disubstituted derivatives (Shuman *et al.*, 1973), a kind of negative synergyistic effect has been observed. The corresponding singly-substituted derivatives are good substrates for the phosphodiesterase, while the related disubstituted analogues are completely resistant to hydrolysis. Taken together the data indicate that the *purine* and the *phosphate* rings are important binding sites for the cyclic nucleotide phosphodiesterases.

C Inhibition of Cyclic Nucleotide Phosphodiesterases by Cyclic Nucleotide Analogues

In addition to mimicking the physiological action of the naturally occurring cyclic nucleotides, the analogues can also diminish the rate of breakdown of cAMP and/or cGMP within the cells by inhibiting the various cyclic nucleotide phosphodiesterases. The structure–activity relationships between the position

Table 6.5 Hydrolysis of cyclic nucleotide analogues containing substitutions or modifications of the *ribose* or *cyclic phosphate* ring by cyclic nucleotide phosphodiesterases

class of analogue	relative rate of hydrolysis	tissue	references
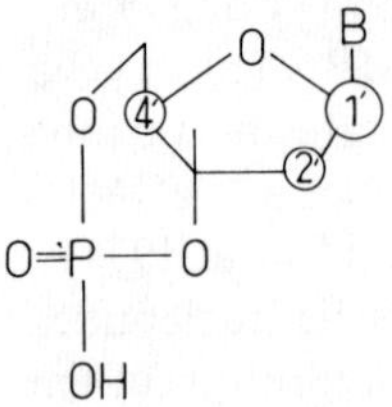 B=Ade, Gua, Hyp			
1′-derivative			
L-cAMP	<.005		Holy, 1972; Michal *et al.*, 1974
2′-O-acyl-cNMP derivatives			
	0.06–1.2	bovine brain, heart; porcine brain; rabbit kidney, lung	Michal *et al.*, 1970, 1974; Miller *et al.*, 1973c; 1976a
2′-deoxy-cNMP derivatives			
	0.2–12.	bovine brain, heart; dog heart; rabbit kidney, lung; rat adipocytes	Drummond *et al.*, 1963; Michal *et al.*, 1970, 1974; Miller *et al.*, 1973c, 1976a
2′-arabinosyl-cNMP derivatives			
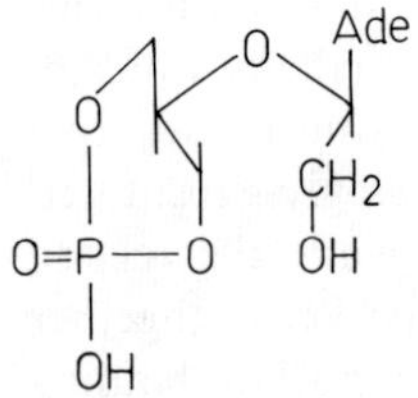	0.17–0.53	bovine brain, heart; rabbit kidney, lung	Miller *et al.*, 1973c, 1976a
	0	bovine heart	Scheit, personal communication
4′-derivatives			
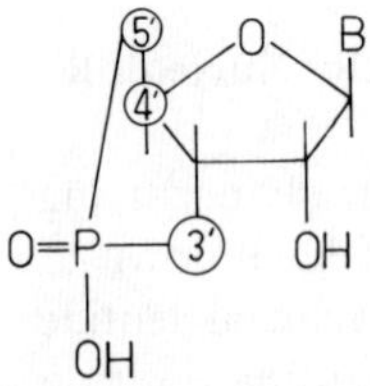 B = Ade, Gua, Hyp	0.42–0.95	bovine heart; rabbit kidney	Anisuzzaman *et al.*, 1973

class of analogue	relative rate of hydrolysis	tissue	references
3′-xylosyl-cNMP derivatives			
	<0.01–0,04	rabbit brain, kidney	Drummond and Powell, 1970
3′-deoxy-3′-substituted-cAMP derivatives			
3′-CH_2-	0		Drummond and Powell, 1970;
3′-NH-	0.28	bovine heart; rabbit brain	Panitz *et al.*, 1975
5′-deoxy-5′-substituted-cAMP derivatives			
5′-CH_2-, -NH-	0.04–0.06		Drummond and Powell,
5′-S-	0.29	bovine heart; rabbit brain; kidney	1970; Shuman *et al.*, 1973; Panitz *et al.*, 1975
phosphate-substituted-cAMP derivatives			
	0.007–0.02		Jastorff and Bär, 1973; Eckstein *et al.*, 1974

of single or multiple substitutions on or modification to cyclic nucleotides and their relative phosphodiesterase inhibitory properties will not be dealt with here. There exist large, yet unsystematic, variations in the inhibitory properties of different substitutents at the same position of the cyclic phosphate. In addition, with certain analogues, there are large variations in the ability of enzymes from different sources to be inhibited (Boswell *et al.*, 1973; Harris *et al.*, 1973; Meyer *et al.*, 1973b, 1975a; Miller *et al.*, 1973a,b,c; 1975a,b,c,d; Christensen *et al.*, 1975; Khwaja *et al.*, 1975; Uno *et al.*, 1976a,b). A further complication is the ability of certain analogues (most notably cGMP and cIMP derivatives) to activate some phosphodiesterases at low cyclic nucleotide concentrations (Michal *et al.*, 1974). Derivatives of cAMP, in virtually every case which has been examined, have been shown competitively to inhibit various cyclic nucleotide phosphodiesterases. This type of inhibition has been reported for derivatives with modifications in the purine, ribose and cyclic phosphate rings (Harris *et al.*, 1973; Jastorff and Bär, 1973; Miller *et al.*, 1973a; Shuman *et al.*, 1973; Eckstein *et al.*, 1974; Christensen *et al.*, 1975; Meyer *et al.*, 1975a; Gillen and Nagyvary, 1976).

Two fluorescent cAMP analogues (1,N^6-etheno-2-aza-cAMP and 2-NH_2-cNMP) have been utilized to develop new assays for cAMP phosphodiesterase. The assay using the former compound utilizes the fluorescence of the resulting 1,N^6-etheno-2-aza-Ado formed after phosphatase treatment of the 5′-phosphate, to quantify the amount of product formed (Tsou *et al.*, 1974). The assay using 2-NH_2-cNMP utilizes an ultraviolet spectral difference between the cyclic phosphate substrate and the 5′-phosphate product (Scheit, 1974).

III EFFECTS OF CYCLIC NUCLEOTIDE ANALOGUES ON INTACT CELLS

An examination of the structure–activity relationships of cyclic nucleotide analogues and their ability to induce cAMP or cGMP-mediated functions when added exogenously to intact cells provides additional insight into the physiological actions of these analogues not provided by studies using isolated enzymes. Information can be gained concerning (*i*) the ability of analogues to penetrate cell membranes; (*ii*) the sensitivity of analogues to phosphodiesterases (possibly the 'low K_m' enzymes) at the 'physiological concentration' of the analogues; and/or (*iii*) the ability of the analogues to stimulate protein kinases *in vivo*.

A Induction of β-Galactosidase Synthesis in *Escherichia coli*

In *E. coli*, the synthesis of a number of enzymes is induced by an increase in the intracellular concentration of cAMP (Pastan and Perlman, 1970). Table 6.6 summarizes the reported effect of some cyclic nucleotide analogues on β-galactosidase synthesis in *E. coli*. The results of Anderson *et al.* (1972) on β-galactosidase synthesis have been duplicated by Krakow (1975) who studied the binding of the cAMP-receptor protein to poly d(I–C).

7-Deaza-cAMP has been found to be as effective as cAMP. The results with 8-substituted- and 2′-*O*-substituted-cAMP derivatives indicate that the type of substitution determines the degree of inducing activity. In the 8-substituted series, 8-HO- and 8-MeO-cAMP have significant activity and 8-NH_2-cAMP has slight activity, while 8-Br-cAMP is completely inactive. Comparing the 2′-derivatives, 2′-acyl-, deoxy- and ara-cAMP are inactive, while 2′-*O*-Me-cAMP has activity approaching that of cAMP.

The conclusions drawn concerning the structure–activity relationships between cAMP analogues and the cAMP receptor (Anderson *et al.*, 1972; Pačes and Smrž, 1973) require some modification when the results of all investigators are compared. Apparently, only certain alterations to the purine and ribose rings are tolerated by the cAMP receptor. It may well be that a cAMP analogue with the 'proper' modification in the cyclic phosphate ring will also retain activity.

A recent report on a specific cGMP-binding protein from *Caulobacter crescentus* shows that Bt_2-cGMP and cIMP are able to inhibit the binding of cGMP by 50% when present in a fifty-fold molar excess (Sun *et al.*, 1975). The function of this protein is as yet unknown.

Table 6.6 Effect of cyclic nucleotide analogues on β-galactosidase synthesis in *E. coli*

analogue	relative activity	type of assay	reference
cAMP	1		Pastan and Perlman,
7-Deaza-cAMP	1	*in vitro*	1970; Anderson *et al.*,
8-Br-, 8-HS-cAMP	0	and	1972
N^6-Bt-cAMP, 6-HO-cNMP	0	*in vivo*	
2′-O-Bt-, 2′-O-succinyl-cAMP	0		
2′-deoxy-, ara-cAMP	0		
3′deoxy-3′-CH_2-cAMP	0		
5′-deoxy-5′-CH_2-cAMP	0		
2′-*O*-Me-cAMP	0.5	*in vivo*	Pačes and Smrž, 1973
2-H_2N-cNMP	0.08		
8-Br-cAMP	0		
8-HO-cAMP	0.1	*in vivo*	W. Epstein, personal communication
8-MeO-cAMP	0.1		
8-H_2N-cAMP	0.03		
N^6-Me-cAMP	0		
N^6-$(CH_3)_2C{=}CHCH_2$-cAMP	0		

B Cell Growth

The levels of cAMP and cGMP and/or the ratio of their concentrations determine in part whether cells proliferate or not (Shields, 1974). In general, cAMP inhibits and cGMP initiates growth. When examining the effects of cyclic nucleotide analogues on cell growth and related processes *in vitro* it is important to keep in mind the question of whether the cyclic nucleotide or one of its metabolites is exerting the observed effects on growth, especially when considering the cyclic phosphate derivatives of known cytotoxic nucleosides (e.g., ara-cAMP, 6-HS-cNMP and 6-MeS-cNMP (Table 6.7). With ara-cAMP, it has been demonstrated that the cyclic nucleotide is probably converted to the 5′-nucleotide intracellularly (LePage and Hersh, 1972). In fact, these workers have suggested the general use of derivatives of toxic nucleotides as carcinostatic agents because it is not necessary for the target cell to phosphorylate the nucleoside. Hong *et al.*, (1975) have compared cytotoxic effects of the nucleosides, 5′-nucleotide and 3′,5′-cyclic nucleotide of a number of N^6-substituted cAMP derivatives. In the case of N^6-Ph-cAMP, only the cyclic phosphate is cytotoxic. Examples have also been found where the nucleoside or 5′-nucleotide is more active than the cyclic nucleotide. Plunkett and Cohen (1975) have suggested the use of cyclic phosphates to replace toxic nucleosides when the solubility and/or membrane permeability of a compound needs to be improved. That ara-cAMP, 6-HS-cNMP and 6-MeS-cNMP are hydrolysed intracellularly to the 5′-nucleotide is to be expected from the reported ability of these cyclic phosphates to be hydrolysed by cyclic nucleotide phosphodiesterases (Meyer *et al.*, 1972; Miller *et al.*, 1973c).

Table 6.7 Effect of cyclic nucleotide analogues on cell growth

analogue	cell type	effect measured	reference
ara-cAMP	L-1210 leukaemia cells	inhibition of DNA synthesis	Hughes and Kimball, 1972
ara-cAMP 6-HS-cNMP, 6-MeS-cNMP	Human lymphocytes	inhibition of DNA synthesis	Lepage and Hersch, 1972
ara-cAMP ara-cGMP	L-cells, KB cells, HeLa cells, sarcoma 180 cells, RK-13 cells	Cytotoxicity, decrease in cell viability	Mian *et al.*, 1974;
Bt_2-cAMP, N^6-Bt-cAMP 8-(HO-, HS-, MeS-, *p*-ClPhS-, Br-)cAMP	Reuber H35 hepatoma cells human prostatic epithelial cells (MA-160), L-929 cells, 3T3-4 cells, NRK cells	inhibition of growth	Johnson *et al.*, 1971; VanWijk *et al.*, 1972; Wicks *et al.*, 1976; Carchman *et al.*, 1974
6-(HS-, MeS-)cNMP 8-(H_2N-, $HOCH_2CH_2HN$-)cAMP	Reuber H35 hepatoma cells	cell death	Koontz and Wicks, 1976
N^6-$PhCH_2$-cAMP	Reuber H35 hepatoma cells	inhibition of growth	Koontz and Wicks, 1976
N^6-($[CH_3]_2C{=}CHH_2C$-, $PhCH_2$-, $CH_3[Cl]C{=}CHCH_2$-, [2-furfuryl]CH_2-, Ph-, *p*ClPhNH(O)C-)cAMP	Nc-37 cells RPMI-6410 cells L-1210 cells	decrease in cell viability	Hong *et al.*, 1975
8-($BrH_2N[H_2N]CSe$-, EtSe-, *p*NO_2PhCH_2Se-)cGMP	L-5178Y cells	decrease in cell viability	Chu *et al.*, 1975
N^6,2′-*O*-diacetyl-cAMP methyl ester	Ehrlich ascites cells, CHO cells chick embryofibroblasts	inhibition of growth	Nagyvary *et al.*, 1973

When only inhibition of growth is observed, especially when it is known that the inhibition is reversible, it is possible that the analogue is exerting its effect as the cyclic phosphate on a cyclic nucleotide-sensitive metabolic step in the cell. This can to some extent be predicted by the resistance of these compounds to phosphodiesterases. Examples of these types of analogues are Bt_2-cAMP, 8-HO-, 8-HS and 8-MeS-cAMP, N^6-alkyl- and N^6-aryl-cAMP and 8-alkylseleno-cGMP-dervatives.

The situation with N^6-2′-*O*-diacetyl-cAMP methyl ester may be that the compound is more stable to hydrolysis and/or is more permeable to cells than is cAMP, and is deacetylated to yield cAMP within the cell (Nagyvary *et al.*, 1973).

1 Immunocompetent Cells

Evidence is accumulating which indicates that cyclic nucleotides act as intracellular mediators of the expression of antigen-sensitive cells (Watson *et al.*, 1973). Strom *et al.*, (1973) have shown that agents which increase cAMP levels in antigen-sensitized rat spleen cells cause a decrease in the cytotoxicity of these cells. On the other hand, 8-Br-cGMP or agents which increase cGMP levels cause an enhancement of this lymphocyte-mediated cytotoxicity. Luckasen *et al.* (1974) have reported that cAMP and Bt_2-cAMP inhibit the antigen binding of murine lymphocytes. This inhibition can be largely prevented by 8-Br-cGMP, even though this analogue does not by itself inhibit or enhance binding.

8-Br-cGMP has also been found to be mitogenic in spleen cells from five different mouse strains, and has been shown to be a specific B-cell mitogen (Chambers *et al.*, 1974; Weinstein *et al.*, 1974, 1975). By comparison, Bt_2-cAMP exhibits an antimitogenic effect and inhibits the immunological response to antigen (Watson *et al.*, 1973; Weinstein *et al.*, 1974). Weinstein *et al.* (1975) have also found that some other 8-substituted-cGMP derivatives have mitogenic activity, the order of effectiveness being 8-Br-cGMP $>$ Bt_2-cGMP $\simeq$ 8-HS-cGMP $>$ 8-MeS-cGMP. In addition, Bt_2-cGMP induces phosphoribosyl pyrophosphate synthetase in murine splenic lymphocytes concomitant with the increase in DNA synthesis (Chambers *et al.*, 1974). Bt_2-cAMP inhibits the mitogenic actions of concanavalin A in these cells (Chambers *et al.*, 1974).

2 Antiviral Effects

One effect that the inhibition of cell growth can have on virus-infected cells is the inhibition of virus replication. A number of cyclic nucleotide derivatives of known toxic nucleosides (ara-cAMP, ara-cIMP, and ara-cCMP, 6-HS-cNMP and MeS-cNMP) has demonstrated significant antiviral activity (Long *et al.*, 1972; Sidwell *et al.*, 1973a,b; Mian *et al.*, 1974; Revankar *et al.*, 1975). As with the effects noted above on cell growth, most probably these compounds exert their activity after the hydrolysis of the cyclic phosphate. In contrast, derivatives of cyclic nucleotides which are stable to the action of phosphodiesterase may well exert their antiviral effects as the intact cyclic nucleotides (Long *et al.*, 1972; Sidwell *et al.*, 1974a,b). In addition, it has been found that a number of 8-substituted derivatives of cAMP exerts an antiviral effect which is host-cell specific, indicating either a differential sensitivity to the cAMP analogues among different cell types or a variation in the extent of hydrolysis of the cyclic phosphate within the cells. Lastly, Allen *et al.* (1974) has shown that 6-HS-cNMP enhances the antiviral action of interferon in L-cells.

C Action of Cyclic Nucleotide Analogues in Lipocytes, Adrenal Cells and Hepatocytes

A large number of cyclic nucleotide analogues has been examined for their abilities to activate lipolysis in rat epididymal fat cells and also to activate

Table 6.8 Activation of steroidogenesis and lipolysis by cyclic nucleotide analogues

class of analogue	relative activity	
	adrenal cells	lipocytes
cAMP	1	1
Bt_2 cAMP	35	17
N^1-substituted-cAMP	~1	< 0.9–16
2-substituted-cAMP and 2,6-disubstituted-cNMP derivatives	0–2	< 1–17
N^6-substituted-cAMP and 6-substituted-cNMP derivatives	< 1–30	6–170
8-substituted-cAMP derivatives	< 1–19	< 1–140
8-substituted-cIMP derivatives	< 1–7	2–10
8-substituted-cGMP derivatives	0–< 1	0–< 3
N^6,8-disubstituted-cAMP and 6,8-disubstituted-cNMP derivatives	< 1–28	3–89
2′-OMe-, deoxy- or ara-cAMP	0	0
2′-acyl, 8-disubstituted:	1–5	16–65
N^6,8-trisubstituted-cAMP; and 2′-, acyl,6,8-trisubstituted-cNMP derivatives 2′-ara,8-disubstituted-cAMP derivatives	0	0
7-deaza-cAMP		~10

Data taken from Braun *et al.*, 1969; Blecher *et al.*, 1971; Free *et al.*, 1971, 1972. The relative activity is the ratio of the A_{50} (concentration which results in half maximal stimulation) for cAMP to the A_{50} for the analogue.

steroidogenesis in rat adrenal cells. The available data is summarized in Table 6.8. N^1-Substituted and 2-substituted-cAMP derivatives are poor activators, the exceptions being 1-PhCH$_2$O-cAMP in adrenal cells, and 2-Ph-cAMP and 2-(2-pyridy)-cAMP in both lipocytes and adrenal cells. It is interesting to note that all three analogues contain an aromatic substitution, which may facilitate transport or impart resistance to phosphodiesterase.

6-Substituted-cNMP derivatives are more active than the N^6-substituted-cAMP derivatives in both cells. In addition, the former analogues show a greater relative activity in the adrenal cells, while the latter do not demonstrate this specificity.

8-Substituted-cAMP derivatives are also more active in the adrenal cells, but most of the derivatives are able to stimulate cells to the same level of activity as produced by either noradrenaline or ACTH in the lipocytes, or ACTH in the adrenal cells (Free *et al.*, 1971). The 8-alkylthio derivatives are generally more active than the 8-alkylamino compounds, while 8-substituted derivatives of cGMP and cIMP are inactive. These findings are consistent with the fact that the processes are cAMP-mediated.

Those 2′-derivatives (or 2′,6,8-derivatives) which contain a hydrolysable 2′-*O*-acyl group show significant activity. All other 2′-modified analogues (for example, 2′-*O*-Me- and 2′-ara or 2′-deoxy derivatives) are completely inactive.

From these studies of structure–activity relationships it is clear that the ability of cyclic nucleotide analogues to stimulate cellular processes is qualitatively related to their ability to stimulate protein kinases. A few cyclic nucleotide derivatives have been found to inhibit the ability of hormones to stimulate lipocytes and adrenal cells. 2′-*O*-Palmitoyl-cAMP, 1,N^6-etheno-cAMP and 8-*n*-Pr(O)C-cGMP inhibit hormone-stimulated lipolysis, and 1,N^6-etheno-cAMP is, at the same time, able to stimulate steroidogenesis. N^6-HO-cAMP interferes with hormone-stimulated steroidogenesis, but has no effect on fat cells.

Kneer *et al.* (1974) have reported the effects of cGMP on glucose metabolism in hepatocytes. cGMP, Bt_2-cGMP, 8-Br-cGMP, 8-$PhCH_2S$-cGMP, and 8-*p*-ClPhS-cGMP stimulate glucose synthesis, and inhibit lactate and pyruvate formation from galactose. The activity of the derivatives ranges from slightly more potent than cGMP to twice the activity of the parent cyclic nucleotide. Consistent with these data are the findings of Bauer *et al.* (1971), who showed that the protein kinase-stimulating potency of a number of 8-substituted derivatives of cAMP is proportional to the ability of these analogues to stimulate glycogenolysis in rat liver slices. In addition, Levine and Washington (1970) reported that Bt_2-cAMP, 7-deaza-cAMP, cIMP and 2′-*O*-Bt-cIMP demonstrate significant stimulation of glycogenolytic activity in perfused rat liver.

D Effects of Cyclic Nucleotide Analogues on Differentiation and Differentiated Functions

Kram *et al.* (1973a) have helped to provide a model for the long-recognized fact that, in general, growth and differentiation are mutually exclusive. cAMP induces differentiation (pleiotypic control) and cGMP may be considered to stimulate replication at the expense of differentiation. The effect of cAMP analogues on cells (as discussed in Section IIIB) may be due to the induction of differentiation. In most cases, a differentiated function has not been assayed, but rather cell growth has been considered to be the end point of the experiment. Many of the analogues which induce cell growth have been found to be effective inducers of differentiation. In most cases when differentiation has been examined, derivatives of cGMP either have no effect on differentiation or antagonize the induction of differentiation by cAMP analogues or other inducers. Those differentiation functions which are induced or stimulated by cyclic nucleotide analogues will be examined separately.

1 Enzyme Induction

One differentiation function of cells is the production of enzymes which are specific to each type of cell. The effect of cyclic nucleotide analogues on the induction of enzymes in various cell types is, as with differentiation, cAMP-specific and only elicited by cAMP analogues. Analogues of cGMP are usually without effect.

Two separate studies have attempted to correlate enzyme induction with protein kinase activation. In Reuber H35 hepatoma cells, only those analogues which are active as tyrosine transaminase inducers are capable of stimulating phosphorylation of the serine-37 residue of the endogenous f_1 histone in intact cells (Wagner *et al.*, 1975; Wicks, *et al.*, 1975). The induction of the same enzyme in rat liver *in vivo* by a large number of 8- and N^6-substituted derivatives of cAMP and 6-substituted derivatives of cNMP has been correlated by linear regression analysis with the ability of the analogues to activate rat liver protein kinase and to be hydrolysed by rat liver phosphodiesterase. For nineteen analogues, a correlation coefficient of 0.920 has been obtained, supporting the idea that protein kinase activation is part of the mechanism of enzyme induction (Miller *et al.*, 1975a).

2 Secretory Processes

The secretion of various substances from cells is another differentiated function which can be mimicked by cyclic nucleotide analogues. The effects of numerous cAMP analogues on the release of hypothalamic hormones *in vitro* have been studied extensively (Cehovic *et al.*, 1968, 1970, 1972, 1976b; Giao *et al.*, 1974; Posternak *et al.*, 1969; Posternak and Cehovic, 1971). It was found that N^6-*n*-Bu-cAMP is significantly more active than cAMP, as is Bt_2-cAMP, while other N^6-alkyl-cAMP derivatives examined are equally or only slightly more active than cAMP in stimulating TSH release *in vitro*. Of a number of mono- and di-butyryl-derivatives of 8-substituted-cAMP analogues which have been tested as inducers of hypothalamic hormone release, the N^6-Bt- and Bt_2-derivatives of 8-SH- and 8-OH-cAMP are the most active inducers of GH release.

The most active inducers of prolactin release, in contrast, are the Bt_2-derivatives of 8-Br- and 8-$PhCH_2$-S-cAMP, while the former compounds are relatively inactive by comparison (Cehovic *et al.*, 1976b).

As stimulators of α-amylase release from rat parotid gland, Bt_2-cAMP, N^6-Bt-cAMP, 8-HS-cAMP and 8-$PhCH_2$-S-cAMP are approximately equal in activity, while 8-HO-, MeS- and Br-cAMP are significantly less active (Butcher *et al.*, 1976). Also, it is interesting to note that derivatives of cAMP and cGMP have opposite actions on lysosomal enzyme release from human neutrophils; actions which mimic the effects of cAMP and cGMP (see Ignarro, this volume).

IV EFFECTS OF cNMP ANALOGUES IN WHOLE ANIMALS

Just as the effects of cyclic nucleotide analogues on whole cells allow quantitation of the transport and metabolic properties of these analogues, the effects of the derivatives on whole animals allow analysis of their pharmacokinetic properties. Compared to studies with isolated enzymes, and cells or tissues *in vitro*, the available *in vivo* data are scarce. Only in a few cases have enough derivatives been examined to allow the formulation of even qualitative structure–activity relationships.

A Blood Levels of Glucose and Various Hormones

Intraperitoneal injection of rats with 2′-derivatives of cAMP has little or no effect on blood glucose, but increases plasma steroid levels (Imura *et al.*, 1965).

A large number of N^6- and 8-substituted-cAMP derivatives and 6-substituted-cNMP derivatives found is equal to or better than cAMP. In general, in causing hyperglycaemia the 8-substituted derivatives are less effective than the N^6- and 6-substituted analogues. In examining 8-substituted derivatives of either cAMP, cIMP or cGMP the order of effectiveness in producing hyperglycaemia is: cAMP analogues > cIMP analogues > cGMP analogues. This would be expected from the protein kinase data presented above. In contrast, the order of effectiveness of these derivatives in increasing plasma steroid levels is: cIMP analogues $\simeq$ cGMP analogues > cAMP analogues. The reason for this difference is not known. The analogues which produce the greatest increases in blood glucose, which are approximately twice that of cAMP, are N^6-PhCH$_2$-cAMP, N^2-(4-MePh)CH$_2$-cAMP, N^6-(2,4-[OMe]$_2$-Ph)H$_2$C-cAMP, 6-morpholino-cNMP, 6-piperidino-cNMP and 8-(4-MePh)CH$_2$HN-cIMP. The greatest increases in plasma steroid levels are produced by Bt$_2$-cAMP, 8-Br-cAMP, and 8-PhCH$_2$-cIMP. Consistant with the generally superior activity of N^6-substituted cAMP derivatives is the hyperglycaemic activity of N^6-(Δ^2-isopentenyl)-cAMP and N^6-(*trans*-Cl-butenyl)-cAMP.

Cehovic *et al.* (1976a) have examined a number of 8-substituted-cAMP derivatives, including 8-HO-, 8-H$_2$N-, 8-N$_3$-, 8-HS-, 8-MeS-, 8-PhCH$_2$S- and 8-Br-cAMP, which may also contain either 2′-*O*-Bt-, N^6-Bt or N^6-2′-*O*-Bt$_2$-substitutions, for their effects on serum thyroxine levels. All of these analogues are as active and many are more active than Bt$_2$-cAMP. Introduction of one or two butyryl groups into an 8-substituted-cAMP derivative alters the biological activity of the resulting analogues, but no structure–activity relationships can be deduced from the analogues' effect on serum thyroxine levels and their degree of butyrylation. Moderately active analogues, such as 8-HS-cAMP or 8-HO-cAMP, are made more active and less toxic by addition of one or two butyryl groups. In contrast, very potent analogues, such as 8-MeS-cAMP or 8-H$_2$N-cAMP, demonstrate decreased activity after butyrylation. It has been generally observed that N^6-Bt-8-substituted cAMP analogues are more active than the corresponding dibutyryl derivatives. These results suggest that the butyryl groups have an effect other than, or in addition to, increasing the membrane transport of the analogues.

B The Cardiovascular System

Significant reductions in blood pressure and heart rate have been found by various N^6-substituted-cAMP analogues. With the most active analogue in this latter group, N^6-(2,4[OMe]$_2$Ph)CH$_2$-cAMP, the blood pressure and heart rate thirty-six minutes after an intravenous dose of 20 mg/kg are 33 and 22%, respectively, of the control values. In comparison, 8-substituted analogues of cAMP, cIMP and cGMP, and 2-substituted analogues of cIMP have little or no effect on these cardiovascular parameters. Rubin *et al.* (1971) tested a number of 8-substituted-cAMP derivatives for their relaxant activity in rat portal vein. In contrast, these workers showed that 8-PhCH$_2$S-cAMP and 8-N$_3$-cAMP are the most active analogues of the group and are approximately equal in activity to Bt$_2$-cAMP. Dietmann *et al.* (1972) have reported that Bt$_2$-cAMP and N^6-Bt-cAMP also have little effect on heart rate or blood pressure in conscious, resting dogs but produce a dose-dependent increase in coronary blood flow. The authors reported that Bt$_2$-cGMP has no effect on any of the above parameters. These data are consistent with an earlier report that Bt$_2$-cAMP accelerated, and Bt$_2$-cGMP slowed, the beating of cultured rat heart cells (Krause *et al.*, 1972).

C The Pulmonary System

Van Winkle (personal communication) has examined a number of analogues of cAMP and cGMP as inhibitors of reaginic passive cutaneous anaphylaxis in rats and of histamine-induced bronchoconstriction in guinea pigs. The N^6-and 8-substituted analogues of cAMP demonstrate only slight or no activity. 8-Br-cGMP and 8-PhCH$_2$S-cGMP reduce the cutaneous anaphylactic response by 80–85% when using a intraperitoneal dose of 200 mg/kg. 8-PhCH$_2$HN-cGMP at the same dose gives only a 25% reduction. After receiving the same dose of 8-Br-cGMP or 8-PhCH$_2$S-cGMP, guinea pigs are protected from collapse resulting from a histamine aerosol for between four and seven times longer than the untreated controls. Similar results with 8-Br-cGMP have been reported by Szaduykis-Szadurski and Berti (1972).

A number of cyclic nucleotide analogues has been tested for their relaxant activity on guinea-pig tracheal rings *in vitro*. Of a series of 8-substituted analogues of cAMP the N$_3$-, HS-, MeS- and PhCH$_2$S-derivatives are most active (Rubin *et al.*, 1971). 8-PhCH$_2$S-cAMP is ten-times more potent than Bt$_2$-cAMP. In general, the order of activity is 8-thio-derivatives $>$ 8-oxo-derivatives $\simeq$ 8-Br-cAMP $>$ 8-amino-derivatives. In similar experiments *in* cNMP. The 8-NH$_2$-substituted cGMP and cIMP derivatives are approximately *vitro* the 8-NH$_2$-substituted cGMP and cIMP derivatives are approximately equal in activity to Bt$_2$-cAMP and Bt$_2$-cGMP (Szaduykis-Szadurski *et al.*, 1972). 8-Br-cGMP and N^2-Ph(O)C-cGMP are the most active of all the compounds examined. Treatment with 8-Br-guanosine results in contraction of the tracheal rings, suggesting that the effect of the analogues may be mediated by the intact cyclic nucleotide and not by a metabolic product.

In contrast to the above data, Kaliner *et al.* (1972) have reported that 8-Br-cGMP stimulates the antigen-induced, IgE-dependent secretion of histamine and slow-reacting substance of anaphylaxis in human lung fragments. The observation by Kaliner and Austen (1974) of a similar effect of 8-Br-cGMP in three distinct effector systems of inflammation indicates that the effect of 8-Br-cGMP is not due to an artifact. The polymeric amine-induced release of histamine from rat peritoneal mast cells is inhibited by Bt_2-cAMP. This suggests that cAMP and cGMP have opposing effects in the lung.

D The Nervous System

The role of cAMP in the nervous system is only beginning to be understood. Little data on the *in vivo* effects of cyclic nucleotide analogues are available. 6-piperidino-cNMP has a sedative effect in rats when given intraperitoneally, while cAMP and Bt_2-cAMP, under the same conditions, are ineffective (Vargin and Spano, 1971). Consistent with these results is the report of Siggins and Henriksen (1975) that N^6- or 8-substituted analogues of cAMP are able to inhibit the spontaneous firing of rat cerebellar Purkinje neurones, with 8-H_2N-, 8-$PhCH_2S$-and 8-*p*-Cl-PhS-cAMP derivatives demonstrating greater activity than 8-Br-, 8-MeS-, 8-*i*-PrS-cAMP, N^6-Bt-cAMP or Bt_2-cAMP. 8-MeHN-cAMP, 2′-*O*-Bt-cAMP and 2′-deoxy-cAMP are essentially ineffective in depressing firing and, in fact, do excite the neurones. The ability of these analogues to inhibit neuronal firing correlated with their ranked protein kinase-activating activities ($r = 0.78$) (Meyer and Miller, 1974; Siggins and Henriksen, 1975).

Hoffer *et al.* (personal communication) have observed that superfusion of 8-Br-cGMP or 8-*p*-Cl-PhS-cGMP produces excitations of rat hippocampal pyramidal neurones which had been grafted onto the anterior chamber of the eye. These cGMP derivatives mimic the effect of acetylcholine in this same system.

This mutually antagonistic action of cAMP and cGMP analogues has also been seen in studies on their effects on frog-skin short-circuit currents (Seridan *et al.*, personal communication) Bt_2-cAMP increases while 8-Br-cGMP decreases this current. Higher concentrations of 8-Br-cGMP, in contrast, produce a cAMP-like effect, presumably due to its interaction with a cAMP receptor at the higher concentrations.

In this respect, 8-substituted derivatives of cAMP, as well as Bt_2-cAMP, cause hyperpolarization of the rabbit superior cervical sympathetic ganglion, mimicking the action of dopamine (McAfee and Greengard, unpublished results). In the same system, Bt_2-cGMP depolarizes the ganglionic neurones, while 8-$PhCH_2S$-cGMP is ineffective.

Contrasting with the above results, N^6-Bt-cAMP and 8-MeS-cAMP, as well as 8-Br-cGMP, enhance the electrically stimulated outflow of noradrenaline and dopamine-β-hydroxylase from the isolated perfused cat spleen (Cubeddu *et al.*, 1975).

E Effects of Cyclic Nucleotide Analogues on Tumour Growth In Vivo

The role of cyclic nucleotides in carcinogenesis has been reviewed by Schultz and Gratzner (1973). The rationale for using cyclic nucleotide analogues as antitumour agents is the inhibition of cellular growth caused by cAMP. This topic has been briefly considered previously — see Section IIIB. Reports of tumour growth being inhibited by administration of cyclic nucleotide analogues are summarized in Table 6.9.

Gohil *et al.* (1974) have suggested that the cAMP phosphotriesters (cAMP-P-O-alkylesters) may be easily transported into cells and when there may liberate cAMP. Gillen *et al.* (personal communication) have observed that treatment of Walker 256 tumours in rats with cAMP-ethyl ester reduces metastases and increases tumour cAMP concentrations.

The antitumour effects of Bt_2-cAMP have not been seen in all systems studied. For example, Cho-Chung and Gullino (1974a) have isolated both Bt_2-cAMP-sensitive and resistant cell populations from Walker 256 carcinoma. In addition, Pastan and his coworkers (personal communication) have observed that Bt_2-cAMP has no effect on L-1210 leukaemia cells, Lewis lung carcinoma, Bl melanoma or polyoma-induced sarcoma (Py 89).

Table 6.9 Effect of cyclic nucleotide analogues on tumour growth *in vivo*

analogue	tumour	animal	reference
cAMP ethyl ester, N^6,2′-*O*-diacetyl-cAMP methyl (or ethyl)ester	Ehrlich ascites sarcoma 180	Mice	Nagyvary *et al.*, 1973; Cotton *et al.*, 1975
Bt_2-cAMP, 8-(Br-, MeS-)cAMP	mammary Walker 256 or WTW9	Rats	Cho-Chung, 1974; Cho-Chung and Berghoffer, 1974; Cho-Chung and Gullino, 1974a,b; Cho-Chung and Clair, 1975
8-H_2N(BrH_2N)CSe-cGMP	leukaemia	Mice	Chu *et al.*, 1975
8-(MeSe-, *n*-PrSe-, PhCH_2Se-, *p*-NO_2-PhCH_2Se-)-cGMP 8-(Br-, EtSe-, *n*-BuSe-)cGMP			
6-HS-cNMP	Ridgeway osteogenic sarcoma sarcoma 180	Mice	Stock and Tarnowski, personal communication
ara-cAMP	leukaemia L-1210	Mice	Long *et al.*, 1972

The most toxic of a group of 8-alkylseleno-cGMP derivatives is 8-$H_2N(BrH_2N)CSe$-cGMP. Since cGMP is known to promote growth, it would be expected that these derivatives are acting in some other manner by mimicking the effects of cGMP.

The activity of 6-HS-cNMP is presumably due to its conversion to toxic nucleotides after hydrolysis of the cyclic phosphate to yield the 5′-nucleotide. LaPage (personal communication) has compared the pharmacokinetic properties of 6-HS-cNMP and 2′-*O*-Bt-6-HS-cNMP. The latter is excreted much more slowly when administered at the same dose. Also, the butyrylated compound attains higher tissue levels than does the 6-HS derivative.

The significant antileukaemic activity of ara-cCMP is most intriguing in light of a report by Bloch *et al.* (1974) in which they state that cCMP initiates leukaemia L-1210 cell growth *in vitro*. cCMP has been isolated from L-1210 cells (Bloch, 1974) and a cytidylate cyclase is apparently present in these cells (Bloch and Leonard, 1976). In addition cCMP has been found in the urine of leukaemic patients, but not in that of normal subjects (Bloch *et al.*, 1975). cUMP, which has been identified in rat liver extracts (Bloch, 1975), delays the onset of growth of staged L-1210 cells, an effect opposing that of cCMP. It may be that cCMP and cUMP make up a ‘cyclic nucleotide see-saw’ (Shields, 1974) with interrelationships similar to those observed for cGMP and cAMP in many systems. Only a few pyrimidine cyclic phosphate analogues (cTMP, ara-cCMP, ara-cUMP, 4-HS-cUMP, 4-MeS-cUMP, 6-aza-cUMP and 3,N^4-etheno-cCMP), have been reported, (Holy *et al.*, 1965; Long *et al.*, 1972; Panitz, 1974). Except for ara-cCMP, the effects of these analogues on leukaemia or on tumour growth are yet to be examined.

F Effects of cAMP Analogues in the Treatment of Psoriasis

In psoriasis, the hyperproliferative epidermis contains elevated levels of cGMP and decreased levels of cAMP compared to normal skin (Duell and Voorhees, this volume). A few limited and inconclusive clinical studies have been carried out to determine the possible use of cAMP analogues in the treatment of psoriasis.

The reports concerning Bt_2-cAMP are varied. Intramuscular injections of Bt_2-cAMP (75 mg) for from two weeks to two months result in varying degrees of improvement including a complete clearing of the lesions (Anonymous, 1974). That these results may not be due to a direct effect of the intact cyclic phosphate is indicated by the fact that 5′-AMP injected intramuscularly produces a similar effect (Harrell and Voorhees, personal communication). Topical application of Bt_2-cAMP alone results in no improvement or even a worsening of the lesions (Auerbach, 1974), while modest improvement has been observed using a cream containing Bt_2-cAMP together with theophylline. Posternak *et al.* (1976) have compared the effects of intralesional injections of Bt_2-8-HS-cAMP, theophylline, Bt_2-8-HS-cAMP plus theophylline, or cAMP plus theophylline. Significant improvement was seen with all treatments, but the differences between the various treatments are small.

V CYCLIC NUCLEOTIDE ANALOGUES AS ANALYTICAL TOOLS

Immobilized cAMP analogues can be used in the isolation and purification of the regulatory subunit of protein kinase. It is clear that the type of spacer group and its position of attachment to the cAMP molecule affects the ease with which the regulatory subunit can be eluted from the cAMP–gel. Dills *et al.* (1976) have examined the binding of rabbit muscle regulatory subunit to immobilized cAMP derivatives with spacer groups of varying lengths in the 2-, N^6- and 8-positions of cAMP. The stronger binding of the N^6-substituted analogues to the regulatory subunit allows removal of other proteins before elution of the regulatory subunit with cAMP. Severin *et al.* (1974) used an analogue which binds to the regulatory subunit only weakly — 8-$HOOC(CH_2)_3S$-cAMP — to purify porcine brain regulatory subunit to homogeneity in one step. These results suggest that strong binding is not necessarily required to accomplish the purification.

Analogues of cAMP containing substitutions which can undergo photocatalysed covalent binding to cAMP-specific sites on enzymes present the opportunity to study and characterize these sites. Antonoff and Ferguson (1973) have reported the photocatalysed covalent binding of cAMP itself in extracts of testis and adrenal cortex. The nature of the reaction resulting in the covalent attachment is unknown. In the case of phosphofructokinase, after covalent attachment of the analogue, the enzyme is still sensitive to inhibition by ATP, but is no longer significantly reactivated by cAMP (Brunswill and Cooperman, 1971, 1973, Cooperman and Brunswick, 1973). Other enzymes which have been labelled include cAMP-dependent protein kinase in red blood cell ghost membranes (Guthrow *et al.*, 1973; Haley, 1975) and bovine brain cAMP-dependent protein kinase (Pomerantz *et al.*, 1975).

Immunization of rabbits with 2′-*O*-succinyl-cAMP (or cGMP) linked to protein results in the production of antibodies with a high affinity for the unmodified cyclic nucleotide (Steiner *et al.*, 1972). These antibodies do, however, exhibit a much higher affinity for cyclic nucleotide containing a 2′-*O*-substitution than for the unsubstituted cyclic nucleotides. Methods have been devised to convert cyclic nucleotides in tissue extracts to their 2′-*O*-succinyl- (Cailla *et al.*, 1972) or 2′-*O*-acetyl- (Harper and Brooker, 1973) derivatives. These methods allow the detection of 10^{-15} mole of the cyclic nucleotides.

VI CONCLUSIONS

The data presented in this review show that a great deal of work remains to be done before an analogue of a cyclic nucleotide will become a viable drug entity. Those analogues which are more active as protein kinase stimulators are, in many cases, also more active in whole animals.

To develop an analogue which is tissue-specific when administered to whole animals will require additional chemical modification of the more active analogues. Hopefully, these 'second generation' analogues will demonstrate the required specificity (*i*) by specifically interacting with a specific tissue receptor, (*ii*) by accumulating preferentially in a specific tissue, or (*iii*) by being produced from a prodrug form only in specific tissue. The last process has great potential in the treatment of cancers and hyperproliferative skin diseases where a constant supply of analogue would be necessary to arrest growth.

One possibility would be a derivative such as a 2′-*O*-ester of an active analogue, which by virtue of its acyl group would be easily transported into cells. Once inside the target cells, the acyl group would be enzymically removed and the resulting analogue would be retained in the cell to inhibit cell division. It might also be possible to take advantage of tissue-specific acylases, or other hydrolases, to liberate an active cyclic nucleotide analogue from a prodrug only within a specific tissue. The most systematic approach to this type of drug search would be to first find the analogue which most efficiently affects the process under study. Then this analogue would be modified further in the hope of eliminating side-effects and toxicity. Once the structure–activity relationships for the side-effects and toxicity have been ascertained, additional modifications which most effectively optimize the phramacological parameters of the original analogue could then be incorporated.

ACKNOWLEDGEMENTS

A great many individuals generously supplied the author with their data prior to publication. Their efforts helped to make this review more complete and up to date.

CHAPTER 7

Role of Cyclic Nucleotides in Cultured Cells

P. S. Schönhöfer and H. D. Peters

I INTRODUCTION

The presence of cAMP has been demonstrated in almost all cultured mammalian cell types. Stimulation of cAMP formation was obtained by a variety of hormones and other agents while much less is known about the presence and the stimulation of the formation of cGMP. In this chapter only few cell systems will be discussed in detail. This appears justified since a very comprehensive review on the presence and effects of cyclic nucleotides in cultured cells was recently published by Chlapowski *et al.* (1975).

II ENZYMES OF CYCLIC NUCLEOTIDE METABOLISM

A Fibroblasts

Early studies revealed the presence of the enzymes involved in cAMP formation and metabolism in normal and transformed fibroblast cultures. In fibroblast-like BHK 21/13 cells, adenylate cyclase activity was found in the 2000 g particulate fraction (Bürk, 1968). Prostaglandin stimulates adenylate cyclase activity in homogenates from BHK cells as well as from established (3T3 and L-929) or primary embryonic (mouse, rat and hamster) fibroblast cultures (Peery *et al.*, 1971). Glucagon and adrenaline do not enhance adenylate cyclase activity in L-929 cells, but catecholamines and ACTH are able to stimulate the enzyme in cultured human foreskin fibroblasts (Rao, *et al.*, 1971). Glucagon or vasopressin is not effective in these cultures.

Phosphodiesterase activity was found in the soluble as well as in the particulate fractions of L-cells (Heidrick and Ryan, 1971a). This enzyme's activity appears to be regulated by the intracellular level of cAMP. When L-929 fibroblasts are incubated for a day with prostaglandin, cAMP levels rapidly increase to a maximum and then slowly decrease to near basal values. During this time phosphodiesterase activity gradually increases, and this increase continues for two to three days producing a two to three-fold increase in enzyme activity (Manganiello and Vaughan, 1972b). The enhancement of phosphodiesterase activity is also obtained by treatment of 3T3 fibroblast cultures with Bt_2cAMP or theophylline (D'Armiento *et al.*, 1972). Blockade of the rise in phosphodiesterase activity by cycloheximide or actinomycin D indicates that the effect is dependent on the *de novo* synthesis of the enzyme. Induction of phosphodiesterase activity appears to involve a cAMP-dependent protein kinase.

The presence of cAMP-dependent protein kinase activity was demonstrated for HeLa cells and 3T3 fibroblasts by Klein and Makman (1971). The enzyme is stimulated two to three-fold by 1 μM cAMP, with the incorporation of 150–200 pmoles PO_4^{3-} mg protein min^{-1} in 3T3 cells. The presence of a cGMP-dependent protein kinase has not yet been reported for fibroblast cultures, but Tan and Sokol (1974) have found protein kinase activity in African green-monkey kidney cells. This enzyme is stimulated two-fold by 10 μMcGMP, but stimulation in the presence of cAMP is very slight.

Guanylate cyclase activity in Balb/c 3T3 fibroblasts has been shown. The enzymic activity is present mainly in the particulate membrane fraction and is enhanced about ten-fold by treatment of the microsomal fraction with Triton (Rudland *et al.*, 1974c). The activity of the enzyme in the particulate fraction is also increased considerably by fibroblast growth factor. cGMP levels in intact fibroblasts are also elevated by fibroblast growth factor without any changes in cAMP levels. Similar variations in intracellular cGMP levels of fibroblasts have been observed following addition of serum (Seifert and Rudland, 1974a) or prostaglandins (DeAsua *et al.*, 1975) to resting fibroblast cultures.

B Cells of Neural Origin

Cells of neural origin (astrocytoma 1181N1) contain adenylate cyclase, phosphodiesterase and cAMP-dependent protein kinase activity (Perkins *et al.*, 1971). Gilman and Nirenberg (1971a) examined the formation of cAMP in three clonal lines of cultured glial tumours and found an almost 200-fold increase in cAMP levels in the C-6 cell line by addition of noradrenaline or isoprenaline to the cultures. The other clones also respond in a similar manner to catecholamines, but the extent of the response is smaller in these cultures. Several clonal lines of neuroblastoma are not responsive to catecholamines, but show a marked response to prostaglandin E_1 (Gilman and Nirenberg, 1971b).

Attempts have been made to cultivate normal foetal brain cells from rats and to study cAMP metabolism in these cells (Gilman and Schrier, 1972). The foetal brain cultures contain a variety of cell types, such as endothelial, fibroblast-like, glial and neuronal cells, and respond to prostaglandin E_1, catecholamines and adenosine with an elevation of cAMP levels. The responses obtained with foetal brain cultures are similar to those obtained in brain slices. In chick embryo brain cell cultures, variations in the culture conditions have allowed workers to obtain cultures of predominantly glial or neuronal cells. Such differentiation of the cultures demonstrates that guanylate cyclase activity is only localized in neuronal cells (Goridis *et al.*, 1974).

C Hepatocytes and Hepatoma Cells

Liver tissue allows the comparison between normal hepatocytes and hepatoma cells grown in culture. However, conditions of cultivation markedly influence the adenylate cyclase activity in cultures of Chang liver cells (Makman and Klein, 1972). Adenylate cyclase is higher in cells cultured in suspension than those grown as monolayers, when the enzyme is stimulated by noradrenaline, but the opposite situation is found in the case of stimulation by NaF. Rat hepatoma HTC cells are devoid of adenylate cyclase activity, when grown in suspension (Granner *et al.*, 1968), but the same cells show a low, but significant, adenylate cyclase activity, when grown as monolayers (Makman, 1971). changes in the subcellular distribution of adenylate cyclase have been reported for the chemically induced Yoshida rat hepatoma by Tomasi *et al.* (1973). These workers found a decrease in the enzymic activity in the membrane fraction during the tumour growth, accompanied by the appearance of adenylate cyclase activity in the cytoplasmic fraction. In normal hepatocytes the enzyme is exclusively localized in the membrane fraction. Both the membrane-bound and the cytoplasmic enzyme respond to hormonal and NaF stimulation, but the extent of this response is markedly lower than in normal hepatocytes. Such a decrease in adenylate cyclase activity has been consistently reported for hepatoma cells (Allen *et al.*, 1971; Makman, 1971; Granner, 1974).

Studies on protein kinase activity in liver cells show that the relationship between cAMP-dependent and cAMP-independent activity is different in several cell types, although the total protein kinase activity is identical in both normal hepatocytes and tumour cells (Granner, 1974). The difference appears to be related to an alteration in the cAMP binding protein in hepatoma cells, since the binding protein of normal hepatocytes has two apparent bindings sites for cAMP with different affinities (K_d = 0,35 mM and K_d = 0.01 mM). Hepatoma cells only contain binding protein with the lower binding affinity (Granner, 1974). Addition of normal binding protein restores the cAMP-dependency of protein kinase from hepatoma cells. A similar deficiency in cAMP binding protein has been reported in clonal strains of mouse S49 lymphoma cells (Bourne *et al.*, 1975a).

In general, when hepatoma cells are compared with hepatocytes, the hepatoma cells show an apparent decrease in the activities of all enzymes regulating cyclic nucleotide metabolism. The decrease found in phosphodiesterase activity, however, is usually smaller than that in adenylate cyclase or cAMP-dependent protein kinase activity. Therefore, an increase in cAMP levels can only be obtained in the presence of phosphodiesterase inhibitors, even though the activity of this enzyme in hepatoma cells is somewhat lower than in normal hepatocytes (Manganiello and Vaughan, 1972b).

III HORMONES AND DRUGS AFFECTING cAMP LEVELS

A Hormones

Catecholamines (noradrenaline, adrenaline or isoprenaline) have been found to increase cAMP levels in most cell types of neural origin, such as clonal lines of rat glioma (Gilman and Nirenberg, 1971a), rat or human astrocytoma (Browning, *et al.*, 1974b), foetal rat brain cells (Gilman and Schrier, 1972) and rat glioma × fibroblast cell hybrids, when the parent glioma clone shows a sensitivity to catecholamines (Hamprecht and Schultz, 1973).

Studies with adrenergic blocking agents have revealed that the catecholamine receptor of these cells may be classified in most cell types as a β-receptor. Catecholamines are also effective in elevating cAMP levels in several other cell cultures, such as HeLa cells, rabbit lens epithelial cells, human Chang liver cells (Makman *et al.*, 1974), rat liver BRL 30E cells (Gilman and Minna, 1973), mouse S49 lymphoma cells (Daniel *et al.*, 1973a), rat heart cells (Harary *et al.*, 1973) and chick embryonic myoblasts (Kagen and Freedman, 1974). Few fibroblast strains have been found with catecholamine receptors. Human lung WI-38 fibroblasts (Kelly *et al.*, 1974; Kurtz *et al.*, 1974), human foreskin fibroblasts (Manganiello *et al.*, 1972) and human synoviocytes (Newcombe *et al.*, 1975) are all sensitive to catecholamines, while mouse L-929, 3T3 and primary or secondary embryonic mouse fibroblasts do not respond to these neurotransmitters.

Dopamine stimulates adenylate cyclase activity in neuroblastoma cells (Prasad and Gilmer, 1974) and doubles the cAMP levels in foetal rat brain cultures, but has no effect in other neuroblastoma or astrocytoma cells (Gilman and Nirenberg, 1971b). Other cell lines either have not been tested or are insensitive to dopamine. Histamine causes a small rise in cAMP levels in human astrocytoma cells (Perkins *et al.*, 1971), but not in neuroblastoma cells (Gilman and Nirenberg, 1971b). Histamine and 5-hydroxytryptamine have no effect on cAMP levels in primary embryonic mouse fibroblasts (Schönhöfer *et al.*, 1974).

Acetylcholine and cholinergic agents have been shown to stimulate adenylate cyclase from neuroblastoma cells (Prasad *et al.*, 1974), but Gilman and Nirenberg (1971b) did not observe any such stimulatory effect in either neuroblastoma or rat glioma cells by these agents. Neuroblastoma × glioma hybrid cells show a decrease in cAMP levels following additions of cholinergic stimulants, when the intracellular level of cAMP has been raised by pretreatment with prostaglandin E_1 (Gullis *et al.*, 1975a).

Other hormones are also able to elevate cAMP levels in cell cultures derived from target tissues of the hormone. Examples of this are MSH in mouse melanoma cells (Pawelek *et al.*, 1973), LH and FSH in cultured porcine granulosa cells (Kolena and Channing, 1972), LH in isolated Leydig cells and murine Leydig tumour cells (Moyle and Ramachandran, 1973), ACTH in murine adrenal cortical cells (Kowal, 1973) and tumours (Schimmer, 1972). ACTH also stimulates adenylate cyclase activity in human foreskin fibroblasts (Rao *et al.*, 1971), but not in human lung WI-38 fibroblasts (Kelly *et al.*, 1974). Glucagon is ineffective in stimulating adenylate cyclase from Chang liver cells (Makman and Klein, 1972), but is effective in HeLa S3 cells (Makman, 1971). Glucagon also causes a small increase in cAMP levels of human lung WI-38 fibroblasts (Kurtz *et al.*, 1974). Thyrotropin enhances cAMP levels in cultured thyroid cells (Winand and Kohn, 1975).

B Prostaglandins and Other Local Hormones

Prostaglandins (PG) have been found to be the most universal stimulants of cAMP levels in cultured cells. The extent of the cAMP response to PGs, however, is variable, depending on cell type, culture conditions, population density, time elapsed after the last subcultivation by trypsin treatment and the PG used. PGE_1 is generally the most potent agent, followed by PGE_2 and $PGF_{2\alpha}$. Bradykinin is also reported to elevate cAMP levels in primary embryonic mouse fibroblasts (Schönhöfer *et al.*, 1974).

C Steroid Hormones

Incubation of HTC hepatoma cells with dexamethasone for seventy-two hours results in a 25–40% decrease in phosphodiesterase activity and a concomitant increase in basal as well as adrenaline-stimulated cAMP levels (Manganiello and Vaughan, 1972b). Similar results have been obtained using human foreskin fibroblasts. The decrease in phosphodiesterase activity becomes apparent only after exposure of the cells to the steroid for more than ten hours. Brostrom *et al.* (1974) tested a series of glucocorticoids in glial tumour cell cultures and found an increase in both basal and noradrenaline-stimulated cAMP levels as well as in adenylate cyclase. Increases are dependent on the dose and type of glucocorticoid used. An incubation period of forty-eight hours is needed to obtain the maximal effect, but phosphodiesterase activity is not reduced in these cultures compared with untreated cultures. The authors, therefore, concluded that the rise in cAMP levels and adenylate cyclase activity is due to increased enzyme synthesis induced by the glucocorticoids. This situation has also been described for isolated fat cells (Schönhöfer *et al.*, 1972). In H-4-11-E hepatoma cells, hydrocortisone causes increases in adenylate cyclase and protein kinase activities when cells are incubated for four hours (Sahib *et al.*, 1971). In primary embryonic mouse fibroblasts, incubation for ten minutes with prednisolone results in a fall in basal as well as PGE_1-stimulated cAMP levels (Schönhöfer *et al.*, 1974). However, a longer incubation period of twenty-four hours reverses the inhibitory effect on PGE_1-stimulated cAMP levels, indicating an apparent sensitization of the cells to the effect of PGE_1. In this cell type, oestradiol-17β has been found to be an effective stimulant of cAMP, when calls have been exposed to the steroid for five or ten minutes (Schönhöfer *et al.*, 1974).

D Antirheumatic Drugs

Several antiheumatic drugs, such as sodium salicylate, benzydamine, indometacin and phenylbutazone, have been found to decrease basal as well as PGE_1-stimulated cAMP levels in primary embryonic mouse fibroblasts (Schönhöfer *et al.*, 1974; Peters *et al.*, 1975).

E Phosphodiesterase Inhibitors

Phosphodiesterase plays a complex role in the regulation of cAMP levels and is adaptively increased with the elevation of cAMP levels in lymphoma S49 (Bourne *et al.*, 1973), fibroblasts (D'Armiento *et al.*, 1972) and other cell cultures. Phosphodiesterase inhibitors, such as theophylline Ro 20-1724 and (4-(3-butoxy-4-methoxybenzyl)-2-imidazolidinone) reduce the enzymic activity in the cells, but do not prevent the enzyme induction in neuroblastoma cells, (Prasad and Kumar, 1973). The induction of the enzyme is suppressed by cycloheximide or, in the case of fibroblasts by actinomycin D in addition (D'Armiento *et al.*, 1972). Generally, theophylline or the more potent 1-methyl-3-isobutylxanthine potentiates the increase in cAMP levels elicited by a hormone or PGE_1. Other phosphodiesterase inhibitors, such as 2-amino-6-methyl-5-oxy-4-*n*-propyl-4,5-dihydro-S-triazolo-(1,5-a)-pyrimidine (ICI 63-197), have similar affects in cultured human diploid fibroblasts stimulated by isoprenaline or PGE_1 (Franklin and Foster, 1973).

F Drugs Affecting the Secretion of cAMP

The secretion of cAMP from cultured cells appears to be a common phenomenon and has been described for fibroblasts (Franklin and Foster, 1973; Kelly *et al.*, 1974), astrocytoma cells (Clark *et al.*, 1974) and rat glioma cells (de Vellis *et al.*, 1974) after short incubations. The fraction of cAMP secretion increases more than the intracellular level of cAMP when cells are stimulated by PGE_1 or other stimulants. With longer incubation periods the concentration of cAMP is eight to ten-fold higher in the medium than in the cells. The secretion of cAMP into the media is also observed in presence of high medium concentrations of the cyclic nucleotide (1 mM) (Clark *et al.*, 1975). Therefore, secretion appears to involve an active transport process which may constitute an additional system for the inactivation of intracellular cAMP but may also relate to roles of extracellular cAMP in cell-to-cell recognition and interaction. In rat C-6 glioma cells Schlaeger and Köhler (1976) described an external cAMP-dependent protein kinase. Several drugs have been described which affect the secretion of cAMP. Many inhibitors reduce the amount of cAMP secreted by fibroblasts (Kelly *et al.*, 1974) and by rat glioma cells (de Vellis *et al.*, 1974).

Table 7.1 Effect of antirheumatic drugs on the escape of cAMP from primary embryonic mouse fibroblasts treated with 30 μM PGE_1

		pmole cAMP concentration per 10^6 cells	
addition	mM	cells alone	cells with medium
sodium salicylate	0	12.12 ± 0.36	18.36 ± 1.02
	3	10.90 ± 0.65	11.20 ± 1.08
	10	8.15 ± 0.73	8.36 ± 0.72
indometacin	0	6.02 ± 0.34	14.20 ± 1.02
	0.3	4.76 ± 0.47	4.49 ± 0.36
phenylbutazone	0	7.06 ± 0.53	10.50 ± 0.64
	0.1	6.28 ± 0.72	6.13 ± 0.40
	1.0	3.98 ± 0.36	4.08 ± 0.64
mefenamic acid	0	7.35 ± 0.84	11.80 ± 0.66
	0.3	7.25 ± 0.54	7.02 ± 0.31
prednisolone	0	7.15 ± 0.68	14.14 ± 1.42
	0.03 μM	6.54 ± 0.72	13.18 ± 1.39
	0.3 μM	5.18 ± 0.47	10.85 ± 0.82

Cells incubated with 30 μM PGE_1 and drugs for 10 min at 37°C. Values represent the mean ± SEM of at least 5 independent incubations.

Probenecid (Penit *et al.,* 1974), dipyridamole (Clark *et al.*, 1975), PGA_1, valinomycin, oligomycin and uncouplers of oxidative phosphorylation, such as dinitrophenol and carbonyl cyanide *p*-trifluoromethoxyphenylhydrazone, also decrease the secretion of cAMP from rat astrocytoma or glioma cells, again suggesting that the transport of cAMP across the membrane is energy-dependent and regulatory (Mawe *et al.*, 1974; Doore *et al.*, 1975). Several anti-rheumatic drugs also decrease the escape of cAMP from embryonic mouse fibroblasts (Table 7.1), but glucocorticoids have no effect on the release of cAMP into the medium.

G Adenosine and Related Compounds

Adenosine is a potent agent for increasing cAMP levels in brain slices and for potentiating the response to stimulants, such as biogenic amines. A similar effect of adenosine and related compounds, such as 2-chloroadenosine has been reported for astrocytoma cells (Clark and Perkins, 1971), cloned mouse neuroblastoma C-1300 cells (Schultz and Hamprecht, 1973; Gilman, 1974) and foetal rat brain cells (Gilman and Schrier, 1972). Theophylline inhibits the effect of adenosine (Gilman, 1974), but papaverine and 1-methyl-3-isobutylxanthine do not influence the action of adenosine, as was also described in the case of papaverine using human astrocytoma cells (Clark *et al.*, 1974). Schultz and Hamprecht (1973) observed the effect of adenosine on cAMP levels in neuroblastoma cells only when phosphodiesterase inhibitors are added. Gilman (1974) proposed that the effect of adenosine on cAMP levels is mediated by a receptor site on the outer surface of the cell membrane. This is based on the finding that no decrease in the action of adenosine occurs when the uptake of the nucleoside is completely blocked by dipyridamole. A similar regulatory effect of adenosine has been described for cAMP levels in isolated fat cells (Schwabe *et al.*, 1973). The effect of the very potent adenosine analogue, 2-chloroadenosine, was also observed in rat liver cells and human VA2 fibroblasts (Gilman, 1974).

H Cholera Toxin

Cholera toxin has been shown to increase adenylate cyclase activity in almost every tissue tested, when added to whole tissue cells (Pierce *et al.*, 1971). It is, however, without effect in broken-cell preparations. A lag phase of thirty to sixty minutes is necessary before the stimulation of adenylate cyclase is apparent. Cholera toxin has been found to produce the same effect on the accumulation of cAMP in cultured lymphoma S49 cells (Insel *et al.*, 1975), adrenal cortical tumour cells (Wolff and Cook, 1975) and mouse melanoma cells (O'Keefe and Cuatrecasas, 1974). In the last study it was observed that the activation of adenylate cyclase following exposure of the cells to the toxin for four hours is maintained for up to thirteen days, showing a slow but progressive fall with time. The irreversible activation of the enzyme by the toxin in cultured cells resembles the irreversible activation of adenylate cyclase by guanylyl imidodiphosphate (GMP-P(NH)P) in broken-cell preparations (Rodbell *et al.*, 1975), but preliminary experiments with primary embryonic mouse fibroblasts have not revealed any effect of GMP-P(NH)P in cell cultures, even when cells are used immediately after treatment with trypsin.

IV AGENTS AFFECTING BOTH cAMP AND cGMP LEVELS

A Proteases, Serum and Growth Factors

Treatment of cultured cells with proteases is commonly applied for subcultivation of monolayer cells. The digestion leads to the destruction of receptors on the outer surface of the cell membrane and, consequently, to a lack of response to agents affecting the receptors. Moreover, proteases have been found to decrease cAMP levels in monolayer cultures (Sheppard, 1972; Noonan and Burger, 1973). No data are available indicating concomitant changes in cGMP levels.

Addition of serum is necessary to maintain growth in almost all cultured cells. However serum has marked effects on cyclic nucleotide levels in cell cultures. Addition of serum to quiescent, serum-free maintained Balb 3T3 fibroblasts causes a transient decrease in cAMP and increase in cGMP levels (Seifert and Rudland, 1974a; Moens *et al.*, 1975). The changes in cyclic nucleotide levels, in particular the rise in cGMP levels, are linked to the stimulation of growth effected by serum. However, these findings have not been confirmed by a more recent study showing that addition of serum to quiescent Balb 3T3 fibroblasts causes a decrease in both cAMP and cGMP levels, indicating that both cyclic nucleotides may inhibit cell growth (Miller *et al.*, 1975c).

Cells are cultured in the presence of serum, since it contains not only low molecular weight nutrients, but also a variety of macromolecular factors essential for the survival and growth of cells. These growth factors are probably polypeptide hormones, some of which (e.g. nonsuppressible insulin-like activities, somatomedins and epidermal, fibroblast, ovarian and nerve growth factors) have been identified (Holley, 1975). The effect of a fibroblast growth factor from bovine pituitary glands in the serum was studied on cyclic nucleotide levels and adenylate cyclase and guanylate cyclase activity in fibroblasts (Rudland *et al.*, 1974a). Addition of this growth factor to quiescent 3T3 fibroblasts enhances guanylate cyclase activity and raises cGMP levels, while adenylate cyclase activity and cAMP levels are slightly decreased.

B Insulin

Insulin decreases cAMP levels in 3T3, 3T6 and BHK fibroblasts (Sheppard, 1972) and concomitantly increases cGMP levels (DeAsua *et al.*, 1975). Such divergent changes in cyclic nucleotide levels have not been observed in chick embryonic fibroblasts (Hovi *et al.*, 1974) or in human fibroblasts maintained in serum-free medium (Rosenthal and Goldstein, 1975). However, the effect of insulin in lowering cAMP levels and adenylate cyclase activity as well as slightly raising cGMP levels and guanylate cyclase activity in fibroblasts was found to augment significantly the response of the cultures to fibroblast growth factor (Rudland *et al.*, 1974a).

C Steroid Hormones

Dexamethasone does not affect the fall in cAMP and cGMP levels induced by the addition of serum in Balb 3T3 fibroblasts (Miller *et al.*, 1975c), but prednisolone causes a doubling of cGMP levels in primary embryonic mouse fibroblasts, when cultures are incubated with the steroid for thirty or sixty minutes. Dexamethasone and hydrocortisone are 'permissive' factors for the growth-stimulating effect of fibroblast growth factor in fibroblast cultures (Armelin, 1973; Gospodarowicz and Moran, 1974), and hydrocortisone was shown to significantly enhance the rise in cGMP levels brought about by this growth factor (Rudland *et al.*, 1974a).

D Prostaglandins

PGE_1 is one of the most potent agents to elevate cAMP levels in most types of cell cultures. PGs have also been shown to increase cGMP levels in 3T3 fibroblasts (DeAsua *et al.*, 1975). $PGF_{2\alpha}$ is most potent in enhancing cGMP levels at doses which do not affect, or which even decrease, cAMP levels. PGE_1 increases both cyclic nucleotides at low and high concentrations. When PGE_1 is added together with serum to quiescent 3T3 fibroblasts, the decrease in cAMP due to serum is reversed and the enhancement of cGMP reduced, thus counteracting the changes in cyclic nucleotide levels obtained after addition of serum (Rudland *et al.*, 1974b). A concomitant increase in cAMP and cGMP levels is also obtained in PGE_1-treated primary embryonic mouse fibroblasts and neuroblastoma N1E-115 cells (Matsuzawa and Nirenberg, 1975).

E Cholinergic Agents

Carbamoylcholine increases cGMP levels almost 200-fold in cultures of neuroblastoma clone NN1E-115, and at the same time decreases cAMP levels (Matsuzawa and Nirenberg, 1975). The effect of carbamoylcholine is blocked by atropine, but not by d-tubocurarine, indicating a muscarinic receptor. Carbamoylcholine enhances the effect of PGE_1 on cGMP levels, but reduces the effect of PGE_1 on cAMP levels. The opposite effect occurs with carbamoylcholine and adenosine to which the neuroblastoma clone is sensitive. Other clones of neuroblastoma are far less responsive or even insensitive to carbamoylcholine. Similar results have been obtained by Gullis *et al.* (1975a) with the neuroblastoma × glioma hybrid line 108CC15 and four mouse neuroblastoma cell lines. Carbamoylcholine, tetramethylammonium and pilocarpine increase cGMP levels and slightly decrease cAMP levels in these cells. These effects are more pronounced in presence of high concentrations of Na^+ with carbamoylcholine and tetramethylammonium. Pilocarpine at high concentrations of Na^+ elevates cAMP levels whereas cGMP levels are reduced when a low Na^+ concentration is used.

F Morphine and Related Drugs

Morphine antagonizes the increase in cAMP levels effected by PGE_1 in a clonal line of neuroblastoma × glioma hybrid cells; the effect is dose dependent. This effect, however, is not due to an alteration in the release of cAMP from the cells (Traber *et al.*, 1975). Naloxone blocks the action of morphine. Since noradrenaline and acetylcholine are also able to suppress the stimulatory effect of PGE_1 on cAMP levels, Traber *et al.* (1975) explained the action of morphine by a depolarization of the plasma membrane of the cells which then become hyperpolarized by PGE_1. In the absence of PGE_1, low concentrations of morphine (0.1 – 10 μM) increase cGMP levels and decrease cAMP levels in neuroblastoma × glioma hybrids, while the effect is reversed when using millimolar concentrations of the drug (Gullis *et al.*, 1975b). The elevation of cGMP is mimicked by levorphanol, an active congener of morphine, but not by the inactive enantiomer — dextrorphan. The newly found morphine-like peptides, enkephalins, have the same effects on the cyclic nucleotide levels in this cell line as has morphine (Hamprecht, personal communication). Sharma *et al.* (1975) investigated the effect of morphine and other narcotics on basal and PGE_1-stimulated cAMP levels as well as on adenylate cyclase from the same clonal line of neuroblastoma × glioma hybrid cells. The inhibition of adenylate cyclase is stereospecific and reflects the relative affinity of the narcotics for their receptor, the inhibitory effect being blocked by the morphine antagonist — naloxone. Experiments with GMP-P(NH)P indicate that morphine combines with the narcotic receptor and acts like an inhibitory hormone for the adenylate cyclase, since it lowers ten-fold the threshold for adenylate cyclase activation by GMP-P(NH)P. Cells cultured in the presence of morphine for four days show a progressive increase in basal as well as PGE_1-stimulated adenylate cyclase activity with time. These findings indicate a compensatory increase in the synthesis of the enzyme in presence of the inhibitory agent (Klee *et al.*, 1975).

G Phorbol Myristate Acetate

The tumour-promoting agent phorbol myristate acetate was shown to decrease cAMP levels in mouse epidermis (Grimm and Marks, 1974). In fibroblast cultures, this agent increases cGMP levels (Goldberg *et al.*, 1975).

V CULTURE CONDITIONS

Factors which seriously alter the response of the cyclic nucleotide systems in cultured cells are the age and number of passages of the cells, population density, concentration and type of serum used and the pH of the medium. Since so many factors exert an influence on cyclic nucleotide levels in cultured cells, it is often difficult to compare the results of different studies, even when the same cell type has been used.

Treatment of cultures with trypsin is commonly used for subcultivation, but this digestion can alter the outer surface of the cell membrane, destroying or unmasking receptor sites (Kono, 1970). Re-establishment of the normal cell surface is a time-dependent process. Therefore, the responsiveness of cell cultures to hormonal stimulants increases slowly with time to a maximum after subcultivation, as has been shown for the catecholamine receptor in astrocytoma 1181N1 cells (Perkins *et al.*, 1971), or for the catecholamine and PGE_1 responses in human foreskin fibroblasts (Manganiello and Breslow, 1974).

Manganiello and Breslow (1974) observed a slightly higher cAMP response to PGE_1 in 'old' (more than sixty generations in culture) compared with 'young' (less than thirty generations) fibroblast cultures. The catecholamine response, however, is less dependent on population density in 'old' cultures. A similar change has been observed with human lung WI-38 fibroblasts (Haslam and Goldstein, 1974). In 'old' cultures a marked increase in the catecholamine response occurs with increasing population density whereas little change occurs in the response to PGE_1. Thus, the responsiveness of cells to stimulants of the cAMP or cGMP system may vary with the number of passages of the cultures and population density. Differences were also observed in the activity and stimulation of adenylate cyclase from Chang liver cells, when cells were cultured either as monolayers or in suspension (Makman and Klein, 1972).

Cell density influences not only cyclic nucleotide levels in response to hormonal stimulation, but also basal cAMP levels in the activities of adenylate cyclase and phosphodiesterase (Anderson *et al.*, 1973d). During the early phases of growth cAMP levels remain low, while the activities of adenylate cyclase and phosphodiesterase progressively increase with larger population densities. When normal rat renal cells approach confluency, adenylate cyclase activity continues to rise, while activity of phosphodiesterase decreases somewhat and then remains constant. Levels of cAMP rise concomitantly with the changes in enzyme activities reaching the highest values at the density-dependent inhibition of growth. These results have been confirmed for 3T3 fibroblasts (Otten *et al.*, 1972) and L-cells (Heidrick and Ryan, 1971a), showing that confluent, quiescent cultures have higher cAMP levels than logarithmically growing cells. More recent studies, however, indicate that a membrane-mediated event appears to induce the cAMP increase at the establishment of cell–cell contact before confluency, whereas little change in cAMP levels occurs at confluency (Bannai and Sheppard, 1974; Willingham and Pastan, 1975a). In serum-restricted fibroblast cultures, an increase in cAMP levels occurs when the serum concentration is reduced below the minimum necessary to maintain logarithmical growth (Oey *et al.*a, 1974), and cGMP levels decrease concomitantly (Moens *et al.*, 1975). These findings indicate that cAMP and cGMP levels are influenced by cell-to-cell contact and other factors, such as the amount of serum, which determine the state of growth of the cultures.

Serum factors affecting cyclic nucleotide levels in cultured cells have already been discussed (Section III.A). Serum contains a number of hormones in addition to macromolecular factors. The concentrations of steroid hormones and insulin have been studied in serum preparations from different batches of the same animal source or from different animal sources and were found to vary enormously from lot to lot (Esber *et al.*, 1973). Therefore, the hormonal content of the serum can seriously interfere with the effects of any hormones added to the cultures.

Other culture conditions have also been shown to interfere with the growth rate and cAMP production of cells (D'Armiento *et al.*, 1973). Reduction in the pH of the growth medium from 7.7 to 6.6 causes a rise in intracellular cAMP levels from 80 pmole cAMP/mg DNA to 350 pmole cAMP/mg DNA in WI-38 fibroblasts, while the increase in extracellular cAMP is eight-fold. Therefore, small variations in buffer composition and CO_2 gassing of the cultures may result in large changes in cyclic nucleotide levels.

VI SPECIFIC RESPONSES

A General Considerations

The classical second messenger concept describes cAMP as an intracellular messenger which translates the metabolic signal contained in a hormone into a specific metabolic response at the target cell (Sutherland and Rall, 1960). Thus, cAMP-mediated hormonal actions constitute a *specific control* over the activity or concentration of specific enzymes in the target cell (Kram *et al.*, 1973). But hormones known to act *via* cAMP, also affect cell growth, exerting a *pleiotypic control* over a complex, but co-ordinated set of biochemical events related to cell growth. This *pleiotypic programme* includes such mechanistically unrelated parameters as membrane transport, protein, RNA and DNA synthesis, and protein breakdown, which have to take place in a co-ordinated fashion during cell growth and mitosis (Hershko *et al.*, 1971). Pleiotypic control has been studied for cAMP and cGMP in 3T3 fibroblasts (Kram *et al.*, 1973; Kram and Tomkins, 1973). Since cAMP and cGMP affect both the specific and pleiotypic responses of cells and since cell cultures are most often maintained under conditions of cell growth, it is difficult to discuss separately effects of cyclic nucleotides on specific or pleiotypic responses in cultured cells. An attempt has been made, however, to investigate specific cellular metabolic responses in cell cultures. Stimulation of endogenous cAMP production or, more often, addition of cAMP and cGMP or their derivatives is employed in cell cultures to study the actions of cyclic nucleotides on specific metabolic processes contained in the metabolic code of the cell type used.

B Fibroblasts

Bt_2-cAMP increases the amount of sulphate incorporated into glycosaminoglycans secreted by 3T3 fibroblasts (Goggins *et al.*, 1972). The increase reflects an enhanced glycosaminoglycan synthesis effected by the cAMP analogue. PGE_1 was shown to increase cAMP levels in primary embryonic mouse fibroblasts during a short incubation period of ten minutes; increases are dose-dependent. This rise in cAMP levels which lasts for less than one hour is sufficient to trigger an enhanced glycosaminoglycan synthesis in these cells for twenty-four to forty-eight hours (Peters *et al.*, 1974). A similar effect was obtained using bradykinin or oestradiol-17β (Schönhöfer *et al.*, 1974). The regulation of glycosaminoglycans synthesis is antagonistically influenced by cGMP as indicated by the effect of Bt_2-cGMP. Antirheumatic agents and prednisolone reduce either basal or PGE_1-stimulated cAMP levels or the activity of cAMP-dependent and cAMP-independent protein kinase activity respectively. In all instances, this is followed by a reduction in glycosaminoglycan synthesis (Schönhöfer *et al.*, 1974). Cell numbers are little affected under these conditions, indicating that the observed actions of cAMP and cGMP may represent a specific rather than a pleiotypic control.

A similar enhancement of hyaluronic-acid synthesis was observed using cultured human synovial cells (Castor, 1974). Bt_2-cAMP, or high concentrations of cAMP, stimulate hyaluronic-acid synthesis in these cultures and potentiate the effect of connective tissue-activating peptide at sub-stimulatory concentrations, when cells are exposed to the cyclic nucleotide for forty hours. PGE_1 shows a similar effect under these conditions (Castor, 1975a). The effect of Bt_2-cAMP is not blocked by cortisol or cycloheximide, but is nonspecifically inhibited by propranolol, chlorpromazine, imipramine and ethacrynic acid (Castor, 1975b). Cortisol, indometacin and cycloheximide also fail to inhibit the effect of PGE_1 which is blocked by 7-oxa-13-prostynoic acid (Castor, 1975a).

Glycosaminoglycans are formed immediately prior to synthesis of collagen by fibroblasts. Bt_2-cAMP was also found to increase the synthesis of collagen in human embryonic bone fibroblasts (Manner and Kuleba, 1974). The stimulation of collagen synthesis is more pronounced in stationary then in proliferating cultures. However, studies with cells derived from a human giant-cell bone tumour showed that the stimulation of collagen synthesis brought about by Bt_2-cAMP does not correlate with the inhibition of cell proliferation, indicating a specific control by cAMP of collagen synthesis. Bt_2cAMP induces an extensive stimulation of collagen synthesis in epithelial cells, such as Chinese hamster ovary cells (Hsie *et al.*, 1971). Bt_2-cAMP, however, fails to induce the same effect in human epithelial cells derived as primary cultures from skin explants (Manner and Kubela, 1974).

Other specialized synthesis of macromolecules may also be controlled by cAMP in fibroblasts. Interferon synthesis is induced in L-cells by viral infection or by poly I and poly C. Bt_2-cAMP or stimulation of the cells with adrenaline in the presence of theophylline suppresses the interferon synthesis (Dianzani *et al.*, 1972). The effect of Bt_2-cAMP is blocked by cycloheximide, indicating that cAMP may control the synthesis of an inhibitor for interferon synthesis rather than directly affecting interferon synthesis.

C Cells of Neural Origin

In foetal rat brain cell cultures, incubation with Bt-cAMP or Bt_2-cAMP for several days results in a three-fold increase in acetylcholinesterase activity and a reduction of choline acetyltransferase activity by 50% (Shapiro, 1973). Glutamic acid decarboxylase activity is also enhanced (Schrier and Shapiro, 1973). However, these changes in enzyme activity are accompanied by distinct morphological changes in the cultures and by a marked reduction in cell density, indicating that the observed changes in enzymic activity may be related to pleiotypic control rather than to specific actions of cAMP. In primary chick embryonic brain cultures, Bt_2-cAMP causes an increase in the activity of acetyl-cholinesterase and choline acetyltransferase (Werner *et al.*, 1971). In neuro-blastoma cells, addition of Bt_2-cAMP or increase of endogenous cAMP levels by phosphodiesterase inhibitors, catecholamines or PGE_1 causes an increase in the activity of tyrosine hydroxylase (Richelson, 1973), choline acetyltransferase (Prasad and Mandal, 1973) and acetylcholinesterase (Furmanski *et al.*, 1971). The sensitivity of the adenylate cyclase to acetylcholine, catecholamines and PGE_1 is enhanced, when neuroblastoma cells are incubated with Bt_2-cAMP or a phosphodiesterase inhibitor for three days (Prasad and Gilmer, 1974; Prasad *et al.*, 1975). Again this treatment causes 'differentiation' of the tumour cells and may therefore be ascribed to pleiotypic rather than to specific effects of cAMP.

A specific control by cAMP has been demonstrated for glycogen depletion (Opler and Makman, 1972) and increased lactate dehydrogenase activity (de Vellis and Brooker, 1972) in rat astrocytoma cells. Catecholamines, histamine, phosphodiesterase inhibitors and Bt_2-cAMP are able to induce these effects during a short incubation period.

D Hepatoma Cells

The role of cAMP and Bt_2-cAMP in the induction of several liver enzymes has been extensively studied and monolayer cultures of several cell lines of 'differentiated' hepatoma cells have been used to assess the role of cAMP in enzyme induction (Wicks, 1974). Control of enzymic activity is considered to be effected by cAMP at the transcriptional level for serine dehydratase and at the translational level for tyrosine aminotransferase, phosphoenolpyruvate carboxy-kinase (Butcher *et al.*, 1971; Wicks, 1974) and phenylalanine hydroxylase (Haggerty *et al.*, 1973). Glucocorticoids and insulin appear to affect enzyme activity in hepatoma cells as well as in hepatocytes by mechanisms unrelated to the action of cAMP. Studies with different hepatoma cell lines have revealed that Bt_2-cAMP is only able to induce tyrosine aminotransferase and the carboxy-kinase in those cell lines which also show a decrease in DNA synthesis and an increase in doubling time following exposure to Bt_2-cAMP (van Rijn *et al.*, 1974).

The correlation which exists between the effects of cAMP on enzyme induction and growth regulation is related to an altered affinity of the cyclic nucleotide for the cAMP-binding protein found in some hepatoma cell lines (Granner, 1974). However, since the effects of cAMP on the induction of tyrosine aminotransferase, phosphoenolpyruvate carboxykinase and growth occur separately in different hepatoma cell lines, an alteration or deletion of substrate protein for cAMP-dependent protein kinase has to be considered (van Rijn *et al.*, 1974). Thus, studies with different hepatoma cell lines prove to be an useful tool for elucidating the mode of action of cAMP in the induction of enzyme activity. Addition of cGMP or its analogues does not indicate a function of this nucleotide in hepatic enzyme induction (Wicks, 1974).

E Other Cell Types

Cell cultures from different organs have been used in many studies on the regulation of a specific synthetic process by cAMP. Stable cell lines of adrenal cortical cells or adrenal tumour cells have been shown to respond with increased steroid synthesis when cAMP is added to the cultures (Kowal, 1973) or when endogenous cAMP levels are increased by stimulation with ACTH (Buonassisi *et al.*, 1962), cholera toxin or other endotoxic lipopolysaccharides (Wolff and Cook, 1975). Mutants of adrenal tumour cells lacking the ACTH receptor but responsive to cAMP, or which do not respond to ACTH and cAMP, have been used to study the role of the transducer in activating adenylate cyclase (Wolff and Cook, 1975). Mutants have also been used to investigate the role of ions, such as Ca^{2+}, on the action of cAMP (Kuo *et al.*, 1975b) or to elucidate the step, or steps, in the biosynthesis of steroids which is regulated by cAMP (Brush *et al.*, 1974). Interestingly, the rate of growth is only reduced in adrenal tumour cell cultures which show a steroidogenic response to ACTH or cAMP (Masui and Garren, 1971), and where the steroidogenic response is not accompanied by a rise in overall protein synthesis (Kowal, 1973b).

In rat pituitary tumour cells, Bt_2-cAMP increases prolactin production (Tashijan and Hoyt, 1972). In melanoma cells, Bt_2-cAMP produces an increase in pigmentation and reduces growth (Kreider *et al.*, 1973). In primary cultures of Sertoli cells, Bt_2-cAMP stimulates oestradiol-17β synthesis thirty-fold when the medium contains testosterone (Dorrington and Armstrong, 1975). Primary monolayer cultures of thyroid cells reacted to TSH or Bt_2-cAMP with morphological changes to produce more differentiated cells possessing vesicles, which are capable of concentrating iodide and synthesizing iodinated thyroglobulin (Lissitzky *et al.*, 1971). Monolayer cells not exposed to TSH or Bt_2-cAMP can only synthesize a noniodinated material which is bound to antisera against thyroglobulin, but which is metabolically inactive.

F Conclusions

The results obtained with cell cultures indicate that cultured cells are useful tools in studies of the regulatory role of cyclic nucleotides in specific synthetic or metabolic processes. The advantage of cell cultures is that the effects of cyclic nucleotides can be studied without interference form other cell types, as compared to tissue slices or surviving organs. The disadvantage of cell-culture studies is the pleiotypic response to cyclic nucleotides which is obtained in many cultures under the conditions of cell growth. The use of stationary confluent cultures reduces but does not eliminate the possible interferences of pleiotypic effects with the specific effect under investigation.

VII PLEIOTYPIC RESPONSES

A Changes in Cyclic Nucleotide Levels During Cell Growth

Cell cultures, such as nontransformed fibroblasts, normally exist in one of two reversible states of growth. Firstly, there is a state of rapid proliferation, in which varying portions of the cells are in the G_1, S, G_2 and M phase of the cell cycle. Secondly, cells can exist in a state of relative quiescence, where most of the cells are in the G_0 or G_1 phase of the cell cycle, for instance during 'density-dependent inhibition' of growth (Todaro *et al.*, 1965; Smith and Martin, 1973). Normally, cultured cells proliferate to their saturation density and become quiescent (density-dependent regulation of growth). The saturation density varies with cell types and culture conditions, but is primarily determined by the concentration of serum in the medium. The transition between the two states appears to be essentially regulated by growth factors present in the serum (Holley, 1975). Most of these factors seem to act on the cell surface (Pardee *et al.*, 1974). Therefore, the continuation of signals triggering growth probably requires intracellular messengers acting as regulatory signals for growth control. Changes in intracellular cAMP, cGMP and Ca^{2+} concentrations as well as in the influx of essential nutrients into the cell have been suggested as the regulatory signals (Otten *et al.*, 1972; Kram *et al.*, 1973; DeAsua *et al.*, 1974; Goldberg *et al.*, 1975; Dulbecco and Elkington, 1975; Berridge, 1975b).

Cyclic nucleotide concentrations measured in asynchronously growing cell cultures only reflect an average over the entire cell cycle. The average cAMP levels were found to show some correlation with the rate of growth of the cultures, since in rapidly growing cells the G_1 phase is normally shorter than in slowly growing cultures, thus resulting in a higher proportion of cells in the S or M phase (Otten *et al.*, 1972). Analysis of the fluctuation of cyclic nucleotide levels during the cell cycle reveals that in synchronized Chinese hamster ovary cells cAMP levels are high in the early G_1 phase, then fall towards the S phase, being low during the S phase and even lower during mitosis (Sheppard and Prescott, 1972). Similar results were obtained with HeLa cells (Zeilig *et al.*, 1972) and in cultured human lymphoid cells (Millis *et al.*, 1974). These studies show that the highest cAMP levels occur in the G_2 phase. The high cAMP levels in the late G_2 phase may be correlated with increased histone f_1 phosphorylation (Bradbury *et al.*, 1974). In contrast, the amount of phosphate incorporated into nuclear acidic protein parallels the fluctuation in cAMP levels during G_1, S and M phase in synchronized HeLa cells (Karn *et al.*, 1974). The apparent conflicting finding of Makman and Klein (1972) that adenylate cyclase activity in synchronized Chang liver cells is low in the early S phase and then increases, reaching peak values at the M phase can be explained by the findings of Millis *et al.* (1974) in synchronized human lymphoid cells. These workers showed a strong rise in phosphodiesterase activity in the late G_2 and M phase. Thus, the high adenylate cyclase activity during the M phase may be compensated or even overcome by a more marked rise in phosphodiesterase activity.

Fluctuations in cGMP levels have not been observed in the different phases of the cell cycle (Seifert and Rudland, 1974b). A transient increase in cGMP levels was found following the addition of serum to previously serum-restricted, quiescent 3T3 fibroblasts (Seifert and Rudland, 1974a; Moens *et al.*, 1975). The transient stimulation of cGMP levels is only obtained, when cells are specifically arrested in the G_0 phase or when passing through the G_1 phase, but not during the other phases of the cell cycle (Seifert and Rudland, 1974b). The rise in cGMP levels is accompanied by a fall in the cAMP levels. Since the addition of serum stimulates the cells to proceed into the S phase and to proliferate, the changes in cGMP and cAMP levels have been interpreted as triggering events in the initiation of proliferation. A similar rise in cGMP levels is obtained in confluent 3T3 fibroblast cultures stimulated to proliferate by insulin and phorbol (Goldberg *et al.*, 1975), or fibroblast growth factor (Rudland *et al.*, 1974a). These data lead to the concept of a bidirectional control of cell growth which proposes that cGMP may serve as a key regulatory effector in promoting the early events of the pleiotypic programme for proliferation, while cAMP may serve as an inhibitory modulator (Goldberg *et al.*, 1975). However, in a recent study, Miller *et al.* (1975c) were unable to obtain a transient rise in cGMP levels following the addition of serum to serum-restricted 3T3 fibroblasts. They observed a concomitant decrease in cAMP and cGMP levels after addition of serum, even though the cells are stimulated to proliferate by this treatment.

B Influence of Cyclic Nucleotides on Growth

Incubation of lymphoma S49 cells in the presence of high concentrations of Bt_2-cAMP in the medium results in the arrest of growth of the cells in the G_1 phase after completing another cycle of replication in the normal doubling time of seventeen to eighteen hours (Bourne *et al.*, 1975b). Prolonged incubation with Bt_2-cAMP causes cell death after about thirty-six hours. The same results are observed, when intracellular levels of cAMP are elevated by addition of cholera toxin or isoprenaline to the medium. When cells are grown in the presence of Bt_2-cAMP, several clones are obtained which are resistant to the cytotoxic effect of the cyclic nucleotide. Analysis of these mutant cell lines shows that these cells are either devoid of any protein kinase activity or the regulatory subunit of the enzyme has a markedly reduced affinity for cAMP (Insel *et al.*, 1975c). The latter cells are susceptible to a cytotoxic effect of Bt_2-cAMP, when very high concentrations are added to the medium. These findings demonstrate the essential role of cAMP and cAMP-dependent protein kinase in the regulation of growth.

A similar inhibition of growth by cAMP has been obtained with many other cell types, but the arrest does not always occur in the G_1 phase. For example, arrest of growth in the G_1 phase has been observed in 3T3 fibroblasts (Willingham *et al.*, 1972; Kram *et al.*, 1973), and in primary rat fibroblasts (Frank, 1971). In contrast, 3T3 fibroblasts and leukaemic lymphoid cells were also be found to be arrested in the G_2 phase (Willingham *et al.*, 1972; Millis *et al.*, 1974). Reuber hepatoma cells appear to be arrested in the S phase (Wicks, 1974).

Thus high intracellular levels of cAMP or high concentrations of Bt_2-cAMP in the medium normally cause an inhibition of growth in cells with normal growth control mechanisms, the arrest being mostly maintained in the G_0 or G_1 phase of the cell cycle. In most cell types, removal of Bt_2-cAMP from the medium leads to a re-initiation of growth with a pattern similar to that seen after release of cells from serum restriction (Pardee, 1974). In several lines of neuroblastoma cells, however, prolonged incubations with Bt_2-cAMP or agents able to achieve a sustained rise in endogenous cAMP, such as phosphodiesterase inhibitors, induce 'irreversible differentiation' of the cells. Features include formation of neurites, loss of tumourigenicity, elevation of the activities of the enzmes associated with neurotransmission and increased sensitivity of adenylate cyclase to catecholamines or acetylcholine (Prasad, 1972; Prasad and Kumar, 1973; Shapiro, 1973).

Initiation of proliferation can be achieved by a variety of procedures, all resulting in a drop in cAMP levels, as can be obtained by removal of Bt_2-cAMP from the medium (Pardee, 1974). Serum-restriction causes inhibition of growth and raises cAMP levels. Readdition of serum is followed by a sharp drop in cAMP levels and initiation of proliferation (Kram *et al.*, 1973). In stationary density-inhibited quiescent cultures, addition of fresh serum or treatment with pronase initiates another round of cell division, and in each case causes a transient drop in cAMP levels (Otten *et al.*, 1972; Noonan and Burger, 1973; Oey *et al.*, 1974). This drop appears to be related to a triggering event which switches on the pleiotypic programme for the next round of cell division. When Bt_2-cAMP is added to the cultures within five minutes of pronase treatment causing the fall in cAMP, the switch to cell proliferation is suppressed. Addition of the cyclic nucleotide ten minutes after pronase treatment is unable to suppress the cell proliferation (Noonan and Burger, 1973). These results suggest that the programme for cell division is switched on within minutes of the fall in cAMP levels and that a narrow period exists during which an increase in cAMP can prevent the decision of the cell to proliferation.

The role of cGMP in the switching-on of the pleiotypic programme for mitosis is not yet clarified. A transient rise of cGMP levels during stimulation of proliferation by addition of serum (Seifert and Rudland, 1974a,b; Moens *et al.*, 1975) or insulin (Goldberg *et al.*, 1975) has been described, but other workers have found no such increases (Hovi *et al.*, 1974) or even a decrease in cGMP levels (Miller *et al.*, 1975c). Kram and Tomkins (1973) showed that exogenous cGMP is able to counteract the inhibitory effect of Bt_2-cAMP on the earliest metabolic changes induced by the pleiotypic programme, such as increased uptake of uridine, leucine or deoxyglucose. But exogenous cGMP, Bt-cGMP and Bt_2-cGMP have a limited effect on the initiation of proliferation, since only 10–20% of the cells of a quiescent fibroblast culture are induced to proliferate by the nucleotides. Moreover, this effect appears to be rather nonspecific, since Bt-cIMP is as effective as the cGMP derivatives in stimulating growth (Seifert and Rudland, 1974a). Several nucleotides in addition to cGMP have been found to stimulate serum-induced DNA synthesis in 3T3 cells, including cAMP, adenosine, AMP, ADP, ATP and cIMP (Schor and Rozengurt, 1973). This may be caused by nonspecific actions, since serum-induced DNA synthesis is also enhanced by low concentrations of Bt_2-cAMP, whereas high concentrations are strongly inhibitory in the same cell type (Furmanski *et al.*, 1971).

C Mechanisms of Cyclic Nucleotides Influencing Growth Regulation

Once a cell is committed to entering another round of cell division, cAMP is unable to inhibit the read-out of the pleiotypic programme for mitosis (Berridge, 1975b). In cells with normal growth control mechanisms, high cAMP levels prevent the triggering of the mitotic programme. The decision is made at a point in the early G_1 phase of the cell cycle related to the drop in cAMP levels, but the fall in levels is not necessarily the essential signal. A change in ionic fluxes, such as Ca^{2+}, may be involved, since growth in 3T3 fibroblasts is completely dependent on the presence of the cation in the medium (Boynton *et al.*, 1974; Dulbecco and Elkington, 1975).

The mechanism by which elevated cAMP levels prevent cell proliferation is not yet understood. Since the signals which initiate cell proliferation originate from the cell surface, effects of cAMP on the cell membrane may be involved in the growth inhibition elicited by the cyclic nucleotide. Bt_2-cAMP alters the cell shape of Chinese hamster ovarian cells (Hsie *et al.*, 1971), and reduces the motility of fibroblasts (Johnston *et al.*, 1972). Agglutination of the cells by plant lectins is decreased in the Bt_2-cAMP-treated cultures of Chinese hamster ovarian cells (Hsie *et al.*, 1971). A close correlation exists between cAMP levels, organisation of microtubules and microfilaments and cell shape in these as well as in 3T3 or L cells (Borman *et al.*, 1975; Willingham and Pastan, 1975a). Inhibitors of microtubular assembly, such as colcemid and vinblastine, reverse the inhibition of uridine, leucine and deoxyglucose uptake induced by cAMP. This effect of the inhibitors of microtubular assembly occurs without any fall in cAMP levels. Since these transport phenomena represent early events in pleiotypic responses of cells, Kram and Tomkins (1973) suggested that the microtubular apparatus controls the transport processes involved in the pleiotypic programme and that the effects of cAMP, as well as the possibly antagonistic action of cGMP, may be mediated by a modulation of the state of organization of the microtubules. This hypothesis is supported by the finding that morphological changes induced by Bt_2-cAMP are not prevented by inhibitors of RNA or protein synthesis, indicating that changes in cell morphology and cell-membrane structure result from a rearrangement in the assembly of membrane-related material rather than from synthesis of new protein (Johnson *et al.*, 1971b).

Whether cell-surface microstructures, such as microvilli, are also essential features of the pleiotypic programme remains to be clarified. Willingham and Pastan (1975b) observed a decrease in the number of microvilli, when 3T3 fibroblasts were treated with Bt_2-cAMP and kept in the G_1 phase of the cell cycle. Stimulation of cAMP levels by PGE_1 or blockade of phosphodiesterase also causes a retraction of microvilli, while removal of Bt_2-cAMP from the medium or a drop in cAMP levels, which initiates the progression of the cells towards mitosis, is accompanied by an increase in these surface structures. These data indicate that cAMP may exert its inhibitory effect on cell proliferation by maintaining an organisation of the cell membrane specific for the non-proliferating cell.

VIII CYCLIC NUCLEOTIDES IN TRANSFORMED AND MALIGNANT CELLS

A Characteristics

Cells transformed by virus, chemical carcinogen or spontaneously, or cells derived from malignant tumours exhibit morphological and biochemical alterations as compared to the normal parent cells. Morphological alterations included a more rounded and less 'differentiated' cell shape, an increased motility of the cells, a reduced adhesion to the substratum (Johnson *et al.*, 1971b, 1972), a diminished synthesis of macromolecules (Goggins *et al.*, 1972) and an enhanced agglutinability by plant lectins (Hsie *et al.*, 1971; Sheppard, 1972). An essential feature of the malignant or transformed cells appears to be the alteration of growth control, such as density or serum-dependent inhibition of growth. While normal cells stop growing at the saturation density and become quiescent in the $G_0(G_1)$ phase of the cell cycle, virus-transformed cell lines continue to grow to very high densities and are less serum-dependent. As shown with cultures of 3T3, SV–3T3 and SV–3T3–F 101 fibroblasts, an increase in malignancy leads to an increasing loss of growth control, a decreasing dependency on serum factors and an enhanced growth rate (Paul *et al.*, 1974). Density-dependent regulation of growth, however, is not a property of normal cell cultures alone, since some cell cultures derived from malignant tumours have been shown to possess some density-dependent regulation of growth (Glinos and Werrlein, 1972; Kaminskas, 1972).

The altered characteristics of the malignant or transformed cell appear to be related to changes in composition, organization and structure of the cell membrane. The chemical composition of the cell membrane has been extensively studied in normal and transformed cells, indicating that alterations in the content and composition of phospholipid, lipid, carbohydrate and glycoprotein may occur (Wallach, 1972, 1973). Electron-microscopic studies have revealed that the expression of microfilament sheaths is less pronounced in transformed cells (McNutt *et al.*, 1973; Dermer *et al.*, 1974). Striking differences between normal and transformed cells in the organization of filaments have been demonstrated by use of fluorescent actin antibodies in studies of the fluidity of the cell membrane (Pollack *et al.*, 1975). Willingham and Pastan (1975b) found that the transformed state is characterized by an extensive cover of microvilli in L-cells and rat renal cells, and they concluded that this peculiar quality of the transformed cells may be related to the enhanced agglutinability of transformed cells by plant lectins, such as concanavalin A. But others have been unable to confirm an increase in microvilli structures in transformed 3T3 fibroblasts (Collard and Temmink, 1976; Porter *et al.*, 1973). The malignant or transformed cell, therefore, appears to differ from the normal cell due to alterations of the cell membrane, leading to an aberration of the normal mechanisms of growth control.

B Alterations to Cyclic Nucleotide Metabolism in Malignant Cells

The transformation of cells is understood as an introduction of different 'transforming genes' by oncogenic viruses which bring about a change in a critical membrane component—protein, lipid or carbohydrate — leading to a whole cascade of pleiomorphic anomalities (Pastan *et al.*, 1974). Alterations of the cell membrane may affect various components of the system which regulates the formation and metabolism of cyclic nucleotides. Due to the multiplicity of biochemical lesions leading to the transformed state and due to the complex biochemical machinery involved in the formation, metabolism and function of the cyclic nucleotides, the alterations may be expected to vary widely between different cell types.

In accordance with the inhibitory role of cAMP in cell proliferation, basal as well as stimulated adenylate cyclase activity and cAMP levels have been commonly found to be lower in transformed cells than in the respective parent cells (Sheppard, 1972; Anderson *et al.*, 1974). A complete loss of adenylate cyclase activity is obtained in a mutant clone of lymphoma S49 cells by clonal selection techniques (Bourne *et al.*, 1975b). Fluctuation of cAMP levels in transformed cells gives a pattern similar to that in normal cells during the cell cycle, but the extent of the oscillations is reduced. Transformed cells often fail to produce a rise in cAMP levels at saturation density (Otten *et al.*, 1972; Sheppard, 1972). However, there are exceptions to this rule. In some transformed lines from BHK or 3T3 cells, basal and sometimes NaF-stimulated adenylate cyclase activities are higher than those of the parent cells (Bürk, 1968; Peery *et al.*, 1971). When 3T3 cells and a SV40-transformed cell line are transferred into serum-restricted medium, both cell types react with an increase in cAMP levels, even though only the normal cells become quiescent under serum-restriction, while the transformed 3T3 cells continue to grow. However, cAMP levels are lower in the transformed cells under all conditions tested (Oey *et al.*, 1974).

A decrease in cAMP levels is accompanied by elevated cGMP levels in transformed 3T3 fibroblasts, when compared to the normal parent cells (Rudland *et al.*, 1974a; Moens *et al.*, 1975). These findings have not been confirmed in a more recent study showing that transformed cells have both lower cGMP and cAMP levels, and decreased guanylate and adenylate cyclase activities (Nesbitt *et al.*, 1975). Alterations to the membrane structure by transformation may not only affect the catalytic unit of the adenylate cyclase, but also the receptors for stimulants or other steps of the process by which the activation of the catalytic unit is achieved. A loss of responsiveness to ACTH was observed in mutant cell lines of an ACTH-sensitive, steroid-producing adrenocortical tumour cell culture (Schimmer, 1972). Since the mutant cells contain NaF-stimulated adenylate cyclase activity similar to that of the parent cells, the lesion appears to result from a defect in the ACTH receptor or in a step in the coupling between ACTH–receptor complex and catalytic unit of the adenylate cyclase. Little stimulation by glucagon was observed in some hepatoma cultures also displaying low basal adenylate cyclase activity (Granner *et al.*, 1968). Two transformed cell lines derived from PGE_1-sensitive 3T3 cells are unresponsive to PGE_1 and the other PGs tested, but contain basal as well as NaF-stimulated adenylate cyclase activity similar to that of the responsive cell lines (Peery *et al.*, 1971). On the other hand, a much higher increase in cAMP levels was found in transformed WI-38 fibroblasts compared with the parent cells, when cells are stimulated by catecholamines, PGE_1 or adenosine, even though basal cAMP levels are lower in the transformed cells (Kelly *et al.*, 1974).

Several studies indicate possible mechanisms for the altered adenylate cyclase activity and responsiveness in transformed cells. A translocation of adenylate cyclase from the membrane into the cytosol was found in Yoshida hepatoma cells (Tomasi *et al.*, 1973) and transformed hamster fibroblasts (Rethy *et al.*, 1973), thus rendering the enzyme unresponsive to extracellular stimulants. Transformation of chick embryo fibroblasts by two different viruses resulted in one instance in an alteration of the affinity of the enzyme to ATP and variation in its Mg^{2+} dependency (Anderson and Pastan, 1975). In the other case, no alterations to the kinetics of the enzyme were obtained.

Transformation of cells does not increase phosphodiesterase activity. Transformed 3T3 fibroblasts have lower enzyme activity than the normal parent cell (D'Armiento *et al.*, 1972). The same situation was reported for rat kidney cells transformed by Kirsten sarcoma virus (Anderson *et al.*, 1974). Therefore, the lower cAMP levels of transformed or malignant cells cannot be explained on the basis of an enhanced degradation of the cyclic nucleotide by increased phosphodiesterase activity. Moreover, elevated cAMP levels induce the activity of this enzyme in transformed 3T3 fibroblasts (D'Armiento *et al.*, 1972) and in neuroblastoma cells (Prasad and Kumar, 1973), thus indicating an identical mechanism as that in normal cells.

Protein kinase activity is present in malignant cells, such as glial tumour cells (Perkins *et al.*, 1971) or hepatoma cells (Granner, 1972). Viral transformation does not alter the enzyme activity in bovine fibroblasts (Troy *et al.*, 1973). Even though the protein kinase activity is about the same in normal liver and in hepatoma cells, the affinity of the regulatory subunit of the enzyme for cAMP is altered in a variety of hepatoma cell cultures (Granner, 1974 — see also Section II.C). Clonal selection of lymphoma S49 cells yields mutant lines which either have no protein kinase activity or a reduced affinity of the cAMP-binding protein for the cyclic nucleotide. These changes render the cells unable to react to high cAMP concentrations with an induction of phosphodiesterase activity (Daniel *et al.*, 1973b) or an arrest of growth in the G_1 phase of the cell cycle (Bourne *et al.*, 1975a; Insel *et al.*, 1975b). Using a similar clonal selection technique in neuroblastoma cells, Simantov and Sachs (1975) obtained clonal cell lines which have a lower protein kinase activity and a reduced affinity of the binding protein for cAMP, resulting in a substantial decrease in cAMP-dependent enzyme activity in the mutant clones. Elevation of intracellular cAMP levels by PGE_1 or exogenous cAMP fails to cause irreversible differentiation, such as axon formation and induction of enzymes involved in neurotransmission, in the clonal lines but produces these effects in the parent cells.

Transformed or malignant cells differ from their normal parent cells generally by lower basal cAMP levels. The cAMP levels generally show a fluctuation similar to the normal cells during the cell cycle, but often fail to respond to serum restriction with an increase in cAMP levels. Responsiveness to stimulants may be altered. No striking differences occur between normal and transformed cells in regard to phosphodiesterase or protein kinase activities except for a few lines of hepatoma, lymphoma S49, and neuroblastoma cells. Therefore, it appears difficult to establish a causal relationship between lesions in the formation and metababolism of cyclic nucleotides, and the loss of growth control seen in malignant and transformed cells. However, the studies using cell lines defective in protein kinase activity and its regulation reveal that this cAMP-dependent enzyme has a pivotal role in the cAMP-dependent alterations of the cell cycle.

C Effects of Cyclic Nucleotides

The ability of Bt_2-cAMP to induce irreversible differentiation in malignant neuroblastoma cells, characterized by axon formation, inhibition of proliferation, enhanced responsiveness of adenylate cyclase to hormonal stimulation and induction of the enzymes involved in neurotransmission has been discussed (see Section VII.B). Similarly, early studies with transformed BHK cells and 3T3 fibroblasts suggest that Bt_2-cAMP, in the presence or absence of theophylline, reduces the growth rate and restores density-dependent inhibition of growth in a pattern similar to that of the untransformed parent cells (Bürk, 1968; Sheppard, 1972). In addition, Bt_2-cAMP has been shown to alter the morphology of Chinese hamster ovarian cells (Hsie and Puck, 1971) and of transformed fibroblasts (Johnson *et al.*, 1971b; Peterkofsky and Prather, 1974) towards a more normal cell shape, to increase the adhesion of transformed fibroblasts to the substratum (Johnson and Pastan, 1972b), to reduce the agglutinability of the transformed cells by plant lectins (Sheppard, 1972) and to increase the incorporation of sulphate into glycosaminoglycans (Goggins *et al.*, 1972) and of proline into hydroxyproline (Hsie *et al.*, 1971). Transformed 3T3 fibroblasts show a reduction in the growth rate, a more differentiated cell shape and an increased collagen formation when treated with Bt_2-cAMP. Collagen synthesis is not enhanced in the normal parent cells by this treatment, even though these cells are also retarded in their growth rate by the cyclic nucleotide. With Bt_2-cGMP no effect on growth rate or morphology of the transformed or normal cells has been observed (Peterkofsky and Prather, 1974).

The efficacy of Bt_2-cAMP to re-establish morphology, apparent normal regulation of growth and specific synthetic functions in transformed fibroblasts has been contradicted by findings that Bt_2-cAMP does not effectively influence the growth rate in transformed cell lines. Bt_2-cAMP does not prevent these cell lines from growing to high densities (Johnson and Pastan, 1972b; Blat *et al.*, 1973; Carchman *et al.*, 1974). A lack of effect was also observed for Bt_2-cAMP in the growth rate of HTC hepatoma cells (Granner *et al.*, 1968). Also, growth rate and morphology of transformed WI-38 fibroblasts are not affected by elevation of intracellular cAMP levels induced by prolonged incubation with PGE_1 (Chlapowski *et al.*, 1975).

D Conclusions

The data on the effects of cAMP in malignant or transformed cells appear to be contradictory. Transformation causes either a decrease or an increase in cAMP levels, or no apparent change at all. Bt_2-cAMP may or may not reduce growth rate, and normalize morphology or specific cell functions. These conflicting findings may, however, fit into a common scheme, if the pivotal role of cAMP-dependent protein kinase is taken into consideration for the regulation of pleiotypic and specific responses of the cells. Transformation or malignancy is related to the incorporation of transforming genes into the nucleus, leading to — most likely — multiple disturbances of protein synthesis and alterations to the cell membrane, and also to other defects in cellular protein synthesis. Therefore, transformation may affect the cAMP-generating and metabolizing system leading to changes in the cAMP levels as well as the protein kinase and protein kinase-dependent systems leading to alteration or loss of the regulating function of cAMP.

Taking into account these two possible lesions, the apparent conflicting findings in transformed or malignant cells may become more understandable.

(*i*) Lesions in the adenylate cyclase system leading to decreased cAMP levels, no lesions in the protein kinase system: Cells exhibit low cAMP levels which may or may not be altered by stimulants. Elevation of intracellular cAMP or Bt_2-cAMP will be effective in normalizing specific or pleiotypic responses, such as cAMP-dependent inhibition of growth.

(*ii*) Lesions in the adenylate cyclase system leading to increased cAMP levels, no lesions in the protein kinase system: Cells exhibit high cAMP levels. Growth of these cells will be arrested or possibly the cells will be killed as observed in lymphoma S49 cells.

(*iii*) Lesions in the adenylate cyclase system leading to decreased cAMP levels plus a defect in the protein kinase system: Cells exhibit low cAMP levels, but elevation of intracellular cAMP levels or Bt_2-cAMP will be ineffective in reducing growth rate or in restoring specific cell functions.

(*iv*) Lesions in the adenylate cyclase system leading to increased cAMP levels plus a defect in the protein kinase system: Cells exhibit high cAMP levels, but cAMP or Bt_2-cAMP will not restore cAMP-dependent inhibition of growth or specific responses.

(*v*) No lesion in the adenylate cyclase system, but a defect in the protein kinase system: Cells exhibit normal cAMP levels and normal responses of cAMP levels to stimulants, but cAMP or Bt_2-cAMP will be ineffective in restoring cAMP-dependent inhibition of growth or specific responses.

At present time it is difficult to comment on the role of cGMP in transformed or malignant cells, since sufficient information is still lacking.

CHAPTER 8

Cyclic Nucleotides, Calcium and Insulin Secretion

W. Montague

I INTRODUCTION

Many forms of diabetes mellitus appear to result from, or to be accompanied by, an insulin-secretory response which is inappropriate to the requirements of the organism for the maintenance of blood metabolites at their normal concentrations. This, together with the development of techniques for the accurate determination of low concentrations of insulin and for the isolation of functional pancreatic islets of Langerhans in large numbers, has provided the stimulus for numerous studies aimed at defining in molecular terms the mechanism of insulin release.

These studies have revealed a number of features of the insulin-secretory process and its regulation which are common to all tissues which store their secretory products in membrane-lined vesicles prior to release. Furthermore, it has become obvious that the problem of insulin secretion and its alteration in diabetes should not be considered in isolation but in the context of one of the fundamental problems of molecular biology—How do cells respond to stimuli in their environment?

The ability of all mamalian cells to respond to a wide range of external stimuli appears to depend on the existence of signaling systems within the cell. These systems serve to communicate events related to stimulus perception at the plasma membrane to the effector system which mediates the appropriate response within the cell. Ca^{2+} and/or cyclic nucleotides appear to act as the major intracellular signals, or messengers, regulating the activity of most cells. It is the purpose of this review to consider these agents as intracellular signals mediating the effects of various stimuli on the insulin-secretory response of the β-cells of the mammalian pancreatic islets of Langerhans. Mammalian islets of Langerhans represent a heterogeneous cell population composed typically of 70–80% of the insulin-producing β-cells. It is usually assumed that the biochemical characteristics of intact islets reflect those of the predominant β-cells, an assumption that has also been made in this review.

II GENERAL FEATURES OF THE SECRETORY RESPONSE OF THE β-CELL

The secretory response of the β-cell is controlled both on a minute-to-minute basis by fluctuations in the extracellular concentrations of metabolites and hormones, and on a longer-term basis by changes in the physiological status of the animal.

The physiological agents which affect insulin release can be divided into two groups on the basis of their ability to stimulate secretion in the absence of other agents (Malaisse, 1973). Primary stimuli promote insulin release, while secondary stimuli or modifiers alter the response to a primary stimulus but do not directly affect the secretory rate (Figure 8.1). This fundamental difference bet-

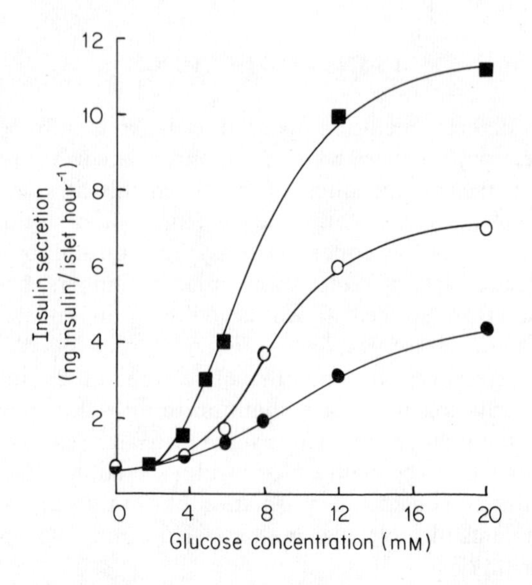

Figure 8.1 Response of the β-cell to variations in the extracellular concentration of a primary stimulus, such as glucose, in the absence (○) and presence of modifiers which increase (■) or decrease (●) the response. In the absence of glucose, modifiers do not affect the rate of release

ween the two classes of stimuli suggests a difference in their mode of action on the release process. The response to all stimuli is rapid, occurring within seconds of the addition of a stimulus, and the control systems which mediate the response appear to be mono-directional, that is they are exercised by a simple on-principle, since the removal of a stimulus results in an immediate cessation of the secretory response (Berridge,1975a). There is a sigmoid relationship between the rate of insulin release and the extracellular concentration of a primary stimulus, such as glucose (Figure 8.1), suggesting that some type of co-operative control may be functioning. In addition, the response of the β-cell to all physiological stimuli depends on the presence of a functional microtubular system within these cells, on the maintenance of intracellular ATP and AMP concentrations, and on the presence of extracellular Ca^{2+}. Thus, the secretory response of the β-cell shows several important features which must be accounted for in any theory of regulation.

III CALCIUM AND THE REGULATION OF INSULIN RELEASE

The release of insulin from β-cells in response to physiological stimuli always requires the presence of extracellular Ca^{2+}. This requirement may simply indicate that Ca^{2+} is a necessary cofactor for the functioning of the secretory mechanism. However, numerous observations suggest that changes in the concentration of Ca^{2+} within the cytoplasmic compartment of the β-cell may trigger the secretory mechanism and serve as an important signal linking events occurring at the plasma membrane, which relates stimulus recognition to the release process (Devis *et al.*, 1975; Hellman, 1975).

The Ca^{2+} concentration in the cytoplasm of the β-cell is determined by its rate of influx and efflux across the plasma membrane, and by its rate of uptake and release from other intracellular compartments, including mitochondria, the endoplasmic reticulum and storage granules (Howell *et al.*, 1975; Montague and Howell, 1976). Efflux of Ca^{2+} from the cytoplasmic compartment across the plasma membrane and into the mitochondria occurs against a concentration gradient and may be related to Ca^{2+}-stimulated ATPase activity. In the nonstimulated β-cell the cytoplasmic Ca^{2+} concentration is likely to be maintained in the region of 10^{-7}–10^{-6}M, whereas the extracellular Ca^{2+} level is normally 10^{-3}M. Thus, there is a concentration gradient favouring the entry of Ca^{2+} into the β-cell. This is augmented by an electrical gradient since the cell interior is electronegative relative to the exterior. This electrochemical gradient favours Ca^{2+} entry into the β-cell, but in the unstimulated cell the rate of influx is usually low since the membrane is relatively impermeable to the cation. In response to a primary stimulus, such as glucose, there is a rapid depolarization of the plasma membrane and a consequent increase in its permeability to Ca^{2+} (Dean and Matthews, 1970), leading to a rapid accumulation of the cation within the cytoplasm of the β-cell. Increased intracellular Ca^{2+} levels then trigger the release mechanism (Malaisse *et al.*, 1973). Similar increases in intracellular Ca^{2+}

associated with stimulation of insulin release can be induced by high extracellular K^+, which also depolarizes the plasma membrane (Henquin and Lambert, 1974), and by the ionophore A-23187, which transports divalent cations across the plasma membrane down the electrochemical gradient (Ashby and Speake, 1975; Sharp, 1975).

In addition to regulation at the plasma membrane the concentration of Ca^{2+} in the cytoplasm is also controlled by its rate of uptake by and efflux from intracellular storage pools. Studies have indicated that uptake into the mitochondrial pool is likely to play the most important role in the regulation of cytoplasmic Ca^{2+} concentrations in the β-cell. This uptake appears to be regulated by cAMP (Howell *et al.*, 1975).

Secondary stimuli, unlike the primary stimuli, do not appear to alter the net uptake of Ca^{2+} into the β-cell. However, these agents may affect the intracellular distribution of the cation within these cells by promoting the efflux of Ca^{2+} from intracellular storage pool (Brisson *et al.*, 1972). Thus, secondary stimuli may increase the response of the β-cell to a primary stimulus by increasing the availability of Ca^{2+} in the cytoplasmic compartment. In the absence of a primary stimulus, such increased availability of Ca^{2+} does not appear to have a marked effect on the release mechanism, possibly because the Ca^{2+} is rapidly transported across the plasma membrane. However, short-lived effects of secondary stimuli have been reported in the absence of primary stimuli, suggesting that transient increases in cytosolic Ca^{2+} sufficient to activate the release mechanism do occur (Brisson *et al.*, 1972). The release of Ca^{2+} from intracelluar storage pools may also account for the ability of the ionophore A-23187 to induce insulin release in the absence of extracellular cation (Karl *et al.*, 1975).

It is thus possible to account for some of the characteristics of the insulin-

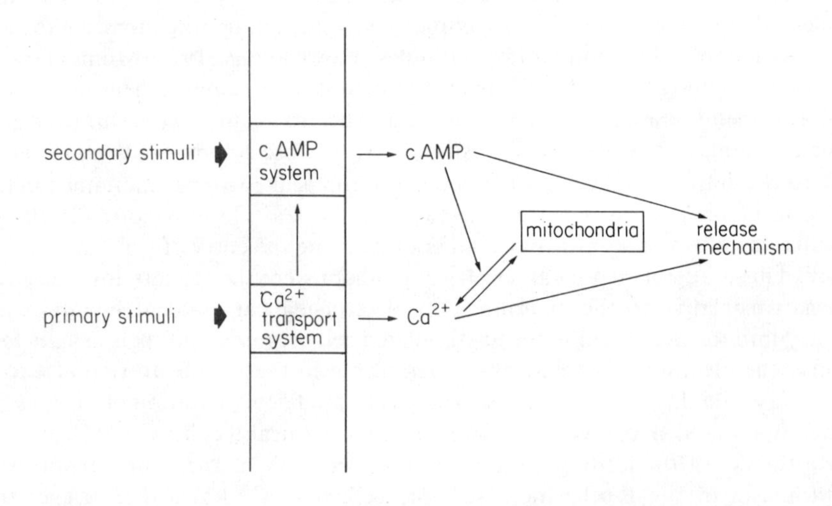

Figure 8.2 Roles of Ca^{2+} and cAMP in the regulation of insulin secretion

secretory response of the β-cell on the basis of a regulatory system which utilizes Ca^{2+} as an intracellular messenger (Figure 8.2). Primary stimuli increase the availability of the messenger by promoting its net transfer across the plasma membrane, while secondary stimuli promote its transfer from intracellular storage pools. Primary stimuli might interact with specific receptors on the plasma membrane of the β-cell to promote net Ca^{2+} uptake. However, it is unlikely that secondary stimuli directly affect the release of calcium from intracellular storage pools since many of these agents are unable to penetrate the β-cell. It is possible that some other regulatory system operates in the β-cell which links the events related to secondary stimulus recognition occurring at the plasma membrane to the release of mitochondrial Ca^{2+}. The cAMP system appears to perform such a function in a variety of tissues and there is considerable evidence for its playing a similar role in the β-cell (Figure 8.2).

IV CYCLIC NUCLEOTIDES AND INSULIN RELEASE

A Short-Term Regulation

Secretory tissues generally respond to a limited number of physiological agents. In many instances, the effect of such an agent on the release process appears to be mediated via the cAMP system. Interaction of the stimulating agent with specific receptor units on the plasma membrane activates adenylate cyclase and produces an increase in the intracellular concentration of cAMP. This, in turn, activates the release mechanism via an effect on cAMP-dependent protein kinase activity. The response in such a system is rapid, specific and quantitatively related to the magnitude of stimulus. A regulatory system of this type appears to operate in a variety of secretory tissues, including the thyroid and anterior pituitary. In view of the possibility that the cAMP system may function in the regulation of insulin release, numerous studies have been performed to investigate the short-term effects of a wide variety of agents on the components of this system in the β-cell. Since these studies have been reviewed recently (Montague and Howell, 1975) the present discussion will include only important new findings.

cGMP may act as an intracellular messenger regulating the activity of specific cell functions in a variety of tissues. There is some evidence that cGMP may play a role similar to that of cAMP in the regulation of insulin release (Howell and Montague, 1974). However, apart from this one study little attention has been given to the role of cGMP in the β-cells and this topic is, therefore, beyond the scope of this review.

1 Effects of Hormones

A variety of hormones alters the responsiveness of the β-cell to primary stimuli and there is considerable evidence to suggest that their effects are mediated via the cAMP system of the β-cell.

Glucagon, pancreozymin and secretin potentiate the response to a primary stimulus. The effects of all these polypeptide hormones on insulin release are related to their ability to increase intracellular levels of cAMP in the β-cell. This is achieved by interaction with specific receptors in the plasma membrane related to the activation of adenylate cyclase (Goldfine *et al.*, 1972b). The effect of glucagon on insulin release and on adenylate cyclase activity is augmented by 5′-guanylyl imidodiphosphate (GMP-P(NH)P) and GTP (Howell and Montague, 1973; Schauder *et al.*, 1975). GTP potentiation of hormonal activation of adenylate cyclase has been observed in a number of tissues and has led to the suggestion that GTP may be an important component of the link between hormone–receptor interaction and the activation of adenylate cyclase (Birnbaumer, 1973).

Many adrenergic agents affect the response of the β-cells to primary stimuli and the effects of these agents are explicable in terms of the presence of both α and β-adrenergic receptors in the β-cell (Ahlquist, 1948), which appear to be linked to adenylate cyclase (Montague and Howell, 1975). The catecholamines, adrenaline and noradrenaline, inhibit the release of insulin in response to all stimuli. This effect appears to be related to their ability to activate α-receptors on the β-cell, since phentolamine, an α-receptor blocker, abolishes the inhibitory effect of adrenaline, while β-receptor blockade with propranolol is without effect (Malaisse *et al.*, 1967). Activation of α-receptors in the β-cell inhibits adenylate cyclase and lowers intracellular cyclic AMP levels. As a consequence the secretory response to any agent is reduced (Howell and Montague, 1973). Such inhibitory effects of catecholamines on insulin release might be physiologically important in maintaining elevated blood glucose concentrations arising from catecholamine-induced glycogenolysis during stress. Isoprenaline has a stimulatory effect on insulin release which is prevented by β-blockade with propranolol but not by α-blockade with phentolamine. β-Receptor activation in the β-cell appears to be linked to adenylate cyclase activation and increased intracellular cAMP (Kuo *et al.*, 1973).

Prostaglandins have no effect on insulin release although they do potentiate the effect of primary stimuli. The actions of prostaglandins on insulin release are paralleled by their ability to activate adenylate cyclase and to increase cAMP levels in the β-cell (Thompson *et al.*, 1973; Howell and Montague, 1973). The effects of prostaglandins on insulin release and on adenylate cyclase activity are increased by GTP (Johnson *et al.*, 1974).

Thus, the short-term effects of a variety of hormones on the secretory response of the β-cell to primary stimuli are mediated via the cAMP system. Polypeptide hormones and prostaglandins activate adenylate cyclase, increase cAMP levels and potentiate the secretory response. In contrast, catecholamines decrease adenylate cyclase activity, lower intracellular cAMP levels and inhibit secretion.

2 Pharmacological and Pathological Agents

Many pharmacological agents alter the responsiveness of the β-cell to a primary stimulus via an effect on the cyclic nucleotide system. Methylxanthines, including theophylline, caffeine and 1-methyl-3-isobutyl-xanthine, inhibit cyclic nucleotide phosphodiesterase activity in the β-cell and increase intracellular cAMP levels, but they do not by themselves stimulate the release of insulin (Charles *et al.*, 1975). However, their relative effectiveness on insulin release in the presence of a primary stimulus parallels their effects on intracellular cAMP concentrations suggesting that the two are related.

Cholera toxin has a dramatic effect on islet cell adenylate cyclase activity and increases intracellular cAMP levels, but in the absence of a primary stimulus it is without effect on insulin release (Hellman *et al.*, 1974).

The mode of action of sulphonylureas on the release process has been the subject of intense study since these agents are widely used in the treatment of some forms of diabetes mellitus. The relative impermeability of the plasma membrane of the β-cell to sulphonylureas has led to the suggestion that their effects may be mediated at the level of the plasma membrane (Hellman *et al.*, 1971a). Sulphonylureas, in general, act as secondary stimuli and at least part of their action appears to be mediated through the cAMP system of the β-cell since they increase the intracellular cAMP concentration (Sams and Montague, 1974) either by activating adenylate cyclase (Kuo *et al.*, 1973) or by inhibiting phosphodiesterase activity (Ashcroft *et al.*, 1972).

The ionophore A-23187 and the sulphydryl inhibitor chloromercuribenzene-*p*-sulphonic acid have been shown to increase cAMP concentrations in the β-cell (Zawalich *et al.*, 1975; Hellman *et al.*, 1974a). These agents under appropriate conditions act as primary stimuli and stimulate insulin release in the absence of other agents. They can also stimulate insulin release in the absence of extracellular Ca^{2+} without increasing cAMP levels. These observations suggest that their effects on release are not mediated directly by increasing cAMP levels.

3 Metabolite Stimuli

Since glucose is the major physiological stimulus to insulin release and since in diabetes mellitus there may be alterations in the response of the β-cell to this stimulus, it is important to determine the mode of action of glucose on the release mechanism. The secretory response to glucose is relatively specific, very rapid and dose-dependent. It has been suggested that the β-cell is equipped with a specific glucose recognition unit which senses and responds to changes in the extracellular glucose concentration. Such a glucose recognition unit need not necessarily respond to the glucose molecule itself but could respond to a metabolite, the concentration of which would be directly related to the glucose concentration. In addition, the recognition unit has not to be located on the plasma membrane since there is a rapid equilibration of intracellular and extracellular glucose concentrations in the β-cell.

Characterization of the glucose recognition unit has not yet been accomplished, although one proposal suggests that it may be part of the adenylate cyclase system (Cerasi, 1975). However, glucose has no short-term effect on islet cell adenylate cyclase (Kuo *et al.*, 1974b; Howell and Montague, 1973) or on phosphodiesterase activity (Sams and Montague, 1972) when these are measured in broken-cell preparations. However, several workers have been able to demonstrate the dramatic effects of glucose in increasing cAMP levels in intact islets.

The most extensive study on cAMP concentrations in islets has been performed by Grill and Cerasi (1974) who used a [^{3}H]adenine prelabelling technique to label intracellular nucleotides and then followed the incorporation of label into cAMP during subsequent incubation periods. Using this technique, glucose was found to have a dramatic effect on the content of cAMP in islet cells. However, the effect appears to be unrelated to the secretory action of glucose, since no significant elevation of cAMP was observed until the glucose concentration was increased to above 3 mg/ml, a concentration at which insulin release is maximally stimulated. Furthermore, an increase in glucose concentration from 3 to 7 mg/ml produces a dramatic increase in cAMP but no further effect on insulin release. Thus, in these experiments, total intracellular cAMP concentrations in response to glucose do not appear to correlate with the secretory response.

There is the possibility that glucose induces an increase in a small labile pool of cAMP and that it is the concentration in this pool that is important in determining the rate of release. Changes in such a small pool might not be detectable when total intracellular cAMP is measured. In a number of tissues the amount of cAMP released from cells has been suggested to reflect changes in physiologically important intracellular pools (Exton *et al.*, 1971). There is a high correlation between the amount of insulin released from the β-cell in response to a variety of stimuli and the extracellular accumulation of cAMP (Cerasi, 1975). The correlation is so striking that it is important to consider the possibility that cAMP might be released as a consequence of insulin release rather than as a reflection of the cAMP concentration in an important regulatory pool.

Evidence in favour of this alternative view is provided by the studies of Leitner *et al.* (1975) who investigated the concentration of various adenine nucleotides released from islets of Langerhans during secretion and in subcellular fractions prepared from islets. The relative amount of adenine nucleotides secreted from islet cells during glucose-stimulated insulin release corresponds to that present in insulin-storage granules. These results suggest that during insulin secretion both cAMP and insulin are released from the storage granules and it is therefore not surprising that the two parameters parallel each other.

Glucose has been found to increase the total intracellular concentration of cAMP in the β-cell in some studies (Charles *et al.*, 1975; Capito and Hedeskov, 1974; Zawalich *et al.*, 1975), although not in others (Hellman *et al.*, 1974; Cooper *et al.*, 1973). The use of starved animals in the latter experiments may have precluded the observation of increased cAMP levels, since starvation appears to decrease islet cell adenylate cyclase activity (Howell *et al.*, 1973). However, in the absence of increased cAMP levels the islets used in these studies still release insulin in response to glucose, suggesting that glucose-induced cAMP elevation is not an essential component of the action of glucose on the release mechanism. Furthermore, in the studies in which increases of cAMP have been observed there is no correlation beween rates of insulin released and cAMP concentrations, or between the time courses in the two parameters.

It is conceivable that changes in cAMP may be secondary to some other effect of glucose. The effect of glucose on cAMP does not appear to be mediated by alterations of cyclic nucleotide phosphodiesterase activity, since the effect is markedly potentiated by the effective phosphodiesterase inhibitor 1-methyl-3-isobutyl-xanthine. Glucose may, therefore, activate adenylate cyclase, although the activation is not typical of that induced by hormones since it is not potentiated by GTP or GMP–P(NH)P (Howell and Montague, 1973; Schauder *et al.*, 1975) and cannot be obtained in broken-cell preparations. Furthermore, it requires the presence of extracellular Ca^{2+} (Charles *et al.*, 1975; Zawalich *et al.*, 1975). These observations suggest that the effect of glucose on adenylate cyclase may be secondary to changes in the transport of Ca^{2+} across the plasma membrane. Evidence in favour of this suggestion comes from the following observations. Glyceraldehyde and dihydroxyacetone, agents which stimulate insulin release and Ca^{2+} uptake, also increase the level of cAMP in the β-cell (Hellman *et al.*, 1974b). In addition, the effects of glucose, glyceraldehyde and dihydroxyacetone on both Ca^{2+} transport and cAMP concentrations are inhibited by mannoheptulose. The ionophore A-23187 and the sulphydryl inhibitor chloromercuribenzene-*p*-sulphonic acid both increase the transport of Ca^{2+} across the β-cell plasma membrane and increase the cAMP concentration (Zawalich *et al.*, 1975; Hellman *et al.*, 1974a). A relationship exists, therefore, between the transport of Ca^{2+} across the plasma membrane and the activation of adenylate cyclase. Changes in cAMP which occur in the β-cell in response to glucose may, therefore, be secondary to the primary effect of glucose in altering the transport of Ca^{2+} across the plasma membrane. Such changes may be important in regulating the intracellular distribution of Ca^{2+} and may form the basis of the co-operative action of secretion which is reflected in the sigmoid dependence of secretion on glucose concentration (Figures 8.1, 8.2).

B Long-Term regulation

In addition to the acute, minute-by-minute response of the β-cell to fluctuations in the extracellular concentration of metabolites and hormones, there exist in the β-cell regulatory mechanisms, responsible for longer-term adaptations of the secretory response. Such mechanisms operate to maintain an enhanced responsiveness of the β-cell to metabolite stimuli during pregnancy, and a decreased responsiveness during starvation. The intracellular regulatory system which permits the β-cell to maintain such altered responsiveness appears to be based on the cyclic nucleotide system.

1 Starvation

During starvation the insulin-secretory response to glucose is drastically reduced. The pattern of response resembles that seen in islets incubated with agents which decrease intracellular cAMP levels (Figure 8.1). The effect of starvation on the secretory response of the β-cell can be prevented by intermittent administration of small amounts of glucose during starvation and can be overcome by intravenous or intraperitoneal administration of glucose or by feeding a high carbohydrate diet again (Grey *et al.*, 1970). Moreover, the response to glucose can be normalized in the presence of agents which increase intracellular cAMP (Hedeskov and Capito, 1975). These observations suggest that the decreased responsiveness of the β-cells to glucose stimulation during starvation might be related to an impairment in the cyclic nucleotide regulatory system. The results of several studies provide experimental evidence in favour of these suggestions. For example, adenylate cyclase activity (Howell *et al.*, 1973) and cAMP levels are reduced in islets from starved animals (Labine *et al.*, 1975), although phosphodiesterase activity is unchanged (Capito and Hedeskov, 1974).

Several observations suggest that glucose, or a glucose metabolite, may have some long-term effects on β-cell adenylate cyclase activity, which may permit the secretory mechanism to adjust to long-term disturbances in the carbohydrate metabolism of the organism. Glucose loading of animals prior to islet isolation markedly increases adenylate cyclase activity of the isolated islets. In addition, a progressive increase in adenylate cyclase activity has been observed when islets were incubated *in vitro* with high extracellular glucose concentrations for up to twenty-four hours (Howell *et al.*, 1973). The mechanism by which glucose increases islet adenylate cyclase activity is unknown. In the *in vitro* experiments of Howell *et al.* (1973), actinomycin D and cycloheximide did not prevent the effect of glucose, suggesting that activation of existing enzyme, rather than induction of new enzyme, might be responsible for the increased activity. The ability of mannose and pyruvate, but not of galactose or 2-deoxyglucose, to mimic the effect of glucose suggests that a metabolite of glucose might be involved (Capito and Hedeskov, 1974). Purine nucleotides, such as GTP, might fulfil a role of this type since they can modulate the basal and hormone-stimulated activities of adenylate cyclase in islets (Johnson *et al.*, 1974; Howell and Montague, 1973).

2 *Pregnancy*

The insulin secretory response of the β-cell to glucose is considerably enhanced during pregnancy. The increased responsiveness is comparable to that normally obtained with glucose and a modulatory factor which increases intracellular cAMP levels, suggesting that the regulatory system involved might be the cAMP system of the β-cell (Figure 8.1). This suggestion is supported by the observation that islets of pregnant rats have a higher adenylate cyclase activity than islets from their non-pregnant littermates resulting in increased intracellular cAMP levels (Green *et al.*, 1973). Thus, part of the mechanism which enables the β-cell to maintain an enhanced response to glucose during pregnancy is an increase in adenylate cyclase activity.

The factors responsible for the increased adenylate cyclase activity appear to be related to the quantity and, in particular, the carbohydrate content of the food eaten during pregnancy since the enhanced responsiveness of the β-cell during pregnancy is diminished when the food intake is reduced. Conversely, responsiveness is enhanced when additional carbohydrate, though not protein, is eaten. These results again suggest an important role for glucose in the long-term regulation of islet cell adenylate cyclase activity, although they do not exclude the possibility that hormonal influences may also be involved.

V MODE OF ACTION OF Ca^{2+} AND CYCLIC NUCLEOTIDES ON INSULIN RELEASE MECHANISM

Insulin released during the first hour of stimulation comes exclusively from the cytoplasmic pool of insulin-storage granules. The release mechanism can, therefore, be considered to have two major components, namely the movement of the insulin-storage granules to the plasma membrane (margination) and the extrusion of the granule contents through an opening created by the fusion of the plasma membrane with the granule membrane (exocytosis).

A Margination

In the unstimulated β-cell, the insulin-storage granules appear to be randomly distributed throughout the cytoplasm, whereas during stimulation there is an increase in the number of granules close to the plasma membrane. Theoretical considerations have led Matthews (1970) to suggest that random collisions of the granules with the plasma membrane should be sufficient to account for the observed rates of insulin release.

However, numerous studies have implicated microtubules and microfilaments in the control of granule movement in the β-cells. It was originally suggested that microtubules may provide a contractile system for granule movement in the β-cell, although it now seems more likely that they provide intracellular pathways guiding the movement of granules (Lacy and Malaisse, 1973). Ca^{2+} may play a role in regulating the interaction between storage granules and microtubules, an interaction which might be necessary to ensure the vectorial transport of granules in the β-cell (Figure 8.3, sites 1 and 2).

Microtubules in a variety of cell types are polymerized structures in a state of dynamic equilibrium with a pool of subunits (tubulin). Recent evidence suggests that changes in the equilibrium between subunits and polymerized microtubules may play an important role in insulin release (Montague *et al.*, 1976). Thus, under conditions in which insulin release is stimulated by glucose and by changes in intracellular cAMP there is a rapid shift in the equilibrium in favour of polymerized microtubules. Conversely, in the absence of extracellular Ca^{2+}, release is inhibited and microtubular polymerization decreased. These effects do not appear to be a consequence of alterations in the rate of insulin release; cAMP is able to increase the conversion of tubulin subunits to microtubules in the absence of Ca^{2+} and cAMP on the release process might be the regulation of microtubule polymerization (Figure 8.3, site 3). The effect of cAMP on

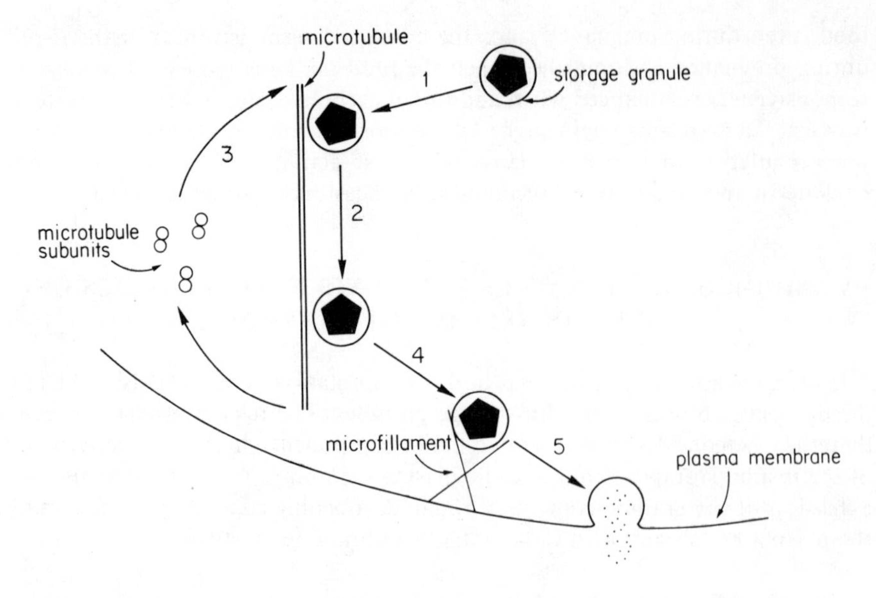

Figure 8.3 Possible sites of action of Ca^{2+} and cAMP on the insulin release process. Site 1: Interaction of storage granules with microtubules. Site 2: Microtubule assisted granule movement. Site 3: Microtubule polymerization. Site 4: Interaction of granules with microfilaments web. Site 5: Fusion of granule and plasma membranes

microtubular polymerization may be a consequence of an increased availability of cytoplasmic Ca^{2+}, although a more direct effect via the activation of cAMP-dependent protein kinase activity seems likely since, in a variety of tissues, phosphorylation of microtubular protein or protein closely associated with microtubules has been shown to occur in response to cAMP (Soifer, 1975; Sloboda *et al.*, 1975).

Microfilaments, organized as a web adjacent to the plasma membrane of the β-cell, appear to be important in insulin release since cytochalasin B, an agent which disrupts the microfilamentous system, has dramatic effects in promoting insulin release (Orci *et al.*, 1972). Such a web system could control the rate at which granules interact with the plasma membrane. Microfilaments are composed of actin-like material (Gabbiani *et al.*, 1974) and Ca^{2+} may play a role in regulating the contractile activity of this system (Figure 8.3, site 4).

B Exocytosis

There is little information available concerning the interaction between granules and plasma membranes due to the difficulties involved in isolating storage granules and plasma membrane fractions from the small amounts of islet tissue which are available. Davis and Lazarus (1975a) have analysed the factors necessary for the release of insulin from storage granules incubated *in vitro* with an impure plasma membrane fraction prepared from mouse islets.

Glucose, ATP and Ca^{2+} are all necessary for the release of insulin. Glucose can be replaced by various glucose metabolites, but release is totally dependent on Ca^{2+} and ATP. These observations suggest that Ca^{2+} plays a role in regulating the interaction between granules and plasma membrane (Figure 8.3, site 5). It has been suggested that an electrical potential barrier exists to prevent the fusion of granules and the plasma membrane, and one function of Ca^{2+} may be to neutralize such a barrier (Dean, 1974). Polar head groups of membrane phospholipids are also a likely barrier to membrane fusion. Ca^{2+}-stimulated phospholipase C activity at the site of fusion would result in the conversion of membrane phospholipids to 1,2-diacylglycerols which could act as fusagenic lipids to promote the fusion of granules and plasma membranes (Allen and Michell, 1975).

cAMP may also play a role in determining the rate of membrane fusion by inducing changes in the level of phosphorylation of specific proteins in granule and plasma membrane (Figure 8.3, site 5). The existence of substrates for cAMP-dependent protein kinase in islet-cell plasma membranes has recently been demonstrated by Davis and Lazarus (1975b), although the roles of these proteins have not been defined yet.

Thus, while no definite conclusions can be drawn concerning the sites of action of Ca^{2+} and cAMP on the insulin release mechanism, several potential sites of action have been described and these will undoubtedly be the subject of future investigation.

VI SUMMARY AND CONCLUSIONS

On the basis of the results summarized in this review several important conclusions can be made concerning the control system which regulates the secretory activity of the β-cell.

The β-cell responds, on a minute-to-minute basis, to fluctuations in the extracellular concentrations of metabolites and hormones, the magnitude of the response being related to the physiological status of the animal. The regulatory system operates to ensure that the insulin-secretory response is sufficient to maintain plasma metabolites at a concentration which is appropriate to the varying needs of the organism. Ca^{2+} and cAMP act as signal molecules in the β-cells to link events related to stimulus recognition to the activation of the release mechanism (Figure 8.2). Since two intracellular messenger systems operate in the β-cell it is of interest to determine which, if either, plays the major role in mediating the effect of glucose. Such interest is of more than academic importance since the response of the β-cell to glucose may be altered in diabetes mellitus. Two schools of thought currently exist, one of which suggests that Ca^{2+} mediates the effect of glucose (Malaisse, 1973) and the other that cAMP is the signal molecule (Cerasi, 1975). The weight of experimental evidence as reviewed in this article suggests that the primary effect of glucose on the insulin-secretory apparatus is not mediated by an increase in intracellular cAMP, although the modulatory effects of a variety of agents on the release process are mediated via the cAMP system. It seems likely, therefore, that changes in intracellular cAMP induced by glucose and other metabolites are secondary to other effects of the agents.

The intracellular concentration of Ca^{2+} in the cytoplasmic compartment of the β-cell appears to play a major role in determining the activity of the secretory mechanism. The concentration of Ca^{2+} presumably determines the rate at which it reacts with functional protein components of the secretory machinery. Such proteins might be part of the microtubular or microfilamentous system, or might be located on the granule or plasma membranes. Glucose and other primary metabolite stimuli appear to promote the transport of Ca^{2+} into the β-cell thereby triggering the release mechanism. Such an effect of glucose might be a direct one on the plasma-membrane transport system (Malaisse, 1973; Hellman *et al.*, 1971b) or secondary to an effect of glucose in promoting the interaction of Ca^{2+} with the secretory apparatus (Montague and Howell, 1976; Hellman, 1975).

Whatever the mechanism of the increased Ca^{2+} transport is, it cannot by itself trigger the insulin-release mechanism since in the absence of adequate intracellular cAMP concentrations release is not stimulated. The concentration of cyclic AMP in the β-cell is controlled in the short term by secondary stimuli and in the long term by metabolites and hormones. cAMP is important in regulating the subcellular distribution of calcium in the β-cell and in controlling the level of phosphorylation and, hence, activity of specific functional proteins of the release mechanism. The effects of cAMP on the release mechanism are also not sufficient to trigger release since in the absence of metabolite-stimulated calcium uptake, the release mechanism is not activated.

Thus, for a secretory response, the β-cell needs the combined information from two intracellular messenger systems which control the cascade of events leading from stimulus recognition to the release of insulin (Figure 8.2).

CHAPTER 9

Cyclic Nucleotides and Gastric Secretion

H. J. Ruoff and K.-F. Sewing

I INTRODUCTION

The investigation of biochemical processes at the cellular level requires a homogeneity of the organ to be studied. In cases of heterogenous tissue, experimental techniques must be used to enrich the structures of interest. The mammalian gastric mucosa contains at least four different cell types of which three are necessary for the secretion of the components of the gastric juice: the parietal cells for secretion of water, electrolytes and intrinsic factor; the chief cells for secretion of proteolytic enzymes and the mucous cells for the secretion of mucus. The function of the last cell type, the enterochromaffin-like cells, is still obscure. Although all cell types are localized in special layers of the gastric mucosa, it is very difficult to isolate single-cell populations.

This is the main difficulty encountered when investigating the role of cyclic nucleotides in the action of gastric secretagogues. The reason for looking for such a mechanism stems from two discoveries. Firstly, gastric acid secretion is in part hormonally controlled and, secondly, Orloff and Handler (1962) were able to stimulate Na^+ transport in the toad bladder by cAMP. Their work provided the first evidence that ionic transport might be associated with the cAMP system. The role of cyclic nucleotides in gastric secretion has been reviewed by Kimberg (1974).

II AMPHIBIAN GASTRIC MUCOSA

The main problem in the isolation of single-cell populations may be overcome by use of a relatively simple gastric mucosal model which is almost uniformly

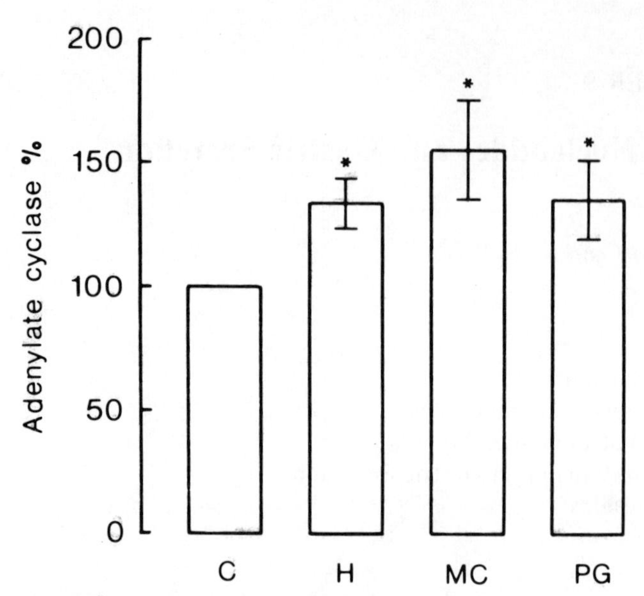

Figure 9.1 Effects of 10^{-4}M histamine (H), 10^{-4}M methylcholine (M) and 5×10^{-5}M pentagastrin (*PG*) in comparison with controls (C) on adenylate cyclase activity of the light membrane fraction from bullfrog oxyntic cells. $N=4$. Values: $x \pm s$, $=p<0.05$ (Data from Ray and Forte, 1974)

organized. A suitable system appears to be the amphibian gastric mucosa which has been studied extensively.

It was demonstrated by Nakajima *et al.* (1971) in *Necturus*, and by Ray and Forte (1974) in bullfrog, that the adenylate cyclase present in the gastric mucosa of these species is stimulated by histamine, pentagastrin and methylcholine (Figure 9.1). Charters *et al.* (1973) demonstrated that stimulation of the frog gastric mucosal adenylate cyclase by pentagastrin results in an increase in cAMP concentrations accompanied by an acid response. Previously, Harris and Alonso (1965), and Way and Durbin (1969) had described how exogenously administered cAMP stimulates H^+ secretion in the gastric mucosa of *Rana pipiens in vitro.*

More indirect evidence for the role of cyclic nucleotides in gastric acid secretion comes from investigations of gastric mucosal phosphodiesterase. From studies in man, it has been known for a considerable time that methylxanthines stimulate gastric acid secretion (Katsch and Kalk, 1925; Roth and Ivy, 1944; Krasnow and Grossman, 1949). Similar effects were observed in *in vitro* preparations of the gastric mucosa of *R. pipiens* and *R. catesbiana* (Alonso and Harris, 1965; Alonso *et al.*, 1965), suggesting that the inhibition of phosphodiesterase by methylxanthines and the subsequent rise in cAMP concentration represent the mechanism of action as shown in other systems by Sutherland and Rall (1958), and by Butcher and Sutherland (1962). This view was supported by Harris *et al.* (1969) who demonstrated that methylxanthines raise the cAMP concentration before the onset of secretion. On the other hand, imidazole, which stimulates phosphodiesterase (Butcher and Sutherland, 1962), suppresses the stimulatory effect of aminophylline and Bt_2-cAMP in the isolated gastric mucosa of *Necturus*. However, the effect of pentagastrin on acid secretion is not suppressed, suggesting that different stimulatory mechanisms are operating (Nakajima *et al.*, 1970). The gastric mucosal phosphodiesterase of *Necturus*, which is preferentially localized in the parietal cell region (Sung *et al.*, 1972), is a soluble enzyme unable to be activated by histamine, pentagastrin and methylcholine. Thus, the available data favour the view that in amphibians cAMP plays a physiological role in gastric-acid secretion.

III MAMMALIAN GASTRIC MUCOSA

The number of relevant studies in rodents is much greater. There is good evidence for the existence of a gastric mucosal adenylate cyclase in rat, guinea pig and rabbit (Bersimbaev *et al.*, 1971; Holian *et al.*, 1973; McNeill and Verma, 1974; Ohkura and Hattori, 1975; Mangla *et al.*, 1974; Ruoff and Sewing, 1975a; Perrier and Laster, 1969; Dousa and Code, 1974; Sung *et al.*, 1973). The enzyme

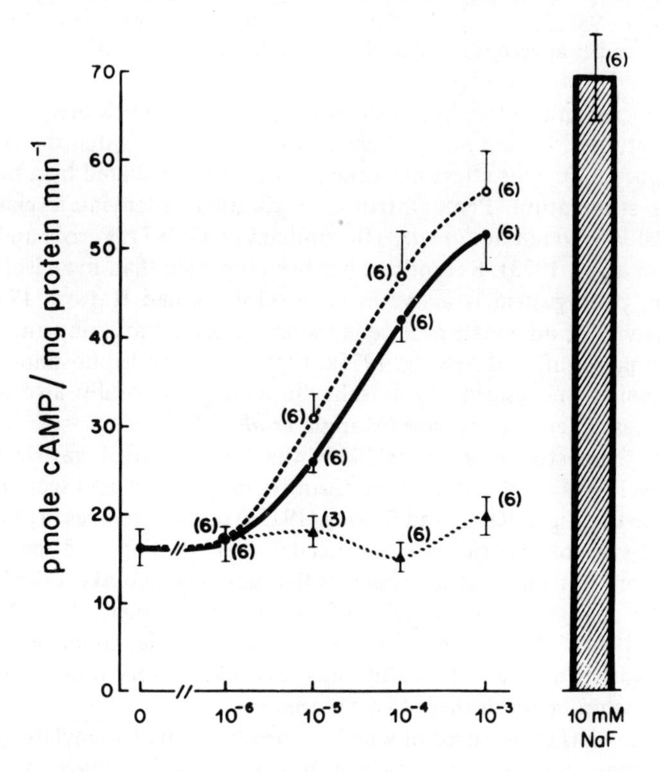

Figure 9.2 Effect of histamine (●), N-methylhistamine (O) and NaF ( ) on guinea-pig gastric mucosal adenylate cyclase activity. Values: $\bar{x} \pm s_{\bar{x}}$. Numbers in parentheses are the numbers of experiments (Data from Dousa and Code, 1974)

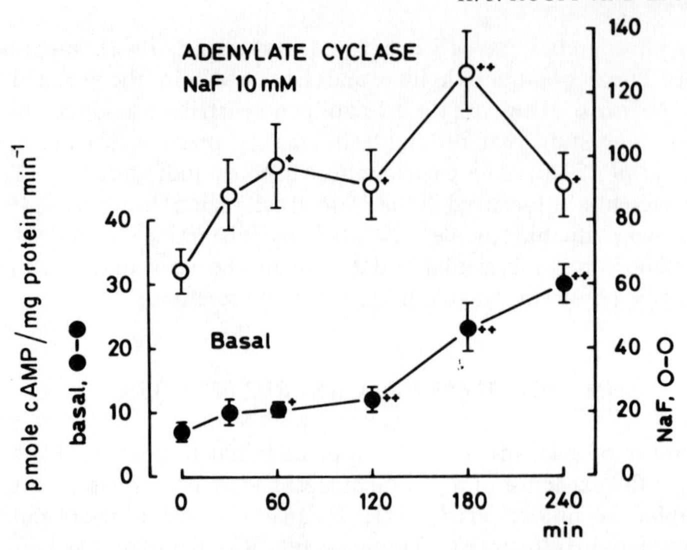

Figure 9.3 Gastric mucosal basal (●) and NaF-stimulated (O) adenylate cyclase activity of rats starved for 48 hr after refeeding. $N=7$. Values: $\bar{x} \pm s_{\bar{x}}$. $+p<0.05$, $++p<0.01$ in comparison with values at zero time (Data from Ruoff and Sewing, 1975a)

has been detected in cell-free homogenates and in isolated cell preparations. In all of these studies, histamine — when it was tested — activated the adenylate cyclase (Figure 9.2). This effect has been shown to be mediated by a histamine H_2-receptor stimulation. Pentagastrin does stimulate adenylate cyclase when administered *in vivo* but not *in vitro* (Bersimbaev *et al.*, 1971; Perrier and Laster, 1969; Holian *et al.*, 1973). Recently, it has been reported that, in an isolated cell preparation, pentagastrin is active *in vitro* (Ohkura and Hattori, 1975). The enzymic activity of adenylate cyclase is reduced during starvation and restored after refeeding (Ruoff and Sewing, 1975a; Figure 9.3), while phosphodiesterase activity remains unchanged. Acetylsalicylic acid given orally also activates gastric mucosal adenylate cyclase (Mangla *et al.*, 1974).

From all these studies it seems likely that activation of gastric mucosal adenylate cyclase is an essential event during stimulation of acid secretion. This view, and the finding of Ruoff and Sewing (1975a) that no changes in phosphodiesterase activity of gastric mucosa occur after feeding and pentagastrin administration, is in contrast to proposals that gastric mucosal cAMP levels are regulated by changes in the activity of this enzyme (Amer, 1972; Amer and McKinney, 1972). Domschke *et al.* (1972) described an identical circadian rhythm for gastric acid secretion and mucosal cAMP levels in rats indicating a close relationship between these two parameters.

The question has to be raised of whether an activation of adenylate cyclase by gastric secretagogues elevates gastric mucosal cAMP concentrations. In rodents, the most reliable gastric secretagogue, which elevates gastric mucosal cAMP levels *in vivo* and *in vitro*, is histamine acting on H_2-receptors (Domschke *et al.*, 1973; Karppanen *et al.*, 1974; Narumi and Maki, 1973; Ruoff

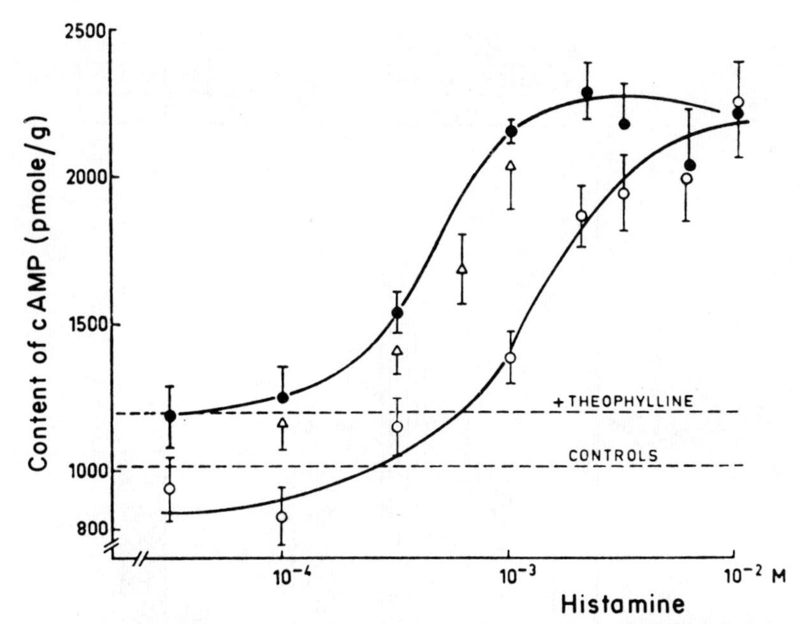

Figure 9.4 Effect of 3.3×10^{-4} burimamide (O) and 3.3×10^{-4}M diphenhydramine (Δ) on histamine-stimulated (●) rise in cAMP concentration in the presence of theophylline (10^{-3}M) in minced guinea-pig stomach. $N = 10$. Values: $\bar{x} \pm s_{\bar{x}}$ (Data from Karppanen and Westermann, 1973)

and Sewing, 1974). Tetra or pentagastrin seems to be active *in vivo* (Ruoff and Sewing, 1973a, 1974; Narumi and Maki, 1973), while reports on the pentagastrin action *in vitro* are controversial (Karppanen *et al.*, 1974). Cholinergic stimulation elevates cAMP levels only in *in vivo* experiments (Narumi and Maki, 1973; Ruoff and Sewing, 1974, 1975b). H_2-Receptor blockade by burimamide and metiamide inhibits the stimulatory effects of histamine *in vitro* (see Figure 9.4) (Karppanen and Westermann, 1973; Dousa and Code, 1974). Ruoff and Sewing (1975b) found that metiamide inhibited the histamine and pentagastrin-stimulated increase in cAMP concentration *in vivo*.

According to Thompson and Jacobson (1975) a gastric mucosal guanylate cyclase is inhibited in rats by histamine and carbachol (in the presence of Ca^{2+}). In contrast, this cyclase is activated by secretin. Pentagastrin is without any effect.

A detailed investigation has been carried out concerning the role of adrenergic mechanisms in relation to gastric mucosal cyclic nucleotide levels (Ruoff and Sewing, 1975a). In this study, adrenaline has been shown to elevate the concentration of both cAMP and cGMP in rat gastric mucosa. When during stimulation with adrenaline the α-adrenergic receptor is blocked by phenoxybenzamine the increase in cAMP levels is potentiated and the elevation of the cGMP concentration abolished (Figure 9.5). When the β-adrenergic receptor is blocked by propranolol a potentiation of the rise in cGMP concentration is observed while the increase in cAMP is inhibited (Figure 9.6). These results

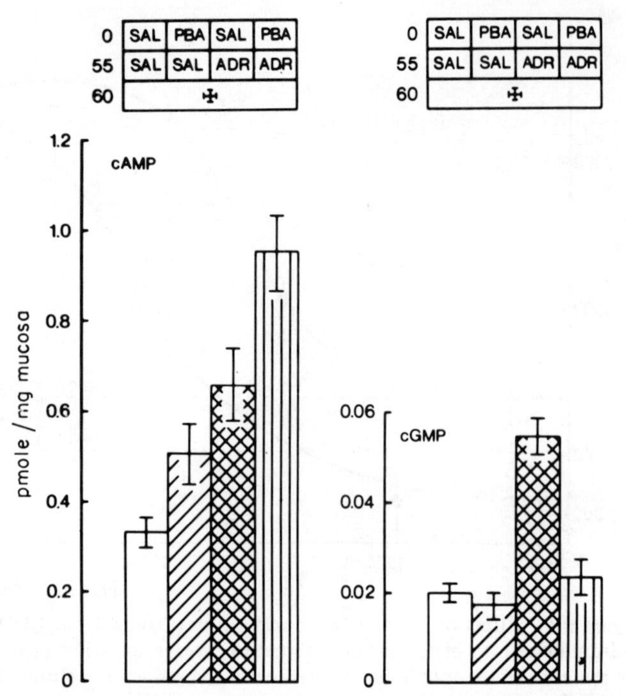

Figure 9.5 Effect of phenoxybenzamine (PBA) (4 mg/kg) on rat gastric mucosal cAMP and cGMP levels. Rats were treated (i.p.) with saline (Sal) or adrenaline (ADR) (2 mg/kg). $N = 5$–6. Values: $\bar{x} \pm s_{\bar{x}}$. The figures on top indicate the time intervals in minutes

suggest that the rise in cAMP concentration is mediated by the β-adrenergic receptor and that in cGMP concentration by the α-receptor. These findings demonstrate that the cAMP and cGMP systems of the gastric mucosa are not only susceptible to gastric secretagogues, but also to mechanisms quite obviously not involved in the secretory process.

This raises the question of whether it is possible to distinguish between a cyclic nucleotide response elicited by adrenergic mechanisms and the one associated with stimulation of gastric acid secretion. If the time course of cyclic nucleotide levels is measured after administration of pentagastrin, histamine or carbachol to rats, for up to thirty minutes none of the gastric acid stimulants increase the cGMP content above control levels. However, cAMP concentrations are elevated for different lengths of time by these agents, resulting in an increased ratio of cAMP/cGMP. This effect in comparison to the effect of adrenaline may indicate that catecholamines are involved only to a very small extent, if at all, in the response of the cyclic nucleotide system to gastric acid stimulants.

When summarizing all the *in vitro* and *in vivo* findings one has to conclude that the question of if the gastric secretory process is mediated by cyclic nucleotides in rodents cannot be finally settled until techniques have been developed which allow specific investigation of parietal cells. A step forward has

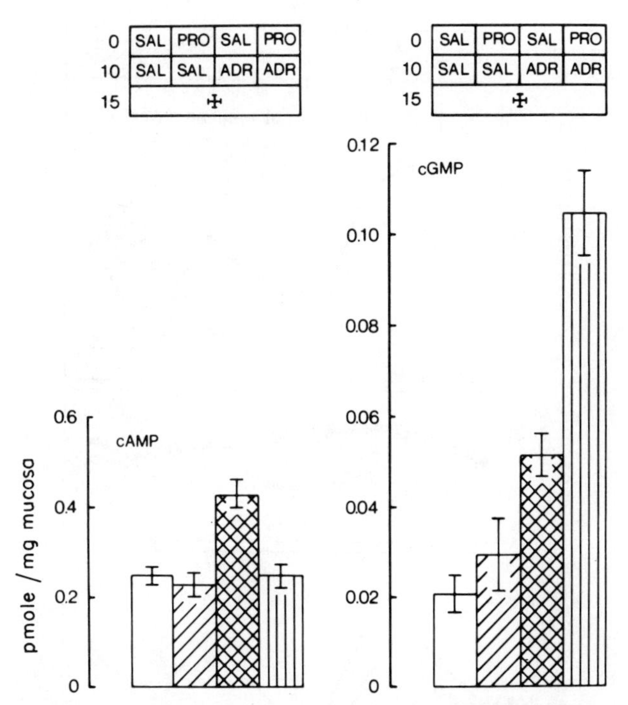

Figure 9.6 Effect of propranolol (PRO) (4 mg/kg) on rat gastric mucosal cAMP and cGMP. Rats were treated (i.p.) with saline (Sal) or adrenaline (ADR) (2 mg/kg). $N=6$. Values: $\bar{x} \pm s_{\bar{x}}$. The figures on top indicate the time intervals in minutes

been made by Glick *et al.* (1974) who studied the cAMP concentrations in different layers of glandular stomach of fasted and fed rats (Figure 9.7). In fasted rats, a peak of cAMP concentration was found in the area of the highest density of parietal cells. Feeding does not change this peak quantitatively or qualitatively. However, a second peak was found in the area of the highest chief-cell density suggesting that in this area the adenylate cyclase is activated by feeding.

It is unlikely that cGMP in rodents acts as an intracellular mediator of gastric secretion since the concentration of this nucleotide is never raised by application of various secretagogues. It is also difficult to evaluate whether changes in the ratio of cAMP/cGMP can cause intracellular processes related to gastric acid secretion.

Other studies have shown that gastric mucosal cAMP induces an activation of protein kinase and carbonic anhydrase. Narumi and Kanno (1973) studied the effect of gastric acid stimulants and inhibitors on the gastric mucosal Mg^{2+}-ATPase and carbonic anhydrase, and found that pretreatment of rats with tetragastrin, histamine or carbachol increases the activity of the enzymes in a gastric mucosal homogenate and in the mitochondrial fraction. The increase in enzyme activity of the mitochondrial fraction is prevented by pretreatment with

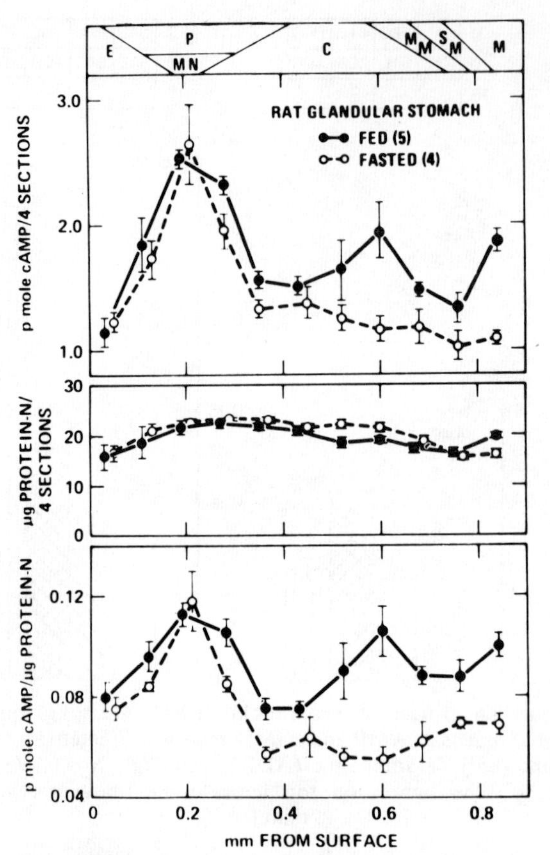

Figure 9.7 Quantitative histological distribution of cAMP in the body of the glandular stomach of fed and twenty-four hour-fasted rats. The points are means from experiments on separate stomachs of rats, the number of which is given in parentheses. Values: $\bar{x} \pm s_{\bar{x}}$. Histological zones are indicated by E = epithelial, P = parietal, C = chief, MN = mucous neck cells, MM = muscularis mucosa, SM = submucosa, M = muscle. Circular fresh frozen microtome sections (4 mm ∅, 16 μm thick, 0.2 μl volume) were used for analysis (Data from Glick *et al.*, 1974)

acetazolamide or atropine. *In vitro* addition of the gastric-acid stimulants to the enzyme preparations has no effect. An activation of the rat gastric mucosal carbonic anhydrase can also be obtained *in vivo* by Bt_2-cAMP or *in vitro* by adding cAMP to a carbonic anhydrase preparation from cytosol (Narumi and Maki, 1973). Recently, Narumi and Miyamoto (1974) found that in bovine erythrocytes the activation of carbonic anhydrase is associated with a phosphorylation of the enzyme protein by a cAMP-dependent protein kinase. That

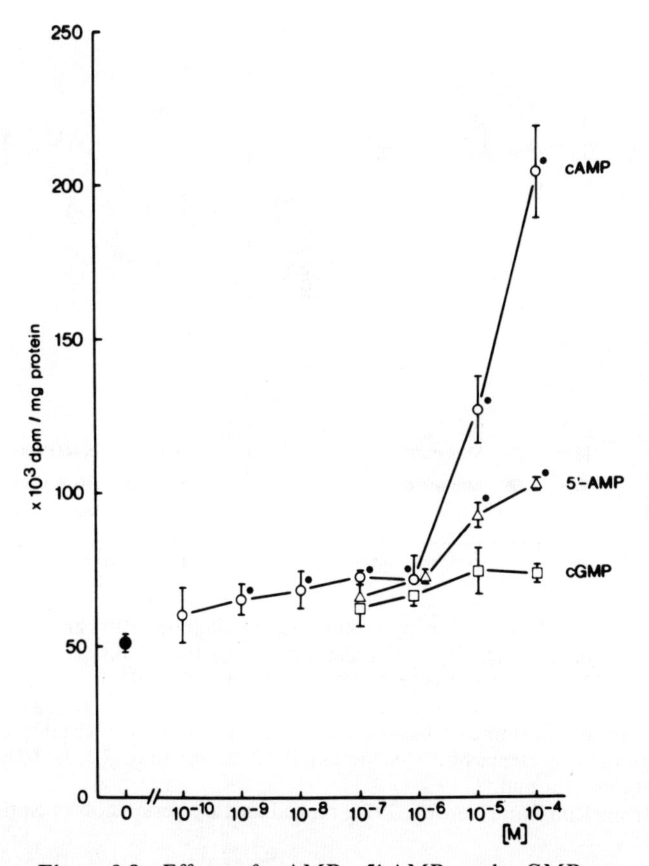

Figure 9.8 Effect of cAMP, 5′-AMP and cGMP on protein synthesis (as measured by incorporation of [C^{14}]leucine) in an isolated system of rat gastric mucosal deoxycholate polysomes. $N = 5-10$. Values: $x \pm s_{\bar{x}}$. $\cdot\ p = < 0.01$

the same sequence of events also operates in the gastric mucosa is likely since Wollin *et al.* (1976) found a protein kinase in fundic strips of the guinea-pig gastric mucosa. This kinase can be activated by histamine but not by the biologically inactive histamine derivative 1,4-dimethylhistamine.

A new aspect was raised when the effect of cyclic nucleotides on the protein biosynthesis of the gastric mucosa of rats was studied. Preliminary data had shown that starvation and refeeding of rats changes gastric mucosal adenylate cyclase activity, cAMP levels and protein synthesis as measured by [^{14}C]-leucine incorporation in an almost parallel fashion. Although in a crude gastric mucosal homogenate cAMP fails to stimulate protein synthesis (Sewing *et al.*, 1976) an attempt has been made to study the effect of cyclic nucleotides on protein synthesis of isolated rat gastric mucosal polysomes (Legler and Sewing, 1976). In deoxycholate polysomes — a preparation consisting entirely of ribonucleo-

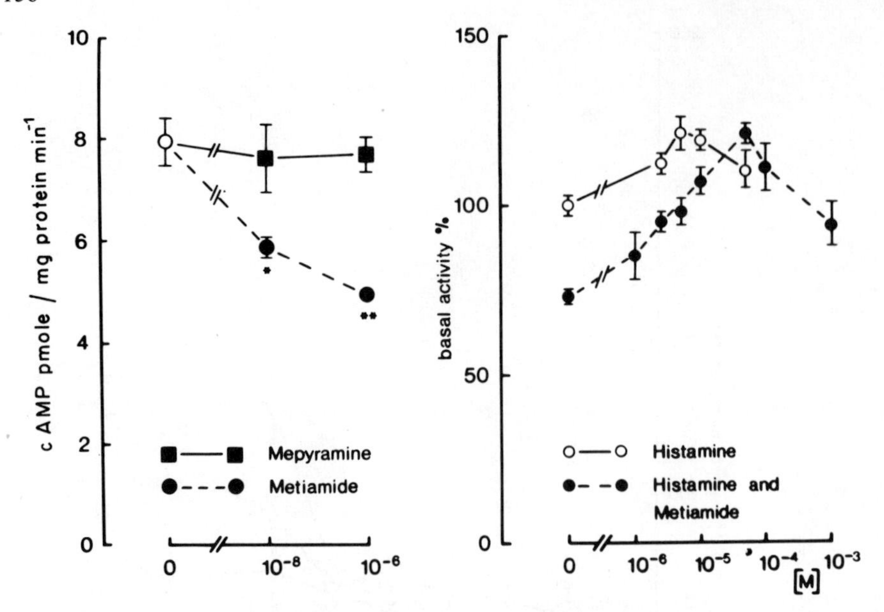

Figure 9.9 Left: Effect of mepyramine and metiamide (10^{-8} and 10^{-6} M) on basal adenylate cyclase activity of the dog gastric mucosa. $N = 5$. Values: $\bar{x} \pm s_{\bar{x}}$. $+ = p < 0.025$, $++ \; p < 0.001$

Right: Effect of histamine on basal adenylate cyclase activity of the dog gastric mucosa without or in the presence of 10^{-5} M metiamide. $N = 10$, values: $\bar{x} \pm s_{\bar{x}}$. 100% = 9.4 pmole cAMP/mg protein min.$^{-1}$

(Data from Ruoff and Sewing, 1976; reproduced by permission of Springer-Verlag)

protein particles — cAMP stimulates the incorporation of [^{14}C]-leucine by up to 300% at 10^{-4}M (Figure 9.8), the effect is concentration-dependent. cGMP is inactive and 5′-AMP only slightly active. The effect of cAMP on protein synthesis requires the intact bond between polysomes and a cAMP-binding protein complex which can be washed off from the particles with ammonium chloride solution. These data support the hypothesis that in rat gastric mucosa cAMP may be involved in regulating protein biosynthesis.

Although there is no doubt about the presence of an adenylate cyclase in the canine gastric mucosa, attempts to demonstrate an activation of the enzyme by gastric secretagogues, in particular by histamine, have failed (Mao *et al.*, 1972; Cooke *et al.*, 1974). This may be caused either by a complete insensitivity of the enzyme to histamine or by the experimental conditions employed.

The data of Ruoff and Sewing (1976) favour the latter view since the activation of *in vitro* canine gastric mucosal adenylate cyclase by histamine in a concentration-dependent manner has been achieved (Figure 9.9), when the histamine H_2-receptor antagonist metiamide is present in the incubation medium. Metiamide reduces the basal adenylate cyclase activity, indicating that endogenous histamine may contribute to what is measured as 'basal adenylate cyclase activity'.

Table 9.1 Cyclic GMP content of canine gastric mucosa

	Fundus (pmol g^{-1})	*N*	*P*
Baseline	28 ± 2		
Vagal stimulation*	66 ± 9	8	< 0.001
Baseline after atropine	39 ± 5		
Vagal stimulation after atropine	35 ± 6	4	< 0.3
Baseline	58 ± 1		
ACh infusion*	104 ± 5	4	< 0.001
Baseline	30 ± 5		
Insulin infusion*	53 ± 10	4	< 0.02
Baseline	35 ± 4		
Insulin infusion, vagi cut	31 ± 4	4	> 0.2
Baseline	30 ± 2		
Pentagastrin infusion*	29 ± 2	4	> 0.5
Baseline			
Vagal stimulation, fundus out			

*Acid secretion (reduced gastric mucosal pH).

Mao *et al.* (1972) have been able to demonstrate in the gastric mucosa of dog a phosphodiesterase which is inhibited by theophylline and papaverine *in vitro*. These data show that both enzyme systems responsible for cAMP synthesis and for degradation are present. Histamine and pentagastrin do not bring about any rise in cAMP concentration in the dog gastric mucosa *in vivo* (Mao *et al.*, 1973). In a study using Heidenhain pouch of dogs, Bieck *et al.* (1973) found that during stimulation of gastric acid secretion with intravenous infusions of pentagastrin and histamine, the output of cAMP into the gastric juice is elevated for a short time. This cAMP output is outlasted by the acid-secretory response. These findings led Mao *et al.* (1973) to regard the cAMP output merely as a washout phenomenon.

So far no positive evidence has been provided that, in dogs, cAMP acts as an intracellular mediator for gastric acid secretion. This view is supported by the finding that both Bt-cAMP and cAMP inhibit gastric acid secretion stimulated by histamine and gastrin (Levine and Wilson, 1971). cGMP seems to be involved in cholinergic stimulation (Eichhorn *et al.*, 1974; Table 9.1). Electrical and insulin stimulation as well as intraarterial infusions of acetylcholine elevate mucosal cGMP levels in the fundus, but not in the antrum. This effect can be blocked by atropine and vagal denervation. There is no response of cGMP after pentagastrin administration.

Preliminary studies indicate that adenylate cyclase is also present in the human gastric mucosa. A number of investigations favour the assumption that phosphodiesterase is also present and its inhibition can add to gastric acid secretion, since methylxanthines stimulate gastric-acid secretion (Katsch and Kalk, 1925; Roth and Ivy, 1944; Krasnow and Grossman, 1949). These findings were confirmed when Mertz ((1970) and Ottenjann *et al.* (1971) found that gastric-acid secretion stimulated by betazol or pentagastrin is enhanced by theophylline. This effect of theophylline has also been observed in patients with duodenal ulcers in whom gastric-acid secretion is maximally stimulated by pentagastrin (Bittner *et al.*, 1972). When Domschke *et al.* (1974) tried to establish a correlation between gastric mucosal cAMP and gastric acid secretion in man they failed. Their results showed that when acid secretion is maximally stimulated by subcutaneous administration of 6 μg/kg pentagastrin, gastric mucosal cAMP concentrations do not exceed basal values. Even in two patients with extremely high serum gastrin levels and basal hypersecretion the cAMP values were normal.

IV ROLE OF CALCIUM IN GASTRIC SECRETION

This report remains incomplete as long as the role of Ca^{2+} in relation to gastric secretion and cyclic nucleotides is not discussed. The relationship between Ca^{2+} and gastro-intestinal secretion has been extensively reviewed by Case (1973). It is apparent from this review that, in contrast to other exocrine organs the role of this cation in gastric-acid secretion is poorly understood. One major reason for this is that extensive species differences complicate any uniform interpretation of the experimental observations. In some animals and in man, removal of Ca^{2+} by chelators from the extracellular fluid reduces gastric-acid secretion (Hotz *et al.*, 1971a,b; Hertin and Steenbok, 1955; Ward *et al.*, 1964). No such uniformity was found in the effect of hypercalcaemia on gastric-acid secretion in different species. Therefore, it may be that a lack of Ca^{2+} at least disturbs the normal gastric secretory function. It would be interesting to know where Ca^{2+} has its place in the sequence of events between stimulation and secretion of gastric juice, and whether it is connected with cyclic nucleotides. Berridge (1975b) suggested three models for stimulus–secretion coupling which may be operative in different organs. The coupling in the adrenal medulla, synaptic nerve ending, neurosecretory cells and exocrine pancreas in the case of amylase release, is Ca^{2+}-dependent, while coupling is Ca^{2+}-dependent and cAMP-modulated in the exocrine pancreas platelets and mast cells. In the mammalian and *Calliphora* salivary glands and the exocrine pancreas (fluid secretion), stimulus–secretion coupling is both Ca^{2+} and cAMP-dependent.

For the stomach one can only speculate about what system is involved since only one report (Kasbekar, 1973) describes that, in amphibians, H^+ secretion is completely inhibited after the removal of Ca^{2+}. Subsequent addition of Bt_2-cAMP stimulates gastric acid secretion in a dose-dependent fashion, suggesting that Ca^{2+} plays a permissive role at a site before cyclic nucleotides may act.

V CONCLUSIONS

In summary it seems likely that in amphibians cAMP acts as a second messenger for transforming a stimulation into a secretory process. In all other species, including man, the role of cyclic nucleotides is as yet poorly understood although we have evidence that in rodents and dogs gastric-acid secretagogues have an influence on nucleotide cyclases and/or levels of cyclic nucleotides. The place of Ca^{2+} in this process is entirely unknown.

ACKNOWLEDGEMENT

Studies were supported by grants from the Deutsche Forschungsgemeinschaft and the Alfred Teufel Foundation.

Note added in proof

The present report has shown that histamine stimulates gastric mucosal adenylate cyclase of amphibians, rodents and dogs. Whether this activation is due to parietal or other gastric mucosal cell origin is not known as yet. The activity of adenylate cyclase was measured in human biopsies of fundic, antral and duodenal mucosa from normal subjects, duodenal ulcer and antrectomized patients (Billroth II resection). Histamine (10^{-5} M) stimulated the basal activity in the fundic mucosa of normal subjects by 200% its effect on antral (40%) and duodenal mucosa (15%) was much less pronounced. In the fundic mucosa of duodenal ulcer patients histamine activated the basal rate of cAMP formation by 256%, whereas in the same tissue of Billroth II patients it was without any effect. The data strongly support the view that histamine stimulates predominately adenylate cyclase of the parietal cell, since the histamine effect was limited to the secretory part of the stomach, enhanced in the parietal cell rich mucosa of duodenal ulcer patients and abolished in parietal cell poor mucosa of Billroth II patients. The function of the antrum and the different pattern of gastrin release in these patients may play an essential role in this process.

CHAPTER 10

Cyclic Nucleotides in the Immune Response

C. W. Parker

I INTRODUCTION

A question of considerable current interest in immunology and cell biology as a whole is the mechanism whereby lymphocytes are activated by antigens and other stimuli to undergo differentiation and division. Since an individual antigen stimulates only a small percentage of cells, regardless of whether animals have been immunized or not, detection of the biochemical changes in the tissue as a whole may be difficult, or even impossible. In attempts to elucidate activation mechanism or mechanisms most of the emphasis has been on responses to mitogenic plant lectins, such as phytohaemagglutinin (PHA), concanavalin A (con A) and pokeweed mitogen (PWM). These agents produce responses apparently similar in time course and biochemical mechanism to those produced by antigens, but a much higher percentage of the cells are activated. Since the lymphocyte surface contains receptors for these lectins and a response is produced by preparations of lectin rendered insoluble on beads, interfering with their entry into the cell (Greaves and Bauminger, 1972), it seems likely that the activation process is initiated at the cell surface. This suggests the need for a cyclic nucleotide, a critical cellular nutrient, such as Ca^{2+}, or some other second messenger to transmit the stimulus from the outside to the inside of the cell (Parker *et al.*, 1974b).

Because lymphocyte activation is a long and complex process requiring up to seventy-two hours for completion and involves changes in every aspect of cell metabolism, an attempt has been made to concentrate primarily on the early activation events. Alterations in Ca^{2+}, K^{+}, nucleoside and monosaccharide transport, phosphatidylinositol turnover and cAMP levels are observed within a few minutes of exposure of lymphocytes to lectins. Changes in histone acetylation and amino-acid transport occur a short time later (Wedner and Parker, 1976). Primary emphasis is given to a summary of recent studies of human lymphocytes including attempts to elucidate the role of cyclic nucleotides, Ca^{2+}, microfilaments and microtubules in the activation process. A working hypothesis of lymphocyte activation is also presented which it is felt can reconcile most of the existing, and at times seemingly contradictory, data on the role of cyclic nucleotides in this process. More comprehensive reviews of the modulatory role of cyclic nucleotides in the immune response have been presented previously (Parker, 1974b; Wedner and Parker, 1976; Parker *et al.*, 1974a). This subject is also discussed in the chapter by Keirns *et al.* in this volume (see Chapter 23).

II METHODS

Peripheral lymphocytes have been separated from heparinized venous blood of normal human volunteers by dextran sedimentation and isopycnic centrifugation over a Ficoll–Hypaque mixture (Eisen *et al.*, 1972). On average, 70% of the original lymphocyte cell number are recovered. Based on morphological criteria 92–98% of the nucleated cells are lymphocytes. Some of the lymphocytes (20–25%) become fluorescent when exposed to a polyvalent mixture of fluorescein-labelled rabbit anti-human γ-globulins and are therefore considered to be B-cells. Platelets normally numbered less than one per lymphocyte. In selected experiments, essentially pure lymphocytes (<1% contaminating nucleated cells) are obtained by passing Ficoll–Hypaque purified lymphocytes over nylon-wool columns (Eisen *et al.*, 1972). Essentially pure T-cells (< 2% of the cells stained with fluorescein-labelled rabbit anti-human γ-globulin) are obtained by use of longer columns. Preparations enriched in B-cells (approximately 80%) are obtained by exposing mixtures of T and B-lymphocytes to sheep erythrocytes followed by centrifugation to remove the rosette-forming cells, which are almost exclusively T-cells. Suitable controls with purified erythrocytes, polymorphonuclear leucocytes and platelets, as well as with lymphocytes purified additionally by passage through a nylon-wool column, as described above, establish conclusively that in each of the systems described below, the changes are taking place in lymphocytes.

In most experiments involving measurements of DNA synthesis, 1×10^6 lymphocytes were suspended in sterile TC199 (1 ml) supplemented with heated foetal calf serum (15%, by volume) and penicillin–streptomycin (100 U/ml) and incubated in the presence and absence of stimulatory agent for seventy-two hours at 37° C in a humidified atmosphere (Eisen *et al.*, 1973). Radioactive thymidine was present during the final five hours of the culture. The cells were then collected on glass-fibre filters, washed extensively with phosphate-buffered saline and trichloroacetic acid, and counted by liquid scintillation. PHA and con A produce fifty to 200-fold increases in thymidine uptake in this system. In selected experiments, DNA synthesis was measured in microtitre plates by a modification of the method of Luckasen *et al.* (1974b). PHA and con A produce eight to forty-fold increases in thymidine uptake under the experimental conditions employed.

Uptake of Ca^{2+} was measured by a modification of the method of Whitney and Sutherland (1972b), as previously described (Parker, 1974a).

During the first hour of the response PHA and con A produce two to ten-fold increases in Ca^{2+} uptake, with the most marked changes occurring during the first few minutes of stimulation. The data of Whitney and Sutherland (1973) indicate that Ca^{2+} is transported in lymphocytes by facilitated diffusion and that the stimulatory effect of lectin is on the transport K_m.

Aminoisobutyric acid transport was measured by a modification of the techniques of Mendelsohn *et al.* (1971), and of Van den Berg and Betel (1971). Lymphocytes (5×10^6) in a final volume of 0.5 ml Eagle's minimal essential medium were incubated, in the presence and absence of stimulatory agent, for one to seven hours at 37° C. Radioactive aminoisobutyric acid (0.2 mM final concentration) was added and the cells were incubated for an additional ten to thirty minutes. Ice-cold phosphate-buffered saline (2 ml) containing unlabelled amino acid was added. After centrifugation and washing in the same medium the cell pellets were sonicated and counted by liquid scintillation. Aminoisobutyric acid is a nonmetabolizable amino acid transported by Na^+-dependent mechanisms. PHA and con A produce two to four-fold increases in the amino acid uptake after three hours with a response first appearing after thirty to sixty minutes and continuing for at least eighteen hours. The increase in transport in response to lectin is mediated through changes in the V_{max} of the transport (Mendelsohn *et al.*, 1971; Van den Berg and Betel, 1971).

Phosphatidylinositol turnover was studied by a modification of the methods of Fisher and Mueller (1968) and of Maino *et al.* (1975b). Cells were preincubated with gentle shaking for ninety minutes at 37° C in phosphate-free Eagle's medium supplemented with vitamins, amino acids and heat-inactivated foetal calf serum. After separation by centrifugation, cells were resuspended in phosphate-free Eagle's medium containing foetal calf serum. Aliquots of 12.5×10^6 cells in a total volume of 2.5 ml were incubated in the presence or absence of stimulatory agents for various time periods at 37° C; $^{32}PO_4$ was added and the incubation was continued for an additional five to forty-five minutes. The cells were washed, treated with trichloroacetic acid, sonicated, extracted with ethanol–chloroform and sonicated again. The organic solvent mixture was extracted with HCl and the chloroform fraction was counted by liquid scintillation.

Optimal concentrations of PHA produce up to a four-fold overall increase in incorporation of ^{32}P into phospholipid with maximal changes at forty-five to sixty minutes. In accordance with earlier studies using thin layer chromatography, 70–90% of the early increase in $^{32}PO_4^{3-}$ incorporation was found to be in phosphatidylinositol. The increase in phosphatidylinositol specific activity is six to twelve-fold and is due primarily to an increase in phosphatidylinositol turnover rather than to net phosphatidylinositol synthesis.

cAMP was studied in lymphocytes by previously described methods (Smith *et al.*, 1971b; Steiner *et al.*, 1969). Lymphocytes ($1–3 \times 10^6$) were suspended in Gey's solution or Eagle's medium containing AB-serum and incubated for up to two hours with various stimulants. The samples were then rapidly centrifuged, the supernatants decanted, and the pellets were stored at –80° C until assay. cAMP was measured by radioimmunoassay (Steiner *et al.*, 1969; Parker, 1971; Steiner *et al.*, 1972). cGMP, and in some experiments cAMP as well, was determined in extracts obtained by boiling and sonication or by treatment of the entire lymphocyte suspension with 0.4 M perchloric acid, followed by heating to 80° C, neutralization with 3 M TRIS, absorption with neutral alumina, chromatography on AG 2×100 resin (BioRad) and radioimmunoassay. The validation of these assays in extracts of human lymphocytes has been described in detail elsewhere (Smith *et al.*, 1971b; Wedner *et al.*, 1975b).

In measurements of protein phosphorylation, lymphocytes were incubated in phosphate-free medium for one hour followed by a second incubation for one hour in medium supplemented with $^{32}PO_4^3$ (Wedner *et al.*, 1975a; Wedner and Parker, 1976). After subsequent exposure to stimulatory agent or buffer for one to thirty minutes, the cells were disrupted by Dounce homogenization or addition of non-ionic detergent (NP–40). The soluble non-nuclear fraction was boiled and analysed by SDS–polyacrylamide gel electrophoresis.

For measurements of lectin binding, con A and WGA were labelled with ^{125}I by the method of Hunter and Greenwood (Greenwood *et al.*, 1963) at a final chloramine T concentration of 0.33 mg/ml. Unconjugated radio-iodine was removed by gel filtration on Biogel P–10 (BioRad). The iodinated con A was further purified by chromatography on Sephadex G–25 (Sigma). Lectin binding to lymphocytes was studied by the method of Phillips *et al.* (1974). Con A or WGA was incubated with cells in Eagle's minimal essential medium supplemented with heat-inactivated human serum or Gey's solution. Cell suspensions were incubated for varying times and then layered over heated foetal calf serum formed in Beckman microfuge tubes and centrifuged. Cell-associated radioactivity was determined by γ-scintillation.

III CYCLIC NUCLEOTIDES

During extensive studies, a number of experimental observations suggesting the possible importance of cAMP in the activation of human lymphocytes by lectins has been made.

In the first place, PHA and con A both produce small but consistent increases in cAMP. Changes are seen within one minute, thus, preceding other known metabolic alterations in lectin-stimulated cells. A maximal response is obtained after five to fifteen minutes (Parker *et al.*, 1974a; Smith *et al.*, 1971b; Lyle and Parker, 1974) followed by a decline to control values after one to two hours. Other investigators have confirmed these findings (Webb *et al.*, 1974; Krishnaraj and Talwar, 1973).

Secondly, in many experiments, a biphasic curve is seen with stimulation maxima at high ($> 30\ \mu g/ml$) and low ($1-10\ \mu g/ml$) PHA or con A concentrations (Parker *et al.*, 1974a). When the lymphocytes are fractionated by nylon-fibre chromatography (to remove the B-cells) the high dose–response is largely or entirely eliminated, but a response is still seen in the low dose range. When most of the lymphocytes forming rosettes with sheep erythrocytes are eliminated (decreasing the number of T-cells) only the high response is seen. Thus, it appears that both T and B-cells respond to PHA and con A, but that at lectin concentrations optimal for mitogenesis most of the response is in T-cells. In this connection it is of interest that while T and B-lymphocytes contain approximately equal numbers of binding sites, under the usual stimulation conditions T-cells appear to account for most or all of the mitogenic response to PHA and con A. On the other hand, there is evidence that if lectin is rendered insoluble or if present in high concentration at least some response in B-cells is possible (Wedner and Parker, 1975).

The cAMP response to con A is specifically blocked or reversed by low (1–10 mM) concentrations of α-methyl-D-glucoside or α-methyl-D-mannoside (Lyle *et al.*, 1974). Increases in cAMP have been obtained with con A covalently bound to polylysine–Sepharose beads. Taken together these observations suggest that the increase in cAMP is initiated through specific *carbohydrate receptors* on the lymphocyte surface.

Also, cells exposed to PHA and con A appear to accumulate cAMP in or near the lymphocyte surface as indicated by cAMP-immunofluorescence studies using rabbit anti-cAMP antibody and fluorescein-treated goat anti-rabbit IgG (Wedner *et al.*, 1972; Bloom *et al.*, 1973). Localized patches of fluorescence have been observed in or near the external cell membranes.

PHA and con A have been shown to activate lymphocyte adenylate cyclase directly in crude lymphocyte homogenates and in plasma membrane-rich subcellular fractions (Parker *et al.*, 1974a; Smith *et al.*, 1971b). However the response is small and not seen in every experiment. It may not necessarily be concluded that this is the major mechanism by which increases in cAMP are obtained in intact cells.

cAMP and Bt_2-cAMP stimulate mitogenesis although the magnitude of change is only two to three-fold and much less than the fifty to hundred-fold response seen with PHA and con A (Smith *et al.*, 1971a; Hirschhorn *et al.*, 1970). Higher concentrations of cAMP ($10-100\ \mu M$) as well as theophylline, isoprenaline and PGE_1 in concentrations producing sustained increases in cAMP markedly inhibit the DNA synthetic response to PHA and con A. Amino-isobutyric acid uptake is also inhibited.

Transient increases in phosphorylation are seen in proteins with molecular weights in the range 40,000–110,000, one to three minutes after exposure of cells to PHA or monobutyryl-cAMP (Wedner *et al.*, 1975a; Wedner and Parker, 1975). However, the response to PHA differs in that it is sustained for a longer period and in its later phases the size of the phosphorylated proteins progressively declines. None of these changes is seen with PGE_1 or theophylline.

In contrast to cAMP no consistent alterations in cGMP concentrations have been seen in response to lectin; cGMP and 8-bromo-cGMP, over a concentration range of 10^{-12}–10^{-3}M, either inhibit or have no effect on DNA synthesis (Parker, 1974b; Wedner *et al.*, 1975b). Moreover, neither cGMP nor 8-bromo-cGMP stimulates non-nuclear protein phosphorylation.

Based on the above observations, a model for lymphocyte activation by lectin has been suggested (Parker *et al.*, 1974a; Wedner and Parker, 1976) involving: (*i*) binding of lectin to specific receptors on the lymphocyte surface; (*ii*) activation of adenylate cyclase due either to physical perturbation of the membrane by lectin or secondary to changes in transport; (*iii*) stimulation of an increase in cAMP in a functionally distinct pool of cAMP in or just beneath the plasma membrane; (*iv*) selective activation of one or a group of protein kinases under the control of the cAMP pool; (*v*) activation of membrane metabolism and transport; (*vi*) increased uptake of Ca^{2+} and other essential nutrients into the cytoplasm; (*vii*) conceivably, the release from the plasma membrane of a cAMP-dependent protein kinase — or some other modulatory enzyme — with a sufficiently narrow spectrum of reactivity to exert a selective action in the cytoplasm or nucleus (Parker, *et al.*, 1974a); and (*viii*) widespread changes in intracellular metabolism, developed and co-ordinated at least in part by the influx of Ca^{2+} and other regulatory molecules into the cytoplasm.

According to this scheme cAMP will have different, and even antagonistic, metabolic actions depending on the part of the cell in which it is formed, and only the cAMP that is generated in the appropriate area of the external cell membrane will be able to promote mitogenesis effectively. Suggestive, if not conclusive, support for this model has been obtained in cAMP-immunofluorescence studies in which cells responding to PHA, PGE_1 and isoprenaline show different staining patterns, suggesting that cAMP is accumulating in different parts of the cell (Bloom *et al.*, 1973). Some evidence has also been obtained by studies of adenylate cyclase in which different subcellular fractions respond selectively to the different stimuli (Parker *et al.*, 1974a). Additional evidence bearing on this model is presented below.

IV CALCIUM

Ca^{2+} is important in a wide variety of cellular processes including cell growth and division. Human lymphocytes show a marked diminution in their DNA-synthetic response to mitogenic lectins when the medium contains no Ca^{2+} or when enough ethylene-bis-(oxyethylene-nitrilo)-tetraacetic acid (EGTA) has been added to chelate all or nearly all of the extracellular Ca^{2+} (Alford, 1970; Whitney and Sutherland, 1972a). In addition, lymphocytes exhibit rapid increases in $^{45}Ca^{2+}$ uptake in response to PHA and con A. Lectin induced increases in cell-associated Ca^{2+} are sustained over a period of many hours and appear to involve a direct effect of lectin on the K_m of transport (Whitney and Sutherland, 1973). Nonlectin mitogens, such as Hg^{2+}, Zn^{2+}, periodate and trypsin, also produce rapid increases in Ca^{2+} uptake, whereas six related nonmitogenic lectins and divalent cations fail to do so (Parker, 1974a). The alterations in Ca^{2+} uptake in response to these agents are sufficiently rapid to justify speculation on the possible importance of Ca^{2+}, either alone or in combination with cAMP or cGMP, in very early activation events. Studies to elucidate further the importance of Ca^{2+} early in the response have recently been conducted (Greene and Parker, 1975).

Purified human peripheral blood lymphocytes are stimulated by the Ca^{2+} ionophore, A-23187, which transports Ca^{2+} into and out of cells, or by mitogenic lectins at various extracellular Ca^{2+} concentrations (<1–>1000 μM). A number of biochemical parameters known to be altered during activation by mitogenic lectins have been studied including: (*i*) aminoisobutyric-acid transport, (*ii*) phosphatidylinositol turnover, (*iii*) cyclic nucleotide accumulation and (*iv*) Ca^{2+} uptake.

The ionophore A-23187 produces stimulatory effects in all of these systems with the changes similar to those produced by PHA and con A, both with respect to magnitude and time course. In accordance with results in animal lymphocytes (Luckasen *et al.*, 1974b; Maino *et al.*, 1975a), A-23187 also produces five to twelve-fold increases in DNA synthesis measured two to three days after exposure of the cells. The responses to A-23187 are largely, or entirely, inhibited when Ca^{2+} is absent from the medium. These observations suggest that Ca^{2+} is important during the early as well as the later phases of lymphocyte activation. However, short time-course studies of early metabolic responses to lectin in Ca^{2+}-deficient media have failed to provide convincing evidence of this Ca^{2+}-dependency. In a 'Ca^{2+}-free' medium which contains 1–3 mM EGTA giving an estimated final Ca^{2+} concentration of <1 μM, PHA and con A continue to produce substantial increases in aminoisobutyric acid transport, phosphatidylinositol turnover and cAMP accumulation. While the absolute responses in these systems are somewhat diminished, concurrent changes occur in control values resulting in stimulation indices 70–90% of those obtained in the presence of Ca^{2+}. A greater than 50% inhibition of the lectin response is seen only in cells preincubated in Ca^{2+}-free, EGTA-containing medium for at least thirty minutes prior to stimulation with lectin, which may be deleterious to overall cellular economy. Thus, in the absence of an exogenous source of Ca^{2+} early activation events appear to be better sustained than the later events, making the stimulatory effects of A-23187 early in the response somewhat difficult to interpret.

There are several possible ways of reconciling the Ca^{2+}-ionophore experiments, which suggest that Ca^{2+} is critically involved in the early phases of activation, and the contradictory experiments with lectin in Ca^{2+}-deficient media.

Since lectins produce increases in intracellular cAMP, even when exogenous Ca^{2+} is absent, cAMP and Ca^{2+} may be alternative modulators of activation.

This would be analogous to the situation in skeletal muscle in which increases in phosphorylase *b* kinase activity are produced by increases either in intracellular cAMP or intracellular Ca^{2+} (Robison *et al.*, 1971). cAMP and its lipophilic derivatives are, at best, weak inducers of lymphocyte activation, but high concentrations of these substances are inhibitory. This may be because cAMP is not being delivered to the appropriate subcellular region.

Lectins may initiate the release of stored intracellular Ca^{2+} permitting activation even in the absence of the exogenous cation. Much of the Ca^{2+} in mammalian cells is sequestered in intracellular organelles. In skeletal muscle a portion is released into the cytoplasm in response to cAMP (Gergely, 1964; Borle, 1974), suggesting a possible model for cAMP action in lymphocytes.

The third possibility is that the ionophore, in the presence of extracellular Ca^{2+} is acting primarily as a membrane perturbant rather than by promoting the transport of Ca^{2+} into the cell. This is suggested by the ability of the ionophore to produce rapid, several-fold increases in intralymphocytic cAMP concentrations, by the marked cytotoxicity of the ionophore at concentrations of less than one order of magnitude above those leading to maximal stimulation, by a lack of complete Ca^{2+}-dependency in these responses. When the ionophore becomes complexed with Ca^{2+} it loses its carboxylate charge, possibly permitting new pharmacological actions of the agent independent of its ability to alter Ca^{2+} fluxes. Loss of carboxylate charge also occurs when Mg^{2+} ionophore complexes are formed. These complexes are only weakly stimulatory but they are also less stable and have a different fluorescence activation spectrum (Pfeiffer *et al.*, 1974), providing for a possible difference in pharmacological action, exclusive of alterations in divalent cation transport. Ionophore effects on cellular cAMP concentrations have not been previously described, except in one study in unpurified human leucocytes, where an increase in cAMP was also seen (Lichtenstein, 1975) and in a recent report in isolated rat pancreatic islets where several-fold increases in cAMP were observed (Karl *et al.*, 1975).

In summary, it seems likely that Ca^{2+} has a significant role in the early as well as in the late phases of lymphocyte activation but at present the evidence is by no means conclusive. Regardless of what the eventual answer is, it is surprising that the Ca^{2+} ionophore has been used to study so many different cellular processes with so little attention being given to a possible direct action at the level of the plasma membrane. The ionophore does indeed transport Ca^{2+} into cells, and there is little doubt that this is an important part of its action. However, conclusions as to whether this is necessarily its primary mode of action in all of the diverse biological responses it produces are far from certain. Possibly modulatory effects at the level of the plasma membrane deserve careful study.

V OTHER MITOGENS

Apart from PHA and con A, a number of other stimuli are capable of inducing DNA synthesis in human lymphocytes, including a wide variety of specific antigens, other lectins (pokeweed mitogen, *Phaseolus luminaris, Lens culinaris* and *Wistaria fluoribunda*), heavy metal ions (Hg^{2+} and Zn^{2+}), calcium ionophore (see Section IV), antibodies directed towards immunoglobulins or other antigenic determinants on the cell surface, bacterial products including endotoxin, oxidizing agents, such as periodate, and proteolytic enzymes (Wedner and Parker, 1976).

Matters would be simplified if all these agents acted through a common mechanism. On the basis of measurements of early changes in Ca^{2+} uptake this may to some extent be true since four different classes of mitogens (lectins, heavy metal ions, proteolytic enzymes and periodate) as well as Ca^{2+} ionophore A-23187 have been found to stimulate Ca^{2+} uptake at ten or sixty minutes, whereas seven related nonmitogenic substances produced little or no change (Parker, 1974a). In addition, as discussed above, when A-23187 and PHA or con A are compared in regard to cAMP accumulation, aminoisobutyric acid uptake and phosphatidylinositol turnover marked similarities are observed. Protein-phosphorylation studies indicate that early increases in phosphorylation are produced by the ionophore also. On the other hand, when the same experimental parameters were examined with several of the other mitogens some interesting differences emerged. Hg^{2+} and Zn^{2+} produce little or no change in protein phosphorylation, less consistent or no increase in cAMP, and an inhibition of both aminoisobutyric acid uptake and phosphatidylinositol turnover. Periodate failed to alter cAMP convincingly and neither inhibited nor stimulated aminoisobutyric acid uptake or phosphatidylinositol turnover. Thus, there appear to be at least two classes of mitogens. One class brings about increases in cAMP, uptake of the amino acid, protein phosphorylation and phosphatidylinositol turnover. Included in this class are PHA, con A, A-23187 and cytochalasin B in combination with PHA or con A. A second class, including periodate and Hg^{2+}, inhibits or fails to stimulate these responses.

While the explanation for these interesting differences is not clear at present it seems necessary to assume that different activation mechanisms are involved. Because the changes in cAMP, phosphatidylinositol turnover, protein phosphorylation and uptake of the amino acid tend to occur together it is tempting to speculate that cAMP is directly involved in the other three responses as is sometimes the case in other tissues. Hg^{2+} and Zn^{2+} might be acting at a later stage of the activation sequence, by-passing the cAMP-requiring step. Unfortunately, as discussed above, with the exception of the protein phosphorylation response, which can be partially reproduced by Bt-cAMP it has not been possible to obtain direct evidence for cAMP dependency in these responses, perhaps because of an inability to deliver cAMP to the appropriate area of the cell membrane.

VI MICROFILAMENTS

In addition to Ca^{2+} and the cyclic nucleotides, the possible role of microtubules and microfilaments in the regulation of lymphocyte function must be considered. On the basis of recent ultrastructural studies, most or all eukaryotic cells, including lymphocytes, contain a rich array of microtubules and microfilaments in close association with the plasma membrane. Under favorable circumstances, apparent insertions of these structures into the inner portions of the plasma membrane can be visualized (Bryan, 1974; Wessels *et al.*, 1971) raising the possibility that they are somehow involved in maintaining the integrity of the plasma membrane. Furthermore, since morphological studies suggest that microtubules and microfilaments can extend from the plasma membrane into the cytoplasm either or both structures could be a part of a mechanical transduction system through which perturbations exerted at the lymphocyte surface are conveyed to the interior of the cell.

One approach to studying the participation of microfilaments in various cellular processes is by the use of a family of small molecular weight fungal metabolites, the cytochalasins, which appear to alter normal microfilament structure and function in intact cells (Malawista *et al.*, 1971). This is achieved through an ability to interact directly with actin (Spudich, 1972) or myosin (Puszkin *et al.*, 1973). These agents modulate a variety of processes known to be initated at the cell surface, including random motility, chemotaxis (Becker *et al.*, 1972), secretion (Schofield, 1971), phagocytosis (Davis *et al.*, 1971), pinocytosis (Wagner *et al.*, 1971), cytoplasmic streaming (Wessels *et al.*, 1971), and adhesion and capping (Taylor *et al.*, 1971; Loor *et al.*, 1972) suggesting that microfilaments play an important role in cell-surface regulation. Cytochalasins also display inhibitory effects in systems with no obvious requirement for microfilament function, including glucose (Zigmond and Hirsch, 1972a) and nucleoside transport (Plagemann and Erbe, 1974), mucopolysaccharide synthesis (Sanger and Holtzer, 1972) and glycolysis (Zigmond and Hirsch, 1972b). Nonetheless at low concentrations they appear to provide a powerful tool to probe the role of microfilaments in a variety of membrane-associated phenomena. Cytochalasins have recently been utilized by the author in an effort to elucidate the importance of microfilaments or related cytochalasin-sensitive structures in lymphocyte activation (Greene and Parker, 1975).

A DNA Synthesis

The ability of cytochalasin-sensitive structures to modulate DNA synthesis in unstimulated and lectin-stimulated human lymphocytes has been studied. Cells were incubated in the presence and absence of lectin with varying concentrations of cytochalasin B (CB). Basal DNA synthesis is unaffected by CB but the DNA-synthetic response to E–PHA is markedly enhanced. The potentiation of lectin stimulation is seen in the high nanomolar to low micromolar CB range with higher and lower concentrations, respectively, inhibiting or failing to affect the response. CB also augments DNA synthesis in cells stimulated with con A or P–PHA. The magnitude of the increase varies with each of the three mitogens. Enhancing effects of CB on lectin-stimulated DNA synthesis have also been reported in rat lymphocytes (Yoshinaga *et al.*, 1972) and in rabbit lymphocytes (Ono and Hozumi, 1973). Augmentation of DNA synthesis is seen at lectin concentrations above, at and below the mitogenic optimum, suggesting that CB does not act by altering lectin binding. This is in accordance with the results of the direct-binding experiments presented below. Cytochalasin B effects on DNA-synthetic responses to lectin are seen as early as twenty-four hours after treatment with maximal increases at three days. At this stage the response to lectin alone is at its peak. When CB is added after the response to lectin is already underway a less marked effect is seen. Even a delay of a few hours in the addition of CB decreases the enhancement, and after twenty-four hours no enhancement is seen. Thus, the modulatory effect of CB on DNA synthesis appears to be initiated at a relatively early time in the activation sequence.

Since CB is hydrophobic and may be acting nonspecifically to enhance DNA synthesis, other cytochalasin analogues have been studied. Cytochalasin A (CA), cytochalasin D (CD) and cytochalasin E (CE) all act similarly to CB with the exception that maximal effects are obtained at ten or even hundred-fold lower concentrations than with CB. Two cytochalasin derivatives, dihydrocytochalasin B and the γ-lactone derivative of CB, with little or no biological activity in other cell types do not augment lectin-induced DNA synthesis. The relationship between structure, effects in other biological systems and enhancement of DNA synthesis strongly suggests that microfilaments or closely related structures are involved in the latter response.

B Calcium Uptake

In short incubation experiments lasting for up to two hours, cells treated with a combination of CB and P–PHA undergo up to a four-fold greater increase in $^{45}Ca^{2+}$ uptake compared with cells treated with P–PHA alone. The effect is very rapid with changes taking place within a few minutes. Three other cytochalasins, CA, CD, and CE, produce similar responses.Amplification of the P–PHA response is seen in the same cytochalasin concentration range which enhances DNA synthesis. In the absence of lectin, modest (30% or less) stimulation of $^{45}Ca^{2+}$ uptake is occasionally seen but in most experiments there is no effect at all. E–PHA and con A responses are also enhanced when CB is present. The con A response is largely, or entirely, eliminated by 50 mM α-methyl-D-mannoside indicating that CB cannot exert its effect unless con A binds to receptors on the cell surface. Another mitogenic agent whose Ca^{2+} uptake response is enhanced by CB is periodate. Cells treated with the nonmitogen wheat germ agglutinin (WGA) together with CB do not accumulate any more Ca^{2+} than cells treated with WGA alone, strengthening the impression that CB is acting by modulating a response to a positive stimulus rather than by having a direct stimulatory effect of its own (Greene *et al.*, 1975).

In contrast to the mitogenic lectins and periodate, in the case of the heavy metal ion mitogens Hg^{2+} and Zn^{2+}, the early Ca^{2+} response is not enhanced by CB. While the basis for this difference is not clear, it is of interest that the heavy metal mitogens also differ from PHA and con A in inhibiting rather than stimulating phosphatidylinositol turnover and aminoisobutyric acid uptake, and in failing to increase protein phosphorylation convincingly. Taking these observations together, it appears that Hg^{2+} and Zn^{2+} activate lymphocytes through a different mechanism compared with other mitogens studied (for further discussion, see above).

C Aminoisobutyric Acid Transport

Mitogenic lectins produce several-fold increases in the transport of aminoisobutyric acid a nonmetabolizable amino acid which is transported by Na^+-dependent mechanisms. Effects are first apparent after thirty to sixty minutes and progressively increase over several hours. CB amplifies the response to con A and PHA producing up to a three-fold increase in amino acid uptake compared with the response to lectin alone. Again the effect is analogous to that seen in the DNA-synthetic response in that: (*i*) other cytochalasins (CA, CD and CE) produce similar changes; (*ii*) dihydrocytochalasin and the γ-lactone of CB are much less effective; (*iii*) a response is obtained at cytochalasin concentrations in the range of 0.01–10 μM; (*iv*) enhancement is seen over a broad range of lectin concentrations; and (*v*) the time course of lectin stimulation is the same, regardless of whether cytochalasin is present or not. Moreover, as in the DNA-synthetic experiments, the enhancement of amino acid uptake is maximal only when CB is present during the entire period of stimulation by lectin. Interestingly, for reasons that are not clear, when cells are preincubated with CB for thirty minutes prior to lectin stimulation the enhancement of lectin responsiveness is largely eliminated. CB alone produces a small but consistent increase in transport not seen with control solutions containing DMSO.

In previous investigations of aminoisobutyric acid transport kinetics in human (Mendelsohn *et al.*, 1971) and rodent lymphocytes (Van den Berg and Betel, 1973), mitogenic lectins have been shown to act by altering the velocity (V_{max}) rather than the affinity of the transport system. In the presence of CB, the usual increase in the V_{max} seen with lectin alone is magnified whereas the K_m of transport is unaffected. Thus, CB appears to alter the magnitude but not the mechanism of the transport response to lectin.

D Phosphatidylinositol Turnover

Mitogenic lectins have been shown to stimulate selectively several-fold increases in phosphatidylinositol turnover in human lymphocytes with changes appearing within five to ten minutes and reaching a maximum after thirty to sixty minutes (Fisher and Mueller, 1968; Maino *et al.*, 1975b). When effects of cytochalasins A, B, D and E on P–PHA-stimulated phosphatidylinositol turnover are evaluated, each of these agents enhance the turnover response compared with cells treated with lectin alone. In accordance with the binding experiments presented below and with what is normally seen in other stimulation systems (amino-acid uptake and DNA synthesis), CA, CD, and, in particular, CE are more effective than CB in producing a response, whereas the γ-lactone of CB and dihydrocytochalasin B are inactive. When lipid extracts from stimulated cells are fractionated by thin layer chromatography, most or all of the increase in $^{32}PO_4^{3-}$ radioactivity is in the phosphatidylinositol area of the chromatogram; a qualitatively similar response to that obtained with lectin alone. Thus, the cytochalasins augment the usual increase in phosphatidylinositol turnover in response to lectin rather than by exerting a less selective effect on phospholipid metabolism. As in the aminoisobutyric-acid transport experiments, an alteration in the intensity but not in the mechanism of the response seems to be taking place.

E Cyclic Nucleotides

As discussed above, increases in cAMP in human lymphocytes in response to mitogenic lectins begin within one minute, reach a maximum after five to fifteen minutes and then slowly decline over several hours. In lymphocytes exposed simultaneously to a mixture of CB and con A, P–PHA or E–PHA and incubated for two minutes, a several-fold enhancement of the early cAMP response to lectin is seen; CB alone is without effect. The major effect of CB is on the time course rather than on the magnitude of the response although in about 50% of experiments the response to CB and lectin in combination after two minutes considerably exceeds the response to lectin alone, even at later times. Thus, CB appears to increase the rate and, to a more variable extent, the magnitude of the cAMP response to mitogenic lectins. After ten minutes, the increase in cAMP accumulation is no greater than in cells stimulated with lectin alone. Indeed, the normal delayed fall in cAMP occurs earlier when CB is present. CB and lectin, either together or separately, fail to alter significantly cGMP levels.

F Protein Phosphorylation

Preliminary evidence indicates that CB considerably enhances the early increase in protein phorphorylation in response to PHA.

G Lectin Binding

Possible effects of CB on [^{125}I]con A binding were studied over a broad range of CB concentrations and incubation times. Binding was also evaluated after adding partially inhibitory concentrations of α-methyl-D-mannoside as a means of screening for subtle alterations in binding affinity. No effect on con A binding was observed. While it is difficult to exclude entirely the possibility that only a small segment of the total con A-receptor population is involved in the activation process and that this particular interaction is being selectively affected by CB, the failure to detect any alterations in overall con A binding makes it unlikely that CB effects are being exerted at this level.

H Cytochalasin B-Binding Studies

To characterize the interaction of cytochalasins with lymphocytes, [^{3}H]CB was prepared by reduction of CA with [^{3}H] sodium borohydride (Lin *et al.*, 1974). Binding studies were carried out with lymphocytes which had been purified in the usual way and then filtered through a short nylon-wool column to remove the few remaining polymorphonuclear leucocytes and monocytes; both on the basis of cell number and cell protein these cells contain more binding sites for CB than lympocytes. Over a wide concentration range, CB binds very rapidly to the filtered lymphocytes both at 22 and 37° C with equilibration being essentially complete within a minute. At 4° C binding is not maximal until at least fifteen minutes have elapsed. In accordance with results reported previously with radiolabelled CB or CD in other cell types (Lin *et al.*, 1974; Tannenbaum *et al.*, 1974), a Scatchard plot of CB binding suggests the presence of two classes of binding sites. By extrapolation of the high-affinity region of the binding curve, the cells were estimated to contain 15 ± 4.5 pmole of high-affinity CB-binding sites/10^7 cells with an average association constant of 1.6×10^6 LM^{-1}. This is comparable to the association constant of 10^7 LM^{-1} reported for high-affinity CB-binding sites in erythrocytes (Lin *et al.*, 1974). However, when different cytochalasins (CA, CB, CD and CE) are compared with respect to their ability to inhibit [^{3}H]CB binding an interesting difference between erythrocytes and lymphocytes is seen. In erythrocytes, CB is a considerable better inhibitor of binding than the other cytochalasins but in filtered lymphocytes, CB is the poorest of the four compounds, the order of inhibition being CE > CA > CD > CB. The relative effectiveness of cytochalasins as inhibitors is the same in cells preincubated for up to sixty minutes with unlabelled cytochalasins before adding [^{3}H]-CB. Thus, differences in rates of cytochalasin entry into the cells cannot explain the different inhibition patterns.

Biologically ineffective cytochalasin analogues, such as dihydrocytochalasin B and cytochalasin γ-lactone are ineffective or weak inhibitors of cytochalasin binding. While it has to be kept in mind that nonspecific binding sites may appear to be specific by many of the criteria classically used to describe receptors, including saturability, specificity, reversibility and high affinity (Cuatrecasas, 1975), results using different cytochalasin derivatives provide strong presumptive evidence for the existence and biochemical significance of specific cytochalasin receptors in these cells.

When the effect of D-glucose on CB binding is compared with erythrocytes and lymphocytes yet another interesting difference emerges. In lymphocytes, concentrations of D-glucose as high as 50 mM fail to inhibit CB binding, whereas in erythrocytes glucose produces a dose-related inhibition of CB binding in accordance with earlier studies (Lin and Spudich, 1974a). Thus, high-affinity CB-binding sites in lymphocytes may be less closely coupled to components of the membrane involved in glucose transport than appears to be the case in erythrocytes.

The effect of sulphydryl reagents on CB binding in lymphocytes was investigated. Increases or decreases in CB binding are seen in cells preincubated with iodoacetamide, *p*-mercuriphenylsulphonate, *p*-mercuribenzoate, mercaptoethylamine or cysteine. Enhancement of CB binding is particularly marked with cysteine (2 mM and higher). Since thiols have been reported to promote and to accelerate lymphocyte responses to lectin (Dabrowski *et al.*,

1974; Fanger *et al.*, 1970) as well as to eliminate the need for glass-adherent cells in immune responses to foreign erythrocytes (Chen and Hirsch, 1972a; Chen and Hirsch, 1972b; Click *et al.*, 1972) their ability to alter CB binding is of considerable interest. Quite possibly the stimulatory effects of thiols are exerted through cytochalasin-sensitive structures.

The location of cytochalasin-binding structures in lymphocytes has been investigated by autoradiography. Experiments were carried out at low [^{3}H]CB concentrations using long development times of up to forty days so that most CB radioactivity would be localized at high-affinity sites. At least 80% of the autoradiography grains were seen to have a superficial location on the cell. The involvement of plasma membranes in cytochalasin binding in lymphocytes also is indicated in experiments with purified plasma membranes, although there is a high level of binding in other subcellular fractions making it difficult to conclude that plasma membranes are the sole, or even predominant, site of binding. A plasma-membrane localization for at least some of the cytochalasin-sensitive structures is further supported by the ability of *p*-hydroxymercuriphenyl sulphonate, which penetrates poorly into cells, to alter CB binding markedly.

I Further Comments

Cytochalasins inhibit the induction of capping of membrane receptors by lectins and other cross-linking agents in lymphocytes. As described above, they also enhance the ability of mitogenic lectins to stimulate Ca^{2+} and aminoisobutyric-acid uptake, phospholipid turnover, early increases in cAMP accumulation, protein phosphorylation and mitogenesis in these cells. These effects do not appear to involve changes in lectin binding. They are seen using a variety of biologically active cytochalasins at very low nanomolar or even subnanomolar concentrations making it likely that specific cytochalasin-binding structures are involved. Generally, effects in the different metabolic systems are remarkably similar in regard to optimal cytochalasin dose, order of effectiveness among the different cytochalasins, an absolute requirement for lectin and an apparent effect on the magnitude, but not the mechanism, of the response.

The site of cytochalasin action is a question of considerable interest. While cytochalasins can affect energy utilization and transport, it seems probable that many of the changes are being exerted at the level of the plasma membrane. In cells exposed simultaneously to lectin and cytochalasin, either with or without glucose present, cytochalasin effects on cAMP metabolism and Ca^{2+} transport occur within two minutes. This finding is more consistent with a direct plasma-membrane action rather than a remote effect on intracellular metabolism (Greene and Parker, 1975). The most convincing evidence that plasma membranes contain cytochalasin-binding structures comes from human erythrocytes where the ghosts contain virtually all of the cytochalasin-binding activity present in intact erythrocytes and also cytochalasins exert marked effects on glucose transport (Lin and Spudich, 1974b). In lymphocytes, autoradiography, subcellular fractionation and organic mercurial inhibition data provide strong presumptive evidence for the existence of high affinity cytochalasin-binding sites at this location, but final proof is not yet available.

A possible model for cytochalasin action in lymphocytes has previously been suggested based on the assumption that filaments containing actin and myosin in combination, or filaments of actin alone, insert directly onto the plasma membrane and help to modulate cell-surface movement in response to external stimuli (Greene and Parker, 1975). Since both actomyosin and actin are sensitive to Ca^{2+} concentrations and uptake of the cation is increased during exposure of the cells to mitogenic lectin, one of the presumed early effects of lectin binding would be to activate either actin or actomyosin contraction. In addition to a response mediated by local increases in Ca^{2+} concentration, mechanical alterations in the membrane itself may serve as the stimulus for contraction. Once filament contraction occurs it may help to restore the cell surface to its original unperturbed configuration serving as a negative feedback mechanism to decrease the response (Durham, 1974). In the presence of CB, which inhibits microfilament function, the disorganization of membrane structure normally associated with lectin binding might be qualitatively altered or accentuated thereby increasing the response. An alternative possibility, that CB is acting solely through its ability to promote Ca^{2+} transport, leading to an augmented accumulation of Ca^{2+} in the cell, is not entirely excluded. Neither explanation may be fully satisfactory since certain of the enhancing effects of cytochalasin on lectin responsiveness continue to be seen, even in the absence of extracellular Ca^{2+}.

In contrast to the stimulatory effects of the cytochalasins, all of the responses studied are inhibited at high concentrations of these agents. Since the concentrations at which inhibition occurs are in the 0.1–1 mM range nonselective effects on cell metabolism are difficult to exclude. However, in as much as lymphocytes appear to contain both high and low-affinity binding sites for cytochalasins, both effects may involve specific cytochalasin-binding proteins. The very presence of cytochalasin-binding structures in lymphocytes suggests that naturally occurring cytochalasin analogues may exist in lymphoid tissues and help in modulating lymphocyte function.

VII MICROTUBULES

As discussed above, a modulatory role for microtubules in lymphocyte activation must be considered. Despite considerable investigation, studies into the role of microtubules in lymphocyte activation have largely been confined to a description of effects on membrane-receptor movement (Edelman *et al.*, 1973) and inhibition of lectin-induced DNA synthesis (Medrano *et al.*, 1974; Wang *et al.*, 1975b). Recent studies have confirmed that agents which dissociate microtubules inhibit DNA synthesis. Such results further indicate that some, but not all, of the early metabolic responses to lectin are inhibited also.

A DNA Synthesis

The effects of colchicine and vinblastine which both depolymerize tubulin, but interact at different sites on the tubulin molecule have been studied. Human lymphocytes were incubated for seventy-two hours at optimal concentrations of PHA or con A, in the presence and absence of microtubular agents, with [^{3}H]thymidine present during the final five hours of the culture. Colchicine and vinblastine both markedly inhibit lectin-stimulated DNA synthesis with comparable levels of inhibition being seen with the two lectins. Both agents produce significant inhibition of mitogenesis at concentrations as low as 10^{-8} M; at concentrations of 10^{-6} M and above, greater than 75% decreases in [^{3}H]thymidine uptake are obtained. Similar inhibitory effects of colchicine and vinblastine on lectin-induced DNA synthesis in lymphocytes have been reported by Medrano *et al.* (1974) and Wang *et al.* (1975b). The inhibition is most marked when the inhibitors are present throughout the incubation period although up to 25% inhibition is seen when additions of colchicine or vinblastine are made at forty-eight hours.

B Calcium Uptake

In view of the suspected role of Ca^{2+} in the stimulation of lymphocyte transformation by lectin, the effect of vinblastine and colchicine on early Ca^{2+} uptake was studied. Neither agent significantly alters the $^{45}Ca^{2+}$ uptake response to con A, although apparent modest inhibition was observed in several individual experiments at high concentrations of these agents. Nor do cells preincubated with colchicine or vinblastine for thirty minutes show any changes. Thus, while delayed effects on Ca^{2+} transport are not excluded, altered uptake of Ca^{2+} is probably not the basis for the inhibition of lymphocyte transformation by these agents.

C Aminoisobutyric Acid Transport

In murine lymphocytes, early changes in transport of the amino acid appear to be correlated closely with the ultimate development of a DNA-synthetic response (Van den Berg and Betel, 1974). While a cause-and-effect relationship has not been established conclusively, amino acid transport does represent a potential target in colchicine inhibition of DNA synthesis. Colchicine and vinblastine inhibit the amino acid's response to con A, P–PHA, and E–PHA. In contrast to the other systems studied, this transport is inhibited even in the absence of lectin (10–30% inhibition compared with 40–60% inhibition in the presence of lectin). The inhibition by microtubular agents does not appear to involve an effect on cell viability. Trypan blue dye exclusion and diacetyl fluorescein deacylation studies reveal no differences in cell viability in lymphocytes exposed to lectin or colchicine or combinations of these agents for periods of up to three hours. Lumicolchicine, an inactive or weakly active congener, is much less effective in producing inhibition than colchicine, although modest inhibition has been observed in some experiments.

Convincing effects of con A on amino-acid transport are first seen at one to two hours. A similar time dependency was observed for colchicine inhibition of transport in the absence of lectin with changes first appearing at one hour and persisting for at least eighteen to twenty hours. A similar time course of inhibition was seen in the presence of lectin.

As discussed above, mitogenic lectins increase aminoisobutyric acid transport through changes in V_{max} rather than K_m. Thus, a quantitative rather than a qualitative change in the response is taking place.

The possibility that colchicine is releasing an inhibitory substance from contaminating cells present in the usual Ficoll–Hypaque preparation of lymphocytes (3–5% polymorphonuclear leucocytes, 1–2% monocytes) was also considered. Essentially pure (> 99%) lymphocytes were obtained by filtration of the cells through nylon-wool columns. Colchicine continues to produce marked inhibition of lectin-stimulated amino acid transport in the more highly purified cells indicating that it is acting directly on lymphocytes rather than through non-lymphocytic cells. Moreover, since the filtered cells contained less than 2% B-lymphocytes it is apparent that the colchicine is acting on an essentially pure population of T-cells.

D Phosphatidylinositol Turnover

Preliminary phosphatidylinositol turnover studies have failed to suggest any effect of colchicine or vinblastine on the early turnover response to lectins measured at one hour. While experiments in which cells are preincubated with colchicine or vinblastine for thirty to sixty minutes before the addition of lectin are necessary to exclude more slowly evolving effects, taken together with the results of the Ca^{2+}-uptake experiments in which preincubation experiments were carried out, it appears that the microtubular agents are exerting their action after the activation process is well underway.

E Cyclic Nucleotides

In cells incubated for thirty minutes with lectin, colchicine and vinblastine produce up to a three-fold enhancement of the cAMP response. Combinations of colchicine and vinblastine are no more effective in increasing the cAMP response. With colchicine or vinblastine alone, cAMP levels do not differ significantly from control cells suggesting that a modulation of the response to lectin, rather than a direct effect on cellular metabolism, is taking place. This has been, in large measure, confirmed by studies with adrenaline and PGE_1. A modest increase is seen in the PGE_1 response when 10^{-5} M colchicine is present, but much larger changes are seen when PGE_1 is combined with theophylline, a known inhibitor of the lymphocyte cAMP phosphodiesterase. Moreover, the response to adrenaline is completely unaffected by colchicine. Thus, the ability of colchicine to modulate cAMP accumulation is proportionately greater in cells stimulated by lectin than in cells stimulated by PGE_1 or adrenaline.

When the early time course of colchicine enhancement of the cAMP response to lectin is studied, enhancement is usually maximal at sixty minutes, but some increase in the response persists for several hours. No enhancement occurs prior to 15 minutes incubation. Vinblastine produces identical effects to colchicine in regard to the time course of cAMP accumulation. Again, irrespective of the length of the incubation, colchicine and vinblastine fail to alter cAMP levels when lectin is absent.

Lumicolchicine fails to alter the con A-induced cAMP response strongly suggesting that microtubules or closely related structures mediate the response. No consistent alterations of cGMP levels have been observed with vinblastine or colchicine, either alone or in combination with lectin.

F Lectin Binding

[^{125}I]Con A binding was measured in lymphocytes incubated in the presence and absence of colchicine and vinblastine. No inhibition of binding was observed. In the same experiments, binding was markedly reduced at all times by α-methyl-D-mannoside, a specific inhibitor of con A binding (So and Goldstein, 1967). Thus, neither of the microtubular agents demonstrably affect overall lectin binding.

G Colchicine Binding

Studies in progress indicate that colchicine-binding sites exist in intact purified T-cells. In contrast to observations with cytochalasin B, uptake of [^{3}H]colchicine is slow and after one hour at 37° C the level of cell-associated colchicine is still increasing. Even at that time the quantity of [^{3}H]colchicine bound is small. Mercuric chloride markedly inhibits, while vinblastine stimulates, colchicine uptake, in accordance with their effects on colchicine binding by isolated tubulin. These observations suggest that when colchicine uptake is measured in intact lymphocytes a substantial proportion of the binding probably involves microtubules or microtubule-like proteins. However, because of the slow rate of colchicine uptake and its depolymerizing effect on tubulin — gradually increasing the number of available colchicine-binding sites — the usefulness of this approach is limited.

H Further Comments

It is apparent that both vinblastine and colchicine can affect early metabolic responses to mitogenic lectins, such as aminoisobutyric-acid transport and cAMP accumulation, beginning as early as thirty minutes after addition, and that the concentrations involved are similar to those which inhibit DNA synthesis. While other early responses, such as Ca^{2+} uptake and phosphatidylinositol turnover, are unaffected by these agents, nonetheless it seems probable that the effect of microtubular reagents on DNA synthesis is initiated at an early stage in the activation sequence. The fact that colchicine is maximally effective in inhibiting lectin-stimulated DNA synthesis only when it is present throughout the culture is consistent with this view (Wang *et al.*, 1975a).

While the basis for the inhibition is poorly understood, the failure of vinblastine and colchicine to affect lectin binding, early Ca^{2+} transport or phosphatidylinositol turnover suggests that the membrane structure is not grossly disorganized and that the inhibition occurs at a later step in the activation process than the positive response to cytochalasins. One possible locus of action could be through the ability of vinblastine and colchicine to prolong and to potentiate the early cAMP response to lectin. As discussed above (see Section VIIA), while low concentrations of exogenous cAMP have weak stimulatory effects on DNA synthesis in human lymphocytes (Smith *et al.*, 1971b), high concentrations inhibit amino-acid transport and DNA synthesis. Moreover, prolonged exposures to high concentrations of PGE_1, isoprenaline and theophylline, all of which raise cAMP in these cells, reduce aminoisobutyric acid uptake and DNA synthesis. Thus, too marked or too sustained an increase in cAMP is associated with an inhibition of lymphocyte activation, suggesting a possible basis for the action of vinblastine and colchicine.

In addition to an effect involving cAMP, several alternative explanations for the inhibitory effects of colchicine and vinblastine appear possible.

Firstly, if the concept of stimulatory and inhibitory sites on the lymphocyte surface is accepted as true (see below), microtubular disruption might eliminate the physical restraints which normally separate these functionally distinct regions. In rabbit leucocytes, very few if any of the plasma-membrane sites involved in nucleoside and amino acid transport are introduced into the cell interior during phagocytosis as opposed to a large proportion of con A-binding sites which is suggesting that there are distinct segments of the membrane which selectively participate in phagocytic vacuole formation (Oliver *et al.*, 1974). However, when colchicine is present these relationships are altered, suggesting that a randomization of membrane organization has occurred. Another observation consistent with an important role of microtubules in normal membrane assembly is that colchicine interferes with the ability of con A to alter nonlectin receptor distribution in lymphocytes (Edelman *et al.*, 1973). Secondly, since the effect of vinblastine and colchicine takes some time to evolve nonspecific interference with energy utilization cannot be excluded, and thirdly, microtubular disruption may interfere with an as yet undefined mechanical transducing mechanism for the transmission of the stimulus to the cell interior.

The mechanism by which these agents sustain the cAMP response is equally unclear. The early increase in cAMP in response to lectin is normally followed by a progressive fall in cAMP, usually to control levels by 120–180 minutes (Smith *et al.*, 1971a). The slow decline in the cAMP response is not associated with changes in overall lectin binding or a loss of responsiveness to PGE_1 or adrenaline. Thus, the decline may be analogous to the delayed reduction in cAMP responsiveness observed during sustained hormonal stimulation in a variety of cell systems which can involve either a de-activation of adenylate cyclase or an activation of phosphodiesterase (Robison *et al.*, 1971). As discussed above, the lymphocyte cAMP response to lectin is considered to be highly compartmentalized with enzymes which metabolize and mediate the action of cAMP being positioned in or near the plasma membrane close to the site of cAMP formation (Wedner *et al.*, 1975a; Wedner *et al.*, 1972). If this is the case, the sustained cAMP response to lectin in the presence of colchicine and vinblastine could be due to a disorganization of membrane structure bringing about a change in the distribution of cAMP within the membrane and interfering with the local negative feedback mechanism which normally limits the response.

VIII NEGATIVE MODULATION OF LYMPHOCYTE FUNCTION

A Inhibitory Effects of Wheat Germ Agglutinin

While there has been much emphasis given to the stimulatory effects of lectins on lymphocyte function the possibility of negative modulatory responses has received relatively little attention. A variety of lectins have recently been studied to assess their ability to inhibit responses normally stimulated by PHA and con A, and to obtain evidence that wheat germ agglutinin (WGA), a nonmitogenic lectin, and, to a lesser extent, *Agaricus* mushroom agglutinin diminish both early and late responses to mitogenic lectins (Parker *et al.*, 1976; Parker, 1976; Greene *et al.*, 1976).

1 DNA Synthesis

The combination of PHA or con A with WGA, which is a protein of molecular weight 22,000 with binding specificity for *N*-acetylglucosamine, and con A or PHA, produces less increase in DNA synthesis than does PHA or con A alone. WGA alone is not consistently inhibitory although unequivocal inhibition has been seen in some experiments. Inhibition of mitogenic lectin stimulation was obtained at WGA concentrations of 5 or 10 μg/ml. No evidence for the stimulation of DNA synthesis by WGA was obtained even at relatively high concentrations (100 μg/ml). Karsenti *et al.* (1975) have also observed that WGA reduces DNA synthesis in rat thymocytes.

2 Calcium Uptake

In short incubation experiments lasting an hour or less, neither WGA nor mushroom agglutinin consistently alter Ca^{2+} uptake either by themselves or in the presence of PHA or con A. When cells are preincubated for ten to twenty minutes with WGA before the addition of mitogenic lectin and Ca^{2+}, again no inhibition of Ca^{2+} uptake is observed. While delayed effects of WGA on Ca^{2+} uptake are not excluded, the failure of WGA to inhibit in this system appears to be in contrast to its inhibitory effects on several early activation events described below.

3 Aminoisobutyric Acid Uptake

In contrast to PHA and con A, WGA not only fails to stimulate transport of the amino acid but produces a marked inhibition of the response. The inhibition occurs over a broad range of WGA concentrations and incubation times, indicating that the failure to stimulate is not due to an arbitrarily chosen set of incubation conditions. When lymphocytes are exposed to a combination of WGA and con A or PHA, the effect of mitogen is markedly diminished. Inhibition is obtained even when WGA is added up to two hours after the mitogen.

The decrease in amino acid uptake is almost completely blocked or reversed by 50 mM *N*-acetylglucosamine, with partial inhibition at concentrations as low as 1 mM; other monosaccharides are poorly inhibitory. Over the same concentration range of *N*-acetylglucosamine there is a graded reduction in binding of [^{125}I]WGA to cells, suggesting that WGA must bind to the cell surface before it can initiate its inhibitory effects.

Since cross-linking of surface receptors is thought to be important in the activation of lymphocytes by mitogenic lectins, a possible inability of WGA to undergo multivalent interactions with cells was considered to explain the inhibitory or nonstimulatory properties of WGA. However, the protein is normally bivalent in neutral aqueous solution, and both caps and agglutinates lymphocytes (Nagata and Burger, 1974). Moreover, after cross-linking with glutaraldehyde to form tetramers or larger units, WGA continues to be inhibitory whereas malelyated-con A (Young, 1974), which is divalent, and unmodified con A, which is primarily tetravalent, (Sharon and Lis, 1972) are both stimulatory.

The effects of WGA on the kinetics of aminoisobutyric acid transport have been studied. WGA produces small but consistent increases in the transport K_m of the amino acid, suggesting an alteration in the mechanism of transport, as may be produced by a direct interaction of WGA with components of the transport system, rather than a simpler effect on the transport rate. Because of the inhibition of amino acid uptake by WGA the possibility that it might be cytotoxic to lymphocytes was evaluated. Using conventional cytochemical criteria, no evidence of impaired cell viability has been obtained, even in cells exposed for several hours to the lectin.

4 Phosphatidylinositol Turnover

When used alone the nonmitogenic lectin WGA significantly diminishes phosphatidylinositol turnover; levels of inhibition range from 10 to 30%. Control experiments demonstrated that the WGA is not contaminated with inorganic phosphate thereby diluting the specific activity of the $^{32}PO_4^{3-}$ label. In addition, WGA has been shown not to interfere with the uptake of $^{32}PO_4^{3-}$ into cells. WGA also alters the increase in phospholipid turnover seen in the presence of PHA or con A. In cells incubated with WGA and PHA or con A for one hour at 37 °C the increase in phospholipid radioactivity obtained with mitogenic lectin alone is decreased by as much as 50%, to about a two-fold one. Modest inhibition was also seen in four out of seven experiments with mushroom agglutinin.

5 Protein Phosphorylation

Studies in progress by Wedner indicate that when intact lymphocytes are preincubated with $^{32}PO_4^{3-}$ and exposed to WGA there is a rapid decrease in the level of ^{32}P in non-nuclear proteins of high molecular weight. Interestingly, these proteins have similar or identical molecular weights to those which exhibit increased ^{32}P incorporation when PHA or con A is present. Combinations of WGA and PHA give intermediate responses. The negative effect of WGA on phosphorylation is seen between two and ten minutes, thus following a similar time course to the increase in phosphorylation seen with PHA and con A. Therefore the response to WGA is essentially a mirror image of the responses to PHA and con A.

6 Cyclic Nucleotide Measurements

WGA, con A and mushroom agglutinin all produce significant increases in intracellular cAMP with a maximal response occurring between five and twenty minutes. There is little or no change in cGMP. The cAMP responses to WGA are blocked by *N*-acetylglucosamine but not by α-methyl-D-mannoside, indicating that the responses are modulated through specific cell-surface receptors. The cAMP response is obtained at concentrations of WGA which produce inhibition of WGA uptake. Combinations of con A with WGA or mushroom agglutinin produce approximately additive increases in cAMP, whereas responses to mushroom agglutinin with WGA are no greater than mushroom agglutinin or WGA alone. While a variety of interpretations needs to be considered, one possibility is that WGA and mushroom agglutinin are acting through the same or overlapping areas on the lymphocyte surface presumably through activation of a common adenylate cyclase, whereas con A is acting through different areas and through a different cyclase.

7 Lectin Binding

The physical relationship of binding sites for WGA and con A to one another was studied by determining if unlabelled WGA inhibits the binding of [^{125}I]con A. Little or no alteration in con A binding was observed, even in cells preincubated with high concentrations of WGA either in soluble form or coupled to 30 nm latex particles. While additional studies are necessary, it appears that WGA and con A are indeed interacting with different receptors and that the average distance of separation is probably at least 30 nm.

B Inhibitory Effects of Polystyrene Latex Beads

Polystyrene latex particles have been reported to be bound to highly purified human peripheral blood lymphocytes and to produce a rapid dose-related increase in intracellular cAMP levels by as much as forty-fold (Atkinson *et al.*, 1975). Increases in cAMP occur within two minutes, reach a maximum at five to ten minutes and completely disappear by sixty minutes; a similar time course is followed by PHA and con A. The accumulation of cAMP is primarily in areas of the cell adjacent to the site of particle attachment, as indicated by immunofluorescence studies (see above). Because of the ability of these latex particles to modulate lymphocyte cAMP levels, their effects on various parameters of lymphocyte activation were studied. No consistent effect on DNA synthesis was observed over a wide range of particle/cell ratios. However, aminoacid transport and phosphatidylinositol turnover responses to PHA and con A are consistently inhibited by latex both in the presence and absence of lectin. Other preliminary studies suggest that the latex particles also inhibit protein phosphorylation. Thus, the particles partially mimic the effects of WGA on human lymphocytes in producing both transient rises in cAMP as well as inhibition of amino-acid uptake, phosphatidylinositol turnover and protein phosphorylation. Interestingly, while latex particles have no apparent effect on cAMP levels in human platelets, human neutrophiles, rabbit alveolar macrophages and rat peritoneal mast cells (Atkinson *et al.*, 1976)., in the rat thyroid, latex not only raises cAMP levels but also produces small increases in I^- trapping, phospholipid turnover, protein and RNA synthesis, and amino-acid uptake (Kowalski *et al.*, 1972a; Kowalski *et al.*, 1972b). Similar changes are produced by thyroid-stimulating hormone, which presumably is acting through cAMP, and exogenous cAMP itself suggesting that in at least one cell type latex particles can mimic the action of a more conventional extracellular stimulus.

C Further Comments

While the basis for the inhibitory effects of WGA on aminoisobutyric-acid uptake, DNA synthesis and phosphatidylinositol turnover are not altogether clear, the prevention or reversal of the WGA inhibition by *N*-acetylglucosamine suggests that WGA must interact selectively with cell-surface receptors before it can produce its effects. While a mechanism in which the WGA aggregates cell-surface receptors, is taken into the cell by endocytosis and produces its inhibitory effects somewhere in the cell interior cannot altogether be excluded, it is tempting to assume that the inhibitory effects of WGA are being modulated at the cell surface. Presumably, the less striking inhibitory effects of mushroom agglutinin and latex beads are also initiated at this level. Perhaps the major argument in favour of a cell-surface action is that the inhibitory effects of WGA on protein phosphorylation are seen within a relatively short time, making it doubtful that the response is modulated through changes in protein synthesis or other remote effects on cellular metabolism. Moreover, the ability of all three agents to increase whole lymphocyte cAMP concentrations provides additional presumptive evidence for a local action at the level of the plasma membrane. Since high concentrations of exogenous cAMP inhibit amino acid uptake and DNA synthesis, it seems very possible that local changes in cAMP are involved in the inhibition. The protein-phosphorylation studies suggest a more explicit model for the modulation of lymphocyte function; WGA and PHA (or con A) acting through cAMP, exert opposing effects on protein phosphorylation presumably through selective effects on nearby phosphoprotein phosphatase or protein kinase molecules in the membrane.

If the argument that WGA is producing its inhibitory effects at the lymphocyte surface is valid, then some explanation for the differing responses to WGA and PHA or con A is needed. While the lectins differ from one another in valence, overall charge and other molecular properties, their different carbohydrate-binding specificities provide a more likely basis for their differing actions. Since WGA does not compete with con A for binding, it seems reasonable to assume that WGA and con A are binding to different sites on the cell surface. Moreover, work on animal and human lymphocytes (Wray and Walborg, 1971; Jansons and Burger, 1973; Madyastha *et al.*, 1975; Green *et al.*, 1976) indicates that WGA receptors do not co-cap with con A receptors. It will be of interest to determine whether receptors for negative modulatory lectins, such as WGA or mushroom agglutinin, or positive modulatory lectins, such as con A or E–PHA, are randomly distributed or fall within common micro-anatomical domains (Greene *et al.*, 1976). While a random distribution would be consistent with recent lipid bilayer models for membrane structure (Singer and Nicolson, 1972), a mosaic model with partial segregation of membrane components into anatomically and functionally distinct regions is not excluded. In this connection, Abbas *et al.* (1975) have recently reported that cell-surface immunoglobulins appear to be nonrandomly distributed in murine lymphocytes despite efforts to exclude artifactual redistribution.

Although the biological significance of the presence of negative modulatory sites on lymphocyte surfaces is a matter of conjecture, several lines of speculation are possible.

(*i*) Increases in cAMP and a diminution in aminoisobutyric-acid uptake occur when various nontransformed cell lines undergo contact inhibition in tissue culture (Sheppard, 1974; Foster and Pardee, 1969). Since the response is apparently triggered by cell-to-cell contact and direct cell-surface perturbation may be important in its initiation, the participation of surface receptors analogous to those involved in inhibition of aminoisobutyric acid uptake in lymphocytes seems worth considering.

(*ii*) Under the usual stimulatory conditions tetravalent con A produces maximal stimulation of lymphocyte mitogenesis; lesser degrees of stimulation occur at higher concentrations. Divalent con A is also stimulatory but fails to display any inhibition at high doses (Wang *et al.*, 1975c). Thus, there appear to be stimulatory and inhibitory portions of the lectin dose–response curve which can be manipulated independently depending on the state of aggregation of the con A preparation being used. One possible explanation is that the stimulatory and inhibitory signals are manipulated by two different sets of receptors with high and low affinities, respectively, for con A. Co-operative binding effects would permit tetravalent con A to interact with low as well as high-affinity receptors, provided that the lectin is present at high concentrations, whereas divalent con A would be unable to do so, explaining the presence or absence of inhibition at high doses. Wang *et al.*, (1975c) have independently suggested a similar model as one of several mechanistic possibilities for the inhibition observed at high doses. While alternative explanations are not excluded, the very fact that WGA produces an inhibitory signal strengthens the argument for an inhibitory signal in the con A system as well.

(*iii*) The immune response is susceptible to inhibition at high doses of antigen. It is possible that this inhibition is modulated through low-affinity receptors attached to negative modulating sites on these cells, analogous to the argument presented for con A.

Obviously the hypothesis that con A and WGA are acting in microanatomically and functionally different areas of the plasma membrane is speculative and requires extensive substantiation. A combination of ultrastructural and biochemical techniques will have to be utilized before final conclusions can be drawn; in ultrastructural studies high levels of resolution may be necessary. At the moment, this model appears to provide the most attractive explanation for the interesting differences in lymphocyte responsiveness to concanavalin A and wheat germ agglutinin.

IX CONCLUSIONS

While the evidence is still incomplete, it is becoming increasingly clear that cAMP has an important, if still poorly understood, role in lymphocyte activation with both positive and negative modulatory effects. In addition to the work in human lymphocytes, cAMP appears to be implicated in effects of crude and purified thymic extracts on T-cell differentiation or proliferation (Bach and Bach, 1973). Moreover, work in progress in several laboratories indicates that cAMP is also involved in certain of the stimulatory or inhibitory effects on lymphocytes and other cells of products (lymphokinines) given off by stimulated lymphocytes (Kishimoto *et al.*, 1975).

cGMP is also under intensive investigation and while the evidence that it is important in the early phases of human T-cell activation (Hadden *et al.*, 1972) is controversial (Wedner *et al.*, 1975b), a somewhat better argument can be made for a modulatory role of cGMP in responses of murine B-lymphocytes to liposaccharides (Watson, 1975; Weinstein *et al.*, 1975). However, even in this system the action of cGMP may require a nondialysable factor released from macrophages rather than a direct effect of cGMP on lymphocytes (Diamantstein and Ulmer, 1975). Clearly, a great deal of additional work is needed before the relative importance of cGMP and cAMP in lymphocyte activation and their relationship to one another and to Ca^{2+} is fully clarified.

Microfilaments and microtubules also appear to be implicated in the modulation of lymphocyte responses but their precise role and the extent to which their actions are controlled by cyclic nucleotides and Ca^{2+} is largely a matter of conjecture.

In addition to its effects on lymphocytes, cAMP has been shown markedly to affect cell growth and differentiation in a wide variety of nonlymphocytic tissues. A number of transformed cell lines has been studied in some detail and an alteration in cAMP metabolism has usually been demonstrated (see the review by Chlapowski *et al.*, 1975 and Chapter 6, this volume). Frequently lowered cAMP levels have been demonstrated in transformed cells leading to speculation that a deficiency in cAMP is responsible for the failure to control cell growth. However, other transformed cell lines have been shown to have increased levels of cAMP making it difficult to construct a coherent hypothesis. In view of the evidence presented above that cAMP can have stimulatory as well as inhibitory effects on cell growth in lymphocytes, depending on the particular pool of cAMP which is stimulated, it is suggested that the same situation may pertain in nonlymphocytic cells. According to this scheme, malignant transformation could involve a loss of inhibitory cAMP, an excess of stimulatory cAMP or some combination of the two, explaining the apparent dichotomy in experimental results in the various cell lines. Thus, a derangement of cAMP metabolism would be of central importance in the altered growth behaviour of neoplastic cells, but the exact nature of the alteration depends on whether a positive or negative modulatory site is involved.

ACKNOWLEDGMENTS

This work was supported by USPHS Programme Project (AI12450) and Medical Science Training Programme (GM02016) (WCG) Grants. The author would like to thank Deborah Noakes for her secretarial assistance.

CHAPTER 11

Cyclic Nucleotides and Lysosomal Enzyme Secretion

L. J. Ignarro

I INTRODUCTION

The principal objective of this chapter is to review the experimental evidence that cyclic nucleotides, and the pharmacological agents which bring about their accumulation in tissues, play a biological role in the regulation of lysosomal enzyme secretion. Lysosomal enzyme secretion from neutrophils (polymorphonuclear leucocytes) will be considered primarily, although some attention is also given to macrophages (mononuclear phagocytes) and platelets, as the latter cells also contain granule-bound lysosomal enzymes which are secreted into the extracellular environment. In addition to discussing enzyme secretion, several other physiological or pathophysiological functions of these cells will be considered with respect to cyclic nucleotides. This is because, in many instances, several different functions of a given cell occur concomitantly in order to facilitate the overall function of the cell. Cyclic nucleotides may help to regulate and modulate such multiple but complementary cell functions.

Secretion of lysosomal enzymes from neutrophils into the surrounding extracellular environment occurs as a consequence of contact between neutrophils and certain macromolecular substances which are immunological or immune reactants. Immunological reactants, which provoke lysosomal enzyme secretion from neutrophils, can occur in soluble or particulate form. Particulate material of suitable size can be introduced by phagocytosis into neutrophils. This cellular function is almost always accompanied by lysosomal-enzyme discharge. Large, immobilized, immune reactants, which do not enter the cell by phagocytosis, also provoke enzyme secretion upon making contact with the neutrophil plasma membrane. In addition, certain soluble components of complement are capable of triggering lysosomal enzyme secretion. Immune reactants can provoke lysosomal enzyme release from other cells, such as macrophages and platelets. The interaction of neutrophils and the other cells with immunological reactants has been reviewed recently (Becker and Henson, 1973).

The pathophysiological consequence of this discharge is the destruction of the surrounding connective and supporting vascular tissue. Breakdown of tissue occurs because of the presence within the lysosomes of enzymes capable of degrading connective tissue constituents, such as the proteoglycan matrix, collagen and elastin. This subject has been reviewed by Ignarro (1975, 1976). Neutrophils migrate towards and accumulate at sites of tissue injury, come into contact with immune reactants and discharge their lysosomal contents. Both tissue macrophages and peripheral blood mononuclear phagocytes perform functions similar to those of neutrophils in response to tissue injury. During platelet aggregation platelets discharge lysosomal enzymes and other granular contents, a serious response to tissue injury resulting in haemostasis and thrombosis (Mustard and Packham, 1970).

Because of the severity of the detrimental effects resulting from lysosomal enzyme secretion research into the development of drugs designed to curtail enzyme secretion from inflammatory cells, such as neutrophils and macrophages, has been stimulated (Ignarro, 1974c, 1976). Successful development of effective inhibitors of lysosomal enzyme secretion necessitates extensive knowledge of the biological mechanisms responsible for regulating and modulating enzyme secretion as well as other complementary functions. cAMP has been recognized as an important mediator of cell function before the discovery of cGMP as a natural constituent of mammalian tissues. The possible importance of endogenous cGMP in biological functions was discovered by George *et al.* (1970), who showed that the decrease in cardiac contractility mediated by acetylcholine is associated with a concomitant and rapid accumulation of cardiac cGMP.

cGMP may find its importance in the regulation of biological function by virtue of its capacity to mediate or signal cellular processes which are antagonistic or opposite in direction to those mediated by cAMP (Goldberg *et al.*, 1974). The original experiments reported by George *et al.* on rat heart (1970) established the basis for the *dualism* hypothesis. This model suggests that cellular control mechanisms in bidirectionally regulated systems are mediated through the opposing actions of cGMP and cAMP (Goldberg *et al.*, 1973). Such a mechanism has been suggested for a variety of peripheral blood and related cells, such as neutrophils, lymphocytes, platelets, monocytes, basophilic leucocytes and tissue mast cells (Ignarro, 1976; Hadden *et al.*, 1974; Kaliner and Austen, 1974). Many studies on cyclic nucleotides and biological control mechanisms have indicated the importance of Ca^{2+} as a component of the overall system. For example, Ca^{2+} appears to be required for certain functions of neutrophils (Woodin and Wieneke, 1963; Ignarro and George, 1974b). Ca^{2+} plays a definite, although poorly understood, role in the metabolism of cGMP. Like many cholinergic effects on cellular function, cholinergically induced accumulation of tissue cGMP may be dependent on the presence of Ca^{2+} (Schultz *et al.*, 1973). Consistent with this is the reported stimulation of guanylate cyclase activity by Ca^{2+} (Hardman *et al.*, 1971).

Hormones of various types are well known to alter tissue concentrations of cAMP and cGMP. Recent studies with neutrophils, macrophages and platelets have revealed similar hormonal effects. The findings that certain hormones stimulate whereas other hormones inhibit the functions of these blood cells has uncovered a new and potentially important biological mechanism for the regulation of peripheral blood cell function. For example, a variety of hormones, including neurohormones and tissue hormones, have been found to influence markedly lysosomal enzyme secretion from, and other functions of, human neutrophils. These hormones include autonomic neurohormones (adrenergic and cholinergic), prostaglandins (E, A and F series), glucocorticosteroids and other tissue hormones (histamine).

II FUNCTIONS OF NEUTROPHILS, MACROPHAGES AND PLATELETS

The inflammatory process constitutes numerous responses of cells and tissues to injury. One of the most prominent responses is the discharge of lysosomal contents from neutrophils, macrophages and platelets. Lysosomes from these cells contain, in addition to acid hydrolases and cartilage-degrading proteases, other enzymic and nonenzymic proteins capable of eliciting local acute inflammatory reactions, such as vasodilation and enhanced permeability of the microcirculation. The discharge of lysosomal contents from as well as other functions of phagocytic blood cells are intended to defend the host against noxious stimuli and thereby preserve the viability of the tissues. When the inflammatory responses proceed through a symmetrical cascade of events wound healing ensues. However, intensive and uncontrolled inflammatory responses result in extensive lysosomal enzyme discharge and further tissue injury. Connective tissue degradation by lysosomal enzymes is a characteristic feature of chronic immune complex disorders, such as rheumatoid arthritis.

Neutrophils are the principal phagocytes in peripheral blood. These cells engulf by phagocytosis a variety of noxious particulate materials in an attempt to protect the host. Phagocytosis is usually accompanied by intracellular digestion of the ingested material and by lysosomal enzyme secretion. Neutrophils also adhere to large nonengulfable complexes and discharge their lysosomal contents. Chemotaxis and motility are complementary functions which enable neutrophils to seek and to adhere to immunological reactants. Macrophages, which perform functions similar to those of neutrophils, play an important physiological role in the development and progression of the immune response. These phagocytes ingest and digest a wide spectrum of particulate and soluble materials, including antigenic substances. Lysosomal enzyme discharge accompanies macrophage phagocytosis and can provoke degradation of the surrounding connective tissues. Certain soluble factors released from lymphocytes are capable of activating macrophages, thereby enhancing the capacity of these cells to engulf material and discharge their lysosomal contents.

Platelets, which are much smaller than either neutrophils or macrophages and which possess no nuclei, play a vital role in the response of tissue to injury. Haemostasis and thrombosis are two important consequences of platelet function. Injury to blood vessels usually results in exposure of basement membranes, collagen, elastin and other connective tissue components to the circulating blood platelets. Such exposure then leads to adherence of platelets to these components, thereby triggering a sequence of platelet reactions resulting ultimately in the formation of either a beneficial haemostatic plug or a harmful intravascular thrombus. Platelet adherence provokes the platelet reaction, in which the contents of at least two types of granules are discharged into the extracellular environment. Some of the released constituents, such as ADP, provoke platelet aggregation and further release of granule contents. Among the substances discharged are lysosomal enzymes, similar in properties to those released from phagocytic leucocytes. Vaso-active amines are also discharged from platelets during the release reaction.

Lysosomal enzyme release from neutrophils, macrophages and platelets can be regarded as a secretory process because of the requirement for Ca^{2+} and because of the lack of release of soluble cytoplasmic constituents (Ignarro, 1976). As with other secretory processes, Ca^{2+} probably functions as the link in the stimulus–secretion coupling mechanism which governs lysosomal enzyme release. Consequently, the availability of extracellular Ca^{2+} to these secretory cells may determine the magnitude of lysosomal enzyme secretion. It follows that pharmacological agents known to affect the availability of Ca^{2+} to the cell interior should influence the magnitude of lysosomal enzyme secretion.

III BIOREGULATION OF NEUTROPHILIC FUNCTION

Extensive progress has been made recently in the elucidation of biological factors affecting the functions of neutrophils. A significant portion of this progress can be attributed to the discovery of the mechanisms by which cAMP mediates and regulates cellular function.

The possibility that cAMP plays a role in influencing neutrophilic function first became apparent about six years ago when adrenaline and PGE_1 were shown to increase cAMP formation by human leucocyte homogenates (Scott, 1970). Similarly, human leucocyte adenylate cyclase activity is stimulated by noradrenaline and isoprenaline, but not by phenylephrine or glucagon (Bourne and Melmon, 1971). Associated with a stimulation of cAMP accumulation by prostaglandins E_1 and E_2 is the inhibition of intracellular killing of *Candida albicans* by human phagocytic leucoctyes (Bourne *et al.*, 1971). Thus, the basis of the involvement of cAMP in human neutrophilic function was established by the identification of hormone-responsive synthetic processes for cAMP in these cells.

Lysosomal enzyme secretion from human neutrophils was the first function of these cells reported to be affected by cyclic nucleotides. cAMP (May *et al.*, 1970; Weissmann *et al.*, 1971a; Ignarro, 1974a) and agents found to stimulate the accumulation of this nucleotide in neutrophils (Ignarro and George, 1974a; Ignarro *et al.*, 1974b) inhibit lysosomal enzyme secretion. On the other hand, cGMP and agents which stimulate the accumulation of cGMP in neutrophils accelerate lysosomal enzyme secretion (Ignarro, 1974a; Ignarro and George, 1974a,b; Zurier *et al.*, 1974).

The contrasting effects of the two nucleotides on lysosomal enzyme secretion from human neutrophils illustrate that the dualism hypothesis of biological control mechanisms may apply to neutrophils. Indeed, this concept may hold true for other neutrophilic functions. For example, cGMP and cAMP have been reported to stimulate and to inhibit, respectively, such functions as phagocytosis, motility and antibody-dependent cellular cytotoxicity (to be discussed in a subsequent section).

A Lysosomal Enzyme Secretion

Experimental evidence indicates that the accumulation of cGMP and cAMP in human neutrophils is associated with stimulation and inhibition, respectively of lysosomal enzyme secretion. A variety of hormones, drugs and other pharmacological agents has been shown to influence the cyclic nucleotide levels in, and lysosomal enzyme secretion from, human neutrophils. Some of the enzyme systems associated with the metabolism and possible actions of cGMP and cAMP in neutrophils have been identified and partially characterized. As with secretory processes in other tissues, stimulation and enhancement of lysosomal enzyme secretion were found to require, in most instances, the presence of Ca^{2+}.

Working with a mixed leucocyte fraction from human peripheral blood, Weissmann and coworkers (May *et al.*, 1970; Weissmann *et al.*, 1971a) discovered that cAMP, dibutyryl-cyclic AMP (Bt_2-cAMP), theophylline and PGE_1 inhibit the immunological discharge of lysosomal enzymes. The stimulus for lysosomal enzyme discharge is serum-treated particulate material of a size suitable to be taken up by the leucocytes by phagocytosis. In fact, phagocytosis does occur and may have provided at least one mechanism by which lysosomal contents could be extruded. This mechanism has been described by Weissmann *et al.* (1971a) as regurgitation during feeding. In view of the possibility that inhibitors of phagocytosis could also inhibit the release of lysosomal contents, selective lysosomal enzyme secretion might best be studied in the absence of particle ingestion. Therefore, Zurier *et al.* (1973a) employed human polymorphonuclear leucocytes treated with cytochalasin B, a fungal metabolite known to interfere with the function of microfilaments. Microfilaments play a key role in heterophagic vacuole formation from the invaginating plasma membrane and are therefore necessary for phagocytosis. Thus, cytochalasin B is a useful pharmacological tool in the study of lysosomal enzyme secretion provoked by particulate immune reactants in the absence of phagocytosis. Likewise, the influence of drugs on enzyme secretion can be analysed independently of their possible additional influence on phagocytosis. Inhibition of lysosomal enzyme release from cytochalasin B-treated human granulocytes by cAMP, Bt_2-cAMP, theophylline and various prostaglandins was reported from Weissmann's laboratory (Zurier *et al.*, 1973a; 1973; Goldstein *et al.*, 1973). In addition, or as an alternative, to the use of cytochalasin B to inhibit phagocytosis, soluble immune reactants have been used to stimulate lysosomal enzyme release from neutrophils (Goldstein *et al.*, 1973a). Untreated (Ignarro and George, 1974b) or cytochalasin B-treated (Weissmann *et al.*, 1975) human neutrophils mixed with particulate-free, zymosan-treated serum, which contains activated, soluble complement components, results in stimulation of lysosomal enzyme secretion. cAMP or Bt_2-cAMP inhibits this stimulation.

Studies from the author's laboratory have revealed that cAMP, Bt_2-cAMP and other 8-substituted analogues of cyclic AMP inhibit the release of lysosomal enzymes (β-glucuronidase and neutral protease) from human polymorphonuclear leucocytes which have been provoked by particulate immune reactants. cAMP and its analogues were found also to inhibit lysosomal enzyme release from human neutrophils in a model system not envolving phagocytosis in which the cells were allowed to adhere to immobilized immune reactants (Ignarro, 1974, 1975). In addition to cAMP, several hormones and pharmacological agents that are known to elevate the concentration of cAMP in various tissues have been reported to inhibit lysosomal enzyme release from human polymorphonuclear leucocytes in both phagocytic and nonphagocytic systems (Weissmann *et al.*, 1971a; Zurier *et al.*, 1973b; Ignarro, 1974a; Ignarro and George, 1974a, 1974b). In most cases, the pharmacological agents which inhibit lysosomal enzyme release also elevate concomitantly the levels of cAMP in purified human neutrophils (Ignarro and George, 1974a,b). These data form the basis of the hypothesis that intracellular cAMP is responsible for mediating intracellular processes associated with inhibition of lysosomal enzyme secretion (Ignarro and George, 1974a).

Cyclic GMP, Bt_2-cGMP and certain cholinergic agents (muscarinic agonists) were found to enhance the magnitude of lysosomal enzyme release initially provoked by particulate immune reactants. Muscarinic agonists are well known to stimulate the accumulation of cGMP in many mammalian tissues. Indeed, this effect is believed to account for many of the physiological and pharmacological actions of acetylcholine. Similarly, acetylcholine was found to elevate neutrophilic levels of cGMP along a time course which strongly suggests that intracellular cGMP is responsible for mediating lysosomal enzyme secretion (Ignarro and George, 1974a,b). Other muscarinic agonists were also found to stimulate concomitantly cGMP accumulation in, and lysosomal enzyme release from, human neutrophils (Ignarro and George, 1974b; Zurier *et al.*, 1974). These findings, that the effect of cGMP and muscarinic agonists are opposite to that of cAMP and catecholamines, provides the basis for the hypothesis that cGMP and cAMP mediate antagonistic actions on lysosomal enzyme secretion from human neutrophils (Ignarro and George, 1974a).

In vitro stimulation of lysosomal enzyme release from human neutrophils by particulate, immobilized or soluble immunological reactants requires, in most instances, the presence of a divalent cation, specifically Ca^{2+}, in the extracellular medium (Ignarro, 1974a; Ignarro and George, 1974b). Similar data had been reported earlier for rabbit neutrophils (Woodin and Wieneke, 1963). Also, stimulation of enzyme release by a divalent cation ionophore requires the presence of extracellular Ca^{2+} (Smith and Ignarro, 1975). EDTA inhibits the stimulation of enzyme release by immune reactants and muscarinic agonists, respectively (Ignarro and George, 1974b). These findings, namely that calcium may control, modulate or mediate the exocytosis of granule contents, are very similar to those reported for many other secretory systems.

B Phagocytosis, Motility and Other Functions

Cyclic nucleotides and agents which alter levels of cyclic nucleotides in the tissue influence functions of neutrophils in addition to lysosomal enzyme secretion. cAMP, its analogues and agents which elevate neutrophilic levels of cAMP have been reported to inhibit phagocytosis of particulate immunological reactants by neutrophilic leucocytes (Bourne *et al.*, 1971; Cox and Karnovsky, 1973; Ignarro and Cech, 1976). This has been demonstrated in polymorphonuclear leucocytes from man and guinea pig. A variety of prostaglandins is found to inhibit leucocyte phagocytosis (Cox and Karnovsky, 1973), and this effect correlates with the capacities of the prostaglandins to stimulate the accumulation of cAMP in human neutrophils (Ignarro, 1975). Catecholamines have also been demonstrated to inhibit phagocytosis in purified human neutrophils, an effect which correlated with the capacities of the catecholamines to elevate neutrophilic levels of cAMP (Ignarro and George, 1974a,b).

In an initial study, cGMP and muscarinic agonists were not observed to alter phagocytosis to any significant extent. However, in a subsequent study (Ignarro and Cech, 1976), cGMP, acetylcholine and pilocarpine (0.1–1 μM) were observed to enhance the magnitude of phagocytosis in human neutrophils. The contrast in results is attributed to differences in the ratio of particles to cells used in the two studies. In the former study, where a higher particle/cell ratio of 60/1 was used, the possibility exists that maximal rates of phagocytosis prevail such that further stimulation would be difficult to observe. On the other hand, when the particle/cell ratio was reduced to 5/1 significant enhancement of the magnitude of particle ingestion by cGMP and muscarinic agonists was observed. These data indicate that acetylcholine, and perhaps other endogenous tissue hormones, possess the capacity to enhance phagocytosis by neutrophils during the early stages of cell confrontation with particles.

Motility is another important function of polymorphonuclear leucocytes which is influenced by cyclic nucleotides. The leucotactic response of granulocytes was found to be stimulated by cGMP and pharmacological agents which provoke the accumulation of cGMP in these cells (Estensen *et al.*, 1973). In contrast, cAMP and agents which cause cAMP to accumulate in polymorphonuclear leucocytes were reported to inhibit motility (Estensen *et al.*, 1973; Rivkin *et al.*, 1975). Hill *et al.* (1975) showed that various pharmacological agents reported to elevate the levels of cGMP and cAMP in human polymorphonuclear leucocytes enhance and inhibit, respectively, the chemotactic response of these cells to a bacterial factor.

The process of adhesion of granulocytes to solid surfaces and the relationship of this process to cAMP were studied by Bryant and Sutcliffe (1974). These investigators found that Bt_2-cAMP, PGE_1, histamine and theophylline each inhibit the adhesion of neutrophils, eosinophils and basophils to glass surfaces. However, the inhibitory effects on the neutrophils are much smaller than those on the other two granulocyte types. The effects of cGMP and cholinergic agents on cell adhesion were not examined. Unpublished observations from the author's laboratory indicate that adhesion of purified human neutrophils to glass surfaces or immobilized immunological reactants (that is, cellulose-fibre discs coated with heat-aggregated human IgG) is not influenced by cGMP, 8-bromo-cGMP or carbamylcholine.

Another function of polymorphonuclear leucocytes influenced by cyclic nucleotides is a special form of cellular cytotoxicity which is mediated by human neutrophils and which is dependent on antibody. cGMP was found to facilitate, whereas cAMP inhibits, this neutrophil-mediated cytotoxic response (Gale and Zighelboim, 1974). Moreover, pharmacological agents capable of elevating cGMP levels, such as acetyl-β-methylcholine and carbamylcholine, facilitate whereas agents capable of elevating cAMP levels, such as adrenaline, isoprenaline and aminophylline, inhibit this cytotoxic response.

The above data indicate that both nucleotides are associated closely with facilitatory and inhibitory influences on at least four independent functions of human neutrophils. Such opposing effects have been elicited not only by the cyclic nucleotides or their analogues directly, but also by tissue hormones known to elevate the tissue levels of the appropriate cyclic nucleotide. The antagonistic actions of cGMP and cAMP support the dualism hypothesis of biological control in a bidirectional regulated system (Goldberg *et al.*, 1973a).

IV PHARMACOLOGICAL MODULATION OF LYSOSOMAL ENZYME SECRETION AND OTHER NEUTROPHILIC FUNCTIONS

Numerous pharmacological agents from diverse chemical classes have been reported to modify or influence the principal functions of polymorphonuclear leucocytes. The present discussion is limited to those agents known to or believed to elicit their effects by altering the cyclic nucleotide levels in neutrophils. A discussion of the effects of other drugs has appeared elsewhere (Ignarro, 1976).

In the previous section, experimental evidence for the direct effects of the cyclic nucleotides on several neutrophilic functions was discussed. In addition, some mention was made of the effects of tissue hormones and other chemical agents on both cell function and cyclic nucleotide levels. The relationships between cell function and cyclic nucleotide concentrations with regard to the actions of autonomic neurohormones, prostaglandins, glucocorticosteroids, Ca^{2+} and other pharmacological agents are presented below.

A Cyclic Nucleotides

cGMP and cAMP elicit direct positive and negative effects, respectively, on lysosomal enzyme secretion and several other functions of polymorphonuclear leucocytes. In the presence of an appropriate immunological stimulus which initiates a particular cellular response, such as lysosomal enzyme release, cGMP facilitates and cAMP inhibits that particular response. Numerous pharmacological agents have been reported to influence neutrophilic function similarly and to elicit concomitant and corresponding changes in the cellular concentration of the appropriate cyclic nucleotide.

B Autonomic Neurohormones

Sympathomimetic and parasympathomimetic or cholinergic agonists elicit a multitude of biological actions that are, for the most part, antagonistic in direction. Similarly, certain catecholamines and cholinergic agonists have been found to elicit opposing influences on neutrophilic function. Adrenaline, isoprenaline and, in certain cases, noradrenaline inhibit human neutrophilic functions, including lysosomal enzyme secretion, phagocytosis, motility and antibody-dependent cellular cytotoxicity (Estensen *et al.*, 1973; Gale and Zighelboim, 1974; Ignarro, 1974a; Zurier *et al.*, 1974; Hill *et al.*, 1975; Rivkin *et al.*, 1975; Ignarro and Cech, 1976). The inhibitory effects of the catecholamines appear to be mediated via an interaction with specific receptor sites located, most probably, on the extracellular surface of the plasma membrane. The receptors involved are classified as β-adrenergic on the basis of selective blockade of receptor interactions with known β-adrenergic receptor antagonists. Thus propranolol (a known β-adrenergic blocker) but not phentolamine (an α-adrenergic blocker), blocks the inhibitory effects of the catecholamines on the functions of neutrophils.

Catecholamines also bring about the accumulation of cAMP in human polymorphonuclear leucocytes (Ignarro and George, 1974a,b; Zurier *et al.*, 1974; Rivkin *et al.*, 1975). The mechanism by which the catecholamines elevate neutrophilic cAMP levels is most likely by stimulation of membrane-bound adenylate cyclase (Scott, 1970; Bourne and Melmon, 1971). It is of interest that the stimulation by noradrenaline, adrenaline and isoprenaline of adenylate cyclase activity is blocked by propranolol but not by phentolamine. Similarly, phenylephrine (a selective α-adrenergic agonist) does not stimulate adenylate cyclase (Bourne and Melmon, 1971). These data indicate that human neutrophilic adenylate cyclase is associated very closely with the β-adrenergic receptor. Moreover, the findings that catecholamines stimulate cAMP accumulation and inhibit several neutrophilic functions concomitantly, and that both types of effect are blocked selectively with β-adrenergic antagonists, suggest strongly that the pharmacological effects of the catecholamines on neutrophils are mediated by cAMP. The enhancement of both the physiological and biochemical effects of the catecholamines by phosphodiesterase inhibitors, such as theophylline and aminophylline, further supports the role of cAMP as a second messenger in human neutrophilic leucocytes.

Acetylcholine and structural analogues possess the capacity to stimulate or enhance certain functions of neutrophils, such as lysosomal enzyme secretion, phagocytosis, chemotactic responsiveness or motility, and antibody-dependent cellular cytotoxicity (Gale and Zighelboim, 1974; Ignarro, 1974a; Ignarro and George, 1974a; Zurier *et al.*, 1974; Hill *et al.*, 1975; Ignarro and Cech, 1976). The stimulatory effects of these cholinergic agents appear to be mediated via an interaction with muscarinic receptors, which are most likely to be located on the plasma membrane. Evidence for this is that atropine blocks these stimulatory effects in most cases. Lysosomal enzyme secretion from human neutrophils provoked by immune reactants is not blocked with atropine. However, the enhancement of secretion by muscarinic agonists is blocked selectively by atropine but not by either hexamethonium (a ganglionic blocker) or d-tubocurarine (a neuromuscular blocker) (Ignarro, 1974a; Ignarro and George, 1974a; Zurier *et al.*, 1974). The same findings apply to the effects of atropine on phagocytosis (Ignarro and Cech, 1976). Muscarinic agonists also provoke further accumulation of cGMP in human neutrophils (Ignarro and George, 1974a,b; Zurier *et al.*, 1974); an effect which, like lysosomal enzyme secretion, is blocked by atropine but not by hexamethonium or d-tubocurarine (Ignarro and George, 1974a,b). These data suggest that the muscarinic effects on lysosomal enzyme secretion are mediated by neutrophilic cGMP.

C Prostaglandins

In recent years the prostaglandins have received widespread attention in their role as endogenous modulators of the inflammatory process. Prostaglandins are capable of eliciting both pro-inflammatory and anti-inflammatory effects *in vitro* and *in vivo*. Neutrophilic function is affected profoundly by these local tissue hormones. Prostaglandins E_1, E_2, A_1, A_2, $F_{1\beta}$ and F_2 possess the capacity to inhibit lysosomal enzyme secretion from numan polymorphonuclear leucocytes in both phagocytic and nonphagocytic model systems (Weissmann *et al.*, 1971a; Goldstein *et al.*, 1973; Zurier *et al.*, 1973a,b; Ignarro, 1975). In certain instances, the presence of theophylline is required in order to observe significant effects. Also, prostaglandins inhibit phagocytosis by polymorphonuclear leucocytes (Cox and Karnovsky, 1973). Enhancement of the inhibitory actions of the prostaglandins by theophylline suggests that cAMP may mediate these effects. Indeed, certain prostaglandins are capable of stimulating human neutrophilic adenylate cyclase activity (Scott, 1970; Bourne and Melmon, 1971; Bourne *et al.*, 1971; Polgar *et al.*, 1973). In agreement with these data are the findings that prostaglandins promote the accumulation of cAMP in purified human neutrophils during cell contact with particles capable of being taken up by phagocytosis (Ignarro, 1975).

Prostaglandin $F_{2\alpha}$ elicites a biphasic effect on the nonphagocytic secretion of lysosomal enzymes from human neutrophils (Ignarro, 1975). At high concentrations (10^{-6}–10^{-5} M) $PGF_{2\alpha}$ accelerates whereas at low concentrations (10^{-8}–10^{-7}M) it inhibits lysosomal enzyme secretion. Theophylline reduces the accelerating action and enhances the inhibitory action of $PGF_{2\alpha}$ Elevations in the level of cAMP and cGMP in neutrophils are associated with inhibitory and accelerating influences, respectively, of this prostaglandin. Thus, once again inhibition and acceleration of lysosomal enzyme secretion from neutrophils are associated with elevations in the cellular levels of cAMP and cGMP, respectively.

Adhesion and motility of polymorphonuclear leucocytes are also affected by prostaglandins. Adhesion of human neutrophils to glass surfaces is inhibited by high concentrations (10^{-6}–10^{-4}M) of PGE_1; the effect is enhanced by theophylline (Bryant and Sutcliffe, 1974). PGE_1 and PGA_1 inhibit chemotaxis of human neutrophils and elevate the cellular levels of cAMP concomitantly (Rivkin *et al.*, 1975). Thus, these data on cAMP levels are in agreement with those of Ignarro (1975), in that PGE_1 and PGA_1 stimulate the accumulation of cAMP in human neutrophils. Hill and coworkers (1975) reported that PGE_1 (10^{-8}–10^{-7} M) inhibit the chemotactic motility of human neutrophils to a bacterial factor. PGE_2 is also inhibitory, but only at 10^{-6}M. PGF_2 inhibits motility at 10^{-6}M but markedly enhances motility at lower concentrations, namely 10^{-8} and 10^{-7}M. None of the effects of the prostaglandins on neutrophilic function or cyclic nucleotide levels are inhibited by autonomic receptor antagonists, suggesting that separate prostaglandin receptors exist on the neutrophil.

D Glucocorticosteroids

An adequate review of the effects of glucocorticosteroids on neutrophil function and the inflammatory process is beyond the scope of this limited chapter—see Ignarro (1976) for a more comprehensive review and bibliography. Glucocorticoids have been reported to inhibit several neutrophilic functions, such as lysosomal enzyme secretion, phagocytosis, chemotaxis and motility, and antimicrobial activity. The present discussion is limited to the relationship between glucocorticoids and cyclic nucleotides as they pertain to lysosomal enzyme secretion. Several glucocorticosteroids have been found to inhibit Ca^{2+}-dependent lysosomal enzyme secretion from purified human neutrophils in both nonphagocytic (Ignarro, 1974b) and phagocytic model systems (Ignarro and Cech, 1976). Mineralocorticosteroids are inactive. Figure 11.1 illustrates the inhibitory effect of methylprednisolone sodium

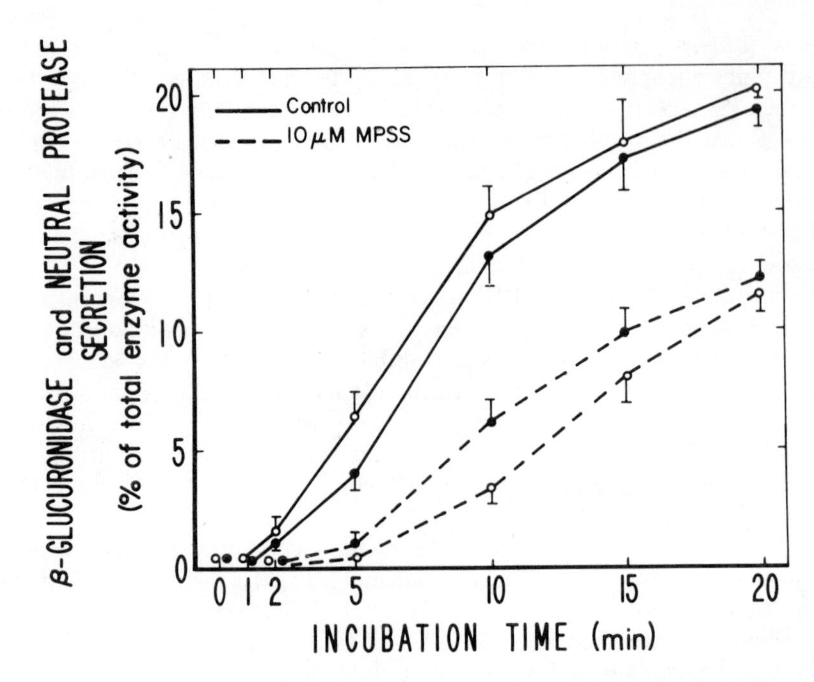

Figure 11.1 Effect of methylprednisolone on lysosomal enzyme secretion from human neutrophils in contact with serum-treated zymosan. Neutrophils (5×10^6) were preincubated with 10 μM methylprednisolone sodium succinate (MPSS) at 37° C for up to 20 min: β-glucuronidase (●) and neutral protease (o) activities, respectively, in lysates of 5×10^6 neutrophils were 234–277 μg phenolphthalein liberated from phenolphthalein glucuronide in 18 hr at 37° C, and 83–102 μg tyrosine equivalents liberated from haemoglobin in 18 hr at 37° C. $N = 4$. Values $= \bar{x} \pm s_{\bar{x}}$.

succinate on lysosomal enzyme secretion from phagocytic human neutrophils. Triamcinolone, paramethasone and hydrocortisone elicit a similar effect. These glucocorticoids, but not the mineralocorticoids deoxycorticosterone and aldosterone, also inhibit phagocytosis.

The inhibitory influence of glucocorticoids on neutrophilic function is similar to the inhibitory influence of cAMP and pharmacological agents which elevate cAMP levels in these cells. However, neutrophilic levels of cAMP have been found not to be elevated by any of the glucocorticoids. Instead, the immunologically-provoked accumulation of cGMP, which is believed to be the intracellular signal for secretion, is significantly inhibited (Figure 11.2). Methylprednisolone (Figure 11.1) and several other glucocorticoids therefore increase the concentration ratio of cAMP to cGMP, and this is associated with an inhibitory effect on secretion. Similarly, other inhibitors of secretion, such as catecholamines, prostaglandins and theophylline, increase the cAMP/cGMP concentration ratio in neutrophils.

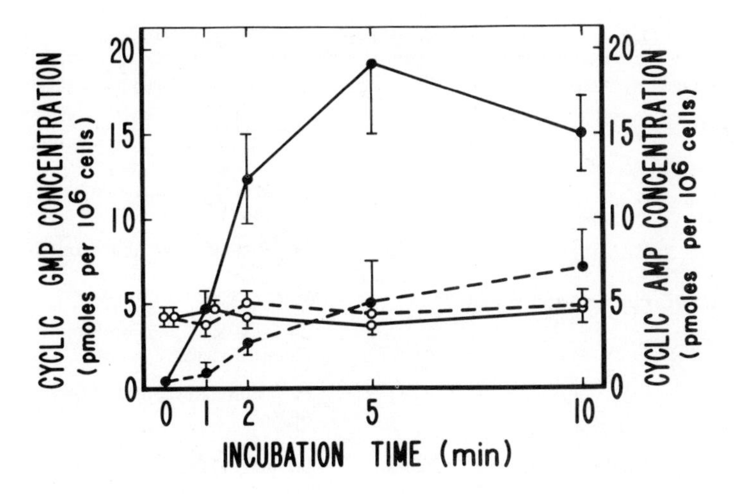

Figure 11.2 Effects of methylprednisolone on accumulation of cGMP and cAMP in human neutrophils. Neutrophils (5×10^6) were preincubated with 10 μM methylprednisolone sodium succinate at 37°C for 10 min in Hanks' glucose solution, prior to addition of 4×10^8 serum-treated zymosan particles. Samples were further incubated for up to 10 min. Incubation media were immediately frozen and analysed for cGMP and cAMP (Ignarro and George, 1974a). cGMP concentration (●); cAMP concentration (O); control (—), methylprednisolone present (— — —). $N = 4$–5. Values $= x \pm s_{\bar{x}}$

The mechanism by which methylprednisolone decreases the accumulation of cGMP in neutrophils is not clearly understood. It is reasoned that inhibition of synthesis and/or stimulation of degradation of cGMP should account for the glucocorticoid effect. cGMP formation from GTP by guanylate cyclase has been studied. The presence of guanylate cyclase has been demonstrated in purified human neutrophils (Ignarro and George, 1975). Some of the properties of this enzyme are listed in Table 11.1. Human neutrophilic guanylate cyclase activity from the soluble fraction is not inhibited significantly by the glucocorticoids tested. Particulate enzyme activity is not inhibited either. Thus, direct inhibition of guanylate cyclase activity cannot account for the action of the glucocorticoids. These negative data, however, do not exclude the possibility of steroid-induced indirect inhibition of enzyme activity or blockade of activation of guanylate cyclase by divalent cations. Influx of extracellular Ca^{2+} into human neutrophils has been found to be associated temporally with immunologically induced lysosomal enzyme secretion (Smith and Ignarro, 1975). Moreover, acetylcholine and a divalent cation ionophore (A-23187) each stimulate Ca^{2+} influx, cGMP accumulation and lysosomal enzyme secretion. Methylprednisolone inhibits not only cGMP accumulation and secretion, but also Ca^{2+} influx (Figure 11.3). Several other glucocorticoids, but not mineralocorticoids, inhibit Ca^{2+} influx into human neutrophils provoked by a soluble

Table 11.1 Some properties of human neutrophilic guanylate cyclase

subcellular distributions	23% of total activity* is particulate**
	80% of total activity is soluble
K_m	66 μM GTP for particulate fraction
	14 μM GTP for soluble fraction
divalent cation requirement***	Mn^{2+} or Ca^{2+}, but not Mg^{2+}
	2–3 mM Mn^{2+} is 2.5-fold more potent than Ca^{2+}
	Ca^{2+} stimulates 2-fold in presence of Mn^{2+}
effect of inhibitors on soluble fraction	0.1 mM ATP–53% inhibition
	0.1 mM CTP–41% inhibition
	0.1 mM P_2O_7–48% inhibition
	0.1 mM Au–75% inhibition
	0.1 mM Ascorbate–51% inhibition

*Total activity signifies detergent-treated whole homogenate activity
**Particulate signifies detergent-treated sediment from a 100,000 *g* spin for 60 min
***Data apply to both the soluble and particulate fractions

immune reactant and a divalent cation ionophore. These data suggest that glucocorticosteroids inhibit lysosomal enzyme secretion, and perhaps phagocytosis, by inhibiting the intracellular accumulation of Ca^{2+}, thereby blocking the signal for these neutrophilic functions.

E Other Pharmacological Agents

Several substances other than tissue hormones have been shown to influence neutrophilic function probably by altering the cellular levels of cGMP and/or cAMP. The stimulatory effect of a divalent cation ionophore, in the presence of Ca^{2+}, on lysosomal enzyme secretion (in the absence of an additional immunological reactant) has been discussed. Phorbol myristate acetate, a tumour-promoting agent, provoked the release of lysosomal contents from polymorphonuclear leucocytes (Repine *et al.*, 1974; Weissmann *et al.*, 1975). Alterations in cyclic nucleotide levels of neutrophils have not been reported, but phorbol myristate acetate does stimulate cGMP accumulations in platelets (White *et al.*, 1973) and fibroblasts (Estensen *et al.*, 1974). Accumulation of cGMP in neutrophils could account for the stimulatory effect of this tumour promoter on lysosomal enzyme release.

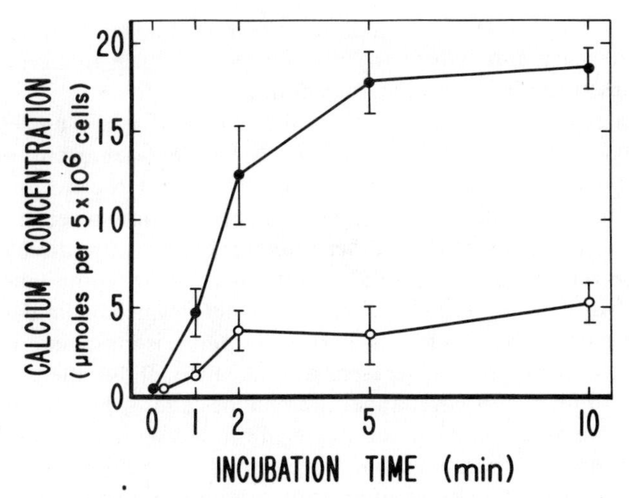

Figure 11.3 Effect of methylprednisolone on Ca^{2+} accumulation in human neutrophils during cell contact with zymosan-treated serum. Neutrophils (5×10^6) were preincubated with $10\,\mu M$ methylprednisolone sodium succinate at 37°C for 10 min in Hanks' glucose solution, prior to addition of $^{45}CaCl_2$ and zymosan-treated serum (Smith and Ignarro, 1975). Samples were then further incubated for up to 10 min. Incubations were terminated by rapid micropore filtration and the filters, containing the cells, were analysed for $^{45}Ca^{2+}$ (Smith and Ignarro, 1975). Control (●); methylprednisolone present (O). $N = 5$; values $= \bar{x} \pm s_{\bar{x}}$

V REGULATION OF LYSOSOMAL ENZYME SECRETION FROM MACROPHAGES AND PLATELETS

CAMP, Bt_2-cAMP and high concentrations of cGMP have been reported to inhibit lysosomal enzyme release from purified mouse peritoneal macrophages (Weissmann *et al.*, 1971b). Phagocytosis is also inhibited by high concentrations of cAMP, but lower concentrations (10^{-10}M) of cAMP and cGMP enhanced phagocytosis (Weissmann *et al.*, 1971b). Phagocytosis by peritoneal macrophages has been reported to be stimulated by imidazole and levamisole (Lima *et al.*, 1974), agents which reduce cAMP but increase cGMP levels in certain tissues (Hadden *et al.*, 1975a). cGMP levels in macrophages are increased by 5-hydroxytryptamine and carbamylcholine, both agents which also enhanced macrophage chemotaxis (Sandler *et al.*, 1975). A confusing problem in the study of macrophages is that although cGMP and cAMP stimulate and inhibit secretion, respectively, cell contact with certain particles causes cAMP, but not cGMP, accumulation—see Ignarro (1976b) for discussion. A subject which deserves extensive investigation is the regulation of macrophage function by soluble protein factors discharged from activated lymphocytes (Ignarro, 1976). Lymphocyte factors, known as lymphokinines, inhibit the motility and induce the proliferation of macrophages, and these effects may be mediated by cyclic nucleotides.

Numerous reports have appeared on the effects of cyclic nucleotides and pharmacological agents, which affect cyclic nucleotide levels, on platelet function including secretion—see Haslam (1975) for a review. In summary, cAMP and Bt_2-cAMP inhibit the platelet release reaction, a secretory process whereby lysosomal enzymes and other granular contents are discharged from the cell. By the inhibition of the secretion of these compounds platelet aggregation is also inhibited. Chemical agents which stimulate platelet adenylate cyclase activity, such as PGE_1, glucagon, isoprenaline and adenosine, inhibit platelet secretion, whereas agents which inhibit adenylate cyclase, including thrombin, noradrenaline, adrenaline and 5-hydroxytryptamine, stimulate or provoke secretion and aggregation. However, careful analysis of the data and further experimentation have provided the basis for the argument that whereas cAMP almost certainly mediates inhibition of platelet secretion, a decrease in cAMP does not, in itself, mediate platelet secretion (Haslam, 1973).

In view of the evidence that cGMP mediates secretion of granular contents from neutrophils and histamine-containing cells, Haslam and McClenaghan (1974) studied the influence of promoters of platelet secretion on cGMP levels. Collagen increases platelet cGMP and this cellular event precedes and is related quantitatively to platelet aggregation. Similar effects were reported by others for collagen, ADP and adrenaline, and, in addition, guanylate cyclase has been identified in platelets (Ignarro, 1976). These data support the hypothesis that cGMP mediates platelet secretion, while they somewhat weaken the view that a fall in cAMP levels accounts for this action (Haslam, 1975). The importance of Ca^{2+} in controlling cGMP accumulation and the platelet release reaction has been discussed recently (Haslam, 1975; Ignarro, 1976). Ca^{2+} may serve as an important link in the stimulus–secretion coupling mechanism which regulates the platelet release reaction.

VI SUMMARY AND CONCLUSIONS

The experimental data discussed indicate that cyclic nucleotides, and pharmacological agents which cause them to accumulate within the cells, play a biological role in regulating lysosomal enzyme secretion from neutrophils and, perhaps, other blood cells. In addition, these chemical agents may influence other functions of blood cells, such as phagocytosis, motility and interactions between cells. cGMP and cAMP are associated with and may mediate antagonistic effects on lysosomal enzyme secretion and other neutrophil functions. Thus, cGMP and agents which cause its accumulation in neutrophils accelerate, whereas cAMP and agents which cause its accumulation inhibit, lysosomal enzyme secretion, phagocytosis, chemotactic motility and antibody-dependent cellular cytotoxicity. It appears that cGMP and cAMP are associated with opposing actions in macrophages and platelets also. These findings support the dualism concept of cellular control mechanisms in bidirectionally regulated biological systems, whereby cGMP and cAMP mediate opposing or antagonistic influences on cell function (Goldberg *et al.*, 1973a).

Lysosomal enzyme secretion from human neutrophils is associated closely with Ca^{2+} influx and cGMP accumulation. In fact, Ca^{2+} influx and elevation of cGMP levels in neutrophils precede any measurable discharge of lysosomal enzymes (Smith and Ignarro, 1975). Ca^{2+} is required in most experimental systems in order that lysosomal enzyme secretion may occur. Indeed, Ca^{2+} is an absolute requirement for numerous secretory processes. As is true for other secretory processes, Ca^{2+} may function as a link in the stimulus–secretion coupling mechanism. For example, contact between the neutrophilic plasma membrane and an immunological reactant triggers an increase in membrane permeability to Ca^{2+}, which then enters the cell. Increased intracellular calcium levels signal the secretory process by raising the intracellular concentration of cGMP. Consistent with this view is the reported activation of guanylate cyclase by Ca^{2+} (Hardman *et al.*, 1971). The finding that a divalent cation ionophore, in the presence of Ca^{2+}, provokes Ca^{2+} influx, cGMP accumulation and secretion in neutrophils suggests that immunological reactants which promote secretion function as divalent cation ionophores.

Although considerable information has accumulated on the modulation of cyclic nucleotide levels in neutrophils and other blood cells, the mechanisms by which these second messengers influence cell function are not understood. It is speculated that protein kinases activated by cAMP and phosphoprotein phosphatases activated by cGMP exist in the neutrophil and that these enzyme systems govern lysosomal enzyme secretion by modulating the state of phosphorylation of certain key protein macromolecules within the cell (Smith and Ignarro, 1975). For example, stimulation of secretion may be associated with phosphoprotein dephosphorylation, whereas inhibition of secretion may be associated with protein phosphorylation. cAMP-activated protein kinase activity, as well as phosphoprotein phosphatase activity, has been identified recently in human neutrophils (Tsung *et al.*, 1975). Macrophages and platelets also have cAMP-activated protein kinase activity.

The ultimate mechanism by which cyclic nucleotide-mediated intra-cellular events actually alter lysosomal enzyme secretion from neutrophils and other blood cells is not understood. Weissmann and coworkers (1975) have attributed the inhibitory effect of cAMP on lysosomal enzyme release to an interference with microtubule assembly, thereby reducing the translocation of primary lysosomes to, and their subsequent fusion with, the neutrophil plasma membrane. Presumably, stimulation of secretion by cGMP is a consequence of the promotion of microtubule assembly and therefore the acceleration of flow of the lysosome to and merger with the plasma membrane. Alternatively, or in addition, intracellular cyclic nucleotides may alter the properties of lysosome granule membranes to an extent which would influence the movement of the granules and/or their fusion with other membranes (Ignarro and George, 1974a). Clearly, the effects of cyclic nucleotide-modulated protein kinases and phosphoprotein phosphatases on microtubules, lysosomes and the plasma membrane of neutrophils and other lysosomal enzyme-secreting cells must be assessed in order to more fully comprehend the secretory mechanisms involved.

An interesting concept emerging from the analysis of the effects of tissue hormones on the functions of neutrophils, macrophages and platelets is that these blood cells possess specific receptor sites on their surface which are capable of responding to specific hormones. Neutrophils are capable of responding *in vitro* to muscarinic agonists, β-adrenergic agonists, histamine, prostaglandins, glucocorticosteroids and other tissue hormones. The fact that physiological concentrations of most of these naturally occurring substances affect neutrophil function suggests that these cells may act similarly *in vivo*. Unfortunately, no clear evidence for this is presently available. Thus, any suggestion of *in vivo* modulation of neutrophil function by tissue hormones is purely speculative. However, it appears reasonable to speculate, in view of the available *in vitro* data, that at least one aspect of the inflammatory process, namely neutrophilic function, is modulated by a variety of local and circulating hormones.

ACKNOWLEDGEMENT

Research was supported partially by USPHS Grant No. AM 17692, the Arthritis Foundation and the Edward G. Schlieder Educational Fund. The author receives a Research Career Development Award (USPHS Grant No. AM 00076).

CHAPTER 12

Cyclic Nucleotides in Adipose Tissue

J. N. Fain

I INTRODUCTION

This chapter reviews the current status of cyclic nucleotides as regulators of fat-cell metabolism. Emphasis is placed on recent findings which have provided new insights into the regulation of fat-cell lipolysis by cyclic nucleotides. A previous review has emphasized the effects of a wide variety of drugs and hormones on cAMP metabolism in fat cells (Fain, 1973a).

II cAMP AND LIPOLYSIS

cAMP is now recognized as the most important factor involved in the regulation of triglyceride lipolysis by the fat cells of adipose tissue. The key role of cAMP in the regulation of lipolysis by catecholamines is shown in Figure 12.1. The model suggests that adenosine and fatty acids inhibit catecholamine activation of adenylate cyclase. Methylxanthines are postulated to increase cAMP accumulation by inhibiting both cAMP phosphodiesterase and adenosine uptake by fat cells. These aspects are covered in later sections of this review.

The initial step in catecholamine action is binding with a β-1 receptor. The structure–activity relationships for activation of lipolysis by catecholamines and for inhibitors of catecholamine action fits a β-1 receptor model (Fain, 1973a).

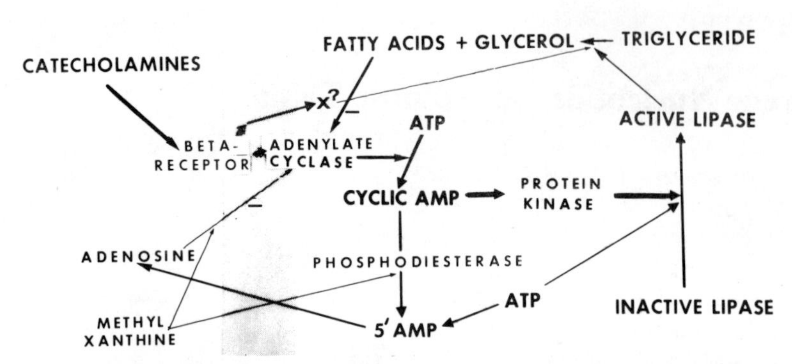

Figure 12.1 Model for activation of lipolysis by catecholamines and its modulation by methylxanthines, fatty acids and adenosine. Cathecholamines interact with the β-adrenergic receptor resulting in adenylate cyclase activation and possibly release of a messenger (X). Adenylate cyclase increases cAMP formation which activates protein kinase. Inactive trigylceride lipase is converted to its active form by protein kinase in the presence of ATP. Triglycerides are hydrolysed; free fatty acids formed act as product inhibitors of the lipase and feedback inhibitors of adenylate cyclase. Adenosine release to the medium is regulated by 5′-AMP formed by hydrolysis of cAMP or breakdown of ATP. Methylxanthines inhibit adenosine action and cAMP phophodiesterase activity. The – signs refer to inhibition of adenylate cyclase by adenosine and fatty acids

The link between catecholamine binding to receptors and adenylate cyclase activation is unknown. There have been problems in identifying the β-adrenergic receptors in any tissue due to a large amount of non-specific binding of catecholamines (Cuatrecasas *et al.*, 1974). Pairault and Laudat (1975) reported that much of the non-specific binding of catecholamines can be abolished by incubating fat-cell plasma-membrane vesicles in the presence of 1 mM EGTA or 0.5 mM pyrocatechol. The activation of adenylate cyclase of fat-cell membranes by 5 μM noradrenaline is unaffected while catecholamine binding is inhibited by 98% in the presence of 0.5 mM pyrocatechol. Propranolol (1 μM) can inhibit the residual binding of 0.5 μM catecholamine in the presence of 0.5 mM pyrocatechol by about 50% (Pairault and Laudat, 1975).

The evidence for cyclic AMP involvement in the lipolytic action of catecholamines is impressive: (*i*) Catecholamines activate adenylate cyclase in broken-cell preparations, and addition of cAMP to these preparations activates protein kinase and triglyceride lipase. (*ii*) β-Adrenergic antagonists block the elevation of cAMP and of lipolysis due to catecholamines. (*iii*) Agents such as methylxanthines and adenosine deaminase which elevate intracellular cAMP also activate lipolysis. (*iv*) Cholera toxin also activates lipolysis and cAMP accumulation in fat cells but only after a lag period. (*v*) Prostaglandin E_1, adenosine and cyclic carboxylic acids, such as nicotinic acid, are inhibitors of both cAMP accumulation and lipolysis.

Rodbell (1967) reported that catecholamines activate adenylate cyclase in lysed fat-cell preparations known as ghosts. Isolation of hormonally-responsive rat fat-cell ghosts involves lysis of fat cells with hypotonic buffer in the presence of ATP (Rodbell, 1967). Subsequent work has demonstrated that the presence of ATP during rupture of fat cells and preparation of plasma-membrane fractions is required in order to maintain a maximal response to hormones (Combret and Laudat, 1972; Sahyoun and Cuatrecasas, 1975; Rodbell, 1975). Birnbaumer and Rodbell (1969) found that a variety of hormones activate the same adenylate cyclase enzyme. There appear to be different hormone-specific sites (receptors) which bind catecholamines, ACTH, TSH and glucagon. It is not yet clear how a variety of hormones activates the same enzyme.

Rodbell (1975) has shown that nucleotides can both inhibit and stimulate the activity of fat-cell adenylate cyclase. 5′-Guanylyl imidodiphosphate — GMP-P(NH)P — is an analogue of GTP resistant to hydrolysis by nucleotide phosphohydrolases in membranes. GMP-P(NH)P activates adenylate cyclase in a variety of eukaryotic cells (Londos *et al.*, 1974). The analogue also activates fat-cell adenylate cyclase at high pH (8.5) and high Mg^{2+} concentrations (50 mM), but is an inhibitor of the enzyme if the concentration of Mg^{2+} and the pH are lowered (Rodbell, 1975). These results suggest that the regulation of adenylate cyclase is a complex function of the concentrations of both free activating nucleotide (guanine or adenine), protonated substrate and the affinities of various states of the enzyme for these ligands.

Corbin *et al.* (1969) demonstrated that protein kinase activity of adipose tissue homogenates is activated by cAMP. Maximal activation is seen with only 100 nM cAMP. In subsequent studies it was shown that the concentration of cAMP required for activation of protein kinase depends on the concentration of kinase (Swillens *et al.*, 1974), and possibly the protein inhibitor concentration. Beavo *et al.* (1974) have found that 300 nM cAMP gives half-maximal stimulation if the concentration of protein kinase is 9 nM. In contrast, 1.5 μM cAMP is required for the same stimulation if the protein kinase concentration is increased to 150 nM.

The muscle protein kinase concentration is about 150 nM while that of cAMP is about 250 nM. Since the concentration of protein kinase is nearly equal to that of cAMP, a large portion of basal cAMP should be bound to the kinase. Conceivably, agents could alter the protein kinase activity by affecting the equilibrium between cAMP and protein kinase. One factor could be the ATP concentration since Haddox *et al.* (1972) and Beavo *et al.* (1974) both found that the concentration of Mg^{2+}–ATP affects cAMP binding to protein kinase.

Corbin *et al.* (1970) and Huttunen *et al.* (1970) reported that triglyceride lipase preparations from rat adipose tissue are activated by a cAMP-dependent protein kinase. Activation of the lipase is associated with a transfer of phosphate from the γ-position of ATP to the enzyme protein. There is no evidence for participation of an intermediate enzyme analogous to phosphorylase *b* kinase in the activation of triglyceride lipase (Khoo *et al.*, 1973, 1974).

Khoo and Steinberg (1974) found that a partially purified triglyceride lipase

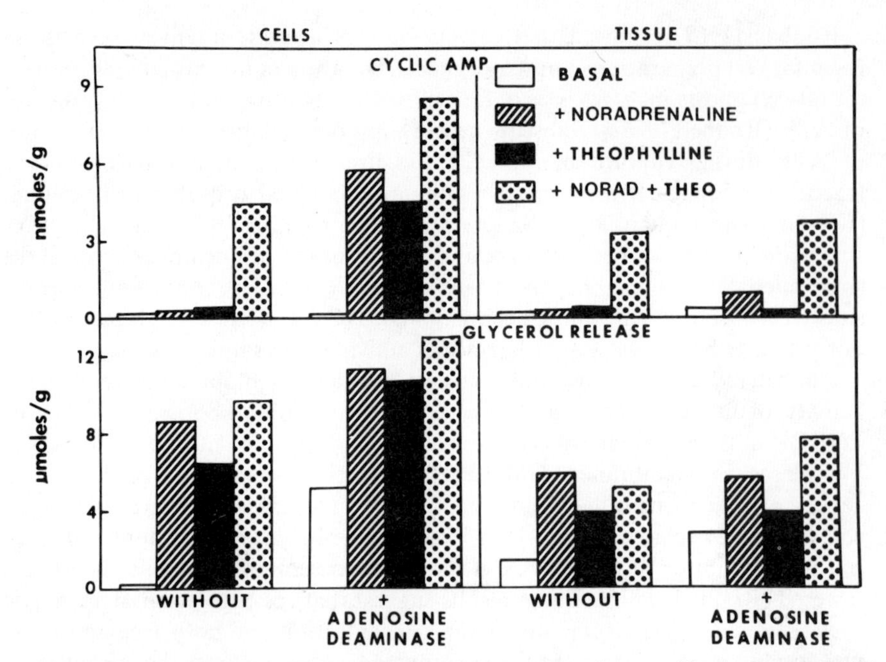

Figure 12.2 Comparison of adenosine deaminase as an activator of lipolysis and cAMP accumulation in adipose tissue and fat cells. Adipose tissue from 3 rats was minced. Aliqucts were incubated for 1 hr in buffer (4 ml) with or without collagenase (2 mg). Adipose tissue pieces (32–50 mg/tube) were then rinsed twice with fresh buffer and added to tubes. Fat cells (38 mg/tube) were also rinsed twice and added to tubes. Incubation was for 20 min with or without adenosine deaminase (0.5 μg/ml). Noradrenaline concentration = 1.5 μM. Theophylline concentration = 100 μM. $N = 4$. (Data from Fain, unpublished results)

from chicken adipose tissue is activated by addition of cAMP and muscle protein kinase; activation is four to ten-fold. The fully activated enzyme is de-activated by dialysis but can be re-activated by the addition of cAMP and Mg^{2+}–ATP. The reversible de-activation cycle can be repeated several times and is presumed to reflect the presence of a lipase phosphatase.

There is a reciprocal relationship between the activity of the hormone-sensitive triglyceride lipase and lipoprotein lipase. Robinson and Wing (1970) and Patten (1970) suggested that this occurs by a mechanism similar to that for regulation of glycogen synthetase and phosphorylase. The reciprocal relationship of the two lipases may reflect changes in cAMP with the same enzyme responsible for activation of triglyceride lipase and inactivation of lipoprotein lipase. This would make sense as lipoprotein lipase is activated by insulin and hydrolyses plasma lipoproteins enabling fatty acids to be taken up for storage into triglycerides of the fat cells. In contrast, the hormone-sensitive lipase which hydrolyses triglycerides to fatty acids during starvation is inactivated by insulin. However, it is unlikely that insulin affects the activity of

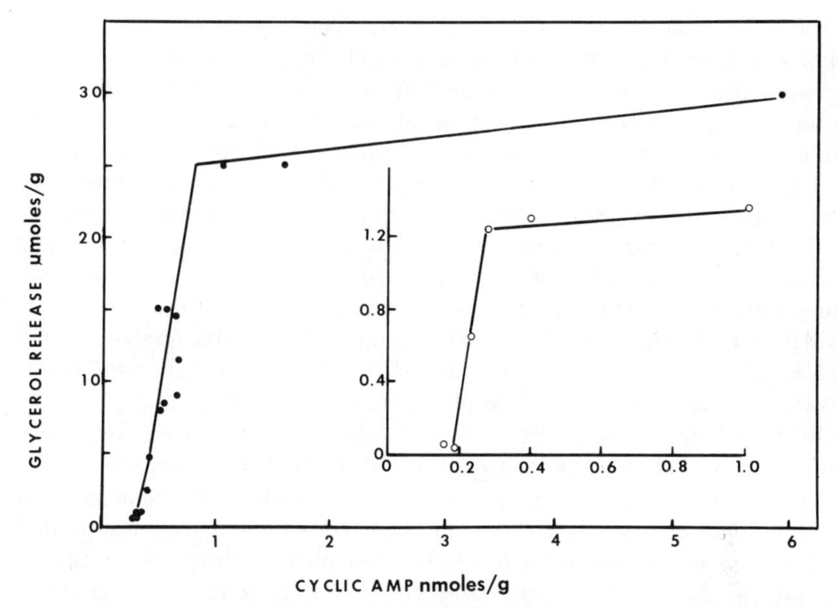

Figure 12.3 Correlation between cAMP accumulation and lipolysis. Data in the main figure are based on 3 paired studies in which fat cells (20 mg/ml) were incubated in 3% albumin containing glucagon (50 mg/ml) and varying concentrations of theophylline or 1-methyl-3-isobutyl xanthine. cAMP was measured after incubation for 10 min and glycerol after 60 min at 37° C. Data in the insert are redrawn from a figure in Butcher and Baird (1969), and are for 5 experiments in which incubated adipose tissue (ca. 50 mg/ml) was incubated for 24 min at 37° C in buffer without glucose or albumin. Tissues incubated for 24 min at 37°C in fresh buffer without glucose or albumin but with varying concentrations of catecholamines and caffeine. cAMP levels in the tissue and glycerol release to the medium were then determined and are in the same units as the main part of the figure

either lipase through changes in cAMP levels. Furthermore, Steinberg and Khoo (1976) showed that lipoprotein lipase is not affected by the cAMP-dependent protein kinase and that the two lipases are distinct enzymes in adipose tissue. Passage of fat-cell extracts through heparin-sepharose affinity columns removes all the lipoprotein lipase activity. The lipase activities against triolein, diolein, monolein and cholesterol oleate have been inseparable to date and are all enhanced by cAMP-dependent protein kinase (Steinberg and Khoo, 1976).

The major evidence against cAMP being the sole mediator of catecholamine-activated lipolysis is the lack of correlation between levels of cAMP and lipolysis. Incubation of fat cells or adipose tissue with either noradrenaline or theophylline alone stimulates lipolysis but does not noticeably increase cAMP levels (Figure 12.2). Adenosine deaminase also increases lipolysis without elevating cAMP levels in fat cells. In the presence of theophylline or adenosine deaminase, there is a marked accumulation of cAMP due to catecholamine.

The correlation between cAMP accumulation and lipolysis occurs over such a small concentration range that it is difficult to see. Figure 12.3 has been prepared from a series of experiments in which cAMP accumulation at ten minutes in the absence and presence of varying concentrations of methylxanthines was correlated with lipolysis over one hour. Near maximal activation of lipolysis is seen by increasing the cAMP concentration from 0.3 to 0.6 nmoles/g. These data support the hypothesis that cAMP is involved in lipolysis if a small increase over the basal level is all that is necessary for maximal activation of lipolysis. This was first pointed out by Butcher and Baird (1969) whose data are shown in the insert of Figure 12.3. In this case, a rise in fat-pad cAMP from 0.18 to only 0.40 nmole/g maximally activates lipolysis. Butcher and Baird (1969) suggested that values for cAMP greater than 0.4 nmole/g represent redundant cAMP of no physiological significance.

Other evidence suggests that either an unknown factor acts synergistically with cAMP or that there are multiple pools of cAMP. If a substantial fraction of the basal cAMP content of fat cells is in physiologically inactive compartments then one could increase the active pool of cAMP without detecting a rise in the basal level. The large amounts of cAMP (equivalent to about 25% of the total amount of ATP), which can be seen in the presence of a millimolar concentration of methylxanthines and pharmacological concentrations of hormones, probably have no physiological relevance (Fain, 1973a; Stock and Prilop, 1974). Concentrations of hormones or other agents which elevate lipolysis to the half-maximal level may produce no detectable rise in cAMP. Conversely, drugs or hormones, such as insulin, which lower cAMP may inhibit lipolysis by mechanisms unrelated to their effects on cyclic nucleotides.

The difference between the anti-lipolytic action of insulin and that of an inhibitor of cAMP accumulation, such as adenosine, is shown in Figure 12.4. N^6-Phenylisopropyladenosine was used since it is a derivative of adenosine which is more potent an inhibitor of cAMP accumulation than adenosine itself and cannot be deaminated (Fain, 1973b). In fat cells, N^6-phenylisopropyl-adenosine lowers cAMP to a greater extent than does insulin yet the derivative is not anti-lipolytic (Figure 12.4). N^6-Phenylisopropyladenosine also lowers both lipolysis and cAMP accumulation in incubated adipose tissue. Insulin, on the other hand, markedly decreases lipolysis in fat cells but reduces cAMP to a lesser extent than the adenosine analogue. In intact adipose tissue, insulin lowers cAMP but does not affect lipolysis. The lack of anti-lipolytic action by insulin in incubated adipose tissue is well established (Fain, 1973a). The opposite effect was seen with N^6-phenylisopropyladenosine, which lowers noradrenaline induced lipolysis in intact adipose tissue but not in fat cells. These results illustrate that the total cAMP values do not correlate well with lipolysis. It could be argued that it is inappropriate to expect a correlation between cAMP values measured at the end of a twenty-minute incubation and lipolysis over the same period. However, similar results have been reported with regard to the effects of insulin and adenosine on fat cells in which cAMP was measured after two minutes (Fain, 1973b).

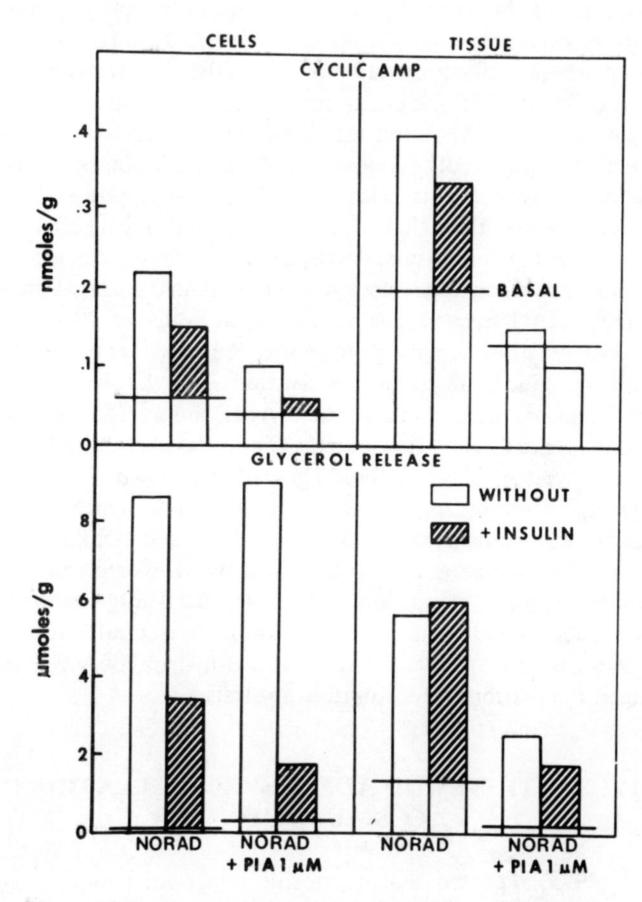

Figure 12.4 Comparison of N^6-(phenylisopropyl)adenosine and insulin as inhibitors of lipolysis and cAMP accumulation in adipose tissue and fat cells. Pooled adipose tissue from 3 rats was minced. Aliquots were incubated for 1 hr in buffer (4 ml) with collagenase (2 mg) or with buffer without collagenase. After the hour incubation tissue was rinsed twice with fresh buffer and added to tubes. The fat cells were also rinsed twice and added to tubes. The fat cells (38 μmoles of triglyceride per tube) and the tissue (32–50 μmoles of triglyceride per tube) were incubated for 20 min. with noradrenaline (NORAD) (1.5 μM) in the absence or presence of 1 μMN^6-phenylisoprophyl-adenosine (PIA) either without (open bars) or with insulin (120 μU/ml) (stippled bars). The values of lipolysis and cAMP in the absence of noradrenaline are indicated by the line across the bars either in the absence or presence of PIA. The values are the means of four paired experiments. (Data from Fain, unpublished results)

Westermann *et al.* (1969) showed that *in vitro* N^6-phenylisopropyl-adenosine at concentrations of between 1 and 10 nM inhibits lipolysis due to low concentrations of catecholamine. Unlike a β-blocker, the maximal inhibition of lipolysis due to N^6-phenylisopropyladenosine is 50% (Westermann and Stock, 1970; Stock and Prilop, 1974). Similar results are seen with PGE_1. For example, Wieser and Fain (1975) found that N^6-phenylisopropyladenosine, PGE_1 and nicotinic acid all markedly reduce cAMP accumulation by rat fat cells in the presence of lipolytic agents. However, the anti-lipolytic action of these compounds is no greater than that of insulin, which has little effect on cAMP accumulation. Possibly adenosine, PGE_1 and nicotinic acid are all potent inhibitors of lipolysis when adenylate cyclase is activated to only a small extent. However, when adenylate cyclase is maximally activated, a 99% inhibition of adenylate cyclase may not be enough to reduce the cAMP concentration below levels which produce maximal activation of lipolysis. Adenosine, prostaglandins and nicotinic acid all lower cAMP in intact fat cells but do not inhibit adenylate cyclase in cell-free systems. The reason for the failure to see any inhibition of adenylate cyclase by these agents is unknown.

Another group of inhibitors of cAMP accumulation, which appears to act like prostaglandins, adenosine and nicotinic acid, are short-chain acids, such as lactic and β-hydroxybutyric acids. Both lactate and β-hydroxybutyrate inhibit lipolysis under certain conditions. Lactate markedly inhibits cAMP accumulation under conditions in which lipolysis is unaffected (Fain and Shepherd, 1976). Similar results are seen with β-hydroxybutyrate except that after prolonged incubation it does inhibit lipolysis.

III METHYLXANTHINE AND ADENOSINE REGULATION OF cAMP ACCUMULATION

Fain *et al.* (1972) reported that adenosine is a potent inhibitor of cAMP accumulation and lipolysis by rat fat cells. Adenosine (73 μM) inhibits the activation of fat-cell ghost adenylate cyclase due to 200 μM adrenaline by about 22%. In contrast, 2′,5′-dideoxyadenosine (5 μM) inhibits adenylate cyclase by 50%; but N^6-phenylisopropyladenosine is inactive as an inhibitor of adenylate cyclase even at a concentration of 73 μM. However, as little as 0.05 μM N^6-phenylisopropyladenosine inhibits cAMP accumulation by 50% in intact cells while 2 μM 2′,5′-dideoxyadenosine gives only a 30% inhibition of cAMP accumulation (Fain, 1973b). These data suggest that the mechanism by which adenosine inhibits cAMP accumulation in intact cells is different from that for inhibition of adenylate cyclase in ghosts. This conclusion is based on the differences between the effects of N^6-phenylisopropyladenosine and 2′,5′-dideoxyadenosine, and the finding that 100 to 500 times as much adenosine is required to inhibit adenylate cyclase activity of fat-cell ghosts by 50% as is required to reduce cAMP accumulation by 50% in intact cells. The mechanism by which adenosine inhibits cAMP accumulation in intact fat cells is not known, but it is similar to that for nicotinic acid and prostaglandins of the E series which do not inhibit adenylate cyclase in cell-free systems (Fain, 1973a).

Evidence that adenosine might regulate fat-cell cAMP accumulation came from the work of Schwabe *et al.* (1973). Previously Schwabe and Ebert (1972) had noted that if a small number of fat cells (20,000 cells/ml, equivalent to about 4 mg of fat cells) are incubated with isoprenaline, there is a large accumulation of cAMP. Accumulation is not enhanced by the presence of 1 mM theophylline. In contrast, if a high level of fat cells is incubated (100,000 cells/ml) there is little accumulation of cAMP due to isoprenaline alone but theophylline is now able to markedly potentiate cAMP accumulation. Schwabe and Ebert (1972) suggested that an inhibitory factor is released into the medium whose action is blocked by methylxanthines.

Schwabe *et al.* (1973) subsequently reported that the inhibitory factor is adenosine. If a high enough concentration of fat cells is incubated, adenosine rapidly accumulates in the medium, which then inhibits the rise in cAMP accumulation due to lipolytic agents. Schwabe *et al.* (1973) found no evidence that adenosine release is stimulated by catecholamines.

The net accumulation of adenosine in the medium depends on the balance between release from cells and its disappearance from the medium. The main factor regulating adenosine release is the concentration of fat cells. Disappearance depends on two factors. The first is uptake and incorporation of adenosine into nucleotides by fat cells, which is probably related to the concentration of cells. The second factor is deamination of adenosine to inosine. Chicken fat cells are capable of deaminating much more adenosine than are rat fat cells. Thus, much of the conflicting data with respect to the time of onset, maximal value and the subsequent drop in cAMP concentration may result from species differences in the accumulation of adenosine.

If adenosine released during the incubation of fat cells is important in feedback control, then its removal by the addition of adenosine deaminase to the medium should enhance cAMP accumulation and lipolysis. Adenosine deaminase has been shown to be a potent stimulator of rat fat-cell lipolysis at concentrations as low as 10–100 ng/ml. (Fain, 1973b; Schwabe and Ebert, 1974; Fain and Wieser, 1975). Adenosine deaminase also markedly increases the level of cAMP when activators of adenylate cyclase are added to fat cells (Figure 12.2). The effects of adenosine deaminase are blocked by an equimolar concentration of coformycin, an extremely potent inhibitor of the enzyme (Fain and Wieser, 1975). Furthermore, N^6-phenylisopropyladenosine reverses the effects of adenosine deaminase.

From data in Figure 12.1, it can be seen that adenosine is an extracellular inhibitor of cAMP accumulation whose action is antagonized by methylxanthines. The rate of adenosine release to the medium may be dependent on the concentration of 5′-AMP and the activity of 5′-nucleotidase.

Angel *et al.* (1971) and Cushman *et al.* (1973) have shown that ATP levels are decreased if high levels of intracellular free fatty acids are obtained by incubating fat cells with lipolytic agents in the absence of albumin and glucose. Presumably these conditions would accelerate adenosine release.

Methylxanthines are potent inhibitors of cAMP phosphodiesterase activity in fat cells (Fain, 1973a), and they are also potent activators of fat-cell lipolysis. There is a good correlation between the effects of over sixty xanthine compounds on lipolysis and cAMP hydrolysis (Beavo *et al.*, 1970).

Dipyridamole and papaverine are much more potent than theophylline as inhibitors of cAMP phosphodiesterase in fat-cell homogenates (Schwabe *et al.*, 1972; Fain, 1973b). However, there is only a very small potentiation of the rise in cAMP due to catecholamines in intact fat cells in the presence of dipyridamole or papaverine (Schwabe *et al.*, 1972; Fain, 1973b). There are several possible explanations for this discrepancy, one being that papaverine and dipyridamole do not affect the metabolism of intact fat cells. However, this appears unlikely since papaverine actually inhibits lipolysis in fat cells (Fain, 1973b). Another possibility is that the cAMP phosphodiesterase activity in broken-cell preparations at cAMP concentrations in the range of 0.1 to 1 μM does not reflect the activity of this enzyme in intact cells. It is more likely that papaverine and dipyridamole have multiple effects on fat cells. It should be noted that caution must be exercised in extrapolating from effects of substances on isolated enzymes to the situation in the intact cell, as in the case of F^-. This anion is a potent stimulator of adenylate cyclase in broken-cell preparations but inhibits cAMP accumulation in intact cells.

In chicken fat cells, both theophylline and adenosine deaminase are relatively ineffective as activators of cAMP accumulation and lipolysis (Malgieri *et al.*, 1975). Chicken fat cells accumulate less adenosine in the medium during incubation with lipolytic agents than do rat fat cells. This probably is the result of the finding that chicken fat cells are capable of deaminating five to ten times more adenosine to inosine than are rat fat cells. In other words, there may be so much endogenous adenosine deaminase activity in chicken fat cells that there would be very little further effect by adding deaminase to the medium. The large stimulation of cAMP accumulation due to lipolytic agents in chicken fat cells probably results from the absence of sufficient amounts of adenosine to inhibit the adenylate cyclase. However, cAMP accumulation in chicken fat cells can be readily inhibited by the addition of N^6-phenylisopropyladenosine which is resistant to deamination. An increase in cAMP accumulation due to methylxanthines in chicken fat cells can only be seen in the presence of added adenosine.

Is the antagonism between methylxanthines and adenosine with regard to cAMP accumulation secondary to the inhibition of adenosine uptake by fat cells? We found that methylxanthines markedly inhibited adenosine uptake by isolated rat fat cells. Both theophylline (200 μM) and 1-methyl-3-isobutylxanthine (4 μM) inhibit adenosine uptake by rat fat cells by 50%. The failure of methylxanthines to have much of an effect on cAMP accumulation in chicken fat cells involves more than a lowered net accumulation of adenosine by chicken compared with rat fat cells. Methylxanthines are less effective as inhibitors of the soluble cAMP phosphodiesterase activity of homogenized chicken fat cells as compared with rat fat cells. Thus, the combination of less net adenosine accumulation by chicken fat cells and less inhibition of cAMP phosphodiesterase activity results in fat cells which respond poorly to methylxanthines.

Adenosine is not the sole feedback regulator of cAMP accumulation released during incubation of rat fat cells with lipolytic agents. Schwabe and Ebert (1974) found that even in the presence of adenosine deaminase the rise in cAMP due to noradrenaline plus theophylline returns to basal values within one hour. Fain and Shepherd (1975) found a potent inhibitor of cAMP accumulation whose release into the medium is seen only in the presence of lipolytic agents and which cannot be inactivated by adenosine deaminase. This inhibitor is also different from adenosine in that it is non-dialysable.

IV FEEDBACK REGULATION OF ADENYLATE CYCLASE BY FATTY ACIDS

Ho and Sutherland (1971) and Manganiello *et al.* (1971) reported that incubation of rat fat cells with lipolytic agents results in the formation of antagonists or feedback regulators which block the ability of cells to elevate cAMP in response to a second addition of hormone. The formation of the antagonists is enhanced by incubation with adrenaline, ACTH, glucagon or Bt_2-cAMP. The antagonists which accumulate in the presence of adrenaline block the subsequent response to ACTH and glucagon as well as that to the catecholamine.

The response of cells which have been incubated with lipolytic agents is restored by washing the cells and then incubating them in fresh buffer (Ho and Sutherland, 1971; Manganiello *et al.*, 1971). This suggests that the antagonist is released into the medium. The stimulation of cAMP due to catecholamines is markedly reduced if fat cells are incubated in medium which has previously been exposed to fat cells and catecholamines. Subsequent work has confirmed that adenylate cyclase inhibitors are released to the medium by rat fat cells incubated with lipolytic agents (Fain and Shepherd, 1975). The antagonists do not act by increasing cAMP phosphodiesterase activity (Pawlson *et al.*, 1974) which is in agreement with the initial report of Manganiello *et al.* (1971). There is some disagreement about the chemical nature of the feedback regulators of adenylate cyclase which are released into the medium during incubation of rat fat cells with lipolytic agents. Prostaglandins, adenosine, free fatty acids and a compound of unknown structure have all been suggested as possible feedback regulators of adenylate cyclase.

Illiano and Cuatrecasas (1971) postulated that prostaglandins were physiologically important feedback regulators of lipolysis based on studies with indometacin. However, other workers have been unable to find any potentiation of lipolysis or cAMP accumulation in fat cells by indometacin (Fain *et al.*, 1973; Dalton and Hope, 1973). Indometacin is a drug which effectively blocks prostaglandin formation in rat fat cells (Dalton and Hope, 1974). Studies with prostaglandin antagonists have been difficult to interpret and their specificity of action remains to be established. Illiano and Cuatrecasas (1971) observed that SC-19220, an oxazepine derivative, enhances the lipolytic action of catecholamines. However, Radzialowski and Rosenberg (1973) found no effect of SC-19220 on the adrenaline-induced lipolysis but they found that it does antagonize the action of PGE_2.

Further evidence against the prostaglandin theory of feedback regulator was the report by Fredholm and Rosell (1970). These workers showed that the amounts of prostaglandins belonging to the E series released during and after stimulation of nerves supplying canine subcutaneous adipose tissue are lower than the amounts required for inhibition of lipolysis. Bowery and Lewis (1973) found that indometacin inhibits prostaglandin formation and functional vasodilation in rabbit adipose tissue but does not affect lipolysis during ACTH infusion. The only recent report suggesting that prostaglandins are feedback regulators is based on chronic treatment of rats with a new inhibitor of prostaglandin biosynthesis, L8027 (pyridylindolyl ketone) (Lefebvre and Luyckx, 1974). Whether the results are indeed due to inhibition of prostaglandin biosynthesis is not known, as acute treatment of rats with indometacin actually decreases the plasma free fatty acid values two or five hours later (Deby and Bacq, 1972). These data suggest it is unlikely that prostaglandins are important feedback regulators of adenylate cyclase in fat cells.

Gorman (1975) has suggested a modification of this hypothesis with prostaglandin endoperoxide intermediates acting as feedback regulators. Gorman *et al.* (1975) reported that 0.28–28 μM PGH_2 (15-hydroxy-9-peroxidoprosta-5,13-dienoic acid) inhibits the adenylate cyclase activity of fat cell ghosts. However, the real question is whether indometacin also inhibits the formation of the endoperoxides. In tissues in which this has been tested, indometacin blocks formation of endoperoxides as well as that of prostaglandins (Hamberg and Samuelson, 1974). Furthermore, endoperoxides in platelets have functions which are opposite to those of the prostaglandins of the E series.

The feedback regulators of Ho and Sutherland (1971) and of Fain and Shepherd (1975) inhibit adenylate cyclase as do prostaglandin endoperoxides. In contrast PGE_1 and PGE_2 have never been found to inhibit adenylate cyclase (Fain, 1973a). The addition of arachidonic acid to the medium should result in a large increase in endoperoxide formation. However, when this acid is added to incubated fat cells it is no more effective than an equimolar concentration of oleate or any other long chain fatty acid (unpublished results). Dalton and Hope (1973) found that 40 μM arachidonic acid does not affect basal or hormone-stimulated lipolysis by rat fat cells when incubated for one hour.

Possibly, exogenous addition of arachidonic acid does not result in endoperoxide formation. Thus, there is still a chance that endoperoxides formed from endogenous arachidonic acid are feedback regulators of adenylate cyclase. Christ and Nugteren (1970) reported that arachidonic acid accounts for about 5% of the fatty acids released during hormone-stimulated lipolysis in incubated adipose tissue. Dalton and Hope (1974) found that release of PGE_2 is about 0.1 nmole/g of fat cells during incubation of the cells with lipolytic agents. Basal release is about one-tenth of this value. Dalton and Hope (1974) suggested that the PGE_2 comes from the cAMP activation of a phospholipase, since arachidonic acid accounts for 18% of phospholipid fatty acids (Dalton and Hope, 1974). However, arachidonic acid could have arisen during triglycerid hydrolysis since 2.4 μmoles of arachidonic acid would have been accumulated if the amount of fatty acid released in the presence of hormones was 240 μmoles per gram over one hour and 1% of triglyceride fatty acids were accounted for as arachidonic acid. The prostaglandins formed from arachidonic acid released during lipolysis most likely act as vasodilators, increasing blood flow through adipose tissue. This may enhance lipolysis by preventing saturation of plasma albumin with fatty acids.

During the activation of lipolysis by hormones, such as catecholamines and ACTH, the fatty acids released into the medium appear to be the major feedback regulators of adenylate cyclase. The important factor is the ratio of free fatty acid to albumin within the medium. Rodbell (1965) first demonstrated that lipolysis by isolated fat cells in the presence of hormones virtually ceased when the ratio of free fatty acids/albumin is greater than three. What was not appreciated then was that if the primary binding sites on albumin in the medium were saturated, a further increase in free fatty acids would inhibit adenylate cyclase.

The first report to show inhibition of cAMP accumulation by free fatty acids was that of Burns *et al.* (1975). After incubating human fat cells with medium containing enough sodium oleate to give a free fatty acid/albumin ratio of 2.5 to 3, they found marked inhibition of cAMP accumulation. Furthermore, there is little stimulation of cAMP accumulation in human fat cells by lipolytic agents in the absence of albumin from the medium. Unlike rat fat cells (Fain and Shepherd, 1975), there is no direct inhibition adenylate cyclase from human fat cells by free fatty acids (Burns *et al.*, 1975). However, in ghosts prepared from human fat cells previously incubated with high fatty acid/albumin ratios there is a reduced sensitivity to adrenaline but not to F^- (Burns *et al.*, 1975).

Medium in which rat fat cells have previously been incubated with lipolytic agents or added free fatty acids markedly inhibits cAMP accumulation by a fresh batch of fat cells (Fain and Shepherd, 1975). The medium from cells incubated with lipolytic agents also inhibits adenylate cyclase of fat cell ghosts. The addition of sufficient sodium oleate to raise the free fatty acid/albumin ratio above two inhibits adenylate cyclase; maximal inhibition is seen at a ratio of five.

Further support for the hypothesis that free fatty acids may be important feedback regulators of adenylate cyclase comes from studies with chicken fat cells (Malgieri *et al.*, 1975). In chicken fat cells, glucagon alone produces a large rise in cAMP which is maintained for up to one hour. Catecholamines are ineffective as activators of lipolysis or cAMP accumulation in chicken fat cells. Furthermore, cAMP accumulation in chicken fat cells is little affected by the addition of theophylline (Malgieri *et al.*, 1975) or adenosine deaminase.

The addition of enough sodium oleate to raise the free fatty acid/albumin ratio to eight does not affect the level of cAMP in chicken fat cells under conditions in which cAMP accumulation by rat fat cells is markedly reduced. Ratios as high as twelve do not inhibit adenylate cyclase activity of chicken fat-cell ghosts (Malgieri *et al.*, 1975). In both chicken and rat fat cells, there is product inhibition of triglyceride lipase by free fatty acids, but inhibition of chicken fat cell lipase requires higher free fatty acid/albumin ratios than are necessary for rat fat cells. Chicken fat cells differ from the rat cells in that free fatty acids are not feedback regulators of adenylate cyclase. There is no significant inhibition by fatty acids of protein kinase or cAMP phosphodiesterase activities of either chicken or rat fat cells by low concentrations of free fatty acids in the absence of albumin (Malgieri *et al.*, 1975).

A perplexing problem at the moment is the relationship of free fatty acids to the feedback regulator of adenylate cyclase originally described by Ho and Sutherland (1971). The simplest explanation is that they are the same substance. The feedback regulator binds tightly to albumin and is nondialysable, but can be extracted with a lipid solvent (80% ethanol). This situation is what one would expect of free fatty acids. However, Ho *et al.* (1975b) has claimed that their feedback regulator could be separated from free fatty acids by partition chromatography on Sephadex LH–20 columns using a mobile phase of acetone:ethanol:heptane:water (4:50:1:9). The chromatographically purified regulator was like fatty acids in that it gave opaque suspensions which were heat stable. However, it was acid-labile since 90% of the antagonist activity was lost upon exposure to pH 1 for only ten seconds (Ho *et al.*, 1975a). One explanation is that the feedback regulator described by Ho is an artifactual inhibitor formed from free fatty acids during isolation and purification. This would account for the appearance of the regulator during the stimulation of fat cells by lipolytic agents, including Bt_2-cAMP (Ho *et al.*, 1975b). Perhaps there was no appearance of feedback regulator in the absence of lipolytic agents, since fat cells were incubated in albumin freed of fatty acids by extraction with petroleum ether and acetic acid (Ho *et al.*, 1975b,c).

In contrast, Fain and Shepherd (1975) found that a potent inhibitor of cAMP accumulation could be obtained by the original procedure of Ho and Sutherland (1971) after incubation of fat cells in the absence of lipolytic agents with non-defatted albumin in the medium. In fact, the presence of fat cells is not necessary to obtain an inhibitor of cAMP accumulation, as it is sufficient to dialyse albumin extensively against distilled water and then to lyophilize the albumin (Fain and Shepherd, 1975). Whether the inhibitor was formed from fatty acids present in the albumin is not known. What is clear is that artifactual inhibitors of cAMP accumulation can be readily formed by treatment of albumin. Albumin is the only known acceptor for the fatty acids released during lipolysis and there is little lipolysis or cAMP accumulation by fat cells incubated in its absence. Therefore, the problem cannot be overcome by the omission of albumin.

The feedback regulator described by Ho *et al.* (1975b,c) could be a physiologically important compound quite distinct from either free fatty acids or adenosine. Since this regulator is apparently a lipid-like substance of low molecular weight, its relationship to prostaglandins is of interest. Ho and Sutherland (1971) claim that the feedback regulator differs from prostaglandins in that it is not inactivated by 15-hydroxyprostaglandin dehydrogenase, is more stable than PGE_1, can be separated from PGE_1 using column chromatography and inhibits fat-cell adenylate cyclase. However, the feedback regulator could be one of the endoperoxides or thromboxanes formed from arachidonic acid rather than a prostaglandin of the E series. Gorman *et al.* (1975) have shown that, at fairly high concentrations, endoperoxides formed from arachidonic acid are inhibitors of fat cell adenylate cyclase. Arachidonic acid released during lipolysis could be converted to a more potent inhibitor of adenylate cyclase than the parent compound is. However, no effect of indometacin on the accumulation of feedback inhibitors of cAMP accumulation has been observed during incubation of rat fat cells with lipolytic agents (Fain, unpublished results).

Schimmel (1974) has looked at feedback regulation of cAMP accumulation in incubated segments of rat adipose tissue. Results obtained are different from those using isolated fat cells. No antagonists are released into the medium from tissue segments during incubation with lipolytic agents. Fatty acids released into the medium do not accumulate in amounts which can inhibit cAMP accumulation since the maximal fatty acid/albumin ratio for the released fatty acids is one. Furthermore, it is unlikely that appreciable amounts of adenosine accumulate in the medium. Studies comparing intact adipose tissue with incubated fat cells carried out in the author's laboratory have shown little response in intact adipose tissue to adenosine deaminase (Figure 12.2). One difference between rat fat cells and intact tissue is the presence in the latter of substantial amounts of capillary endothelial cells and erythrocytes. These cells appear to deaminate adenosine much more rapidly than do lipocytes.

Schimmel (1974) found that rat adipose tissue which had been exposed for one hour to ACTH or adrenaline responds poorly with respect to cAMP accumulation to a second addition of either agent even after transferring the tissue to fresh medium. Exposure for one hour to a concentration of theophylline which does not appreciably elevate cAMP levels but gives an increase in lipolysis equivalent to that seen with either ACTH or adrenaline, fails to reduce the subsequent rise in cAMP due to adrenaline. Furthermore, the inhibition due to prior exposure to adrenaline is reversed if PGE_1 or nicotinic acid is also present. These data suggest that prior activation of adenylate cyclase in intact adipose tissue is responsible for the poor response to a second addition of activators of this enzyme.

V ACTIVATION OF LIPOLYSIS AND ADENYLATE CYCLASE BY CHOLERA TOXIN

After a lag period of about an hour, cholera toxin activates adenylate cyclase irreversibly by a process independent of protein synthesis. There is nothing unique about the activation of adenylate cyclase activity in fat cells compared with other cells. Cholera toxin appears to universally activate adenylate cyclase in mammalian cells.

The initial step in cholera toxin's action involves rapid binding to a monosialoganglioside (GM_1) on the cell surface (King and van Heyningen, 1973). The toxin is composed of two major subunits; the larger subunit of 54,000 daltons binds to GM_1. A lag period appears to be required for the release of the smaller subunit of about 32,000 daltons, which is hydrophobic and thus insoluble in aqueous solutions, except in the presence of detergent. Alternatively, the small subunit passes through the plasma membrane to bind with adenylate cyclase, probably indirectly. In membranes from tumour cells or pigeon erythrocytes, the 32,000 subunit appears to be cleaved into further subunits of 25,000 and 7,000 daltons which are held together by disulphide bonds (Gill, 1975; Bitensky *et al.*, 1975). The subunit of about 7,000 daltons is probably the link between the 25,000 and the 55,000 subunits since thiolytic cleavage of intact cholera toxin releases an insoluble protein of 25,000 daltons (Sattler and Wiegand, 1975). Cleavage of the 55,000 subunit from the 32,000 one can be accomplished by detergents or high concentrations of urea, indicating that the subunits are bound by non-covalent bonds. A hypothesis to explain cholera toxin action is shown in Figure 12.5. In the model, the 32,000 subunit dissociates from the toxin–ganglioside complex and diffuses through the lipid phase of the plasma cell membrane to its inner surface. At the inner surface of the membrane, the 25,000 subunit is released into the medium where it binds to and activates adenylate cyclase.

The nature of the activated state of fat-cell adenylate cyclase was investigated by Bennett *et al.* (1951a). They found that in the particulate fraction obtained after Polytron homogenization of fat cells which have been exposed to cholera toxin for at least an hour there is a marked enhancement of adenylate cyclase activity. The adenylate cyclase activity of membranes from toxin-treated fat cells is more responsive to stimulation by GTP than is the enzyme from membranes of control cells. The basal activity of adenylate cyclase is five to tenfold higher in membranes from cells exposed to toxin and the apparent affinity for stimulation of the cyclase by isoprenaline, ACTH, glucagon and vaso-active intestinal polypeptide is increased by a factor of two (Bennett *et al.*, 1975a).

Sahyoun and Cuatrecasas (1975) found that addition of cholera toxin to particulate preparations obtained from fat-cell homogenates results in increased adenylate cyclase activity after a lag period of about twenty-five minutes. This effect is seen only if the particulate fraction is incubated during this time in the presence of ATP and a regenerating system for the ATP. Actually, the toxin prevents the loss of activity which occurs during the two-hour incubation

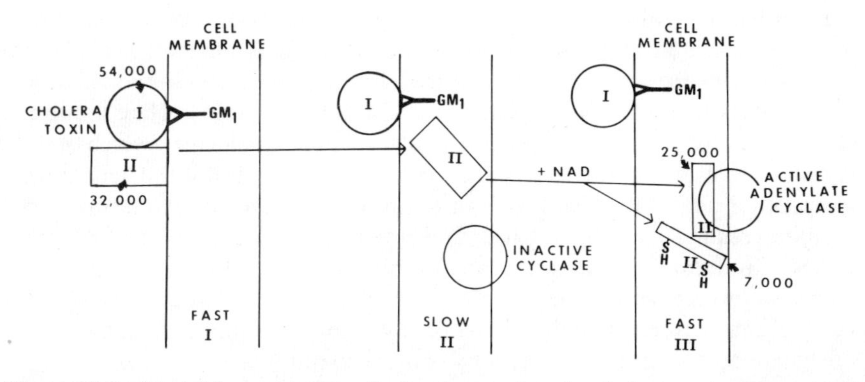

Figure 12.5 Model for activation of adenylate cyclase by cholera toxin. The model is based on the structure and molecular weight of the components of cholera toxin as proposed by Sattler *et al.* (1975). The intact toxin approaches the cells and rapidly binds via its protein I to the ganglioside receptor (GM_1) on the outer face of the cell membrane. The long lag period in toxin action is necessary for transfer of the protein II to the inside of the lipid bilayer. The lag period is abolished when the toxin is added to broken-cell preparations. Protein II is hydrophobic which may facilitate its migration through the lipid bilayer of the plasma membrane. The last step in activation of cyclase is conversion of protein II to a form which binds irreversibly to adenylate cyclase and activates the enzyme. NAD or thiol reducing agents appear to be required for this step which may invoke thiolytic reduction of the disulphide bridges linking the two protein subunits of protein II

(Sahyoun and Cuatrecasas, 1975). The effect of the toxin is blocked with exogenous GM_1 or choleragenoid.

Another line of investigation has shown that the 'active' subunit dissolved in 0.05% Lubrol–PX activates adenylate cyclase when added to intact fat cells. The lag period is reduced to about fifteen minutes and the stimulation is not inhibited by choleragenoid or GM_1 (Sahyoun and Cautrecasas, 1975). After toxin activation, substantial precipitation of adenylate cyclase activity occurs with antisera against the active subunit (Sahyoun and Cuatrecasas, 1975; Bennett *et al.*, 1975a,b). These data support the hypothesis that the hydrophobic active subunit of cholera toxin binds to adenylate cyclase, resulting in activation of the enzyme and increased sensitivity to its activators. The work of Gill (1975) and Bitensky *et al.* (1975) has suggested that this process involves cleavage of the 32,000 subunit to give a protein which binds to adenylate cyclase.

Bennett *et al.* (1975b) have proposed that cholera toxin binds to cell-surface gangliosides and that there is mobility in the plane of the membrane which leads to patching and capping of cholera toxin–ganglioside complexes. This mobility in the lateral plane is thought to account for the lag period. Sayhoun and Cuatrecasas (1975) suggested that lateral mobility is required in order to obtain multivalent binding of toxin with membrane gangliosides for release of the active subunit into the membrane. If this occurs it is not by a process which involves formation of microfilaments or microtubules since colchicine, colcemide, vincristine, vinblastine and cytochalasin B do not inhibit cholera toxin as measured by the activation of fat-cell lipolysis (Bennett *et al.*, 1975a). The best evidence against the lateral mobility hypothesis for action of cholera toxin comes from the data of Craig and Cuatrecasas (1975). They found that concentrations of vinblastine, vincristine and colchicine which inhibit lymphocyte capping due to cholera toxin by 55–60% inhibit the activation of lymphocyte adenylate cyclase due to toxin by less than 11%. Concanavalin A inhibits capping by 80% but was said not to inhibit activation of adenylate cyclase. Sodium azide inhibits capping by more than 96% but only inhibits adenylate cyclase activation by 55%. These data suggest that there is very little correlation between capping induced by cholera toxin and activation of adenylate cyclase in lymphocytes by cholera toxin.

VI LIPOLYTIC ACTION OF GROWTH HORMONE AND GLUCOCORTICOIDS

Cholera toxin, in common with growth hormone and glucocorticoids, shares the ability to activate lipolysis and increase cAMP accumulation after a lag period of one to two hours. One key difference, however, is that the lipolytic action of cholera toxin is not blocked by inhibitors of RNA or protein synthesis, such as puromycin, actinomycin D or cycloheximide (Cuatrecasas, 1973). Similar results are shown in Table 12.1 with respect to cycloheximide, a potent inhibitor of protein synthesis. Cycloheximide is able to block the increased lipolysis due to the combination of growth hormone with glucocorticoid but not that due to cholera toxin (see Table 12.1).

Much less is known of the processes involved in activation of lipolysis by growth hormone or glucocorticoid as compared to cholera toxin. Incubation of fat cells with either growth hormone or glucocorticoid for 3.5 hours prior to preparation of ghosts results in a 40 and 20% increase, respectively, in the stimulation of adenylate cyclase by maximal concentrations of noradrenaline (Fain and Czech, 1975). There is no significant effect of prior incubation with growth hormone or glucocorticoid on basal, ACTH or F^--stimulated adenylate cyclase activity (Fain and Czech, 1975). In fat cells incubated with cholera toxin, adenylate cyclase activity of washed particulate fractions is greatly increased along with the sensitivity to all hormones (Bennett *et al.*, 1975a).

Growth hormone increases glycogen phosphorylase activity and cAMP accumulation by rat fat cells, and these effects are blocked by puromycin and cycloheximide (Moskowitz and Fain, 1970). Similar increases of cAMP were noted in other studies after incubation with growth hormone, but the presence of dexamethasone does not result in any further rise in cAMP accumulation (Fain *et al.*, 1971; Fain and Saperstein, 1970). These results support the hypothesis hat the lipolytic action of growth hormone involves synthesis of a protein via DNA-dependent RNA synthesis which increases the sensitivity of adenylate cyclase to activation by hormones (Fain, 1973a).

Table 12.1 Failure of cycloheximide to block lipolytic action of cholera toxin

Addition	Basal	Δ due to cholera toxin (0.2 μg/ml)	Δ due to Dex (0.016 μg/ml) + GH (1.0 μg/ml)
		Glycerol release (mole/g)	
None	4.0	+ 25.5	+ 5.0
Cycloheximide (2 μg/ml)	6.5	+ 22.0	+ 0.3

Fat cells (30 mg/tube) were isolated and then incubated in 1 ml of 3% albumin buffer for 3 h. The values are the means of 2 paired experiments (Data from Fain, unpublished results).

The permissive effect of glucocorticoids on lipolysis occurs after a lag period and requires continuing protein synthesis. Since glucocorticoids do not increase cAMP accumulation it appears that they may act by potentiating the action of cAMP. Lamberts *et al.* (1975) have shown an increase in cAMP-dependent protein kinase activity in homogenates of fat-cell suspensions after incubation of fat cells with glucocorticoids. The increase in lipolysis due to glucocorticoids is also seen during maximal inhibition of cAMP phosphodiesterase activity by theophylline or Bt_2-cAMP (Lamberts *et al.*, 1975).

VII INSULIN AND cAMP PHOSPHODIESTERASE

The involvement of cAMP in the action of insulin on fat cells appears minimal. However, there is ample evidence that insulin can reduce cAMP both by direct inhibition of adenylate cyclase and by activation of cAMP phosphodiesterase. Butcher *et al.* (1966) originally reported that insulin can lower cAMP in adipose tissue. Jungas (1966) shortly afterwards reported that pretreatment of adipose tissue with insulin reduces adenylate cyclase and glycogen phosphorylase and increases glycogen synthetase activity of the tissue homogenates. Other investigators have shown direct inhibitory effects of insulin on adenylate cyclase activity of fat-cell ghosts (Hepp and Renner, 1972; Illiano and Cuatrecasas, 1972; Renner *et al.*, 1974).

There is substantial evidence that the cAMP phosphodiesterase activity of fat-cell homogenates is enhanced if the cells are exposed to insulin prior to homogenization (Loten and Sneyd, 1970; Manganiello and Vaughan, 1973; Zinman and Hollenberg, 1974; Sakai *et al.*, 1974; Solomon, 1975; Pawlson *et al.*, 1974; Kono *et al.*, 1975). However, no direct effects of insulin on the enzyme have been observed in cell-free preparations. The effect of insulin is primarily on the phospho-diesterase activity of particles derived from the endoplasmic reticulum rather than from the plasma membrane (Kono *et al.*, 1975).

One curious finding is that prior exposure of fat cells to adrenaline or other lipolytic agents, including Bt_2-cAMP, mimicks the action of insulin (Zinman and Hollenberg, 1974; Solomon, 1975; Pawlson *et al.*, 1974; Kono *et al.*, 1975). Even more unusual is the report by Solomon (1975) that 0.1 mM propranolol or 2 μM adenosine also activates cAMP phosphodiesterase activity. The only substances which are known to be increased by both adrenaline and insulin are cGMP and cytosol Ca^{2+}. Activation of cAMP phosphodiesterase possibly results from increases in cytosol Ca^{2+} and cGMP. The use of adenosine or high concentrations of propranolol as activators of cGMP has not been assessed, but carbachol, noradrenaline and insulin all elevate cGMP (Fain and Butcher, 1976).

There are some contradictions between various reports since Pawlson *et al.* (1974) reported that the stimulatory effects of ACTH and insulin are additive. However, Kono *et al.* (1975) found that the combined effect of noradrenaline and insulin is much less than that due to either agent alone. Solomon (1975) also reported that the combination of noradrenaline with either propranolol or adenosine gives less activation of the enzyme than is seen with any of the three agents alone.

What relationship does the inhibition of adenylate cyclase and activation of cAMP phosphodiesterase by insulin have to its physiological effects? While some reports have clearly shown an inhibition of net cAMP accumulation by fat cells in the presence of insulin (Butcher *et al.*, 1966; Siddle and Hales, 1974; Kono and Barham, 1973; Desai *et al.*, 1973), others have observed that under conditions in which insulin is just as antilipolytic there is no decrease in the cAMP content (Fain and Rosenberg, 1972; Khoo *et al.*, 1973; Knight and Iliffe, 1973). Wieset and Fain (1975) compared the antilipolytic action of insulin with those of PGE_1, nicotinic acid and adenosine. Although all four agents are antilipolytic, there is no inhibition of cAMP accumulation by insulin; the other agents markedly reduce cAMP levels. The available evidence indicates that an antilipolytic action of insulin can be seen under conditions in which total cAMP accumulation is unaltered. There is some interaction between insulin and cAMP in that the stimulation of glucose metabolism and inhibition of lipolysis by insulin is reduced under conditions in which cAMP is elevated (Wieser and Fain, 1975; Fain and Wieser, 1975).

Fain (1975) found that incubation of fat cells with insulin and lipolytic agents actually enhances the ability of fat cells to elevate cAMP accumulation after a second addition of adenylate cyclase activators. If fat cells are incubated with lipolytic agents under conditions in which the free fatty acids/albumin ratio exceeds three cAMP accumulation due to a second hormone addition is markedly reduced. If insulin is also present this ratio remains below three and there is no inhibition of cAMP accumulation. Incubation of fat cells with PGE_1 has quite different effects as it inhibits both lipolysis and the rise in cAMP accumulation (Fain, 1975).

Pawlson *et al.* (1974) pointed out that activation of cAMP phosphodiesterase by lipolytic agents is an unlikely explanation for feedback regulation of cAMP accumulation. The increase in cAMP phosphodiesterase due to lipolytic agents is transient and disappears after incubation for twenty minutes. This is the time at which feedback regulation of adenylate cyclase is maximal. Furthermore, insulin by inhibiting the activation of lipolysis by hormones prevents the accumulation of feedback regulators of adenylate cyclase and thus enhances, rather than inhibits, cAMP accumulation. Therefore, the relative effect of insulin on cAMP accumulation depends on a balance between its ability to lower cAMP by activating phosphodiesterase and its acceleration of feedback regulators.

VIII cGMP AND FAT-CELL LIPOLYSIS

Illiano *et al* (1973) reported that insulin and carbachol elevate cGMP in rat fat cells. Fain and Butcher (1975) and Vydelingum *et al.* (1975) found that insulin increases cGMP levels in fat cells. Fain and Butcher (1976) confirmed that carbachol, a long-lasting cholinergic agent, elevates cGMP. Furthermore, cGMP levels are also elevated in fat cells by the divalent cation ionophore A-23187, noradrenaline and oleate (Fain and Butcher, 1975).

The problem in involving any role of cGMP in the regulation of lipolysis is the lack of correlation between the nucleotide and lipolysis. Chlouverakis (1963) first reported that cholinergic agents do not affect fat-cell lipolysis. This was confirmed by Fain and Butcher (1976). Ionophore A-23187 has no inhibitory effect on lipolysis while noradrenaline activates lipolysis. There is no apparent correlation between lipolysis and cGMP since the latter is increased by agents which either activate, inhibit or have no effect on lipolysis (Fain and Butcher, 1976).

The increases in cGMP accumulation due to noradrenaline, carbachol, insulin, oleate and A-23187 are markedly inhibited by incubation of fat cells with the Ca^{2+} chelator EGTA (Fain and Butcher, 1975). However, lipolysis due to catecholamines is unaffected by incubation in Ca^{2+}-free buffer containing EGTA (Khoo *et al.*, 1973; Fain and Butcher, 1976). Vydelingum *et al.* (1975) observed that the increase in cGMP due to insulin is abolished in Ca^{2+}-free buffer as is that of leucine incorporation into protein. Their results suggest that insulin-induced protein synthesis may require the presence of extracellular Ca^{2+} and invoke cGMP accumulation. However, the antilipolytic action of insulin is not affected by omission of extracellular Ca^{2+} (Fain and Butcher, 1976).

The role of Ca^{2+} in the regulation of fat-cell metabolism has been difficult to establish. Berridge (1975a) suggested that a rise in cGMP reflects an increase in free cytosol Ca^{2+}. If this is actually the case it might explain the rise in cGMP seen after exposure of fat cells to a variety of agents. The elevation of cGMP in fat cells is clearly dependent on the presence of extracellular Ca^{2+}. However, lipolysis is little affected by the absence of extracellular Ca^{2+}.

The regulation of adipose tissue phosphorylase by cAMP is thought to be mediated through phosphorylation of a phosphorylase *b* kinase by protein kinase. The role of Ca^{2+} in the regulation of phosphorylase *b* kinase activity was originally considered to be of little consequence (Khoo *et al.*, 1973) but more recently it has been shown that adipose tissue phorphorylase kinase requires micromolar amounts of free Ca^{2+} for optimal activity (Khoo, 1976). Khoo *et al.* (1973) originally saw an activation of glycogen phosphorylase and triglyceride lipase after incubation of cells with adrenaline. However, there was no activation of phosphorylase kinase. Possibly the increased activation of phosphorylase could have resulted solely from an increase in intracellular Ca^{2+}. If an increase in cytosol Ca^{2+}. is involved in catecholamine activation of phosphorylase, it is unlikely to result from increased uptake of extracellular Ca^{2+}. This is based on the fact that catecholamine activation of fat-cell cAMP, total phosphorylase or lipase is unaffected by incubation in Ca^{2+}-free buffer containing 1 mM EGTA (Khoo *et al.*, 1973).

Kissebach *et al.* (1975) have proposed that insulin facilitates a rise in intracellular Ca^{2+} of fat cells. Ca^{2+} is postulated to be the missing second messenger from insulin action. The major problem with this proposal is that an increase in cytosol Ca^{2+} should activate phosphorylase (Khoo, 1975). However, insulin inhibits phosphorylase. Thus, it appears unlikely that the multitude of changes in fat-cell metabolism elicited by insulin results from an increase in intracellular Ca^{2+}. A rise in cytosol Ca^{2+} is more likely one of the many secondary effects of insulin which result from an unknown first event. This could well be an alteration in plasma-membrane function. While Larner (1972) has postulated an unknown second messenger for insulin there is no proof of its existence or even of the necessity for insulin action to be transmitted through the cytosol by a second messenger similar to, but different from, cAMP or cGMP.

CHAPTER 13

Cyclic Nucleotides and Heart

E.-G. Krause and A. Wollenberger

I INTRODUCTION

The function of the heart as a pump depends on the co-ordination of electric, mechanical and metabolic activities of its cells. These activities are subject to neural and hormonal regulation, mainly by the autonomic nervous system and its transmitter substances, noradrenaline (in the mammalian heart) and acetylcholine, and by circulating catecholamines, chiefly adrenaline. In addition, other hormones and hormone-like compounds, such as those of the thyroid and the adrenal cortex, glucagon, histamine and prostaglandins are capable, either directly or by interaction with catecholamines and perhaps with acetylcholine, of exerting modulatory effects on cardiac function and metabolism. All these regulatory influences appear to be realized by modulating the role of Ca^{2+}, the prime regulator of cardiac activity. In many instances these regulatory influences seem to involve the cyclic nucleotides, cAMP and cGMP as their intracellular mediators.

The remarkable interplay between Ca^{2+} and the cyclic nucleotides, especially cAMP, in a wide variety of biological processes has been summarized by Rasmussen and Nagata (1970) and has been discussed in a previous chapter by Berridge.

According to a widely accepted theory (Greengard and Kuo, 1970) cAMP, as intracellular mediator of neural and hormonal signals, exerts its effects in the target cell through the activation of cAMP-dependent protein kinases. These enzymes catalyse the transfer of the terminal phosphate of ATP to protein substrates, leading to a modification of the properties of these proteins. cGMP, which has emerged more recently as another intracellular mediator of extracellular signals, may fundamentally act in the same way by activating cGMP-dependent protein kinases. In the performance of their regulatory functions these two cyclic nucleotides may act either independently in monodirectional control or they may constitute a pair of antagonists in 'bidirectional systems of dual control'. As will be shown, the molecular events of adrenergic and cholinergic regulation in heart cells at the level of cAMP and cGMP may reflect this dualism.

A number of reviews provides a detailed description of the mediator function of cAMP in the adrenergic regulation of glycogenolysis and contractility in muscle tissues (Haugaard and Hess, 1965; Sutherland *et al.*, 1968; Epstein *et al.*, 1971; Sobel and Mayer, 1973; Entman, 1974; Perry, 1975). In the reviews of Wollenberger (1975) and Williamson (1976), the interrelationship of cAMP and Ca^{2+} in the regulation of cardiac muscle is discussed at length. An extensive bibliography of work in this area published prior to 1967 can be found in the review by Drummond (1967). Biological regulation through the opposing influences of cGMP and cAMP on cardiac muscle and other tissues is an important topic of the review of Goldberg *et al.* (1975).

II ENZYMES OF CYCLIC NUCLEOTIDE SYSTEMS IN THE MYOCARDIUM

According to the current concept of hormonal action via the second messenger cAMP the biochemical reaction sequence involved comprises the following steps: stimulation of an adenylate cyclase sensitive to a hormonal or other extracellular signal → intracellular accumulation of cAMP → cAMP-induced phosphorylation of enzymes and other proteins by protein kinases → biological effect(s). In contrast to this relatively well-documented sequence, the analogous cascade for cGMP, namely, hormonal or other stimulation of guanylate cyclase → formation of cGMP → cGMP-enhanced protein phosphorylation → physiological effect(s), has only partly been confirmed. Its existence, however, is within the realms of possibility. It must be added that enzymatic dephosphorylation of phosphoprotein and enzymic degradation of the two cyclic nucleotides serve as means of terminating the cellular effects of cyclic nucleotides and of re-establishing cellular homeostasis.

A Adenylate Cyclase

In the myocardium, as in other tissues, the enzymes of the cAMP and cGMP systems are located in, or associated with, different subcellular structures and compartments. Hormone-sensitive adenylate cyclase is located mainly in the plasma membranes of the myocardium (Drummond and Duncan, 1970; Wollenberger and Schultz, 1976). Some activity may also reside in the membranes of the sarcoplasmic reticulum (Katz *et al.*, 1974). Particulate adenylate cyclase preparations from the hearts of several species are stimulated by a variety of hormones and neurohumours. These substances include catecholamines — in the order of potency: isoprenaline > adrenaline *D* noradrenaline — glucagon, histamine, thyroid hormones and some prostaglandins (see Drummond and Severson, 1974). The stimulatory action of catecholamines is inhibited by β-adrenergic blocking agents, which do not, however, prevent the action of glucagon and the other agents mentioned. This fact is indicative of the concept that cardiac adenylate cyclase is associated with the β-adrenergic receptor (Robison *et al.*, 1967). There is also evidence for an association of cardiac adenylate cyclase with specific receptors for some of the other agents. The sensitivity of the enzyme to catecholamines and glucagon appears in the human heart at different stages of embryonic development (Palmer and Dail, 1975).

Membrane lipids play a crucial role in the process by which hormones stimulate adenylate cyclase. Treatment of myocardial enzymic preparation with detergents or phospholipase leads to a loss of sensitivity of the enzyme to hormones. Addition of exogenous lipids to these preparations partially restores the hormonal response of the enzyme. For example, phosphatidylserine restores glucagon and histamine responsiveness, while monophosphatidylinositol restores responsiveness to catecholamines but not to histamine or glucagon (Levey, 1972). The activity of adenylate cyclase is modulated in the cell by Mg^{2+}, Mn^{2+}, Ca^{2+}, ATP, pyrophosphate (Drummond *et al.*, 1971), GTP (Lefkowitz *et al.*, 1973b, Lefkowitz, 1975) and possibly by an inhibitory feedback regulator (Ho *et al.*, 1975c) which has not yet been isolated from myocardium.

B cAMP-Dependent Protein Kinase Activity

This enzyme was first demonstrated in heart muscle by Brostrom *et al.*, (1970) and by Kuo *et al.* (1970). The kinase exists in the heart muscle in different forms in the cytosol (Corbin *et al.*, 1975) and is also found associated with other cellular structures, such as the myofibrils (Reddy *et al.*, 1973), microsomal fractions enriched in sarcoplasmic reticulum (Katz and Repke, 1973) and fragments of the plasma membrane (Krause *et al.*, 1973; Dowd and Schwartz, 1975). Transfer of cAMP-dependent protein kinase from cytosol to membranous structures has been suspected to occur in uterine muscle (Korenman *et al.*, 1974) and could be of regulatory importance. Other authors (Keely *et al.*, 1975a), however, think that such transfer may be an *in vitro* artifact, occurring only after homogenization of the tissue.

cAMP activates protein kinase by inducing dissociation of the enzyme into cAMP-binding and cAMP-independent catalytic subunits. This effect has been demonstrated repeatedly *in vitro* and also appears to occur in the myocardium *in vivo* (Corbin *et al.*, 1975). Only few data are available on the physiological substrate(s) involved with the different protein kinases in the myocardium. Partially purified bovine heart kinase catalyses the phosphorylation of such substrates as protamine, arginine-rich histone and lysine-rich histone at relative rates of 1, 0.5, and 0.2, respectively. Table 13.1 lists a number of muscle enzymes and proteins phosphorylated *in vitro* by cAMP-dependent protein kinase from heart and/or skeletal muscle. Possible functional consequences of the protein phosphorylations are also listed in Table 13.1. The criteria proposed by Krebs (1973) for judging whether cAMP exerts its action via protein phosphorylation must be met before the participation of the proteins listed in Table 13.1 in a cardiac response to neurotransmitters and hormones can be confirmed. Only in the case of troponin I (see Table 13.1), a subunit of troponin, which is regulatory

Table 13.1 Known protein substrates of cAMP-dependent protein kinase of heart or skeletal muscle

protein substrate of kinase(s)	probable functional response(s)	demonstration of phosphorylation *in vivo*	
1. *Enzymes of energy metabolism*			
phosphorylase kinase	(i) increased rate of phosphorylase $b - a$ conversion and of glycogenolysis	–	*a*
	(ii) phosphorylation of troponin I with unknown consequence	–	*b, c*
	(iii) phosphorylation of 95,000 dalton protein of sarcoplasmic reticulum; increased Ca^{2+} transport in vesicles of sacroplasmic reticulum.	–	*d*
glycogen synthetase D	inhibition of glycogen synthesis	–	*e*
lipase	increased rate in lipolysis	–	*f*
2. *Proteins of contractile apparatus*			
actin	increase in Ca^{2+} sensitivity of actomyosin ATPase after reconstitution from components	–	*g*
actomyosin, cardiac muscle	shift in Ca^{2+} dependence of ATPase towards lower concentrations	–	*h*
troponin I	modification of the binding of actin and thus, the activity of actomyosin ATPase	–	*i, j*
3. *Proteins of membrane systems*			
22,000 dalton protein (phospholamban) of sarcoplasmic reticulum cardiac muscle	increased Ca^{2+} transport in vesicles of sacroplasmic reticulum	–	*k*
24,000 dalton protein of cell surface membrane fragments	shift in Ca^{2+} affinity of high-affinity binding site(s) towards lower concentrations	–	*l*
sarcolemmal fragments of skeletal muscle	increased Ca^{2+} accumulation and Ca^{2+} stimulated Mg^{2+}–ATPase	–	*m*

a. Krebs (1972)
b. Huang *et al.* (1975)
c. Moir *et al.* (1974)
d. Schwartz *et al.* (1976)
e. Schlender *et al.* (1969)
f. Steinberg and Huttunen (1972)
g. Pratje and Heilmeyer (1972)
h. Rubio *et al.* (1975)
i. Cole and Perry (1975)
j. England (1975)
k. Tada *et al.* (1975)
l. Krause *et al.* (1975b)
m. Sulakhe and Drummond (1974)

protein of the contractile systems, have two of these criteria (phosphorylation *in vitro* and *in vivo*) been met for the myocardium. In addition, a third criterion has been partly fulfilled in that troponin I phosphorylation in the isolated perfused heart in response to adrenaline correlates well with the increase in the strength of contraction (England, 1975). Attempts to demonstrate increased phosphorylation associated with hormonal activation of phosphorylase *b* kinase *in vivo* have been unsuccessful (Meyer and Krebs, 1970).

C Guanylate Cyclase

In contrast to the membrane-associated adenylate cyclase, this enzyme has been found to exist both in particulate and soluble forms. Each form has distinct enzymic properties (Kimura and Murad, 1974). Myocardial guanylate cyclase activity requires Mn^{2+} and may be dependent on the Mn^{2+}/Ca^{2+} ratio. Evidence as to neurohumoral effects on particulate and soluble guanylate cyclase is contradictory. George *et al.* (1970) observed a correlation between cholinergic effects on heart function and increased cGMP levels. Evidence for a stimulation of guanylate cyclase activity in broken-cell preparations by acetylcholine was found by White *et al.* (1973) and Sulakhe *et al.* (1975), while Kimura and Murad (1974), and Limbird and Lefkowitz (1975) were unable to obtain such a stimulation.

D cGMP-Dependent Protein Kinase

Recently, evidence for the occurrence of cGMP-dependent protein kinase(s) in mammalian heart has been obtained with the help of a heat-stable modulator protein in the assay of the enzyme (Kuo, 1974). Such modulator protein(s) may be important in the regulation of both cGMP and cAMP-dependent protein

Table 13.2 Hypothetical reaction sequence for action of β-adrenergic agonists in heart muscle at the levels of the sarcolemma and the sarcoplasmic reticulum

sarcolemmal membrane	sarcoplasmic reticulum*
Agonist	Agonist
Receptor binding	Receptor binding
Adenylate cyclase activation	Adenylate cyclase activation
Increased cAMP	Increased cAMP
Phosphorylation of membrane protein(s)	Phosphorylation of phospholamban
Changed membrane permeability	–
Increased influx of Ca^{2+} into myoplasma during action potential plateau	Stimulation of Ca^{2+} transport by sarcoplasmic reticulum
Increased intracellular Ca^{2+}	–
Augmented contraction	–
Increased Ca^{2+} extrusion (after repolarization)	–
Enhanced rate of relaxation	Accelerated rate of relaxation

* as proposed by Katz *et al.* (1975).

kinases (Kuo, 1975a). A natural substrate for cGMP-dependent protein kinase in the heart has not yet been identified.

E Cyclic Nucleotide Phosphodiesterase

Multiple forms of cyclic nucleotide phosphodiesterase exist in many tissues. These forms may differ in cellular location, substrate specificity, substrate affinity and other kinetic properties (Appleman *et al.*, 1973). At least three enzyme activities, which hydrolyse cAMP and cGMP to the respective 5′-nucleotides at different rates, have been separated from extracts of the heart (Appleman and Terasaki, 1975).

1 cAMP–cGMP Phosphodiesterase

This enzyme is activated in a positive co-operative manner by very low concentrations of cGMP ($<10^{-6}$M). It has similar affinities for both nucleotides ($K_m = 2.6 \times 10^{-5}$M).

2 cAMP Phosphodiesterase

The activity of this enzyme is regulated by its own substrate which acts as a negative co-operative ligand.

3 cGMP–cAMP Phosphodiesterase

This phosphodiesterase is dependent for its activity on a thermostable protein, Mg^{2+} and Ca^{2+}. It possesses high affinity for cGMP ($K_m \simeq 10^{-6}$M) and also a low affinity for cAMP ($K_m = 6 \times 10^{-5}$M) (Kakiuchi *et al.*, 1975a). The activity of this enzyme accounts for more than 80% of the total phosphodiesterase activity in the myocardium. It has been suggested that the thermostable protein activation factor of this phosphodiesterase, which requires the presence of Ca^{2+}, plays a role similar to that of troponin in the function of the actomyosin complex. Half maximal activation occurs at 2 — 4 μM Ca^{2+}, which lies within the range of the intracellular free Ca^{2+} concentration during cardiac activity ($5 \times 10^{-8} - 1 \times 10^{-5}$M). Phosphodiesterase activity in myocardial fibres may, therefore, be modulated by changes in the intracellular Ca^{2+} concentration during the heart cycle.

Whether cardiac phosphodiesterases, like the enzymes in adipose tissue and liver, may have their content, activity or both regulated by the action of hormones, such as insulin and thyroxine, remains to be clarified. However phosphodiesterases are the chief target of drugs which affect cellular function modulated by cyclic nucleotides. For example, in coronary vasodilation by smooth muscle relaxing agents, the efficiency correlates well with the extent of inhibition of cAMP degradation (Kukovetz and Pöch, 1970a). Furthermore, 1-methyl-3-isobutylxanthine, an inhibitor of phosphodiesterase (Beavo *et al.*, 1970), causes an increase in the level of cAMP and an activation of cAMP-dependent protein kinase in the heart perfused with a constant submaximal concentration of this inhibitory agent (Keely *et al.*, 1975b). These effects are the same as those known to occur after administration of hormones, such as adrenaline and glucagon, which stimulate the synthesis of cAMP. The prospect of a drug-mediated, selective increase in the level of cAMP (and probably cGMP) in a given tissue is an attractive one for clinical medicine.

F Phosphoprotein Phosphatase

Activity can be regarded as an additional means of terminating the action of cyclic nucleotides. For a time, phosphoprotein phosphatase was studied in dephosphorylation reactions of substrates, such as histones, phosphorylase *a*, and glycogen synthetase I. More recently, enzymic activity in canine heart extracts, which is capable of dephosphorylating [^{32}P]phosphohistone, has been separated into four fractions with distinct properties (Li, 1975). Modulation of heart phosphoprotein-phosphatase activity by various metabolites has been reported. cAMP inhibits dephosphorylation of histone by this enzyme. This effect is inhibited by an, as yet, unidentified cAMP antagonist isolated from liver (Wasner, 1975). Evidence has been obtained using fat cells that this cAMP antagonist may have a more general role as an intracellular feedback regulator of cAMP-mediated processes at different stages of metabolism (Ho and Sutherland, 1975).

III CYCLIC NUCLEOTIDES AND NORMAL HEART FUNCTION

The role of cAMP in normal heart function has been the subject of many pharmacological, physiological and biochemical studies. As an introduction to a discussion of the possible mechanisms responsible for the enhancement of myocardial function, particularly contractility, by cAMP-stimulating agents, some general remarks about the biochemical basis of the heart beat may be useful.

Contraction of myocardial myofibrils is initiated by depolarization of the cell-surface membrane (sarcolemma), with sequential and temporary increases in membrane permeability to Na^+, Ca^{2+} and K^+. The initial rapid inward Na^+ current is followed by a slow inward Ca^{2+} current during the plateau phase of the action potential (Reuter, 1967). The action potential in some way triggers the release of Ca^{2+} stored in the junctions of the sarcoplasmic reticulum (subsarcolemmal cisternae and lateral cisternae). It is believed that in the mammalian myocardium the amount of releasable Ca^{2+} in this store can account for 50–70% of the total rise in the intracellular free Ca^{2+} concentration (Morad and Goldman, 1973). Intracellular free Ca^{2+} concentration increases from a value of less than 10^{-7}M in the resting muscle to approximately 10^{-5}M during the activation phase of contraction. The 'activator' Ca^{2+} diffuses to the myofibrils and there combines with a subunit of the regulatory contractile protein, troponin. This results in an alteration in the conformation of the troponin–tropomyosin complex which in turn leads to a removal of the inhibition of actomyosin–ATPase, combination of F-actin with myosin and contraction. Contraction strength is determined by the total Ca^{2+} made available from the inflowing extracellular Ca^{2+} and Ca^{2+} released from internal stores. The normal twitch produces only about 10–20% of the tension which the myofibrils are capable. This is equivalent, in terms of Ca^{2+} movement, to the transfer of 4.5–9 pmole/mg myocardium for each beat. Trans-sarcolemmal Ca^{2+} movement is probably in the range of 1.0–3.6 pmole/mg myocardium for each beat as calculated from electrophysiological studies and ^{45}Ca influx data (Grossman and Furchgott, 1964; Morad and Goldman, 1973).

Relaxation is initiated by theATP-dependent sequestration of Ca^{2+} from the myoplasma by network elements of the sarcoplasmic reticulum and by extrusion of the ion across the sarcolemma to the extracellular space. Alternatively, Ca^{2+} may be removed to superficial cell-membrane binding sites. This removal occurs when the intracellular Ca^{2+} concentration is reached at which the troponin–tropomyosin complex exerts an inhibitory effect on actomyosin ATPase. It is believed that the outward transport of Ca^{2+}, in the case of cardiac muscle, occurs chiefly by way of Ca^{2+}/Na^{+} exchange. However, there is also evidence suggesting the existence of an ATP-dependent pump mechanism. The molecular basis of the processes of muscular contraction and relaxation has been reviewed by Ebashi and Endo (1968), Katz (1970), Langer (1973), and Morad and Goldman (1973).

Figure 13.1 gives a model of the Ca^{2+} cycle in cardiac muscle. According to this scheme there is an external and an internal Ca^{2+} cycle. These cycles interact through the myoplasmic Ca^{2+} pools. The proposed interaction of the Ca^{2+} cycles with the cAMP-dependent processes is illustrated in the diagram and will be discussed in detail in the following sections.

A Adrenergic Regulation

1 Cyclic Nucleotides and the Inotropic Response

The adrenergic catecholamines are the most important external chemical regulators of contractile activity of the myocardial fibres in higher animals. It was once thought that the β-adrenergic receptor and cAMP-mediated activation of the myocardial phosphorylase system, and the ensuing acceleration of glycogenolysis and of glycolytic energy production, might be directly related to the increase in myocardial contractile force. Later it was shown that small doses of catecholamines could have positive inotropic effects on the myocardium without increasing the phosphorylase *a* level (Mayer *et al.*, 1963). Using higher doses the peak of the positive inotropic response precedes the maximum activation of phosphorylase (Øye, 1965; Drummond *et al.*, 1966). Moreover, the rise in cAMP occurs slightly earlier than the increase in contractility (Cheung and Williamson, 1965). It thus became clear that activation of phosphorylase is not a prerequisite for the positive inotropic action. Neither is the increase in contractile force a necessary condition for the phosphorylase-activating effect of the catecholamines. The hypothesis that cAMP mediates the inotropic response to adrenergic amines was supported by

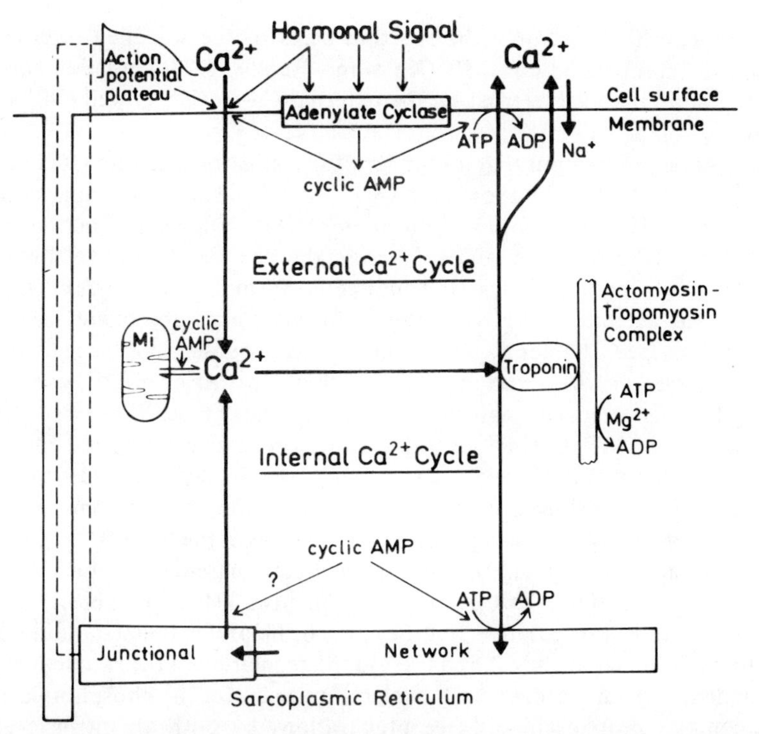

Figure 13.1 Model for the control of the Ca^{2+} cycle, and its interaction with cAMP-dependent processes in cardiac muscle. The junctional sarcoplasmic reticulum comprises the subsarcolemmal cisternae and the terminal cisternae apposed to the T-tubules. Mi: Mitochondrium

the finding that β-adrenergic blockers which inhibit the cardiac effect of β-adrenergic agents also inhibit the increase in cAMP. There is, furthermore, a positive correlation between the increase in cAMP levels and the inotropic effect of other agents which, like glucagon and histamine, increase cAMP by stimulating adenylate cyclase (La Raia and Reddy, 1969) or which, like papaverine, inhibit phosphodiesterase (Kukovetz *et al.*, 1975).

The stimulatory effect of cAMP on Ca^{2+} uptake by cardiac microsomes was first described by Entman *et al.* (1969). Catecholamines and other hormones which stimulate the formation of cAMP in the myocardium may promote the transfer of Ca^{2+} from the myoplasm and their binding sites on troponin to the Ca^{2+}-storage organelles of the sarcoplasmic reticulum network, as has been proposed by Katz and Repke (1973). Such an effect can explain the acceleration of relaxation which is an important component of the action of catecholamines on the cardiac contraction cycle (Morad and Rolett, 1972). With augmented Ca^{2+} replacement of storage sites, more Ca^{2+} can be set free by excitation during the succeeding heart beats, thus leading to increased development of tension. It now appears that not only the reaction steps of the so-called internal Ca^{2+} cycle are modulated by cAMP (Katz and Repke, 1973), but that this modulatory action is also exerted on the external Ca^{2+} cycle as well (Williamson, 1976).

It must be pointed out that a number of investigators has presented evidence against a direct participation of cAMP in the positive inotropic action of catecholamines. They favour an immediate, cAMP-independent action by these amines on the movement of Ca^{2+} across the cell membrane. In this theory, cAMP only supports contraction indirectly by making more ATP available through the activation of glycogen phosphorylase and subsequent acceleration of glycolysis (see Chapter 5). This concept was proposed by Ebashi (1969), and by Rasmussen (1970). Experimental results of whether or not the contraction-strengthening effect of catecholamines on the myocardium is mediated primarily by cAMP, by Ca^{2+}, or by both have been inconclusive and opinions on this question are still divided — for a review see Wollenberger (1975).

If cAMP is the link connecting the catecholamines with the augmented contractile response, one can propose, on the basis of the two Ca^{2+} cycles model of Morad and Goldman (1973) and the protein kinase theory of cAMP action (Greengard and Kuo, 1970), the two sequences listed in Table 13.2. These sequences are still speculative, but they may be helpful in understanding the true relationship between cAMP and Ca^{2+} in the regulation of heart function. There is evidence from cardiac and skeletal muscle for a phosphorylation of sarcolemmal proteins in cell-free preparations by both an intrinsic and an added, extrinsice cAMP-dependent protein kinase (Krause *et al.*, 1973; Sulakhe *et al.*, 1974). A phosphorylated cardiac membrane fragment which exists as 'inside-out' vesicle and is largely derived from the cell surface (Lüllmann *et al.*, 1975), accumulates Ca^{2+} in the presence of oxalate and ATP at a greater rate than do control membrane vesicles. The same effect has been demonstrated for phosphorylated membranes of skeletal muscle (Sulakhe *et al.*, 1975). According to Lüllmann *et al.* (1975) Ca^{2+} uptake by inside-out vesicles may represent active Ca^{2+} transport outwards across the myocardial cell surface. The data allow the tentative suggestion that the phosphorylation of sarcolemmal protein (see Table 13.1) may have a role in the regulation of this Ca^{2+} outward movement. This transport may assist the sarcoplasmic reticulum in its function of lowering myoplasmic Ca^{2+}.

Protein components of the sarcoplasmic reticulum can serve, too, as substrates for cAMP-dependent protein kinases (Kirchberger *et al.*, 1972; La Raia and Morkin, 1974; Wray *et al.*, 1973) as well as for phosphorylase *b* kinase (Schwartz *et al.*, 1976). In the case of the reaction catalysed by cAMP-dependent kinases, phosphate incorporation occurs mainly in a 22,000 dalton protein (see also Table 13.1). Incorporation is associated with a two to three-fold acceleration of the oxalate-dependent Ca^{2+} uptake into the microsomal vesicles and with a corresponding increase in Ca^{2+}-dependent Mg^{2+}-ATPase activity. An increased rate of Ca^{2+} uptake by sarcoplasmic reticulum may also lead to an increased level of filling of this internal Ca^{2+} store.

The hypothetical model of the mechanism of the inotropic action of catecholamines presented in Table 13.2 draws support from the fact that Sutherland's four criteria for judging if cAMP mediates a given hormonal effect have been met for catecholamines and cardiac contraction. Sutherland *et al.* (1968) demonstrated that: (*i*) myocardial adenylate cyclase (see Chapter II) in broken-cell preparations is stimulated by low, inotropically effective concentrations of catecholamines; (*ii*) the inotropic effect is associated with an increase in cAMP; (*iii*) it is potentiated by inhibition of cAMP breakdown; and (*iv*) the hormonal effect may be reproduced by added cAMP or one of its derivatives.

The N^6,2′-O-dibutyryl derivative of cAMP (Bt_2-cAMP) was found (Ahren *et al.*, 1971; Kukovetz and Pöch, 1970b; Skelton *et al.*, 1970) to provoke a positive inotropic cardiac response and to mimic the inotropic action of adrenaline both with respect to changes in certain parameters of contractility (Morad and Rolett, 1972; Skelton *et al.*, 1970) and with respect to the influence of the extracellular Ca^{2+} concentration on its magnitude and time course (Meinertz *et al.*, 1973a). Whereas adrenaline preferentially increases contractile force in comparison to activation of phosphorylase, Bt_2-cAMP does not do so (Ahren *et al.*, 1971), or may even show the reverse preference (Øye and Langslet, 1972). An important parallel between adrenaline and Bt_2-cAMP was established by the finding that the latter, like adrenaline, increases both the Ca^{2+} inward current (Tsien *et al.*, 1972) and the rate of $^{45}Ca^{2+}$ uptake in cardiac cells (Meinertz *et al.*, 1973b). Further evidence for a cAMP-mediated increase in Ca^{2+} influx was presented in experiments on K^+-arrested isolated hearts (Watanabe and Besch, 1974) and on K^+-arrested myocardial cells in culture (Warbanow *et al.*, 1975). In both instances Bt_2-cAMP, like adrenaline, restores the spontaneous contractions. Introduction of cAMP by the 'cut-end' method into calf ventricular muscle trabeculae was found to produce positive inotropic effects (Tsien and Weingart, 1974). The ability of cAMP to mimic the relaxation effect of adrenaline was demonstrated in arrested heart muscle preparations and in skinned cardiac cells (Fabiato and Fabiato, 1975). Preliminary findings have been presented which suggest that cAMP-promoted phosphorylation of a protein of cardiac sarcoplasmic reticulum may enhance the release of Ca^{2+} from the cellular storage site of the ion (Katz *et al.*, 1975).

A cAMP-dependent protein kinase and its substrate have been found to be present in isolated liver mitochondria (Kleitke and Wollenberger, 1976). However, it is not yet known, if the activity of this enzyme is related to the Ca^{2+} efflux which cAMP has been shown to stimulate in isolated mitochondria from liver and other organs, including heart.

cAMP-dependent protein kinases and phosphorylase *b* kinase can phosphorylate components of the contractile apparatus (see Table 13.1). Since the discovery that troponin is a substrate for cAMP-enhanced phosphorylation investigators have been aware of the possibility that the control function of troponin in muscle contraction might be modified by its phosphorylation. This in turn may depend on the tissue level of cAMP. A specific role has not yet been found for phosphotroponin (Lallemant *et al.*, 1975). Theoretically, agents such as adrenaline which alter myocardial contractility could do so without the participation of cAMP by either directly increasing the affinity of troponin for Ca^{2+} or by promoting the activating effect of Ca^{2+} on the contractile system (Katz, 1970).

2 Cyclic Nucleotides and the Chronotropic Response

There have been few attempts to apply biochemical techniques to the elucidation of the possible role of cAMP in the positive chronotropic action of the catecholamines, mainly on account of the small dimension of the pacemaker region of the heart and because of difficulties in localizing these regions. There is little information as yet on the level of cAMP and on adenylate cyclase activity in cardiac pacemaker cells and tissue; none exists on phosphodiesterase.

Adenylate cyclase activity is stimulated by adrenaline to about the same extent in particulate preparations of the new-born rat heart as it is in particles prepared from cell cultures of new-born rat hearts (Karczewski, personal communication). In the latter preparation the majority of myocardial cells has probably regained the facility of autorhythmicity (Wollenberger, 1964). The ability of the adenylate cyclase system in cultured heart cells to respond to adrenaline was shown as a 290% increase in the cAMP level after exposure to 3×10^{-7}M adrenaline. This elevation is enhanced five-fold in the presence of 1-methyl-3-isobutylxanthine. These changes can be prevented by the β-blocker (–)-propranolol, but not by the α-blocker phentolamine. Propranolol also blocks the stimulatory action of positive chronotropic concentrations of adrenaline on adenylate cyclase in homogenized Purkinje fibres of the adult dog heart (Dhalla *et al.*, 1973). It was suggested that the adenylate cyclase system is involved in the β-adrenergic regulation of automacity. N^6-Monobutyryl-cAMP (Bt-cAMP) when added to superfused, electrically driven calf Purkinje fibres resembles noradrenaline in its effect on the shape of the action potential (Tsien *et al.*, 1972). The same effect is also observed following addition of Bt_2-cAMP to spontaneously contracting dog heart Purkinje fibres (Danilo *et al.*, 1972). Bt-cAMP was also found to resemble adrenaline in its ability to restore spontaneous activity in arrested cardiac pacemaker tissue (Tuganowski *et al.*, 1973; Warbanov *et al.*, 1975). Tsien (1973) found that iontophoretic injection of cAMP into spontaneously active calf Purkinje fibres causes changes similar to those produced by adrenaline, while 5′-AMP is without effect. This was also observed by Yamasaki *et al.* (1974) in rabbit sino-atrial nodal cells.

The findings have been taken as evidence for a mediator function of cAMP in the electrical effects of adrenaline on the heart. An involvement of cAMP in the positive chronotropic cardiac action of catecholamines had been suggested earlier by Krause *et al.* (1970), and by Kukovetz and Pöch (1970b). Chiba *et al.* (1972) and Wildenthal (1973), on the other hand, were unable to observe a positive chronotropic effect of Bt_2-cAMP.

3 cAMP and Glycogenolysis

There is no doubt that the most prominent metabolic effect of catecholamines on the heart is the acceleration of glycogen breakdown. Acceleration is mediated in the myoplasma of cardiac cells by cAMP in the presence of Ca^{2+}.

B Cholinergic Regulation

Parasympathetic innervation of the cardiac ventricle is sparse compared to that of the sino–auricular node and the atria. The restraining influence of the vagus on the heart, which is continuously counteracted by sympathetic or equivalent adrenergic influences, is exerted predominantly by decreasing heart rate, atrio–ventricular transmission speed and atrial contractile force. Ventricular contraction can indirectly be weakened by these two processes (Braunwald *et al.*, 1967). However, biochemical studies of the effect of acetylcholine and other parasympathomimetic substances have mainly been carried out on ventricular muscle.

1 cAMP

Murad *et al.* (1962) reported that adenylate cyclase in cell-free heart muscle particles is moderately depressed by acetylcholine and other choline esters. Atropine prevents this depression. These effects offered an explanation for the lowering of myocardial phosphorylase *a* levels by vagal stimulation and by administering acetylcholine (Hess *et al.*, 1962), and for the phosphorylase *a*-stimulating effect of atropine (Kukovetz, 1962). It was shown later that carbamylcholine in isolated electrically driven atria produces pronounced reductions in contractile tension, adenylate cyclase activity and cAMP levels, whereas phosphodiesterase activity remains unaffected (La Raia and Sonnenblick, 1971). There is no definite evidence, however, that a decrease in myocardial cAMP levels caused by cholinergic agents is responsible for the decrease in phosphorylase *a*, the weakening of cardiac contractility and the slowing of Ca^{2+} influx into the cells (Grossman and Furchgott, 1964) — effects that occur in cardiac tissues exposed to these agents. Neither is much known about the biochemical changes underlying the increase in K^{+} permeability which occurs in pacemaker and atrial tissues exposed to acetylcholine (Trautwein, 1963).

2 cGMP

This cyclic nucleotide has been suspected of being involved in cholinergic neurotransmission (Ferrendelli *et al.*, 1970). Indirect support for such involvement was provided by George *et al.* (1970, 1973) who showed that acetylcholine markedly increases the level of cGMP in the heart, while the level of cAMP is moderately decreased or remains unchanged. The increase in cGMP is closely correlated with a decrease in the rate of tension development when the heart rate is kept constant. A correlation between the magnitude of increase in cGMP levels and the slowing of the heart has not been established (George *et al.*, 1970). A rise in cGMP levels in beating atria following exposure to acetylcholine precedes the depression of contractility by two to three seconds (Goldberg *et al.*, 1975). In quiescent ventricular slices, acetylcholine causes a very pronounced rise in cGMP levels (Kuo *et al.*, 1972). The negative chronotropic effect of carbamylcholine is mimicked by Bt_2-cGMP.

These results may be considered as evidence that cGMP is involved in cholinergic neurotransmission. Whether cGMP functions chiefly as a second messenger of cholinergic agents or whether, in addition, it may act via phosphodiesterase stimulation by lowering the cAMP level in a special cellular pool is a question which remains to be resolved.

The present state of knowledge concerning a possible cholinergic stimulation of guanylate cyclase has been discussed elsewhere. In the light of the work of Schultz *et al.* (1973), who studied the changes in cGMP levels in rat ductus deferens to response to cholinergic agents, Ca^{2+} may be the most important effector of guanylate cyclase. It is not unreasonable to expect that those effects of acetylcholine which are possibly mediated by cGMP are brought about by influencing the Ca^{2+} cycle (Figure 13.1) in a manner analogous to the mode of action of cAMP, but in the opposite direction. Casnellie and Greengard (1974) reported that a membrane fraction from smooth muscle catalyses the phosphorylation of two endogenous proteins, and that this phosphorylation is stimulated by physiological levels of cGMP. A ten-times greater concentration of cAMP is necessary to bring about the same increase in phosphorylation. A cGMP-dependent protein kinase has not yet been discovered in cardiac membrane fractions. Furthermore, a physiological role for cGMP-dependent protein kinases (Kuo, 1975b) has not yet been established.

C Influence of Various Hormones

A number of hormones and hormone-like compounds which influence cardiac activity has been reported to stimulate cardiac adenylate cyalse in cell-free preparations and to increase cAMP levels in intact heart tissue. These substances include glucagon, thyroid hormones, histamine and prostaglandins (Epstein *et al.*, 1971; Robison *et al.*, 1971; Williamson, 1976).

The β-adrenergic blocker propranolol, at concentrations which abolish the biochemical effects of adrenaline, has no effect on the cAMP responses of the heart to glucagon. These data support the view that the catalytic moiety of adenylate cyclase is not the β-receptor and that in heart muscle two different classes of hormone-binding sites participate in the activation of adenylate cyclase by catecholamines and by glucagon. These binding sites appear to be connected to the same cyclase, evidence is provided by the fact that combined maximal doses of the two hormones do not produce an effect on adenylate cyclase activity greater than the effect produced by the hormones separately (Levey and Epstein, 1968).

Other evidence, however, suggests that the heart may contain more than one adenylate cyclase. For example, 10^{-6}M L-triiodothyronine, in the presence of a maximal stimulatory dose of catecholamine, produces a further increase in the activity of the enzyme (Levey and Epstein, 1969). As has been mentioned, GTP is a positive effector of adenylate cyclase, possibly making the enzyme sensitive to hormones and, in some cases, affecting the binding of hormones to the receptor sites. The stimulatory effect of thyroid hormones on cardiac adenylate cyclase activity is exerted at lower hormone concentration in the presence of the GTP analogue 5′-guanylyl imidodiphosphate (Will-Shahab *et al.*, 1975). In experiments with slices of rat myocardium, glucagon, L-triiodothyronine, insulin and ouabain have no significant effect on cAMP synthesis (La Raia and

Reddy, 1969). However, other authors (Mayer *et al.*, 1970; Lee *et al.*, 1971) have reported that glucagon does increase cAMP levels in heart muscle.

It has been postulated that the positive inotropic and chronotropic effects of histamine on the heart are mediated by cAMP. Adenylate cyclase activity is stimulated by histamine in particulate preparations from cardiac muscle (Klein and Levey, 1971a); an increase in cAMP content (Kukovetz and Pöch, 1972) is associated with the inotropic action of histamine on the isolated heart. Cytochemical examination of the effect of histamine on adenylate cyclase in myocardial tissue raises the question whether the histamine-sensitive adenylate cyclase activity is concentrated in the endothelial cells of the coronary capillaries (Wollenberger *et al.*, 1973). It has been suggested that cAMP formed in capillaries of the heart exposed to histamine is a second hormonal messenger in the endothelial cells. Also, by acting on neighbouring myocardial cells cAMP may function as an intracellular messenger, thus acting as a local heart hormone.

The significance of the action of prostaglandins on the heart is difficult to assess in view of highly contradictory results concerning contractility, metabolic, and coronary vasodilation responses in heart of different species (Sobel and Mayer, 1973; Glaviano and Masters, 1971; Levey and Killebrew, 1971; Block *et al.*, 1975).

Some hormones may exert a regulatory influence on the heart by allowing for an expression of the effect of other hormones in this organ. Such a permissive role has been demonstrated for cortisol in the control of the cAMP-mediated action of adrenaline in carbohydrate metabolism. Miller *et al.* (1971a) reported that the glycogenolytic response to suboptimal doses of adrenaline is impaired in perfused hearts of adrenal-deficient rats. This defect can be overcome not only by *in vivo* administration of cortisol but also by raising the Ca^{2+} concentration in the perfusate. Namm *et al.* (1968) reported that the phosphorylase kinase step is blocked when Ca^{2+} is omitted from the medium. An impairment of Ca^{2+} uptake in cardiac atria isolated from adrenalectomized rat has been observed (Gerlach and Van Zwieten, 1969). A permissive role of thyroid hormones in β-adrenergic effects on the heart may explain several features of the thyroid hormone–catecholamine interaction in this organ (see Waldenstein, 1966). The nature of such a permissive influence and the site at which it is exerted in the cardiac cAMP system is at present under investigation. Based on work conducted on adipose tissue, this effect may be exerted by suppression of cAMP phosphodiesterase activity (Armstrong *et al.*, 1974; Correze *et al.*, 1974). On the other hand, there is also evidence that adenylate cyclase stimulation may be involved (Guttler *et al.*, 1975; Will-Shahab *et al.*, 1975).

D Cyclic Nucleotides and Coronary Vascular Tone

The hypothesis that cAMP participates together with Ca^{2+} in the regulation of smooth muscle contractility has been extensively tested (see Andersson *et al.*, 1975; Somlyo *et al.*, 1972). cAMP has been assumed to be a mediator of β-

adrenergic actions on the vascular smooth-muscle membrane potential. β-adrenergic agents under appropriate conditions hyperpolarize the vascular smooth-muscle membrane and relax the muscle through mechanisms which are not necessarily controlled by the membrane potential. Bt_2-cAMP has been shown to mimic β-adrenergic hyperpolarization of vascular smooth muscle (Somlyo *et al.*, 1972). Efforts to correlate adrenergic effects on vascular smooth muscle with changes in cAMP levels have not produced uniform results. Increased cAMP levels have been found to be associated with both contraction and relaxation. The increase in cAMP following α-adrenergic stimulation is believed to be secondary to the increase in contractile tone. This is probably caused by the action of Ca^{2+} on cAMP metabolism which consists in a direct or indirect inhibitory effect on cAMP degradation through a rise in cGMP levels (Schultz *et al.*, 1975). The increase in cAMP levels following β-adrenergic stimulation, on the other hand, precedes the relaxation of smooth muscle. The initiation and the maintenance of this relaxation is presumably the consequence of a cAMP-promoted sequestration of Ca^{2+} by sarcoplasmic reticulum. Uptake of Ca^{2+} by microsomal fractions from the aorta (Baudouin-Legros and Meyer, 1973) and colon (Anderson *et al.*, 1975) are increased by cAMP.

Cyclic nucleotide levels in the coronary arteries can be influenced by stimulation or inhibition of their synthesis and their degradation. Many drugs which alter the tone and contractility of the smooth muscle of the coronary arteries are inhibitors of phosphodiesterases. These drugs include caffeine, theophylline, diazoxide, dipyramidol, prenylamine and papaverine. Positive correlations have been established between inhibition of phosphodiesterase and the mechanical response of coronary arterial strips. The inhibitory effects on phosphodiesterase and the resulting increase in cAMP levels precede the onset of smooth muscle relaxation (Kukovetz and Pöch, 1970a; Pöch and Kukovetz, 1972). It appears likely that inhibition of phosphodiesterases is an important biochemical effect of these drugs and that their cardiovascular effects are partly due to increases in the cellular levels of cAMP.

The physiological role of cGMP in the action of some agents which affect the contraction of smooth muscle is still unclear and has not yet been investigated in coronary arteries. The intracellular concentration of cGMP is increased by numerous agents which promote smooth muscle contraction, such as cholinergic and α-adrenergic agonists, histamine, and 5-hydroxytryptamine. The increased formation of cGMP induced by these agonists is apparently secondary to an increase in the cytoplasmic Ca^{2+} concentration (Schultz *et al.*, 1975). It has been suggested that cGMP and cAMP play opposing roles in regulating the tone of venous smooth muscle (Dunham *et al.*, 1974).

E Cyclic Nucleotide Oscillations during the Contraction–Relaxation Cycle

In 1973 two groups of authors independently discovered oscillations in the level of cAMP in the frog heart during the contraction–relaxation cycle (Brooker, 1973; Wollenberger *et al.*, 1973). The fluctuations in cAMP level are

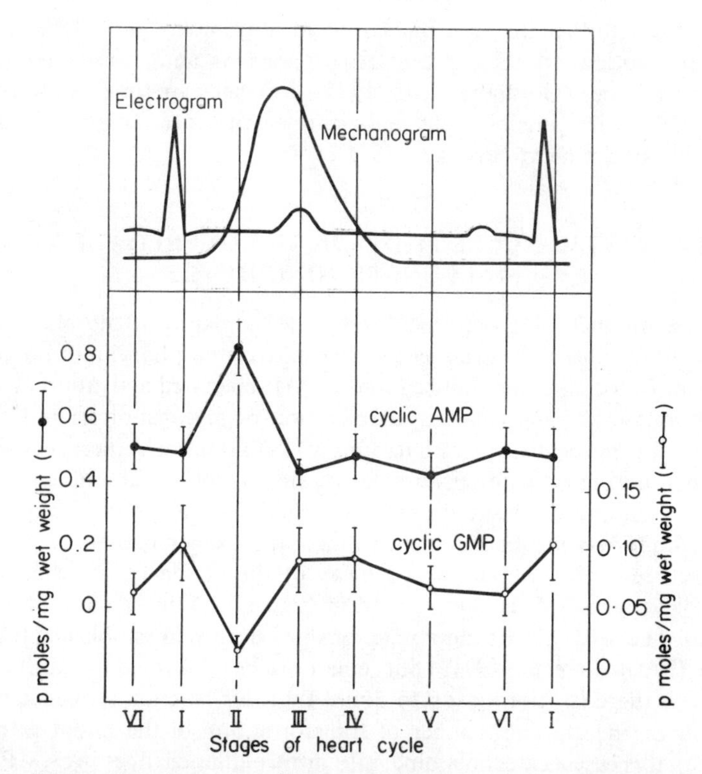

Figure 13.2 Levels of cAMP and cGMP in the frog ventricle at different stages of a cardiac cycle. Duration of the heart cycle: 3 sec. (Data from Wollenberger *et al.*, 1973)

accompanied by variations in the level of cGMP (Wollenberger *et al.*, 1973), which decrease while those of cAMP increase (Figure 13.2). These reciprocal changes in cyclic nucleotide levels may result from simultaneously occurring changes in the activities of adenylate and guanylate cyclase. These changes are conceivably brought about by pulse-synchronous discharges of cardiac sympathetic and parasympathetic neurotransmitters and/or transient local changes in the concentration of Ca^{2+} to which these enzymes are sensitive. The changes in Ca^{2+} concentrations can also affect cGMP phosphodiesterase which is activated by this ion.

The cAMP fluctuations during the heart beat are abolished in frogs pretreated with 6-hydroxydopamine, which causes depletion of the cardiac adrenaline stores, or pretreated with sotalol, a β-adrenergic blocker. The rapid rise in cGMP toward the diastolic level at the beginning of systole is slowed (Krause *et al.*, 1975a) in the presence of atropine. These findings indicate the participation of autonomic neural transmitters in the observed fluctuations in cyclic nucleotide levels during the cardiac cycle. These cyclic fluctuations can be regarded as another instance of reciprocal movements in the level of the two cyclic nucleotides in their presumed capacity as antagonistic regulators of cellular activities (Goldberg *et al.*, 1975). Further experiments are needed to determine their possible physiological significance in the beat-to-beat regulation of the heart (Brooker, 1975).

IV CYCLIC NUCLEOTIDES IN THE OVERLOADED AND MALFUNCTIONING HEART

Almost all the ATP required for the performance of muscle function in aerobic heart muscle is produced by oxidative phosphorylation in the mitochondria. Catecholamine-induced and cAMP-mediated activation of glycogen phosphorylase and the resulting acceleration of glycogenolysis and glycolysis may help the muscle to replenish its supply of ATP in such emergency situations as sudden coronary insufficiency, paroxysmal tachycardia and acute physical exertion. These are all situations in which the availability of oxygen to the myocardium may not be sufficient to meet the oxygen demands.

The current view about the pathway of the action of catecholamines on muscle glycogenolysis is summarized in Figure 13.3 in the form of a reaction cascade. The sequence of these reactions has been well established for skeletal muscle (Brostrom *et al.*, 1971a) but remains to be confirmed for cardiac muscle. However, there is little reason to doubt that this reaction sequence does take place in the heart. The manner of transformation of the initial extracellular signal, in this case a catecholamine, into intracellular changes in cAMP level has been discussed earlier in this chapter. Two of the later steps illustrated in Figure 13.3 deserve comment. Firstly, the phosphorylation of non-activated phosphorylase kinase and glycogen synthetase I by the same cAMP-dependent

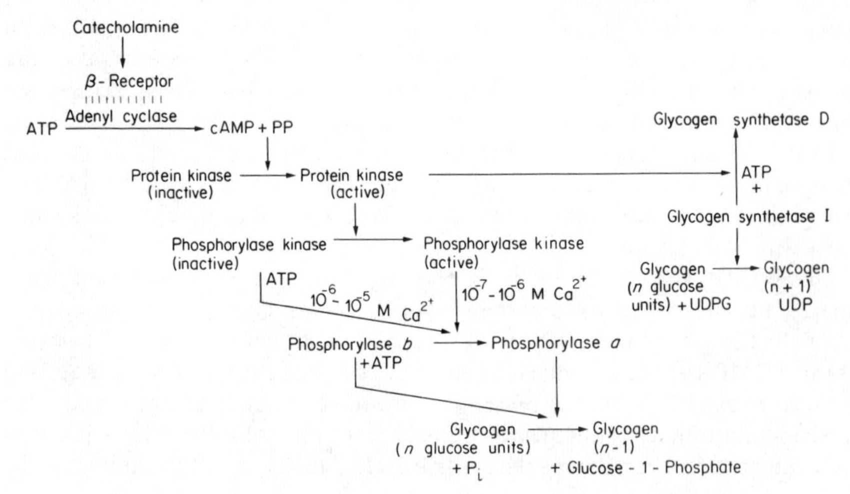

Figure 13.3 The catecholamine–phosphorylase cascade in muscle

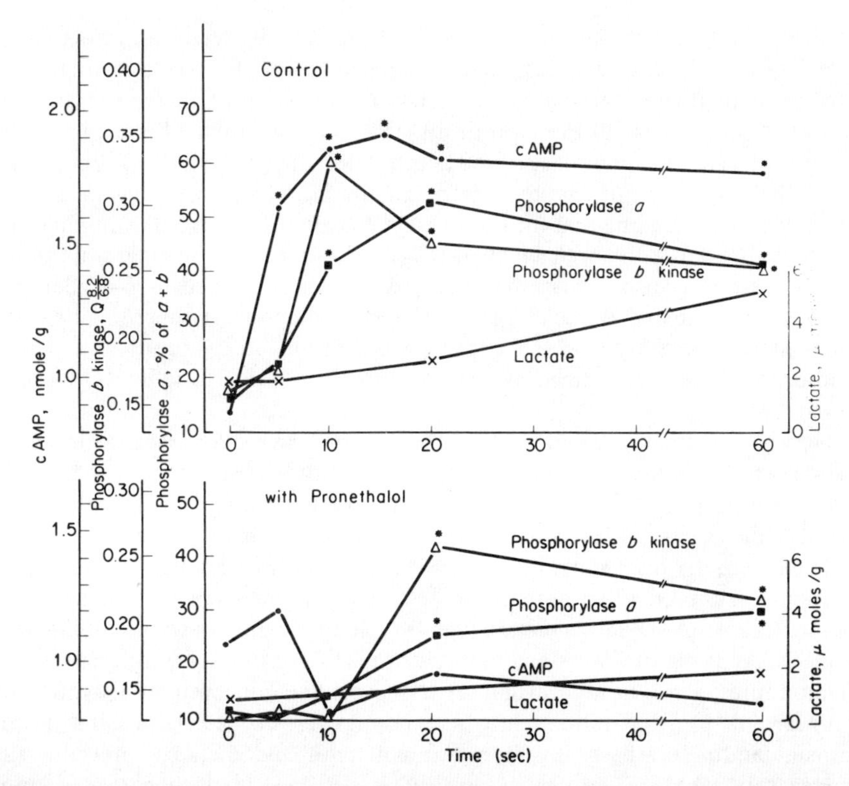

Figure 13.4 Changes in the levels of cAMP, activated phosphorylase kinase, phosphorylase *a* and lactate following arrest of blood flow in the left ventricle of the dog. (Data from Wollenberger and Krause, 1973)

protein kinase converts the kinase into the active and the synthetase into the inactive (glycogen synthetase D) forms thus favouring the breakdown of glycogen. Secondly, the catalytic activities of both the activated and non-activated forms of phosphorylase kinase are dependent on the presence of Ca^{2+}. Whereas the activated phosphorylase kinase requires Ca^{2+} concentrations of the order of 10^{-7} to 10^{-6}M, higher concentrations in the range of 10^{-6} to 10^{-5}M are necessary for the activity of the 'non-activated' form of the enzyme. As has been described these ionic concentrations are reached in the myoplasma of cardiac cells at the beginning of the contraction cycle. Thus, it appears that release of Ca^{2+} into the myoplasma is a common factor for the triggering of both glycogenolysis and contraction (Brostrom *et al.*, 1971a).

There is evidence that both the cAMP-mediated activation of phosphorylase kinase and the Ca^{2+}-dependent increase in the activity of this enzyme occur in heart muscle. Some of the biochemical events involved in the activation of phosphorylase kinase in dog myocardium after sudden arrest of coronary flow are depicted in Figure 13.4. After such an event, mobilization of endogenous cardiac noradrenaline was observed (Shahab *et al.*, 1969), which appears to be a direct effect of acute myocardial ischaemia and not the result of increased cardiac sympathetic nervous activity. The mobilization of the catecholamine results in an increase in the level of cAMP, a rise in activated phosphorylase kinase, a rise in phosphorylase *a*, and acceleration of glycolysis (Wollenberger *et al.*, 1969). The predominantly adrenergic initiation of these events is also indicated by the finding that the rise in cAMP is abolished when the animals are pretreated with β-adrenergic blockers (Figure 13.4). Under the same conditions the increase in phosphorylase *a* is delayed but not abolished by β-adrenergic blockers (Krause and Wollenberger, 1967), indicating the presence of a cAMP-independent, possibly Ca^{2+}-dependent, acceleration of glycogenolysis. These findings have been confirmed by Dobson and Mayer (1973) and by Rabinowitz *et al.*, 1975).

George and Buttusil (1975) have observed that in cardiac tissue the levels of cGMP are significantly elevated during anoxia. Treatment with methylprednisolone prevents this increase in the cGMP level and at the same time reduces the release of lysosomal enzyme from anoxic heart muscle.

It has recently been shown that cAMP rises sharply within the first minute of coronary arterial ligation both in the ischaemic and non-ischaemic areas of the myocardium. cAMP levels remain high for a longer period in the non-ischaemic regions. Acute cardiac overload induced by aortic constriction produces, within five minutes, a rise in myocardial cAMP levels and concomitant activation of myocardial prostaglandin synthetase (Limas *et al.*, 1974). Inhibition of prostaglandin synthesis by pretreatment with indometacin prevents the increase in cAMP. The rise in cardiac cAMP level in the overloaded non-ischaemic muscle may play a regulatory role in metabolic processes other than glycogenolysis, possibly macromolecular synthesis which forms the molecular basis of compensatory cardiac hypertrophy (see Rabinowitz and Zak, 1972). In preliminary experiments with isolated cardiac mitochondria, cAMP brings about an increase in RNA synthesis (Wollenberger and Kleitke, 1973). Ischaemia fails to produce a rise in cardiac cAMP immediately after the arrest of blood flow in animals with cardiac hypertrophy due to aortic constriction or renal hypertension (Wollenberger *et al.*, 1973). This failure may be due to the severe depletion of noradrenaline (Chidsey *et al.*, 1964) and the resulting inability to release this catecholamine on arrest of the coronary flow (Wollenberger *et al.*, 1973). It is unlikely that the failure is due to a loss of the ability of adenylate cyclase to be stimulated by catecholamines (Henry *et al.*, 1971).

Very little is known at present about the enzymes of the cAMP system in the malfunctioning heart. Aberrations in cAMP and cGMP metabolism have been observed in hearts and blood vessels of hypertensive animals and may provide a biochemical basis for some of the pathogenetic factors in hypertension, as discussed by Amer (see Chapter 21). Gold *et al.* (1970) reported that cardiac adenylate cyclase looses its ability to be activated by glucagon in experimental heart failure, whereas stimulation by adrenaline is not affected. These results have been confirmed in work with particulate adenylate cyclase preparations from the papillary muscles of patients with congestive heart failure (Goldstein *et al.*, 1971). In these preparations, there is a lack of response of adenylate cyclase to glucagon which is associated with an absence of a positive inotropic response to this hormone. These observations might explain the lack of efficacy of glucagon in the treatment of some patients with chronic cardiac failure. Activation of isolated particulate adenylate cyclase by adrenaline is markedly reduced and the hormone-independent basal adenylate cyclase activity is also reduced in the isolated perfused heart failure caused by substrate deficiency (Dhalla *et al.*, 1972). A diminution in basal and F -stimulated adenylate cyclase activity and loss of responsiveness to adrenaline were also observed in particulate fractions from hearts of genetically myopathic hamsters at the advanced stage of heart failure (Sulakhe and Dhalla, 1972).

A release of cAMP from the myocardium has been observed in isolated perfused heart preparations (O'Brien and Strange, 1975). Figure 13.5 summarizes the findings of Böhm *et al.* (1976). cAMP levels were measured in coronary sinus and arterial blood of patients with congestive heart failure before

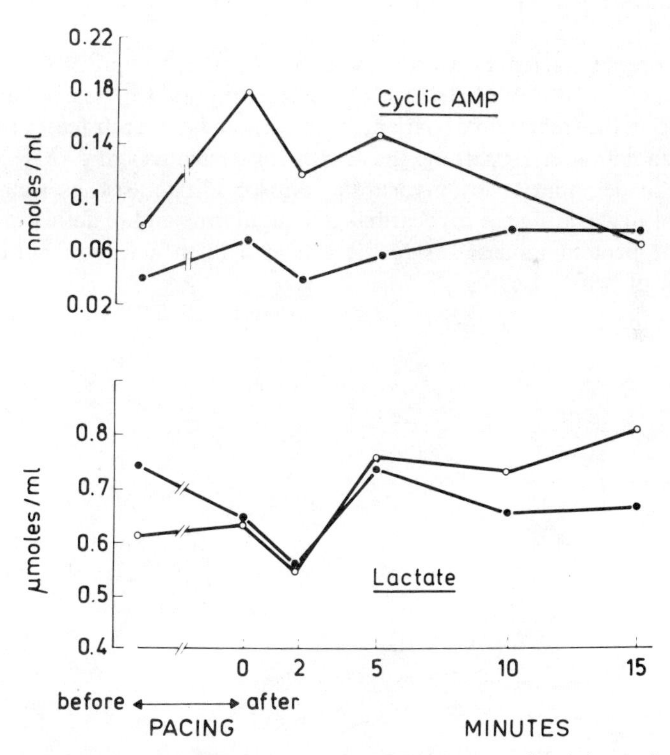

Figure 13.5 Changes in plasma cAMP and lactate levels in aortic (O) and coronary sinus (●) blood samples in a patient before and after electrical pacing of the heart

and during electric pacing of the heart to the point of symptoms of angina pectoris. At this time the venous–arterial cAMP difference is at least doubled. It can be assumed that this increase is due to an increased efflux of the nucleotide from the heart.

Rabinowitz *et al.* (1974) found that in some patients with acute myocardial infarction the cAMP levels in coronary sinus blood exceed those in arterial blood. According to these authors and to Strange *et al.* (1974) a slight increase in the level of plasma cAMP occurs in patients with acute myocardial infarction during the first day of onset. The correlation noted between plasma cAMP level, haemodynamics and prognosis of myocardial infarction may be of clinical significance. Hamet *et al.* (1973) reported that labile hypertensive patients receiving an infusion of isoprenaline excrete more cAMP in their urine than do normotensive patients. It is not known, whether the excreted cAMP originated in the blood plasma or the kidney.

Therapeutic Applications

The effects of methylxanthines and many other drugs on the myocardium and coronary vessels are due at least in part to their ability to inhibit cyclic nucleotide phosphodiesterases. The detailed action of these compounds on heart has been discussed.

It has been reported by Gertler *et al.* (1971) that lipophilic derivatives of cAMP, such as Bt_2-cAMP have been successfully used in severe cardiogenic shock and in the treatment of patients with cardiac failure refractory to digitalis and catecholamines. Hopefully, the search for derivatives of cAMP which are resistant to degradation by myocardial phosphodiesterases, as well as being capable of penetrating the myocardial cell membrane and of activating cAMP-dependent protein kinases inside the cell, will open new possibilities in the treatment of heart diseases.

CHAPTER 14

Cyclic Nucleotides in Regulation of Renal Function

T. P. Dousa and L. D. Barnes

I INTRODUCTION

The primary function of the kidney is to maintain fluid and electrolyte balance. Therefore, the major cellular processes bringing about the basic renal function involve mostly active or passive transport of solutes and fluids across the epithelial cells, resulting finally in the formation of urine. Although the kidney is innervated and neural regulation certainly plays an important role in the control of some aspects of kidney function, such as renal blood flow, basically the kidney can function in a denervated state and the main actions are regulated by hormones or other humoral agents.

It was suggested that cAMP plays a fundamental role in the cellular action of vasopressin (VP) (Orloff and Handler, 1962). Subsequently, cyclic nucleotides have been implicated in the action of many other hormones on the kidney. During recent years investigations into the role of cyclic nucleotides in the cellular action of hormones on the kidney and other renal functions have progressed considerably. In this chapter the following points will be reviewed: (*i*) evidence for participation of cAMP and cGMP in the action of humoral agents of the kidney; (*ii*) the possible role of cyclic nucleotides in renal growth; and (*iii*) renal handling of cyclic nucleotides—a unique feature of this organ, which is useful in the assessment of hormonal action on renal tissue as well as on other organs.

II ROLE OF cAMP IN THE ACTION OF HUMORAL AGENTS

A Vasopressin

1 Basic Mechanism of Cellular Action

Vasopressin (VP) acts on the kidney at the collecting ducts and at the adjacent portion of the distal convoluted tubules to increase relative water permeability with the formation of concentrated urine (Grantham, 1974; Dousa and Valtin, 1976). It has never been convincingly demonstrated that VP at physiological doses affects electrolyte excretion in mammalian kidney (Grantham, 1974). The view that cAMP is an intracellular mediator of VP is supported by a variety of experimental data. Adenylate cyclase stimulated specifically by vasopressin has been described in kidney tissue of all species studied, including man (Dousa and Valtin, 1976; Dousa, 1976) and VP-sensitive adenylate cyclase has been localized in the medullo-papillary portion of the renal medulla (Dousa and Valtin, 1976). It has also been demonstrated that activation of adenylate cyclase by VP is closely related to the binding of the hormone to its receptor (Jard *et al.*, 1975). In recent microdissection studies, VP-sensitive adenylate cyclase was assayed in many segments of tubules, and it was demonstrated that the highest enzyme activity is in collecting ducts and tubules. Some VP-sensitive adenylate cyclase activity is also present in the ascending limb of Henlé's loop (Imbert *et al.*, 1975).

cAMP phosphodiesterase is not influenced by VP either *in vivo* or *in vitro* (Dousa and Valtin, 1976), although the enzyme plays an important role in the control of intracellular levels of cAMP. It has been demonstrated in many experiments that stimulation of adenylate cyclase by VP results in accumulation of cAMP in renal medulla (Grantham, 1974; Dousa and Valtin, 1976). The notion that VP elicits its functional response by increasing the level of cAMP is supported by the finding that exogenous cAMP and its analogues can mimic the action of VP in isolated microperfused tubules (Grantham, 1974). Also, cAMP phosphodiesterase inhibitors can mimic or potentiate the VP effects. However, neither cAMP nor its analogue or cAMP phosphodiesterase inhibitors clearly mimic the VP effect if they are administered into systemic circulation or into the renal artery. This is probably due to the stimulatory effect of these compounds on other segments of the nephron or due to systemic effects (Dousa and Valtin, 1976; Dousa, 1976).

The present experimental evidence suggests that the adenylate cyclase sensitive to VP is localized on the peritubular (basilar) side of the plasma membranes of tubular cells (Dousa and Valtin, 1976). On the other hand, the luminal plasma membrane of tubular cells is the barrier for water permeability (Grantham, 1974).

The mechanism by which the cAMP, which is increased in cells by VP, elicits its final response—an increase in the water permeability of the luminal plasma membrane—is not yet clear. It is still only a working hypothesis that cAMP acts on the luminal plasma membrane by increasing phosphorylation of a specific protein which in turn can influence water-permeability properties of this structure (Dousa and Valtin, 1976; Dousa, 1976). It has been shown that components of cAMP-dependent protein phosphorylation are present in the renal medulla, and that cAMP-dependent protein kinase can be activated in renal medullary cells *in situ*. This activation is proportional to the dose of VP and is related to the elevation in the tissue level of cAMP (Dousa, 1976).

However, to date there is no direct evidence that luminal plasma membrane protein(s) of the collecting ducts are phosphorylated in intact tissue and that such phosphorylations are associated with a large water flux.

Another intracellular component probably playing a critical role in the action of VP is a system of cytoplasmic microtubules. Although integrity of microtubules is necessary for normal action of VP and their function is probably critical for the steps in the cellular action of VP subsequent to cAMP generation, the exact mechanism of their involvement is not clear. Contrary to all expectations, elevation of the renal medullary cAMP in response to VP is not followed by increase in urinary cAMP excretion. Although the findings differ widely from one laboratory to another, in most of the studies no consistent increase in the urinary cAMP level has been found after administration of VP (Dousa, 1974, 1976; Murad, 1973; Broadus *et al.*, 1970).

2 Factors Influencing Vasopressin Response at the Cellular Level

cAMP formation or degradation of the VP-responsive cells of the kidney can be influenced in several ways. Only those factors which are likely to have their effect in intact cells *in situ* are mentioned in this review.

a. Other Hormones

There is some evidence that adrenal steroids can modulate the effect of VP on cAMP metabolism. Adrenalectomy results in decreased activation of adenylate cyclase by VP (Rajerison *et al.*, 1974). Glucocorticoids probably are required for the coupling between the VP receptor and adenylate cyclase within the plasma membrane, while aldosterone probably increases the binding of VP to its receptor in the membrane (Rajerison *et al.*, 1974). Defects due to adrenalectomy may be restored by the administration of glucorticoids and mineralocorticoids. In amphibian urinary bladder, in which VP acts on water permeability in a similar way as in mammalian kidney, it was shown that aldosterone can greatly potentiate the effect of VP through a decrease in activity of cAMP phosphodiesterase (Stoff *et al.*, 1972). It is not known whether such an effect of aldosterone is also manifest in mammalian kidney.

Thyroid hormones apparently have a marked influence on VP-dependent cAMP metabolism. In the hypothyroid state, renal medullary adenylate cyclase activities, both basal and VP-stimulated, are markedly reduced while cAMP phosphodiesterase activity is unchanged. We found that treatment of hypothyroid animals with thyroxin restores the responsiveness of adenylate cyclase to VP and decreases the activity of cAMP phosphodiesterase.

b. Other Cellular Modulating Factors

Other humoral agents besides hormones in the classical sense, have been implicated in modulation of the effect of VP on cAMP metabolism.

Prostaglandins have been studied extensively in both mammalian and amphibian VP-responsive epithelial systems (Dousa and Valtin, 1976; Dousa, 1976). At present the only general conclusion appears to be that prostaglandins (specifically of the E-series) may modulate the VP-dependent cAMP metabolism in some way, but there are numerous controversies on this topic. Some studies have shown that prostaglandins can inhibit VP-stimulated adenylate cyclase, while in other studies, and under different conditions, prostaglandins actually stimulate renal medullary adenylate cyclase and increase tissue levels of cAMP. Moreover, in most of the experiments carried out using prostaglandins, PGE_1, which is not a natural occurring compound in mammalian kidney, was employed. Thus, another question may be raised as to how these results relate to the physiological cellular response to VP (Dousa and Valtin, 1976).

Noradrenaline has been reported to inhibit both the VP response *in vivo* and the VP-dependent accumulation in renal medullary cells (Dousa, 1976; Kurokawa and Massry, 1973). One specific feature of VP-sensitive cAMP system is that it operates in the renal medulla, a tissue with a high osmolarity which fluctuates widely depending on the functional state of the organ (Jamison, 1974). Solutes can profoundly influence the VP-dependent adenylate cyclase, with lower concentrations stimulating and higher concentrations inhibiting its activity (Dousa, 1972). Thus, it is likely that it is the osmolarity of the medulla which markedly influences the VP effect on cAMP accumulation and the ultimate functional response in this specific tissue.

3 Abnormalities in Vasopressin Responsiveness

Recent experimental evidence suggests that hypo or hyper-responsiveness to VP, which is either inherited or induced, may be due to the alteration of VP-dependent cAMP metabolism (Dousa, 1974). A decreased or absent response to VP, such as in hereditary nephrogenic diabetes insipidus in mice, in hypercalcaemic states or caused by drug action, may be due, at least in part, to decreased formation of cAMP in response to VP (Dousa and Valtin, 1976; Dousa, 1974; Singer and Forrest, 1976). On the other hand, hyper-responsiveness to VP after administration of chlorpropamide may be explained by sensitization of adenylate cyclase to VP or inhibition of cAMP phosphodiesterase (Dousa and Valtin, 1976; Dousa, 1974; Singer and Forrest, 1976). In other renal concentrating defects, the cellular action of VP is impaired probably in the steps after cAMP generation (Dousa and Valtin, 1976; Dousa, 1974).

B Parathyroid Hormone

1 Basic Mechanism of Cellular Action

As with vasopressin, there is a strong experimental evidence indicating that the action of parathyroid hormone (PTH) on the kidney tubules is mediated by cAMP (Aurbach and Heath, 1974). However, the picture is more complex because PTH brings about several functional responses in renal transport processes (Aurbach and Heath, 1974; Diaz-Buxo and Knox, 1975); the most prominent is inhibition of phosphate reabsorption resulting in phosphaturia and decreased urinary Ca^{2+} excretion (Diaz-Buxo and Knox, 1975). It is not clear if both of these major effects of PTH on renal function are mediated by cAMP, but it seems likely.

Adenylate cyclase stimulated by PTH has been found in renal cortical sections of all mammalian species so far studied (Dousa, 1976; Aurbach and Heath, 1974; Aurbach *et al.*, 1972). Moreover, very active PTH-sensitive enzyme has also been found in some birds and reptiles (Dousa, 1976). PTH elicits an increase in cAMP levels in cortical tissue *in situ* or in slices incubated with hormone (Dousa, 1976; Aurbach and Heath, 1974; Diaz-Buxo and Knox, 1975).

In experiments to examine the location of adenylate cyclase in various tubular segments, the highest activation of adenylate cyclase by PTH appears to be in cortical convoluted tubules with a lower degree of stimulation occurring in the proximal straight tubules. However, appreciable stimulation of adenylate cyclase has also been observed in certain distal segments of the nephron. On the other hand, no stimulation of adenylate cyclase by PTH is detected in the collecting ducts (Chabardes *et al.*, 1975) but PTH-senstitive adenylate cyclase is present in glomeruli (Sraer *et al.*, 1974).

As with vasopressin, it appears that PTH-sensitive adenylate cyclase is located on the antiluminal side of the plasma membrane of tubule cells, but the site of cAMP action is likely to be at the luminal brush-border side of the cells, at least in case of proximal tubules (Aurbach and Heath, 1974).

Phosphaturic and hypocalcaemic effects of PTH were mimicked by exogenous cAMP or a cAMP phosphodiesterase inhibitor (Rasmussen *et al.*, 1975). cAMP also mimicks the inhibitory effect of PTH on phosphate reabsorption in micropuncture studies (Aurbach and Heath, 1974; Diaz-Buxo and Knox, 1975). The mechanism by which cAMP, which is increased in the tubular cells by PTH, elicits its effect on the transcellular phosphate and Ca^{2+} fluxes is unknown. As in the case of VP, it appears to be a reasonable hypothesis that protein kinase activation and protein phosphorylations are involved in the actions of the cyclic nucleotide. Hypothetical protein substrates for the protein kinase would, of course, be specific proteins associated with phosphate and Ca^{2+} transport, and entirely different from proteins supposedly related to the water-permeability control in collecting ducts (Dousa and Valtin, 1976; Dousa, 1974). There are several clues which suggest that cAMP-dependent phosphorylation occurs on luminal side of tubular cells. Immunoreactive cAMP appears to be located at the luminal site of the proximal tubule cells and increases after stimulation with parathyroid hormone, according to our findings. cAMP-dependent protein kinase is also associated with luminal aspects of the renal cortical plasma membranes while extensive binding of cAMP has been found in brush border membranes (Insel *et al.*, 1975a).

A prominent feature of action of PTH on cAMP metabolism in the renal tubules is the finding that in most species cAMP escapes from the cell and appears in urine in considerable quantity (Dousa, 1976; Murad, 1973; Aurbach and Heath, 1974). Therefore, the renal excretion of cAMP after administration of PTH is markedly increased and, as will be discussed below, this increase can be used as an indication of PTH interaction with adenylate cyclase in the intact kidney. Also, it has been shown that cAMP is transported from the peritubular space into the lumen by a similar mechanism compared to that for other organic acids (Coulson and Bowman, 1974), but it does not seem that the bulk of the increase in urinary cAMP after addition of PTH is due to a stimulation of transcellular transport. For instance, it has been found that in presence of probenecid the increase in urinary cAMP after addition of PTH is unaltered (Dousa, 1976).

2 Factors Influencing Parathyroid Hormone Response

Several factors have been tentatively identified as affecting the PTH-dependent cAMP metabolism in the mammalian kidney. It has recently been reported that in vitamin D-deficient rats the response of renal cortical adenylate cyclase to PTH is diminished markedly as is the urinary cAMP excretion (Forte *et al.*, 1976). This decrease can be restored by administration of vitamin D. The defect appears to be at the PTH receptor–adenylate cyclase coupling since basal adenylate cyclase activity is not different in vitamin D-deficient animals (Forte *et al.*, 1976). In contrast to VP-responsive adenylate cyclase, the PTH-sensitive system does not apear to be influenced by adrenal steroids (Rajerison *et al.*, 1974).

Thyroxin is another hormone which may possibly modulate the effect of PTH. In hypothyroid animals, basal adenylate cyclase activity in renal cortex is unaltered, but the response to PTH is markedly decreased. cAMP phosphodiesterase activity is unchanged in the hypothyroid rat. The reduced response of cortical adenylate cyclase to PTH can not be restored by short-term treatment with thyroid hormones. This contrasts with the restoration of the adenylate cyclase response to VP in hypothyroid rats treated with thyroid hormones.

The possible effect of prostaglandins on the action of PTH on cAMP metabolism in kidney is as unclear as it is in the case of VP.

Ca^{2+} has been reported to modulate the action of PTH on renal cortical cAMP metabolism in several studies (Dousa, 1976; Aurbach and Heath, 1974). It has been consistently found that PTH stimulation of adenylate cyclase as well as the basal activity of adenylate cyclase is suppressed by addition of Ca^{2+} in high concentration to the cell-free system (Dousa, 1976). There is no indication that physiological variations in Ca^{2+} levels modulate PTH-dependent cAMP metabolism.

3 Abnormalities in Parathyroid Hormone Responsiveness

Some diseases with decreased functional responsiveness of the kidney to PTH appear to involve alterations in the cAMP system. A prominent feature of pseudo-poparathyroidism type I is the absence of any increase in urinary cAMP after injection of PTH, in contrast to other syndromes, such as phosphaturia, in which the kidney fails to respond to PTH (Murad, 1973; Aurbach and Heath, 1974). On the other hand, in pseudohypoparathyroidism type II the increase in urinary cAMP after PTH is normal (Drizer *et al.*, 1973). It is unclear which step in the cAMP system is impaired in pseudohypoparathyroidism type I. Possible sites are cAMP synthesis, breakdown or transport out of tubular cells (Aurbach and Heath, 1974; Murad, 1973).

From previous comments, it appears that diminished PTH-dependent cAMP formation may be involved in pathogenesis of renal PTH hyporesponsiveness in hypovitaminosis D (Forte *et al.*, 1976), pathological hypercalcaemia (Beck *et al.*, 1974), or hypothyreosis.

C Calcitonin

The exact role of calcitonin in the physiological regulation of renal function is not yet firmly established. If administered acutely it can produce phosphaturia (Aurbach and Heath, 1974). On the other hand, there is definite experimental evidence that calcitonin can stimulate cAMP formation and accumulation in the mammalian kidney (Dousa, 1976; Aurbach and Heath, 1974). It is assumed that the effect of calcitonin on the kidney is mediated by cAMP. Calcitonin-sensitive adenylate cyclase was found in the kidney, with the highest activity located in the outer medulla (Aurbach and Heath, 1974). Calcitonin increases tissue levels of cAMP (Dousa, 1976; Aurbach and Heath, 1974). Urinary excretion of cAMP after calcitonin administration has only been noted in some studies and is not as prominent as with PTH (Dousa, 1976; Murad, 1973).

D Glucagon

Glucagon is not, in general, considered to be a primary agent regulating excretion of water and electrolytes. However, this hormone has been shown on many occasions to increase the excretion of phosphorus, Na^+ and water probably by the inhibition of electrolyte and water reabsorption from the tubular ultrafiltrate. Glucagon has been found to stimulate adenylate cyclase prepared from both renal cortical and medullary tissue (Dousa, 1976). This glucagon-sensitive adenylate cyclase appears to be different from that stimulated by other agents, such as VP or catecholamines. It is not known whether glucagon-dependent cAMP metabolism in the kidney is related to the functional effects of glucagon. Glucagon increases markedly urinary cAMP excretion, but all this cyclic nucleotide is derived from extrarenal sources (Broadus *et al.*, 1971; Dousa, 1976; Murad, 1973).

E Catecholamines

Catecholamines have been implicated in the formation of urine as well as in the mediation of renin release. There is almost complete agreement that β-adrenergic agents can stimulate cAMP formation and that adenylate cyclase which is under the influence of β-adrenergic agents is not directly related to other hormone-sensitive adenylate cyclases (Dousa, 1976). Isoprenaline-sensitive adenylate cyclase is localized in segments of tubules different from those in which cAMP formation is stimulated predominantly by VP or PTH (Chambardes *et al.*, 1976). The highest activity has been found in certain segments of the distal convoluted tubules and collecting tubules. Noradrenaline has been shown on several occasions to inhibit the action of VP on the adenylate cyclase in the kidney tissue (Dousa, 1976).

III OTHER CYCLIC 3′,5′-NUCLEOTIDES IN HORMONAL ACTION

After the initial discovery of the presence of cGMP in urine (Ashman *et al.*, 1963), the possible role of this cyclic nucleotide in hormonal action in the kidney and related tissues has been investigated in several studies. Reports suggest that cGMP can stimulate active Na^+ transport across amphibian urinary bladder, increase water permeability and inhibit the response to oxytoxin. However, these findings have not been reproduced by other workers (Dousa, 1975). In studies which examined the effect of hormones on guanylate cyclase in a cell-free system or on GMP levels in kidney tissues slices, no effect of VP, PTH or other humoural agents was detected by Helwig *et al.* (1975) and in our laboratory. It is of considerable interest that the specific activity of guanylate cyclase is much higher in glomeruli than in tubules (Dousa, 1976). This is in accord with our immunocytochemical studies (Dousa, Barnes, Ong and Steiner; unpublished data) and the fact that in human kidney cortex most of the guanylate cyclase is membrane-bound while in the rat most of the enzyme activity is in the cytosol (Kim *et al.*, 1976b).

Acetylcholine has recently been reported to increase the level of cGMP in the kidney cortex, an effect similar to that observed in many other tissues (DeRubertis *et al.*, 1976). The relationship of these increases in tissue cGMP levels by acetylcholine and the effects of this biogenic amine on renal function are not known. It was also noted in this study that increased cGMP levels produced by acetylcholine are accompanied by an increased Na^+ excretion. In micropuncture studies, both an increase in Na^+ reabsorption and an increase in fluid excretion were reported in response to exogenously applied cGMP (Osswald and Jacobs, 1974).

IV ROLE OF cAMP AND cGMP IN RENAL GROWTH

Renal tissue has an outstanding potential for compensatory hypertrophy after unilateral nephrectomy. Also there is regeneration of acutely necrotic tubule cells. The kidney is also affected by benign and malignant tumours originating in the renal tubule endothelium. There is some evidence that both cAMP and cGMP may play a role in normal or tumourous renal growth.

Enzymes of cAMP and cGMP metabolism have been studied both in human renal cell carcinoma (Kim *et al.*, 1976a,b) and in experimental kidney tumours (Criss *et al.*, 1976). Adenylate cyclase in renal cell carcinoma—tumours originated in proximal tubule cells—does not differ in terms of basal activity and activities stimulated by NaF or 5′-guanylyl imidodiphosphate from adenylate cyclase from normal renal cortical tissue. However, adenylate cyclase from renal cell carcinoma is completely unresponsive to stimulation by PTH and VP. Stimulation by PGE_1 and PGE_2 is lower than in normal cortex. Two major types of renal cell carcinoma, the 'clear cell' and 'dark cell' types, differ in sensitivity of adenylate cyclase to glycogenolytic agents (Kim *et al.*, 1976b). While renal cell carcinoma of clear cell type is completely unresponsive to stimulatory doses of glucagon and isoprenaline, the adenylate cyclase prepared from dark cell type tumour is highly stimulated by both substances. Characteristics of this stimulation do not differ essentially from stimulatory effects of these two agents on a normal human renal cortical tissue. cAMP phosphodiesterase and cAMP-dependent protein kinase activities from renal cell carcinoma and normal renal cortical tissue are not substantially different (Kim *et al.*, 1976a; Kim *et al.*, 1976b). Thus, it appears that the neoplastic transformation of renal tubule cells is associated with changes in hormonal control of cAMP metabolism, although it is not possible to judge at present whether these alterations in hormonal responses are primary or secondary consequences of the neoplastic transformation. In experimental adenoma of rat cortex, an increase in the tissue level of cGMP but less activity of guanylate cyclase has been reported recently (Criss *et al.*, 1976). In spontaneous renal adenocarcinoma of men, on the other hand, the subcellular distribution of guanylate cyclase was reversed in comparison with normal cortex; the activity of cGMP phosphodiesterase was markedly reduced in human clear cell renal adenocarcinoma (Kim *et al.*, 1976b). In general, it appears that accelerated renal growth may be associated with reciprocal changes in cGMP and cAMP levels.

V URINARY EXCRETION OF cAMP AND cGMP

It was established relatively early in the course of investigations on cyclic nucleotides, that cAMP and cGMP are the only nucleotides excreted in substantial quantities in urine (Ashman *et al.*, 1963). Considerable interest has been devoted to studies of renal handling of cyclic nucleotides as well as the excretion of cyclic nucleotides originating from renal tissue, since such investigations provide a convenient, noninvasive tool for the indirect evaluation of physiological and pathological processes involving cyclic nucleotides in kidney and other tissues.

In studies on human subjects and animals, it has been established that renal clearance of cAMP and cGMP represents a substantial portion of the total body clearance of both nucleotides (Broadus *et al.*, 1971; Murad, 1973; Jard, 1975). Renal clearance of these cyclic nucleotides has represented about 20% of total body clearance depending on the species studied (Broadus *et al.*, 1971; Murad, 1973; Jard, 1975). It appears that both cAMP and cGMP are freely filtered through glomeruli to the tubuler ultrafiltrate. There are several factors besides the filtered load of cyclic nucleotides which may determine the quantity excreted in the final urine. In man, the urinary cAMP excretion exceeds the filtered load (Broadus *et al.*, 1971), indicating that a certain portion of the nucleotide excreted in the final urine is added from tubules into tubular fluid. For practical purposes the portion of urinary cAMP of renal origin is referred to as 'nephrogenous cAMP' (Broadus *et al.*, 1971). It is likely that most of the cAMP added to the glomerular filtrate originates from *de novo* synthesis in the renal tubule cells. However, it has recently been demonstrated that the kidney is capable of taking up cAMP from the peritubular capillary network; this uptake is blocked by probenecid (Coulson and Bowman, 1974; Coulson *et al.*, 1974). This would suggest that cAMP is secreted by renal tubules through the transport system for organic acids, such as *p*-aminohippurate. It also appears that a substantial portion of cAMP taken up by tubuli is catabolized in the tissue by a phosphodiesterase, but certain portions appear in the urine (Forte *et al.*, 1976).

It should be noted that renal handling may vary from species to species. While in man urinary excretion of cAMP always exceeds filtered load (Broadus *et al.*, 1971), in dog a small amount of the cAMP is added from the kidney (Blonde et al., 1974). In the rat, the clearance of cAMP has occasionally been found to be even lower than the glomerular filtration rate (Jard, 1975).

Another feature which should be taken into account is the possibility that excretion of cAMP, which is a weak acid, can be influenced in some segments of tubules by urinary pH. It appears that, in accordance with such a consideration, increasing the pH of urine increases the excretion of cAMP, while acidification promotes 'diffusion trapping' by tubules (Czekalski *et al.*, 1974). cGMP appears also to be filtered completely by glomeruli, but there is no major difference between the filtered load of cGMP and urinary excretion in man (Broadus *et al.*, 1971). On the other hand, renal clearance of cGMP exceeds the filtered load in dog (Blonde *et al.*, 1974).

All these features should be considered in evaluation of any study in which urinary cyclic nucleotide excretion is used as a measure of its metabolism in the kidney or extrarenal tissues. The basic distinction should be made between urinary cAMP of renal origin, the so-called 'nephrogenic cAMP', and cAMP which is filtered from plasma by glomeruli (Broadus *et al.*, 1971). The most accurate measurement of the nephrogenous component of cAMP would be to estimate simultaneously the clearance of cAMP and of inulin in man or creatinine in dog (Blonde *et al.*, 1974). Both inuline and creatinine are used as measures of the glomerular filtration rate. In a simplified form, urinary cAMP excretion is often expressed in relation to total excretion of endogenous creatinine. It is based on the general observation that urinary creatinine excretion approximates to creatinine clearance provided that plasma levels of creatinine are not changed. Although the measurement of urinary cAMP excretion in relation to creatinine excretion is better than a simple measurement of total cAMP excretion (Murad, 1973), it should be borne in mind that the changes in this ratio are useful only if plasma levels of either creatinine or cAMP are unchanged. This may only be a reasonable assumption under certain experimental circumstances. In any case, to be able to assess accurately nephrogenous or extrarenal cAMP if the changes are small and variability high, simultaneous measurement of inulin and cAMP clearance is essential (Murad, 1973).

Determination of urinary excretion of cAMP, either the total or the 'nephrogenous' amount, was used in numerous studies by Murad (1973) when testing the renal response to PTH. PTH was acutely infused or abnormal levels of the hormone were suspected. Also, it appears that urinary excretion of cAMP after administration of glucagon may be a good indicator of the glucagon effect on hepatic cAMP generation. Other agents acting on cyclic nucleotide metabolism either in kidney or extrarenal tissues, so far have produced relatively small changes in urinary excretion of the nucleotide and have not yet been established as reliable tools for diagnosis or investigation. However, it is quite possible that with the development of new or improved testing techniques which can correct for the factors influencing urinary nucleotide excretion, urinary cAMP excretion may be more accurately and more reliably used to estimate humoral interactions involving abnormal cAMP metabolism.

cGMP excretion into urine has been studied less extensively and its potential remains to be established (Murad, 1973). However, recent observations which showed increased urinary excretion of cGMP in animals bearing experimental tumours suggest that measurement of urinary cGMP may be particularly useful in assessment of cGMP turnover. Measurement may also be used in the study of conditions characterized by accelerated tissue growth, either in the kidney or in extrarenal organs (Criss *et al.*, 1976).

VI CONCLUSIONS

It can be stated that cAMP plays a key role in the actions of a number of hormones on the kidney, and that it is also likely that both cAMP and cGMP may participate in regulation of kidney-tissue growth. The participation of cAMP in the cellular action of VP and PTH in the kidney has been firmly established. In spite of uncertainties in the mechanism by which cAMP elicits the final functional response, there is little doubt that it serves as a mediator in the action of these two hormones. There is also consistent evidence for the view that β-adenergic agents, calcitonin, glucagon and prostaglandins influence the cAMP metabolism in the kidney. However, in case of these agents, it is not known at present how their effects on cAMP metabolism relate to their ultimate functional responses in the kidney.

cAMP and cGMP levels and activities of enzymes involved in cyclic nucleotide metabolism appear to change in the course of accelerated renal growth which is in general agreement with observations from other organs. There is, however, no knowledge as to which change in the metabolism of the two cyclic nucleotide is the primary one and whether the changes are related to growth as a cause or an effect.

Kidney is an unique organ as urinary excretion of cAMP and cGMP serves as one of the major outlets for cyclic nucleotides from body fluids in general. Cyclic nucleotides from the kidney can contribute markedly to the urinary content of cyclic nucleotide. This feature of kidney function provides a unique opportunity to use measurements of the excretion of cyclic nucleotides as a noninvasive technique to assess cyclic nucleotide dynamics in the kidney and in other organs in response to a variety of stimuli.

ACKNOWLEDGEMENTS

This work was supported by the USPHS Research Grant AM-16105, by Grant-in-Aid from the American Heart Association, the Minnesota Heart Association and the Mayo Foundation. Dr. T. P. Dousa is an Established Investigator of the American Heart Association and Dr. L. D. Barnes is a Senior Mayo Research Fellow.

The authors thank Ms. Ardith Benjegerdes for her secretarial assistance.

CHAPTER 15

The Thyroid Gland as a Model for the Study of Cyclic Nucleotide Metabolism and Action

J. E. Dumont and J. M. Boeynaems

I INTRODUCTION

The main function of the thyroid gland is to secrete thyroid hormones. All the other specialized functions of the gland are subordinate. Subsidiary functions include: the synthesis of thyroglobulin; exocytotic transport of thyroglobulin to the follicular lumeni; the active pumping of plasma iodide; the generation of hydrogen peroxide and its reduction by a peroxidase coupled with the conversion of iodide to an activated form of unknown structure; the incorporation of active iodine into tyrosyl residues of thyroglobulin; and the coupling of iodinated tyrosines into iodothyronines, which constitute the effective thyroid hormones, triiodothyronine (T3) and thyroxine (T4) (Figure 15.1) (De Groot and Stanbury, 1975; Greer and Solomon, 1974). Thyroid exhibits a unique characteristic among the endocrine glands: namely, the existence of a follicular structure, the lumen of which is filled with thyroglobulin. Thyroglobulin exerts at least two functions. It is a storage form of iodine and an efficient template for hormonal biosynthesis. The scarce and irregular availability of iodine within the environment constitutes an evolutionary explanation for such a complex system. Due to the thyroid gland's particular storage form, the process of secretion of thyroid hormones is also unique. In the first step, iodinated thyroglobulin is engulfed at the apex of the thyroid cells, and after this endocytosis, it is then hydrolysed in the lyosomes to release T3 and T4. These hormones can then enter the plasma (De Groot and Stanbury, 1975; Greer and Solomon, 1974).

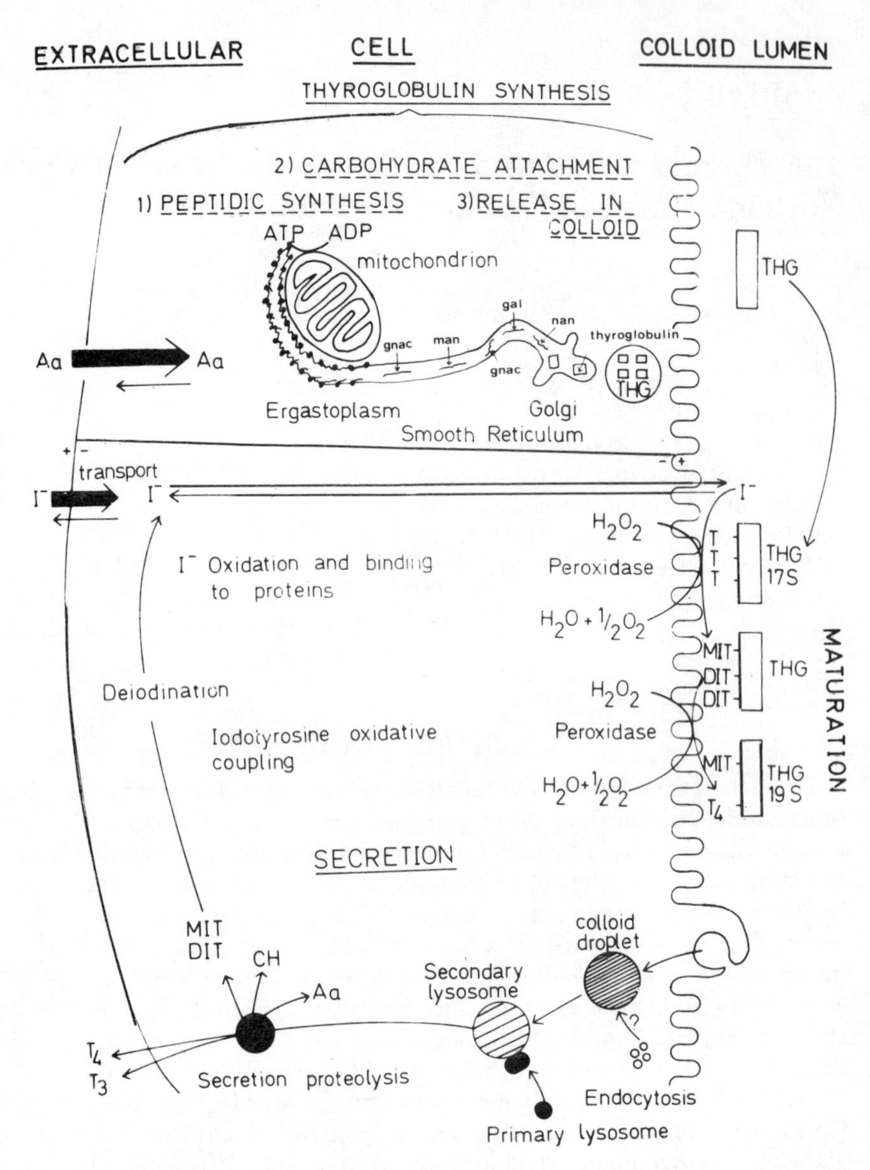

Figure 15.1 Integrated picture of the various specialized functions of the thyroid follicular cells: Thyroglobulin (THG) biosynthesis, its iodination and the secretion of thyroid hormones — triiodothyronine (T_3) and thyroxine (T_4)

All vertebrates require a constant supply of thyroid hormones, for brain maturation and body growth, and for maintaining a normal level of intermediary metabolism and specialized functions in all tissues. The constancy of thyroid hormones plasma level is controlled by a thyroid–hypophysis feedback loop. In this control system, thyrotropin (TSH) stimulates the secretion of thyroid hormones, and circulating thyroid hormones inhibit the secretion of TSH. It appears that the stimulatory action of TSH is mediated, at least partially, by cAMP (Dumont, 1971; Field, 1975).

Basically, the situation in the thyroid seems to be a typical unidirectional system (Goldberg *et al.*, 1975) which expresses a single pattern of functional and metabolic activity — to synthetize and secrete hormones — in response to a single positive regulator — TSH (Dumont, 1971; Field, 1975). However, it now appears that thyroid is also controlled by a variety of extra and intra-cellular regulatory molecules, other than TSH and cAMP, among which are adrenergic and cholinergic agents (Cannon and Smith; 1922; Melander *et al.*, 1974), iodide supply itself (Ingbar, 1972), prostaglandins (Shenkman *et al.*, 1974), Ca^{2+} and cGMP (Van Sande *et al.*, 1975). Thyroid was, and still is, a good illustration for the model of the second messenger (Sutherland, 1972). Also, it provides a good experimental model for studying the question of the respective roles and complex interrelationships between cAMP, Ca^{2+} and cGMP in regulating cell function and growth.

II cAMP AS THE SECOND MESSENGER OF TSH

It is now well accepted that most acute effects of TSH on the thyroid are mediated by cAMP (Dumont, 1971; Field, 1975). The classical criteria proposed by Sutherland (1972) in order to validate his hypothesis have been satisfied, especially for dog thyroid. (*i*) TSH activates directly thyroid adenylate cyclase in cell-free preparations; (*ii*) TSH promotes the accumulation of cAMP in intact thyroid tissue; (*iii*) The acute stimulatory effects of TSH on the thyroid, mostly on secretion, iodination of thyroglobulin and pentose-phosphate pathway are mimicked by exogenous cAMP and by its analogue Bt_2-cAMP, and these effects of TSH are reproduced by agents which raise cAMP levels in the thyroid, such as pgE_1 and cholera toxin; and (*iv*) The effects of low TSH concentrations are potentiated by the methylxanthines which inhibit cAMP phosphodiesterase.

Adenylate cyclase and phosphodiesterases hydrolysing cAMP and cGMP have been characterized in the thyroid (Pastan and Katzan, 1967; Wolff and Jones, 1971; Pochet *et al.*, 1974; Verrier *et al.*, 1974; Bastomsky *et al.*, 1971; Szabo and Burke, 1972; Erneux *et al.*, 1976). Their basic properties are remarkably similar to those of other tissues. It should be noted that some effects of TSH are clearly not mediated by cAMP, in particular, the enhancement of phosphatidylinositol turn-over (Scott *et al.*, 1970). The physiological significance of this phospholipid does not appear any clearer in the thyroid than in numerous other tissues where it has been studied (Hokin and Hokin, 1953; Michell, 1975).

Kuo and Greengard (1969) suggested that cAMP exerts its effects through an activation of protein kinases. Several cAMP-dependent protein kinase activities have been demonstrated in the thyroid (Rappaport *et al.*, 1971; Rappaport and

De Groot, 1972; Yamashita and Field, 1972a; Roques *et al.*, 1973; Kozireff and De Visscher, 1974). TSH activates protein kinase in the intact thyroid (Spaulding and Barrow, 1974; Field *et al.*, 1975), it greatly enhances the phosphorylation of some specific proteins in intact tissue (Lamy and Dumont, 1974). Histone f_1 and a protein belonging to the contractile apparatus of the cell, which is possibly a component of troponin (C unit) have been shown to be phosphorylated. The latter protein could be involved in the endocytosis of thyroglobulin and, thus, the secretory effect of TSH, which is known to require the integrity of microtubules and microfilaments (Temple *et al.*, 1972; Neve *et al.*, 1972). Histones f_1 phosphorylation, which has been suggested to play a role in the activation of transcription (Langan, 1971), could be related to the well-known growth-promoting action of TSH on the thyroid (De Groot and Stanbury, 1975; Greer and Solomon, 1974; Dumont, 1971; Field, 1975). The identification of these natural substrates of cAMP-dependent protein kinases will certainly help in reducing the huge gap which exists between the knowledge of the cAMP system and the molecular mechanisms of the terminal actions of thyrotropin.

III QUANTITATIVE ANALYSIS OF THE cAMP SYSTEM IN THE THYROID

Some quantitative aspects of the cyclic AMP system and its hormonal stimulation have been carefully studied in the thyroid gland and integrated into a model which can be applicable to other systems. In particular, a comparative study of the quantitative relationships between the concentration of TSH and the amplitude of three causally related events (adenylate cyclase activation, cAMP accumulation and hormone secretion) has been carried out.

Thyroid adenylate cyclase activation by TSH is negatively co-operative. However, the response becomes positively co-operative in the presence of the nucleotide ITP (Pochet *et al.*, 1974). This characteristic which is not restricted to thyroid (Hanoune *et al.*, 1975), provides a potential mechanism by which the target cell could regulate its own sensitivity to its activating signal. A variety of hormones has now been shown to interact with their receptors in a negatively co-operative way (De Meyts *et al.*, 1973; Frazier *et al.*, 1974; Limbird *et al.*, 1975).

Adenylate cyclase activation in cell-free preparations appears more sensitive to TSH than cAMP accumulation in intact thyroid tissue; a half-maximal stimulatory effect is obtained at a lower TSH concentration. The factor of stimulation for cAMP accumulation is much higher than for adenylate cyclase activation. While the activation of this enzyme appears to be negatively co-operative, cAMP accumulation is positively co-operative (Table 15.1) (Boeynaems *et al.*, 1974). It can be theoretically shown that these discrepancies could be accounted for by the existence of at least two phosphodiesterases differing both in K_m and V_{max} values. Such a situation has been observed in

Table 15.1 Parameters characterizing the response of horse thyroid to TSH

	stimulation factor	TSH concentration at half-maximal effect (M)	Hill's coefficient
adenylate cyclase activation	4	1.3×10^{-9}	0.76
cAMP accumulation	437	6.5×10^{-9}	1.30

Values are the mean of 3 experiments.

most tissues studied (Russell *et al.*, 1972) including the thyroid (Szabo and Burke, 1972; Erneux *et al.*, 1976).

The secretory response of the thyroid to TSH is positively co-operative, exactly to the same extent as the cAMP accumulation. Thus, the thyroid cells are extremely sensitive to very slight variations in TSH concentration. Moreover, the secretory response to TSH occurs at ten-fold lower concentrations than those at which cAMP accumulation is observed (Figure 15.2). This discrepancy is certainly not an argument against a causal role for cAMP secretion in thyroid secretion and can be explained by the existence of a non-linear saturable step between cAMP and secretion, for instance protein kinase activation. A similar situation, with positive co-operation and 'spare receptors', has been observed in two steroidogenic glands.

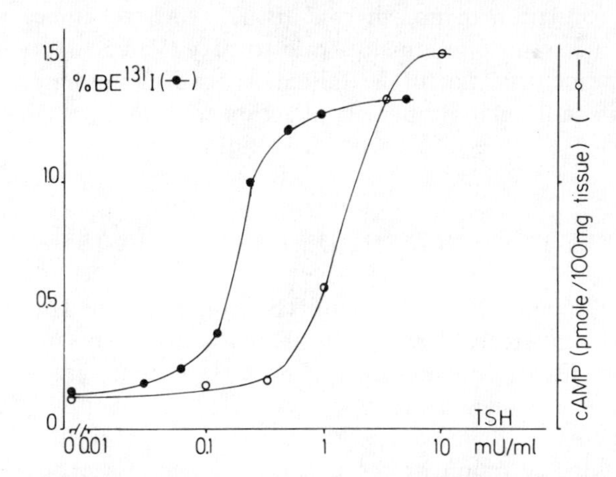

Figure 15.2 Comparison of the dose–response curves for cAMP accumulation and of thyroid hormones by canine thyroid slices, stimulated by TSH. Secretion was estimated by the release of butanol-extractable ^{131}I (BE^{13}I) from glands prelabeled *in vivo*

Similar behaviour was described by Rodbard (1974) for the adrenal cortex and by Dufau *et al.* (1973) for the Leydig cells.

IV INHIBITION BY IODINE OF THE cAMP SYSTEM

Thyroid metabolism seems to be geared to take up and use iodide most efficiently. It is, therefore, to be expected that a regulatory mechanism should exist in order to prevent thyrotoxicosis occurring in cases of excess iodine intake. In fact, a battery of such mechanisms exists in the thyroid, (Ingbar, 1972) and include inhibition at high iodide concentrations of iodide binding to proteins (Wolff–Chaikoff effect), inhibition by iodide of the thyroid blood flow and decreased susceptibility of highly iodinated thyroglobulin to lysosomal digestion. One target of the iodide intra-thyroid feedback loop is the cAMP system (Figure 15.3) (Sherwin and Tong, 1975; Van Sande *et al.*, 1975; Rapport *et al.*, 1975).

Iodide *in vitro* inhibits the TSH enhancement of cAMP accumulation. The stimulatory effect of PGE_1 is also inhibited. Iodide has to be trapped inside the thyroid cells and oxidized in order to exert its inhibitory effect. The active molecular species may be either an oxidized form of iodide or an iodinated organic compound of unknown structure. In accordance with the second messenger model, iodide also inhibits cAMP-dependent effects of TSH, such as hormonal secretion (Van Sande *et al.*, 1975). The decrease in cAMP accumulation is due to adenylate cyclase inhibition rather than to activation of cAMP phosphodiesterases (Rapoport *et al.*, 1975). Iodide is devoid of any effect on cAMP accumulation in non-thyroid tissues, such as kidney or parotid glands, where it has no physiological role to play (Van Sande *et al.*, 1975). Nevertheless, the identification of the iodinated organic compound as well as the elucidation of its molecular mechanism of action could add considerably to the available knowledge concerning cyclic nucleotides in general.

V POSSIBLE INVOLVEMENT OF CALCIUM IN TSH ACTION

The secretory response of the thyroid to TSH is not abolished by the complete removal of extracellular free Ca^{2+} (Willems *et al.*, 1971). This behavior is rare among secretory tissues. For example, if no Ca^{2+} is added to the incubation medium of adrenal medulla neurohypophysis (Douglas, 1968) or β-cells of the endocrine pancreas (Malaisee, 1973) secretion is blocked. Under these conditions, enzyme secretion by the exocrine pancreas is preserved, but the addition of a Ca^{2+} chelator will suppress it completely (Williams and Chandler, 1975). Thyroid secretion is also unique in that it involves an endocytosis instead of an exocytosis. It could be speculated that this unusual feature is related to the lack of a requirement for extracellular Ca^{2+}. cAMP accumulation is also independent of extracellular Ca^{2+} (Van Sande and Dumont, 1973). However,

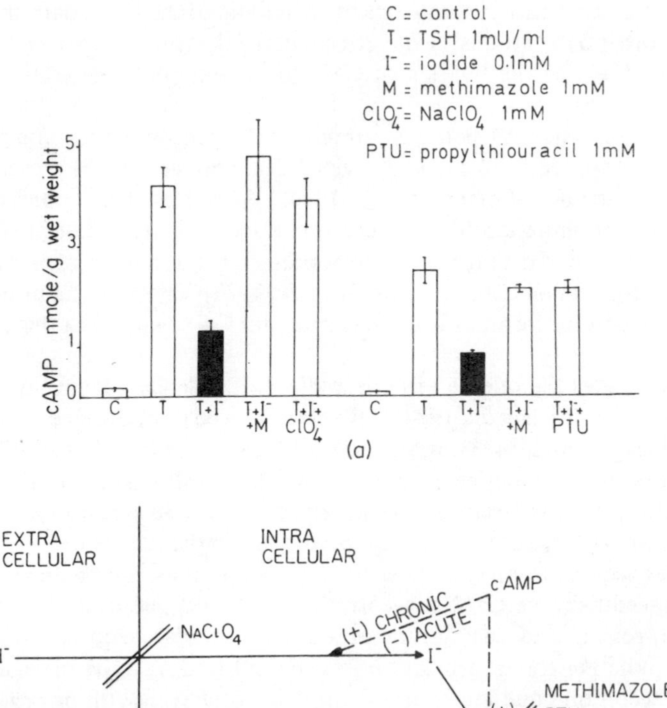

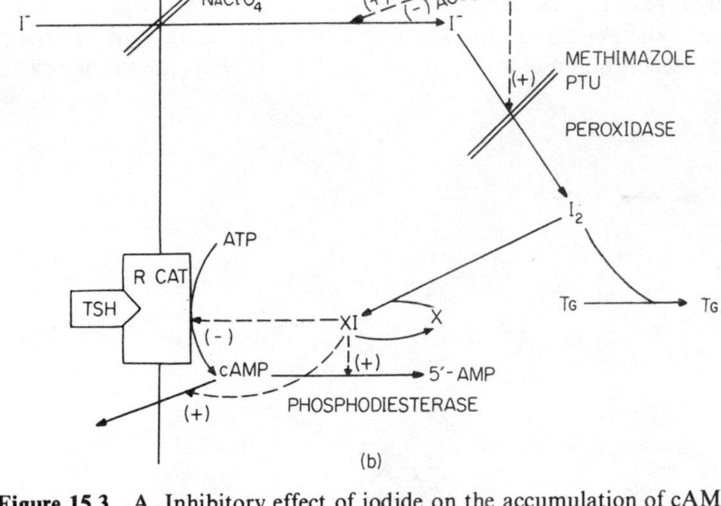

Figure 15.3 A. Inhibitory effect of iodide on the accumulation of cAMP in canine thyroid slices in response to TSH. The inhibition is relieved by $NaClO_4$, which prevents iodide trapping in the thyroid cells, and by propylthiouracil and methimazole, which inhibits the oxidation of iodide by peroxidase. B. Model for the mechanism of the inhibition of iodide of the thyroid cAMP system

two other acute effects of TSH, namely stimulation of the pentose-phosphate pathway and of iodide binding to proteins, are blocked when no extracellular Ca^{2+} is available (Willems *et al.*, 1971). Moreover, a slight increase in the extracellular Ca^{2+} concentration stimulates protein iodination. These data show that two of the three most studied acute effects of TSH require either an influx of extracellular Ca^{2+} or the binding of Ca^{2+} to the external face of the plasma membrane.

The ionophore A-23187 potently stimulates the pentose-phosphate pathway and binding of iodide to proteins, provided that enough Ca^{2+} is present in the extracellular medium (Grenier *et al.*, 1974). At a high Ca^{2+} concentration (10^{-3}M), the ionophore inhibits the secretory effect of TSH and of Bt_2-AMP. It should be remembered that Ca^{2+} also supresses the secretion of parathyroid hormone and glucagon (Leclercq-Meyer *et al.*, 1973). Thus, two acute effects of TSH are reproduced by an influx of extracellular Ca^{2+}, whereas a third effect is inhibited.

TSH accelerates the release of nonrapidly diffusible Ca^{2+} from thyroid cells (Figure 15.4) (Rodesch *et al.*, 1974). This effect can be reproduced by exogenous cAMP. Glucagon and cAMP in the islets of Langerhans (Malaisse, 1973), and acetylcholine in the exocrine pancrease (Williams and Chandler, 1975) exert similar actions which seem to be involved in stimulus–secretion coupling (Douglas, 1968). The TSH effect may correspond to the release of Ca^{2+} from an intracellular sequestrating site to the cytosol, and be associated with an increase of the intracellular free Ca^{2+} concentration. This interpretation has not been definitely proven, and the nature of the intracellular sequestrating site is unknown. Whether the translocation of Ca^{2+} induced by TSH plays a role in stimulus—secretion coupling or in other effects of TSH is still not clear.

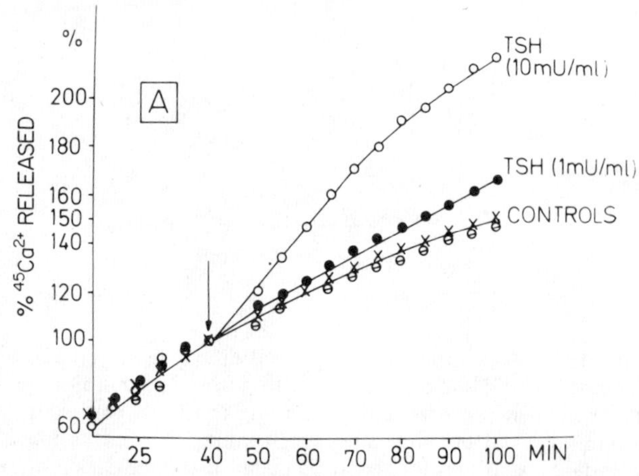

Figure 15.4 Canine thyroid slices were loaded with ^{45}Ca and then washed in order to remove rapidly diffusible Ca^{2+}. Efflux of ^{45}Ca into the incubation medium was measured. ^{45}Ca efflux is expressed in percentage of radioactivity released after 40 min. The arrow indicates the addition of TSH

In conclusion, extracellular Ca^{2+} is required for and can mimic some effects of TSH. TSH, by a cAMP-mediated action, alters the distribution of Ca^{2+} in thyroid cells. Thus, there is some evidence that Ca^{2+} can be an important regulator of thyroid cell functions and play a role in TSH action.

VI cGMP, CALCIUM AND cAMP

Acetylcholine promotes the accumulation of cGMP in the thyroid, as in several other tissues (Yamashita and Field, 1972b; Van Sande *et al.*, 1975). This

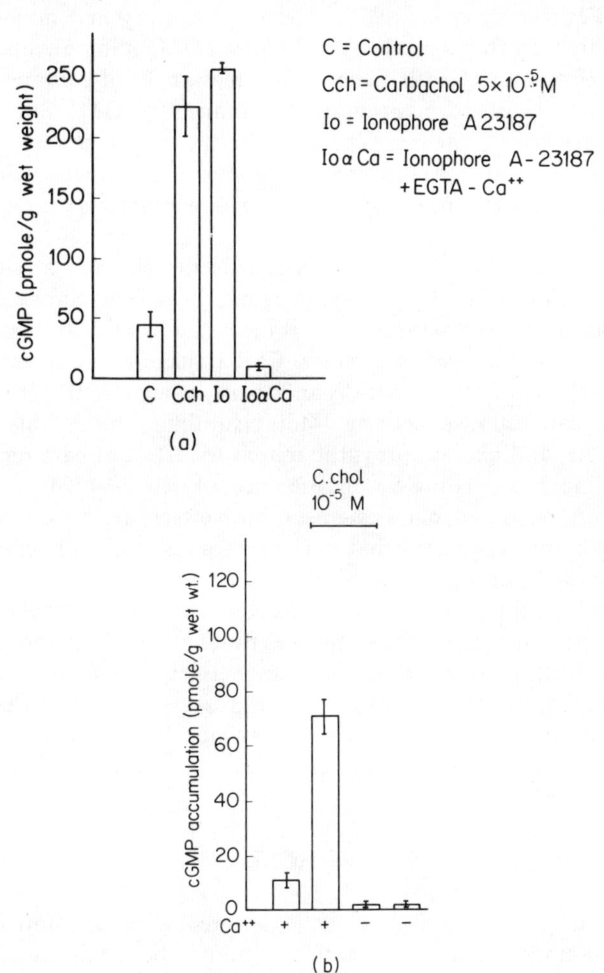

Figure 15.5 Accumulation of cGMP in canine thyroid slices in response to carbamylcholine and ionophore A-23187, in a normal Krebs–Ringer-bicarbonate buffer (A) or in a Ca^{2+}-free medium, containing EGTA 1 m (B)

stimulatory effect is abolished in the absence of extracellular Ca^{2+} and is mimicked by the Ca^{2+} ionophore A-23187 (Van Sande *et al.*, 1975) (Figure 15.5). These data support the hypothesis that the primary factor regulating cGMP levels in the cell is the intracellular concentration of Ca^{2+} (Schultz *et al.*, 1973a), through the control of a Ca^{2+}-sensitive guanylate cyclase (Chrisman *et al.*, 1975). It is not known if carbamylcholine promotes an influx of extracellular Ca^{2+} into the thyroid cells, but it has been shown that it induces a translocation of intracellular Ca^{2+} (Rodesch *et al.*, 1974). Like iononphore A-23187, carbamylcholine stimulates the pentose-phosphate pathway and the binding of iodide to proteins in the thyroid, and inhibits the secretion of thyroid hormones (Van Sande *et al.*, 1975). These stimulatory and inhibitory effects could be mediated either directly by Ca^{2+} or by cGMP. Some evidence suggests that Ca^{2+} rather than cGMP is the second messenger of carbamylcholine for these three effects. Indeed, exogenous cGMP and Bt_2-cGMP fail to reproduce these three effects of carbamylcholine. On the other hand, in a medium containing Mn^{2+} (10^{-3}M) but no exogenous Ca^{2+}, carbamylcholine still promotes an accumulation of cGMP, but no longer inhibits the secretory effect of TSH.

Carbamylcholine, like ionophore A-23187, inhibits the accumulation of cAMP in response to TSH (Champion *et al.*, 1974; Van Sande *et al.*, 1975). Several findings suggest that this effect is mediated by cGMP, through a direct activation of cAMP phosphodiesterase. Carbamylcholine increases the rate of disappearance of cAMP in intact thyroid tissue, while cGMP activates thyroid cAMP phosphodiesterase activity. Methylxanthines block this stimulatory effect of cGMP and also suppress the inhibitory effect of carbamylcholine on cAMP accumulation. However in presence of Ro 20-1724, a non-methylxanthine inhibitor of phosphodiesterases, both effects are preserved (Erneux *et al.*, 1976). It is thus suggested that in the thyroid, some cholinergic effects are mediated by Ca^{2+} and others by cGMP.

The physiological significance of these actions of acetylcholine is unclear and certainly the primary regulators of thyroid function are TSH and iodide, which have positive and negative effects, respectively, and not acetylcholine. Nevertheless, the investigation of the action of acetylcholine on the thyroid has revealed some aspects of the interrelationships between Ca^{2+}, cGMP and cAMP, which may be of general interest.

VII CONCLUSION

The thyroid gland is subject to at least three types of control of varying physiological importance. TSH activates and iodide inhibits all aspects of specialized function and intermediary metabolism associated with the thyroid, whereas acetylcholine inhibits some of these functions and stimulates others. The mixed effects of acetylcholine do not allow the classification of the thyroid as either a uni or bidirectional system. Control of thyroid function by TSH is a good illustration of Sutherland's concept of cAMP as second messenger of hormone action. One mechanism of the inhibitory action of iodide on thyroid function is to block the activation of the cAMP system by TSH. Some effects of acetylcholine seem to be mediated by Ca^{2+}, but, for at least one of these, Ca^{2+} may act by a direct effect on cGMP. The thyroid certainly constitutes an original and useful experimental model in order to clarify the intricate relationships between cAMP, Ca^{2+} and cGMP and their respective roles in cell regulation.

ACKNOWLEDGEMENT

This work has been carried out under contract of the Ministère de la Politique Scientifique (Actions Concertées). It was supported by a grant of the Fonds de la Recherche Cancérologique de la Caisse Générale d'Epargne et de Retraite.

The authors would like to thank Mrs D. Leemans for the preparation of the manuscript.

CHAPTER 16

Cyclic Nucleotides and Hypothalamic Hormones

T. Kaneko

I INTRODUCTION

The existence of a regulatory system between the release of adenohypophyseal hormones and the hypothalamic releasing or inhibitory hormones is generally accepted.

A relationship between hypothalamus and anterior pituitary function had been suspected since the identification of the hypothalamo–hypophyseal portal vessel (Popa and Fielding, 1930) and the establishment of the direction of blood flow in this vessel (Green and Harris, 1947). The first experimental demonstration of neurohumoral control in the release of adenohypophyseal hormones was provided by Benoit and Assenmacher (1952), who have shown that sectioning of the portal vessels in the drake results in the atrophy of testis and thus have demonstrated the presence of a gonodotropic regulatory substance in the portal blood.

During the last two decades, many investigators tried to extract and purify hypothalamic releasing or inhibitory substances. First, Guillemin and coworkers (Burgus *et al.*, 1969) and Schally's group (Folkers *et al.*, 1969), purified TSH-releasing hormone (TRH) from sheep and porcine hypothalami, and its structure was determined as pyro–Glu–His–Pro–NH_2. When chemically synthesized, this peptide was shown to possess identical biological activity to that of purified natural TRH, confirming the tripeptide amide structure. Soon after the discovery of TRH, the primary structures of LH/FSH-releasing hormone (LH–RH/FSH–RH) (Matsuo *et al.*, 1971) and GH-release inhibitory hormone (somatostatin) (Brazeau *et al.*, 1973) have been determined. The synthetic hormones exhibited full biological activities. Thus, when sufficient amounts of the pure synthetic preparations became available further investigations of these hormones were made possible. Elucidation of the mechanism of action of the hypothalamic hormones on a molecular basis had thus become a most attractive subject for many endocrinologists.

By this time, cAMP had been widely accepted as the mediator of action of various hormones (Sutherland and Robison, 1966). The first observation suggesting a relationship between cAMP and hypothalamic hormones was reported by Schofield (1967). He demonstrated that theophylline, a phosphodiesterase inhibitor, stimulates GH release from the bovine adenohypophysis *in vitro*. It was subsequently found that release of TSH from adenohypophysis is also stimulated by theophylline (Wilber *et al.*, 1968). On the other hand, TSH, GH and ACTH release is stimulated by cAMP or Bt_2-cAMP (Wilber *et al.*, 1969; Gagliardino and Martin, 1968; Fleischer *et al.*, 1969). Wilber *et al.* (1969) showed that natural TRH stimulates cAMP formation in the adenohypophysis, and Zor *et al.* (1969, 1970) reported that ovine hypothalamic extracts increased both cAMP concentration and LH release in rat adenohypophysis *in vitro*. These results strongly suggest that hypothalamic hormones may control the release of adenohypophyseal hormones through the stimulation of adenylate cyclase in adenohypophyseal cells. However, the physiological regulation of secretion of these hormones seems to be more complicated and many problems still remain to be solved in order to clarify the regulatory mechanism of hormonal release.

Based on the results of experiments on TRH, LH–RH and somastatin, using highly purified synthetic materials, this paper will review the possible role of cyclic nucleotides in the release of adenohypophyseal hormones induced by hypothalamic hormones. In addition, some comments will be made on the role of Ca^{2+} with reference to the function of cyclic nucleotides in the secretory process, also, the effects of hypothalamic hormones on cAMP-dependent protein kinase activity in the adenohypophysis will be discussed.

II HYPOTHALAMIC RELEASING HORMONES

It was proposed by Sutherland and Robison (1966) that four criteria should be satisfied before it can be concluded that the action of a hormone in its target cell is mediated by cAMP.

The first criterion is activation of adenylate cyclase by the hormone in the plasma membrane of the target cell. Zor *et al.* (1969) showed that hypothalamic extracts stimulate adenylate cyclase activity in rat anterior pituitary homogenate after short incubation periods. Similar findings with crude hypothalamic extracts were described later (Steiner *et al.*, 1970; Jutisz *et al.*, 1972). On the other hand, Poirier *et al.* (1972) reported that synthetic TRH induces a slight but significant stimulation of adenylate cyclase activity in a plasma membrane fraction of bovine anterior pituitary gland. However, Jutisz *et al.* (1972) did not observe any significant activation of adenylate cyclase by pure

LH–RH in the homogenate or in plasma membranes of the adenohypophysis. These workers did find an activation of adenylate cyclase in whole anterior pituitary gland or hemipituitary glands, but only when incubated with a high dose of LH–RH. Similar observations were made with the response of testes to HCG (Dufau *et al.*, 1971) and of fat cell membranes to glucagon (Jarett *et al.*, 1971). Derry and Howell (1973), however, reported that synthetic LH–RH has little effect on the activity of adenylate cyclase in the adenohypophysis, but in the presence of *GTP*, LH–RF produces a marked stimulation of adenylate cyclase activity in the homogenate. The reason for the difficulty in demonstrating activation of adenylate cyclase by pure synthetic hormones in the homogenate or in plasma membrane fractions of the adenohypophysis is not understood. These observations may be explained, at least in part, by the small population of TSH or LH-secretory cells in the adenohypophysis or by the instability of the adenohypophyseal membrane fraction during homogenization and/or fractionation. The observation by Derry and Howell (1973) seems to suggest that GTP may modulate the response of the adenylate cyclase to a hormone in the adenohypophyseal homogenate similar to the activation by glucagon of liver parenchymal cell adenylate cyclase (Rodbell *et al.*, 1971a).

The second criterion for the participation of cAMP in a hormonal response is an increase in the concentration of cAMP in the target cell prior to the physiological response of the cell to the hormone. In contrast to the first

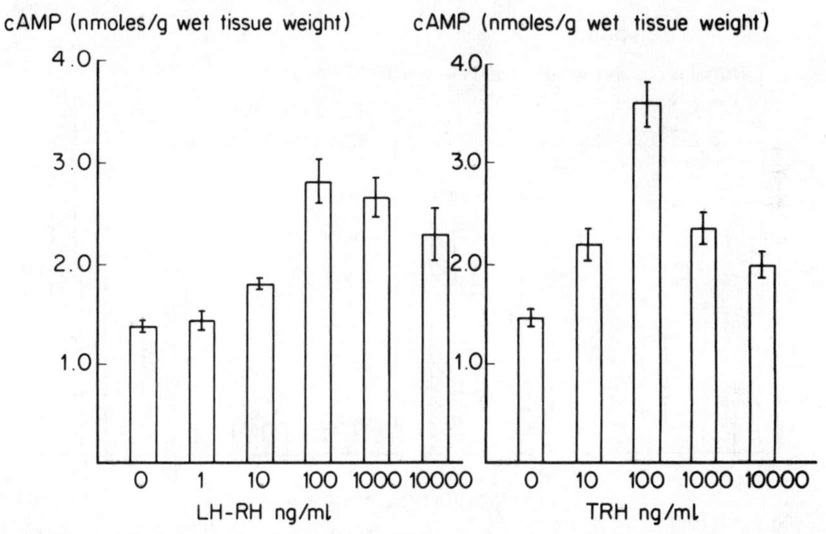

Figure 16.1 The effect of synthetic LH–RH and TRH on cAMP concentration of rat anterior pituitary *in vitro*. The tissue was incubated for 15 min in Krebs–Ringer bicarbonate buffer containing glucose (1 mg/ml), bovine serum albumin (1 mg/ml), theophylline (10^{-2} M) and the hormone. Values are means ± S.E.M. Three whole intact anterior pituitaries used in each experimental group. Triplicate determinations were performed for cAMP assay. (Data from Kaneko *et al.*, 1973c. Reproduced by permission of Excerpta Medica.)

criterion, several experiments have proven this phenomenon in the case of hypothalamic hormones.

As shown in Figure 16.1 cAMP formation in rat adenohypophysis is stimulated by synthetic TRH and LH-RH *in vitro* (Kaneko *et al.*, 1973b). The elevation of cAMP levels in the tissue is maximal at 100 ng/ml TRH or LH-RH and decreased when higher doses are used. This suggests that, under physiological conditions the response of adenohypophysis to the stimuli is regulated not only by the well-known feedback mechanism of peripheral hormones, but also by a mechanism which regulates cell's reactivity to the stimuli, or by an intracellular autoregulatory mechanism for the activation of adenylate cyclase or phosphodiesterase. The concentration of cAMP in the rat adenohypophysis is significantly increased by synthetic LH-RH within an incubation period of one minute, while the release of LH from the tissue into the incubation medium is not markedly increased until incubation for fifteen minutes (Figure 16.2). Examination of the effects of various synthetic analogues of LH-RH on the release of LH and cAMP production, reveals a parallel relationship between these activities in the adenohypophysis. The analogues of LH-RH which enhance LH release into the medium show a stimulatory effect on cAMP production in the glands, while those lacking LH-releasing activity do not (Kaneko *et al.*, 1973b, 1974a). The elevation of cAMP concentrations in the tissue prior to the final step of the adenohypophyseal cell response to LH-RH satisfies the second criterion and strongly supports the hypothesis that cAMP

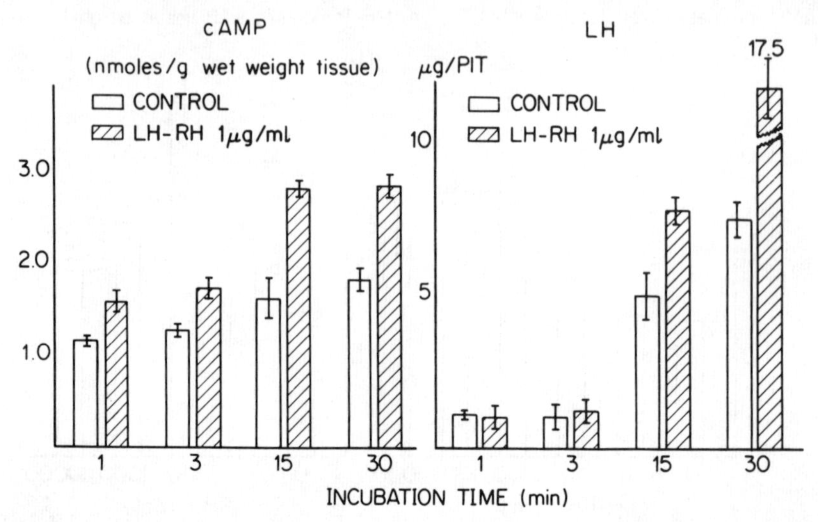

Figure 16.2 The effect of synthetic LH–RH on cAMP concentration and LH release into the incubation medium from rat anterior pituitary. The bars represent means ± S.E.M. Five whole intact anterior pituitaries were used in each experimental group. Triplicate determinations and duplicate determinations were performed for cAMP and LH, respectively. (Data from Kanko *et al.*, 1973c. Reproduced by permission of Exerpta Medica.)

might play an important role in the action of hypothalamic releasing hormones.

Stimulation of hormonal release from the adenohypophysis by cAMP or Bt_2-cAMP *in vitro* has been described for TSH (Wilber *et al.*, 1969), FSH (Jutisz and de la Llosa, 1970), LH (Ratner, 1970), ACTH (Gagliardino and Martin, 1968), GH (Macleod and Lehmeyer, 1970) and prolactin (Lemay *et al.*, 1972). It has also been reported that theophylline increases cAMP formation in the adenohypophysis (Fleischer *et al.*, 1969; Zor *et al.*, 1969; Steiner *et al.*, 1970) and stimulates the release of TSH (Wilber *et al.*, 1969; Steiner *et al.*, 1970), GH (Steiner *et al.*, 1970) and ACTH (Milligan and Kraicer, 1974). Although the release of LH and FSH is not stimulated by the phosphodiesterase inhibitor, the inhibitor does potentiate the stimulatory effect of synthetic LH-RH on release of FSH (Jutisz and de la Llosa, 1969; Jutisz and Kraicer, 1975). These observations meet the third and fourth criteria for the participation of cAMP in the process, as proposed by Sutherland and Robison (1966), which are the potentiation of a hormonal effect by a phosphodiesterase inhibitor and the reproduction of a hormonal response by cAMP or Bt_2-cAMP. The effect of hypothalamic releasing hormones on cAMP formation is specific for the adenohypophysis and has not been found in other tissues as indicated in Table 16.1.

Table 16.1 Specificity of THR stimulation of cAMP production

tissue	cAMP production (nmole/g wet weight tissue) control	THR (10µg/ml)	TSH (10mU/ml)	glucagon (10µg/ml)
adenohypophysis	1.2	2.8	1.1	
neurohypophysis	2.1	1.9	1.8	
thyroid gland	1.6	1.4	12.7	
liver	2.2	2.1		18.7

Rat tissue was incubated in Krebs–Ringer bicarbonate buffer containing glucose (1mg/ml), bovine serum albumin (1mg/ml), theophylline (10^{-2} M) and appropriate substance to be tested for 10 min at 37°C.
Values represent the mean of two experiments.

In response to a peptide hormone, the first step in the reactions taking place at the target cells is the binding of the hormone to its receptor on the cell surface. Using [^{3}H]TRH, two receptor sites specific for TRH in the adenohypophysis have been found (Grant *et al.*, 1973a), one with an affinity constant of 2×10^{-8} M and the other of 5×10^{-7}M. The presence of only one TRH-receptor which has an affinity constant of 3.5×10^{-8}M has also been reported by Barden and Labrie (1973). Grant *et al.* (1973b) showed that the biological activities of TRH and TRH analogues are parallel with their affinity constants for the TRH receptor. The value of this constant for TRH in the case of the cell-membrane fraction was ten times higher than that calculated from its biological activity to stimulate TSH release from the adenohypophysis. This suggests that only partial saturation of the binding sites on the cell surface gives rise to the maximal biological action as reported for of other tissues, or that the naturally occurring TRH is somewhat different from the synthetic tripeptide, as suggested for LH-RH by Fawcett *et al.* (1975). The presence of two kinds of binding sites in the adenohypophysis has also been reported for LH-RH. One has an affinity constant of about 1×10^{-9}M and is thought to be the site at which LH-RH binding can be replaced by an antagonist of LH-RH. The other site affinity constant of about 2×10^{-8}M) is rather nonspecific, although it does not bind TRH (Grant *et al.*, 1973a). The evidence which has been accumulated so far indicates that the action of at least TRH and LH-RH on the release process in the adenohypophysis is intimately associated with changes in the intracellular cAMP concentration.

By satisfying the criteria proposed by Sutherland and Robison (1966), these two hypothalamic releasing hormones can be accepted as belonging to a group of hormones whose actions on their target cells are mediated by cAMP. In the author's laboratory neither significant changes in cGMP concentration nor inhibition of phosphodiesterase activity have been observed in the adenohypophysis in response to TRH or LH-RH.

III HYPOTHALAMIC INHIBITORY HORMONES

Hypothalamic inhibitory, as well as releasing, hormones regulate the secretion of some adenohypophyseal hormones. Attention has been focused mainly on prolactin-release inhibitory hormone, GH-release inhibitory hormone (somatostatin) and MSH-release inhibitory hormone. More recently, the presence of a LH-release inhibitory hormone has also been demonstrated (Johansson *et al.*, 1975).

The structures of the prolactin-release inhibitory and the LH-release inhibitory hormones have not yet been determined. Somatostatin is a tetradecapeptide with a disulphide bridge (cyclic form); the linear form is also biologically active.

Since somatostatin is a peptide hormone like TRH or LH-RH, the first stage in its action is most likely its binding to a specific receptor site on the surface of GH-secretory cells. The hypothalamic releasing hormones (TRH and LH-RH), as already discussed, activate adenylate cyclase through binding to their receptors, resulting in augmented release of the respective hormone from the adenohypophysis. It is conceivable that somatostatin affects adenylate cyclase by competitively binding to the receptor of GH-releasing hormone in the target cell.

Kaneko *et al.* (1973a) and Borgeat *et al.* (1974) have demonstrated *in vitro* the inhibitory effects of somatostatin on cAMP formation in the rat adenohypophysis and on GH release into the incubation medium. Somatostatin decreases

Table 16.2 Effects of synthetic somatostatin on cyclic nucleotide levels in rat anterior pituitary gland in vitro

	somatostatin (ng/ml)	cAMP (pmole/g tissue)	cGMP (pmole/g tissue)
Experiment 1			
	0	1120 ± 25	21.8 ± 1.3
	1	996 ± 58	39.6 ± 4.1
	10	844 ± 25	40.9 ± 1.7
	100	727 ± 30	62.1 ± 2.5
	1000	680 ± 41	68.6 ± 3.1
	10000	635 ± 17	108.4 ± 3.0
Experiment 2			
	0	1200 ± 28	12.4 ± 0.8
	100	1033 ± 10	22.6 ± 0.9
	5000	830 ± 24	83.5 ± 1.2

Anterior pituitary glands were incubated in Krebs–Ringer bicarbonate buffer containing glucose (1mg/ml), bovine serum albumin (1mg/ml) and theophylline (10^{-2} M) with or without somatostatin for 10 min at 37°C.
Values represent the mean ± SEM of triplicate determinations.
(Data from Kaneko *et al.*, 1974b. Reproduced by permission of Academic Press)

not only the stimulatory effect of prostaglandins on cAMP formation and GH release, but also the basal level of cAMP in the adenohypophysis.

Hall *et al.* (1973) reported that somatostatin inhibits not only the release of GH, but also the release of TSH induced by TRH in man. Somatostatin also inhibits *in vitro* the stimulation of cAMP production by TRH in the rat adenohypophysis (Kaneko *et al.*, 1974b). Efforts to gain direct evidence of competition between somatostatin and hypothalamic GH-releasing factor (GH-RH) at the receptor site on GH-secretory cell membrane, have so far been unsuccessful. Since a generally accepted method of purification of GH-RH has not yet been evolved. Bowers and Folkers's group recently reported that two compounds, A–GH-RH and B–GH-RH, purified from porcine hypothalami, stimulate the release of immunoassayable GH *in vitro*. The stimulatory effect is inhibited by somatostatin (Currie *et al.*, 1974; Johansson *et al.*, 1974).

cGMP has attracted attention as another cyclic nucleotide which plays an important role in hormonal action. Recent data implicate cGMP in certain biological systems as a chemical mediator which acts antagonistically to the metabolic effects of cAMP. It has been reported that insulin provokes a transient increase in cGMP concentrations in rat liver slices and in isolated fat cells with no effect on cAMP levels (Illiano *et al.*, 1973). The effect of somatostatin on both cAMP and cGMP concentrations using rat whole anterior pituitary glands has been examined *in vitro*. As can be seen from Table 16.2 synthetic somatostatin at the minimum effective dose of 10 ng/ml raises cGMP concentrations in the adenohypophysis while decreasing cAMP concentrations (Kaneko *et al.*, 1974b). Prostaglandins are well-known stimulators of cAMP production and GH release in rat adenohypophysis.

Table 16.3 Inhibitory effect of synthetic somatostatin on thyrotropin-releasing hormone stimulation of cAMP production of rat anterior pituitary *in vitro*

	cAMP (pmole/g tissue)	cGMP (pmole/g tissue)
control	667 ± 53	42.3 ± 1.9
TRH (1μg/ml)	1463 ± 95	45.3 ± 1.9
TRH (1μg/ml) + somatostatin (5μg/ml)	790 ± 86	64.0 ± 1.5

Anterior pituitary glands were incubated in Krebs–Ringer bicarbonate buffer containing glucose (1mg/ml), bovine serum albumin (1mg/ml), theophylline (10^{-2} M) and appropriate substance(s) to be tested for 10 min at 37°C.
Values represent the mean ± SEM of triplicate determinations.
(Data from Kaneko *et al.*, 1974b. Reproduced by permission of Academic Press).

Synthetic somatostatin slightly, but significantly, inhibits the increase in adenohypophyseal cAMP induced by prostaglandins. The effect of TRH in increasing cAMP level in the tissue is also inhibited by somatostatin. It has, however, been observed that synthetic somatostatin also stimulates the production of cGMP in the adenohypophysis (Table 16.3 and 16.4).

Available data indicate that guanylate cyclase is located mainly in the soluble fraction of cells, thus differing from adenylate cyclase, although the presence of guanylate cyclase in the particulate fraction has been demonstrated. The production of cGMP induced by somatostatin in the adenohypophysis occurs within a very short incubation period. One may, therefore, speculate that guanylate cyclase is located very close to the adenylate cyclase system within the plasma membrane. Alternately, the enzyme may be loosely bound to the

Table 16.4 Inhibitory effect of synthetic somatostatin on prostaglandin E_1 stimulation of cAMP production in rat anterior pituitary *in vitro*

	cAMP (pmole/g)	cGMP (pmole/g)
control	1140 ± 55	33.3 ± 1.1
PGE_1 (1μg/ml)	37800 ± 1270	36.0 ± 2.4
somatostatin (10μg/ml)	820 ± 24	57.3 ± 1.1
PGE_1 (1μg/ml) + somatostatin (10μg/ml)	28000 ± 1604	65.0 ± 0.9

Anterior pituitary glands were incubated in Krebs–Ringer bicarbonate buffer containing glucose (1mg/ml), bovine serum albumin (1mg/ml), theophylline (10^{-2} M) and appropriate substance(s) to be tested for 10 min at 37°C.
Values represent the mean ± SEM of triplicate determinations.
(Data from Kaneko *et al.*, 1974b. Reproduced by permission of Academic Press)

membrane, so that the binding of somatostatin to its receptor (which may be the same as or different from that for GH-RH) affects both adenylate cyclase and guanylate cyclase resulting in the regulation of GH secretion from the adenohypophysis. This hypothesis is also supported by the observation of Rodbell *et al.* (1971a) who showed that GTP, the substrate for guanylate cyclase, regulates adenylate cyclase activity in liver and in the thyroid gland, apparently as an allosteric effector (Wolff and Cook, 1973). Additional support is provided by the finding of Kuo and Kuo (1973) that guanylate cyclase in rat lung is controlled by both β-adrenergic and muscarinic cholinergic receptors.

Secretion of GH from the adenohypophysis appears to be controlled by the releasing and inhibitory hormones of the hypothalamus. Peake *et al.* (1972) reported that cGMP is a potent GH secretagogue and can serve as an intracellular mediator of stimulation of GH release, while Labrie *et al.* (1973) demonstrated that a purified fraction of porcine hypothalami containing GH-RH stimulates cAMP production in the adenohypophysis. The author's results show that synthetic somatostatin at a dose which inhibits GH release prevents cAMP production and stimulates cGMP accumulation in the adenohypophysis simultaneously. These observations suggest that GH release from the adenohypophysis is controlled by both cAMP and cGMP in the gland and that somatostatin inhibits GH release by changes in the concentration of the cyclic nucleotides in the target cells.

In addition to the inhibitory effect of GH release, somatostatin is known to inhibit the increased release of TSH by TRH from the adenohypophysis and to supress the secretion of gastrin (Bloom *et al.*, 1974b), insulin (Alberti *et al.*, 1973), glucagon (Gerich *et al.*, 1974) and secretin (Boden *et al.*, 1975). It has been shown that cAMP concentrations in isolated rat islets are lowered by somatostatin (unpublished data).

The fact that somatostatin inhibits the increased release of TSH induced by TRH as well as the release of GH in the adenohypophysis may indicate a conformational similarity between the receptors for somatostatin and for TRH, or a common origin from which somatostatin and TRH receptor cells have developed.

Schultz *et al.* (1973a) demonstrated that Ca^{2+} is important for the control of cGMP levels in the ductus deferens of the rat. The effect of somatostatin on influx or intracellular redistribution of Ca^{2+} and interaction between cGMP and Ca^{2+} should also be clarified in order to determine the mechanisms of action of hypothalamic inhibitory hormones.

IV CALCIUM AND cAMP IN THE ADENOHYPOPHYSIS

Release of a hormone from the adenohypophysis induced by various kinds of secretagogues *in vitro* requires Ca^{2+} in the medium. Removal of extracellular Ca^{2+} depresses hormone release from the adenohypophysis evoked by hypothalamic extract. Addition of excess Ca^{2+} to this system restores the release process. These results suggest an interrelationship between Ca^{2+} and cAMP in the secretory process. cAMP may stimulate this process by enhancing the mobilization of Ca^{2+}, or cAMP. Alternatively, Ca^{2+} may independently affect the secretory process. In either case, both factors are essential for the hormonal secretion and both are controlled by physiological stimuli. In the absence of Ca^{2+}, stimulation of cAMP formation by hormones has been observed in the insect salivary gland (Rasmussen and Tenenhouse, 1968), dog thyroid gland (Dekker and Field, 1970) and cat adrenal cortex (Carcham *et al.*, 1971), as well as in the rat adenohypophysis (Zor *et al.*, 1970; Steiner *et al.*, 1970).

Assuming that there is a physiological relationship between Ca^{2+} and cAMP, it is necessary to determine if cAMP modulates Ca^{2+} metabolism, for example, by altering Ca^{2+} permeability of the cell membrane and/or by affecting the intracellular distribution of Ca^{2+}. There is also the possibility that the physiological stimulation modulates Ca^{2+} metabolism through an alternative mechanism without the participation of cAMP.

The adenohypophyseal hormone release *in vitro*, which is induced by a wide variety of secretagogues, is suppressed by removal of Ca^{2+} from the incubation medium. According to Eto *et al.* (1974), K^+ (57 mM) or ouabain (1×10^{-3}M) can increase the total Ca^{2+} level in the pituitary and the release of ACTH, GH and TSH. However, verapamil (5×10^{-5}M), an agent which inhibits Ca^{2+} influx, prevents the accumulation of $^{45}Ca^{2+}$ and the release of these hormones. Addition of verapamil does not influence ACTH release stimulated by crude hypothalamic extract, Bt_2-cAMP or theophylline, TSH release evoked by TRH or the GH release produced by hypothalamic extract. From these results, it is concluded that hypothalamic releasing hormones do not affect the process of 'stimulus–secretion coupling' by promoting a specific Ca^{2+} influx, and therefore differ from K^+ and ouabain.

Milligan *et al.* (1972) studied the effect of various secretagogues on Ca^{2+} influx by examining the uptake of $^{45}Ca^{2+}$ and [^{3}H]mannitol into rat adenohypophysis. They demonstrated that whereas a purified fraction of hypothalamus possessing GH-releasing activity increases the influx of Ca^{2+}, the augmented release of hormones induced by other secretagogues, including a crude hypothalamic extract, purified hypothalamic fraction possessing LH-RH activity, Bt_2-cAMP and theophylline, is not associated with any increased influx of the cation. These data seem to establish a critical role of Ca^{2+} in the release process of the adenohypophyseal hormones. Although physiological stimulators, such as hypothalamic releasing hormones, appear to affect both Ca^{2+} influx and the intracellular redistribution of Ca^{2+}, the latter might be the more crucial event in the process of hormone release.

It was reported by Lefkowitz *et al.* (1970) that the binding of ACTH to the adrenal receptor does not require Ca^{2+} and is inhibited by high concentrations of the cation. In the adenohypophyseal plasma membrane, [^{3}H]TRH binding is also inhibited by Ca^{2+} (Labrie *et al.*, 1972). These findings appear to eliminate an important role of Ca^{2+} in the ACTH and TRH–receptor interaction.

It has been proposed that cAMP activates phosphorylation of microtubular protein followed by Ca^{2+} activation of the phosphorylated protein moiety (Goodman *et al.*, 1970). Gillespie (1971) suggested that Ca^{2+} enhances the release of the hormones by promoting the formation of microtubules from available subunits. The intracellular action of the cation in the release reaction however, has, not yet been established.

Depending on the tissue used, Ca^{2+} may act in the opposite direction as in the case of parathyroid-hormone secretion and histamine release. Such co-operative or inhibitory actions of cAMP and Ca^{2+} have not been found in the release process in the adenohypophysis, while it has been observed that Ca^{2+} inhibits both the activity of adenylate cyclase and of cAMP-dependent protein kinase in the adenohypophysis. Therefore, Ca^{2+} may play another role as an intracellular modulator in the hormone-release reaction by regulating the activities of both adenylate cyclase and cAMP-dependent protein kinase.

V PROTEIN KINASES IN THE ADENOHYPOPHYSIS

The activation of cAMP-dependent protein kinase triggers the cell to respond to a hormone. Recently, the presence of a cAMP-dependent protein kinase has been demonstrated in bovine adenohypophysis. According to these investigations approximately 80% of the protein kinase is present in the cytosol fraction, although this enzyme is widely distributed throughout all the subcellular fractions of the adenohypophysis (Labrie *et al.*, 1971a; Lemaire *et al.*, 1971; Labrie *et al.*, 1971b). The addition of histone to a total adenohypophyseal homogenate leads to a four-fold stimulation of protein kinase activity, suggesting that the substrate may be the rate-limiting factor under conditions of maximal activation of the enzyme by cAMP. In the particulate fraction, little or no cAMP dependency of the protein kinase is observed in the presence of the endogenous substrate. cAMP dependency in all fractions is recovered when sufficient amounts of exogenous substrate (histone) are added.

The protein component of the secretory granules in bovine anterior pituitary gland is also phosphorylated by a protein kinase, but cAMP does not stimulate, and sometimes even decreases, the endogenous protein kinase activity of the secretory granules. These phenomena are assumed to be due to the presence of a maximal level of cyclic nucleotide in the particulate fraction or secretory granules. It was reported that a substantial amount of cAMP is present in fractions containing microsomes and nerve endings and fractions which possess high protein kinase activity (Lust and Goldberg, 1970; Maeno *et al.*, 1971). The partial recovery of the cAMP dependency in the presence of histone may suggest that components of the isolated secretory granules and other particulate fractions are phosphorylated by cAMP-independent protein kinase, while a cAMP-dependent enzyme, which is bound in some way to these components, is responsible for the increased phosphorylation in the presence of histone. Basal secretion of hormone in the gland seems to be controlled by this cAMP-independent protein kinase in the particulate fraction. It is assumed that the activation of protein kinase by the cyclic nucleotide is caused by a cAMP-induced dissociation of the receptor–catalytic complex within the cytosol followed by release of the active catalytic subunit and its binding to the various subcellular substrates, including protein components of the secretory granules.

Labrie *et al.* (1971a) reported that cAMP (2.5×10^{-8} M) causes the half-maximal stimulation of adenohypophyseal cAMP-dependent protein kinase. However, the concentration of cAMP induced by hypothalamic hormones reaches much higher levels than that corresponding to the apparent K_m value for cAMP of protein kinase.

The inhibitory effect of Ca^{2+} on endogenous phosphorylation of plasma membranes was demonstrated by Lemay *et al.* (1974), who showed that the phosphorylation occurs specifically in eleven of the thirty-six protein bands obtained using sodium dodecyl sulphate — gel electrophoresis of adenohypophyseal plasma membranes. The properties of these protein moieties are unknown. Therefore, it is suggested that protein kinases which are activated by cAMP, may affect the membrane functions of the cell and of the secretory granules or influence microtubule-associated protein components in the adenohypophysis.

Korenman *et al.* (1974) discussed the possibility of intracellular translocation of a catalytic subunit of protein kinase to the particulate fraction in uterine muscle treated with isoprenaline. It was suggested that such translocation is responsible for the effect of this agent on smooth muscle contraction. Keely *et al.* (1975a) considered the intracellular translocation of the catalytic subunit of protein kinase, at least in heart muscle, to be physiologically insignificant. This was based on the finding that the catalytic subunit of protein kinase binds not only to the particulate fraction but also to denatured proteins, and that the binding of the catalytic subunit to the particulate materials can be prevented or reversed by the addition of 150 mM KCl. On the other hand, Kuo (1975) speculated that a protein kinase modulator may play a physiological role in controlling the opposing effects of cAMP and cGMP by the regulation of the activities of their respective protein kinases.

No evidence has yet been presented to show that protein kinase activity in the adenohypophysis is altered by hypothalamic hormones. Investigations are currently being carried out using rat hemipituitary gland to determine how protein kinase activity is influenced by hypothalamic hormones *in vitro*. After incubation with or without the hypothalamic hormone to be tested, the tissue is homogenized and centrifuged. Protein kinase activity in the 20,000 *g* supernatant is assayed by the method of Kuo and Greengard (1970c) using calf thymus histone in the presence or absence of exogenous cAMP. Synthetic LH-RH increases the incorporation of ^{32}P into protein of the hemipituitary after ten minutes regardless of the presence or absence of exogenous cAMP. On the other hand, when the tissue is incubated with somatostatin for ten minutes, the protein kinase activity in the 20,000 g supernatant is reduced (see Table 16.5). These results suggest another possible action of hypothalamic hormones by which the availability of a cAMP-dependent protein kinase in the cell or the binding of

Table 16.5 The effect of hypothalamic hormones, prostaglandin E_1 and Bt_2-cAMP on protein kinase activity in the rat adenohypophysis

	protein kinase activity (pmole ^{32}P incorporated/ mg protein 30 min^{-1})	
	− cAMP	+ cAMP
control	292 ± 4	693 ± 7
LH-RH (1μg/ml)	652 ± 8	1293 ± 13
control	264 ± 3	594 ± 6
somatostatin (10μg/ml)	189 ± 3	285 ± 4
control	307 ± 1	638 ± 20
PGE_1 (1μg/ml)	367 ± 6	487 ± 2
control	284 ± 5	532 ± 8
Bt_2-cAMP (10^{-3} M)	489 ± 8	514 ± 4

Anterior hemipituitary was incubated in Krebs–Ringer bicarbonate buffer containing glucose (1mg/ml), bovine serum albumin (1mg/ml) and theophylline (10^{-2} M) with or without substance to be tested for 10 min at 37°C. Tissue was then homogenized and centrifuged. Protein kinase activity of the 20,000 *g*-supernatant was assayed by the method of Kuo and Greengard (1970) in the presence and absence of exogenous cAMP.
Values represent the mean ± SEM of triplicate determinations.

cAMP to the enzyme is increased. The protein kinase activity of the 20,000 *g* supernatant which was incubated with Bt_2-cAMP for ten minutes is almost independent of cAMP and its total activity, in the presence of exogenous cyclic AMP, is approximately equal to the control. No significant effect of TRH on the protein kinase activity in 20,000 *g* supernatant of hemipituitary has been observed in the presence or absence of exogenous cAMP under the same experimental conditions. Further investigations are necessary to evaluate the effect of hypothalamic hormones on protein kinase activity.

Somatostatin increases cGMP concentrations and decreases cAMP formation in the adenohypophysis *in vitro*, but the elevated cGMP levels do not reach high enough values to activate the cAMP-dependent protein kinase. Although cGMP-dependent protein kinases present in several mammalian tissues have been purified and partially characterized, the physiological substrate and the function of cGMP-dependent protein kinase (as well as the corresponding cAMP-dependent enzyme) have not yet been established in the adenohypophysis. It is, however, possible that cGMP interacts with cAMP, or that cGMP-dependent protein kinase interacts with the cAMP-dependent enzyme functionally. Both interactions may modulate the subsequent release process in the adenohypophysis.

VI CONCLUSION

On the basis of the available data, it may be concluded that cAMP and/or cGMP play an important role in the mechanism of action of hypothalamic releasing and/or inhibitory hormones and that these nucleotides regulate at least in part the secretory process of adenohypophyseal hormones.

Although the intracellular action of these nucleotides may be exerted through activation of cyclic nucleotide-dependent protein kinases, there still remains many difficult problems to be overcome before the exact physiological role of the cyclic nucleotides in regulation of the secretion of the adenohypophyseal hormones by hypothalamic hormones can be elucidated. Clarification is also requied of the role of the 'switch-off' mechanisms, involving phosphodiesterases and phosphoprotein phosphatases, in the regulation of the secretory process of the adenohypophysis.

CHAPTER 17

Regulation of cAMP Formation in Brain Tissue by Putative Neurotransmitters

P. Skolnick and J. W. Daly

I INTRODUCTION

cAMP has been firmly established as mediating the actions of a variety of hormones in target tissues outside the central nervous system (CNS) (Robison *et al.*, 1971). However, attempts to establish a second messenger role for cAMP in the actions of putative neurotransmitters and neurohormones in the CNS have been confounded by lack of knowledge of the precise functional roles of these neurotransmitters and neurohormones. The complex architectural and functional organizations of the brain have been major obstacles to a clear understanding of neurotransmitter function as well as understanding the role(s) that cAMP plays in regulating neurophysiological processes.

Several lines of investigation have been employed in an attempt to define better the role of cAMP in neuronal processes. These include: (*i*) enzymologic studies of the components of cAMP systems, including adenylate cyclases, phosphodiesterases, protein kinases and phosphoprotein phosphatases; (*ii*) studies of cultured cells derived from nervous tissue; (*iii*) *in vitro* studies of intact brain tissue, most notably using tissue slices; (*iv*) studies of specific neurones in intact brain, primarily using electrophysiological techniques; and (*v*) studies with intact animals in order to assign behavioural roles to cAMP or to assess the effects of drugs and neurohormones on *in situ* levels of cAMP.

Within the past two years, there have been several reviews dealing with these aspects of research on the role of cAMP in the CNS (Drummond and Ma, 1973; Von Hungen and Roberts, 1974; Daly, 1975a). The present review will be limited to the progress made towards an understanding of the regulation of cAMP formation by putative neurotransmitters in brain slices.

The brain slice technique, despite the obvious limitations of the heterogeneity of the tissue, even when taken from discrete anatomical regions, the possibility of tissue injury leading to artifacts and the difficulties in extrapolating the results obtained from tissue slice studies to physiological function, has proved to be a valuable system for elucidation of the role of cAMP in nervous function. It is particularly useful when used in conjunction with other neurochemical, electrophysiological or behavioural studies.

II METHODS

Studies on cAMP in the nervous system have been greatly facilitated by the numerous techniques now available for measurement of this cyclic nucleotide. The most widely used assay methods (see *Advances in Cyclic Nucleotide Research*, Volume 2) include protein-binding techniques (Gilman, 1972; Brown *et al.*, 1972), radioimmunoassay, and prelabelling techniques with radioactive adenine or adenosine (Shimizu *et al.*, 1969).

The technique for preparation of brain slices is as follows. The animal is rapidly killed, the brain is removed and placed in ice-cold Krebs-Ringer bicarbonate buffer, containing glucose (Elliot, 1955). The brain is then dissected and the blocks of tissue are chopped into uniform slices (200–300μm). Tissue slices are transferred to incubation medium at 37° which has been equilibrated with 95% O_2:5% CO_2. When a prelabelling technique is utilized, incubation with low concentrations (4–20 μM) of radioactive adenine (Shimizu and Daly, 1972a) is carried out for thirty minutes to one hour. After a suitable preincubation (usually 40 to 90 min), tissue is divided into portions and transferred to individual vessels for incubation with agonists. Incubations are terminated by addition of trichloroacetic acid, perchloric acid, or hydrochloric acid, or by boiling, and cAMP is determined.

In most brain-slice systems, for example, rat cerebral cortex (Perkins and Moore, 1973a) guinea-pig cerebral cortex (Shimizu *et al.*, 1970a; Schultz and Daly, 1973a) and rat hypothalamus (Weissman *et al.*, 1975a) a good correlation exists between endogenous and accumulation of radiolabelled cAMP. A cursory review of the literature on the effects of putative neurotransmitters on the formation of cAMP in brain slices will reveal that wide ranges for both basal and hormone-stimulated cAMP levels have been reported. There are several possible explanations for such differences; for example, different strains of animals may have been used. Skolnick and Daly (1974a) have demonstrated that significant differences in basal and stimulated levels of cAMP exist in several inbred strains of rats. The frequently used Sprague–Dawley-derived rat is an outbred strain; in brain slices from these rats the basal and hormone-stimulated values for cAMP vary considerably. Also, methodological differences in preparation and incubation procedures, cAMP and protein assays can influence the apparent levels of cAMP reported for brain slices (see Skolnick, *et al.*, 1975). Attempts to minimize inter-laboratory variation in observed values for levels of cAMP in brain slices by comparing percentage increases or Δ values for a stimulatory agent have not been any more satisfactory. Thus, in respect to the brain-slice preparations, only qualitative or semi-quantitative agreement of cAMP levels can be expected from different laboratories.

Phosphodiesterase Inhibitors

Mammalian brain contains a very high level of phosphodiesterase (Weiss and Costa, 1968), perhaps indicative of an important role for this enzyme in the regulation of steady-state levels of cAMP. Its activation and inhibition by chemical agents has been extensively studied both in crude and in purified preparations of phosphodiesterases from various sources (Amer and Kreighbaum, 1975). In brain slices, a number of phosphodiesterase inhibitors have been employed.

1 Theophylline

The earliest studies using theophylline in brain slices described a potentiation of the response of cAMP-generating systems to histamine in the presence of theophylline (Kakiuchi and Rall, 1968a,b). In a subsequent study, no effect on the histamine response was noted (Kakiuchi *et al.*, 1969). The inability of theophylline at concentrations of 0.5–1.0 mM to enhance consistently the accumulation of cAMP elicited in brain slices by putative neurotransmitters (Forn and Krishna, 1971; Sattin *et al.*, 1975; Schultz and Daly, 1973b) and its marked ability to antagonize the accumulation of cAMP brought about by adenosine in brain slices (Sattin and Rall, 1970) cast doubt on the usefulness of this methylxanthine as a phosphodiesterase inhibitor in brain-slice preparations. Theophylline induces large accumulations of cAMP in the cerebrospinal fluid of the intact rat (Kiessling *et al.*, 1975), but this methylxanthine has been shown to inhibit transport of cyclic nucleotides into the isolated choroid plexus (Hammers *et al.*, 1977).

2 1-Methyl-3-isobutyl-xanthine

This methylxanthine is a much more potent inhibitor of phosphodiesterases than theophylline (Amer and Kreighbaum, 1975), and is capable of enhancing amine-elicited accumulations in brain slices. However, like theophylline, 1-methyl-3-isobutyl-xanthine antagonizes accumulations of cAMP due to adenosine (Schultz and Daly, 1973b).

3 Benzylimidaxolidinones

One member of this class of phosphodiesterase inhibitors 4-(3-butoxy-4-methoxybenzyl)-2-imidazolidinone (Ro 20-1724), has been studied extensively with brain-slice preparations (Schultz, 1974b). Ro 20-1724 increases basal levels of cAMP and potentiates amine and adenosine-stimulated accumulation of cAMP in brain slices even at relatively low concentrations. However, it also inhibits the uptake of adenosine (Mah and Daly, 1976). Its effects on cAMP levels would appear mainly to be due to an enhanced net efflux of adenosine from slices rather than to inhibition of phosphodiesterases.

4 Pyrazolo-(3,4b)-pyridines

1-Ethyl-4-(isopropylidenehydrazino)-1H-pyrazalo-(3,4b)-pyridine-5-carboxylate ethyl ester (SQ 20009), the most studied member of this group of phosphodiesterase inhibitors, has a K_1 value of about 2 μM for the enzyme in rat-brain homogenates (Chasin *et al.*, 1972). In view of this very high high potency, it is surprising that in brain slices at 100–250-fold greater concentrations it has virtually no effect on amine-stimulated accumulations of cAMP (Schultz, 1974; Mah and Daly, 1976) or on levels of cGMP (Ohga and Daly, 1976). Forn *et al.* (1974) have reported that 1 mM SQ 20009 does enhance dopamine-elicited accumulation of cAMP in slices prepared from rat caudate nucleus.

5 Other inhibitors

Phenothiazines are potent inhibitors of Ca^{2+}-dependent phosphodiesterases in brain homogenates, but are much less effective against other isoenzymes (Weiss *et al.*, 1974). In brain-slice preparations, phenothiazines depress rather than potentiate accumulations of cAMP due to catecholamines (Uzunov and Weiss, 1971; Palmer *et al.*, 1972). This antagonism is probably due to either inhibition of adenylate cyclase (Uzunov and Weiss, 1971; Palmer and Manian, 1974b) by the phenothiazines or blockade of catecholamine receptor sites (Palmer and Manian, 1974a). At very high concentrations, phenothiazines enhance basal levels of cAMP through interference with the formation of ATP, leading to enhanced adenosine production (Huang and Daly, 1972, 1974a).

Papaverine has been used as a phosphodiesterase inhibitor in brain slices, but much of its effect on levels of cAMP appears to be due to inhibition of uptake of adenosine (Huang and Daly, 1972, 1974b). Similarly, diazepam, a particularly potent agent for potentiation of amine responses in brain slices, has been shown to inhibit uptake of adenosine (Mah and Daly, 1976).

Thus, not only do the various phosphodiesterase inhibitors exhibit a multiplicity of effects in brain slices, but questions of penetration and differential effects on isoenzymes of brain phosphodiesterases can lead to remarkable differences between their efficacies in brain-slice and cell-free preparations. Clearly, the indiscriminate use of agents which inhibit purified preparations of phospho-diesterase is not as yet warranted in brain-slice preparations.

III CATECHOLAMINE RESPONSES

The pioneering experiments of Rall and associates (Kakiuchi and Rall, 1968a,b), which demonstrated the noradrenaline-stimulated formation of cAMP in rabbit cerebral cortex, have led to numerous studies on the nature and regulation of cAMP formation by adrenergic agents in brain slices of many different species (Shimizu *et al.*, 1969; Forn and Krishna, 1971) including man (Berti *et al.*, 1972; Kodama *et al.*, 1973). Many attempts have been made to classify and to characterize the adrenergic receptors which regulate cAMP-generating systems in the CNS. In peripheral systems sensitive to catecholamines, stimulation of the β-adrenergic receptor is associated with an increase in cAMP formation, while stimulation of the α-adrenergic receptor is associated with no change or even a decrease in formation of the nucleotide (Rall, 1973, Robison *et al.*, 1971; Cramer *et al.*, 1973; Cramer and Lindl, 1974). In the initial studies by Kakiuchi and Rall (1968b), noradrenaline was shown to stimulate cAMP formation in rabbit cerebellar slices by interaction with a β-adrenergic receptor. However, Chasin *et al.* (1971, 1973) subsequently demonstrated that adrenaline stimulates the formation of cAMP in guinea-pig cerebral cortical slices through interaction with an α-adrenergic receptor. The adrenergic character of the cAMP-generating system in brain-slice preparations ranges from responses which appear to be primarily β-adrenergic in nature found, for example, in rabbit and rat cerebellum (Kakiuchi and Rall, 1968a,b), via the mixed α and β-responses found in rat cerebral cortex (Perkins and Moore, 1973a; Schultz and Daly, 1973c; Skolnick and Daly, 1975b) to the primarily α-adrenergic responses as found in guinea-pig cerebral cortex (Chasin *et al.*, 1973). α-Adrenergic antagonists effective in brain slices include phentolamine, dibenamine, phenoxybenzamine and dihydroergotamine. β-Adrenergic antagonists include sotalol, propranolol and dichlorisoprenaline.

A Cerebral Cortex

1 Primates

Noradrenaline has only a marginal (Shimizu and Daly, 1972a) or no effect (Forn and Krishna, 1971) on cAMP formation in cerebral cortical slices from the rhesus monkey. In contrast, in studies using the squirrel monkey, noradrenaline elicits a four to five-fold increase in cAMP formation in slices from the polysensory, auditory and visual cortices (Skolnick *et al.*, 1973). Two groups have investigated catecholamine-stimulated formation in human cerebral cortex obtained by biopsies. Noradrenaline elicits a three to four-fold increase (Berti *et al.*, 1972) or a twelve to forty-fold increase (Kodama *et al.*, 1973) in cAMP formation. Adrenaline and isoprenaline are more effective in increasing cAMP levels than noradrenaline. The noradrenaline response is inhibited by β-antagonists. Thus, the data from human cortical slices is suggestive of a β-adrenergic response.

2 Rabbit

In cerebral cortical slices from rabbits, noradrenaline has been reported to bring about either no stimulation or less than a two-fold increase in cAMP levels (Kakiuchi and Rall, 1968a; Shimizu *et al.*, 1970a; Forn and Krishna, 1971; Schmidt and Robison, 1971). No data on the nature of the adrenergic receptor have been reported. Schmidt and Robison (1971) observed a marked stimulation of cAMP by noradrenaline in slices from frontal cortex of new-born rabbits. The magnitude of the response reaches a maximum nine to fourteen days postpartum and then declines to finally disappear in the adult rabbit.

3 Guinea-pig

Noradrenaline produces either no stimulation or only a small increase in levels of cAMP in guinea-pig cortical slices (Kakiuchi *et al.*, 1969; Shimizu *et al.*, 1969; Sattin and Rall, 1970, Schultz and Daly, 1973a). The stimulatory effect of adrenaline is blocked by α-antagonists but is unaffected by β-antagonists (Chasin *et al.*, 1971). Furthermore, isoprenaline, a pure β-agonist is only marginally effective in stimulating the formation of cAMP. These data are compatible with the presence of primarily or solely α-receptor-mediated responses to catecholamines in guinea-pig cortex.

Schultz (1975a) has recently demonstrated that while noradrenaline has little or no effect on cAMP levels under standard incubation conditions, it elicits a significant accumulation of cAMP in slices incubated in a Ca^{2+}-deficient medium. EGTA is used to reduce the extracellular Ca^{2+} concentration. However, it has been demonstrated in the authors' laboratory that theophylline or adenosine deaminase prevents the stimulatory effect of noradrenaline in guinea-pig cortical slices incubated in medium containing EGTA. The latter result suggests that reduction of extracellular Ca^{2+} by EGTA leads to an enhanced efflux of adenosine, and that the released adenosine potentiates the response to noradrenaline. The observation that adenosine deaminase reduces the response to noradrenaline but not that to histamine or the adenosine analogue N^6-phenylisopropyladenosine suggests that adenosine plays a critical role in the action of noradrenaline in guinea-pig cortex.

4 Rat

The response of cAMP-generating systems to catecholamines in rat cerebral cortical slices appears to exhibit both α and β-adrenergic characteristics (Perkins and Moore, 1973a; Palmer *et al.*, 1973; Schultz and Daly, 1973c; Huang *et al.*, 1973a; Skolnick and Daly, 1975b). Thus isoprenaline, although a more potent β-agonist than noradrenaline, brings about accumulations of cAMP significantly lower than those due to noradrenaline. The accumulation of cAMP due to noradrenaline can be blocked only partially by α or β-antagonists, while combinations of α and β-antagonists completely block the response (Perkins and Moore, 1973a; Skolnick and Daly, 1975b). The apparent β-adrenergic component of the response to noradrenaline appears to represent somewhat more than half of the total response (Perkins and Moore, 1973a;

Huang *et al.*, 1973a; Schultz and Daly, 1973c) but the proportion varies with different strains of rat (Skolnick and Daly, 1974a). Current studies utilizing both β_1 and β_2-agonists and antagonists suggest that both β_1 and β_2-adrenergic receptors are involved in the regulation of cAMP formation in rat cerebral cortical slices.

Classical α and β-adrenergic antagonists are not, however, the only agents capable of antagonizing the catecholamine-stimulated formation of cAMP in rat cerebral cortical slices. Uzunov and Weiss (1971) have shown that the phenothiazines, trifluoperazine and chlorpromazine are capable of completely blocking the response to noradrenaline in slices from rat cerebrum. However, at the ver high concentrations employed phenothiazines inhibit both basal and F^- stimulated adenylate cyclases in brain homogenates (Uzunov and Weiss, 1971; Palmer and Manian, 1974b). Clinically active phenothiazines and their metabolites do partially inhibit noradrenaline-stimulated formation of cAMP in cerebral cortical slices at concentrations of only 1–100 μM (Palmer and Manian, 1974a).

Skolnick and Daly (1975b) found that clonidine (2,6-dichlorophenylamino-2-imidazolidine), which is considered to be an α-agonist in both the periphery and in the CNS (Hoefke *et al.*, 1975a), is capable of inhibiting the α-adrenergic component of the noradrenaline-stimulated accumulation of cAMP. In the presence of clonidine, noradrenaline brings about accumulations of cAMP equal in magnitude to those due to isoprenaline. In the presence of clonidine, the α-antagonist phenoxybenzamine does not further antagonize the noradrenaline-stimulated accumulations of cAMP, while a combination of clonidine and the β-blocker, propranolol completely blocks the response to noradrenaline. Clonidine completely blocks the response of the cAMP-generating systems to the α-agonist, methoxamine (see below), and has no effect on the response to maximal concentrations of isoprenaline.

Certain other α-agonists, such as oxymetazoline, naphazoline and tetrahydrazoline, inhibit noradrenaline-stimulated accumulations of cAMP in rat cortical slices (Skolnick and Daly, 1975b). However, the α-agonists, phenylephrine and metaraminol, are ineffective in inhibiting the α-adrenergic component of noradrenaline action. It should be noted that the α-agonists which effectively block the α-adrenergic component of noradrenaline-stimulated formation of cAMP in rat brain slices all contain an imidazole moiety.

α-Adrenergic agonists, including clonidine, oxymetazoline, phenylephrine, naphazoline and tetrahydrazoline, potentiate rather than inhibit the response of cortical cAMP-generating systems to submaximal concentrations of isoprenaline (Skolnick and Daly, 1975b, 1976), while they have no significant effect on responses to maximal or supramaximal concentrations of the β-agonist isoprenaline. However, Schultz (personal communication) has found that neither clonidine nor phenylephrine potentiates the isoprenaline-elicited accumulation of cAMP in cortical slices from Sprague–Dawley rats. In the latter study, 0.1 μM isoprenaline did not stimulate cAMP accumulation in contrast to the results of Skolnick and Daly (1975b, 1976). The potentiative effect of clonidine on responses to low concentrations of isoprenaline would appear to reflect a clonidine-induced increase in the affinity of isoprenaline for the β-receptor–adenylate cyclase complex. It is not yet completely clear if the inhibition of the α-component of the noradrenaline response and the potentiation of the response to isoprenaline by clonidine occur at the same or at different loci. Both effects are still manifest in slices from rats in which presynaptic noradrenergic terminals have been selectively destroyed by pretreatment with 6-hydroxydopamine (Skolnick and Daly, 1976). Thus both actions of clonidine apparently occur at a postsynaptic locus.

The α-adrenergic agonist, methoxamine, has been found to produce a significant accumulation of cAMP in rat cortical slices (Skolnick and Daly, 1975c). The response to methoxamine is blocked by α-antagonists and is unaffected by β-antagonists. Surprisingly, the response to a combination of methoxamine and noradrenaline is additive. It has been proposed that cerebral cortical tissue contains a set of α-adrenergic receptors which are activated by methoxamine but not by noradrenaline. In the presence of adenosine, however, the methoxamine-sensitive adrenergic α-receptors can be activated by noradrenaline.

The accumulation of cAMP stimulated by methoxamine in Sprague–Dawley rats has been confirmed by Schwabe and Daly (1976), who demonstrated net increases of approximately 15 pmole/mg protein over the basal level. In contrast, Schultz (personal communication) has been unable to show an increase in cAMP over basal in the presence of methoxamine in cortical slices using a Sprague–Dawley strain from German sources, but methoxamine competitively inhibited the noradrenaline effect. Methoxamine did, however, bring about an accumulation of cAMP in the presence of adenosine.

The presence of adrenaline and the enzyme responsible for its formation, phenethanolamine-*N*-methyltransferase, has been demonstrated in rat brain (Ciaranello *et al.*, 1969). However, stimulatory effects of this catecholamine on cAMP formation in rat cerebral cortex have not been extensively studied. Schultz and Daly (1973c) have demonstrated that adrenaline produces an accumulation intermediate to that elicited by noradrenaline and isoprenaline, which is very similar to the effects of catecholamines in rat superior cervical ganglia (Cramer *et al.*, 1973). The response in brain slices is partially blocked by α or β-antagonists.

Prostaglandins of the E series at relatively high concentrations, but not of A, B, or F series, have been shown to stimulate the formation of cAMP in rat-brain slices (Berti *et al.*, 1972; Kuehl *et al.*, 1972; Dismukes and Daly, 1975a); maximal accumulations are reached at approximately 85 μM. Using 3 μM or less, PGE_1 has no effect on cAMP formation in rat-brain slices (Palmer *et al.*, 1973). Prostaglandin-stimulated formation of cAMP appears to be unique in several respects, and may involve an interaction with the receptors which mediate catecholamine-sensitive formation of cAMP. Although the accumulation of cAMP elicited by prostaglandin is not blocked by either classical α or β-adrenergic antagonists, combinations of adenosine, isoprenaline or noradrenaline with prostaglandin E_1 do not have additive effects on formation of cAMP. This may indicate either a common locus of stimulatory action or inhibition of adenosine and catecholamine-responses by the prostaglandin (Dismukes and Daly, 1975a).

5. *Mouse*

The effect of catecholamines on formation of cAMP in mouse cerebral cortical slices has not been studied extensively. Schultz and Daly (1973c) have reported a five to seven-fold increase in cAMP formation in an outbred mouse strain using noradrenaline, adrenaline and isoprenaline. The response to isoprenaline is significantly less than the responses to the other catecholamines. The latter responses are partially blocked by both α and β-antagonists, with the β-antagonists being more effective.

B Cerebellum

1. *Rabbit*

In rabbit cerebellar slices, the accumulation of cAMP brought about by noradrenaline is blocked by a β-antagonist but unaffected by an α-antagonist (Kakiuchi and Rall, 1968a,b). Chlorpromazine, at a relatively high concentration, partially antagonizes the response to noradrenaline. Isoprenaline produces an accumulation of cAMP similar in magnitude to that due to noradrenaline.

2. *Guinea Pig*

Accumulations of cAMP stimulated by noradrenaline in guinea-pig cerebellar slices are blocked by propranolol, pronethalol and dichloriso-prenaline, while α-antagonists, such as phentolamine and phenoxybenzamine, are ineffective in reducing this response (Chasin *et al.*, 1971).

3. *Rat*

Basal levels of cAMP as high as 300–400 pmole/mg protein have been reported in Sprague–Dawley and F-344 rat cerebellar slices (Palmer *et al.*, 1973, Skolnick *et al.*, 1976). In contrast, levels in cortical slices are only 10–30 pmole/mg protein. Theophylline greatly reduces the high levels of cAMP observed in the cerebellar slices, providing evidence for the involvement of adenosine. In another rat strain (Carworth CFN), basal levels of cAMP in cerebellar slices are only about 50 pmole/mg protein and are reduced to about 10 pmole/mg protein in the presence of theophylline (Hoffer *et al.*, 1976). Noradrenaline produces only a two-fold increase in cAMP levels in rat cerebellar slices in the absence of theophylline, compared with a nearly six-fold increase after reduction in the basal levels by theophylline (Skolnick *et al.*, 1976). Under these conditions the potency of isoprenaline is increased to approximately that of noradrenaline. Responses to noradrenaline and isoprenaline are blocked by sotalol. The neuroleptics, fluphenazine and α-flupenthixol, at concentrations of 100 μM or less, inhibit catecholamine-elicited accumulations of cAMP by 30–60% in cerebellar slices (Skolnick *et al.*, 1976; Hoffer *et al.*, 1976). Clinically inactive phenothiazines and thioxanthenes, such as promethazine and β-flupenthixol, have no effect. Uzunov and Weiss (1971) had previously reported the inhibition of responses to noradrenaline by phenothiazines in rat cerebellar slices.

The blockade of catecholamine-elicited accumulations of cAMP by clinically active neuroleptics, but not by inactive analogues, is mirrored in their blockade of catecholamine-induced inhibition of Purkinje-cell firing (Skolnick *et al.*, 1976). In view of the total inhibition of noradrenaline-stimulated accumulations of cAMP by sotalol and the only partial inhibition by the neuroleptics, it has been postulated that there are at least two populations of cells in the cerebellum containing catecholamine-sensitive cAMP-generating systems. The receptors of one population, including the cerebellar Purkinje cells, may be blocked by both β-blockers and clinically active neuroleptics. This hypothesis has been tested with rats whose late-maturing cerebellar elements (granule, basket and stellate cells) had been destroyed by neonatal X-irradiation. The early-maturing Purkinje cells are relatively unaffected by this treatment. Fluphenazine has been found to completely antagonize the noradrenaline-stimulated formation of cAMP in cerebellar slices from these irradiated rats. These data strongly support the proposal that only the receptors of certain elements of the cerebellum, including Purkinje cells, are sensitive to blockade by neuroleptics. The accumulation of cAMP stimulated by noradrenaline is reduced in the irradiated animals compared with non-irradiated controls (Hoffer *et al.*, 1976).

The demonstration that neuroleptics inhibit a noradrenaline-stimulated accumulation of cAMP in slices from cerebellum, the brain area, which has no known dopaminergic innervation should renew interest in the interaction between neuroleptics and noradrenergic pathways *in vivo*. Neuroleptics have been previously reported to inhibit noradrenaline-stimulated formation of cAMP in slices from cerebral cortex (Palmer and Manian, 1974a) and limbic cortex (Blumberg *et al.*, 1975). However, the argument could be made that noradrenaline stimulates dopamine receptors present in these areas and that the neuroleptics inhibit only this component of the noradrenaline-response. It should be emphasized that dopamine itself does not elicit significant accumulations in slices from these brain regions (Blumberg, *et al.*, 1975; Dismukes and Daly, 1974).

4 Mouse

In mouse cerebellar slices, noradrenaline and adrenaline both stimulate cAMP levels approximately forty-fold (Ferrendelli *et al.*, 1975). This response is almost completely blocked by either α or β-adrenergic antagonists. Isoprenaline brings about a much smaller accumulation of cAMP in mouse cerebellar slices than do either adrenaline or noradrenaline. The α-adrenergic agonist, phenylephrine brings about a significant accumulation of cAMP. These observations indicate that adrenergic receptors controlling cAMP-generating systems in this tissue differ significantly from those in cerebella of other species.

C Hypothalamus

1 Primates

Noradrenaline produces a five-fold increase in levels of cAMP in hypothalamic slices from the rhesus monkey (Forn and Krishna, 1971).

2 Rabbit

In adult rabbit hypothalamus, noradrenaline elicits less than a two-fold increase in cAMP levels (Kakiuchi and Rall, 1968a,b; Forn and Krishna, 1971; Schmidt and Robison, 1971). In contrast, a seven to eight-fold stimulation by noradrenaline has been observed in hypothalamus from immature rabbits.

3 Rat

A seven-fold increase in levels of cAMP is produced by noradrenaline in hypothalamic slices from adult rats (Palmer *et al.*, 1972, 1973). Adrenaline and isoprenaline do not elicit such large accumulations of the nucleotide. The classical α-agonists, phenylephrine and metaraminol, produce small but significant stimulation of cAMP levels in rat hypothalamic slices. Despite the fact that this brain region contains both dopaminergic cell bodies and terminals (Fuxe and Hökfelt, 1969), dopamine has no effect on cAMP formation in hypothalamus (Palmer *et al.*, 1973). Both α and β-antagonists are capable of completely blocking the response to noradrenaline in this preparation (Palmer *et al.*, 1973). Blockade of the responses to noradrenaline by either α or β-antagonists in hypothalamic slices has been confirmed by two other laboratories (Gunaga and Menon, 1973; Weissman *et al.*, 1975a). A variety of psychoactive compounds, such as chlorpromazine, prochlorperazine, imipramine, Li^+, and *p*-chloramphetamine, inhibit noradrenaline-elicited accumulations of the nucleotide in rat hypothalamus (Palmer *et al.*, 1971; 1972).

Oestrogenic compounds are capable of elevating cAMP levels in the hypothalamus both *in vitro* and *in vivo* and this effect appears to be mediated *via* catecholamine release (Gunaga and Menon, 1973; Gunaga *et al.*, 1974; Weissman and Skolnick, 1975; Weissman *et al.*, 1975a). Also, Gunaga and Menon (1973) reported that oestradiol-17β produces a significant formation of cAMP in whole hypothalami from immature female rats.

The increases in cAMP due to oestrogens differ in several respects from those elicited by putative neurotransmitters. The responses to putative neurotransmitters occur rapidly, being significant within one minute of their introduction. In contrast, oestrogens, such as oestradiol-17β and diethylstilboestrol, stimulate the formation of cAMP only after prolonged incubation, significant accumulations occurring only after forty to fifty minutes. This delayed response strongly indicates an indirect mode of action. Both α and β-adrenergic blocking agents are capable of blocking the effect of oestradiol (Gunaga and Menon, 1973), leading to the proposal that oestrogens stimulate cAMP accumulation via adrenergic mechanisms. The magnitude of stimulation by oestrogenic compounds (Weissman *et al.*, 1975a; Weissman and Skolnick,

1975) appears to correlate to their *in vivo* potency as oestrogens and their ability to bind to cytoplasmic oestrogen receptors. Clomiphene, the oestrogen antagonist, inhibits the response to diethylstilboestrol *in vitro*, but only if the antagonist is added together with the oestrogen. Further studies support the intermediacy of catecholamines in the response to oestrogens in hypothalamus since combinations of oestradiol and noradrenaline elicit accumulations of cAMP no greater than those due to noradrenaline alone. Destruction of intra-hypothalamic neural networks by chopping the slices abolishes responses to oestradiol and to low concentrations of diethylstilboestrol, but does not affect the response to noradrenaline (Weissman *et al.*, 1975a).

In toto, the data indicate that oestrogens stimulate the formation of cAMP in the rat hypothalamus by initially interacting with cytoplasmic sites in cell bodies. A time-dependent transport process which may involve axoplasmic flow of protein and/or catecholamines leads to an enhanced release of catecholamines from the terminals and stimulation of catecholamine-sensitive adenylate cyclases at postsynaptic sites.

More recent investigations (Gunaga *et al.*, 1974) have demonstrated similar stimulatory effects of oestrogens *in vivo*. After intraperitoneal injection of oestradiol benzoate into immature female rats, an enhanced formation of cAMP was noted in the hypothalamus. This increase can be prevented by pretreating the animals with either clomiphene or adrenergic blockers. The involvement of oestrogen–catecholamine–cAMP mechanisms in the feedback control by oestrogens on gonadotrophin secretion at the level of the hypothalamus requires further study.

D Basal Ganglia

The effects of catecholamines on the formation of cAMP in elements of the basal ganglia have been a source of controversy because of the lack of responses to dopamine in tissue-slice preparations. This contrasts to the presence of an adenylate cyclase capable of being stimulated by dopamine in cell-free preparations of the caudate nucleus—see review by Iversen (1975). Noradrenaline does elicit a significant accumulation of cAMP in slices of caudate nucleus, the most extensively studied portion of the basal ganglia.

More recently, Forn *et al.* (1974) have observed that dopamine can approximately double the cAMP formation in slices of rat caudate nucleus in the presence of high concentrations of a phosphodiesterase inhibitor, 1-methyl-3-isobutylxanthine. The dopaminergic agonist, apomorphine, has similar effects. Noradrenaline and isoprenaline, under the same conditions, bring about three to four-fold increases in the nucleotide concentration. Propranolol inhibits the responses to noradrenaline and isoprenaline, but it has no effect on the response to dopamine. Fluphenazine blocks the response to dopamine and slightly inhibits the response to noradrenaline. These results are consistent with the presence of cAMP-generating systems controlled by dopaminergic and by β-adrenergic receptors in caudate nucleus. The lack of response to dopamine in the absence of a phosphodiesterase inhibitor suggests that the dopamine-sensitive compartment has high levels of phosphodiesterase activity.

Using slices of rat striatum prelabelled with radioactive adenine, Harris (1976) reported results seemingly at variance with those of Forn *et al.* (1974). Noradrenaline, adrenaline and isoprenaline are much more potent than dopamine in eliciting accumulation of radioactive cAMP. Dopamine and *d*-noradrenaline display similar concentration–response curves. Harris (1976) found no additive effects with combinations of dopamine and noradrenaline or with dopamine and isoprenaline. The response to dopamine is antagonized by β-blockers, but is unaffected by trifluoperazine and chlorpromazine. These results suggest that dopamine stimulates accumulation of labelled cAMP by interaction not with a dopamine receptor but rather with a β-adrenergic receptor. It is possible that radioactive adenine selectively labels ATP compartments associated with β-adrenergically controlled cAMP systems as found in rat superior cervical ganglia (Lindl *et al.*, 1975). Further experiments are needed to clarify the factors involved in dopaminergic control of cAMP-generating systems in brain.

E Limbic System

There have been several studies conducted on various elements of the limbic system. Noradrenaline elicits a five to six-fold increase in cAMP levels in cingulate gyrus from the squirrel monkey (Skolnick *et al.,* 1973). In hippocampal slices from neonatal rabbits, noradrenaline produces a four to five-fold increase in levels of the nucleotide (Schmidt and Robison, 1971). This response in rabbit hippocampus is virtually absent by day 22 *post partum*. Adrenaline elicits accumulations of cAMP in slices prepared from amygdala and hippocampus (Chasin *et al.*, 1973). These responses are inhibited by α but not by β-antagonists. A three-fold increase in cAMP is produced by noradrenaline in slices of rat hippocampus (Palmer *et al.*, 1973) and a four to five-fold increase in cAMP occurs in slices from 'limbic forebrain', which includes amygdala, pre-optic area, olfactory tubercle and portions of the nucleus accumbens, nucleus interstitialis and stria terminalis (Blumberg *et al.*, 1975). Although dopamine is ineffective in limbic forebrain, pimozide, a specific dopaminergic antagonist, effectively inhibits the response to noradrenaline. Clozapine, an antipsychotic with little or no anti-dopaminergic properties, also effectively inhibits the noradrenaline response. These observations emphasize the need for further investigations into the action of neuroleptics in both cell-free and brain-slice preparations.

F Brain stem

Noradrenaline brings about a two-fold increase in cAMP in slices of rabbit brain stem obtained from animals pretreated with reserpine (Kakiuchi and Rall, 1968b). Several groups have reported noradrenaline-elicited accumulations of cAMP in rat brain stem. Uzunov and Weiss (1971) reported a two to three-fold increase which is completely blocked by high concentrations of chlorpromazine or trifluoperazine. The response to noradrenaline in rat brain-stem slices is significantly inhibited by prochlorperazine, lysergic acid diethylamide and 2-bromolysergic acid diethylamide (Palmer *et al.*, 1972, 1973).

IV HISTAMINE RESPONSES

A Cerebral Cortex

Histamine has only marginal effects or no effect at all in cortical slices from two species of monkey (Forn and Krishna, 1971, Shimizu and Daly, 1972a; Skolnick *et al.*, 1973), and only a marginal effect in human cortical slices (Shimizu *et al.*, 1971). Responses to histamine were initially reported in guinea-pig cerebral cortical slices by Kakiuchi *et al.* (1969), and have subsequently been studied in this species in some detail (Daly, 1975). Early investigations indicated that relatively high concentrations of classical H_1-histaminergic antagonists, such as chlorpheniramine and pyrilamine, completely block histamine responses in guinea pig-cerebral cortical slices (Chasin *et al.*, 1971; 1973). Subsequent investigations with both H_1 and H_2 antagonists and agonists indicate that, in guinea-pig cerebral cortex, histaminergic receptors exhibit both H_1 and H_2-characteristics (Rogers *et al.*, 1975; Dismukes *et al.*, 1976), and that the H_1-receptors appear to predominate. Histamine has either a marginal (Sattin and Rall, 1970; Shimizu *et al.*, 1970) or no (Forn and Krishna, 1971; Palmer *et al.*, 1973) effect on cAMP in rat cerebral cortical slices. However, in the presence of phosphodiesterase inhibitors, a significant accumulation of cAMP is elicited by histamine and this response is antagonized by H_2-antagonists or by high concentrations of H_1-antagonists (Dismukes *et al.*, 1976). In chick cerebral slices, histamine responses are blocked effectively only by the H_2-antagonist metiamide (Nahorski *et al.*, 1974).

B Cerebellum

In rabbit cerebellar slices, histamine produces a three to ten-fold increase in cAMP levels (Kakiuchi and Rall, 1968a,b; Shimizu *et al.*, 1970a). Histamine has no effect in guinea-pig (Ohga and Daly, 1976) or mouse (Ferrendelli *et al.*, 1975) cerebellar slices.

C Limbic System

Histamine brings about very large increases in cAMP in guinea-pig hippocampal slices (Chasin *et al.*, 1973; Rogers *et al.*, 1975). H_1-Antagonists, such as brompheniramine, can partially block the response to histamine, as can H_2-antagonists including metiamide. Combinations of H_1 and H_2-antagonists have additive inhibitory effects on responses to histamine. Recently, clonidine has been found to stimulate cAMP formation in hippocampal slices via histaminergic receptors.

V 5-HYDROXYTRYPTAMINE RESPONSES

In slices of polysensory cortex from squirrel monkey, 5-hydroxytryptamine (5-HT) elicits a two-fold stimulation of cAMP levels (Skolnick *et al.*, 1973). This region is rich in serotonergic nerve terminals. A similar response to 5-HT has not been reported from studies with slices from other species and regions of the brain (Daly, 1975). For example, even in the presence of a phosphodiesterase inhibitor, 5-HT has no effect on cAMP levels of this nucleotide in slices of rat cerebral cortex (Dismukes and Daly, 1974). The reason for lack of stimulation of cAMP formation in brain slices by 5-HT is reminiscent of the situation with dopamine and merits further investigation.

VI ADENOSINE RESPONSES

Adenosine has become established as an important endogenous stimulant of cAMP-generating systems in brain slices from virtually all species and regions studied. Adenosine-stimulated accumulations of cAMP have been studied most extensively using guinea-pig cerebral cortical slices, but similar studies have also been carried out using cerebellar slices from guinea pig (Zanella and Rall, 1973), rabbit cerebral cortical slices, rat cortical, cerebellar and midbrain–striatal slices (Sattin and Rall, 1970; Schultz and Daly, 1973c; Skolnick and Daly, 1974a), mouse cortical and cerebellar slices (Sattin and Rall, 1970; Skolnick and Daly, 1974b; Ferrendelli *et al.*, 1975) and cortical slices from monkey (Skolnick *et al.*, 1973) and humans (Berti *et al.*, 1972, Kodama *et al.*, 1973). In contrast, no response to adenosine has been detected in chick brain slices 1974).

These studies are based on the observation that electrical stimulation of guinea-pig cortical slices produces dramatic increases in the levels of cAMP (Kakiuchi *et al.*, 1969). In the search for a neurotransmitter or neuromodulator released by electrical stimulation and capable of stimulating cAMP formation, it was discovered that adenosine and adenine nucleotides are effective endogenous stimulants of cAMP-generating systems in brain slices (Sattin and Rall, 1970). Subsequent investigations have been concerned with the following problems: (*i*) the mechanism by which adenosine stimulates cAMP formation; (*ii*) the role of adenosine in the accumulation of the nucleotide brought about by depolarizing agents and electrical stimulation; (*iii*) the potentiating interaction of adenosine with the responses of cAMP-generating systems to biogenic amines; and (*iv*) the physiological role of adenosine–cAMP mechanisms in brain function.

A Mechanism

In guinea-pig cerebral cortical slices, adenosine significantly increases the level of cAMP. The effect can be competitively blocked by theophylline (Sattin and Rall, 1970). Adenine nucleotides, such as 5′-AMP, ADP and ATP, stimulate cAMP formation, but it appears that dephosphorylation by nucleotidases to adenosine is a prerequisite for the stimulation of the brain-slice cAMP-generating systems.

Structure–activity relationship studies for the stimulation of cAMP formation by adenosine analogues have been carried out using guinea-pig cerebral cortical slices (Mah and Daly, 1976). Alterations in the ribose portion of adenosine result in analogues which have no agonist activity but are instead antagonists, for example, 2′-deoxyadenosine and adenine xylofuranoside. Certain analogues modified in the purine portion such as 2-chloro-, N^6-phenylisopropyl-and N^6-benzyl-adenosine do retain agonist activity, while other purine-modified analogues are inactive, including inosine, 8-bromoadenosine, and 1-methyladenosine. The specificity of the response to adenosine indicates that the adenosine receptor is highly discriminatory.

Evidence that such a receptor is situated extracellularly has been provided by studies with agents which inhibit the uptake and incorporation of adenosine into brain slices (Huang and Daly, 1974b). Thus, when uptake of submaximal concentrations of adenosine is inhibited by dipyridamole, hexobendine, papaverine or 6-(*p*-nitrophenylthio)guanosine, the adenosine-elicited accumulation of cAMP is potentiated two to three-fold. At high concentrations of adenosine, these inhibitors of uptake have little effect on either incorporation of adenosine or on formation of the cyclic nucleotide. Further evidence for an extracellular site for the action of adenosine has been provided in studies with cultured foetal rat-brain cells (Sturgill *et al.*, 1975). 2-Chloroadenosine, a potent stimulant of cAMP formation in both brain slices and cultured brain cells, is not significantly incorporated into the brain cells using a concentration at which it produces a large accumulation of cAMP. In addition to stimulating cAMP formation through interaction with an extracellular receptor, exogenous adenosine serves as a precursor of ATP which is subsequently converted to the cyclic nucleotide. The contribution of exogenous adenosine to accumulations of cAMP has been reported as to be from less than 20% of the total cAMP in guinea-pig (Schultz and Daly, 1973a) to as high as 40% in mouse (Skolnick and Daly, 1975a) cerebral cortical slices.

B Role in Depolarization-Elicited Formation of cAMP

Initial studies by Kakiuchi *et al.* (1969) indicate that release of biogenic amines probably does not play a major role in the accumulation of cAMP brought about by electrical stimulation of guinea-pig cortical slices. In contrast, the partial or complete blockade by theophylline, 2′-deoxyadenosine or adenosine deaminase of accumulations of cAMP due to electrical stimulation or depolarizing agents provides strong evidence that adenosine is involved in depolarization-elicited formation of cAMP (Huang *et al.*, 1973b; Mah and Daly, 1976). The greater than additive response to combinations of biogenic amines and depolarizing conditions is also compatible with the intermediacy of adenosine; the potentiation of the amine responses by adenosine is well established. Finally, electrical stimulation or incubation of brain cortical slices with depolarizing agents, such as batrachotoxin, veratridine, ouabain or high concentrations of K^+ has been shown to cause an enhanced efflux of adenosine from the slices (Shimizu *et al.*, 1970c; Pull and McIlwain, 1972a,b;). Although adenosine has been strongly implicated in responses of cAMP systems to various depolarizing conditions, clearly all such conditions are not equivalent in terms of stimulation of cAMP systems and blockade by adenosine antagonists (Zanella and Rall, 1973; Mah and Daly, 1976). For example, 40 mM K^+ stimulates accumulations of cAMP in guinea-pig cerebral cortical slices, but not in cerebellar slices; electrical stimulation produces responses in both regions (Zanella and Rall, 1973). Responses to K^+ and ouabain in cortical slices are only partially blocked by theophylline, while responses to electrical stimulation, batrachotoxin or veratridine are nearly completely blocked (Mah and Daly, 1976; Shimizu and Daly, 1972b). Combinations of 40 mM K^+ and a maximal stimulatory concentrations of adenosine have greater than additive effects on cAMP levels (Shimizu *et al.*, 1970b; Huang *et al.*, 1971). Depolarizing agents elicit maximal accumulations of cAMP which are significantly greater than those induced by adenosine (Shimizu *et al.*, 1970c). Clearly, the last two observations indicate that factors other than adenosine are also involved in depolarization-stimulated accumulations of cAMP in brain tissue and point to the need for further research in this area.

C Potentiation of Amine Responses

Adenosine combined with biogenic amines, such as noradrenaline, histamine or 5-HT, often produce greater than additive effects on the accumulation of cAMP in brain slices. Indeed, biogenic amines which in a particular brain tissue may have only marginal or no stimulatory effect alone, induce significant accumulations of the nucleotide when combined with adenosine or when tested under depolarizing conditions (Shimizu *et al.*, 1970b,c; Huang *et al.*, 1971; Schultz and Daly, 1973c,d). In rat and mouse cortical slices, 5-HT has no effect even in the presence of adenosine. Dopamine responses are not detected in the presence of adenosine or depolarizing conditions in either guinea-pig cerebral cortical slices (Shimizu *et al.*, 1970b,c) or in slices from either rat cortex or midbrain–striatum.

The potentiative interaction between catecholamines and adenosine appears to involve mainly α-adrenergic receptors. In rat cortex which contains both α and β-receptors, a combination of noradrenaline and adenosine elicits a much greater than additive response, while the response to a combination of isoprenaline and adenosine is only slightly greater than additive (Schultz and Daly, 1973c). The response to noradrenaline and adenosine is much greater than additive in guinea-pig cortex (Schultz and Daly, 1973d), a tissue which contains predominantly α-receptors. In a tissue which contains primarily β-receptors, such as the cerebellum, there is only an additive effect (Sattin *et al.*, 1975). The EC_{50} for histamine and noradrenaline responses appear to be reduced by the presence of adenosine in guinea-pig cortical slices (Huang *et al.*, 1971). A ten-fold decrease in the EC_{50} for the H_1-agonist 2-aminoethylthiazole was observed in guinea-pig hippocampal slices in the presence of adenosine (Dismukes *et al.*, 1976). In addition 4-methylhistamine, a H_2-agonist, appears to stimulate both H_1 and H_2-receptors in the presence of adenosine, while being specific for H_2-receptors in the absence of adenosine.

Alterations in the nature of adrenergic receptor in the presence of adenosine have been proposed in order to account for the responses to combinations of methoxamine, noradrenaline and adenosine in rat cerebral cortical slices (Skolnick and Daly, 1975c). Adrenergic and adenosine-elicited accumulations of cAMP in rat cortical slices can be proposed to consist of a three component-system: (*i*) stimulation of α and β-receptor-linked cyclases by noradrenaline; (*ii*) stimulation of adenosine-receptor-linked cyclases; and (*iii*) stimulation of 'silent' α-receptors which are activated by noradrenaline only in the presence of adenosine.

Sattin *et al.* (1975) have proposed a similar hypothesis of adenosine-*independent* and adenosine-*dependent* receptors in guinea-pig cerebral cortex. In this tissue, the α-adrenergic response to catecholamines is virtually absent except in the presence of adenosine. The α-adrenergic response in this tissue is thus linked to an adenosine-dependent or silent receptor. Similarly, the greater than additive response to combinations of histamine and noradrenaline are consonant with a silent α-receptor requiring histamine for activation by noradrenaline. α-Adrenergic blockers prevent the response to adrenaline or noradrenaline in the presence of histamine, while β-blockers are ineffective (Schultz and Daly, 1973d). The synergism between histamine and noradrenaline may, however, have an adenosine component since methylxanthines partially antagonize the response (Schultz and Daly, 1973b).

Adenosine responses in rat cerebral cortex develop about five days *post partum*, while a response to catecholamines is not detected until eleven to twelve days after birth (Perkins and Moore, 1973b). However, as early as three to four days *post partum* a synergistic response to a combination of adenosine and noradrenaline occurs. Thus, it appears that the silent adrenergic receptors dependent on adenosine develop before the adenosine-independent adrenergic receptors.

Adenosine–amine interactions are apparently important for the maintenance of amine receptors in responsive states in brain-slice preparations. Amine receptors involved in regulation of cAMP formation can become refractory to activation after first being stimulated with, for example, histamine or histamine–noradrenaline in guinea-pig cortical slices (Schultz and Daly, 1973e) or with histamine or noradrenaline in rabbit cerebellar slices (Kakiuchi and Rall, 1968b). This phenomenon has been studied in detail in guinea-pig cortical slices, where histamine or histamine–noradrenaline can produce only one stimulation of cAMP. However, in the presence of adenosine, multiple stimulations by the amines are observed and co-addition of the amine with adenosine to a 'refractory' slice preparation results in a synergistic response (Schultz and Daly, 1973e). It has been proposed that the presence of adenosine is required in order to maintain the amine-sensitive cAMP-generating system in a responsive state. Since addition of a phosphodiesterase inhibitor, 1-methyl-3-isobutylxanthine, results in a partial restoration of the amine response and absence of extracellular Ca^{2+} prevents the receptors being in a refractory state (Schultz, 1975a), it has been speculated that adenosine may prevent or reverse the activation of phosphodiesterase. A possible mechanism, however, whereby adenosine might have such an effect on phosphodiesterase is unclear. The role of adenosine in the multiple restimulation of cAMP-generating systems by biogenic amines is further complicated by the observation that in rat cortical slices, noradrenaline and the α-adrenergic agonist methoxamine are capable of restimulating cAMP formation (Skolnick *et al.*, 1975).

At the present time, two possible explanations for the lack of restimulation by biogenic amines in guinea pig and the presence of restimulation in rat would appear possible. Firstly, the release of adenosine in rat slices may be sufficient to maintain the cAMP systems in a responsive state to amines. Secondly, there may be fundamental differences in the control of receptor responses in the two species.

D Physiological Role

It is apparent that adenosine, either alone or together with biogenic amines, probably plays an imortant role in cAMP-dependent mechanisms in the central nervous system. Thus, not only can adenosine activate cAMP-generating systems in brain, but it can lower the threshold and potentiate the response to putative neurotransmitters, such as noradrenaline, histamine and 5-HT. Adenosine may be derived from ATP released with the amines during neuro-exocytosis or as a by-product of ATP metabolism during intense neuronal activity. Adenosine has been shown to be released during electrical stimulation of brain slices (Pull and McIlwain, 1972a,b), as well as during the *in situ* stimulation of cat cerebral cortex (Sulakhe and Phillis, 1975). In olfactory tract–olfactory cortex preparations, adenosine inhibits the transsynaptic generation of action potentials after stimulation of the innervating tract and elicits a concommitant accumulation of cAMP (Kuroda and Kobayashi, 1975; Okada and Kuroda, 1975). Theophylline can block both the electro-physiological and biochemical effects of adenosine in this preparation. Adenosine does not inhibit postsynaptic potentials in the optic tract–superior colliculus preparation.

Adenosine and adenosine precursors will perhaps be found to have a variety of physiological roles in the CNS. At present, one can only speculate on direct neurotransmitter roles involving cAMP formation and 'permissive' neuro-modulatory roles in which adenosine greatly potentiates the response of cAMP systems to biogenic amine neurotransmitters.

VII ALTERED RESPONSIVENESS AFTER CHANGES IN SYNAPTIC INPUT

A Noradrenaline

The catecholamine-sensitive cAMP-generating systems appear to be an integral part of the adrenergic receptor (Robison *et al.*, 1971). The investigation of catecholamine-sensitive cAMP-generating systems in brain slices may be an important technique for examining the changes which occur in postsynaptic receptor mechanisms when presynaptic input is chronically altered.

Treatment of animals with 6-hydroxydopamine provides a selective method of destroying presynaptic noradrenergic terminals, and thereby preventing any transsynaptic input to postsynaptic noradrenergic receptors. Weiss and Strada (1972) reported that treatment of rats with 6-hydroxydopamine results in an increased response of the cAMP-generating systems to 5 μM but not to 50 μM noradrenaline in cerebral cortical slices 4 weeks later. Palmer (1972) reported enhanced responses to 1–10 μM noradrenaline in slices from rat cerebrum, hypothalamus and brain stem after treatment with 6-hydroxydopamine. The response to noradrenaline in cerebellum is not enhanced, but may be obscured by the very high basal levels of cAMP found in rat cerebellar slices.

The effects of treatment with 6-hydroxydopamine on rat brain cAMP-generating systems, have subsequently been studied in more detail (Kalisker *et al.,* 1973; Huang *et al.*, 1973a). The enhanced responsiveness to noradrenaline in cortical slices appears to consist of two components. Within two days of administration of 6-hydroxydopamine, an apparent decrease in the EC_{50} for noradrenaline occurs (Kalisker *et al.*, 1973). This has been shown to be due to a reduction in the presynaptic uptake of noradrenaline due to the destruction of the nerve terminals. In the presence of cocaine, an uptake inhibitor, dose–response curves for noradrenaline are identical in control and 6-hydroxydopamine-treated animals. The response to isoprenaline, which is not taken up into terminals, is unaltered two days after treatment with 6-hydroxydopamine. The second component appears within 4 days of treatment with 6-hydroxydopamine and consisted of a nearly two-fold increase in the maximal response to either noradrenaline or isoprenaline. Both the α and β-adrenergic portions of the noradrenaline-response are increased after 6-hydroxydopamine (Huang *et al.*, 1973a). The response to an adenosine–noradrenaline combination is increased, while that to adenosine alone is only marginally stimulated. Thus, destruction of the central noradrenergic terminals appears to be followed by a rather specific adaptive increase in the responses of postsynaptic cAMP-generating systems to noradrenaline. Parenteral administration of 6-hydroxydopamine to new-born rats prevents adrenergic innervation of cortical structures and markedly increases the numbers of adrenergic terminals in subcortical regions, such as the midbrain (Sachs *et al.*, 1974). Noradrenaline elicits an enhanced response of cAMP-generating systems in cortical slices of young (Palmer and Scott, 1974) or adult (Dismukes and Daly, 1975b) rats treated as neonates with 6-hydroxydopamine. Responses to adenosine are not affected. In the midbrain, responses to noradrenaline are elevated, but this probably does not represent a denervation hyper-responsiveness since noradrenergic innervation is increased in this brain region after neonatal 6-hydroxydopamine and since hyper-responsive noradrenaline-sensitive cAMP-generating systems do not develop in midbrain of adult rats treated with 6-hydroxydopamine.

Presynaptic terminals and hence input can also be destroyed by lesions introduced into the noradrenergic tracts. Thus, lesions of the medial forebrain bundle destroy fibres rostral to the lesion and result in a 50–60% decrease in noradrenaline levels in the ipsilateral cortex; the contralateral cortex serves as a paired control. The response of cAMP systems to noradrenaline is enhanced in ipsilateral cortex and hippocampus, but is unaffected in midbrain and cerebellum, regions in which the innervating noradrenergic tracts are caudal to the lesion (Dismukes *et al.*, 1975). An enhanced response to noradrenaline occurs in cortex within two days of lesioning and is fully developed within nine days. Thus, development of hyper-responsiveness of cAMP-generating systems precedes the relatively slow decline in levels of cortical noradrenaline and is correlated to disruption of synaptic input rather than to decreases in noradrenaline levels. The medial forebrain bundle contains serotonergic and histaminergic tracts in addition to noradrenergic tracts. Lesions produce an enhanced response to histamine, and perhaps 5-HT, in rat cortical slices (Dismukes *et al.*, 1975).

The amount of noradrenaline 'available' to the postsynaptic receptor can be pharmacologically altered without any destruction of the presynaptic nerve terminals. Thus, reserpine by depleting noradrenaline can reduce its synaptic availability, while amphetamine by releasing noradrenaline and other compounds due to inhibition of their uptake can increase synaptic availability of noradrenaline. The effects of such pharmacological alterations in synaptic noradrenergic input on responsiveness of cAMP-generating systems have been investigated.

After treatment of rats with resperine for four days noradrenaline produces an enhanced response of the cAMP systems in slices of cerebrum, hypothalamus and hippocampus, but not in slices of cerebellum, midbrain and brain stem (Palmer *et al.*, 1973). Reserpine causes a nearly four-fold increase in responses to noradrenaline in cortical slices, and treatment does not result in significant responses to dopamine in any brain region. After treatment with reserpine for only 2 days, a less than two-fold increase in noradrenaline-response occurs in rat cortical slices (Dismukes and Daly, 1974). An enhanced responsivess to noradrenaline in rat cortical slices is not manifest five hours after reserpine, but develops fully within two days. It is maintained for at least 9 days and disappears within 16 days. Reserpine treatment does not have a significant effect on the EC_{50} for noradrenaline-responses in cortical slices, although French *et al.* (1975a) have described a decrease in the EC_{50} of noradrenaline after reserpine. In contrast to increases in both α and β-responses after 6-hydroxydopamine, reserpine treatment apparently results in a preferential increase in β-

adrenergically controlled responses in rat cortical slices (Dismukes and Daly, 1974). After eight to twelve days of treatment of rats with reserpine, Williams and Pirch (1974) reported an enhanced response to noradrenaline in whole brain 'slices' prepared with a Harvard tissue press fitted with a 1.5 mm sieve. Caffeine was also present in the incubation medium. In addition, Vetulani and Sulser (1975) have reported enhanced responses to noradrenaline in slices of limbic forebrain from rats treated with reserpine.

Chronic increases in noradrenergic transsynaptic input result in a 'subsensitivity' or hypo-responsivensss of noradrenaline-sensitive cAMP-generating systems in brain slices. Thus, treatment of mice with amphetamine, a noradrenaline-releasing agent, results in a diminished response to noradrenaline in cerebral cortical slices, to the extent of about 20–30% (Martres *et al.*, 1975). This diminished response occurs within five hours of chronic ingestion of amphetamine and is maintained for at least ten days of treatment. Only the magnitude of the response to noradrenaline is reduced, with no apparent alteration in the EC_{50}. Responses to dopamine, 5-HT and adenosine are not altered after amphetamine treatment. After treatment of rats with antidepressants, such as desipramine and iprindole, for one to two months responses of cAMP-generating systems to noradrenaline in slices of limbic forebrain are reduced by nearly 75% (Vetulani and Sulser, 1975). These drugs inhibit reabsorption of noradrenaline into presynaptic terminals and thus, presumably, increase synaptic availability of noradrenaline. Frazer *et al.* (1974) had previously reported reduced responses of cAMP systems to noradrenaline in cortical slices of rats treated with imipramine, another inhibitor of noradrenaline uptake. It has been observed that treatment of rats with either imipramine or chlorpromazine for a minimum of six days results in a marked reduction in the response to noradrenaline using cortical slices (Schultz, 1976). Other centrally active drugs, such as phenobarbital or diazepam, have no effect on adrenaline or noradrenaline responses (Schultz, personal communication).

Recent experiments with guinea-pigs (Dismukes *et al.*, 1976) have raised important questions regarding the universal nature of adaptive hyper-responsiveness in catecholamine-sensitive cAMP-generating systems following chronic decreases in transsynaptic input. Treatment with reserpine and with 6-hydroxydopamine, and medial forebrain bundle lesions have been used to reduce input to the postsynaptic adrenergic receptor in guinea-pig cerebral cortex. These procedures increase the responses of cAMP-generating systems to catecholamines in rat cerebral cortex but, although effective in reducing the catecholamine content of guinea-pig brain, such approaches do *not* result in enhanced responses to noradrenaline in guinea-pig cerebral cortex. Curiously, the response to adenosine is only enhanced in reserpine-treated guinea pigs. Histamine and 5-HT responses are not altered in guinea-pig cortex after lesions in medial forebrain bundle, again in contrast to the enhanced responses resulting from such lesions in rat. An understanding of the lack of alterations in responsiveness in the guinea-pig CNS is critical to a full assessment of the significance of such adaptic changes.

Alterations in the responsiveness of cAMP-generating systems during chronic ethanol ingestion and during withdrawal has led French and associates (French and Palmer, 1973; French *et al.*, 1974; French *et al.*, 1975b,) to propose that ethanol dependence is associated with reduced sensitivity of cAMP-generating systems to catecholamines while increased sensitivity results during withdrawals. Chronic ingestion of ethanol in rats leads to a reduced sensitivity of cAMP systems in cortical slices to noradrenaline. This is reflected in a four-fold increase in the EC_{50} for the catecholamine. Maximal accumulations of cAMP elicited by noradremaline are not altered. This is in marked contrast to the adaptive changes in cAMP-generating systems after other treatments where only the maximal response to noradrenaline is altered. Ethanol ingestion causes a sustained increase in turnover of noradrenaline in brain (Hunt and Majchrowicz, 1974; Porohecky, 1974) which may lead to reduced sensitivity of noradrenaline sensitive cAMP-generating systems. It has been proposed by French and associates that the diminished turnover of noradrenaline which occurs in brain during withdrawal from ethanol, coupled with the presence of subsensitive noradrenergic receptors, leads to seizures. During withdrawal from ethanol, an increased sensitivity of the cortical cAMP-systems to noradrenaline develops after three days and is manifest in a two-fold decrease in the EC_{50} for the catecholamine. The maximal accumulations of cAMP brought about by noradrenaline are not altered. Not only is the response to submaximal concentrations of noradrenaline enhanced in rats withdrawn from alcohol, but the responses to histamine and 5-HT are also enhanced. This general increase in sensitivity of cAMP-generating systems has been interpreted in terms of a non-specific postsynaptic supersensitivity, perhaps involving partial depolarization. The relevance of adenosine mechanisms to this general enhanced sensitivity to amines warrants investigation.

Physiological stresses, such as shock applied to the foot pad and restraint, which increase the turnover of noradrenaline and dopamine in brain (Bliss *et al.*, 1968) significantly increase cAMP levels in septum and hippocampus of rat brain (Delapaz *et al.*, 1975), indicating that an increased turnover of neuro-transmitter in the limbic system can result in stimulation of cAMP formation.

B Dopamine

Effects of altered transsynaptic input on responsiveness of dopamine-sensitive cAMP systems have been investigated in brain for the dopaminergic nigro-striatal pathway. Mishra *et al.* (1974) have reported that electrolytic lesions or injection of 6-hydroxydopamine to the substantia nigra results in enhanced responses of adenylate cyclases to dopamine in cell-free preparations of the caudate nucleus. Such an enhanced response is compatible with the apparent 'denervation supersensitivity' to dopaminergic agonists which pertains *in vivo* in such animals. These enhanced responses of dopamine-sensitive adenylate cyclases were measured ten to thirty-six days after electrolytic lesions and 100–120 days after administration of 6-hydroxydopamine. However, other laboratories (Von Voigtlander *et al.*, 1973; Krueger *et al.*, 1976) have not observed enhanced responses of adenylate cyclases to dopamine using homogenates after electrolytic or chemical lesioning. It is uncertain whether differences in protocol, time after lesioning or strain differences account for this discrepancy. Krueger *et al.* (1976), although observing no effect of the lesions on the activity of dopamine-sensitive adenylate cyclase in homogenates, have reported a reduction in the EC_{50} for responses of the cAMP-generating system to dopamine and apomorphine in slices of caudate nucleus. The maximal response to dopamine is unchanged. It has been proposed that lesioning and degeneration of dopaminergic terminals in the caudate nucleus results in an enhanced accessibility of postsynaptic receptors to the catecholamines.

Chronic treatment of animals with neuroleptics or reserpine does *not* produce the expected enhanced activity of dopamine-sensitive adenylate cyclase in cell-free preparations of rat caudate (Von Voigtlander *et al.*, 1975; Rotrosen *et al.*, 1975). However, one group has reported that acute or chronic treatment of rats with haloperidol does increase the activity of the dopamine-sensitive adenylate cyclase (Iwatsubo and Clouet, 1975). Clearly, further studies are necessary to resolve the conflicting reports on the effects of alterations in transsynaptic input on responses of dopamine-sensitive cAMP-generating systems (see also Chapter 19).

C Mechanism

The biochemical bases for alterations in the responsiveness of cAMP-generating systems in brain slices after changes in synaptic input are not yet defined. Changes in any one of many factors can result in enhanced or diminished responses of such a system. Activity and hormonal-responsiveness of adenylate cyclases, activity of associated phosphodiesterases, availability of substrate ATP and other cofactors and amounts of protein kinases must all be considered as potential causative factors in this system.

The obvious experimental paradigm is the study of the adenylate cyclase and phosphodiesterase activities in purified or cell-free preparations from the relevant brain region. This has been done for the dopamine-sensitive adenylate cyclase system of caudate nucleus and the results are conflicting as has been previously discussed. Unfortunately, noradrenaline produces only marginal activation of adenylate cyclases in cell-free preparations of the CNS (Daly, 1975). Nonetheless, studies of noradrenaline-dependent adenylate cyclases after treatment with 6-hydroxydopamine and reserpine might provide valuable data. With regard to phosphodiesterase, it is unlikely that small changes in the enzyme specifically associated with noradrenergic receptors would be detectable in homogenates containing all of the phosphodiesterase activity of a particular brain region.

Kalisker *et al.* (1973) did demonstrate a small decrease in phosphodiesterase activity in cerebral homogenates from rats pretreated with 6-hydroxydopamine, while Breckenridge and Johnson (1969) reported no change after lesions of the medial forebrain bundle. Noradrenaline elicits an enhanced response of cAMP-generating systems in cortical slices from 6-hydroxy-dopamine and reserpine-treated rats even in the presence of a phosphodiesterase inhibitor (Huang *et al.*, 1973a; Dismukes and Daly, 1974), suggesting that the enhanced response is due to increased adenylate cyclase rather than decreased phosphodiesterase activity. However, complete inhibition of phosphodiesterase in slice preparations cannot be produced with the inhibitors presently available (Schultz and Daly, 1973). Thus, a firm conclusion as to the mechanism underlying adaptive changes in cAMP-generating systems in brain is not yet possible.

Several fundamental questions are important for a clear understanding of the significance of adaptive changes in neurohormone-sensitive cAMP-generating systems and their relationship to homeostasis and plasticity of the central nervous system.

The first pertains to the general nature of the phenomenon. Thus, in guinea-pig cortex, no adaptation to reduction in catecholamine has been observed (Dismukes *et al.*, 1976), nor does the cAMP-generating system of rat midbrain undergo adaptive changes after a reduction in noradrenergic input (Dismukes and Daly, 1975b). Responses to catecholamines in cortical slices of several strains of rats after pretreatment with 6-hydroxydopamine have recently been examined. These results indicate that cAMP systems in those strains in which noradrenaline already produces a large accumulation of the cyclic nucleotide in cortical slices from control rats do not become hyper-responsive after 6-hydroxy-dopamine. However, the systems do become hyper-responsive in rats which normally display low responses to noradrenaline.

A second question concerns the fundamental differences in the adaptive alterations of noradrenaline-sensitive cAMP-generating systems observed after 6-hydroxydopamine and reserpine and after ethanol withdrawal. Treatment with the former drugs produces an increase in maximal responses, while ethanol withdrawal apparently increases the affinity of noradrenaline for the receptor. Clearly, studies on the relationship between synaptic input and adaptive changes in postsynaptic cAMP-generating systems remains a fertile area for future research.

VIII FUNCTIONAL ROLES OF cAMP

While elucidating the roles of cAMP in regulation of physiological phenomena in peripheral tissues has progressed greatly, it has proved difficult to establish a causal relationship between changes in cAMP levels, and the biochemical or physiological sequelae in a complex organ such as brain (Daly, 1975). At present, the primary role of cAMP in all tissues appears to be activation of protein kinases. The physiological effect depends on the nature of the structural and/or functional protein or proteins phosphorylated by the cAMP-dependent kinase. It has been proposed that cAMP activation of protein kinase is an integral component of inhibitory neuronal transmission by catecholamines (Miyamoto *et al.*, 1969a,b; 1971; Siggins and Henricksen, 1975). Release of catecholamine from the presynaptic terminal enhances the formation of cAMP in postsynaptic structures by way of activation of a catecholamine–receptor-linked adenylate cyclase. Activation of protein kinase by cAMP ensues, followed by phosphorylation of a membrane protein. The phosphorylated membrane protein is probably responsible for the inhibitory hyperpolarization of the postsynaptic structure. cAMP in the CNS has been shown to modify a number of model systems including glycogenolysis, stabilization of neurotubules and activation of tyrosine hydroxylase. It is tempting to speculate that cAMP plays a pivotal role in altering such functions as behaviour and cardiovascular control as an inhibitory second messenger in central postsynaptic neuronal structures.

A Behavioural Alterations

The study of the effects of putative neurotransmitters on cAMP, and their relevance to the role of these neurotransmitter in all aspects of behaviour, including agressiveness, motor activity, exploratory behaviour and learning, although still in the developmental stage, would appear to be a valuable approach to a more fundamental understanding of the role of neurotransmitters in all aspects of behaviour. There have been many reports on the behavioural effects of cAMP or its derivatives when injected directly into the CNS. These effects range from increased spontaneous motor activity to catatonia, aggressive behaviour and convulsions (Daly, 1975). However, these reports must be interpreted with caution as discussed by Myers (1974) since injection of relatively large amounts of compound might be expected to have broad effects on many brain areas which may have opposing functions.

Possible relationships between behavioural parameters and neurotransmitter-dependent accumulations of cAMP in brain slices have been initiated (Skolnick and Daly, 1974a). The response of cAMP-generating systems to various agents has been determined in cerebral cortical and midbrain–striatal slices in rat strains exhibiting various levels of spontaneous behavioural activity (Segal *et al.*, 1972). An inverse relationship has been observed between tyrosine hydroxylase activity and the noradrenaline-stimulated formation of cAMP in midbrain–striatal slices of these inbred rat strains and also between spontaneous behavioural activity and noradrenaline-stimulated formation of cAMP in cortical tissue. However, a direct relationship between spontaneous behavioural activity and cAMP formation elicited by noradrenaline has been obtained in midbrain–striatal slices. The correlations mentioned are all highly significant, with values of greater than —0.9 or +0.9, respectively. The correlations are only obtained using noradrenaline. Other agents tested include isoprenaline, adenosine, veratridine, dopamine and an adenosine–noradrenaline combination.

This study supports the hypotheses of Segal *et al.* (1972) demonstrating an inverse correlation between spontaneous behavioural activity and midbrain–striatal tyrosine hydroxylase activity. They postulated that a low tyrosine hydroxylase activity is compensated for by an increase in the number or functional activity of adrenergic receptors which is ultimately reflected as a relatively high spontaneous behavioural activity. Conversely, in animals endowed with a high intrinsic activity of the tyrosine hydroxylase, this is compensated for by fewer adrenergic receptros which results in a relatively low spontaneous behavioural activity. These mechanisms are in agreement with the concept of ahomeostatic control in the CNS. The inverse correlation between responses to noradrenaline in cortical slices and behaviour in the four rat strains (Skolnick and Daly, 1975c) may reflect the inhibitory role of cortical noradrenergic pathways in control of behaviour.

Preliminary behavioural experiments with clonidine strongly support a role for the α-component of noradrenaline-responses in control of behaviour. Preliminary experiments indicate that pretreatment of animals with clonidine effectively antagonizes the increases in spontaneous behavioural activity observed when noradrenaline is infused into the lateral ventricles of rats (Segal, personal communication).

Williams and Pirch (1974) have observed a striking positive correlation between increases in motor activity and the noradrenaline-stimulated formation of cAMP in rats treated with reserpine for prolonged periods. It has been proposed that enhanced noradrenergic-receptor responsiveness combined with return of noradrenergic input during chronic reserpine reatment is responsible for hyperactivity observed in such rats.

IX CONCLUSION

Much insight has been gained during the past seven years regarding the regulation of cAMP formation in the CNS by putative neurotransmitters. The brain-slice preparation has proved to be a useful tool for the understanding of these phenomena. Further investigation is, however, necessary to assess the physiological role which cAMP plays in the brain. Undoubtedly, the progress made in this respect will parallel advances made with regard to the functions of the neurotransmitters themselves.

CHAPTER 18

Electrophysiological Effects of Cyclic Nucleotides on Excitable Tissues

G. R. Siggins

I INTRODUCTION

Research over the last decade has demonstrated that excitable tissue constitutes a major store of cyclic nucleotides (Sutherland *et al.*, 1962; Klainer *et al.*, 1962; Goldberg *et al.*, 1969). The ensuing analysis of the physiological roles for these nucleotides has been studied by many workers. The first stage was the observation that several mammalian neurotransmitters, especially the monoamines, and hormones are capable of stimulating cAMP production (Klainer *et al.*, 1962; Robison *et al.*, 1971). Since the best known function of neurotransmitters is the modulation of the electrical activity of neurones and muscle, the next logical step in determining the role of cyclic nucleotides was to assess their effects on the excitability of these cells. This chapter is devoted primarily to an analysis of the electrophysiological effects of cyclic nucleotides and their putative neurohormones or first messengers on excitable tissue. No attempt will be made to cite all the literature pointing to the actions of cyclic nucleotides in a given function since this subject has been recently surveyed by Daly (1975) and by Bloom (1975), as well as by other authors in this volume. Nor will the techniques employed be explored in detail — see the review of iontophoresis techniques by Hoffer and Siggins (1975). Instead, significant concepts and implications of second-messenger mediation of humoral and synaptic or junctional events will be considered, especially within the central nervous system (CNS). The methods used will only be discussed where variations have resulted in artefacts or controversies. However, a short introduction on what bioelectrical events need to be measured and what methods are used to administer first and second messengers will greatly facilitate the subsequent discussion of the electrophysiological effects of these messengers.

II PARAMETERS OF EXCITABILITY

A Intracellular Recording

The most direct measure of cell excitability is obtained by recording the potential difference across the intact cell membrane, usually by inserting a glass microelectrode filled with electrolyte into the cell (Ling and Gerard, 1949; Katz, 1966). Using this approach, changes in cell excitability generally reflect changes in the resting transmembrane potential, with increased potentials (hyperpolarization) usually indicating reduced excitability and moderately diminished potentials (depolarization) increased excitability. Moderate depolarization tends to increase the probability of the abrupt regenerative reversals of potential, known as action potentials or spikes, while hyperpolarization reduces the likelihood of action potentials by shifting the membrane potential away from the threshold for spike generation. This subject has been fully discussed by Katz (1966).

Intracellular recording offers an added powerful advantage in that, by using appropriate techniques for intracellular stimulation, the relative permeability (conductance) of the membrane to ions can be assessed. Since changes in membrane ionic conductance are usually the cause of changes in membrane potential and excitability, measurement of conductance provides further information about ionic mechanisms of the excitability change. Experimental manipulation of the intra or extracellular concentrations of various ions then allows evaluation of the specific ion (or ions) responsible for the change in excitability. Determination of the ionic basis of excitatory and inhibitory postsynaptic potentials is a classic example of the use of measurements of membrane conductance (Eccles, 1964; Katz, 1966). However, it is not clear what changes in ionic conductance, if any, would appear if ionic pumps are the cause of excitability changes (Thomas, 1972). Such 'electrogenic' pumps are thought by some to account for certain 'slow' synaptic potentials in sympathetic ganglia (Nishi and Koketsu, 1967).

B Gross Recording of Cell Populations

Intracellular recording with micropipettes is convenient for large, easily isolated cells which are relatively uninjured by the impaling electrode. However, many excitable cells are too small to withstand penetration by the microelectrode and require less direct measures of excitability. One such approach is the sucrose-gap method of recording from either single large fibres or the total activity of a large number of longitudinally arrayed and electrically isolated cells (Stämpfli, 1954). In this technique, the potential difference is measured between one end of the fibre or cellular array placed in a suitable medium and the other end bathed in concentrated KCl and consequently depolarized. Electrical shunting between the two ends along extracellular pathways is kept to a minimum by bathing the middle section in a non-conducting sucrose solution. The potential measured is essentially 'intracellular'.

C Extracellular Recording of Single Units

A disadvantage of the sucrose-gap, or any gross recording, method is that it is necessary to isolate *in vitro* a large number of nearly homogenous cell types, preferably arranged longitudinally. This is virtually impossible for most mammalian central neurones. However, many such neurones spontaneously discharge action potentials. By recording these action potentials extracellularly with appropriate glass or metal microelectrodes, another indirect index of cell excitability may be obtained if it is assumed that a high rate of discharge reflects membrane depolarization and a reduced rate of firing, hyperpolarization. If the neurone under study does not fire spontaneously it can often be activated to generate evoked spikes by stimulating the excitatory inputs to the cell. Changes in excitability will then be reflected in changes in the response to the excitatory input. If specific excitatory inputs cannot be activated then an alternate, though less acceptable, method of making the cell discharge artificially is to apply an excitatory drug, such as glutamic acid, directly to the cell by iontophoresis (Bloom, 1974, 1975).

III MODES OF DRUG ADMINISTRATION

A Perfusion and Superfusion

In the case of those tissues which can be safely isolated and which are largely composed of homogeneous cell types not electrically or synaptically connected, *in vitro* perfusion of agents is the best method. Artifacts which arise are the minor correctible ones of tissue or ion movement (due to fluid flow), temperature and pH. Alternatively, perfusion, whether through an intact vascular system or in a bath, is often used in studies of isolated populations of heterogeneous cell types. Care must be exercised in these cases to allow for the different cell types involved and their possible interconnections, since drug effects may be exerted at a site remote from the actual cell under study. For example, inhibition of neuronal activity indicated by hyperpolarization could arise from the excitation by perfused drug of a remote neurone which sends inhibitory terminals to the cell under study; the conclusion that the agent is inhibitory would then constitute an interpretive error.

Superfusion of drugs to *in situ* exposed structures, such as surfaces of cerebral or cerebellar cortex, or brain tissue transplanted *in oculo* (Hoffer *et al.*, 1975, 1976b), requires similar controls. The intact vascular supply, which may carry the drug to remote but connected brain areas, constitutes an additional source of error in these studies. Removal of major afferent pathways to the cell under study, for example by surgery (Hoffer *et al.*, 1975, 1976), X-irradiation

(Woodward *et al.*, 1974), genetic mutations (Siggins *et al.*, 1976b) or chemical denervation (Hoffer *et al.*, 1971b) can often overcome these errors.

B Microiontophoresis

By far the most widely used method of drug administration for mammalian central neurones in the last decade is the local application of drug into the immediate environment of the recorded cell by iontophoresis (Curtis, 1964). This technique has several advantages since the area of brain (and numbers of cells and cell types) affected by the drug are substantially reduced, thus enabling the study of single neurones in intact brain. Also, drugs may be applied directly

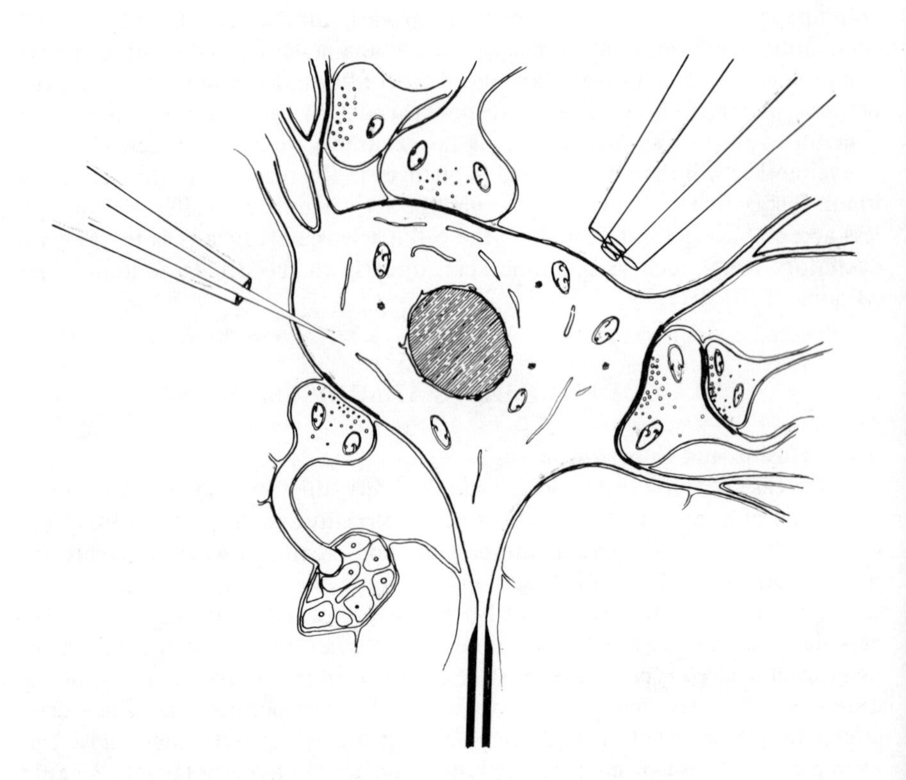

Figure 18.1 Diagrammatic representation of a single neuronal unit, with two types of iontophoretic drug delivery and recording assemblies, drawn to approximate scale. The extracellular assembly on the right has five barrels, one for recording neuronal discharge, one for current control and three for ejection of ionized drugs. The assembly on the left has an intracellular microelectrode arranged with its tip in advance of a single or multiple-barrelled iontophoretic electrode; the optimum assembly in this system may be obtained with two extracellular electrodes, one containing ionized drug for ejection and one for current neutralization. Also depicted are several synaptic arrangements constituting possible sources of indirect drug actions in iontophoretic experiments. (Modified from Salmoiraghi and Bloom, 1964)

to a particular cell in quantities too low to reach the systemic circulation and cause cardiovascular changes and subsequent mechanical or metabolic alterations in neuronal firing. Furthermore, drugs that do not penetrate pial surfaces or the blood–brain barrier can be studied, even in deep brain structures.

This method can be adapted for recording either extracellular neuronal activity, or in the case of large cells, transmembrane potentials. The review by Hoffer and Siggins (1975) provides details of the construction and use of the most common types of microelectrodes. The most widely used ones are those for extracellular recording, consisting of five to seven pipettes fused together and pulled to a fine open tip (2 – 10 μm overall tip diameter). The central barrel is usually filled with 3 M NaCl and is used for recording neuronal discharge (Figure 18.1). Another barrel is also usually filled with electrolyte and is used as a current control or for current neutralization (Salmoiraghi and Weight, 1967). The remaining barrels are filled with concentrated solutions of ionized drugs. Current of the same polarity as the ionized drug is applied to these barrels to eject the drugs into tissue. Since the glass pipettes are fused together, drugs can be applied at the exact site of neuronal recording and drug-evoked changes in neuronal discharge can be recorded simultaneously. However, many artifacts are encountered with this technique, especially regarding negative results and current effects; the reader interested in these artifacts should consult the review of techniques by Bloom (1974).

Obviously, the best method of electrophysiological recording is obtained by intracellular electrodes. Methods are available for recording intracellularly while applying drugs extracellularly by iontophoresis (Figure 18.1) — see Hoffer and Siggins (1975). However, since cyclic nucleotides are only sparingly permeable across cell membranes and their presumed site of action is intracellular (Kuo and Greengard, 1969), the best approach would be to administer these compounds intracellularly. Although this has been accomplished for some large non-neuronal cells (Tsien, 1973) it is a formidable task to inject nucleotides into the smaller mammalian central neurones without injuring them.

IV EFFECTS OF CYCLIC NUCLEOTIDES AND RELATED FIRST MESSENGERS

Analysis of the electrophysiological actions of cyclic nucleotides assumes functional importance when compared to the electrophysiological effects of the specific neurotransmitters and other humoral substances known to elevate levels of cyclic nucleotides in excitable tissue. Mimicry between the physiological actions of humoral substances and cyclic nucleotides is a criterion in order to prove the mediation by a nucleotide, as a second messenger, of a specific humoral response (Robison *et al.*, 1971; Bloom, 1975).

Of all the neurotransmitters and humoral substances known to elevate levels of cyclic nucleotides in excitable tissues, the catecholamines (particularly adrenaline, noradrenaline and dopamine) seem to be the most universal in their ability to stimulate cAMP formation. The β-receptor is most often involved in such actions, although dopamine receptors and occasionally α-receptors may also stimulate cAMP production. For this reason, most of the electrophysiological studies described below have compared the effects of β-receptor agonists and cAMP. While the link between acetylcholine and cGMP as the mediator of muscarinic-receptor stimulation has been well documented for certain peripheral tissues (Goldberg *et al.*, 1973a), it is less well understood in the CNS, since acetylcholine has such diverse effects in many areas of the CNS and many non-cholinergic agonists are capable of elevating cGMP levels (Ferrendelli *et al.*, 1974; Mao *et al.*, 1974b). Moreover, relatively few studies have appeared on the electrophysiological effects of cGMP. Another potential modulator, adenosine, has provoked some interest due to its ability to stimulate adenylate cyclase in the CNS. Therefore, the following account highlights primarily the comparisons between catecholamines, adenosine and cAMP, and between acetylcholine and cGMP.

A Muscle

1 Skeletal Muscle

In spite of early studies which showed elevation of cAMP levels in skeletal muscle by β-receptor stimulation (Klainer *et al.*, 1962), little is known of the functional role of either catecholamines or cAMP in this tissue. In several types of vertebrate and invertebrate skeletal muscle, β-receptor stimulation has been reported to result in hyperpolarization associated with decreased membrane conductance (Hidaka and Kuriyama, 1969; Kuba, 1970; Kuba and Tomita, 1971; Ito *et al.*, 1971). Both isoprenaline (a selective β-receptor agonist) and cAMP have been reported to hyperpolarize the 'slow' or tonic skeletal muscle fibres of the pigeon (Somlyo and Somlyo, 1969).

Catecholamines also affect motor nerve terminals of some preparations, but generally such effects have been characterized pharmacologically as α-responses. Stimulation of these receptors restores junctional transmission after fatigue (Breckenridge *et al.*, 1967) and facilitates the frequency of miniature endplate potentials (Jenkinson *et al.*, 1968; Goldberg and Singer, 1969; Kuba, 1970; Kuba and Tomita, 1971). Such potentials are thought to result from spontaneous release of small packets of neurotransmitter from the nerve terminals. The potentials produced in the muscle cell are too small to result in action potentials. However measurement of their frequency can indicate the excitability of the motor nerve terminal (Katz, 1967). In several of these studies the presynaptic effects of exogenous cAMP are similar to those of the catecholamine (Goldberg and Singer, 1969; Takamori *et al.*, 1973).

2 Cardiac Muscle

In cardiac muscle, catecholamines act at a β-receptor, leading to increased membrane conductance and a more rapid pacemaker depolarization. Similar inotropic effects have been seen using external application of cAMP or one of its butyrylated derivatives (Tsien *et al.*, 1972). The physiological result of both catecholamine and cAMP administration is an acceleration of pacemaker activity and an elevation of plateau height and duration (Reuter, 1974a).

The proposition that the cAMP produced intracellularly mediates the physiological action of the catecholamine (Robison *et al.*, 1971; Wollenberger, this volume) is strengthened by the finding that cAMP has a more potent action in mimicking catecholamines when iontophoretically applied into myocardial cells (Tsien, 1973); catecholamines injected into the cell are less effective than when applied externally (Tsien, 1973; Reuter, 1974a; Yamasaki *et al.*, 1974). Reuter (1974a) suggested that both the catecholamines and Bt_2-cAMP specifically increase the membrane conductance to Ca^{2+} in ventricular myocardial preparations.

While cGMP has been strongly implicated in acetylcholine-induced tachycardia, (Goldberg *et al.*, 1973b; George *et al.*, 1970; Wollenberger, this volume), no studies on the electrophysiological effects of this cyclic nucleotide have been published to date.

3 Smooth Muscle

The actions of catecholamines on membrane properties of smooth muscle is complicated by the fact that varying ratios of α and β-receptors may be present. For example, Bülbring and Tomita (1969) found that noradrenaline (a mixed-receptor agonist) hyperpolarizes the smooth muscle of guinea-pig *Taenia coli* by increasing conductance mainly to K^+, whereas isoprenaline hyperpolarizes the membrane without changing the membrane conductance. Isoprenaline also hyperpolarizes myometrial fibres (Diamond and Marshall, 1969; Kroeger and Marshall, 1973). In support of a role for cAMP in these β-responses are the findings that Bt_2-cAMP hyperpolarizes smooth muscle cells of the *Taenia coli* (Takagi *et al.*, 1971; Kroeger and Marshall, 1973).

In vascular smooth muscle isoprenaline has been shown to produce hyperpolarization of the membrane accompanied by muscular relaxation (Somlyo and Somlyo, 1969; von Loh, 1971). The mechanism of the relaxation of smooth muscle produced by isoprenaline is still not clear. Indeed, relaxation can result without any significant change in membrane potential (Daniel *et al.*, 1970). Some authors have suggested that an electrogenic Na^+ pump may be involved (Somlyo and Somlyo, 1969) while others have speculated that a Ca^{2+} pump, which reduces intracellular free Ca^{2+} levels (Magaribuchi and Kuriyama, 1972) may be important. cAMP has also been suggested to play a role in β-receptor-mediated vasodilation (Robison *et al.*, 1971; Amer, this volume). However, a few studies have been reported on the effects of cAMP on electrophysiology of vascular smooth muscle. Somlyo and Somlyo (1969) and Somlyo *et al.* (1970) reported that Bt_2-cAMP induces hyperpolarizations of the smooth muscle of the rabbit pulmonary vein which are inversely dependent on extracellular K^+ levels, as are the hyperpolarizing responses to isoprenaline. The fact that these hyperpolarizing responses to both agents are potentiated by theophylline (an inhibitor of phosphodiesterase) further supports a second messenger role for this cyclic nucleotide in the effect of isoprenaline.

B Liver

Although not usually classified as an excitable tissue, liver cells do exhibit substantial resting membrane potentials, and respond to catecholamines and glucagon with increases in cAMP levels (Robison *et al.*, 1971). Furthermore, hyperpolarization of liver cells has been reported using catecholamines and glucagon in perfused rat liver (Friedman *et al.*, 1971), catecholamines in superfused guinea-pig liver slices (Haylett and Jenkinson, 1969) and iontophoresis of catecholamines in short-term tissue cultures of guinea-pig liver parenchymal cells (Green *et al.*, 1972). It is of great interest, therefore, that both cAMP and cGMP hyperpolarize perfused rat liver cells. These hyperpolarizations are preceded by an increase in Ca^{2+} efflux, paralleled by an increase in K^+ efflux and blocked by tetracaine (Friedman *et al.*, 1972). The similarity of action of glucagon, isoprenaline and cAMP greatly supports the hypothesis that cAMP is the second messenger in a variety of hormonal effects produced by catecholamines and glucagon in liver (Robison *et al.*, 1971).

C Fat Cells

Fat cells represent another tissue thought to utilize cAMP as a second messenger to a variety of hormones (Robison *et al.*, 1971; Fain, this volume). Intracellular recordings from brown fat cells have revealed that noradrenaline, whether superfused *in vitro*, perfused *in situ* or released by stimulation of the sympathetic nerves, depolarizes the membrane (Girardier *et al.*, 1968; Horowitz *et al.*, 1969; Krishna *et al.*, 1970). This depolarization is accompanied by a decrease in membrane resistance, (Horowitz *et al.*, 1971) and is blocked by the antilipolytics insulin and propranolol.

Although early reports indicated that cyclic nucleotides have no effect on membrane potentials of brown fat cells (Girardier *et al.*, 1968; Horowitz *et al.*, 1969; Krishna *et al.*, 1970), a later study by Williams and Matthews (1974a, b) showed pronounced depolarization of 11 – 28 mV with high concentrations of Bt_2-cAMP (2 mM) or theophylline. It has been suggested that the earlier negative findings might result from poor intracellular penetration of the nucleotide. The ionic mechanism involved in these depolarizations is unclear, although both Na^+ and Cl^- may be implicated. Moreover, Williams and Matthews (1974a, b) point out that it has not yet been determined whether the depolarizing effect of catecholamines and cAMP results from a direct or an indirect action on the membrane, since the fatty acid octanoate is also able to depolarize fat cells.

D Invertebrate Salivary Glands

There is pharmacological evidence that cAMP mediates at least part of the secretory response of blowfly salivary glands to 5-hydroxytryptamine (5-HT) (Berridge and Prince, 1971, 1972b; Berridge, this volume). However, when the transepithelial potential is measured using a liquid paraffin-gap technique, the electrical responses to 5-HT and cAMP are completely different. 5-HT causes a negative luminal potential (with respect to the bathing medium) while cAMP (and theophylline) makes the lumen more positive (Berridge and Prince, 1972b).

However, when 5-HT is administered in the presence of theophylline or in a Cl^--free medium, a positive luminal potential results. Berridge and Prince (1972b) therefore concluded that 5-HT has two actions: (*i*) elevation of anion transport (perhaps mediated by Ca^{2+} fluxes); and (*ii*) stimulation of cAMP production which results in enhanced cation transport.

Work by Hax *et al.* (1974) suggests that cAMP may have yet another function in isolated salivary glands of the larvae of *Drosophila hydie.* cAMP and Bt_2-cAMP hyperpolarize these cells with a decrease in the ionic permeability of non-junctional membranes and increased permeability of the specialized low-resistance junctions between cells. These responses are mimicked by theophylline and by ecdysterone (a hydroxylated derivative of the insect-moulting hormone ecdysone) both of which elevate intracellular cAMP levels. Thus, it is proposed that cAMP plays a role in passive electrical communication between cells.

E Pineal Gland

A definite role for cAMP in pineal gland is the conversion of 5-HT to its *N*-acetyl derivative, by induction of the enzyme 5-HT *N*-acetyltransferase (Klein *et al.*, 1970). Since it had been known for some time that noradrenaline stimulates cAMP production in the pineal (Weiss and Costa, 1968), it was of some interest when Sakai and Marks (1972), and Kakiuchi and Marks (1972) found that noradrenaline and isoprenaline evoked a rapid hyperpolarization of pinealocytes. Subsequently, Parfitt *et al.* (1975), also utilizing intracellular microelectrode recording, noted that cAMP or Bt_2-cAMP, as well as noradrenaline, induces a small but significant hyperpolarization of rat pinealocytes. Furthermore, ouabain or a high extracellular K^+ concentration blocks the hyperpolarizing effects of noradrenaline and cAMP. Since ouabain does not prevent the stimulation of cAMP generation by noradrenaline, but does block the induction of the *N*-acetyltransferase, Klein and Parfitt (1975) propose that hyperpolarization of the pineal cell may serve to establish an intracellular ionic environment more conducive to the cAMP induction of the *N*-acetyltransferase. However, since membrane conductance was not measured, it is difficult to determine which ions may be involved in this process and which produce the hyperpolarizations. Indeed, the involvement of an electrogenic pump as the source of the hyperpolarization cannot be ruled out.

F Invertebrate Neurones

There is scant information in the literature on a putative electrophysiological effect of cAMP in invertebrate neurones, even though several neurotransmitter candidates, such as dopamine, octopamine and 5-HT, are known to elevate cAMP levels in invertebrate nervous tissue — see review by Nathanson and Greengard (1976). This situation arises partly because cAMP has no apparent effect on the electrical membrane properties of many invertebrate neurones (Gerschenfeld, personal communication). The most comprehensive paper on this subject (Takeuchi *et al.*, 1975) reports that neither cAMP nor cGMP have any effect on two types of dopamine-sensitive neurones in the suboesophageal ganglion of the African giant snail.

However, interpretation of negative results in these invertebrate preparations is hazardous for several reasons: (*i*) in many cases, it has not yet been proved that the neurotransmitter candidate, that is the first messenger, studied is actually a neurotransmitter for the cell from which recordings are being made; (*ii*) stimulation of cAMP formation by the first messenger has not been shown for the specific cells studied electrophysiologically; and (*iii*) most studies have not used greater concentrations of cAMP than 1 mM, yet other experiments have shown that cAMP must be applied at levels several orders of magnitude greater than that of the first messenger in order to be effective (Bloom, 1974). This arises because the first messenger has its receptor on the external surface of the membrane while the relatively impermeable cyclic nucleotide most likely must reach the intracellular receptor, the protein kinase. It is suggested that further work on invertebrates proceed by localizing cAMP generation in specific neurones, for example by histochemical techniques. Electrophysiological studies should be carried out using intracellular injection of cAMP, extracellular perfusion of relatively high concentrations of cAMP (up to 5 mM), or using derivatives known to be more potent or permeable than cAMP itself (Siggins and Henriksen, 1975).

G Photoreceptors

Cyclic nucleotides have been implicated in the role of modulation of photoreceptors (Bitensky *et al.*, 1973). Studies by several groups (Miki *et al.*, 1973; Chader *et al.*, 1974; Goridis *et al.*, 1973; Goridis and Virmaux, 1974) indicate that the phosphodiesterase of retinal outer segments is stimulated by light while the adenylate cyclase is unaffected. Moreover, the phosphodiesterase is more specific for cGMP than for cAMP (Goridis and Virmaux, 1974). The implication, therefore, is that the most important nucleotide in photoreception is cGMP. Unfortunately, only one preliminary report has appeared on the electrophysiological effects of cGMP on photoreceptors (Miller, 1973). Although the results of this study are inconsistent, it is interesting that intracellular injection of both cAMP and cGMP in several experiments produced large increases in membrane and receptor potentials. However, the exact role of the cyclic nucleotides in photoreception, if any, still remains unclear.

H Vertebrate Nervous Tissue

By far the greatest proportion of literature on the electrophysiology of cyclic nucleotides concerns vertebrate nervous tissue. The primary reason for this is the well-documented effect of neurotransmitter candidates on cAMP or cGMP levels in vertebrate nervous tissue (Daly, 1975; Bloom, 1975; Skolnick and Daly, this volume). Since almost all the reports to date have described a postsynaptic action of the nucleotides, the following account is concerned only with the effect of nucleotides, and their presumed first messengers, at this level of integration.

1 Peripheral Nervous System

The vertebrate sympathetic ganglia are a favourable model for the study of the electrophysiology of cyclic nucleotides at synapses for the following reasons: (*i*) the principal neurones are large enough to be impaled by microelectrodes; (*ii*) the tissue is easily isolated and well organized making it suitable for sucrose-gap recording; (*iii*) the synaptic anatomy is fairly simple, with one or two inputs and only one output pathway; (*iv*) repetitive stimulation of the inputs produces a unique slow inhibitory postsynaptic potential (sIPSP) and a slow excitatory postsynaptic potential (sEPSP) distinct from the classical fast excitatory postsynaptic potential which normally triggers the action potential (Eccles and Libet, 1961; Nishi and Koketsu, 1967); (*v*) repetitive stimulation elevates the levels of both cAMP and cGMP in sympathetic ganglia (McAfee *et al.*, 1971; Weight *et al.*, 1974); (*vi*) superfused catecholamines or histamine also evoke increases in ganglionic cAMP (Kebabian and Greengard, 1971; Cramer *et al.*, 1973; Cramer and Lindl, 1974; Lindl *et al.*, 1975); and (*vii*) immunohistochemistry shows that increases in cAMP and cGMP in response to superfused dopamine and acetylcholine, respectively, appear predominantly in the postganglionic neurones (Kebabian *et al.*, 1975).

The last three points and other pharmacological data are taken to implicate cAMP as the intracellular postsynaptic mediator of the sIPSP, and cGMP as the second messenger for the sEPSP. Since responses to exogenous dopamine and acetylcholine mimic the respective slow postsynaptic potentials, the first messengers for these systems in mammals are thought to be dopamine for the sIPSP and acetylcholine for the sEPSP. However, some findings have implicated noradrenaline in the sIPSP of rat ganglia (Lindl and Cramer, 1975). The cellular source of the dopamine in mammals is thought to be the small intensely fluorescent cell. It has been suggested that this cell functions as an interneurone interposed between the preganglionic input fibres and the principal ganglio-neurones. Intracellular recordings show that sIPSP and sEPSP are generally accompanied by an increase in membrane (input) resistance which is indicated by a decrease in ionic conductance. A recent controversy centres on whether the slow postsynaptic potentials arise as a result of activation of electrogenic pumps (Nishi and Koketsu, 1967; Kobayashi and Libet, 1968), or whether they result from a decreased conductance of the membrane to Na^+ and K^+, respectively (Weight and Votava, 1970; Weight and Padjen, 1973b).

The first electrophysiological studies of cyclic nucleotides in sympathetic ganglia showed that both cAMP and dopamine hyperpolarize the ganglioneurones while cGMP produces mainly a depolarizing response (McAfee and Greengard, 1972); theophylline potentiates the dopamine response. These data are taken as evidence that cAMP postsynaptically mediates the sIPSP. However, since this study utilized the sucrose-gap method of recording, changes in membrane resistance to ionic conductance were not recorded. Therefore, it is not yet known if the responses to cAMP and cGMP exactly mimic the sIPSP and sEPSP, respectively, and what ionic mechanisms are involved. A recent report (Libet *et al.*, 1975) suggests a unique function for the cyclic nucleotides, in addition to the direct effects on membrane potential. Using gross recording methods, these authors noted that extracellular application of dopamine or cAMP to the sympathetic ganglion causes long-term enhancement of the slow depolarizing response to methacholine; addition of cGMP to the bath abolishes this effect. It is, thus, proposed that cAMP may function to produce a 'memory trace' in the ganglion which is 'disrupted' by cGMP.

2 Central Nervous System

a. Cerebellum The electrophysiological effects of cylic nucleotides have been studied in several regions of the CNS. Positive results were first reported for the cerebellum of the rat, in which a concordance of data favours a role for cAMP in synaptic transmission. Furthermore, the nature of the neuronal organization of this region facilitates electrophysiological identification of the only output cell of this structure, the Purkinje cell (Eccles *et al.*, 1967). The principal data implicating a role for cAMP in noradrenergic transmission to the Purkinje cells is as follows: (*i*) biochemically, catecholamines elevate cAMP levels in cerebellar slices *in vitro* (Kakiuchi and Rall, 1968b) via a β-receptor (Chasin *et al.*, 1971); (*ii*) noradrenergic fibres arising from the brain stem nucleus locus coeruleus (LC) terminate on the Purkinje cells (Bloom *et al.*, 1971); (*iii*) stimulation of LC and iontophoresis of noradrenaline inhibit the spontaneous extracellular activity of most Purkinje cells (Hoffer *et al.*, 1971a; Siggins *et al.*, 1971b; Hoffer *et al.*, 1973); (*iv*) ionotophoresis of PGE_1 and PGE_2, nicotinate and adrenergic blockers, such as sotalol and fluphenazine, all agents known to antagonize noradrenaline-evoked elevations of cAMP (Chasin *et al.*, 1971; Dismukes and Daly, 1975; Skolnick *et al.*, 1976), also block the inhibitory responses to LC stimulation or iontophoresis of noradrenaline (Hoffer *et al.*, 1969; Siggins *et al.*, 1971a; Hoffer *et al.*, 1973; Stitt and Hardy, 1975; Freedman and Hoffer, 1975); (*v*) phosphodiesterase inhibitors enhance these inhibitory responses (Siggins *et al.*, 1969; Hoffer *et al.*, 1971; Hoffer *et al.*, 1973); and (*vi*) stimulation of the LC or topical application of noradrenaline elevates immunohistochemically detectable cAMP in Purkinje cells (Siggins *et al.*, 1973).

A major line of evidence linking cAMP to this system is derived from electrophysiological studies. Using extracellular recording of the spontaneous action potentials of identified Purkinje cells of many species, cAMP and several derivatives produce a reduction in the firing rate in most Purkinje cells similar to that produced by noradrenaline and LC stimulation (Siggins *et al.*, 1969; Bloom *et al.*, 1975; Siggins and Henriksen, 1975; Hoffer *et al.*, 1971b). Moreover, using intracellular recording it is shown that stimulation of the LC or extracellular application of cAMP, Bt_2-cAMP or noradrenaline hyperpolarizes Purkinje cells (Figure 18.2) and generally elevates the input resistance (Siggins *et al.*, 1971c). The latter effect is unique since most inhibitory neurotransmitters and

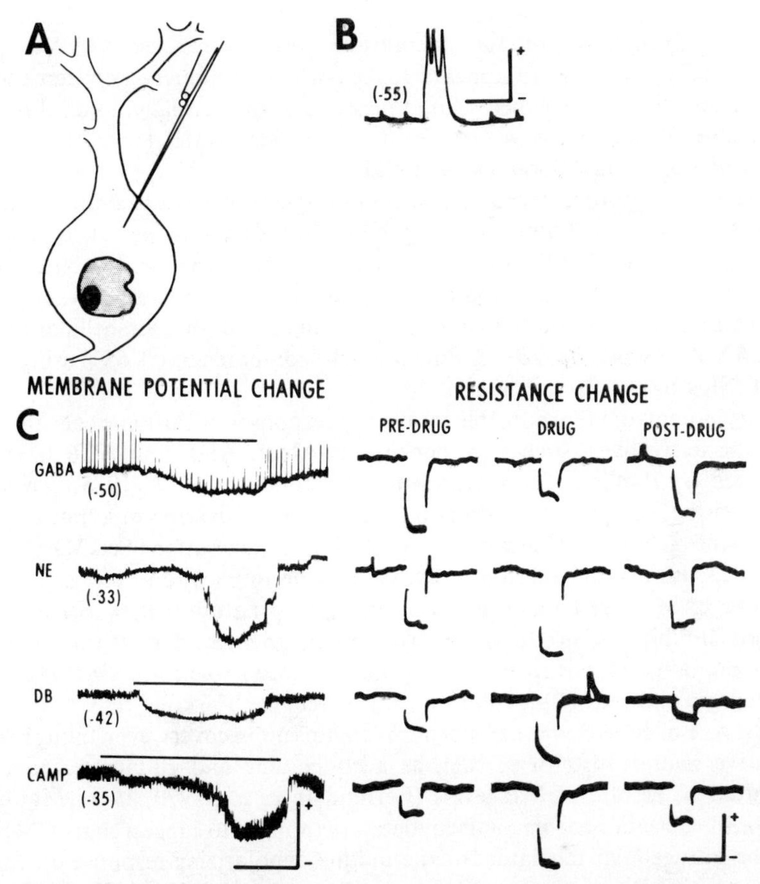

Figure 18.2 Intracellular recordings from rat cerebellar Purkinje cells. A. Schematic representation of a three-barrelled micropipette with a Purkinje cell. The intracellular electrode protrudes beyond the orifices of the two extracellular microelectrophoretic barrels. B. Multispiked spontaneous climbing fibre discharge obtained during intracellular recording from a Purkinje cell. Number in parentheses is resting potential in millivolts (mV); calibration bars are 20 msec and 25 mV. C. Changes in membrane potential and membrane resistance of four Purkinje cells in response to γ-aminobutyrate (GABA), noradrenaline (NE), Bt_2-cAMP (DB) and cAMP. All specimens in each horizontal row of records are from the same cell. Solid bar above each record indicates the extracellular electrophoresis of the indicated drug (100–150 nA). Number in parentheses below each recording is resting potential (mV); calibration bar under membrane potential records is 10 sec and 20 mV for NE, DB, and cAMP, and is 5 sec and 10 mV for GABA. The effective input resistance was judged by the size of pulses resulting from the passage across the membrane of a brief constant current (1 nA) pulse before, during, and after electrophoresis of the respective drugs (1 mV = 1 MΩ). Discontinuities in the fast transients of the pulses result from the loss of high frequencies (>10 kHz) and from the chopped nature of the frequency-modulated magnetic tape recording used. All 'pulse' records have been graphically normalized to the same baseline level. Calibration bar on right indicates 80 msec and 15 mV for all pulse records. (From Siggins *et al.*, 1971d with permission of the American Association for the Advancement of Science. Copyright 1971 by the American Association for the Advancement of Science)

pathways generally reduce membrane resistance (see GABA effect, Figure 18.2). Thus, there appears to be an exact mimicry between the first messenger and the presumed second messenger. Interestingly, iontophoresis of adenosine, which is known to elevate cerebellar cAMP levels, also inhibits Purkinje cell firing (Bloom *et al.*, 1975).

The observation of reproducible inhibitory responses of cultured Purkinje cells to superfusion of high concentrations of cAMP (Gahwiler, 1976) reinforces assumptions that cAMP mimicks the effects of noradrenaline. The latter point is further supported by the fact that iontophoresis of certain derivatives of cAMP possessing a greater ability to activate protein kinase (the intracellular receptor for cAMP) inhibits 80–92% of Purkinje cells compared to 60–65% using cAMP itself (Siggins and Henriksen, 1975).

Very potent and reproducible inhibitory responses of Purkinje cells to several adenine derivatives, such as adenosine, 5′-AMP, ADP and ATP have been reported by Phillis and his associates, (Kostopoulos *et al.*, 1975). They have ascribed these responses in cerebral cortex to the activation of a 'purinergic' or adenosine receptor, which results in postsynaptic generation of cAMP (Phillis *et al.*, 1975; Phillis and Kostopoulos, 1975). The implication here is that intracellular cAMP must therefore be inhibitory in nature if it is to mediate the potent inhibitions produced by the other adenine derivatives. Thus the preponderance of evidence indicates an inhibitory role for cAMP in the noradrenergic and purinergic inhibitions of cerebellar Purkinje cells.

The role of cGMP in the cerebellum remains more covert, even though several putative neurotransmitters, such as acetylcholine and glutamate, have been reported to elevate cGMP levels (Ferrendelli *et al.*, 1970, 1974; Mao *et al.*, 1974a,b). Certain pharmacological data are thought to suggest that cGMP may be the intracellular mediator for the unique depolarizing response of Purkinje cells to activation of climbing fibre inputs (Mao *et al.*, 1974a,b). Here, glutamate may be the neurotransmitter. However, all biochemical studies on the role of cGMP in cerebellum have utilized slices of cerebellum treated with very high concentrations of agonists or intact animals injected systemically. The results of these studies are therefore difficult to interpret because of the multiple sites of possible indirect actions of the agonists. Moreover, early iontophoretic studies in rat cerebellum showed that the responses of Purkinje cells to cGMP are mainly inhibitory (Hoffer *et al.*, 1971b), although a more recent study of normal and mutant (weaver) mouse cerebellum reported a predominance of excitatory, presumed depolarizing, responses (Siggins *et al.*, 1976b). Glutamate, but not acetylcholine, also excites normal mouse Purkinje cells. This finding is more consistent with a role for glutamate as a first messenger than for acetylcholine in mouse. However, weaver Purkinje neurones show more excitations (40%) to acetylcholine. Obviously many more criteria of second messenger mediation need to be satisfied before a synaptic role can be verified for cGMP in cerebellum.

b. Hippocampus. The LC sends noradrenergic fibres to several other cortical areas as well as to the cerebellum. Recent pharmacological and electrophysiological studies of the rat hippocampus, whose pyramidal cells receive a noradrenergic input, suggest that this input is functionally similar to that in cerebellum (Segal and Bloom, 1974a,b). These data show that: (*i*) electrical stimulation of the LC and iontophoresis of noradrenaline inhibit the spontaneous activity of hippocampal pyramidal cells probably invoking a β-receptor; (*ii*) phosphodiesterase inhibitors potentiate both of these inhibitions; (*iii*) agents known to block activation of adenylate cyclase in other systems antagonize the LC and noradrenaline-evoked inhibitions; and (*iv*) biochemical studies show that both noradrenaline (Forn and Krishna, 1971; Palmer and Burks, 1971) and LC stimulation (Guidotti *et al.*, unpublished results) elevate cAMP levels in hippocampus.

However, a major link between cAMP and noradrenaline-evoked inhibition of pyramidal cells is derived from electrophysiological studies which showed that extracellular iontophoresis of cAMP acts in a similar way to noradrenaline and LC stimulation by inhibiting the majority (57%) of pyramidal cells (Segal and Bloom, 1974b). More exact mimicry of responses to noradrenaline and cAMP has been shown in the preliminary intracellular studies of Oliver and Segal (1974). cAMP, but not ATP or 5′-AMP, iontophoretically injected into pyramidal cells produces the same response as LC stimulation or noradrenaline to the outside of the cell — hyperpolarization accompanied by no change or an increase in membrane input resistance. Thus, hippocampal pyramidal cells respond to the LC noradrenergic pathway in the same way as cerebellar Purkinje cells.

A possible electrophysiological role for cGMP in rat hippocampus is also emerging, as the result of studies by Hoffer *et al.* (1976b) using explants of hippocampus transplanted into the anterior chamber of adult rat eyes. This preparation allows the use of superfusion as well as iontophoresis and permits better control of the indirect effects of the two approaches to drug delivery. In this preparation, both acetylcholine and cGMP, or its derivatives, evoke reproducible increases in the spontaneous firing rate of all pyramidal cells, often to the point of seizure (Hoffer *et al.*, 1976b); certain phosphodiesterase inhibitors potentiate the excitations to acetylcholine. These results predict a postsynaptic role for cGMP in mediating the excitatory responses to acetylcholine, a physiologically relevant concept in view of the well known cholinergic input to the hippocampus via the septum.

c. Cerebral Cortex. Although the cerebral cortex of several species responds well to catecholamines in generating cAMP *in vitro* (Kakuichi and Rall, 1968a; Chasin *et al.*, 1971), electrophysiological studies linking the cyclic nucleotide to the effects of catecholamines are not conclusive. One problem has been the inability to determine exactly which type, or types, of cortical neurone receive input from LC. The importance of identifying cell types for obtaining positive responses to cAMP has been described for cerebellum (Siggins *et al.*, 1971c; Siggins and Henriksen, 1975). In general, both noradrenaline and cAMP depress the discharge of identified pyramidal tract cells, with cAMP inhibits about 70% of these cells. The preliminary finding of potentiation of noradrenaline responses by the phosphodiesterase inhibitor papaverine (Stone *et al.*, 1975) adds further evidence that cAMP may be the second messenger for this monoamine in cerebral cortex.

An interesting new development is the possibility that adenosine, or some other non-cyclic adenine derivative, may be first messenger for cAMP in cerebral cortex. Adenosine stimulation of cAMP levels in slices of cortex is well known (Sattin and Rall, 1970; Shimizu and Daly, 1970). It is thought that the cyclic nucleotide is not formed directly from adenosine (as a precursor for ATP and therefore cAMP), but rather from stimulation of an adenosine receptor linked to adenylate cyclase (Daly, 1975). Furthermore, the methylxanthines, in contrast to their ability to elevate cyclic nucleotide levels by inhibition of phosphodiesterase, actually block the adenosine receptor (Kakiuchi and Rall, 1968a,b).

It is of interest, therefore, that Phillis *et al.* (1975), and Phillis and Kostopoulos (1975) have shown that the iontophoresis of adenosine and a variety of adenine derivatives inhibits the spontaneous electrical activity of almost all pyramidal tract and other unidentified cortical neurones. The proposition that this effect is mediated through an adenosine receptor and that intracellular generation of cAMP is strengthened by the finding that iontophoresis of methylxanthines antagonizes the inhibitions produced by adenosine (Phillis and Kostopoulos, 1975). Similar findings have recently been reported for slices of olfactory cortex maintained *in vitro*. Adenosine, 5′-AMP, ATP and ADP ($5 \times 10^{-6} - 5 \times 10^{-3}$ M) all depress the orthodromically-evoked field potential (N-wave), indicative of a population of EPSPs, and enhance the formation of cAMP (Okada and Kuroda, 1975; Kuroda and Kobayashi, 1975). Theophylline inhibits both these effects of the adenine derivatives. Although these authors have reported that cAMP itself has a strong inhibitory effect (50% or more reduction of the N-wave), this inhibition was only seen in four of ten slice preparations (Okada and Kuroda, 1975). It is likely that the poor effects the majority of the preparations reflect the use of relatively low cAMP concentrations (10^{-4} M). The implication derived from all these studies is that intracellular cAMP must have a net inhibitory effect, since the respective first messengers all inhibit these cortical neurones.

The iontophoretic studies of Stone *et al.* (1975) suggest a possible role also for cyclic GMP in rat cerebral cortex. These workers found a strong correlation between excitations of discharge produced by acetylcholine and cGMP in identified pyramidal tract neurones. Approximately 75% of these cells are excited by cGMP. Thus the above findings suggest that cAMP and cGMP may function as reciprocal intracellular second messengers for noradrenaline and acetylcholine, respectively.

d. Caudate Nucleus. Although catecholamines are known to stimulate cAMP formation in the caudate (Kebabian *et al.*, 1972; Walker and Walker, 1973; Miller *et al.*, 1974a), the adenylate cyclase system in this brain structure more closely resembles that of sympathetic ganglia than of the brain cortices. The endogenous catecholamine is predominantly dopamine, derived from cell bodies in the substantia nigra, not noradrenaline. Furthermore, the dopaminergic receptor, as assayed either biochemically or electrophysiologically, is antagonized by neuroleptic agents, such as haloperidol, chlorpromazine and fluphenazine, rather than by β-adrenergic blockers. Inhibitory responses of rat caudate neurones to iotophoresis of dopamine are potentiated by pgE_1 (Siggins *et al.*, 1976a), and not blocked as are noradrenaline responses in the rat cerebellum (Hoffer *et al.*, 1969; Siggins *et al.*, 1971a; Stitt and Hardy, 1975).

In spite of these differences, over 90% of unidentified caudate neurones respond like cerebral and cerebellar cortical neurones to iontophoresis of cAMP, Bt-cyclic AMP and dopamine, with a reduction in spontaneous or amino acid-evoked discharge rate (Siggins *et al.*, 1974, 1976b). The enhancement of dopamine and cAMP-induced inhibitions by phosphodiesterase inhibitors, such as papaverine and 1-methyl-3-isobutylxanthine, further strengthens the link between dopamine and cAMP in caudate neurones at the electrophysiological level.

Adenosine may also play a role in the caudate nucleus. The biochemical studies of Wilkening and Makman (1975) have shown that 2-chloroadenosine stimulates cAMP accumulation in caudate. This derivative of adenosine does not enter the pool of precursors for ATP and cAMP (Sturgill *et al.*, 1975), thus it may function primarily as an agonist for an adenosine receptor. The formation of cAMP in response to 2-chloroadenosine is blocked by the methylxanthines (Wilkening and Makman, 1975). Recent electrophysiological studies reinforce these biochemical findings. Iontophoresis of adenosine and 2-chloroadenosine inhibits the spontaneous firing of nearly all caudate neurones studied (Siggins *et al.*, 1976b). Unlike the inhibitory responses to cAMP and dopamine, those to adenosine are blocked by methylxanthines. Hence, an inhibitory adenosine–adenylate cyclase system distinct from the dopamine–adenylate cyclase system seems likely in rat caudate as well as cerebral cortex.

e. Other Dopamine-Rich Areas. Dopamine-containing fibres, with cells of origin in areas A9 (substantia nigra) or A10, densely innervate several other areas besides the caudate nucleus. Principal among these are the limbic structures such as olfactory tubercle, amygdala and limbic cortex, and the nucleus accumbens. It is, therefore, of some interest that the effects of iontophoretically applied dopamine have been compared to the effects of cAMP in the nucleus accumbens and the olfactory tubercle. The spontaneous activity of about 90% of these (unidentified) cells is inhibited by dopamine and cAMP (Bunney and Aghajanian, 1974). Furthermore, Obata and Yoshida (1973) have noted that iontophoresis of both dopamine and cAMP predominantly inhibits neuronal firing in the entopeduncular nucleus, another region rich in dopamine. Biochemical studies have shown that dopamine can stimulate adenylate cyclase in the nucleus accumbens and olfactory tubercle (Clement–Cormier *et al.*, 1974). However, proof of the intermediation of the inhibitory dopamine responses in these structures by cAMP awaits further electrophysiological tests with agents, such as phosphodiesterase inhibitors, known to influence the dopamine — cAMP system.

f. Brain Stem. An effort has also been made to find a relationship between responses to noradrenaline, histamine and cAMP in unidentified neurones of the cat brain stem (Anderson *et al.*, 1973a). Here, iontophoresis of cAMP depresses about 80% of all neurones studied, including 95% of those units inhibited by noradrenaline and about 90% of those depressed by histamine. Responses to noradrenaline are blocked by nicotinate. However, only three of eleven cells show potentiation of the monoamine effects by the phosphodiesterase inhibitors theophylline and aminophylline, and pgE_1 blocks 50% of the noradrenaline-induced depressions. As a result, Anderson *et al.* (1973a) were reluctant to conclude that cAMP mediates monoamine depressions. However, in view of the lack of histochemical evidence on the presence of noradrenaline synapses on any of the test cells, and the liklihood of diverse isozymes of phosphodiesterase in this brain region (Siggins *et al.*, 1974), the findings of Anderson *et al.* (1973a) do not rule out a link between cAMP and the depressant responses to noradrenaline and histamine in brain stem.

g. Tissue Cultures. A few attempts have been made to study electrophysiological responses of cultured neurones to cyclic nucleotides. These studies have not always produced positive results. Indeed it has often proved difficult to evoke responses even to the first messengers noradrenaline and dopamine, for example in cultured superior cervical ganglion neurones of the rat recorded intracellularly (Obata, 1974). Secondly, responses to the presumed second messengers are not often reproducible, for example, in cultured frog sympathetic ganglion neurones also recorded intracellularly (Siggins *et al.*, unpublished results). It is possible that the culture system itself modifies monoamine receptors or enzymes of the cAMP system.

However, several studies have noted modulation by cAMP of on-going excitability of cultured neurones. Thus, Crain and Pollack (1973) showed that cAMP, or Bt_2-cAMP, restores the excitability of neurones of foetal explant cultures of mouse cerebral cortex and spinal cord after removal of Ca^{2+}. Likewise, Chalazonitis and Green (1974) have reported a greater excitability of neuroblastoma cell lines cultured for long periods in media with high cAMP levels. A similar elevation of excitability was seen with superfusion of Bt_2-cAMP onto exposed kitten cerebral cortex *in situ* (Purpura and Shofer, 1972). However, these studies do not show that cAMP excites these neurones *per se*, but that cAMP so alters the membrane that it is easier to excite by other means (for example, by intracellular injections of current). These data appear to contradict those showing inhibitory effects of cAMP on neurones *in situ*, in fact such an increase in excitability could be produced by slight hyperpolarizing actions accompanied by increased membrane resistance, wherein a constant amount of stimulating or synaptic current would produce a larger potential deflection across the greater membrane resistance. This would permit the evoked potential to approach more closely or to exceed the threshold for spike generation, while spontaneous firing (arising from pacemaker potentials) would be depressed by the hyperpolarization (Hoffer *et al.*, 1975).

A recent study by Gahwiler (1976) on explants of rat cerebellar cortex reinforces the view that cAMP mediates the inhibitory effects of noradrenaline.

In this study, perfusion of the culture chamber with noradrenaline at concentrations greater than 10^{-5} M decreases spontaneous firing of Purkinje cells. In contrast, 10^{-4} M cAMP has no effect and a higher concentration (10^{-3} M) slows the firing of 44% of the cells. Phosphodiesterase inhibitors alone also depress Purkinje cell firing, but at doses below the threshold they also strongly potentiate the inhibitions produced by either cAMP or noradrenaline. Gahwiler (1976) concluded that these data support the hypothesis that cAMP mediates the depressant action of noradrenergic neurotransmission in Purkinje cells.

V CONCLUSIONS AND SPECULATIONS: UNIVERSAL EFFECT AND MECHANISM OF ACTION OF cAMP

Although it is premature to suggest that cAMP has a specific universal electrophysiological effect in all tissues or species, it is tempting to survey the tissues studied to date for possible equivalencies in response to cAMP. Indeed, it can be seen from this survey that with a few exceptions (for example, heart and lipocytes) cAMP, when studied by intracellular methods, predominantly hyperpolarizes the cells studied. The cell types include smooth and skeletal muscle, hepatoctyes, salivary cells, pineal cells, possibly reticulocytes and certain neurones. In addition, those studies utilizing extracellular recording of evoked or spontaneously firing neurones also strongly imply hyperpolarizing effects. However, further studies using intracellular recording techniques and, ideally, the intracellular injection of this cyclic nucleotide are needed to verify a universal action of cAMP.

Determination of the ionic mechanism underlying responses of excitable cells to cAMP also requires further tests, since most studies reported to date neither measured membrane resistance nor changed ionic concentrations during recording. As a result, several mechanisms of action have not been eliminated as possible candidates. These include increases in conductance to ions (K^+, Cl^-), decreases in conductance to other ions (Na^+, Ca^{2+}) and activation of electrogenic pumps. The last mechanism is a candidate in skeletal muscle and fat cells, where active extrusion of ions or activation of an Na^+/K^+-ATPase by catecholamines or cAMP has been observed (Hays *et al.*, 1974; Bressler *et al.*, 1975; Horowitz and Eaton, 1975). It is speculated that there will be little if any change in membrane resistance, with activation of an electrogenic pump (Thomas, 1972).

The molecular events by which cAMP may evoke these membrane changes is another avenue of investigation. It is thought that protein kinase may be involved as the intracellular (or perhaps intramembranous) receptor for most, or all actions, of cAMP (Kuo and Greengard, 1969). Electrophysiologically, this theory is supported by the finding that derivatives of cAMP inhibit Purkinje-cell firing in direct relation to their ability to activate protein kinase (Siggins and Henriksen, 1975). The subsequent steps are still unproven, but there is evidence that the protein kinase activated by cAMP may catalyse the phosphorylation of a membrane-bound protein or activate membrane phosphatase (Maeno *et al.*, 1975). It is speculated that these alterations in the phosphate content of membrane components may result in the observed electrical changes in membrane properties (Greengard and Kebabian, 1974).

However, the production of electrophysiologically detectable events by cyclic nucleotides in a given cell type does not rule out other cyclic nucleotide-dependent physiological events in this cell which might escape detection. Such inapparent effects may be much slower in nature than the electrical events and may involve induction or activation of enzymes, cell growth, differentiation and mitogenesis (Thoenen and Otten, 1975; Parker this volume). Indeed, further investigation could show that the electrophysiological changes in some cells are superfluous or perhaps ontogenic vestiges. A more likely possibility is that these changes bring about ionic conditions in the cell or its membrane which are favourable for initiation of these later, slower phenomena, as suggested by Klein and Parfitt (1975). In this regard, neurones may represent a special case wherein the slow cAMP-evoked hyperpolarizations are utilized both for tonic modulations of electrical excitability and for the optimization of conditions necessary for long-term enzyme induction. The hyperpolarizations in some muscle cells may bring about conditions favorable for sequestation of Ca^{2+} (Tada *et al.*, 1974; Andersson and Mohme-Lundholm, 1970; Wollenberger this volume). In the final analysis, each excitable cell type may be unique in its ultimate use of the cyclic nucleotides.

ACKNOWLEDGEMENTS

I thank my colleagues, Drs. F. Bloom, B. Hoffer, J. Nathanson, D. Taylor and F. Weight for helpful discussions. Particular thanks go to Ms. Patricia Millhouse for typing the manuscript and preparing the bibliography.

CHAPTER 19

Psychotropic Drugs and Cyclic AMP in the Central Nervous System (With Particular Reference to Striatal and Mesolimbic Structures)

F. Sulser and J. Vetulani

I INTRODUCTION

Adenosine 3′,5′-monophosphate (cAMP) was originally shown to be the intracellular mediator of the hepatic glycogenolytic action of adrenaline and glucagon, but the cyclic nucleotide is now also recognized as a second messenger mediating a variety of hormonal effects (Sutherland *et. al*, 1962; Robison *et al.*, 1968, 1971). A great number of hormones have been shown to act by increasing the activity of adenylate cyclase in their appropriate target cells.

Several observations support the view that cAMP also plays an important role in brain function. (*i*) the brain tissue contains the highest activity of both adenylate cyclase and phosphodiesterase in any of the mammalian tissues studied (Sutherland *et al.*, 1962). (*ii*) studies on the subcellular distribution of the enzyme suggest that brain adenylate cyclase may be located in synaptic membranes (De Robertis *et al.*, 1967); extrapolation from studies using homogenates of the pineal gland (Weiss and Costa, 1967), and using slices from cerebral cortex (Kalisker *et al.*, 1973) and limbic forebrain (Blumberg *et al.*, 1976; Vetulani *et al.*, 1976), suggests that the enzyme is localized at a postsynaptic site. (*iii*) Several agents thought to function as neurotransmitters are capable of affecting adenylate cyclase from several tissues including brain (Klainer *et al.*, 1962). A number of biogenic amines are capable of increasing the level of cAMP in brain slices from several species (Kakiuchi and Rall, 1968,a,b; Shimizu *et al.*, 1971; Palmer *et al.*, 1973; Perkins and Moore, 1973). (*iv*) Electrical stimulation of brain slices leads to large increases in the levels of cyclic AMP (Kakiuchi *et al.*, 1969). (*v*) Introduction of Bt_2-cAMP into the cerebral ventricles of several species has been reported to elicit excitatory behavioural responses (Gessa *et al.*, 1970). (*vi*) A cAMP-dependent protein kinase (Miyamato *et al.*, 1969b; Maeno *et al.*, 1971), endogenous substrates for the kinase (Johnson *et al.*, 1971a; Ueda *et al.*, 1973) and a phosphoprotein phosphatase (Maeno and Greengard, 1972) have been found to be present at high levels in synaptic-membrane fractions from brain. (*vii*) Using a specific immunofluorescent histochemical method, Siggins *et al.* (1973) have demonstrated that cAMP can be generated postsynaptically in the central nervous system (CNS) in response to noradrenergic stimuli.

Although over the last ten years, cAMP has been the focus of biological regulatory effectors, it has now become clear that cGMP may play a separate role in regulating cellular metabolism and function (Goldberg *et al.*, 1969; Ferrendelli *et al.*, 1970). cGMP has been shown to occur as a natural constituent of brain (Ferrendelli *et al.*, 1970; Ishikawa *et al.*, 1969; Goldberg *et al.*, 1969) and guanylate cyclase activity has been detected in all mammalian tissues examined to date (Goldberg *et al.*, 1973a). Unlike mammalian adenylate cyclase, which is considered to be totally particulate, the major portion of guanylate cyclase activity resides after homogenization of tissues in the soluble fraction (Hardman and Sutherland, 1969; Schultz *et al.*, 1969). cGMP phosphodiesterase (Brooker *et al.*, 1968) and a cGMP-dependent protein kinase (Kuo and Greengard, 1970b) have been identified in brain. Moreover, drugs affecting the release and metabolism of biogenic amines in brain tissue also influence the level of cGMP in brain (Ferrendelli *et al.*, 1972; Sulser and Vetulani, unpublished results). A centrally active cholinesterase inhibitor increases the concentration of cGMP in mouse cerebellum (Goldberg *et al.*, 1973a) and also muscarinic cholinergic agonists produce a several-fold increase in cGMP in rabbit brain which is prevented by muscarinic but not nicotinic blocking agents (Lee *et al.*, 1972). These observations provide evidence that a relationship exists between cholinergic action and the accumulation of cGMP. Conversely, muscarinic agonists attenuate the effectivenss of adrenergic agents in producing an increase in cAMP. While the activation of protein kinases by cAMP appears to be the major mechanism, if not the only one, by which this nucleotide carries out its function as a second messenger in the transmission of hormonal and neurohormonal signals — see, for example, the review by Langan, (1973), — the molecular basis of cellular events affected by cGMP has yet to be elucidated. The existence of a cGMP-dependent protein kinase does, however, strengthen the suggested role of cGMP as a biological regulator independent of cyclic AMP.

It is the purpose of this chapter to review the action of major psychotropic drugs on various cAMP-generating systems in the brain, with particular emphasis on striatal and limbic forebrain structures. The involvement of the dopaminergic nigrostriatal system in extrapyramidal motor function is well documented and evidence exists that Parkinson's disease can result from either a deficiency of dopamine in the basal ganglia or from blockade of dopamine receptors (Hornykiewicz, 1966, 1975). The limbic forebrain system receives and integrates sensory inputs from all sensory systems and relays information to the entire hypothalamic and lower reticular fields (Gloor, 1955; MacLean, 1958; Nauta, 1958). According to Morgane and Stern (1972), 'this recurrent system may thereby modulate multimodal inputs and regulate filtering in information channels at several integrative levels in the brain-stem'. Moreover, limbic connections, including those with the thalamus and overlying cerebral cortex, may provide a key integrating system related to selective modulation of emotion and sensory mechanisms of the brain (Powell and Hines, 1974). Neurochemically, the limbic forebrain receives ascending noradrenergic fibres originating in cell bodies of the pons and medulla oblongata, dopaminergic fibres from the A10 region (mesolimbic dopaminergic system) and serotonergic fibres from the anterior raphé complex (Andén *et al.*, 1966; Dahlström and Fuxe, 1964). Recently, by using mass fragmentography and/or enzymic isotopic micromethods, the concentrations of catecholamines and 5-hydroxytryptamine (5-HT) have been measured in discrete nuclei of the limbic system (Koslow *et al.*, 1974; Saavedra *et al.*, 1974) substantiating the location of biogenic amines described earlier in histochemical reports (Andén *et al.*, 1966). For neurophysiological, neurochemical and neuropsychiatric reasons, the limbic system with its complexity of cortical and subcortical projections is a logical focus of interest for studies dealing with the translation of presynaptic stimuli into postsynaptic events by neurotransmitters such as noradrenaline (NA), dopamine (DA) and 5-HT.

II ROLE OF CATECHOLAMINERGIC MECHANISMS IN THE ACTION OF PSYCHOTROPIC DRUGS

Central aminergic mechanisms have been implicated in the mode of action of many psychotropic drugs, including antidepressants, antipsychotic drugs, amphetamine-like stimulants, narcotic analgesics and anti-Parkinson agents. The important aspects of the effects of psychotropic drugs on aminergic mechanisms including active transport of the amino-acid precursors across the blood–brain barrier, sequential biosynthesis, storage and release of monoamines, and termination of their effect by either enzymic breakdown, reabsorption mechanisms or diffusion, have been extensively reviewed (Bloom and Giarman, 1968; Andén *et al.*, 1969; Weiner, 1970; Sulser and Sanders-Bush, 1971; Sulser, 1976; Costa and Garattini 1970; Usdin and Snyder, 1973).

More recently, it has been demonstrated that long-term pharmacological manipulations which facilitate or inhibit catecholaminergic transmission can produce adaptive changes in the levels of tyrosine hydroxylase by varying the rate of new enzyme synthesis (Segal *et al.*, 1974; Mandell, 1975; Kuczenski, 1975). Short-term alterations of nerve excitability in the CNS as a result of psychotropic drug administration have also been implicated. Thus, DA-receptor blockade produced by classical neuroleptics may lead to an immediate compensatory increase in the activity of dopaminergic cells via a neuronal feedback mechanism (Andén *et al.*, 1970). Recently it has been demonstrated that the increased turnover rate of catecholamines, particularly of DA, caused by neuroleptics is associated with a change in the affinity constant of soluble tyrosine hydroxylase. Thus, Zivkovic *et al.* (1974) reported that the K_m for the pteridine cofactor is reduced four to five-fold and the V_{max} for DOPA formation is doubled by a single dose of haloperidol. Similarly, catecholamine-receptor stimulation by amphetamine, for example, may cause neuronally mediated feedback inhibition of catecholaminergic cells.

Bunney *et al.* (1973) have presented direct evidence in support of this hypothesis. Recording single unit activity from noradrenergic cell bodies in the locus coeruleus, they observed that d-amphetamine produces a rapid, reversible decrease in the firing rate of these neurones. Similarly, d-amphetamine decreases the firing rate of nigrostriatal and ventral tegmental (A10) neurones (Bunney and Aghajanian, 1974). The latter decrease can be reversed by chlorpromazine, and the administration of chlorpromazine alone or of haloperidol results in a marked increase in single unit activity. Moreover, tricyclic antidepressant drugs markedly decrease the rate of firing of noradrenergic cells in the locus coeruleus, the order of potency being parallel to their potency in blocking the neuronal reabsorption of NA (Nybäck *et al.*, 1975).

First and second messengers have been implicated in the long-term synaptic regulation of tyrosinehydroxylase in adrenal medulla and superior cervical ganglia. It is generally accepted that the first messenger in transsynaptic induction of the enzyme in the adrenal medulla and the adrenergic neurone terminals is acetylcholine (Thoenen and Otten, 1975). In contrast, the nature of the second messenger is still a matter of debate. While Guidotti *et al.* (1973) maintain that transsynaptic induction of the hydroxylase in adrenal medulla, for example, requires a stimulus-coupled increase in cAMP or an increase in the ratio of cAMP/cGMP, Thoenen and Otten (1975) have provided evidence that a causal relationship between the two processes can be excluded. They showed that an increase in cAMP or in the ratio of cAMP/cGMP is not essential for the subsequent induction of TOH.

III ADENYLATE CYCLASES AS RECEPTOR SITES FOR CATECHOLAMINES

The second messenger concept (Sutherland *et al.*, 1968) has now been extended to include the neurotransmitter function of catecholamines and other biogenic amines (Rall and Gilman, 1970; Daly, 1975). Robison *et al.* (1971) have suggested that the receptors with which many hormones, including the neuro-hormones, combine are closely related to and may even be part of an adenylate cyclase system. Indeed, adenylate cyclase receptors have been proposed for DA in both the caudate nucleus and the limbic forebrain (Kebabian *et al.*, 1972; Clement-Cormier *et al.*, 1974; Miller *et al.*, 1974b) and for catecholamines in the cerebral cortex (Von Hungen and Roberts, 1973; Perkins and Moore, 1973a; Schultz and Daly, 1973c,d). A simple two-component model of the membrane adenylate cyclase has been proposed by Robison *et al.* (1967). This model consists of two subunits: a receptor or regulatory subunit, facing the outside of the cell, and a catalytic subunit facing the cell interior. While the catalytic subunit might be expected to be basically similar in all tissues, including neurones, the structure of the regulatory subunit may vary and impart receptor (hormonal) specificity. Alternative models have been suggested, for example by Birnbaumer and his associates (1970). The characteristics and structure of adenylate cyclase systems have been comprehensively reviewed by Robison *et al.* (1971) and by Perkins (1973). Moreover, in this volume, Birnbaumer deals specifically with adenylate cyclase–hormone receptors and with current views on the transduction of hormone–receptor interactions into activation of adenylate cyclase, known as the coupling process.

IV DOPAMINE-SENSITIVE ADENYLATE CYCLASE SYSTEMS

A Dopamine-Sensitive Adenylate Cyclase and Its Similarity to the Dopamine Receptor

DA-Sensitive adenylate cyclases have been demonstrated in the caudate nucleus of the mammalian brain (Kebabian *et al.*, 1972; Clement-Cormier *et al.*, 1974; Miller *et al.*, 1974b). Kebabian *et al.*, (1972) first showed that an adenylate cyclase in homogenates of the caudate nucleus of the rat brain is sensitive to very low concentrations of DA ($K_a \simeq 4$ μM). Although the maximal stimulation of adenylate cyclase activity elicited by NA is equal to that observed with DA, the K_a value of NA is approximately 28 μM (Figure 19.1). The β-adrenergic agonist isoprenaline has no significant effect on the enzyme's activity even at very high concentrations (Kebabian *et al.*, 1972). This DA-sensitive adenylate cyclase of the caudate nucleus is weakly antagonized by α-adrenergic blocking agents, but is not influenced by classical β-blockers. These findings, and the ability of apomorphine to mimic the actions of DA upon the DA receptor of the caudate nucleus (Andén *et al.*, 1967) and to stimulate the adenylate cyclase activity, suggest that DA-sensitive adenylate cyclase may be the receptor for DA.

There is some disagreement concerning the DA-stimulated accumulation of cAMP in preparations from brain slices (Huang *et al.*, 1973a; Palmer *et al.*, 1973; Forn *et al.*, 1974; Blumberg *et al.*, 1975). Generally, the NA-sensitive adenylate cyclase systems lose their hormonal sensitivity (particularly of the β-

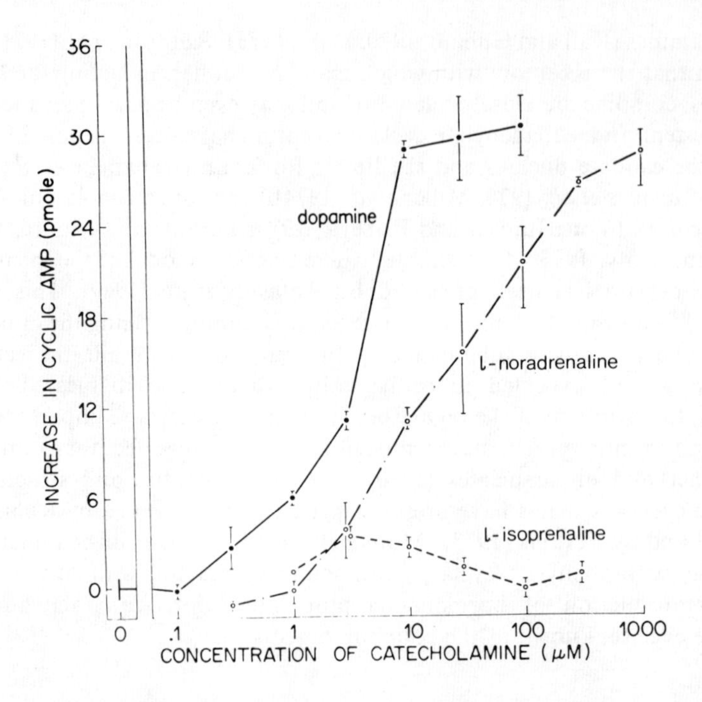

Figure 19.1 Effect of catecholamines on adenylate cyclase activity in a homogenate of rat caudate nucleus. In the absence of added catecholamine, 27.1 ± 1.0 pmol (mean ± S.E.M., $N = 6$) of cAMP was formed. The increase in cAMP above this basal level is plotted as a function of catecholamine concentration. The data give mean values and range for duplicate determinations on each of two to five replicate samples. (From Kebabian *et al.*, 1971)

type) which is best demonstrated in brain-slice preparations. Loss of hormonal sensitivity as a result of homogenization has been observed for a number of adenylate cyclases from a variety of tissues (Øye and Sutherland, 1966; Perkins, 1973). However, homogenization and fractionation techniques yielding cell-free particulate fractions of brain tissue which retain hormonal sensitivity to NA, DA, histamine and adenosine have recently been reported (Von Hungen and Roberts, 1973; Chasin *et al.*, 1974).

DA-sensitive adenylate cyclases with essentially similar pharmacological and biochemical characteristics to those in the caudate nucleus have been identified in the olfactory tubercles and the nucleus accumbens (Clement-Cormier *et al.*, 1974; Horn *et al.*, 1974). These findings are pertinent as a number of studies have suggested that the blockade by antipsychotic drugs of DA receptors in the mesolimbic system may be more closely related to the antipsychotic activity than to the action of these drugs on striatal DA receptors (Andén, 1972; Andén and Stock, 1973).

B Stimulation by Dopamine Analogues and Other Drugs

Because considerable evidence indicates that some anti-Parkinson drugs exert their action as direct or indirect agonists at central DA receptors and that Parkinsonism can be induced by DA antagonists (Hornykievicz, 1975), the structural requirements for DA-receptor agonists and antagonists on the striatal DA-sensitive adenylate cyclase have been extensively studied (Iversen *et al.*, 1975; Sheppard and Burghardt, 1974; Makman *et al.*, 1975). Among the β-phenethylamine analogues of DA, only the *N*-methyl derivative epinine is equipotent to DA (Sheppard and Burghardt, 1974). Non-catecholamines, such as tyramine and amphetamine, and methoxylated DA metabolites are inactive. The same authors also tested compounds in which the side chain of DA is held in a rigid comformation in a second ring system. Apomorphine, a direct-acting DA-receptor stimulant, is a potent agonist of adenylate cyclase in striatal homogenates. It is of interest that the anti-Parkinson drug piribedil, a non-catechol analogue of DA, has been found to be inactive in stimulating the production of cAMP, whereas its metabolite (S584) is a potent stimulator of the adenylate cyclase (Figure 19.2), thus suggesting that the pharmacological

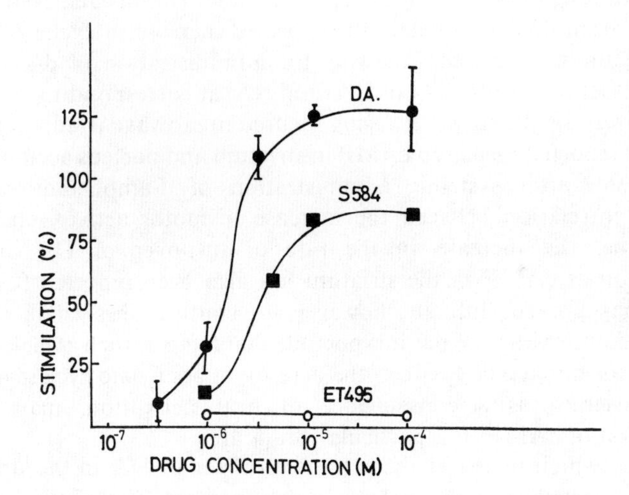

Figure 19.2 Dose–response curves for dopamine, piribedil (ET495) and S584 on cAMP production in rat striatal homogenates. Percentage stimulation refers to stimulation of cAMP production during 2.5 min incubation over control values in absence of added drug. Control values for experiments involving dopamine were 26.1 ± 2.3 pmolecAMP/incubation tube (containing 2 mg wet weight striatal tissue), for ET 495 experiments 34.3 ± 2.6, and for S584 experiments 31.5 ± 3.3 (Mean ± S.E.M., $N = 6$ determinations). Each point is the mean of between 4 and 7 separate incubations, S.E.M. values for ET495 and S584 were less than 10% of the means. (From Miller and Iversen, 1974)

effects of piribedil are mediated *in vivo* through an active catechol metabolite Makman *et al.*, 1975; Kiessling *et al.*, 1975).

Attempts to study the pharmacology of cAMP systems in various brain areas *in vivo* have been hampered by marked changes in the levels of cAMP following sacrifice of the animals by decapitation. The introduction of microwave irradiation as a means of rapid tissue fixation (Jones *et al.*, 1974; Schmidt *et al.*, 1971) has made such studies feasible, although caution in the interpretation of the data must still be exercised. Schmidt *et al.* (1971) studied the distribution of cAMP in various brain areas and found that the amount of this cyclic nucleotide is highest in the cerebellum and brain stem, intermediate in the hypothalamus and midbrain, and lowest in the hippocampus and cortex. The distribution does not appear to correlate with adenylate cyclase activity but a significant relationship exists between areas high in cAMP and low in phosphodiesterase (Weiss and Costa, 1968).

As expected, systemic administration of l-DOPA produces a significant accumulation of cAMP in the caudate nucleus of the rat. Accumulation is prevented by pretreatment with a decarboxylase inhibitor, thus, indicating an *in vivo* activation of DA-sensitive adenylate cyclase (Garelis and Neff, 1974).

We have been unable to detect the expected increase in cyclic AMP in any area of the brain examined following the administration of d-amphetamine which increases the availability of DA (and NA) at corresponding receptor sites (Schmidt *et al.*, 1972). However, using focused microwave irradiation, Carenzi *et al.* (1974) reported increased cAMP in striatum and nucleus accumbens ten to thirty minutes after systemic administration of d-amphetamine. A dose-dependent correlation between the increase in motor activity elicited by d-amphetamine, the increase in the rate of turnover of DA and of the concentration of cAMP in the striatum has also been reported (Costa *et al.*, 1973). It has proved difficult, however, to confirm these data (Blumberg, unpublished observations), but it is possible that factors, such as lack of uniform irradiation, or imbalance between the activity of adenylate cyclase and cAMP phosphodiesterase as a consequence of heat activation, may affect the concentration of cAMP in a particular brain area.

Morphine, which increases the rate of turnover of DA in the striatum has also been reported to cause a long-lasting increase in the concentration of striatal cAMP, which has been interpreted to be the consequence of a persistent stimulation by morphine of postsynaptic DA receptors (Costa *et al.*, 1973). Although results from some studies on the effect of morphine on adenylate cyclase and phosphodiesterase activity are consistent with this view (Puri *et al.*, 1975), conflicting reports have appeared (Iwatsubo and Clouet, 1973).

Van Inwegen *et al.* (1975) have been unable to detect any consistent effect of morphine on adenylate cyclase or phosphodiesterase from any brain area, in response to morphine either injected *in vivo* or added *in vitro*. Recent data indicate that narcotics can bind to 'opiate receptors' in the brain, thus mimicking effects of endogenous ligands ('opioid peptides'). A number of small peptides with morphine-like activity (endorphins) have been isolated from brain tissue

(Goldstein, 1976). The pentapeptide leucine-enkephalin alters cyclic nucleotide levels in neuroblastoma x glioma hybrid cells in a manner similar to that of opiates (Brandt *et al.*, 1976). In rat brain also presynaptic effects of enkephalin have been reported (Taube *et al.*, 1976).

Li^+ has been reported to inhibit NA-induced cyclic AMP increases in slices of the hypothalamus (Palmer *et al.*, 1972) and the cerebral cortex (Forn and Valdecasas, 1971). These effects are difficult to interpret, however, as only very high concentrations of this ion alter the cAMP responses. In cell-free preparations from various regions of the rat brain (including corpus striatum), D-LSD and BOL have been reported to block both DA and NA-induced activation of adenylate cyclase (Von Hungen *et al.*, 1975). Curiously, 5-HT antagonists also consistently stimulate adenylate cyclase activity and activation is blocked by DA-blocking agents (trifluoperazine, thioidazine, chlorpromazine and haloperidol) (Von Hungen *et al.*, 1975). Based on these studies, the authors suggest that LSD causes complex interactions with cerebral receptors for 5-HT, DA and NA.

C Inhibition of Dopamine-Stimulated Adenylate Cyclases

Pharmacological and biochemical studies of various antipsychotic drugs have implicated a blockade of DA-receptor sites in brain in their therapeutic action as well as in the production of extrapyramidal side-effects. The ability and potency of antipsychotic agents to block DA receptors in the striatum parallel their ability to produce disturbances in extrapyramidal function in man. Thus, the potency ratios of various antipsychotic drugs for elevating striatal homovanillic acid (increased turnover of DA as a consequence of DA-receptor blockade) and for blocking amphetamine-induced rotational behavior in animals with unilateral lesions of the substantia nigra (functional DA-receptor blockade) are practically identical (Stawarz *et al.*, 1975). There is, however, no correlation between striatal DA-receptor blockade elicited by antipsychotic drugs and their antipsychotic potency in man.

A number of recent studies has suggested that the blockade by antipsychotic drugs of DA receptors in mesolimbic dopaminergic structures may be more closely related to their antipsychotic activity than to the action of these drugs on striatal DA receptors (Andén, 1972; Andén and Stock, 1973).

It is thus of interest that antipsychotic drugs have been found to be highly potent antagonists of the DA-sensitive adenylate cyclase in the caudate nucleus and in mesolimbic structures, such as the olfactory tubercles and nucleus accumbens (Clement-Cormier *et al.*, 1974; Miller *et al.*, 1974b; Karobath and Leitich, 1974; Lippman *et al.*, 1975; Miller *et al.*, 1975b). The inhibition displays competitive kinetics in all cases (Figure 19.3). Generally, the K_i value of each drug for the enzyme from the caudate nucleus is similar to that for the enzyme from mesolimbic structures even though the drugs vary greatly in their potency (Clement-Cormier *et al.*, 1974).

The question now arises of whether or not a correlation exists between the potency of various antipsychotic drugs in blocking DA receptors in limbic

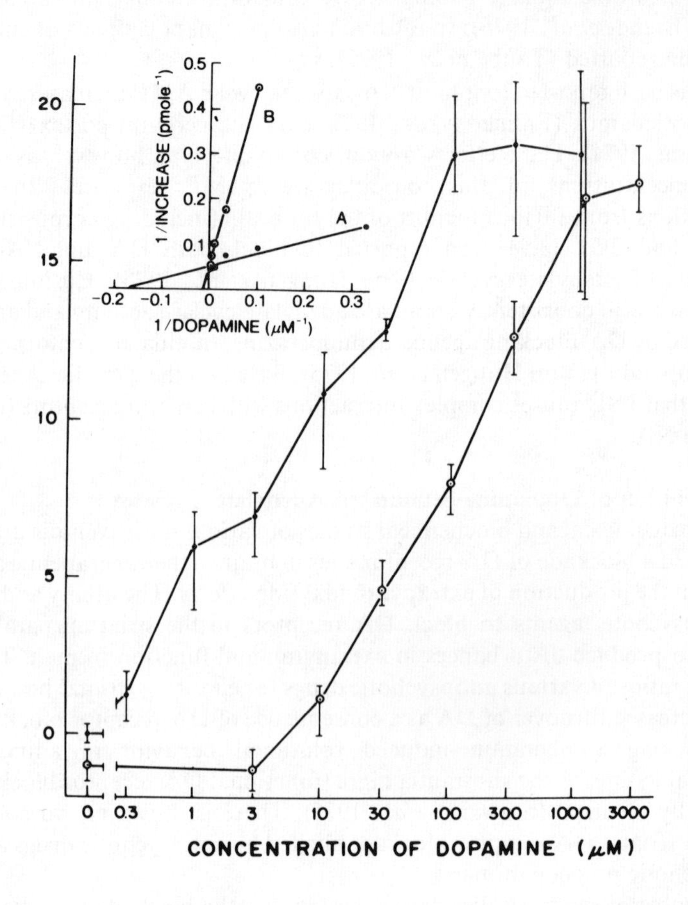

Figure 19.3 Effect of various concentrations of dopamine, alone (•) or in combination with 0.1 μM fluphenazine (⊙), on adenylate cyclase activity in a homogenate of rat olfactory tubercle. In the absence of added DA and fluphenazine, 12.0 ± 0.3 pmole (mean ± S.E.M. = 9) of cAMP was formed; in the presence of 0.1 μM fluphenazine but without added DA, 11.0 ± 0.5 pmole (mean ± S.E.M., $N = 6$) of cAMP was formed. The increase in cAMP above the basal level (the level in the absence of both DA and fluphenazine) is plotted as a function of DA concentration. The data give mean values and ranges for duplicate determinations on each of three replicate samples. Insert: Double-reciprocal plot of cAMP increase as a function of dopamine concentration from 3 μM to 300 μM (A) control; (B) 1×10^{-7} M fluphenazine. (From Clement-Cormier *et al.*, 1974)

structures, their clinical therapeutic potency and their potency as inhibitors of the DA-stimulated cAMP formation. Within a particular group of compounds, for example the phenothiazines, a good correlation exists (Table 19.1). However, discrepancies become apparent if one compares the potency relationships of antipsychotic drugs belonging to different chemical classes. For example, haloperidol and pimozide, which have been shown to be the most potent *in vivo* blockers of DA receptors in both the striatum and limbic forebrain (Stawarz *et al.*, 1975), are relatively weak inhibitors of DA-sensitive adenylate cyclases (Table 19.1). On the other hand, clozapine, which is a very weak *in vivo* inhibitor of striatal DA receptors and does not cause extra-pyramidal side-effects in man, is a relatively potent inhibitor of the DA-sensitive adenylate cyclase in the caudate nucleus. It is possible, of course, that anticholinergic properties of this drug (Snyder *et al.*, 1974) are responsible for its weak *in vivo* action as a DA-receptor blocker and the low incidence of extra-pyramidal side effects in man, although there is little clinical evidence suggesting that clozapine behaves as a central antimuscarinic agent.

Table 19.1 Clinical potency of antipsychotic drugs and their potency as inhibitors of dopamine-sensitive adenylate cyclases

Drug	Caudate nucleus enzyme IC_{50} (μM)	K_i(μM)	Olfactory tubercle enzyme IC_{50} (μM)	K_i(μM)
Phenothiazines and related compounds				
fluphenazine	0.07	0.008	0.10	0.011
trifluoperazine	0.10	0.011	0.15	0.017
promazine	0.35	0.039	–	–
triflupromazine	0.40	0.044	–	–
prochlorperazine	0.50	0.055	–	–
thioridazine	0.50	0.055	0.45	0.050
chlorpromazine	0.60	0.066	0.9	0.100
promethazine	15.0	1.67	–	–
imipramine	27.0	3.00	27.0	3.00
desmethylimipramine	28.0	3.11	25.0	2.77
ethoproperazine	37.0	4.11	–	–
diethazine	100.0	11.11	110.0	12.22
Butyrophenone and related compounds				
haloperidol	2.0	0.22	1.3	0.144
pimozide	11.0	1.22	–	–
Dibenzodiazepine				
clozapine	0.55	0.061	1.0	0.111

Data from experiments in which the concentration of DA is held constant, and the concentration of the test substance is varied. The K_i value is calculated from the relationship $IC_{50} = K_i\,(1 + S/K_m)$, where IC_{50} is the concentration of drug required to give 50% inhibition of the DA-stimulated increase in enzyme activity, and S is the concentration of dopamine (40 μM). The mean value for K_m found in this series of experiments was 5 μM. (From Clement-Cormier *et al.*, 1974).

Among the thioxanthenes, marked differences are known to exist in the pharmacological activities of the *cis* and *trans* isomers. For example, the *cis*-isomer α-flupenthixol is a potent antipsychotic drug whereas the *trans*-isomer β-flupenthixol is devoid of any such activity. Interestingly, Miller *et al.* (1974b) found that α-flupenthixol is an extremely potent inhibitor of the DA-sensitive adenylate cyclase in both striatum and limbic structures ($IC_{50} = 2.2 \times 10^{-8}$ M), while β-flupenthixol is inactive even at very high concentrations ($IC_{50} > 10^{-4}$ M).

Studies on species differences in the rate and pathways of metabolism are necessary, however, before definite conclusions can be drawn on the relationship between a blockade of DA sensitive adenylate cyclases *in vitro* by antipsychotic agents in the mesolimbic system of rats and the antipsychotic activity of these drugs in man.

Studies with a new neuroleptic agent butaclamol, a benzocycloheptapyrido-isoquinolinol derivative, have provided additional evidence for the suggestion that the DA-sensitive adenylate cyclase may be identical to the DA receptor. The pharmacological and biochemical properties of this drug have been observed in the (+)-enantiomer, the (–)-enantiomer being inactive (Bruderlein *et al.*, 1975; Zivkovic *et al.*, 1975). Interestingly, the (+)-enantiomer has also proved to be a potent inhibitor of the DA-sensitive adenylate cyclase in cell-free homogenates of the corpus striatum (Miller *et al.*, 1975b) and of the olfactory tubercle (Lippman *et al.*, 1975), while the (–)-isomer is inactive (Table 19.2). Therefore, there is considerable support for the view that extrapyramidal side-effects and perhaps some of the therapeutic actions of antipsychotic drugs may be related to their ability to block the stimulation by DA of adenylate cyclase activity in the caudate nucleus and in mesolimbic structures, respectively.

Table 19.2 IC_{50} and calculated inhibition constants (K_i) of butaclamol, its enantiomers and fluphenazine for the DA-sensitive adenylate cyclase of rat olfactory tubercle

	Olfactory Tubercle Adenylate Cyclase	
Treatment	IC_{50} (μM)*	K_i (μM)**
butaclamol	0.84	0.168
(+)-enantiomer	0.23	0.046
(–)-enantiomer	>100	>20
fluphenazine	0.15	0.030

Data in the above experiments was obtained by varying the concentration of test substances while the concentration of DA (20 μM) was held constant.

*Drug concentration causing half-maximal inhibition of stimulation of cAMP production by 20 μM DA.

**Inhibition constant calculated from $IC_{50} = K_i(1 + S/K_m)$ where IC_{50} is as above, $S = 20$ μM and the mean value for K_m was 5 μM (From Lippmann *et al.*, 1975).

It still has to be seen, however, how drugs such as metoclopramide and sulpiride fit into the picture. Metoclopramide has been shown to produce extrapyramidal symptoms in man and to block dopaminergic mechanisms in animals (Costall and Naylor, 1974; Robinson and Sulser, 1976), but the drug does not apparently inhibit the activity of striatal DA-sensitive adenylate cyclase (Sarau, personal communication). Sulpiride fails to block DA-sensitive adenylate cyclase in both striatal and mesolimbic structures of the brain (Trabucchi *et al.*, 1975) but nevertheless has been reported to be a potent antipsychotic drug.

An other interesting exception is the finding that molindone, a clinically effective antipsychotic drug with a pharmacological profile similar to that of the phenothiazines (Bunney *et al.*, 1975), is a very weak inhibitor of DA-sensitive adenylate cyclase in homogenates of both striatal and limbic structures (Greengard, 1975). Moreover, the discrepancy between the relatively potent effects of some tricyclic antidepressants, including amitriptyline and doxepin, on DA-sensitive adenylate cyclase *in vitro* and their lack of antipsychotic efficacy *in vivo* has yet to be explained (Karobath, 1975).

Recently, Clement-Cormier *et al.* (1975) have succeeded in demonstrating the stimulation by DA of adenylate cyclase in particulate fractions of the rat caudate nucleus enriched with synaptic membranes. As in unfractionated homogenates, antipsychotic drugs competitively inhibited the DA activation of the enzyme.

D Dopaminergic-Receptor Supersensitivity and Dopamine-Sensitive Adenylate Cyclase

While there is abundant evidence that modification of DA-sensitive adenylate cyclase activity both in striatal and limbic structures is involved in the action of a number of psychotropic drugs, the question of whether DA-sensitive adenylate cyclase contains receptor sites for DA or is identical with the DA receptor has yet to be resolved. One of the characteristics of postsynaptic receptors is the development of supersensitivity to the neurotransmitter following denervation or disuse. The phenomenon of behavioural supersensitivity to DA or other DA agonists has been well established following chronic interruption of nerve impulse transmission at central dopaminergic synapses with 6-hydroxydopamine or α-methyl-p-tyrosine, or after chronic blockade of DA receptors with neuroleptics (Schoenfeld and Uretsky, 1972; Tarsy and Baldessarini, 1974; Gianutsos *et al.*, 1974; Nahorski, 1975; Moore and Thornburg, 1975). Moreover, results from electrophysiological studies have provided evidence at the level of individual caudate neurones for the development of postsynaptic supersensitivity to DA and apomorphine following administration of 6-hydroxydopamine (Siggins *et al.*, 1974) or chronic treatment with the DA-receptor blocker haloperidol (Yarborough, 1975). However, prior destruction of dopaminergic nerve terminals in the striatum by 6-hydroxydopamine or chronic inhibition of the synthesis of DA, both of which produce behavioural effects suggestive of postsynaptically mediated supersensitivity, have been found by most investigators not to be accompanied by increased

Table 19.3 Effect of chemosympathectomy with 6-hydroxydopamine on the cAMP response to noradrenaline, isoprenaline and adenosine in limbic forebrain

	cAMP pmole/mg protein ± S.E.M. control	6-hydroxydopamine	reactivity index[1]
basal level	28.93 ± 1.68 (6)	47.90 ± 8.13 (4)	–
noradrenaline (5 μM)	53.62 ± 5.38 (9)	119.06 ± 6.39 (9)**	288
isoprenaline (5 μM)	44.77 ± 2.56 (9)	69.23 ± 8.30 (9)*	135
adenosine (100 μM)	175.24 ± 17.36 (6)	216.69 ± 26.84 (6)	115

Numbers in parentheses indicate the number of samples.

[1]Calculated with the formula $\frac{\text{stimulated level (drug)} - \text{basal level (drug)}}{\text{stimulated level (control)} - \text{basal level (control)}} \times 100$

*$p < 0.05$
**$p < 0.001$
(From Vetulani *et al.*, 1976b).

responses of the DA-sensitive adenylate cyclase to DA (Von Voigtlander *et al.*, 1973; Kebabian *et al.*, 1975; Iversen, 1975). In contrast, Mishra *et al.* (1974) have reported an enhanced stimulation of caudate adenylate cyclase by DA following lesions of the substantia nigra. The reasons for these conflicting reports are not clear although differences in the handling of tissues may be responsible. Also, the behavioural supersensitivity to d-amphetamine and apomorphine following chronic treatment with neuroleptics is not accompanied by an incrase in the DA-stimulated adenylate cyclase activity (Von Voigtlander, 1974; Rotrosen *et al.*, 1975).

Consequently, it is concluded that the behavioural and neurophysiological postsynaptic super-sensitivity following administration of neuroleptics is not mediated by alterations in the sensitivity of striatal adenylate cyclase to DA. In contrast to the results obtained with DA-sensitive adenylate cyclase preparations in homogenates, the expected 'denervation' supersensitivity to NA can be unequivocally demonstrated in slices from various brain regions containing NA-sensitive adenylate cyclase systems (Table 19.3). It is conceivable that damage of the DA–cyclase system by homogenization is responsible for the observed lack of correlation between increased behavioural and electrophysiological sensitivity to DA and the activity of DA-stimulated adenylate cyclase as measured in homogenates.

V NORADRENALINE-SENSITIVE ADENYLATE CYCLASE SYSTEMS

The presence of a NA sensitive enzyme system in slices from various brain areas was established in the late 1960s — see for example the review by Rall and Sattin (1970). Although the first studies, carried out with slices from rabbit and guinea-pig cerebrum and cerebellum, suggest that histamine rather than NA, is a primary stimulant of cAMP accumulation, subsequent studies using brain slices from rats and mice have revealed that in these species, NA is more potent in stimulating the accumulation of cAMP. Slices of human cerebral cortex also respond to NA (Shimizu *et al.*, 1971).

A Preparation and Properties

In contrast to DA-sensitive adenylate cyclase systems, NA-sensitive systems can easily be detected in slices from the limbic forebrain of the rat, whereas homogenization of the tissue generally leads to a loss of sensitivity. However, under controlled homogenization conditions, adenylate cyclase systems sensitive to a low concentration of NA and adrenaline have been demonstrated in cell-free homogenates of guinea-pig brain (Chasin *et al.*, 1974) and in the rat limbic forebrain (Horn and Phillipson, 1975). It appears, however, that physical damage to the cyclase system is one important factor determining the loss of sensitivity, since the magnitude of the hormonal response is known to vary inversely with the degree of homogenization (Øye and Sutherland, 1966). Whether the integrity of a large part of the cellular membrane is necessary and sufficient to retain the hormonal sensitivity to NA or whether some intracellular structures are also involved in the mediation of the hormonal response remains to be resolved.

The basal levels of cAMP in slices of the limbic forebrain of the rat vary between 20 and 30 pmole/mg protein (Blumberg *et al.*, 1975, 1976; Vetulani *et al.*, 1976b). The level of the nucleotide may rise dramatically if the oxygenation of the medium is insufficient, glucose is omitted, or the pH deviates from 7.4.

The lowest concentration of NA producing a significant accumulation of cAMP varies between 0.5 and 1.0 μM. The maximal stimulation is produced by 10 – 50 μM NA which causes a five to six-fold increase in the level of cAMP. The response is specific to NA; DA and 5-HT do not produce any appreciable increase in the level of the nucleotide (Figure 19.4) even at a concentration of 1 mM (Blumberg *et al.*, 1975). The K_a value of NA is approximately 5 μM.

Adenosine is also capable of increasing the level of cAMP. The K_a value (approxmately 30 μM) and the dose–response and time–response curves for adenosine differ, however, from those for NA (Figure 19.5), suggesting that different cAMP-generating systems are involved (Blumberg *et al.*, 1976).

The specific noradrenergic cyclic AMP-generating system in limbic forebrain slices is also stimulated by isoprenaline (Figure 19.5). The maximal stimulation achieved by the β-adrenergic agent is much lower than that produced by NA, although the K_a value for isoprenaline is approximately the same as that for NA. Responses to NA (5 μM) are blocked by both phentolamine ($IC_{50} = 8.0$ μM) and propranolol ($IC_{50} = 2.7$ μM). In contrast, the response to isoprenaline is only inhibited or blocked by the β-blocking agent propranolol and the adenosine-dependent system is not influenced by either α or β-antagonists (Blumberg *et al.*, 1976). These data suggest that the NA-induced rise of the nucleotide in limbic forebrain slices may be mediated by both α and β-adrenergic receptors.

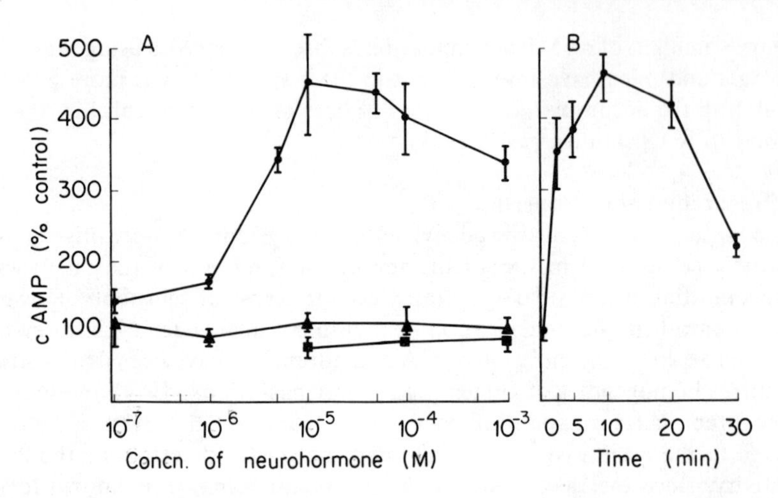

Figure 19.4 (A) Effect of various concentrations of (—)-NA (●——●), DA (▲——▲) and 5-HT (■——■) on the accumulation of cAMP in slices from the limbic forebrain of rats. Basal control values were 26.0 ± 2.0 pmole cAMP/mg protein. (B) Time course of the effect of (—)-NA (5×10^{-5}M) on the accumulation of cAMP in tissue slices from the limbic forebrain. Control values were 46.0 ± 4.9 pmole cAMP/mg protein. All values are expressed as the mean percentage of control values ± S.E.M., $N = 4$. (From Blumberg, *et al.*, 1975)

Whereas the presence of classical β-receptors is fairly well established, the blockade of the NA response by phentolamine does, however, not necessarily imply the presence of classical α-receptors. Thus, the α-agonists clonidine and phenylephrine do not elevate the basal level of cAMP in limbic forebrain slices; they significantly inhibit the response to NA but not to isoprenaline (Figure 19.6; Vetulani *et al.*, 1975). Similar unexpected results using phenylephrine and clonidine in cerebral cortical slices have been published by Skolnick and Daly (1975b).

The noradrenergic cAMP-generating system in slices of the limbic forebrain displays several properties compatible with those of a central NA receptor: specificity, sensitivity and the ability to be blocked by specific antagonists. Another characteristic of a receptor is its ability to develop supersensitivity following prolonged reduction in the neurohormone–receptor interaction (supersensitivity due to denervation or disuse).

The development of supersensitivity of a central NA receptor after treatment with reserpine or 'central chemosympathectomy' using 6-hydroxydopamine has been suggested by several behavioural studies. Thus, behavioural sensitivity to intraventricular NA has been found to be enhanced in supersensitized animals (Geyer and Segal, 1973; Mandell, 1974). An increased responsiveness of the

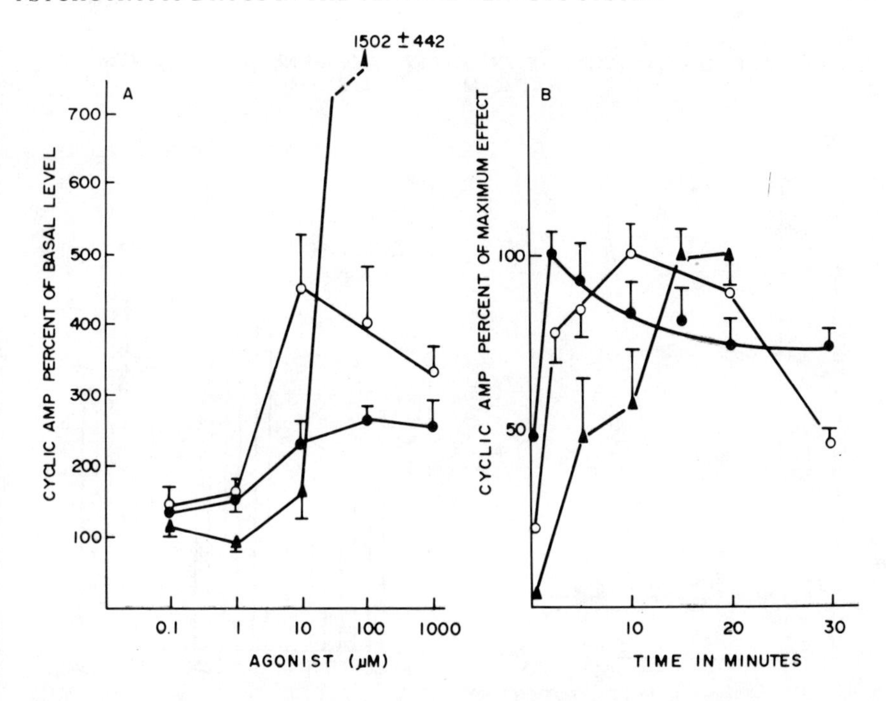

Figure 19.5 (A) Effect of various concentrations of l-noradrenaline (o), l-isoprenaline (•) and adenosine (▲) on the accumulation of cAMP in slices from the limbic forebrain of rats. Basal control values: 26 ± 2 pmole cAMP/mg protein. $N = 4$–8. (B) Time-course of the effect of l-noradrenaline (50 μM), l-isoprenaline (5 μM) and adenosine (100 μM) on the accumulation of cAMP in tissue slices from the limbic forebrain. Symbols are the same as in (A). All values are expressed as the mean percentage of the maximum effect ± S.E.M. $N = 2$–8. (Data from Blumberg *et al.*, 1976)

cAMP-generating system to NA following 6-hydroxydopamine treatment in cortical slices has also been reported (Palmer, 1972; Weiss and Strada, 1972; Huang *et al.*, 1973a; Kalisker *et al.*, 1973). Studies by the authors have shown that the basal level of cAMP in limbic forebrain slices is increased following the destruction of adrenergic nerve terminals with 6-hydroxydopamine. Also the sensitivity of the preparation to exogenous NA is increased approximately four-fold (Vetulani *et al.*, 1976b). It should be noted that the destruction of adrenergic nerve terminals by 6-hydroxydopamine also increases the cAMP response to iso-prenaline in slices from both the cerebral cortex (Huang *et al.*, 1973a) and the limbic forebrain (Table 19.3; Vetulani *et al.*, 1976b), while the responses to adenosine and DA are virtually unaffected. As it is unlikely that the enhanced adrenergic cAMP responses following 6-hydroxydopamine are due to a decrease in the activity of cAMP phosphodiesterase (Huang *et al.*, 1973a;

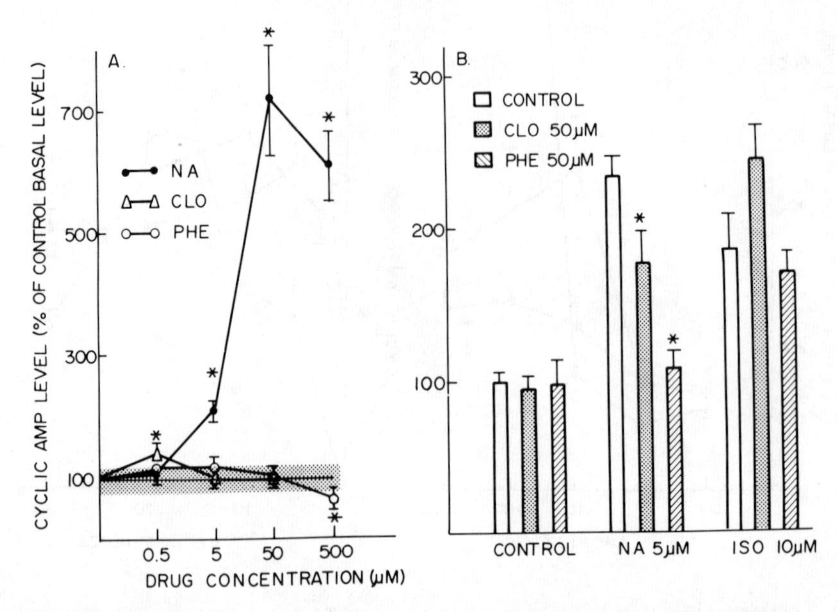

Figure 19.6 (A) Effect of various concentrations of l-NA, clonidine (CLO) and phenylephrine (PHE) on the accumulation of cAMP in slices from the limbic forebrain of rats. Basal control values were 22.6 ± 1.7 pmole cAMP/mg protein. (B) Effect of clonidine (CLO) and phenylephrine (PHE) on the accumulation of cAMP elicited by either 5 μM noradrenaline (NA) or 50 μM isoprenaline (ISO). All data are expressed as the mean percentage of control values ± S.E.M. Basal control values were 23.4 ± 1.4 pmole cAMP/mg protein

Kalisker *et al.*, 1973) and because the response to isoprenaline — which is not taken up by presynaptic nerve terminals (Callingham and Burgen, 1966) — is also enhanced the results reflect, in all probability, changes in the sensitivity of the postsynaptically located NA-sensitive adenylate cyclase–receptor system. Subchronic treatment with reserpine produces similar results; the maximal responsiveness of the preparation is increased with no change in the K_a value.

Provided that the noradrenergic cAMP-generating system is regarded as a receptor-like entity, the classical theory of drug interactions may be applied (Ariëns *et al.*, 1957). The K_a value of NA would reflect the affinity of the neurohormone to the receptor while the maximal response is an expression of its intrinsic activity. It then follows that the supersensitivity resulting from a prolonged deprivation of NA at postsynaptic receptor sites is related to changes in the intrinsic activity rather than the affinity of the neurohormone to the receptor. Assuming that the current model of adenylate cyclase (Robison *et al.*,

1967) applies to the hypothetical central NA receptor, it can be argued that the regulatory subunit on the outer cell surface which recognizes the neurohormone is responsible for both specificity and affinity phenomena. Changes in the activity of the catalytic subunit are reflected in variations in the intrinsic activity. In any event, the elucidation of the molecular mechanism responsible for the change in the receptor sensitivity to NA remains an important topic for future research.

Additional experiments with 6-hydroxydopamine have provided further evidence of the specificity of the limbic cAMP-generating system to NA. Thus, when noradrenergic terminals are protected from the action of 6-hydroxy-dopamine by desipramine no supersensitivity of the cAMP-generating system to NA occurs (Figure 19.7).

Recently, Horn and Phillipson (1975) have reported on a NA-sensitive adenylate cyclase system in homogenates of the limbic forebrain (K_a value of

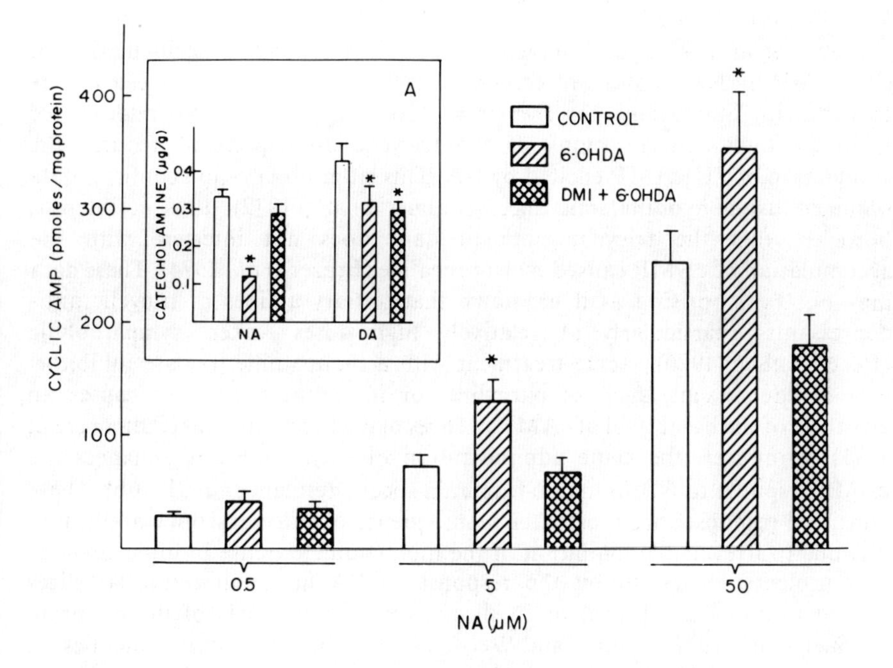

Figure 19.7 Effect of 6-hydroxydopamine (6-OHDA), with and without desipramine (DMT) pretreatment on the cAMP response to noradrenaline (NA) in slices of the limbic forebrain and on the levels of brain catecholamines (insert A). Each bar represents the mean values of cAMP in pmole/mg protein ± S.E.M. ($N = 8$) and of catecholamines in µg/ ± S.E.M. ($N = 11$–16) * < 0.01.
(Data from Vetulani *et al.*, 1976b)

NA = 8.5×10^{-7} M). They showed that active agonists must have a catechol grouping and a β-hydroxy group with the correct stereochemistry. They found DA to be inactive at concentrations of up to 10^{-3} M, as it was in limbic-forebrain slices. The blockade of the stimulation elicited by NA in the presence of adrenolytics is stereoselective. It remains to be seen, however, if this noradrenergic system in homogenates will display similar receptor characteristics to those described for limbic-forebrain slices.

B Effect of Psychotropic Drugs on Noradrenaline-Sensitive Adenylate Cyclase

Although many studies on the effect of psychotropic drugs on cAMP in brain have been concerned with DA-sensitive adenylate cyclase systems, antipsychotic drugs and antidepressants also interfere with noradrenergic cAMP-generating systems in various cortical and subcortical areas of the brain (Uzunov and Weiss, 1971; Palmer *et al.*, 1972, 1973; Von Hungen and Roberts, 1973; Frazer *et al.*, 1974; Blumberg *et al.*, 1975, 1976; Vetulani and Sulser, 1975; Vetulani *et al.*, 1976a,b).

Frazer *et al.* (1974) have shown that the dose-dependent accumulation of [^{3}H]cAMP by NA in slices of cerebral cortex is significantly reduced in rats treated with imipramine for five days. Moreover, the *in vitro* addition of relatively high concentrations of the tricyclic antidepressant reduces the stimulation of [^{3}H]cAMP elicited by NA. This latter observation confirms data obtained using hypothalamic slices (Palmer *et al.*, 1972). It is of interest, however, that the tricyclic antidepressant does not interfere with the accumulation of cAMP caused by isoprenaline (Frazer *et al.*, 1974). These data may not be surprising as it is known that tertiary amines of tricyclic antidepressants — particularly at relatively high doses — exert sympatholytic effects (Sulser, 1976). Acute treatment with a monoamine oxidase inhibitor-type antidepressant, such as pargyline, or its addition *in vitro* causes an elevation of the basal level of cAMP in slices of the hypothalamus (Palmer *et al.*, 1971). Moreover, the acute administration of such inhibitors enhances the cAMP response to NA in limbic-forebrain slices (Vetulani *et al.*, 1976b). These enhanced responses are most likely consequence of a decreased metabolism of NA due to irreversible inhibition of the monoamine oxidase in brain tissue.

Neuroleptic drugs inhibit the response to NA in rabbit cerebellar slices (Kakiuchi and Rall, 1968a) or in slices from various areas of the rat brain (Palmer *et al.*, 1971; Uzunov and Weiss, 1972b). The action of neuroleptics on the response of cAMP to NA may, however, vary according to the brain region considered. Thus, haloperidol has been found to inhibit the cAMP response to NA in slices from the brain stem, but not in hypothalamic slices (Palmer *et al.*, 1971).

The availability of various metabolites of chlorpromazine has provided the opportunity to test further the specificity of phenothiazines in antagonizing the cAMP response to NA in the rat hypothalamus and the brain stem (Palmer *et al.*, 1972). Exposure of brain slices for six minutes to 5×10^{-5} M NA consistently

Table 19.4 Effect of antipsychotic drugs on noradrenaline (NA) and adenosine-induced cAMP responses in the limbic forebrain

Drugs	IC_{50} NA-sensitive cAMP response (μM)	effect of drugs (50 μM) on cAMP response to adenosine (% control ± S.E.M.)
promethazine	inactive up to 10 μM	72 ± 25 (6)
haloperidol	10 (IC_{40})	83 ± 12 (5)
chlorpromazine	9	77 ± 12 (6)
thioridazine	1.2	95 ± 12 (6)
pimozide	0.08	88 ± 13 (6)
clozapine	0.06	74 ± 13 (5)

Tissue slices were incubated at 37° C. The various drugs were added 14 min prior to the addition of 5 μM NA or 100 μM adenosine. The incubation was terminated 10 min later and cAMP determined. Numbers in parentheses represent the number of samples (From Blumberg *et al.*, 1976)

results in a three to five-fold increase in the level of cAMP. Prochlorperazine and 7-hydroxychlorpromazine at a concentration of 10^{-5}M antagonize the increased production of the cyclic nucleotide due to NA in the hypothalamus and the brain stem, while 8-hydroxychlorpromazine and imipramine are only effective in the hypothalamus. The remaining metabolites of chlorpromazine — 7-methoxychlorpromazine, 3,7-dihydroxychlorpromazine, 7-hydroxy-8-methoxychlorpromazine, 8-hydroxy-7-methoxychlorpromazine, 7,8-dihydroxypromazine and chlorpromazine sulphoxide — fail to modify the NA-induced cAMP response. More recent studies by the authors have been concerned with the noradrenergic cAMP-generating system of the limbic forebrain and its modification by antipsychotic drugs (Blumberg *et al.*, 1976). While not altering the basal level of the cyclic nucleotide, clinically effective antipsychotic drugs cause a dose-dependent inhibition of the limbic noradrenergic cAMP response, with clozapine being particularly potent (Table 19.4). In addition, these studies have revealed that pimozide is not a selective DA-blocking agent as it is also a potent antagonist of this noradrenergic response. As has been noted with DA-sensitive adenylate cyclase systems in homogenates from caudate nucleus and nucleus accumbens (Clement-Cormier *et al.*, 1974), the butyrophenone derivative, *haloperidol*, is also a weak antagonist of the NA-induced cyclic AMP response. The action of clinically active antipsychotic drugs on the noradrenergic cAMP-generating system in the limbic forebrain seems to be rather specific, however, as the drugs do not affect the responses brought about by adenosine (Table 19.4). Recent studies with the antipsychotic drug *butaclamol* indicate that its blocking effect on the specific limbic noradrenergic cAMP-generating system resides entirely in the (+)-enantiomer, thus demonstrating stereospecificity for central NA-receptor blockade (Robinson and Sulser, 1976). The availability of a stereochemically specific antagonist of a central NA–adenylate cyclase receptor should provide an important tool to elucidate this system further.

Table 19.5 Inhibition by neuroleptics of noradrenaline-sensitive adenylate cyclase in homogenates of limbic forebrain

Neuroleptics	% Inhibition at 10^{-5} M	Neuroleptics	% Inhibition at 10^{-5} M
promazine	83.5 ± 4.8 (7)	α-chlorprothixene	48.9 ± 14.4 (4)
(+)-butaclamol	73.9 ± 10.4 (3)	β-clopenthixol	44.9 ± 2.8 (4)
clozapine	66.2 ± 2.9 (4)	pimozide	44.6 ± 2.8 (5)
β-chlorprothixene	57.4 ± 13.3 (4)	haloperidol	43.5 ± 3.0 (4)
thioridazine	55.0 ± 5.1 (3)	trifluperazine	39.6 ± 6.6 (4)
chlorpromazine	50.9 ± 0.9 (4)	α-flupenthixol	38.6 ± 2.6 (3)
	$K_i \cong 1.6 \times 10^{-7}$ M	β-flupenthixol	33.9 ± 5.9 (4)
α-clopenthixol	48.9 ± 7.0 (4)	(−)-butaclamol	23.4 ± 3.0 (4)

desimipramine, 7-hydroxychlorpromazine, chlorpromazine sulphoxide were inactive at 10^{-5} M.
EC_{50} = concentration required for half-maximal stimulation by the agonist;
IC_{50} = concentration of antagonist required to half-maximal inhibition of the stimulation produced by 50 μM NA; K_i = inhibition constant.
In experiments using antagonists or neuroleptics the drug was added 15 min prior to the addition of 50 μM NA. Numbers in parentheses indicate the number of determinations. EC_{50} and IC_{50} values were estimated graphically from the means of at least four determinations. (From Horn and Phillipson, 1975).

Interestingly, neuroleptic drugs have also been found to block a NA-sensitive adenylate cyclase in homogenates of the limbic forebrain (Table 19.5). Promazine, clozapine, thioridazine and chlorpromazine are quite active blockers, whereas α-flupenthixol and trifluoperazine, both potent blockers of the DA sensitive adenylate cyclase, are less active (Horn and Phillipson, 1975). Pimozide which is relatively potent in blocking the NA-induced increase in cAMP in limbic-forebrain slices (Blumberg *et al.*, 1975), exerts only weak antagonistic properties in homogenates (Horn, personal communication). Although it would be presumptious to correlate blockade of NA-sensitive adenylate cyclase systems by antipsychotic drugs with their clinical antipsychotic efficacy, these studies nevertheless support the view that a blockade of noradrenergic receptors in limbic structures of the brain may also contribute to the pharmacological profile and perhaps, the therapeutic action of this particular class of drugs.

VI ADAPTIVE MECHANISMS OF THE NORADRENERGIC cAMP SYSTEM

It has been well established that destruction of central adrenergic neurones by 6-hydroxydopamine results in supersensitivity of the cAMP-generating system to NA in cortical slices (Palmer, 1972; Weiss and Strada, 1972; Huang *et al.*, 1973a; Kalisker *et al.*, 1973). Similar supersensitivity develops in the noradrenergic cAMP-generating system in limbic forebrain slices following central chemosympathectomy with 6-hydroxydopamine or subacute treatment with reserpine (see Section VA). The change in receptor sensitivity may be interpreted as an adaptation to a prolonged decrease in the NA-receptor interaction and it probably reflects a postsynaptic counterpart of the presynaptic compensatory increase in the activity of tyrosine hydroxylase which is known to occur following the administration of reserpine (Thoenen *et al.*, 1969; Segal *et al.*, 1974).

Since inhibition of monoamine oxidase is associated with an increased availability of catecholamines at presumptive catecholamine-receptor sites in brain (Strada and Sulser, 1972), it was of interest to study the biochemistry of the noradrenergic cAMP-generating system in the limbic forebrain following acute and chronic administration of monoamine oxidase inhibitors.

While a single dose of nialamide or pargyline results in an enhanced response of the system to NA, treatment for three consecutive days or for up to two weeks does not produce any significant change in the responsiveness. However, the responsiveness of the system to NA is significantly depressed following treatment with these inhibitors for longer than three weeks (Figure 19.8). These

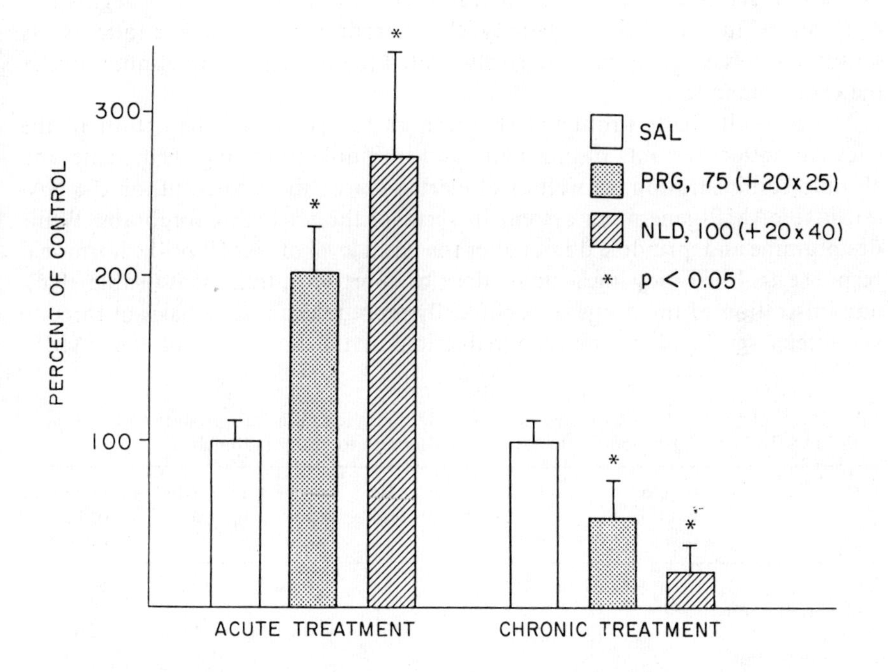

Figure 19.8 Relative cAMP responses to 5 μM noradrenaline in slices of the limbic forebrain of rats following acute and chronic treatment with MAO inhibitors. Animals were killed 18 hr after the last injection. Pargyline (PRG) was administered at a dose of 75 mg/kg, i.p. followed by daily doses of 25 mg/kg for 20 days; nialamide (NLD) was given at a dose of 100 mg/kg, i.p. followed by daily doses of 40 mg/kg for 20 days. SAL = Saline. The data are expressed as a percentage of the control response to NA ± S.E.M.

Table 19.6 Effect of short-term treatment with desipramine or iprindole on the response of the cAMP-generating system in the rat limbic forebrain to noradrenaline

	time of sacrifice[a] (hr)	*N*	basal level of cAMP (pmole/mg protein ± SEM)	cAMP response to NA[b] (pmole/mg protein ± SEM)	percentage of control response
control	1 or 24	23	18.0 ± 1.7	19.4 ± 3.2	100
desipramine HCl	1	8	14.7 ± 1.7	14.6 ± 3.3	75
desipramine HCl	24	28	14.5 ± 1.4	16.6 ± 3.2	85
iprindole HCl	1	8	12.8 ± 1.8	15.6 ± 5.6	80
iprindole HCl	24	17	18.6 ± 2.0	21.1 ± 4.3	108

Desipramine and iprindole (10 mg/kg day^{-1}, i.p.) were injected for 1–2 weeks. The controls received saline.

[a]Time after last injection

[b]Difference in the level of cAMP between the preparation exposed to 5 μM NA and that of the control preparation (corresponding hemisection) (From Vetulani *et al.*, 1976b)

data indicate the ability of the postsynaptic receptor to adapt to a prolonged increase in the NA–receptor interaction and provide evidence for a regulatory mechanism in the CNS, whereby the noradrenergic receptor adapts its sensitivity to NA in a manner inversely related to the degree of its stimulation by the catecholamine.

These results have prompted the authors to investigate the action of the tricyclic antidepressants desipramine and iprindole following their acute and chronic administration, as well as of electro-convulsive treatment, on the NA-sensitive cAMP-generating system in slices of the rat limbic forebrain. While desipramine and iprindole do not alter the basal level of cAMP or the hormonal response to NA following a single dose or short-term treatment (Table 19.6), administration of the drugs on a clinically more relevant time basis of three to six weeks results in a marked reduction in the sensitivity of the cAMP-

Table 19.7 Effect of long-term treatment with desmethylipramine or iprindole on the response of the cAMP-generating system in the rat limbic forebrain to NA noradrenaline

	time of sacrifice[a] (hr)	*N*	basal level of cAMP (pmole/mg protein ± SEM)	cAMP response to NA[b] (pmole/mg protein ± SEM)	percentage of control response
control	1 or 24	15	17.8 ± 2.6	20.4 ± 2.7	100
desmethylipramine HCl	1	12	20.5 ± 2.7	9.9 ± 3.5*	49
desmethylpramine HCl	24	14	16.6 ± 1.6	6.9 ± 2.1***	34
iprindole HCl	1	13	22.3 ± 3.6	9.4 ± 4.7*	46
iprindole HCl	24	15	16.9 ± 1.5	7.9 ± 2.4**	38

Desipramine and iprindole (10 mg/kg day^{-1}, i.p.) were injected for 4–8 weeks. The controls received saline.

*$p < 0.05$; **$p < 0.01$; ***$p < 0.001$ (difference from control response; Student t-test)

[a]Time after last injection.

[b]Difference in the level of cAMP between the preparation exposed to 5 μM NA and that of the control preparation (corresponding hemisection) (From Vetulani *et al.*, 1976a).

generating system to NA (Table 19.7; Vetulani *et al.*, 1976a). It should be stressed that this change in sensitivity is not related to the levels of the drugs in brain. The development of noradrenergic-receptor subsensitivity following prolonged treatment with desipramine may also be due to chronic overexposure of noradrenergic-receptor sites to NA; analogous to the effect elicited by monoamine oxidase inhibitors (Vetulani *et al.*, 1976b). It is known that chronic exposure of various tissues to catecholamines results in reduced β-adrenergic-mediated responses to subsequent catecholamine stimulation (Makman, 1971b; Deguchi and Axelrod, 1973; Mukherjee *et al.*, 1975b). This explanation is not applicable, however, to iprindole as this antidepressant does not block the neuronal uptake of NA, alter its metabolism (Freeman and Sulser, 1972) or change its turnover following chronic administration (Rosloff and Davis, 1974) and thus it does not alter the availability of NA at postsynaptic noradrenergic-receptor sites. The possibility of increased phosphodiesterase activity has to be explored, however, as a cAMP-mediated activation of the enzyme within cerebral cortical slices has been suggested as a consequence of an increased level of intracellular Ca^{2+} (Schultz, 1975a).

Electro-convulsive therapy (ECT), which has been shown to increase the turnover of NA in brain and to lower the high-affinity uptake for NA, reduces the cAMP response to NA. The effect is similar to that observed following chronic treatment with tricyclic antidepressants. However, the onset of action of this altered response is more rapid following ECT. It is interesting to note that eight days following cessation of ECT, the hormonal response to NA is still reduced. Clinically, ECT is preferable in cases of severe depression because the onset of its therapeutic effect is generally more rapid than is that of chemotherapy. The decreased reactivity to NA following ECT may explain the antagonism by ECT of the stimulatory effects elicited by combined treatment with desimpramine and Ro–41284 (Matussek and Ladisich, 1969), or following amphetamine (Papeschi *et al.*, 1974). The stimulatory effects of these drugs are thought to be the consequence of the release of catecholamines onto post-synaptic receptor sites.

The increased noradrenergic cAMP response following central chemo-sympathectomy with 6-hydroxydopamine and its antagonism by ECT has theoretical implications, particulary, since ECT also prevents the hyper-sensitivity of the system to NA in reserpinized rats (Vetulani and Sulser, 1975). Reserpine can precipitate severe depression in man (Bunney and Davis, 1965) and 6-hydroxydopamine has been reported to cause behavioural changes in rhesus monkeys, similar to those exhibited by separated animals (McKinney, 1974). The decreased noradrenergic-receptor function resulting from chemotherapy and ECT suggests that the therapeutic action of antidepressants may be related to postsynaptic changes in the sensitivity of the noradrenergic adenylate cyclase–receptor system rather than to acute action of the drugs on presynaptic sites. These data together with those obtained using brain slices from reserpine-treated animals and from rats treated with 6-hydroxydopamine support the heuristic hypothesis that depression may be associated with increased noradrenergic-receptor responsiveness, particularly in the limbic forebrain.

VII POSSIBLE ROLE OF PRESYNAPTIC RECEPTORS

Presynaptic α-receptors, which inhibit the release of NA when stimulated by the released neurotransmitter, have been postulated. This hypothesis is based predominantly on the findings that low doses of α-agonists inhibit NA release and α-antagonists enhance the liberation of the catecholamine (Häggendal, 1970; Farnebo and Hamberger, 1971a,b; Kirkepar and Puig, 1971; Starke, 1972b; Enero *et al.*, 1972; Starke and Montel, 1973). An additional finding supporting the view that presynaptic inhibitory α-adrenergic receptors may exist in brain has been reported by Svensson *et al.* (1975), who found that iontophoretic injections of NA or clonidine into the locus coeruleus inhibit the firing of noradrenergic neurones. In addition to the prejunctional α-receptor, β-receptors activated by low concentration of NA and leading to an increase in the transmitter output have also been suggested (Langer *et al.*, 1975). Since phosphodiesterase inhibitors and analogues of cAMP have been shown to increase the release of NA due to nerve stimulation (Cubeddu *et al.*, 1974, 1975), it has been suggested that cAMP may also be involved in presynaptic release phenomena with regard to the CNS.

Under phosphorylating conditions cAMP reduces significantly the K_m of tyrosine hydroxylase for the cofactor and increases the K_i for DA (Morgenroth *et al.*, 1975; Goldstein *et al.*, 1975). Since DA activates adenylate cyclase, one would assume that antipsychotic drugs, as blockers of DA-sensitive adenylate cyclase, should decrease the affinity of the tyrosine hydroxylase for the cofactor. However, Zivkovic *et al.* (1974) have reported that antipsychotic drugs alter the kinetic properties of the enzyme in the same direction as cAMP; they reduce the K_m for the pteridine cofactor and increase the K_i for DA. Obviously, these discrepancies need to be resolved in order to understand how DA and cAMP regulate the affinity of tyrosine hydroxylase for the pteridine in DA neurones. The further elucidation of the role of presynaptic inhibitory receptors may prove to be a promising route of investigation.

Several laboratories have reported that cAMP or Bt_2-cAMP stimulates the activity of tyrosine hydroxylase in brain slices (Goldstein *et al.*, 1973c), and in synaptosomal preparations from striatal and mesolimbic structures (Goldstein *et al.*, 1975). Also, in the presence of Mg^{2+} and ATP, the soluble striatal enzyme is stimulated (Goldstein *et al.*, 1975). Protein kinase is probably involved in this activation by cAMP (Morgenroth *et al.*, 1975). Since the activation of protein kinase appears at present to be the major mechanism by which cAMP exerts its action, an activation of the hydroxylase by phosphorylation of the enzyme protein is a distinct possibility. Immunoprecipitation studies, however, appear, not to substantiate such a view (Lovenberg *et al.*, 1975). Although the physiological significance of these results obviously has yet to be established, the data indicate that cAMP and the modification of its level by various drugs may also play an important presynaptic role in the short-term regulation of catecholamine biosynthesis.

VIII EFFECTS ON cAMP PHOSPHODIESTERASE

The activity of cAMP phosphodiesterase plays an important role in the regulation of the level of the cyclic nucleotide as it catalyses the hydrolysis of cAMP to 5′-AMP (Butcher and Sutherland, 1962). The enzyme is distributed unequally throughout the CNS (Weiss and Costa, 1968) and there are multiple forms of the enzyme. Uzunov *et al.* (1974) have suggested that each cell type may have a characteristic number and pattern of these forms. Six distinct forms of the enzyme have been found in the cerebellum (Uzunov and Weiss, 1972a), while only four forms are present in the cerebrum (Uzunov *et al.*, 1974) and a peripheral nervous tissue, the superior cervical ganglion (Lindl *et al.*, 1976). The various forms of phosphodiesterase are differently affected by different inhibitors (Weiss, 1975). For example, in the rat cerebrum, trifluoperazine inhibits peak II to a greater extent than it does peak III, whereas theophylline inhibits peak III more than peak II (Uzunov, *et al.*, 1974).

Methylxanthines are probably the best known phosphodiesterase inhibitors; their action on the enzymes was reported as early as 1962 (Butcher and Sutherland, 1962). The drugs are known as stimulants of the CNS, and it has sometimes been assumed, without specific data, that a cause–effect relationship exists between their central stimulant action and the inhibition of phosphodiesterase. However, methylxanthines have been shown not to increase cAMP levels in slices from guinea-pig cerebral cortex (Kakuichi *et al.*, 1969) and in brain tissue from rabbit and rat (Forn and Krishna, 1971), and not to augment the increase in the levels of the nucleotide produced by either histamine or NA.

These surprising findings might be explained by the ability of methylxanthines to inhibit the cAMP accumulation caused by adenosine (Sattin and Rall, 1970). Recently, Sattin *et al.* (1975) have proposed that guinea-pig cortex contains NA–adenosine-dependent receptors, which must be stimulated simultaneously by NA and adenosine in order to activate the cAMP-generating system. The inhibitory action of methylxanthines on the adenosine part of the receptor may thus preclude the possibility of the receptor being stimulated by NA. The complex interaction between biogenic amines and adenosine with regard to cAMP appears to be unique to brain.

It has been shown that methylxanthines appreciably increase the action of the DOPA and DA agonists, apomorphine and piribedil, respectively, on turning behaviour in animals with unilateral lesions of the substantia nigra, and this action has been attributed to the known effect of methylxanthines to inhibit cAMP phosphodiesterase (Fuxe and Ungerstedt, 1974). Studies by Arbuthnott *et al.* (1974) have shown the difficulty when using phosphodiesterase inhibitors in interpreting pharmacological data in the CNS. Although their results with aminophylline confirm the findings of Fuxe and Ungerstedt (1974), they were unwilling to attribute the observed effects to the actions of the methylxanthines as phosphodiesterase inhibitors. More potent phosphodiesterase inhibitors than the methylxanthines, particularly ICI 63197, do not increase the turning behaviour due to apomorphine in rats with unilateral lesions of the substantia nigra. ICI 63197 actually depresses the action of apomorphine and yet markedly increases the concentration of cAMP in brain. Since aminophylline is the only central stimulant, and all other phosphodiesterase inhibitors tested have some central sedative properties, Fuxe and Ungerstedt (1974) concluded that the central action of the drugs rather than their phosphodiesterase inhibitory-like properties determines their effect on turning behaviour. It should also be remembered that *methylxanthines* exert a number of pharmacological effects unrelated to inhibition of phosphodiesterases — see for example the review by Appleman *et al.*, 1973). Theophylline increases cAMP concentrations in the cerebrospinal fluid of the rat (Kiessling *et al.*, 1975), an effect which may be due largely to an inhibition of nucleotide transport at the cerebrospinal fluid-blood barrier (Hammers *et al.*, 1977) and thus to a probenecid-like action (Cramer *et al.*, 1972).

The *benzodiazepines* are another group of phosphodiesterase inhibitors having psychotropic properties. Both diazepam and chlordiazepoxide are potent phosphodiesterase inhibitors, having IC_{50} values of 33 and 110 μM, respectively (Beer *et al.*, 1972). Benzodiazepines, and their derivatives, elevate the level of cAMP in cerebral cortical slices of the rat and guinea pig (Schultz, 1974a). They also enhance the stimulatory effects of NA, isoprenaline and adenosine in rat cortical slices, as well as the effects of histamine, histamine plus NA and adenosine (Schultz, 1974a). Another potent phosphodiesterase inhibitor is 1-ethyl-4-isopropylidenehydrazino)-1H-pryazolo-[3,4b]-pyridine-5-carboxylic acid (SQ 2000 9) with an IC_{50} of 2 μM (Beer *et al.*, 1972). This compound potentiates the effect of histamine, histamine plus NA on the cAMP accumulation in guinea-pig cortical slices (Schultz, 1974b). Both benzodiazepines and SQ 20009, as well as theophylline, have been reported in animal experiments to reduce anxiety (Beer *et al.*, 1972) and it is possible that their anti-anxiety effect may be correlated with phosphodiesterase inhibition.

Phenothiazine and reserpine derivatives have also been reported to inhibit the activity of phosphodiesterase (Honda and Imamura, 1968), but the physiological and pharmacological significance of these actions has yet to be elucidated. It is worth remembering that no class of drugs has been shown to act physiologically via their effects on cyclic nucleotide phosphodiesterases in either peripheral organs or in the CNS.

IX CONCLUSIONS

The data reviewed in this chapter on psychotropic drugs which act as agonists or antagonists on DA and/or NA-sensitive adenylate cyclase systems in brain support the view that cAMP plays an important role in brain function. Many of the results discussed are consistent with the idea that DA or NA-sensitive adenylate cyclase in the brain are part of a catecholamingeric receptor system which, through the formation of cAMP, may mediate the action of the neurohormones in a particular neurone. The blockade of DA-sensitive adenylate cyclases by antipsychotic drugs in striatum and of DA and NA-sensitive adenylate cyclases in limbic structures, and the observed change in receptor sensitivity of NA-sensitive cAMP-generating systems in the limbic forebrain following chronic administration of antidepressant drugs have provided a new framework for future studies. It is necessary to understand how the drug-induced changes in the availability of cAMP are translated into neurophysiological and, eventually, into behavioural function. The determination of the substrates of cAMP-dependent protein kinase *in vivo* and, more importantly, the elucidation of the function (enzymic or otherwise) of the protein substrates are required to establish the relevance of the mechanisms by which cAMP and/or cGMP bring about any action in the CNS. Studies on modifications of these processes by antipsychotic and antidepressant drugs may eventually lead to an improved understanding of the psychobiology of mental disorders.

ACKNOWLEDGEMENT

The original studies from the authors' laboratories, quoted in this review, were supported by USPHS grant MH 11468.

CHAPTER 20

Cyclic Nucleotides and Skin Disorders

E. A. Duell and J. J. Voorhees

I INTRODUCTION

The skin is composed of three distinct subunits: the avascular epidermis, the dermis containing nerves, blood vessels, exocrine sweat glands, sebaceous glands and hair follicles, and the fatty layer. The epidermal portion of the skin is a highly differentiated tissue composed of four distinct layers as shown in Figure 20.1. The basal layer is the proliferating region of the epidermis. In contrast, the spinous and granular layers do not normally proliferate and have become highly differentiated—containing tonofilaments and keratohyalin granules—while the horny layer, which serves as the protective barrier, is composed of compacted dead cells lacking nuclei and filled with amorphous and fibrous material. Keratinocytes comprise approximately 95% of the volume of these four layers. Since marked stratification does exist in the epidermis, some caution must be exercised in the interpretation of the biochemical data which are an average measurement of all the components present in a given tissue sample. Thus, different biochemical values are obtained if the outer half of the epidermis is compared to the entire epidermis or to the full-thickness skin. These differences may be exaggerated in the case of important regulatory molecules such as the cyclic nucleotides or components of the prostaglandin system.

In contrast to these difficulties encountered in the utilization of the epidermis as a tissue for investigating factors which control proliferation and/or differentiation are its rather obvious advantages of accessibility, the presence of both proliferative and differentiated regions and the occurrence of diseases exhibiting alterations in both proliferation and differentiation. This laboratory has utilized the benign proliferative skin disease, psoriasis, as a model system for

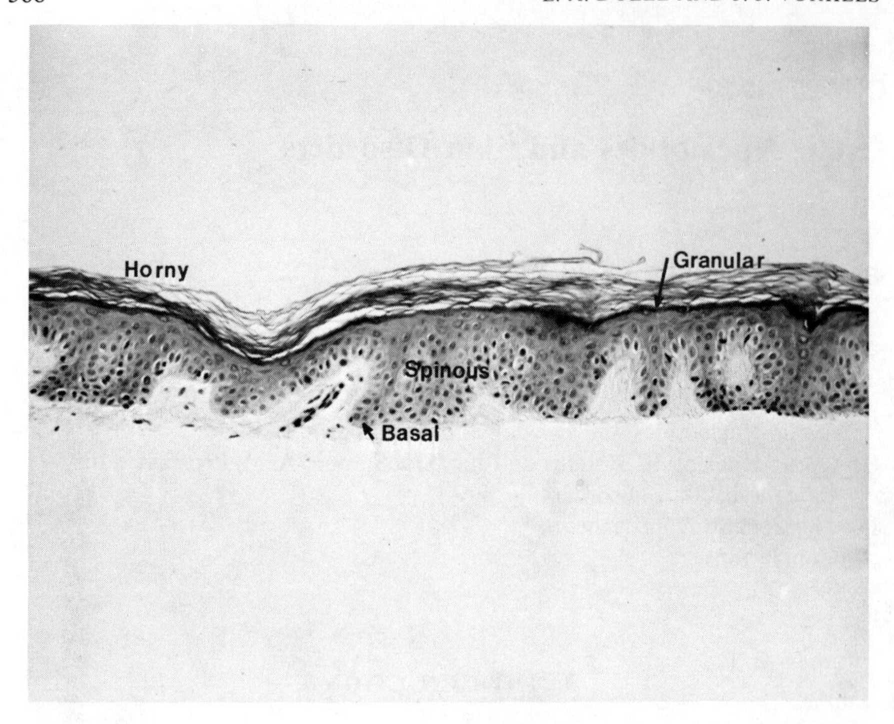

Figure 20.1 Normal epidermis obtained with a keratome set at 0.1 mm. A minimum of dermal contamination is present beneath the dermo–epidermal junction. Stained with hematoxylin and eosin; ×100

investigating the roles of cAMP, cGMP and the components of the arachidonic acid–prostaglandin system as possible regulatory molecules in epidermal proliferation and differentiation. Other skin diseases in which these compounds may play a role as regulatory molecules are the benign proliferative disorder, atopic dermatitis and the malignant disease, melanoma.

II CYCLIC NUCLEOTIDES AND PROSTAGLANDINS IN NORMAL SKIN

Information concerning the cyclic nucleotide system in skin obtained over recent years will be discussed and only a summary of the findings obtained prior to the review of Voorhees *et al.* (1974) will be presented.

A Enzymes

1 Cyclic Nucleotide System

The enzymes, such as adenylate and guanylate cyclases, phosphodiesterase(s) and cAMP-dependent protein kinases, involved in cyclic nucleotide metabolism, have been identified in the epidermis (Voorhees *et al.*, 1974). Epidermal adenylate cyclase is stimulated by catecholamines (Voorhees *et al.*, 1974). PGE_1 and PGE_2 (Voorhees *et al.*, 1973; Mandel *et al.*, 1975; Aso *et al.*, 1975a; Adachi *et al.*, 1975) and adenosine (Voorhees and Duell, 1975). There is some disagreement about the localization of adenylate cyclase activity as determined by electron microscopy. Søndergaard *et al.* (1975) have reported that the adenylate cyclase activity is located in the basal layer and in the lower spinous layer of human epidermis but not in the upper spinous, granular or horny layers. In contrast, Mahrle and Orfanos (1975) using very similar techniques and tissue have localized the enzyme in all layers; increased levels were found in the lower layers after the addition of glucagon. Prior to this report, an increase in cAMP had not been demonstrated in response to glucagon (Voorhees *et al.*, 1974). Some questions still remain concerning the actual techniques used and thus the validity of the results with identified adenylate cyclase activity using electron microscopy. Localization of cAMP in the epidermis by immunofluorescent techniques with a variety of rigorous controls may provide support or may negate the previous ultrastructural experiments

2 *Prostaglandin System*

Prostaglandins and their interrelationship with the cyclic nucleotides have been reviewed by Samuelsson *et al.* (1975), and with particular reference to the skin by Ziboh (1975). Utilizing homogenates of human skin Jonsson and Anggard (1972) demonstrated that the major C_{20}-acid esterified to the phospholipid fraction of this tissue is arachidonic acid, the precursor for the synthesis of both PGE_2 and $PGF_{2\alpha}$. In a homogenate, PGE_2 (mean value 15 ng/g tissue wet weight) is the main prostaglandin component. A microsomal fraction from human skin contains prostaglandin synthetase activity which can be inhibited by the addition of anti-inflammatory drugs, such as indometacin (6 μM inhibits 33%), aspirin (60 μM inhibits 47%) and triamcinolone acetonide (60 μM inhibits 11%) (Ziboh, 1975).

In similar studies utilizing homogenates from human skin, Greaves and McDonald-Gibson (1973) have demonstrated the formation of both PGE_2 and $PGF_{2\alpha}$ from arachidonic acid. In these studies, the addition of 0.56 mM aspirin, 0.28 mM indometacin or 0.5 mM chloroquin results in a statistically significant decrease in PGE_2 synthesis of 19%, 34% or 24%, respectively, while 0.5 mM acetaminophen is ineffective. The inhibition of $PGF_{2\alpha}$ formation under the same conditions produces a different pattern; aspirin is ineffective while inhibition is produced by the other drugs, indometacin (64%), chloroquin (51%) and acetaminophen (45%).

A new set of reactions with arachidonic acid as the precursor has been reported by Hamberg and Samuelsson (1974). Fatty acid cyclo-oxygenase (previously known as prostaglandin synthetase) appears to be the enzyme involved in the formation of cyclic endoperoxide (prostaglandin G_2 and H_2), thromboxane B_2, 12L-hydroxy-5,8,10,14-eicosatetraenoic acid (HETE and 12L-hydroxy-5,8,-10 heptadecatrienoic acid (HHT) from arachidonic acid in platelets. These alternative pathways have also been investigated by Hammarström *et at.* (1975) in human epidermis where very low levels of HETE(<0.05 μg/g wet weight) have been detected. The possible significance of the alternate pathway in skin will be discussed in the section pertaining to psoriasis (see Section IIIA).

B Effects of Cyclic Nucleotides and Prostaglandin E on the Epidermis

1 *cAMP*

Increased levels of cAMP have been shown to be associated with the inhibition of mitosis in the G_2 phase of the cell cycle in mouse epidermis. Also cAMP exhibits a regulatory effect on glycogen metabolism (Voorhees *et al.*, 1974). Several tissue-culture systems have been utilized to obtain additional information about the role of cAMP in epidermal proliferation. Using the outgrowths from human skin in organ culture Flaxman and Harper (1975) investigated the effect of various compounds known to increase cAMP concentrations in the epidermis on the G_2 phase of the cell cycle. At a concentration of 10^{-9}M, L-adrenaline, D, L-isoprenaline and L-noradrenaline inhibit mitosis by 65%, 53% and 31%, respectively. These inhibitory effects are prevented by the addition of a β-adrenergic blocker, such as propranolol. These data are consistent with an increase in the concentration of cAMP mediating the inhibition of mitosis. The mechanism for the inhibition resulting from the addition of millimolar concentrations of ATP, GTP, CTP and UTP is open to speculation at this time since only ATP and GTP can alter cAMP levels by relatively poorly understood mechanisms.

Inhibition of proliferation of primary cell cultures of neonatal mice by the addition of cyclic phosphodiesterase inhibitors to the culture medium has been demonstrated by Voorhees and Duell (1975). Papaverine (10^{-5}M), theophylline (5×10^{-5}M) and 1-methyl-3-isobutylxanthine (10^{-3}M) inhibit the cell cultures by 74%, 60%, and 51%, respectively, as determined by the incorporation of [3H]thymidine into DNA. In these experiments, it is not known at which stage, or stages, of the cell cycle the inhibitory effect is manifest. Similar studies have been conducted with primary cultures of guinea-pig epidermal cells (Delescluse *et al.*, 1974). In addition to the cyclic phosphodiesterase inhibitors, PGE_2 methyl ester (10^{-5}M) produces a 61%, Bt_2-cAMP (10^{-4}M) a 58% and isoprenaline (5×10^{-6}M) a 59% inhibition of cell proliferation. Autoradiographs show that grain densities are similar in both control and Bt_2-cAMP-treated cultures, thus indicating that the inhibition of cell proliferation does not occur in the S phase of the cell cycle. The point, or possibly points, of inhibition in these experiments are at present unknown.

cAMP has been shown to induce differentiation in cultured malignant cells in addition to its effects on cell proliferation (Voorhees *et al.*, 1974). Recently data have been presented (Delescluse *et al.*, 1976) which corroborate the initial hypothesis that cAMP might be expected to promote differentiation in normal keratinizing epithelial cells (Voorhees *et al.*, 1972a) and non-keratinizing epithelial cells (Pratt and Martin, 1975). Primary cultures of guinea-pig epidermal cells were grown for a period of ten days in the absence or presence of 10^{-3} M Bt_2-AMP. The incorporation of radioactive histidine, cystine and arginine into protein during a one-hour terminal labelling period is increased in the presence of the cyclic nucleotide. There is also an increase in the rhodamine B-stainable material—such staining has been used to monitor keratinization histochemically (Delescluse *et al.*, 1976). Thus, evidence is presented which indicates that increases in cAMP levels can promote the synthesis of a measurable component which is part of the differentiation process.

In the non-keratinizing system, rat palatal shelves have been placed in tissue culture at day 14 of gestation with and without 6×10^{-4} M Bt_2-AMP (Pratt and Martin, 1975). Palatal shelves grown in the presence of the derivative of cAMP exhibit decreased proliferation as assessed by reduced [^{3}H]thymidine incorporation, and increased differentiation indicated by stimulated incorporation of [^{3}H]fucose and [^{3}H]glucosamine into glycoproteins, and increased adhesion of the palatal shelves. Thus, cAMP appears to affect proliferation, differentiation and cell adhesion *in vitro*.

2 cGMP

Information about cGMP and its effect on the epidermis is very limited. *In vitro* experiments with hairless mouse epidermal slices indicate that histamine (10^{-3}M) and tetradecanoyl phorbol acetate (75 ng/ml) can increase the concentration of cGMP fifteen-fold after incubation for thirty seconds and less than two-fold after a three-minute incubation, respectively (Voorhees *et al.*, 1974). The addition of 10^{-6}M tetradecanoyl phorbol acetate to the primary cultures of neonatal mouse epidermal cells results in a 240% increase in cell proliferation (Voorhees and Duell, 1975). It is concluded that increased levels of cGMP are associated with induced epidermal-cell proliferation.

Additional circumstantial evidence suggesting that cGMP may be closely related to increased epidermal proliferation has been obtained in experiments conducted *in vivo* using newborn rats (Stegman *et al.*, 1976). Intradermal injections of 5–100 ng of concanavalin A (Con A) were given twenty-four hours prior to sacrifice. The results of these injections are an increase in glycogen deposition throughout the epidermis, a decreased granular layer as determined histologically and by decreased incorporation of [^{3}H]histidine into the granular layer, and increased incorporation of [^{3}H]thymidine into the basal cells. An injection of 0.1 ml of 3 mM α-methyl-D-glucopyranose into the rats previously injected with the lectin can prevent all of these effects from taking place. In these experiments, the concentration of cAMP or cGMP was not assessed following the treatment with Con A. However, on the basis of information available from the literature for the lymphocyte system, it seems possible that the lectin Con A first increases the cGMP levels in the epidermis followed by the alterations in proliferation and differentiation. There is some controversy surrounding lectin-stimulated cyclic nucleotide concentrations. Increases in cAMP levels have been implicated based on the work of Hadden *et al.* (1972), Schumm *et al.* (1974) and Whitfield *et al.* (1974). In contrast, data of Wedner *et al.* (1975b) indicate that cAMP levels are increased, which contradicts the data obtained by Glasgow *et al.* (1975) and DeRubertis *et al.* (1974) in which an elevation of cAMP levels could not be detected. It would be interesting to determine if cGMP levels are in fact increased in epidermal cells.

3 Prostaglandins E_1 and E_2

At the present time, two systems are being investigated to determine what role, if any, the prostaglandins play in the epidermis. Rats which have been maintained on a diet deficient in essential fatty acids develop a scaly dermatosis exhibiting increased proliferation (Ziboh, 1975). The epidermis can be restored to a normal appearance by the topical application of PGE_2. As indicated in an earlier section, the addition of PGE_1 or PGE to epidermal slices from various species including man can increase the level of cAMP in the tissue (Voorhees *et al.*, 1974; Aso *et al.*, 1975b; Adachi *et al.*, 1975). However, it is unknown if cAMP elevation as a result of the topical application of PGE_2 is the mechanism for restoring the scaly lesions to normal tissue.

The role of prostaglandins in the inflammatory response induced by UV light has been investigated (Snyder, 1975; Snyder and Eaglstein, 1974; Eaglstein and Weinstein, 1975). Intradermal injections of PGE_2 can produce areas of erythema and an increased incidence of labelled cells after injection of [^{3}H]thymidine (Eaglestein and Weinstein, 1975), thus implicating PGE_2 in both phenomena. In other studies, the erythema produced by UV irradiation is diminished or blocked by anti-inflammatory drugs such as aspirin, triamcinoline and indometacin (Snyder, 1975). However, in the work of Snyder (1975), keratinocyte death and increased cell proliferation usually associated with UV-damaged skin were not prevented by indometacin. In other systems, it has been shown that UV-irradiation may damage the PGE_2 receptors (Johnson *et al.*, 1973). Thus, it would appear that PGE_2 is probably involved in the inflammatory response, but the levels of PGE_2 have not been measured in the epidermis following UV-irradiation nor have the levels of cAMP and cGMP been determined. Further investigations may provide some answers about the interrelationships between the cyclic nucleotides, prostaglandins and UV-irradiation, and their effects on the epidermis.

III CYCLIC NUCLEOTIDE SYSTEMS AND THE ARACHIDONIC ACID–PROSTAGLANDIN CASCADE IN SKIN DISORDERS

A Psoriasis

Psoriasis is a benign proliferative skin disease found only in humans, which is characterized by increased glycogen accumulation (Braun-Falco, 1958), decreased differentiation (Van Scott, 1972) and an increased rate of cell

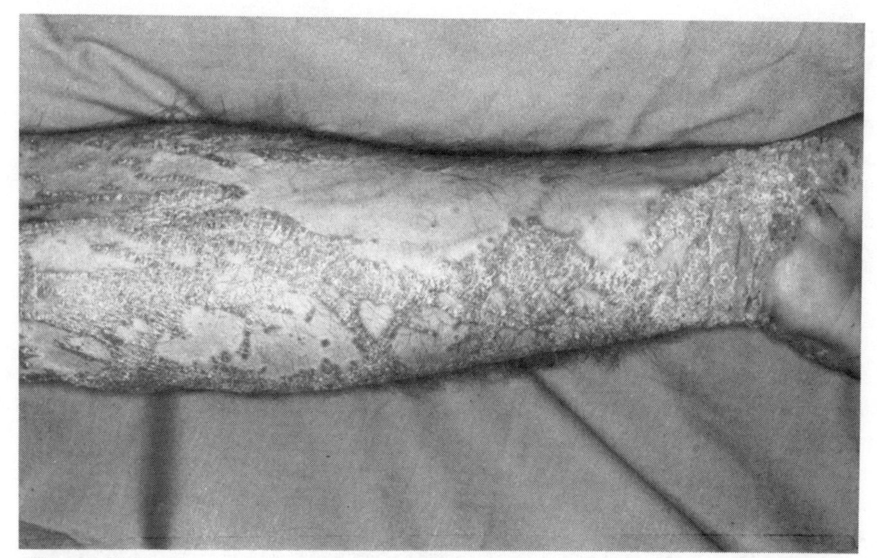

Figure 20.2 The arm of a psoriatic patient with a mild case. Note the sharp margins of the psoriatic lesion

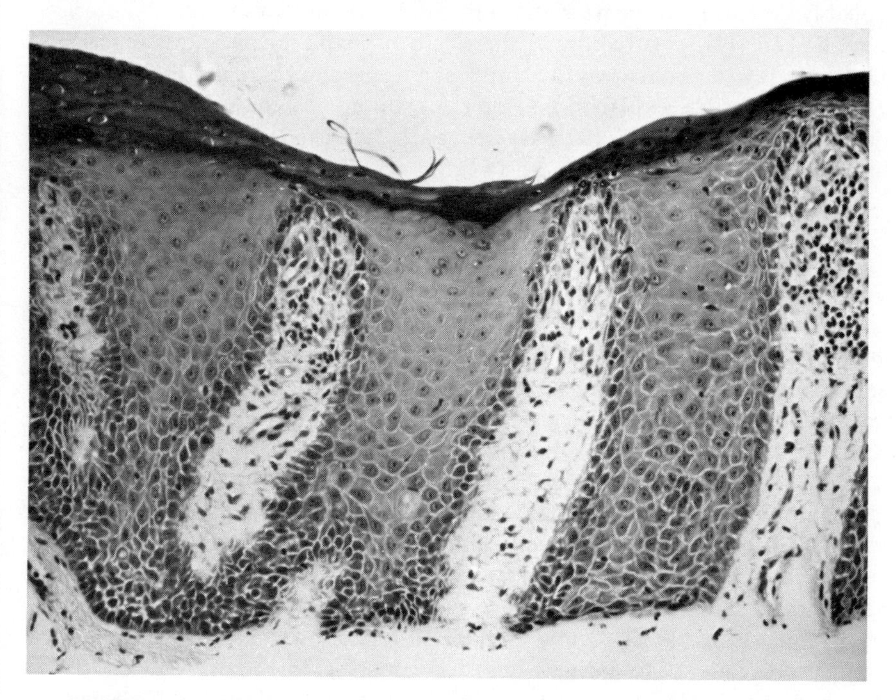

Figure 20.3 Involved psoriatic epidermis obtained with a keratome set between 0.25 and 0.5 mm. The dermal contamination cannot be avoided if one removes the entire epidermis. Stained with hematoxylin and eosin; ×100

proliferation (Weinstein and Frost, 1968). Figure 20.2 shows an example of a psoriatic lesion in a mild form, while Figure 20.3 is a histological section through a psoriatic lesion. The authors have been investigating the role of cAMP in the pathophysiology of psoriasis during the last five years. More recently investigations have been made into the role of cGMP and the importance of the arachidonic acid–prostaglandin cascade in psoriasis. The general topic of cyclic nucleotides and psoriasis was extensively reviewed in 1974 (Voorhees *et al.*); only highlights and new information are included in this review.

1 Cyclic Nucleotide Levels

A controversy still exists as to whether the levels of cAMP are decreased in the lesional areas compared with uninvolved epidermis (Voorhees *et al.*, 1975; Halprin *et al.*, 1975; Voorhees *et al.*, 1976). Also opinions are divided as to whether the β-adrenergic receptor is malfunctioning in the psoriatic lesion. Data have been presented which indicate that the levels of cAMP in the psoriatic lesions of ten patients are the same or slightly elevated in comparison to uninvolved areas (Yoshikawa *et al.*, 1975b). Data from the authors' studies involving fifty patients, indicate that the mean value for cAMP levels is probably decreased in the psoriatic compared with uninvolved areas (Voorhees *et al.*, 1972b, 1973). In contrast, there is a probable mean value increase in cGMP levels in lesional areas in comparison to the control regions (Voorhees *et al.*, 1973). Possible explanations for these differences have previously been presented (Voorhees *et al.*, 1975; Halprin *et al.*, 1975) and will not be discussed in this chapter.

2 Possible Defect in the β-Adrenergic Receptor

Information has been obtained in the authors' laboratory from studies of samples of involved and uninvolved epidermis from twenty-three psoriatic patients which indicates that the cAMP levels in the involved and uninvolved epidermis are similar after incubation for three minutes with 10^{-6} M isoprenaline. These data suggest the presence of a functioning β-adrenergic receptor in the lesional epidermis. Supporting these data are the experiments of Harper *et al.* (1964). The addition of isoprenaline (10^{-6} M) or adrenaline (4.5×10^{-6} M) inhibits the cell proliferation of outgrowths of psoriatic epithelium in culture to the same extent as that of normal human skin.

Using histochemical techniques, Mahrle and Orfanos (1975) suggested that defective β-adrenergic and glucagon receptors exist in psoriatic tissue compared with normal skin. This is the first report of glucagon-stimulating epidermal adenylate cyclase. All previous studies of glucagon's effects on epidermal adenylate cyclase have yielded negative results (Marks and Revien, 1972; Voorhees *et al.*, 1974). Halprin *et al.* (1975) have suggested that the β-adrenergic receptor is defective in the psoriatic lesion. Both the adenine-prelabelling technique and incubation of sliced preparations with catecholamine have been utilized in these studies. It is of considerable interest to note that Halprin *et al.*

(1975) have also reported an increase in glycogen accumulation in psoriatic lesions. They showed that the glycogen is hydrolysed when the slices are incubated with β-adrenergic agonists, such as adrenaline. These data are paradoxical since they show increased lesional cAMP and glycogen levels, and decreased lesional responsiveness of adenylate cyclase to β-adrenergic agonists. These findings require explanation and/or clarification.

3 Arachidonic Acid–Prostaglandin Cascade

In the past two years several laboratories have investigated the concentrations of PGE_2 and $PGE_{2\alpha}$ present in involved and uninvolved psoriatic epidermis. A lower level of PGE_2 has been found in the lesional areas (129 ng/g wet weight) compared with uninvolved epidermis (303 ng/g wet weight) when determined by radioimmunoassay; $PGE_{2\alpha}$ levels are approximately the same in both tissue samples (Aso *et al.*, 1975a). These investigators concluded that decreased PGE_2 synthesis occurs in psoriatic regions. In addition, there appears to be a significant decrease in the responsiveness of adenylate cyclase to added PGE_1 or PGE_2 in the lesional areas in comparison with the uninvolved areas (Aso *et al.*, 1975b). This finding is in contrast to other data which indicate that no significant decrease in the responsiveness of adenylate cyclase to added PGE_1 or PGE_2 occurs in lesional areas (Adachi *et al.*, 1975; Halprin *et al.*, 1975; Mandel *et al.*, 1975).

The levels of PGE_2 (33 ng/g wet weight) in involved epidermis was shown by Hammarstrom *et al.* (1975) to be increased in comparison to the uninvolved areas (24 ng/g wet weight). Similarly the levels of $PGF_{2\alpha}$ (39 ng/g wet weight) are higher in involved that in normal epidermal samples (21 ng/g wet weight) as determined by gas–liquid chromatography and mass spectrometry. Addition of arachidonic acid, the precursor of both $PGE_{2\alpha}$ and $PGE_{2\alpha}$, results in a significantly higher biosynthetic rate of PGE_2 in lesional areas than in uninvolved tissue; no difference in $PGF_{2\alpha}$ synthetic capacity was detected by these same investigators. The alterations in the prostaglandin levels in the lesional psoriatic epidermis are very small when compared to the twenty-five-fold increase in free arachidonic acid observed in the lesional areas with the uninvolved psoriatic epidermis as the reference point. Even more striking is the eighty-one-fold increase in the concentration hydroxyeicosatetraenoic acid (Hammarström *et al.*, 1975), a recently isolated and identified component of the arachidonic acid–prostaglandin cascade (Hamberg and Samuelsson, 1974). The significance of these large increases in free arachidonic acid and hydroxyeicosatetraenoic acid within the lesions is not known. The latter compound may be an important molecule or it may be a 'dead-end' product formed by the system in an attempt to compensate for or to reduce the increased levels of free arachidonic acid. Such increased levels of free arachidonic acid may explain the decrease in PGE_2 synthesis reported (Aso *et al.*, 1975a) since the large amounts of unlabelled arachidonic would dilute the labelled arachidonic acid added as the substrate for the reaction.

All the laboratories investigating a possible role for the cyclic nucleotides and the prostaglandin cascade in the control of epidermal proliferation and differentiation appear to be convinced that these compounds are important in the pathophysiology of psoriasis. However, there is little agreement among the various workers as to the site, or possibly sites, of the defects and the relative levels of the various regulatory molecules. The development of additional techniques and renewed application of those already available may eventually resolve these differences.

B Atopic Dermatitis

Like psoriasis atopic dermatitis is a benign proliferative skin disorder. The disease frequently occurs in individuals with asthma and/or allergic rhinitis. Szentivanyi (1968) originally proposed that a malfunctioning β-adrenergic receptor might be the underlying pathophysiology in bronchial asthma. A similar defect in other tissues, such as the skin, may account for other diseases, such as atopic dermatitis. Since the β-adrenergic receptors are intimately involved in the formation of cAMP, an investigation of the cyclic nucleotide system would seem to be the most appropriate approach for testing this hypothesis.

1 Enzymes Involved in cAMP Metabolism

Punch biopsies (4 mm) from normal volunteers as well as samples from involved and uninvolved areas of patients with atopic dermatitis have been utilized to assay the high K_m cAMP phosphodiesterase (Holla *et al.*, 1972) and adenylate cyclase activities (Mier and Urselmann, 1970). cAMP phosphodiesterase activities appear to be similar in all samples. Likewise, there is no significant change in basal adenylate cyclase activities when expressed in relation to moles of DNA phosphate. Based on wet weight, the activity in lesional areas may be increased. Since the response of adenylate cyclase to β-adrenergic agonists was not assayed, the most important question with respect to Szentivany's proposal—namely is a malfunctioning β-receptor apparent in disease?—was not investigated.

Recently Kumar *et al.* (1975) partially purified protein kinase activity from full-thickness skin obtained from normal individuals and from patients suffering from atopic dermatitis. The basal level of kinase activity is quite comparable in all samples. However, the enzyme from the atopic patients is much less responsive to cAMP; cAMP (5×10^{-6} M) stimulates the protein kinase from normal skin 130% while the enzyme derived from atopic patients is stimulated only 29%. As suggested by Kumar *et al.* (1975) a certain amount of caution should be exercised since alterations in the enzyme may occur during the purification procedure. It is possible that two defects in the cAMP system may occur in atopic dermatitis, one in the cAMP-dependent protein kinase and the other in the β-adrenergic receptor.

A different approach for investigating the site of the defect in atopic dermatitis has been initiated by Carr *et at.* (1971). These workers used the inhibition of mitosis as the criterion for assessing the β-adrenergic receptor in the skin. It was already known from the work of Bullough and Laurence (1961) that adrenaline inhibits mitosis in mouse skin. The data indicate that mitosis is inhibited in normal skin by the addition of the β-agonists adrenaline or isoprenaline, while lesional or uninvolved skin from atopic eczema patients are refractory (Carr *et al.*, 1973). Thus, the data appear to support the hypothesis of Szentivanyi (1968). However, other factors must be considered in evaluating the data. In the studies by Carr *et al.* (1973) the only data base is the surface area, and incorporation of [^{3}H]thymidine into DNA is used to assess the degree of mitotic inhibition. This technique is subject to question due to the variability in the uptake of thymidine and the alterations in pool size which are known to occur as a result of variations in cellular cAMP levels (Roller *et al.*, 1974; Kaukel *et al.*, 1972a).

Attempts to confirm these observations by investigating adenylate cyclase activity were then initiated. The levels of cAMP present in the epidermal samples from five atopic patients and eight normal volunteers after a four-minute incubation period with isoprenaline (10^{-6} M) were determined (Lee *et al.*, 1974). An unexpected observation was made as a result of these incubations. The mean level of cAMP in the skin obtained from the atopic patients is slightly elevated in comparison with the levels in samples from the normal volunteers. Therefore, at the enzymic level the data in support of a defect in the β-adrenergic receptor have not been obtained. While the status of the β-adrenergic receptor in the epidermis of atopic patients is not clear at present, the information available using other tissues does implicate both cAMP and cGMP as important regulatory substances in the release of chemical mediators which are essential to the pathophysiology of asthma.

2 *Knowledge from Other Systems in Support of a Defective β-Adrenergic Receptor in Atopy*

Information from two different lines of research seem applicable. Firstly, leucocytes, lung tissue and nasal polyps have been investigated with respect to their responsiveness to catecholamines. Secondly, the effects of increased cAMP or cGMP levels upon the release of factors involved in the immunological response, such as histamine, slow-reacting substance of anaphylaxis and eosinophilic chemotactic factor of anaphylaxis, have been studied. Regardless of the tissue, increased levels of cAMP appear to inhibit release of these factors. This inhibition has been observed in nasal polyps (Kaliner *et al.*, 1973), lung tissue (Austen, 1974; Okazaki *et al.*, 1975) and leucocytes (Koopman *et al.*, 1970). Also elevations in the levels of cGMP may enhance release as in the lung (Wasserman *et al.*, 1974) and leucocytes (Goldstein *et al.*, 1973a; Strom *et al.*, 1972; Weinstein *et al.*, 1975).

The responsiveness of leucocytes, indicated by increased levels of cAMP following the stimulation of the cells with β-adrenergic agonist, has been investigated in a number of laboratories in order to test for a possible defect at the receptor level. A decrease in responsiveness based on cAMP formation has been observed (Gillespie *et al.*, 1974; Parker and Smith, 1973), while there seems to be no alteration in the binding of adrenaline to the leucocytes from asthmatics (Sokol and Beall, 1975). It seems quite likely, on the basis of available information, that an alteration in the β-adrenergic system does occur in atopy and, with a closer examination of the skin, a similar pattern may emerge in studies on the epidermis.

C Melanoma

Melanoma is a malignant proliferative disease of the skin and is the most difficult of tumours in the skin to control. A number of investigators have utilized a series of melanoma cell lines to determine what control the cyclic nucleotides may exert in this system. When Bt_2-cAMP is added to Cloudman S91 melanomas, cell spreading occurs and the cells appear enlarged and become more pigmented (Johnson and Pastan, 1972c). Even more striking results occur with an amelanotic cell line grown in culture. These cells become highly pigmented if 10^{-7} M melanocyte-stimulating hormone (MSH) is added to the culture (Wong *et al.*, 1974). The system is particularly interesting since MSH binds to the cells only during the G_2 phase of the cell cycle (Wong *et al.*, 1974; Varga *et al.*, 1974; Kerr *et al.*, 1975). In addition to MSH, pgE_1 also increases cAMP levels and stimulates differentiation as measured by increased pigmentation (Kerr *et al.*, 1975; Kreiner *et al.*, 1973a).

In vitro experiments have provided ample evidence to indicate that increasing the concentration of cAMP within the melanoma cells inhibits proliferation as well as induces differentiation (Lincoln and Vaughan, 1975; Kreider *et al.*, 1973; Pawelek *et al.*, 1975). It is known that the increase in pigmentation is due to increased tyrosinase activity. The pathways by which increased levels of cAMP produce the increased tyrosinase activity are unknown as is the mechanism by which cAMP inhibits proliferation. As might be anticipated, the addition of acetylcholine to a test system reverses the effects of MSH, that is, in frog there is a lightening of the skin (Moellmann *et al.*, 1974). While levels of cGMP have not been measured in this system, it seems possible that acetylcholine does increase the cGMP levels since the cholinergic system is known to raise levels of this cyclic nucleotide in tissues.

IV THERAPEUTIC APPROACHES

If one accepts the information which suggests that cyclic nucleotides are important in the control of proliferation and differentiation, then controlling these levels within a lesional area may result in the tissue returning to a more normal condition. The disorders of the skin which appear to be the most amenable to this approach are psoriasis and atopic dermatitis. In both of these disorders a defect apparently exists early in the chain of events which appear to be controlled by the cyclic nucleotides.

Double-blind clinical studies have been carried out to determine if the topical application of compounds known to affect the levels of cAMP in the epidermis can improve psoriatic lesions. Two cyclic phosphodiesterase inhibitors, papaverine (Stawiski *et al.*, 1975) and Ro—1724 have been tested in this way. A cream containing 1% papaverine when applied to forty-five patients produced a statistically significant ($p = 0.011$) improvement in the centre of the lesional areas as compared with lesional areas treated with the cream base alone. A similar study utilizing a cream containing 1% Ro 20–1724 was carried out on twelve patients. A statistically significant ($p < 0.01$) improvement in the lesional areas treated with the Ro 20–1724 cream was observed. Thus, in two separate clinical studies, a statistically significant improvement was noted when the lesional areas were treated topically with compounds which are known to elevate cAMP levels *in vitro*. The use of various combinations of such compounds may lead to a more effective mode of treatment and may also add credence to the contention that altered cyclic nucleotide metabolism is important in the pathophysiology of psoriasis.

V CONCLUSIONS

The cyclic nucleotide systems and the arachidonic acid–prostaglandin cascade appear to be important regulatory systems in the control of proliferation and/or differentiation. Evidence has been accumulating which indicates that alterations in either, or both, of these systems are important in the pathophysiology of diseases, such as psoriasis and atopic dermatitis. Continuing investigation into these systems will be necessary to delineate further the mechanisms by which these controls are exerted. New therapeutic methods based on the alterations of the cyclic nucleotide levels or components of the arachidonic acid–prostaglandin cascade may prove to be effective means of treating diseases, such as psoriasis and atopic dermatitis, with minimum of side-effects being observed.

CHAPTER 21

Cyclic Nucleotides in Vascular Smooth Muscle and in Hypertension

M. S. Amer

I INTRODUCTION

Hypertension afflicts nearly one-fifth of the adult population in the United States (Laragh, 1974). Although elevated blood pressure may be the result of another underlying disease, about 85% of all hypertensive patients are classified as suffering from essential hypertension, indicating that the condition is of unknown cause or aetiology (Hirschman and Herfindal, 1971). In these patients there is frequently elevated peripheral vascular resistance to blood flow due to the decreased capacity of the arterioles. This increased peripheral vascular resistance is primarily due to changes in the structure, tone and sensitivity of the vascular smooth muscles themselves in addition to significant extramyogenic factors. The myogenic changes include (*i*) increased cellular proliferation resulting in an increased wall/lumen ratio, (*ii*) increased vascular smooth-muscle tone and (*iii*) increased sensitivity of vascular smooth muscles to the contractile effects of catecholamines and other information-transferring molecules. Extra-myogenic factors include increased sympathetic tone and/or circulating levels of catecholamines (Roizen *et al.*, 1975; Bevan *et al.*, 1975; Louis *et al.*, 1973; Hashida *et al.*, 1974; Iriuchijima, 1973) and the presence or absence of circulating vasoconstricting or dilating renal factors, respectively.

Since the defects in vascular smooth muscle function in hypertension appear to persist in the absence of extramyogenic factors (Douglas *et al.*, 1975; Albrecht *et al.*, 1975), changes in the metabolism or function of vascular smooth muscles themselves have been strongly suspected (Finch, 1975; Hamilton, 1975). This has led to the study of cyclic nucleotide metabolism in vessels from hypertensive animals since the cyclic nucleotides appear to play an important role in the control of vascular smooth muscle function similar to their roles in almost all other biological systems.

The importance of cyclic nucleotides in vascular smooth muscle function will be discussed first, followed by consideration of the aberration in metabolism in hypertension and the possible importance it may have in the aetiology, maintenance and treatment of the disease.

II CYCLIC NUCLEOTIDES IN VASCULAR SMOOTH MUSCLE

Since their discovery almost two decades ago, the role which cyclic nucleotides play in the control of smooth muscle activity has been under investigation. It now appears that cAMP is important in mediating vascular smooth muscle relaxation whereas cGMP may play a role antagonistic to that of cAMP.

Much of the confusion in this field (Bär, 1974a) can be traced to the lack of data on cGMP and the persistent view that changes in cAMP levels alone should reflect the contractile state of the muscle if these nucleotides are to play their expected mediator roles. The two cyclic nucleotides appear to act in the control of vascular smooth muscle function in a similar way to their action in other biological systems and the determinations of the levels or activity of either cyclic nucleotide are not sufficient to clarify fully the functions of these mediators. Thus vascular smooth muscle does not differ significantly from other smooth muscles and most of the observations made in the latter may be applied to the former.

A Occurrence

The presence of cAMP in vascular smooth muscle has been well demonstrated (Aberg and Anderson, 1972, Amer, 1974 and Somylo *et al.*, 1972). Its levels appear to vary widely in different blood vessels and even in different sections of the same vessel. Comparatively little data on the levels of cGMP in this tissue have been reported (Amer *et al.*, 1974, 1975; Goldberg *et al.*, 1973b).

B Adenylate and Guanylate Cyclases

The presence of an adenylate cyclase capable of synthesizing cAMP in vascular smooth muscle was discovered in Sutherland's laboratory (Klainer *et al.*, 1962). Although the first attempts were unsuccessful, several reports now support the hormonal sensitivity of vascular adenylate cyclase (Amer, 1973; Amer *et al.*, 1974, 1975; Triner *et al.*, 1971, 1972, 1975; Klenerova *et al.*, 1975). As with other tissues, the β-receptor seems to be coupled to cAMP synthesis. Although it is relatively easy to demonstrate cAMP increases in response to adrenaline in intact vascular smooth muscle (Triner *et al.*, 1971, 1972; Volicer and Hynie, 1971; Andersson, 1973b), it is more difficult to demonstrate the hormonal sensitivity in broken-cell preparations. This may be due to the greater need for structural integrity in adenylate cyclase from vascular smooth muscles than from other sources. No such difficulty appears to exist with respect to stimulation by NaF which may act at the level of the catalytic rather than the receptor subunit of the enzyme (Amer and McKinney, 1974), thus needing less structural integrity.

The importance of adenylate cyclase in the normal function of vascular smooth muscle is not clear. Although β-adrenergically mediated vasodilation has been demonstrated, for example in skeletal muscle (Viveros *et al.*, 1968), vascular smooth-muscle tone appears to be controlled primarily by the tonic influence of the sympathetic nervous system mediated by α-adrenergic receptors which trigger increased cGMP synthesis as will be discussed later. Vascular smooth muscle relaxation is effected mainly by the withdrawal of that sympathetic tone. Normally masked active vasodilation (Abboud and Eckstein, 1966) has been demonstrated in a number of systems and may function using a selection of transmitters (Takeuchi and Manning, 1973; Beck, 1965; Tobia *et al.*, 1970; Fox and Hilton, 1958; Horton, 1969). It is conceivable that adenylate cyclase stimulation and cAMP synthesis mediate active vasodilation by these agents. This does appear to be the case in the prostaglandin-mediated bovine and canine venous relaxation (Dunham *et al.*, 1974). It is the loss of the vasodilator response, possibly due to the loss of adenylate cyclase sensitivity, which may underlie the increased sensitivity of the vessels from hypertensive animals to adrenergic stimulation (Tobia *et al.*, 1970).

Although a variety of hormonal factors have been shown to stimulate cGMP production in other systems (Goldberg *et al.*, 1973b) little is known about the *in vivo* activity or sensitivity of guanylate cyclase in vascular smooth muscles. *In vitro* guanylate cyclase sensitivity to hormonal stimulation in vascular smooth muscle and other tissues has been difficult to demonstrate as well. This may be because the enzyme exists in a loosely bound, partially inhibited state due in part to inhibitory concentrations of ATP normally present. Homogenization and/or dilution may result in an enzyme preparation becoming fully stimulated so that its activity cannot be increased further. It is known, however, that cGMP synthesis increases with α-adrenergic (which reflects sympathetic activity at the level of the vascular mooth muscle) and cholinergic stimulation (Goldberg *et al.*, 1973b), with angiotensin (Rosman *et al.*, 1974), and in bovine and canine venous strips with $pgF_{2\alpha}$ (Dunham *et al.*, 1974).

The sensitivity of guanylate cyclase to Ca^{2+} (Schultz and Hardman, 1975) emphasizes the possible importance of cGMP in vascular smooth muscle function in view of the importance of the cation in reactivity of this tissue. It should be realized that evidence is accumulating that guanylate and adenylate cyclases may represent components of the same enzyme system (Amer and Byrne, 1975; Hollenberg and Cuatrecasas, 1975). This restates the importance of the determination of interrelationships between the two cyclic nucleotides.

C Phosphodiesterases

As in other tissue (Amer and Kreighbaum, 1975), the cyclic nucleotide phosphodiesterases of vascular smooth muscles, both in man and experimental animals, appear to exist in at least two forms differing primarily in their affinity for the substrates cAMP and cGMP (Amer, 1973; Amer *et al.*, 1974b, 1975; Wells *et al.*, 1975; Hidaka *et al.*, 1975). These enzymes appear to be the site at which most direct vascular smooth muscle relaxants act (Amer and Kreighbaum, 1975; Lugnier *et al.*, 1972; Berti *et al.*, 1974; Pöch and Kukovetz, 1972; Aporti *et al.*, 1975). The two cyclic nucleotides appear to be hydrolysed by separate sets of enzymes which show different sensitivities to the phosphodiesterase activator and Ca^{2+} (Wells *et al.*, 1975; Hidaka *et al.*, 1975) and possibly also to phosphodiesterase inhibitors (Hess, 1974). Ca^{2+} appear to stimulate selectively cGMP hydrolysis.

D Relation of cAMP and cGMP to Vascular Smooth Muscle Activity

Numerous reports now suggest the possible role of cAMP in mediating vascular smooth muscle relaxation (Andersson, 1973a; Volicer and Hynie, 1971; Bär, 1974a; Seidel *et al.*, 1975). Few reports support the role of cGMP in mediating vascular smooth muscle contraction (Dunham *et al.*, 1974) although its importance in other smooth muscles is more firmly established (Takayanagi and Takagi, 1973; Lee *et al.*, 1972; Schultz *et al.*, 1973; Ham *et al.*, 1974; Puglisi *et al.*, 1972; Amer, 1974; Diamond and Hartle, 1974). The mechanisms by which the two cyclic nucleotides achieve their effects on contractility are largely unknown, but are probably similar to those involved in other effects, including protein phosphorylation, the control of Ca^{2+} availability (Demesy-Waeldele and Stoclet, 1975; Aoki *et al.*, 1974) and membrane effects.

III ABERRATIONS IN HYPERTENSION

In hypertension, the vascular smooth muscles are characterized by defects in cyclic nucleotide metabolism which may be important in mediating the increased peripheral resistance. This increased resistance is apparent in most chronic forms of the disease and seems to be responsible for maintaining the elevated blood pressure. Evidence for the cyclic nucleotide lesion includes the increased cGMP/cAMP ratio, increased activity of cAMP phosphodiesterase and decreased sensitivity of adenylate cyclase to stimulation by catecholamines in the vasculature, in addition to a variable lesion in the heart (Amer *et al.*, 1974b, 1975; Amer, 1973; Kuo and Davis, 1975).

A Decreased Adenylate Cyclase Sensitivity to Stimulation

Reduced stimulation by catecholamines is a typical feature of the hearts and blood vessels from sufferers of the chronic forms of hypertension due to genetic

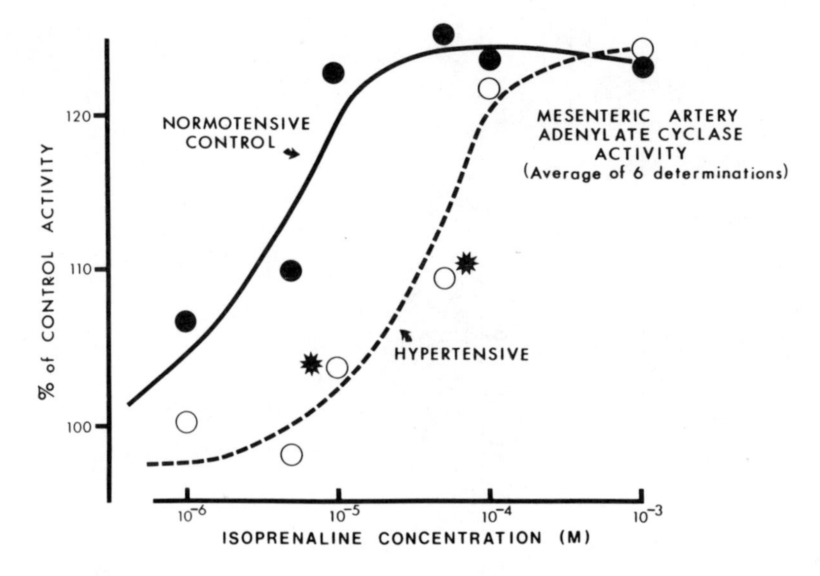

Figure 21.1 The effect of increasing isoprenaline concentrations on adenylate cyclase activity in rat mesenteric arteries from normotensive (•) and spontaneously hypertensive (o) rats. Values are averages of six determinations. $*p < 0.05$

factors, increased Na^+ intake or prolonged stressor an acute form of neurogenic hypertension (Tables 21.1 and 21.2). Higher concentrations of isoprenaline are needed to stimulate adenylate cyclase from the vessels of hypertensive animals to the same extent as those from normotensive animals (Figure 21.1). Using higher concentrations of isoprenaline, no differences have been observed in the adenylate cyclase response as in the study by Klenerova *et al.* (1975) in the heart of spontaneously hypertensive rats. The lesion appears to be limited to the coupling unit since the structural integrity of the receptor and catalytic subunits of the enzyme are apparently preserved. This is shown by the normal response of the adenylate cyclase to NaF and the normal binding of noradrenaline to the vascular smooth muscle receptors (Hubbard *et al.*, 1974). This decreased sensitivity of the adenylate cyclase to stimulation may underlie the decreased response of the hearts from hypertensive animals to isoprenaline (Frohlich, 1974) and their inability to synthesize cAMP (Largis *et al.*, 1973). The factors mediating the loss of adenylate cyclase sensitivity in hypertension are not known but may include increased Ca^{2+} influx and/or unbalanced prostaglandin synthesis.

The studies of Bohr, (1974) and others (Moore *et al.*, 1974; Aoki *et al.*, 1974; Jones and Phelan, 1974; Aoki *et al.*, 1975) have clearly demonstrated what appears to be an alteration in Ca^2 -binding and/or permeability in the vascular smooth muscles from hypertensive animals. Levels of Ca^{2+} are also known to be

Table 21.1 Basal and stimulated adenylate and basal guanylate cyclase activities in the aortas of control and hypertensive rats

		nucleotidate cyclase activity (μmole cyclic nucleotide formed/mg wet tissue min^{-1})			
		basal	adenylate cyclase isoprenaline (10^{-5} M)	NaF (8×10^{-3} M)	guanylate cyclase basal
spontaneous	control	8.33 ± 0.89	24.30 ± 3.94**	31.77 ± 8.52*	4.16 ± 0.73
	hypertensive	9.44 ± 0.58	7.75 ± 1.89	27.33 ± 4.59	8.87 ± 1.03Δ
stress	control	9.03 ± 2.86	19.42 ± 2.13*	23.71 ± 3.03**	5.01 ± 0.93
	hypertensive	11.42 ± 1.93	10.44 ± 1.42	15.44 ± 3.44	2.19 ± 1.01
11-deoxycorticosterone	control	36.20 ± 1.74	49.35 ± 2.51**	83.07 ± 3.81**	24.57 ± 3.13
acetate	hypertensive	55.28 ± 1.06ΔΔ	57.70 ± 10.18	90.61 ± 9.95*	41.99 ± 5.69Δ
neurogenic	control	9.95 ± 1.12	12.32 ± 1.23*	14.67 ± 2.23*	0.99 ± 0.24
	hypertensive	7.04 ± 0.52	7.00 ± 0.42	11.86 ± 1.12*	1.90 ± 0.25Δ

Values are mean of 4–10 experiments ± S.E.M.
Significantly different from respective basal: $^{*}p < 0.05$; $^{**}p < 0.01$.
Significantly different from respective control: $^{\Delta}p < 0.05$; $^{\Delta\Delta}p < 0.01$.

Table 21.2 Basal and stimulated adenylate and basal guanylate cyclase activities in the hearts of control and hypertensive rats

		nucleotidate cyclase activity (μmole cyclic nucleotide formed/mg wet tissue min^{-1})			
		adenylate cyclase basal	adenylate cyclase isoprenaline (10^{-5} M)	adenylate cyclase NaF (8×10^{-3} M)	guanylate cyclase basal
spontaneous	control	15.11 ± 1.19	18.64 ± 0.55*	19.95 ± 0.90*	12.86 ± 0.55
	hypertensive	18.38 ± 2.09	19.12 ± 2.35	24.44 ± 1.86**	10.79 ± 0.50
stress	control	19.73 ± 3.48	25.73 ± 1.64*	26.34 ± 1.73*	13.27 ± 2.18
	hypertensive	20.24 ± 2.69	18.36 ± 2.21	26.15 ± 1.62*	10.23 ± 3.21
DOCA	control	16.90 ± 2.28	20.92 ± 3.80*	24.93 ± 2.31*	7.28 ± 1.85
	hypertensive	21.04 ± 2.78	19.24 ± 1.91	31.03 ± 2.81*	10.44 ± 0.71
neurogenic	control	33.00 ± 0.47	38.55 ± 2.38*	42.47 ± 2.64*	4.75 ± 0.23
	hypertensive	29.79 ± 0.99	28.67 ± 3.08	35.78 ± 1.83*	5.63 ± 0.46

Values are mean of 4–10 experiments ± S.E.M.
Significantly different from respective basal; *$p < 0.05$; **$p < 0.01$.

high in hypertensive blood vessels (Furura *et al.*, 1974). Increased Ca^{2+} flux could underlie the decreased adenylate cyclase sensitivity, and the often increased guanylate cyclase activity, both characteristic of vessels from hypertensive animals.

Several workers (McGiff and Itskovitz, 1973; Aiken and Vane, 1972; Muirhead *et al.*, 1973; Scholkens and Steinbach, 1975) appear to have established a role for PGE_2, not only in the antihypertensive function of the kidney but also in competing with $PGF_{2\alpha}$ for the same vascular smooth-muscle receptors, thus exerting the opposite effects to those of PGE_2 on pulmonary vascular resistance (Kadowitz *et al.*, 1974) and in modulating autonomic transmission (Brody and Kadowitz, 1974). A decreased synthesis of PGE_2 could be important in mediating the decreased cAMP levels observed in the vessels from hypertensive animals. This might be particularly true if the reduced production of PGE_2 is coupled with elevated $PGF_{2\alpha}$ synthesis since the latter promotes the synthesis of cGMP and has prominent hypertensive effects (Ellis and Hutchins, 1974; Sweet *et al.*, 1971). An increased PGF_2/PGE_2 ratio would precipitate the elevated cGMP/cAMP ratio observed in the vessels of hypertensive animals. The observation that PGE_2 and arachidonic acid have strong antihypertensive activities supports that hypothesis. In addition, PGE_2 seems to control regional blood flow and its effects on lowering blood pressure in spontaneously hypertensive rats correlate well with blood pressure. This is consistent with the idea that exogenous PGE_2 may replace deficient endogenous PGE_2 synthesis which may at least partially contribute to the elevated pressure. The lowering of blood pressure by acetylcholine and isoprenaline, for example, does not exhibit this unique relationship.

The importance of the loss in adenylate cyclase hormonal sensitivity may lie in the increased responsiveness of both arteries and veins from hypertensive animals to the contractile effects of catecholamines. This is probably true since elevated cAMP levels are generally associated with vascular smooth muscle relaxation, whereas cGMP may mediate contraction. Thus, the insensitivity of adenylate cyclase to catecholamines would result in relatively more cGMP than cAMP being synthesized in response to either increased sympathetic tone or to circulating catecholamine levels. This imbalance in favour of cGMP synthesis could also occur when circulating catecholamines or sympathetic activity are normal. This appears to be the case in established hypertension and may directly contribute to the increased pressure. Elevated sympathetic tone or circulating catecholamine levels would further aggravate the unbalanced cyclic nucleotide ratio. It is not known at present whether this decreased adenylate cyclase-sensitivity is specific for the catecholamines alone or includes other natural vascular adenylate cyclase stimulants. If it proves to be a nonspecific loss of adenylate cyclase sensitivity, then non-neuronal factors, such as angiotensin-II and the prostaglandins, may also contribute to the imbalance. This is possibly true as in hypertension there is a general increase in sensitivity of the blood vessels from hypertensive animals not only to the catecholamines but also to angiotensin and prostaglandins.

B Increased Activity of Cyclic AMP Phosphodiesterase

In hypertension, the decreased sensitivity of adenylate cyclase to hormonal stimulation is coupled with increased activity associated with the low K_m (and possibly the more important *in vivo* form) of cAMP phosphodiesterase (Table 21.3). The use of high substrate concentrations may obliterate the differences in phosphodiesterase activity (Klenerova *et al.*, 1975). Increased cAMP phosphodiesterase activity appears to be responsible for the decreased basal vascular levels of cAMP and represents the most reproducible cyclic nucleotide-related observation in hypertension. The increased enzyme activity may be related to increased Ca^{2+} concentrations in the vessels from hypertensive animals. Although no studies have so far been carried out on the effects of Ca^{2+} on the phosphodiesterase system of vascular smooth muscle, this cation has been shown to stimulate the enzyme from brain and liver. It is conceivable that the increased Ca^{2+} activity in the vessels from hypertensive animals may result in the activation of cAMP phosphodiesterase in these tissues. The increased ability of hypertensive animals to hydrolyse cAMP may be related to their vessels' reduced ability to relax. However the greater sensitivity of the vessels from hypertensive animals to the relaxant effects of certain phosphodiesterase inhibitors (Stoclet *et al.*, 1971) may be directly associated with the greater dependency of the reduced cAMP levels in the vessels from hypertensive rats on the increased activity of phosphodiesterase.

C Increased cGMP/cAMP Ratios

It appears that loss of adenylate cyclase sensitivity to stimulation and the increased cAMP phosphodiesterase activity account for the increased vascular cGMP/cAMP ratio characteristic of the chronic forms of hypertension (Table 21.4). This increased ratio directly underlies the elevated basal vascular smooth muscle tone and the increased peripheral resistance in hypertension.

The distortion in cyclic nucleotide metabolism could also explain the greater efficacy of some antihypertensive drugs in lowering the blood pressure in hypertensives than in normotensives. As can be seen from Figure 21.2, the major variation in cAMP metabolism in the vessels from hypertensive animals may result from extra-myogenic contributions due to the relative insensitivity of adenyl cyclase to stimulation. Thus, removal of these extra-myogenic influences would be expected to result in cyclic nucleotide metabolism and blood pressure in the hypertensive animals tending towards more normal states. Effects in the normotensive animals would not be so marked. This may explain the greater effects of sympathetic neuronal blockade and anti-angiotensin-II on blood pressure in the hypertensive animals and man than in the normotensives. Removal of extra-myogenic factors, however, may not completely reverse the changes in the hypertensive vascular smooth muscle which may still exhibit a higher tone and a greater sensitivity to noradrenaline irrespective of vascular hypertrophy (Lais and Brody, 1975; Greenberg and Bohr, 1975).

Increased reactivity appears to be a property of the vascular smooth muscles themselves consistent with their changed intracellular cyclic nucleotide

Table 21.3 Phosphodiesterase activity in the aortas of control and hypertensive rats

	cAMP (nmole cyclic nucleotide hydrolysed/5 mg wet tissue 10 min^{-1} at 30°C)				cGMP			
	control		hypertensive		control		hypertensive	
	total activity	% low K_m[a]	total activity	% low K_m	total activity	% low K_m	total activity	% low K_m
spontaneous	6.0 ± 1.2[b]	4.6	10.3 ± 1.4*	12.9*	11.3 ± 2.1	5.4	12.1 ± 1.6	4.9
stress	4.4 ± 0.3	1.95	20.4 ± 2.2**	7.33*	9.7 ± 1.1	4.7	10.1 ± 1.3	5.1
11-deoxycorticosterone acetate	7.3 ± 0.8	2.0	6.4 ± 0.9	4.1**	12.7 ± 1.1	5.4	13.6 ± 1.3	4.7
neurogenic	11.5 ± 1.5	2.6	15.8 ± 4.5	3.8*	12.7 ± 1.1	5.4	13.6 ± 1.3	4.7

[a]Obtained as described previously (Amer, 1973).
Values are mean of 4–10 experiments ± S.E.M.
Significantly different from respective control; $*p < 0.05$; $**p < 0.01$.

Table 21.4 Cyclic nucleotide levels in the aortas of control and hypertensive rats

	cAMP (pmole/mg wet tissue)		cGMP (pmole/mg wet tissue)		
	control	hypertensive	control	hypertensive	index[a]
spontaneous	0.98 ± 0.23	0.45 ± 0.16*	0.07 ± 0.01	0.11 ± 0.02*	3.42†
stress	0.60 ± 0.07	0.36 ± 0.05**	0.08 ± 0.01	0.07 ± 0.01	1.65†
11-deoxycortico-sterone acetate	0.56 ± 0.05	0.55 ± 0.04	0.07 ± 0.01	0.10 ± 0.01*	1.45†
neurogenic	0.38 ± 0.05	0.18 ± 0.01*	0.09 ± 0.03	0.15 ± 0.03*	1.76†

[a] $\dfrac{\text{cAMP control}}{\text{cGMP control}} \Big/ \dfrac{\text{cAMP hypertensive}}{\text{cGMP hypertensive}}$

Values are mean of 3–12 experiments ± S.E.M.
Significantly different from respective controls; $*p < 0.05$; $**p < 0.01$.
Significantly different from 1; $\dagger p < 0.05$.

composition. In addition, hypertrophy of the vascular smooth muscles which is thought to be responsible for the increased wall/lumen ratio in the vessels from hypertensive animals, contributing to the increased peripheral resistance, may also be mediated by the increased cGMP/cAMP ratio. cGMP stimulates and cAMP inhibits cellular proliferation. Furthermore, the increased water and salt contents of hypertensive vessels (Johsson *et al.*, 1975; Nagoaka *et al.*, 1970; Tobian and Binion, 1952) may result from their reduced cAMP levels which normally inhibit the Na^+ pump. The increased protein synthesis in the vessels of hypertensive animals may also be mediated by the increased cGMP/cAMP ratio. cAMP and cGMP appear to promote antagonizing effects on protein

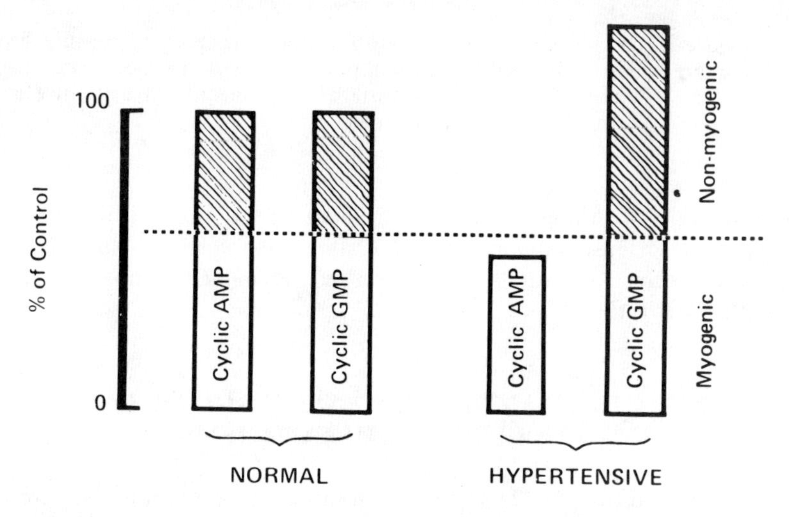

Figure 21.2 Cyclic nucleotide levels in vascular smooth muscles in hypertension

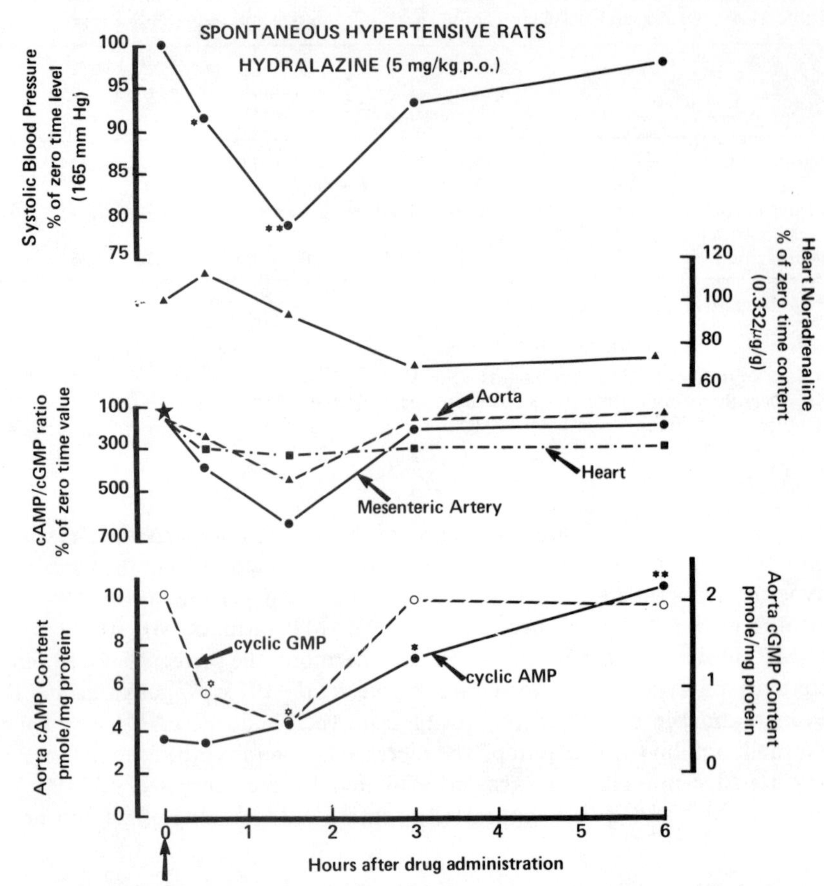

Figure 21.3 Time study of the effects of hydralazine on blood pressure, heart noradrenaline, cAMP/cGMP ratios in the aorta, heart and mesenteric artery, and cGMP and cAMP levels in aorta of spontaneously hypertensive rats. Average of 4–6 experiments $*p < 0.05$ $**p < 0.01$

synthesis. It is of interest to note that the structural changes observed in the vessels of hypertensive animals are reversible upon sustained lowering of the blood pressure, a condition invariably associated with a redressed cGMP/cAMP ratio.

IV EFFECTS OF ANTIHYPERTENSIVE AGENTS ON CYCLIC NUCLEOTIDE METABOLISM

If the elevated peripheral vascular resistance in hypertension is supported, at least in part, by an elevated vascular cGMP/cAMP ratio, drugs which lower the

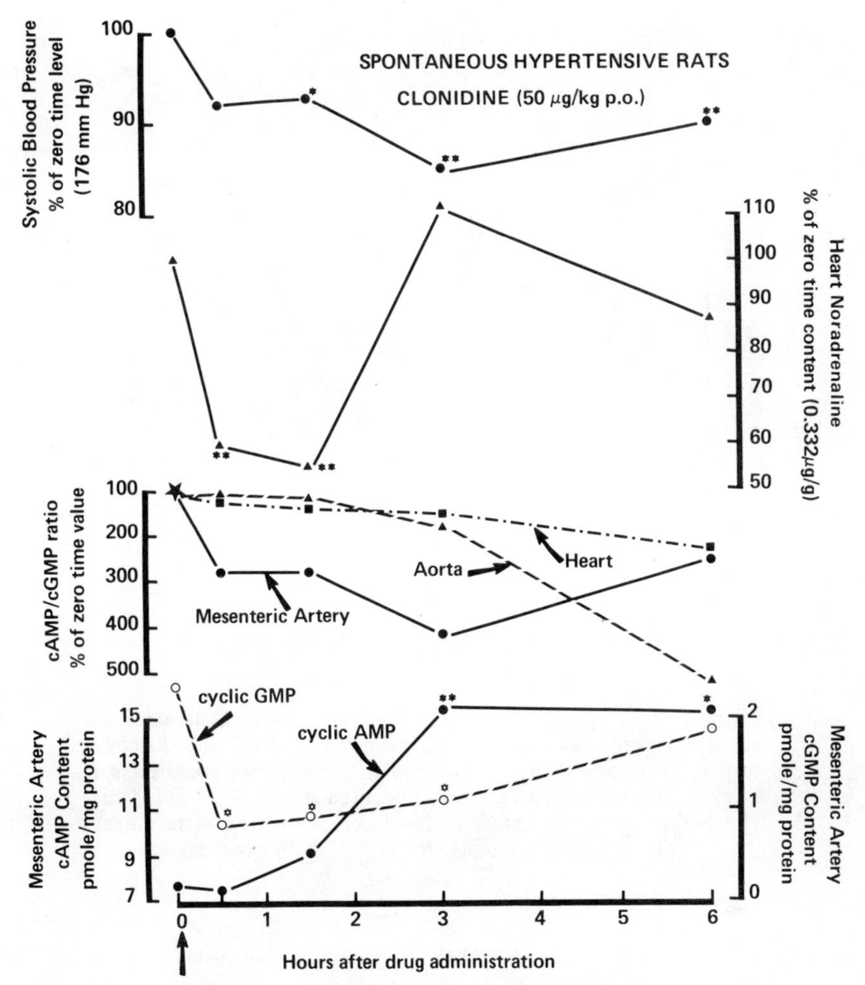

Figure 21.4 Time study of the effects of clonidine on blood pressure, heart noradrenaline, cAMP/cGMP ratios in heart, aorta and mesenteric artery, and cGMP and cAMP levels in the mesenteric artery of spontaneously hypertensive rats. Average of 4–6 experiments $*p < 0.05$ $**p > 0.01$

arterial blood pressure by lowering peripheral vascular resistance should revert the elevated ratio. This has been shown to be the case using a number of drugs acting in several models of hypertension in rats *in vivo*. Examples are shown in Figure 21.3 for a drug which acts by direct dilatation of the arterioles, and in Figure 21.4 for a drug which acts centrally to reduce sympathetic outflow. The effects are dose-dependent and are parallel to the effects on blood pressure as shown in Figure 21.5 for guanethidine. The latter drug appears to act peripherally by sympathetic neuronal blockade. The direct vasodilators appear to act directly by inhibiting vascular phosphodiesterases, with greater effect on

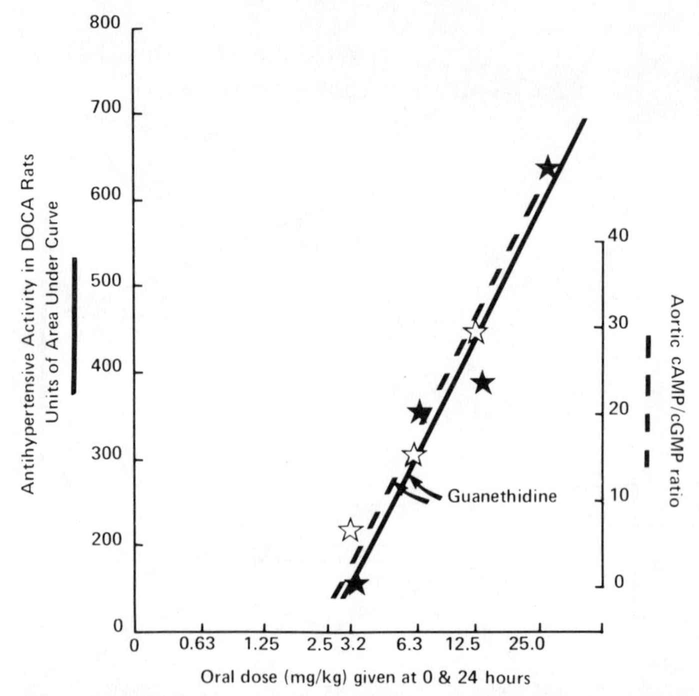

Figure 21.5 Dose–response curves for the effects of guanethidine on blood pressure and aortic cAMP/cGMP ratio of 11-deoxycorticosterone acetate hypertensive rats. The animals were injected at 0 and 24 hours and blood pressure measured at 0, 4, 24, 28 and 48 hrs. The animals were killed at 48 hrs for the determination of their aortic cyclic nucleotide contents. Average of 5–10 experiments

the cAMP enzymes than on the cGMP enzymes. As examples, the results with diazoxide and minoxidil are shown in Table 21.5.

V CYCLIC NUCLEOTIDE METABOLISM IN THE AETIOLOGY OF HYPERTENSION

The changes in both the synthesis and degradation of cyclic nucleotides observed in hypertension may be triggered by an early increase in sympathetic tone and the concomitant increased exposure of the vascular smooth muscle cells to high concentrations of catecholamines. This has clearly been shown to be the case in a neurogenically hypertensive rat model (Amer *et al.*, 1975) which is characterized by increased sympathetic tone. This model has been produced in rats by brain-stem lesions in the nucleus tractus soliterii resulting in central de-afferentation of baroreceptors. The acute loss of adenylate cyclase sensitivity,

Table 21.5 Effect of diazoxide and minoxidil on low K_m phosphodiesterase activity in the aortas of dog, rabbit and rat

	diazoxide			minoxidil		
	K_i (mM) substrate		K_i cGMP / K_i cAMP	K_i (mM) substrate		K_i cGMP / K_i cAMP
	cAMP	cGMP	ratio	cAMP	cGMP	ratio
dog	3.5	7.2	2.1	2.7	7.0	2.6
rabbit	1.7	5.7	3.4	1.9	6.1	3.2
rat	1.4	4.7	3.4	2.3	3.2	1.4

Enzyme assays were carried out on the unfractionated homogenate of fresh tissues obtained immediately after decapitation. The enzyme activity was determined by the method of Thompson and Appleman (Appleman *et al.*, 1973). K_i values were determined at 2 substrate (1 and 2×10^{-6} M) and 4–6 drug concentrations bracketing the K_i value.

which occurs two hours after the lesion in this model is due to catecholamine release and not secondary to the elevated pressure. Thus, the *in vivo* model confirms catecholamine-induced loss of adenylate cyclase sensitivity observed in several systems *in vitro* — see for example Hopkins, 1975; Mukherjee *et al.*, 1975b; R-O'Donnell, 1974.

Although the reversibility of the loss of adenylate cyclase sensitivity observed in these acutely hypertensive rats has not been reported, this acute loss may serve to indicate the type of mechanism which may, when prolonged, result in the permanent loss of enzyme sensitivity which occurs in the more chronic disease. It is conceivable that repeated and sustained increases in sympathetic nerve activity, perhaps driven by heightened states of emotion and stress, may lead to irreversible loss of adenylate cyclase sensitivity characteristic of chronic forms of hypertension. This phenomenon may represent the biochemical progression from the acute and reversible elevation of blood pressure found in the early stages of the disease to the permanent and sustained hypertension. This hypothesis is supported by the apparent significance of the central adrenergic neurones in the initiation of hypertension and their lesser importance in the long-term maintenance of elevated blood pressure. A similar situation may also exist in renal hypertension in which angiotensin-II may be important in the initiation but not in the maintenance of elevated blood pressure. The renin–angiotensin system may produce its effects on adenylate cyclase either directly by modulating cGMP levels in the vasculature (Rosman *et al.*, 1974), or indirectly *via* the sympathetic nervous system.

This hypothesis agrees well with the current concepts of the aetiology of essential hypertension in man. Essential hypertension is thought to be mediated by an early increase in sympathetic tone resulting in increased cardiac output followed by irreversible changes in the peripheral vasculature characterized by increased tone, sensitivity, ion movements and wall thickness. Preliminary studies from this laboratory provide strong support for this hypothesis. Studies indicate that in very young, spontaneously hypertensive rats variations in cyclic nucleotide levels precede the development of elevated blood pressure. Also, the experiments suggest that increased adenylate cyclase activity precedes the development of enzyme subsensitivity characteristic of more established hypertension in older animals. This also appears to be true for other types of hypertension. The well-accepted hypothesis for the development of hypertension in these animals, that is a temporary intermittent increased cardiac output followed by a more lasting increased peripheral resistance, is strongly supported by these studies. These observations are supported by the conclusion that changes in vascular reactivity may reflect an alteration in the smooth muscle and not a response to an increase in arterial pressure. The altered cyclic nucleotide metabolism which may be responsible for the increased peripheral vascular resistance, may also trigger the structural changes which physically support the elevated pressure.

In summary, it is suggested that sustained increases in sympathetic nerve activity or renin–angiotensin levels early in the development of hypertension may lead to the irreversible loss of adenylate cyclase sensitivity to stimulation coupled with increased cAMP phosphodiesterase activity. This leads to an increased cGMP/cAMP ratio which mediates the increased vascular smooth muscle tone and thickness. These events bring about elevated peripheral resistance which maintains the hypertensive state.

CHAPTER 22

Cyclic Nucleotides in Disorders of Carbohydrate and Lipid Metabolism

J. J. Keirns and E. L. Tolman

I INTRODUCTION

The study of the role of cAMP in carbohydrate metabolism began with the discovery of the cyclic nucleotide. The glycogenolytic and lipolytic cascades are the most extensively studied examples of cAMP-mediated hormone action. However, as studies of the involvement of cyclic nucleotides in diseases of carbohydrate and lipid metabolism have been much more limited, results obtained for normal processes are often extrapolated to the situation in the diseased state.

II GLYCOGEN METABOLISM

The cAMP cascade is an efficient mechanism for responding intermittently to glycogenolytic hormones. The synthesis of cAMP in the liver leads to phosphorylation of two proteins with opposite functions, namely glycogen synthetase, which polymerizes glucose in the form of uridine diphosphate-glucose (UDP-glucose) to glycogen, and glycogen phosphorylase, which dismantles the polymer (Robison *et al.*, 1971; Soderling and Park, 1974). This concomitant phosphorylation of proteins and opposing function produces opposite effects on their activities. Synthetase is inhibited, while phosphorylase is activated, thus providing a co-ordinated dual regulation of glycogen

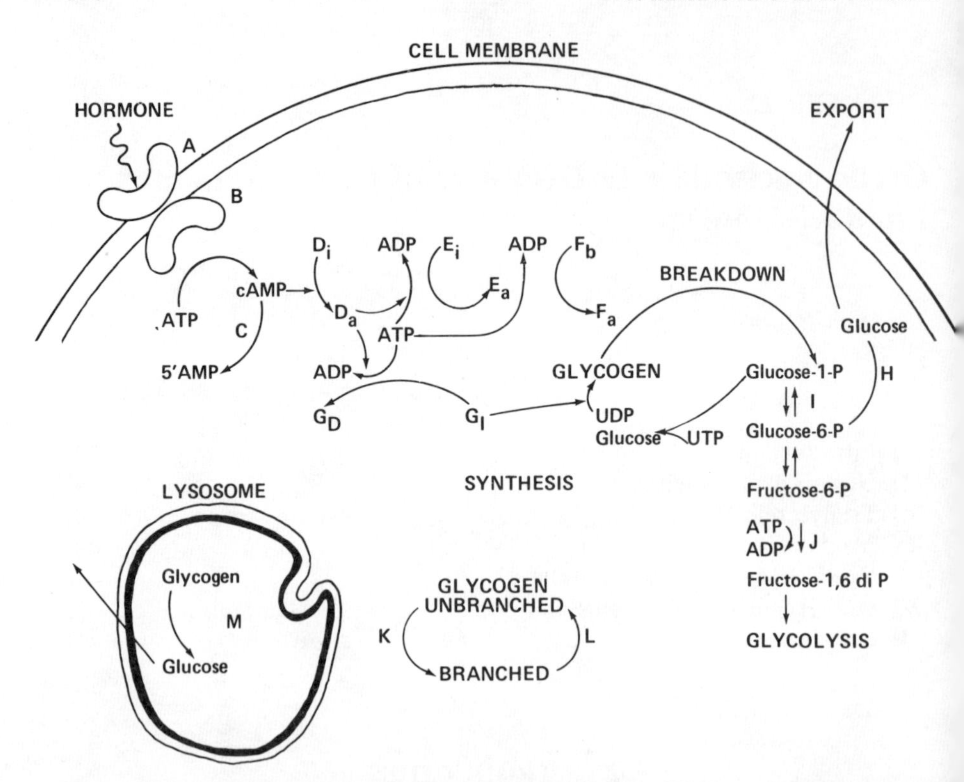

Figure 22.1 Pathways for metabolism of glycogen in liver. The hormone (glucagon or adrenaline) binds to the outward-facing hormonal receptor (A) causing activation of the inward-facing adenylate cyclase (B), which converts ATP to cAMP. cAMP is hydrolysed by phosphodiesterase (C), or is bound to protein kinase converting it from the inactive (D_i) to the active form (D_a). D_a catalyses the phosphorylation of glycogen synthetase, converting it from the active, independent form (G_I) to the inactive glucose-6-phosphate-dependent (G_D) form. D_a can also catalyse the phosphorylation of phosphorylase *b* kinase converting it from an inactive (E_i) to active form (E_a). This enzyme then phosphorylates inactive phosphorylase *b* (F_b) converting it to the active F_a. The phosphorylation reactions can be reversed by phosphoprotein phosphatases. Phosphorylase *a* catalyses the phosphorolysis of glycogen to glucose-6-phosphate by phosphoglucomutase (I). When dephosphorylated by glucose-6-phosphatase (H), glucose can pass into the blood. Alternatively, glucose-6-phosphate enters glycoselysis after conversion to fructose-6-phosphate and fructose-1,6-diphosphate. The latter reation is catalysed by phosphofructokinase (J). Glycogen synthetase and phosphorylase deal with the α-1,4-linkages in glycogen. Synthesis of α-1,6-linkages is accomplished by the branching enzyme (K) and removal of these linkages by the debranching enzymes (L). Following engulfment of glycogen by a lysosome, it is hydrolysed by α-1,4 glucosidase (M)

mobilization. The same seven amino-acid sequence in each enzyme, which contains the serine hydroxyl group, is phosphorylated (Villar-Palasi *et al.*, 1971). Thus, operation of the same 'control panel' in two different enzymes produces opposite effects on their activities.

In Figure 22.1 the pathways of liver glycogen metabolism are divided into five segments: (*i*) the enzymes of the cAMP cascade, which regulate the synthesis and degradation of glycogen in responses to hormones; (*ii*) enzymes involved in producing and removing the branching (α-1,6) linkages; (*iii*) the enzymes of glycolysis; (*iv*) an enzyme which facilitates the transport of glucose from the cell; and (*v*) enzymes involved in the hydrolysis of glycogen in lysosomes. The metabolism of glycogen in heart and skeletal muscle is similar, but there are some exceptions. For example, in skeletal muscle, glucagon does not activate adenylate cyclase and glucose-6-phosphatase is absent from muscle, so there is little transport of glucose from the cell. There is evidence that during muscular contraction phosphorylase *b* kinase may be activated by an increase in the cytoplasmic Ca^{2+} concentration, bypassing the cAMP cascade (Brostrom *et al.*, 1971b).

Glycogen-Storage Diseases

Defects at many different loci in the pathways shown in Figure 22.1 can impair the mobilization of glycogen, thus producing a glycogen-storage disease. In addition to occurring at different enzyme loci, the defect may occur only in one tissue, since some of the enzymes have distinct forms found only in liver or muscle. The different forms are under separate genetic control. In type V glycogenosis there is a defect in muscle phosphorylase, while the liver enzyme is spared. In contrast, in type VI the defect is in liver and not muscle. The different types of glycogen-storage diseases which have been described clinically are listed in Table 22.1, and will be discussed in the following paragraphs.

In von Gierke's disease (type I glycogenosis), glucose-6-phosphatase is absent

Table 22.1 Classification of glycogen-storage diseases

type	disease name	tissue affected	dificient enzyme	reference
I	von Gierke's	liver	glucose-6-phosphatase	Cori and Cori, 1952
II	Pompe's	liver, heart, skeletal muscle	α-1,4-glucosidase (lysosomal)	Hers, 1963
III	Cori's	liver, heart, skeletal muscle	amylo-1,6-glucosidase and/or oligo-1,4–1,4 glucantransferase (debranching enzymes)	Hers, 1960
IV	Andersen's	liver	α-1,4-glucan-6-glucosyl transferase (branching enzyme)	Illingworth and Cori, 1952
V	McArdle's	skeletal muscle	phosphorylase	Mommaerts *et al.*, 1959
Va	–	skeletal muscle	phosphoglucomutase	Satoyoshi and Kowa, 1967
Vb	–	skeletal muscle	phosphofructokinase	Tarui *et al.*, 1965
VI	Hers's	skeletal liver	phosphorylase	Hug and Schubert, 1970
VIa	–	skeletal liver	phosphorylase *b* kinase	Hug *et al.*, 1969
VIb	–	skeletal liver	cAMP-dependent protein kinase	Hug *et al.*, 1970

from liver (Cori and Cori, 1952). This enzyme is not directly involved in the metabolism of glycogen, and it is not regulated by the glycogenolytic hormones. The enzyme is, however, essential if glucose is to be released into the blood for use by other organs, rather than being utilized in the same cell. Normally the phosphatase is absent from muscle and the muscle glycogen stores are used exclusively for intracellular needs. The glycogen accumulation in the liver of patients with von Gierke's disease probably results from mass-action effects on the equilibrium of the phosphorylase reaction, and on the formation of UDP-glucose as well as from stimulation of the dependent form of glycogen synthetase by glucose-6-phosphate. The greatly reduced secretion of glucose produces severe hypoglycaemia, which inhibits the secretion of insulin and stimulates the secretion of adrenaline. Consequently, muscle glycogen and adipose triglycerides are mobilized, and protein synthesis is reduced.

Pompe's disease (type II glycogenosis) is a severe disorder, involving the absence of a lysosomal enzyme. In this disease massive amounts of normal glycogen accumulate in muscle, heart, liver and other organs. The excess glycogen is surrounded by a distended lysosomal enzyme. However, the activities of glycogen-mobilizing enzymes in the cytoplasm are normal (Garancis, 1968). The missing enzyme, α-1,4-glucosidase (Hers, 1963), has optimal activity at pH 4 and sediments with lysosomes. Pompe's disease is prototypical of lysosomal-storage disease. Apparently cytoplasm is constantly ingested by lysosomes. Unless the lysosome is equipped with enzymes to degrade all polymers present in the cytoplasm, thus duplicating any degradative pathways in the cytoplasm, a storage disease results.

Cori's disease (type III) and Andersen's disease (type IV) both involve defects in the enzymes which either remove or form the branches in the glycogen molecule. In type III glycogenosis, the hydrolysis can proceed from the end of a branch only to within a few residues of a branch point resulting in an abnormal, highly branched glycogen (Hers, 1960). Accumulation of this abnormal glycogen results in a fairly mild clinical disease. In type IV glycogenosis there are moderate amounts of largely unbranched glycogen (Illingworth and Cori, 1952). This glycogen is relatively insoluble and provokes a massive fibrotic response and leads to fatal cirrhosis.

Diseases clinically similar to type V glycogenosis (see below) are produced by deficiencies of phosphoglucomutase (Satoyoshi and Kowa, 1967) phosphofructokinase (Tarui *et al.*, 1965) or possibly other enzymes in the glycolytic pathway (Larsson, 1964). In these diseases the accumulation of glycogen in muscle is produced by the mass-action effects mentioned for liver glycogen in von Gierke's disease.

Glycogenolytic hormones bring about glycogenolysis in each of the diseases mentioned above and, with the exception of von Gierke's disease, also produce an elevation in blood glucose levels, since the sequence of reactions from the hormone receptor through phosphorylase is intact. In types V, VI, VIa, and VIb, in which an element of the cAMP-regulatory apparatus is defective, the glycogenolytic response is low or absent. Before discussing these diseases the possible loci for a cAMP-related malfunction should be considered (see Figure 22.1). These include quantitative or qualitative abnormalities of the hormone receptor, the catalytic moiety of adenylate cyclase, phosphodiesterase, cAMP-dependent protein kinase, phosphorylase, glycogen synthetase and phosphoprotein phosphatase. Several of these enzymes consist of many subunits, which are subject to subtle allosteric regulation. Thus, a genetic defect may produce either low or excessively high activity, depending on which subunit is altered or which site within a single subunit is altered. Examples of the results of trauma on regulatory sites are provided by the irreversible conversion from inactive to active forms by limited proteolytic digestion in the case of glycogen synthetase I to D (Belocopitow *et al.*, 1967), phosphorylase *b* to *a* (Huston and Krebs, 1968) and phosphodiesterase (Miki *et al.*, 1975). A lesion in any one of these loci can cause a defect in glycogen metabolism. In the case of phosphodiesterase, glycogen synthetase or phosphoprotein phosphatase, excessively high activity is required to produce a storage disease. Abnormally low activity of one of these three enzymes results in excessive mobilization of glycogen, and leads to insulin-resistant diabetes.

Type V glycogenosis (McArdle's disease) involves a low activity of phosphorylase (Mommaerts *et al.*, 1959). The disorder is limited exclusively to skeletal muscle, while there seems to be no abnormality of glycogen metabolism in myocardium, liver or other tissues. The clinical features of this inborn error of metabolism usually occur during the second decade. An episode of muscular exertion produces muscle pain and myoglobinurea. Prior to and subsequent to the acute episodes the patient is clinically normal, since the principal source of energy for resting muscle if fatty acids. However, during exercise normal muscle obtains extra energy from glycogenolysis and anerobic glycolysis. In the patient with McArdle's disease, this pathway is blocked and thus, muscular exertion cannot be sustained.

Type VI glycogenosis (Hers's disease) is caused by a deficiency in liver phosphorylase (Hug *et al.*, 1969). Symptoms of the disease, which manifests in childhood, include abdominal swelling, mild lethargy and some growth retardation. Liver biopsies reveal increased amounts of normal glycogen. Patients often show a small response to glycogenolytic stimuli. The patients classified as type VI are heterogeneous since it is common to assign inadequately characterized cases of liver glycogenosis to this category. However, the patients collected by Hers showed a reduced activity (about 25% of normal) of glycogen phosphorylase which was often apparent in leucocytes as well as in liver.

Type VIa glycogenosis which is due to the absence of hepatic and sometimes leucocytic phosphorylase kinase, clinically resembles type VI (Hug *et al.*, 1969). In these patients, phosphorylase kinase activity is only about 10% of normal. A single case of glycogenosis due to absence of cAMP-stimulated protein kinase has been described. This patient was also clinically similar to patients with Type VI glycogenosis. It seems likely that in the future further patients with defects in adenylate cyclase or other elements of the cascade and diseases similar to Hers's or McArdle's will be discovered.

III TRIGLYCERIDE METABOLISM

Evidence for the involvement of cAMP in the process of lipolysis in adipose tissue is well documented and has been the subject of extensive reviews (Robison *et al.*, 1971). A series of observations, initially reported in the early 1960s, led to the unequivocal conclusion that cAMP is a major intermediate between the hormones which stimulate lipolysis and the enzymes which eventually catalyse the hydrolysis of fat-cell triglycerides, leading to the release of free fatty acids and glycerol.

Several structurally distinct hormones cause the accumulation of cAMP in adipose tissue. These hormones, including the catecholamines, glucagon, adrenocorticotropin (ACTH) and thyroid-stimulating hormone (TSH) stimulate the release of free fatty acids from rat epididymal fat pads. Sutherland and his co-workers (1962) demonstrated that adipose tissue has catecholamine-sensitive adenylate cyclase activity. Vaughan and Steinberg (1963) have shown that inhibitors of cAMP phosphodiesterase act synergistically with catecholamines to stimulate lipolysis. A more direct relationship between the accumulation of cAMP in fat tissue and lipolysis has been demonstrated by Butcher *et al.*, (1965). When fat pads are incubated with increasing concentrations of adrenaline, there is a progressive rise in both the amount of cAMP within the tissue, and the amount of free fatty acids released from the tissue. The accumulation of cAMP always precedes the stimulation of lipolysis. These observations have been extended to include the other lipolytic hormones and isolated fat cells and fat-cell plasma membranes as well as intact fat pads. Exogenous cAMP, in the form of its dibutyryl derivative (Bt_2-cAMP), stimulates the release of free fatty acids by incubated fat pads and isolated fat cells. Finally, it has been shown that the antilipolytic agents, insulin and pgE_1 act at least in part by preventing the accumulation of cAMP in fat tissue incubated with lipolytic hormones. The mechanisms of the inhibitory effects of both insulin and PGE_1 have been the subject of controversy. The mode of action of insulin will be discussed below.

The very potent anti-lipolytic activity of PGE_1 was reported by Steinberg *et al.* (1963). Paradoxically, PGE_1 blocks the increase in cAMP in the fat pads produced by adrenaline, but by itself PGE_1 elevates cAMP levels in this tissue (Butcher and Baird, 1968). This apparent discrepancy is resolved when the fat cells are separated from the capillaries and other stromal cells. PGE_1 inhibits the lipolytic activities of the catecholamines, glucagon and other hormones by impairing their ability to raise the concentration of cAMP in the fat cell. On the other hand, this prostaglandin elevates cAMP levels in capillary endothelial cells (Wagner *et al.*, 1972), fibroblasts and most other types of cells. The depression of cAMP in fat cells and the elevation in most other cells appears to reflect inhibitory (Butcher, 1970) or stimulatory (Kreiner *et al.*, 1973b) effects of PGE_1 on adenylate cyclase.

The diverse chemical nature of the hormones, which cause the accumulation of cAMP in fat cells and stimulate free fatty acid release, present an interesting

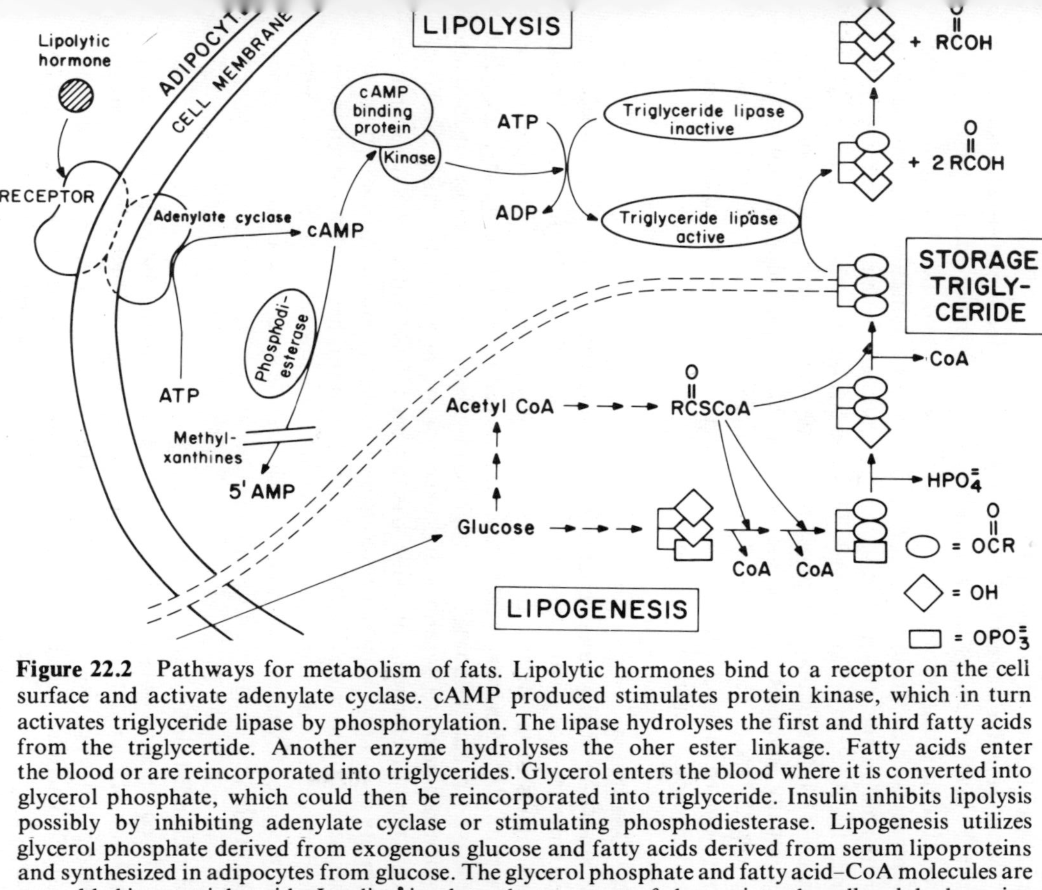

Figure 22.2 Pathways for metabolism of fats. Lipolytic hormones bind to a receptor on the cell surface and activate adenylate cyclase. cAMP produced stimulates protein kinase, which in turn activates triglyceride lipase by phosphorylation. The lipase hydrolyses the first and third fatty acids from the triglycertide. Another enzyme hydrolyses the oher ester linkage. Fatty acids enter the blood or are reincorporated into triglycerides. Glycerol enters the blood where it is converted into glycerol phosphate, which could then be reincorporated into triglyceride. Insulin inhibits lipolysis possibly by inhibiting adenylate cyclase or stimulating phosphodiesterase. Lipogenesis utilizes glycerol phosphate derived from exogenous glucose and fatty acids derived from serum lipoproteins and synthesized in adipocytes from glucose. The glycerol phosphate and fatty acid–CoA molecules are assembled into a triglyceride. Insulin stimulates the transport of glucose into the cell and, by lowering cAMP, stimulates conversion of glucose to acetyl CoA and then to fatty acyl–CoA. (Reprinted from Keirns *et al.* (1974) with permission of the publisher.)

situation. However, the work of Bär and Hechter (1969a), and Rodbell *et al.* (1970) have clarified the point. Their data indicate the presence of a single type of adenylate cyclase within the fat-cell plasma membranes. The membranes however, possess a specific receptor for each hormone. They suggested the existence of a 'transducer' which can carry the message from the receptor to the catalytic subunit of adenylate cyclase.

More recent observations have brought into focus the mechanism of the cAMP involvement. Kuo and Greengard (1969) reported the presence of a cAMP-dependent protein kinase in rat epididymal fat tissue. It has been demonstrated by Corbin *et al.* (1970) that lipolysis is stimulated in adipose tissue incubated in the presence of a partially purified protein kinase (from rabbit muscle), ATP, Mg^{2+} and cAMP. These results indicate that the primary action of cAMP is to stimulate the phosphorylation of an enzyme, or series of enzymes, eventually leading to the activation (probably by phosphorylation) of the triglyceride lipase. The cascade of events, triggered by the binding of a lipolytic hormone to its fat-cell receptor and leading to the net release of free fatty acids and glycerol is outlined in Figure 22.2.

A Obesity

Since cAMP plays analogous roles in glycogen and triglyceride metabolism, it is tempting to postulate defects in the fat-cell cAMP–lipolysis cascade, analogous to the defects of the cAMP–glycogen cascade discussed above (see Section III). Presumably defects in the fat-cell hormonal receptors, adenylate cyclase, protein kinase or triglyceride lipase would cause inadequate mobilization of triglycerides and excessive storage, thus resulting in obesity. This type of biochemical lesion has not been described as a cause of human obesity and is probably not responsible for obesity in the vast majority of such cases. A mobilization defect of this type, however, has been described in obese hyperglycaemic mice in which a glycerol kinase present in the fat cells produces immediate resynthesis of mobilized triglyceride before the component glycerol and fatty acids can reach the blood (Lochaya *et al.*, 1963).

B Lipodistrophy

In contrast to obesity, a defect in the phosphodiesterase or in the regulatory subunit of adenylate cyclase, protein kinase or lipase which results in excessive unregulative activity would produce excessive lipolysis and depleted triglyceride stores. There are rare clinical conditions known as generalized and partial lipodistrophy in which there is a complete lack of stored triglyceride in all or part of the body, respectively. Lipodistrophy may be congenital, but more often follows a febrile illness (Seip, 1971). It is possible that in partial lipodistrophy a neurological defect produces excessive sympathetic stimulation of the adipocyte adenylate cyclase. In generalized lipodistrophy, there has been considerable interest in lipolytic and/or diabetogenic humoral factors (Louis, 1969; Upton *et al.*, 1973), but the possibility of an intrinsic defect in the adipocyte also exists.

IV MODE OF ACTION OF INSULIN

At the level of the whole animal, and also at the level of the perfused organ or isolated cell, insulin is an antagonist of the glycogenolytic and lipolytic hormones. Insulin promotes the synthesis of storage molecules, such as glycogen and triglycerides, while glucagon and adrenaline promote the breakdown of these polymers. Since cAMP has been shown to mediate the glycogenolytic effects of glucagon and adrenaline, and the lipolytic effects of catecholamines and various peptide hormones, it is natural to ask whether insulin antagonizes the effects of these hormones on cAMP metabolism in liver, muscle and adipose tissue. Although this question has been extensively studied during the past fifteen years, no clear answer has emerged. Perhaps it might be more accurate to say that too many answers have emerged. It has been proposed that the primary effect of insulin is one or more of the following: (*i*) stimulation of protein synthesis, especially stimulation of the transcription of certain enzymes; (*ii*) stimulation of cell-membrane transport systems for glucose, amino acids or ions; (*iii*) inhibition of adenylate cyclase; (*iv*) stimulation of cAMP phosphodiesterase; (*v*) elevation of cGMP levels, either by stimulation of guanylate cyclase or by inhibition of cGMP phosphodiesterase; (*vi*) inhibition of protein kinase; or (*vii*) stimulation of phosphoprotein phosphatase. Each of these possibilities will now be considered in detail.

A Protein Synthesis

Application of insulin to intact cells has been shown to increase total protein synthesis, when measured as the net incorporation of amino acids into protein, and also to regulate the activities of a number of enzymes (Krahl, 1972). Some of these studies have been carried out using whole animals. In these cases it is difficult to rule out the possibility that an effect of insulin is mediated by changes in the concentrations of other hormones, especially steroids. However in tissue-cultured cells where this ambiguity does not occur, insulin has been shown to induce the synthesis of hexokinase and tyrosine aminotransferase. Insulin is believed to exert its effects at the level of translation rather than transcription. By this mechanism insulin may increase or decrease the synthesis of an enzyme of the cAMP cascades. Thus, effects of insulin on cAMP may be secondary to effects on protein synthesis. On the other hand, cAMP-stimulated protein kinases are able to stimulate phosphorylation of nuclear proteins (Kish and Kleinsmith, 1974) or ribosomal proteins (Walton and Gill, 1973). By such a mechanism, effects of insulin on protein synthesis may be secondary to effects on cAMP metabolism.

B Transport of Glucose

Insulin stimulates the transport of glucose into fat and muscle cells (Martin and Carter, 1970), but apparently not into liver. The hormone, in particular, stimulates the stereospecific facilitated diffusion of hexoses. In this system, D-glucose and D-galactose, but not L-glucose, are transported across the cell membrane, from a region of high concentration to a one of a lower concentration by a mechanism which is saturable. It has been suggested that binding of insulin to its receptor on the cell membrane may directly stimulate the transport system in much the same way as binding of glucagon directly stimulates adenylate cyclase. However, since insulin is applied to the whole cell, it is possible that a cyclic nucleotide-stimulated phosphorylation of a protein involved in the transport process produces persistant stimulation, or inhibition, of transport. Chang *et al.* (1974) have reported evidence for such a mechanism. They found that a cAMP-stimulated phosphorylation of two membrane proteins is associated with a reversal of the stimulation of hexose transport into fat cells by insulin.

C Transport of Amino Acids

Insulin also stimulates the transport of amino acids into muscle and other tissues (Goldfine *et al.*, 1972a). Like the hexose transport system, the amino acid system is stereospecific and saturable. Unlike the hexoses, amino acids are transported against a concentration gradient, that is from a low concentration to a high one. The driving force for this transport is provided by the 'downhill' cotransport of Na^+. The amino acid uptake does not reflect any intracellular utilization of the amino acid, since the effect is seen with a nonmetabolized synthetic amino acid, α-aminoisobutyric acid. Also, stimulation of amino acid uptake does not require protein synthesis; transport is enhanced by the presence of protein synthesis inhibitors (Fain and Rosenberg, 1972). Finally, this effect of insulin does not appear to be mediated by a decrease in the concentration of cAMP. cAMP or agents which stimulate its synthesis actually mimic the action of insulin in stimulating amino acid uptake.

D Transport of Potassium

Adrenaline, or glucagon, stimulate and insulin inhibit the uptake of K^+ by liver, fat and muscle (Zierler, 1972). It is conceivable that the K^+ taken up or released is dissolved in the water of hydration of glycogen. In this case, K^+ uptake or release may exactly reflect the deposition or mobilization of glycogen. However, this hypothesis is not tenable, since under conditions where the effect of insulin on glycogen metabolism has been completely antagonized by exogenous cAMP, insulin still inhibits the efflux of K^+. Zierler (1972) has suggested that insulin decreases the permeability of the cell membrane to Na^+, thus causing an increase in the membrane potential and the consequent uptake of K^+. As with effects of insulin on hexose and amino acid uptake, the question exists of whether the binding of insulin to its receptor directly affects Na^+ permeability, or whether insulin acts *via* some other process, such as phosphorylation or dephosphorylation.

E Effects of Insulin on Cyclic Nucleotide Metabolism

The remaining proposed mechanisms all suggest that effects of insulin are due to, or are similar to, effects of reducing cAMP levels. The cGMP mechanism can also fall in this category if an increase in the concentration of cGMP is considered to be equivalent to a decrease in the concentration of cAMP (Goldberg *et al.*, 1975). In some of the earlier studies of the effects of insulin on cAMP metabolism, it was found that the hormone reduces the concentration of cAMP in fat cells. (Butcher *et al.*, 1966) while, at the same time, inhibiting lipolysis. This correlation of decreased cAMP levels with the inhibition of lipolysis has been disputed (Fain and Rosenberg, 1972). In this regard, the effects of adenosine are very interesting. Adenosine reduces the concentration of cAMP in fat cells to a much greater extent than insulin does. By itself adenosine has little effect on lipolysis, but it greatly potentiates the inhibitory effect of insulin (Schwabe *et al.*, 1974).

Insulin also decreases the concentration of cAMP in liver. The hormone is more effective in blocking the elevation of cAMP produced by catecholamines than in reducing basal cAMP levels (Park *et al.*, 1972). A lack of insulin, produced either by experimental diabetes or by perfusion of an anti-insulin antiserum, causes a rise in the concentration of cAMP. A reduction in the concentration of cAMP might reflect either reduced synthesis or enhanced degradation. The latter could be due to inhibition of adenylate cyclase or stimulation of phosphodiesterase. Evidence for both of these possibilities has been reported.

F Inhibition of Adenylate Cyclase Activity

Bitensky *et al.* (1972) studied the effects of various hormonal manipulations carried out on the whole animal on the activity of liver adenylate cyclase. If experimental diabetes is induced in a rat by administering the islet poison alloxan or streptozotocin, adrenaline-stimulated adenylate cyclase activity is increased approximately two-fold, while basal and glucagon-stimulated activity remains unchanged. Treatment of the diabetic animals with insulin reduces the adrenaline-stimulated activity to half the value in untreated animals. Bitensky *et al.* (1972) also found that treatment of normal, but not diabetic, animals with glucagon, testosterone or a glucocorticoid reduces adrenaline-stimulated activity by about 50%. These effects of the other hormones as well as the effects of age, sex and adrenalectomy on adrenaline-stimulated adenylate cyclase activity are attributed to secretion of insulin secondary to the other hormonal manipulations. It should be noted that this effect of insulin is only apparent after several days and thus is probably a secondary adaptive phenomenon.

There have been several reports that insulin when applied directly to a membrane suspension inhibits adenylate cyclase. Ray *et al.* (1970) stated that a very high concentration of insulin inhibits both basal and glucagon-stimulated adenylate cyclase activity of liver membranes. Illiano and Cuatrecasas (1972), and Hepp (1972) reported similar inhibitory effects of insulin, but at more physiological concentrations. In contrast, Pilkis and Park (1974), Thompson *et al.* (1973) and Pohl *et al.* (1971b) reported that insulin has no effect on basal or glucagon-stimulated adenylate cyclase activity in liver membranes. Renner *et al.*

(1974) showed that insulin inhibits ACTH, glucagon and catecholamine-stimulated adenylate cyclase in fat cell ghosts, but Fain and Rosenberg (1972) found much smaller effects which they concluded are not significant. In all of these studies the maximal inhibitory effects of insulin were rather small (25 – 50%). To conclude, the question of whether direct inhibition of adenylate cyclase by insulin has any physiological significance or whether there is even a genuine effect is still disputed.

G Effects of Insulin on cAMP Phosphodiesterase Activity

The alternative mechanism for reduction of cAMP levels by insulin has also been extensively studied. Senft *et al.* (1968) reported that cAMP phosphodiesterase activity is reduced in liver, skeletal muscle and fat cells of diabetic rats, and can be restored to normal, or even stimulated above normal by injection of insulin. The effect of insulin on fat cells has been examined in more detail by Loten and Sneyd (1970), Manganiello and Vaughan (1973), and Zinman and Hollenberg (1974). In each of these studies intact fat cells or fat pads were incubated with insulin for a few minutes. Then the cells were lysed and cAMP phosphodiesterase activity examined. Insulin was found to stimulate phosphodiesterase activity by as much as 65%. The stimulatory effect is primarily on the low K_m component of the phosphodiesterase activity, and after cell fractionation is associated with the membrane-bound activity rather than with the soluble phosphodiesterase. This component of the activity may represent a specific enzyme which is bound to the plasma membrane and interacts with an insulin receptor. Fain and Rosenberg (1972) did not demonstrate any effect of insulin on soluble phosphodiesterase from fat cells. House *et al.* (1972), and Thompson *et al.* (1973) also found that insulin stimulates cAMP phosphodiesterase in liver. Thompson *et al.* (1973) reported that the hormonal effect is only apparent when the insulin is administered to the whole animal. Also the effect is specific for a low K_m particulate component of the activity as was found for fat cells.

Except for Senft *et al.* (1968) and House *et al.* (1972), who reported effects when insulin was added directly to the homogenate, all the above workers found that insulin can stimulate phosphodiesterase only if the hormone is added to an whole cell preparation before homogenization and assay. In addition, these studies, with exception of the *in vivo* experiments of Senft *et al.* (1968) and the work of Zinman and Hollenberg (1974), involved very high concentrations of insulin. Finally, other hormones, such as glucocorticoids (Manganiello and Vaughan, 1973), isoprenaline (Zinman and Hollenberg, 1974), and growth hormone (Thompson *et al.*, 1973) which often antagonize insulin, stimulate cAMP phosphodiesterase in a manner similar to that of insulin. Thus, the phenomenon of stimulation of cAMP phosphodiesterase by insulin is more generally recognized than the effect on adenylate cyclase, but several questions remain. Why is insulin only effective when applied to an intact cell? Does this requirement simply reflect the requirement for an intracellular component, such as ATP? Finally, does stimulation of phosphodiesterase by insulin represent a major primary mechanism by which insulin regulates cellular metabolism, or is it a minor modulatory process which is secondary to some other effect of insulin?

H Effect of Insulin on cGMP Metabolism

The possibility that a nucleotide other than cAMP may serve as a second messenger for insulin has been suggested a number of times. The first evidence of this possibility was provided by Illiano *et al.* (1973) who reported that insulin causes a striking increase in the cGMP concentration in liver and fat cells. An insulin-induced increase in the cGMP content of intact cells has also been observed by DeAsua *et al.* (1975). However, Fain *et al.* (1975) and Larner *et al.* (1974) suggested that insulin influences glycogen metabolism in muscle while having no effect on cGMP levels.

As with the effect of insulin on cAMP levels, a change in the concentration of cGMP may reflect regulation of either the synthetic or the degradative enzyme, that is either stimulation of guanylate cyclase or inhibition of cGMP phosphodiesterase. Thompson *et al.* (1973) reported that insulin has no effect on cGMP phosphodiesterase, even when the insulin is administered to the whole animal. As with other cases in which hormones or other agents elevate cGMP levels in intact cells (George *et al.*, 1970; Hadden *et al.*, 1972), it has not been possible to demonstrate any effect on guanylate cyclase in a cell-free system. The elevation of cGMP by insulin is seen only at rather high concentrations of the hormone. It is possible that the effects of insulin on cGMP metabolism are related more to the mitogenic effects of high insulin concentrations than to the metabolic effects of lower concentrations. In any case, a possible link between the large increase in cGMP and the smaller decrease in cAMP is provided by the stimulatory effect of cGMP on cAMP phosphodiesterase (Klotz and Stock, 1972; Sakai *et al.*, 1974). It is possible that this mechanism accounts for the stimulatory effect of insulin on cAMP phosphodiesterase which has been discussed previously.

I Effects of Insulin on Protein Kinases and Phosphatases

A number of studies has suggested that insulin can inhibit lipolysis or stimulate glycogen synthesis without any effect on the concentration of cAMP. One possible mechanism by which insulin might have such effects is by regulating the activity of another protein kinase which is not stimulated by cAMP. Huang *et al.* (1975), and Schlender and Reiman (1975) have presented evidence that glycogen synthetase is regulated by a cAMP-independent as well as a cAMP-dependent protein kinase. They have not, however, shown any effects of insulin on the activity of this enzyme.

Work by Miller and Larner (1973), and by Walaas *et al.* (1973) has shown that insulin, when administered to a whole animal or to a perfused organ, inhibits protein kinase in liver and muscle by converting the kinase from an active, cAMP-independent form to an inactive, cAMP-dependent form. This is the type of change which would result from an insulin-induced decline in the concentration of cAMP. However, Walaas *et al.* (1973) , and Miller and Larner

(1973) reported no measurable decline in cAMP. Walsh and Ashby (1973), and Larner *et al.* (1974) have shown that insulin treatment produces a heat-stable inhibitor of cAMP-stimulated protein kinase in heart or skeletal muscle. Donnelly *et al.* (1973) provided evidence that a heat-stable inhibitor of the cAMP-dependent protein kinase can also stimulate the cGMP-dependent enzyme. It was reported that the activity of this modulator is decreased in adipose tissue from alloxan-diabetic rats, presumably as a result of the absence of insulin (Kuo, 1975).

Bishop (1970), and Gannon and Nuttall (1975) have reported that treatment of a pancreatectomized dog with insulin increases the activity of glycogen synthetase D phosphatase in liver, while glucagon treatment decreases the activity of this phosphatase. Thus, several studies relate effects of insulin on protein kinase or phosphatase activities to its effects on glycogen synthetase. Whether all these studies consider different aspects of the same mechanism and the exact nature of the mechanism remain to be clarified.

Another effect of insulin on protein phosphorylation has been demonstrated by Benjamin and Singer (1975), who found that insulin stimulates the phosphorylation of a fat-cell protein with a molecular weight of 140,000. Adrenaline has no effect on this phosphorylation, so the reaction is probably not mediated by a change in the concentration of cAMP. Possible involvement of this protein in glucose transport or triglyceride metabolism has not yet been demonstrated.

V cAMP IN DIABETES MELLITUS

The prominent role which cAMP plays in the control of intermediary metabolism — see Sutherland *et al.* (1968, 1970) — suggests that abnormalities either in the tissue levels or in the functioning of this nucleotide could lead to serious disturbances in the maintainance of normal metabolism.

Homeostasis of blood glucose, the major energy-supplying substrate, is very heavily dependent on the proper balance of circulating hormones, most of which have direct effects on cAMP metabolism. On the one hand, the catecholamines and glucagon are called forth at times of high energy requirement, during fasting or during stressful situations. These hormones, together with glucocorticoids, raise blood glucose levels by stimulating the synthesis of cAMP in several tissues. In adipose tissue, cAMP stimulates the mobilization of stored triglycerides leading to the release of free fatty acids and glycerol into the circulation. These free fatty acids can serve directly as substrates for energy metabolism in tissues, such as muscle, or can be utilized by other tissues after conversion into ketone bodies by the liver. Glycerol is a substrate for the production of new glucose (gluconeogenesis) in liver. The stimulation of cAMP accumulation in liver, especially by glucagon, results in the activation of glycogenolytic and gluconeogenic enzymes, leading to the release of glucose into the circulation. Proteolysis occurring in skeletal muscle, for example, during long-term starvation, provides amino acids, such as alanine, which are important precursors in gluconeogenesis.

On the other hand, in times of abundance or less immediate need, insulin is elicited to facilitate the removal of glucose from the blood and to promote conversion of glucose to triglycerides and glycogen which are then stored. At least some of the effects of insulin have been implicated with diminished tissue levels of cAMP. These effects include decreased rates of gluconeogenesis and glycogenolysis, and increased rates of glycogen synthesis. In adipose tissue, insulin exerts a potent anti-lipolytic effect and promotes the esterification and storage of fatty acids in the form of triglycerides.

It is evident from the above discussion that maintenance of a proper energy balance depends on the plasma concentrations of certain hormones which produce their opposing effects on intermediary metabolism by affecting tissue cAMP metabolism. The physiological antagonism between two of these hormones, namely glucagon and insulin, is well documented — see for example Mackrell and Sokal (1969). These hormones are now of great interest since an imbalance in their circulating levels has been suggested in the aetiology of diabetes mellitus. Historically, diabetes has been defined as a disease of relative or absolute insulin deficiency. More recent work has suggested that it is also a disease of relative or absolute glucagon overproduction (Unger and Orci, 1975). The net result of excessive glucagon, or insufficient insulin, or both is hyperglycaemia and hyperlipidaemia, possibly leading to the debilitating sequelae of the disease.

While both glucagon and insulin affect tissue cAMP levels, it is of interest to note that cAMP is thought to play a prominent role in the release of insulin. This data is more thoroughly reviewed elsewhere in this volume — see Chapter 8. It is worth noting, however, that it has been suggested that diminished levels of cAMP in the pancreatic β-cell may be the cause of reduction in insulin release in diabetes mellitus. The mechanism of this primary defect is unclear, but it could result from disfunction of the adenylate cyclase associated with the β-cell plasma membrane.

However, though abnormalities in β-cell cAMP metabolism may be an aetiological factor in diabetes, Sutherland and Robison (1969) state '. . . it now seems clear that many of the metabolic derangements associated with this disease are the result of excessive production of cAMP in adipose and hepatic tissue secondary to insulin deficiency.' This conclusion was reached on the basis of both the effects of cAMP on intermediary metabolism discussed above and the results with animal models of diabetes.

It has been shown that livers of rats deficient in circulating insulin, either due to treatment of the animal with alloxan or due to neutralization of plasma insulin by injection of anti-insulin serum, exhibit a net accumulation of cAMP in liver (Jefferson *et al.*, 1968). Treatment of the alloxan-diabetic animals with insulin restores the hepatic cAMP levels to normal. Das (1974), on the other hand, has reported that anti-insulin serum has no effect on blood cAMP levels in rats whose blood sugar levels are at the same time increased three-fold compared with normal animals. Nonetheless, the urinary excretion of cAMP, which is a reflection of its production by several tissues, is higher in diabetics prior to insulin treatment (Tucci *et al.*, 1973). The intravenous administration of glucagon has been shown to cause a rapid increase in the urinary excretion of cAMP and to increase hepatic cAMP levels in both normal and diabetic individuals (Broadus *et al.*, 1970; Taylor *et al.*, 1970).

The data outlined above can be summarized as follows. Hormones which affect energy-substrate availability do so mainly through effects on cAMP levels. This cyclic nucleotide plays an important role in the regulation of normal intermediary metabolism. Excessive levels of cAMP can produce the major disturbances of diabetes, while insufficient levels of cAMP may be responsible for certain types of hypoglycaemia.

VI ORAL HYPOGLYCAEMIC AGENTS AND CYCLIC NUCLEOTIDE METABOLISM

Of the two major classes of oral hypoglycaemic agent, the biguanides and the sulphonylureas, only the latter appears to act by affecting cAMP metabolism. The mechanism of action of the biguanides, such as phenformin, is unclear. Theories have included effects on peripheral glucose uptake, anaerobic metabolism and hepatic production of glucose, or the inhibition of intestinal glucose absorption. No clear evidence exists that effects on cAMP metabolism or actions are involved in the mechanism of phenformin action.

The sulphonylureas, such as tolbutamide, chlorpropamide and glibenclamide, interact with the cAMP-metabolic pathway at a number of sites in several tissues, including the β-cell of the pancreatic islets of Langerhans, the source of insulin. It is accepted that the primary mechanism of action of the sulphonylureas is the stimulation of insulin release from the pancreas. Evidence for this includes the following observations: (*i*) these agents exert no hypoglycaemic effect in pancreatectomized animals, (*ii*) they cause degranulation of the β-cells which is associated with loss of measurable insulin, (*iii*) they increase plasma insulin levels in normal subjects and in maturity onset diabetics, and (*iv*) they directly stimulate insulin release from incubated preparations of pancreas (Mayhew *et al.*, 1969).

Elevated cAMP levels within the pancreatic β-cell are believed to mediate insulin secretion. In isolated rat pancreatic islet cells, for instance, tolbutamide at concentrations as low as 0.1 mM elevates cAMP levels significantly, possibly through the activation of adenylate cyclase (Kuo *et al.*, 1973). Stimulation of the activity of the cyclase by tolbutamide has been observed in mouse islet cells (Bowen and Lazarus, 1974), in a rat pancreatic islet-cell adenoma (Levey *et al.*, 1972) and using subcellular particulate fractions of rabbit and human heart (Levey *et al.*, 1971). However a discrepancy was detected in the mouse islet system in that a twenty-five-fold higher concentration of tolbutamide is needed to stimulate adenylate cyclase activity than is necessary for the initiation of insulin release (Bowen and Lazarus, 1974). In islet cells isolated from the guinea-pig pancreas, glibenclamide has no affect on either adenylate cyclase or guanylate cyclase activity (Howell and Montague, 1974).

In contrast to stimulating adenylate cyclase activity sulphonylurea agents inhibit the activity of cAMP phosphodiesterase, the enzyme responsible for the biological inactivation of the cyclic nucleotide. Aminophylline and theophylline, both potent inhibitors of phosphodiesterase, do not potentiate the stimulation by tolbutamide of insulin secretion (Widstrom and Cerasi, 1973). Such a potentiation would be expected if the sulphonylurea raises cAMP levels by stimulating the adenylate cyclase rather than by inhibiting phosphodiesterase activity. Tolbutamide and glibenclamide inhibit phosphodiesterase activity in islet cells from a gold hamster insulin-secreting tumour (Goldfine *et al.*, 1971) and in normal mouse pancreas (Ashcroft *et al.*, 1972). However, in the latter case tissue levels of tolbutamide as high as 5 mM inhibit the enzyme by 50%, while levels equivalent to those obtained during therapy (0.3 – 1.0 mM) inhibit the activity by only 2 – 10%.

Effects of sulphonylureas on phosphodiesterase activity have been also observed in subcellular preparations of rat lung, kidney, liver and brain, and in human platelets. The doses of tolbutamide and chlorpropamide which effectively inhibit rat kidney phosphodiesterase range from 0.5 to 20 mM (Brooker and Fichman, 1971). It has been suggested that this effect of the sulphonylureas on kidney phosphodiesterase activity explains the vasopressin-potentiating effects of the agents, since the effect of vasopressin on water reabsorption by the collecting tubule is mediated through the activation of adenylate cyclase and resulting elevation of cAMP levels.

While the sulphonylureas act primarily by stimulating insulin release from the pancreas, these agents affect other tissues by interactions with cAMP metabolism. A pertinent example, which also relates the activities of oral hypoglycaemic agents to effects on lipid metabolism, is the inhibition by sulphonylureas of fatty acid mobilization in adipose tissue. In both normal subjects and maturity-onset diabetics, these sulphonylureas decrease the plasma levels of free fatty acids and glycerol. While this action may be attributable in part to the antilipolytic effects of increased plasma insulin, sulphonylureas also directly inhibit lipolysis in isolated adipocytes (Herrera, 1975). Since increased cAMP levels in fat tissue mediate lipolysis, it would be expected that the inhibition by the sulphonylureas should be accompanied by diminished fat-cell cAMP concentrations. This is true for the antilipolytic action of insulin itself. However, 1 mM tolbutamide and 0.1 mM glibenclamide double the basal cAMP levels and potentiate the effects of isoprenaline on cAMP concentrations in isolated rat fat cells (Ebert *et al.*, 1974). As in other tissues, tolbutamide and glibenclamide inhibit phosphodiesterase activity in fat cells (Ebert *et al.*, 1974). The mechanism of the antilipolytic actions of the sulphonylurea agents may be by the uncoupling of oxidative phosphorylation, thereby limiting the availability of ATP for cAMP formation and for the formation of glycerol phosphate (through glycolysis) prior to re-esterification with the fatty acids released during lipolysis. Alternatively, these agents may have effects on intracellular Ca^{2+} concentrations or antagonize the action of cAMP either by inhibiting the binding of cAMP to protein kinase, or by directly inhibiting the lipase.

VII CONCLUSIONS

In this chapter the central role occupied by cAMP, and possibly cGMP, in the regulation of intermediary metabolism by hormones has been emphasized. The interrelated pathways of lipid, carbohydrate and protein metabolism are governed mainly by the relative tissue levels of cAMP and cGMP. Disturbances in these levels result in the disordered metabolism observed in such pathological states as glycogen-storage diseases, diabetes mellitus, obesity and lipodistrophies. Disfunction of cyclic nucleotide metabolism may be either the primary cause of a disease, for example decreased insulin release in diabetes, or the cause of the metabolic derangements associated with the disease, such as the increased gluconeogenesis and lipolysis in diabetes.

CHAPTER 23

Cyclic Nucleotides in Proliferative and Immunological Diseases

J. J. Keirns, J. E. Birnbaum and J. B. Moore, Jr

I INTRODUCTION

A number of topics, which might be considered to be part of the broad subject of cyclic nucleotides in proliferative and immunological disease, has been discussed in considerable detail in other chapters of this volume. Schönhöfer and Peters (Chapter 7) have discussed the role of cyclic nucleotides in cultured cells, especially fibroblasts. Duell and Voorhees (Chapter 20) have considered cyclic nucleotides in the regulation of epidermal cell growth and the treatment of psoriasis. Parker (Chapter 10) has reviewed the role of cyclic nucleotides in the activation of lymphocytes and Ignarro (Chapter 11) has discussed the regulation of phagocytic cell function by cAMP and cGMP. In this chapter an attempt will be made to emphasize aspects which have not been discussed in detail elsewhere. Various aspects of this subject have been treated in the published proceedings of recent meetings (Braun *et al.*, 1974; Brent and Holborow, 1974; Rosenthal, 1975).

II CYCLIC NUCLEOTIDES IN MALIGNANCY AND GROWTH

Investigations in this area have been performed on a plethora of tissues and cell types under diverse experimental conditions. The reader is referred to a number of recent reviews on the subject—see for example Chlapowski *et al.*, 1975; Pastan *et al.*, 1975.

Certain generalizations have emerged from these studies suggesting that in many, but certainly not all, systems cAMP is an inhibitory and cGMP a stimulatory regulator of cell growth and proliferation. In normal growing cells, an inverse relationship between intracellular cAMP levels and proliferation rate has been noted both for cultured cells (Otten *et al.*, 1972; Seifert and Paul, 1972) and for tissues (Marks and Grimm, 1972; Duell and Voorhees, this volume, Chapter 20). cAMP levels rise at confluency in many normal cells, but not in transformed cells which do not cease to grow (Otten *et al.*, 1972; Carchman *et al.*, 1974). Exogenous cAMP derivatives, adenylate cyclase stimulators and phosphodiesterase inhibitors can all block the cell cycle (Rosengurt and Pardee, 1972; Remington and Klevecz, 1973). This may occur at a single site or at multiple sites in the cell cycle. The G_1 and G_2 stages have been implicated as the loci for elevated intracellular cAMP levels. On the other hand, the lowest levels of cAMP are found in mitotic cells (Sheppard and Prescott, 1972; Zeilig *et al.*, 1972). The latter finding may explain the decreased cAMP content observed in rapidly growing cell lines where more cells are undergoing mitosis at a given time.

Agents which stimulate DNA synthesis and cell division, such as serum, trypsin and insulin, produce a fall in the levels of cAMP (Otten *et al.*, 1972). Bt_2-cAMP can prevent the action of some of these agents (DeAsua *et al.*, 1973; Kram *et al.*, 1973). Conversely, serum deprivation of cultured fibroblasts is accompanied by an increase in the levels of cAMP (Seifert and Paul, 1972). This does not appear to be only due to restricted growth, since other nonspecific inhibitors of growth, including disrupters of microtubule assembly, and DNA and RNA-synthesis inhibitors, do not produce such increases (Sheppard and Prasad, 1973).

In malignant cells, or cells transformed by viral or chemical agents, or spontaneously transformed, some or all of the following changes have been seen: (*i*) decreased cAMP or failure of cAMP levels to increase at elevated cell density; (*ii*) loss of contact and density inhibition; (*iii*) loss of the serum requirement for growth; (*iv*) loss of organization; (*v*) morphological alterations; (*vi*) increased agglutination by plant lectins; (*vii*) increased transport of uridine, amino acids and glucose; (*viii*) decreased adhesion; (*ix*) decreased sensitivity to pH changes; and (*x*) tumerogenicity (Pastan *et al.*, 1974, 1975). Some of these changes have also been observed in cAMP-deficient *Escherichia coli* mutants (Pastan and Perlman, 1970). Various alterations, which may be quite specific in nature, have been observed in both adenylate cyclase and phosphodiesterase (Anderson *et al.*, 1973d; Kelley *et al.*, 1974). For example, chick embryo fibroblasts have been transformed with both Bryan high titre strain and the Schmidt–Ruppin strain of Rous sarcoma virus. In the former, the K_m for ATP and the dependence of adenylate cyclase activity on Mg^{2+} concentration are altered whereas in the latter case, the V_{max} of the enzyme is decreased without any change in the K_m for ATP (Pastan *et al.*, 1974). Since phosphodiesterase is an inducible enzyme whose activity in both normal and transformed cells is regulated by changes in intracellular cAMP (D'Armiento *et al.*, 1972), the changes in the activity of this enzyme may be secondary to changes in adenylate cyclase activity.

Some of the transformation characteristics described above can be reversed by exogenous cAMP or Bt_2-AMP (Hsie *et al.*, 1971; Johnson and Pastan, 1972a). Moreover, the effects of cAMP and agents which elevate cAMP may be more pronounced in transformed cells than in normal ones. For example, isoprenaline, Bt_2-cAMP and cAMP greatly inhibit the synthesis of DNA in phytohaemagglutinin-stimulated lymphocytes from patients with chronic lymphocytic leukaemia at concentrations which are without effect on normal lymphocytes (Johnson and Abell, 1970).

Information obtained so far on the role of cGMP in the proliferative process is more limited. Induction of lymphocyte proliferation by concanavalin A or phytohaemagglutinin, and stimulation of fibroblast DNA synthesis and mitosis by serum, trypsin or fibroblast growth factor have been shown to be associated with elevations in cGMP (Hadden *et al.*, 1972; Rudland *et al.*, 1974b). In normal 3T3 fibroblasts, cGMP levels increase when G_0-arrested cells are stimulated to grow or when synchronized growing cells pass through the G_1 stage (Rudland *et al.*, 1974c). cGMP and its dibutyryl analogue can induce DNA synthesis when added to quiescent cells. At least for this type of cultured cells cGMP and cAMP appear to act as positive and negative signals, respectively, in regulating growth. Evidence has been presented that a similar relationship exists in human epidermis and that imbalances in the levels of these nucleotides are associated with the hyperproliferation associated with psoriasis (Duell and Voorhees, this volume).

In contrast to the information outlined above, an increasing number of studies indicates that it may be erroneous to ascribe a generalized role to cAMP and cGMP as inhibitors and stimulators, respectively, of cell growth. Mitogenic activity for cAMP has been observed in thymic lymphocytes (Rixon *et al.*, 1970), peripheral blood lymphocytes (Cross and Ord, 1971) and kidney, liver and haemopoetic cells (Bryant and Sutcliffe, 1974). In attempting to incorporate such disparate findings within a unifying concept of growth control, Berridge (1975a) has suggested that Ca^{2+} is actually the critical intracellular signal. In his model, cAMP acts by *modulating*—either by bringing about an increase or a decrease—the intracellular level of Ca^{2+}. The increases in cGMP associated with cell division in several cell types are considered in this model to be secondary to an increase in intracellular Ca^{2+}.

While results from cell-culture studies indicate that decreases in cAMP levels are characteristic of the malignant transformation of cells, evidence is lacking for a cAMP deficiency within tumour tissue. The levels of this nucleotide, in fact, have been shown to be elevated in a number of animal tumours (Chayoth *et al.*, 1973; Thomas *et al.*, 1973) and in human breast cancer (Minton *et al.*, 1974). This may be a result of enhanced activity of adenylate cyclase as demonstrated for a Morris hepatoma line (Brown *et al.*, 1970), for ethionine-induced hepatocarcinomas (Chayoth *et al.*, 1970), for dimethylamine-biphenyl-induced breast carcinoma (Brown *et al.*, 1969), and for premalignant livers of rats treated with 2-acetyl-aminofluorene (Christofferson *et al.*, 1972) or with 3-methyl-4-dimethylaminobenzene (Boyd *et al.*, 1974). Quantitative and qualitative alterations in the hormonal stimulation of adenylate cyclase have been observed in malignant tissue (Goldfine *et al.*, 1972a; Schorr *et al.*, 1971), as well as possible changes in the subcellular distribution of the enzyme (Tomasi *et al.*, 1973). Decreased activity of cAMP phosphodiesterase, as reported for liver (Chayoth *et al.*, 1973) or adrenal tumours (Goldfine *et al.*, 1972a) might also provide an explanation for the observed elevations in cAMP. In contrast to these findings are reports of unaltered (Butcher *et al.*, 1972) or decreased (Hickie *et al.*, 1974) cAMP levels in several Morris hepatoma lines.

This information need not be regarded as evidence against a role for cAMP in the inhibition of growth of normal tissue; abnormal cells may be defective at a site distal to the generation of cAMP. For example, cAMP does not inhibit cholesterol and fatty acid synthesis in the Morris hepatoma (Bricker and Levey, 1972) as it does in normal liver. This may relate to the absence of cAMP-binding protein which prevents a normal response occurring (Granner, 1972). High levels of cAMP are also known to inhibit the activity of normal rat-liver protein kinase (Ryan and Heidrick, 1974). Moreover, exogenously administered cAMP or its analogues have been found to inhibit the growth of a number of animal tumours (Long *et al.*, 1972; Cho-Chung and Gullino, 1974b). These observations and the presence of different cell types in malignant tissue preclude any generalization on the role of cAMP in the induction of tumours.

The possible involvement of cyclic nucleotides other than cAMP and cGMP in malignant tissue has recently come to light. Cyclic cytosine 3′,5′-monophosphate (cCMP) has been isolated from leukaemia L-1210 cells and from the urine of patients with acute leukaemia, but not from normal subjects (Block, 1975). cCMP is also capable of initiating proliferation of leukaemic L-1210 cells grown to stationary phase and is antagonistic to the growth inhibition by cAMP in these cells. Cyclic uridine 3′,5′ monophosphate (cUMP) has also been found in the urine of leukaemic patients and like cAMP and cGMP, it delays the onset of growth of L-1210 cells (Block, 1975).

III GENETIC ANALYSIS OF S49 LYMPHOMA CELLS

A very interesting series of experiments with the S49 line of cultured lymphosarcoma cells has provided the beginnings of a genetic dissection of the cAMP cascade. The wild-type S49 cells cease to grow and die in the presence of cortisol,

Bt_2-cAMP or isoprenaline. This property of the cells provides a means of isolating mutant lines which have defects in either the system recognizing and responding to steroid hormones, or in the cAMP cascade. The mutant cells are resistant to the cytotoxic effects of either cortisol or cAMP, and thus can be readily separated from the sensitive wild-type cells. In this way, Baxter *et al.* (1971) isolated a number of mutant strains of S49 resistant to cortisol. Most of these mutants lacked the ability to specifically bind [^{3}H]cortisol and are presumably deficient in the glucocorticoid receptor.

Mutants with defects in the cAMP cascade (cA^R) were selected by growing the cells either in the presence of Bt_2-cAMP or isoprenaline. A number of distinct types of mutants was obtained (see Table 23.1). Mutants selected with isoprenaline (Bourne *et al.*, 1975c) and designated I^R have a very low, if any, basal adenylate cyclase activity, which is not stimulated by isoprenaline, PGE_1 or cholera toxin, all of which stimulate the enzyme in wild-type S49 cells. The mutant cells remain sensitive to the cytotoxic effects of Bt_2-cAMP, since the portion of the cascade distal to cAMP is still intact.

Most mutants selected for resistance to Bt_2-cAMP (Coffino *et at.*, 1975a) have defects in the cAMP-binding protein and protein kinase. Mutants designated cA^R have a complete absence of these functions (Daniel *et al.*, 1973a). Other mutants have partial defects so that the affinity (K_m) for cAMP is reduced or,

Table 23.1 S49 Lymphoma mutants for cAMP metabolism

	I^R	cA^R	cA^R (partial 1)	cA^R (partial 2)	cA–G
functional defect	cAMP synthesis	cAMP action	cAMP action	cAMP action	growth inhibition by cAMP
deficient enzyme	adenylate cyclase	cAMP protein kinase (absent)	protein kinase (low affinity for cAMP)	protein kinase (partially absent)	?
stimulation of cAMP synthesis					
isoprenaline	–	+	+	+	+
prostaglandin E_1	–	+	+	+	+
cholera toxin	–	+	+	+	+
induction of phospho-diesterase					
isoprenaline	–	–			+
Bt_2-cAMP	+	–			+
growth inhibition					
isoprenaline	–	–			slight
Bt_2-cAMP	+	–	slight	slight	slight

(Adapted from Bourne *et al.*, 1975b, with the permission of the publisher).

alternatively, so that the affinity is normal but the maximum response (V_{max}) to cAMP is diminished (Insel *et al.*, 1975a).

The cA^R cells are resistant to the cytotoxic effects of isoprenaline and cholera toxin, as well as to Bt_2-cAMP. Another type of mutant designated cA–G is partially resistant to the growth inhibitory effect of Bt_2-cAMP, but has normal cAMP-binding and protein kinase activities. These cells presumably have a defect at a point in the cascade distal to the protein kinase (Bourne *et al.*, 1975b).

During study of the adenylate cyclase activity of the cA^R mutants (Daniel *et al.*, 1973b; Bourne *et al.*, 1973), it has been noted that these cells have higher basal cAMP levels than the wild-type cells and that the increase in cAMP in response to hormones is also much greater. The adenylate cyclase of the cA^R cells is normal, but the cAMP phosphodiesterase activity is greatly reduced. In wild-type S49 cells (and also in the cA–G mutants), elevation of intracellular cAMP produced by isoprenaline PGE_1 or Bt_2-cAMP induces synthesis of phosphodiesterase. In I^R mutants, Bt_2-cAMP induces phosphodiesterase, while the usual stimulators of adenylate cyclase fail to do so. In cA^R mutants, there is no induction of phosphodiesterase by any agent. This induction of the phosphodiesterase by elevated cAMP levels has previously been demonstrated by D'Armiento *et al.* (1972) in fibroblasts, but the studies with the S49 lymphoma clearly show that cAMP-stimulated protein kinase is an essential intermediate in the induction of phosphodiesterase.

These S49 lymphoma studies also provided some insights into the role of cAMP as a mitotic regulator. Using a flow microfluorometer, which can determine the DNA content of individual cells, Coffino *et al.* (1975b) estimated the fraction of cells at different stages of the cell cycle. Cells in G_1 phase have the haploid amount of DNA and those in the G_2 or M phase have the diploid amount, while cells in the S phase have an intermediate amount. Coffino *et al.* (1975b) found the various mutant lines had the same cell-cycle parameters as the wild type. In the presence of Bt_2-cAMP, the sensitive wild-type cells are arrested in G_1. The cA^R mutants progress normally even in the presence of high levels of cAMP, thus providing evidence that regulation of cAMP is not required for normal progression through the cell cycle.

Another interesting point raised by these studies is that genetic regulation of the cAMP cascade appears to occur in complexes rather than in isolated functional units. The I^R mutant is deficient not only in isoprenaline-stimulated adenylate cyclase activity, but also in pgE_1 and cholera toxin-stimulated cyclase activities (Bourne *et al.*, 1975a). The cA^R mutant is deficient in both cAMP-binding and protein kinase activities, although the cAMP-binding site and the protein kinase catalytic activity reside on different polypeptide chains (Daniel *et al.*, 1973a). Finally, as phosphodiesterase is induced and repressed the low K_m and high K_m activities change in parallel (Daniel *et al.*, 1973b). It might be expected that selective hormone-receptor or binding-protein mutants would be found. One possible explanation for the failure to find this type of selective mutation is that synthesis of the hormone receptor–cyclase complex and the binding protein–protein kinase complex is regulated by an operon-type mechanism. Another possible explanation is that mutations in the hormone-receptor or the cAMP-binding protein leave active, unregulated cyclase or kinase catalytic moieties and thus are cytotoxic. Clearly, additional information about the enzymes of cAMP metabolism will be gained through further studies of this experimental system.

IV GROWTH FACTORS

The biological effects of several polypeptide factors which affect growth have recently been related to cAMP and/or cGMP. These compounds include nerve growth factor (NGF), epidermal growth factor (EGF) and fibroblast growth factor (FGF).

A Nerve Growth Factor

NGF, originally prepared from the mouse submaxillary gland, produces sympathetic ganglion hypertrophy and hyperplasia both *in vivo* and *in vitro*. Also, NGF stimulates outgrowth of central and peripheral sympathetic and embryonic sensory neurones *in vitro* (Levi-Montalcini and Angeletti, 1968; Björklund and Steneir, 1972). Its physiological importance is suggested by its widespread distribution in tissues and the rapid degeneration of the sympathetic nervous system following treatment of an animal with anti-NGF antibody (Immunosympathectomy). Biochemical actions include the stimulation of RNA, protein and lipid synthesis, and the uptake and metabolism of glucose. The activities of two neuronal enzymes, tyrosine hydroxylase and dopamine β-hydroxylase, are also increased by NGF (Thoenen *et al.*, 1971).

Treatment of foetal rat, mouse or chick sensory ganglia with Bt_2-cAMP also increases neurite outgrowth (Roisen *et al.*, 1972). Addition of Bt_2-cAMP results in stimulation of tyrosine hydroxylase in superior cervical ganglia and cultured neuroblastoma cells, and stimulation of dopamine β-hydroxylase in superior cervical ganglia (Waymire *et al.*, 1972; Keen and McLean, 1972).

Most of these effects, whether induced by NGF or by Bt_2-cAMP, are blocked by cycloheximide. Moreover, neuronal elongation in response to NGF may be dependent upon microtubule formation (Roisen *et al.*, 1972), a process which is stimulated by cAMP. However, no elevation of the levels of cAMP or of adenylate cyclase activity has been observed in eight-day chick dorsal-root ganglia following exposure to NGF (Frazer *et al.*, 1973). The possible relationship between NGF and cAMP and the potential application of NGF, its analogues, cAMP or cAMP analogues in facilitating nerve regeneration in trauma and disease are of considerable fundamental and clinical importance.

B Epidermal Growth Factor

EGF, which like NGF was isolated from male mouse submaxillary gland, has been shown to stimulate epidermal cell growth *in vivo* (Cohen and Taylor, 1972) and in culture (Cohen and Taylor, 1972), and also to decrease epidermal cAMP levels and adenylate cyclase activity in new-born mice. Biochemical changes in epidermal tissue similar to those produced by NGF in nerve tissue, namely increased RNA and protein synthesis, and glucose transport, have been reported. EGF-induced synthesis of RNA and DNA in human fibroblasts is inhibited by theophylline and/or Bt_2-cAMP, as well as by cholera toxin. The actions of these inhibitors are believed to be mediated through cAMP (Bitensky *et al.*, 1975). Recently, reversal of EGF-enhanced mouse epidermal mitotic activity has been observed for Bt_2-cAMP and agents which stimulate cAMP levels, such as papaverine and isoprenaline, in the G_2 stage of the cell cycle (Birnbaum *et al.*, 1976). Although EGF has now been found in human serum and urine (Starkey *et al.*, 1975), its physiological role in epidermal tissue and elsewhere in the body is not clear. Of obvious interest is its possible implication in the pathogenesis of psoriasis or other proliferative dermatoses.

C Fibroblast Growth Factor

FGF is a polypeptide obtained from the brain and pituitary. Combined with hydrocortisone and a nonspecific carrier protein (bovine serum albumin), it is capable of replacing the requirement of the 3T3 fibroblast for serum in the initiation of DNA synthesis and cell division. Physiological concentrations of FGF produce the same marked transient increases in intracellular cGMP as observed after the addition of serum to quiescent cultures. No significant changes in cAMP levels have been observed (Rudland *et al.*, 1974b). The cGMP increase occurs in G_1 phase and is believed to play a role in mediating other associated serum-like, FGF-induced events in these cells. Such events include the stimulation of amino acid and nucleotide transport, and the increased synthesis of protein, RNA and DNA (Rudland *et al.*, 1974a). Transformed 3T3 cells, on the other hand, do not respond to FGF with an increase in DNA synthesis or cell division, and in these cells the levels of cGMP remain elevated throughout the cell cycle (Rudland *et al.*, 1974c).

D Chalones

A number of other endogenous regulators of cell proliferation has been isolated and classified as chalones. While these are probably a diverse group of soluble substances, chalones have certain common features (Forscher and Houck, 1973). They reversibly inhibit cell proliferation and possibly differentiation by a tissue-specific but not species-specific process. Chalones are produced within their target tissue, but are also found in blood and urine. Chalones are widely distributed, and among the tissues and cells from which chalones have been obtained are epidermis, lymphocytes, erythrocytes, fibroblasts, kidney, lung, liver, uterus, hair follicles, sebaceous glands, melanocytes and thymus. Some of these tissues may actually have multiple chalone activities acting at different points in the mitotic cycle. Epidermal chalones acting at both G_1 and G_2 stages have been identified (Elgjo, 1973). It is widely believed that chalone production by mature cells within a tissue constitutes a means of inhibiting the production of any new cells. None of the chalones has as yet been purified and characterized. However, the similarity of action of many of these substances to that of cAMP suggests their possible mediation through the cyclic nucleotide (Iversen, 1969).

Possible uses of chalones in the treatment of various neoplastic conditions have received considerable attention. They appear particularly attractive for the treatment of humans in light of their tissue selectivity, reversibility (non-cytotoxicity) and some encouraging results in several animal studies (Forscher and Houck, 1973). However, considerable purification and further elucidation of their mechanism(s) of action is needed before chalones can be considered for use in humans.

V MITOTIC OR ANTIMITOTIC EFFECTS OF CYCLIC NUCLEOTIDES

The preceding paragraphs have reviewed the evidence that the levels of cAMP can change at a crucial time when the decision is made to proceed through another cell cycle. The question arises as to how a change in the level of the nucleotide can influence subsequent events. A number of suggestions has been made for mechanisms of action of the cyclic nucleotides, but at present none of these possibilities can be considered to be even as firmly established as the changes in nucleotide concentrations. Possible mechanisms are the stimulation of phosphorylation of chromatin, ribosomes or some other protein, such as RNA polymerase, which then influences the rate of transcription or translation and the regulation of the cytoplasmic Ca^{2+} concentration (Berridge, 1975). Other possibilities are that there is a regulation of the function of microtubules or other contractile elements, either by phosphorylation or by changing the concentration of Ca^{2+}, or that there is regulation of the activity of an enzyme, such as phosphoribosylpyrophosphate synthetase, which plays a crucial role in DNA synthesis. Finally, a cyclic nucleotide may combine with a regulatory protein which binds directly to DNA, such a situation has been found in bacteria (Pastan and Perlman, 1970).

A Stimulation of Transcription by cGMP

Pogo (1972) studied the effects of phytohaemagglutinin on DNA-dependent RNA polymerase activity in isolated lymphocyte nuclei. Three different polymerase activities were distinguished. RNA polymerase I, which requires Mg^{2+}, transcribes ribosomal RNA. RNA polymerase II and III both require Mn^{2+} and are distinguished from each other because only polymerase II is sensitive to inhibition by α-amanitin. Addition of phytohaemagglutinin to intact lymphocytes before isolation of the nuclei causes a rapid stimulation of RNA polymerase I. The time course is similar to that found by Hadden *et al.* (1972b) for increased intracellular cGMP formation. Increases in the activities of RNA polymerases II and III appear much later. Hadden *et al.* (1975) reported that in the presence of 1 mM Ca^{2+}, cGMP stimulates RNA polymerase I activity in isolated nuclei. This stimulation is dependent on the concentration of cGMP. Maximal stimulation occurs at approximately 10^{-8} M cGMP, while higher concentrations of cGMP or cAMP inhibit. Since this enzyme activity is examined in a complex system, namely the nucleus, cGMP may not act directly on the enzyme, but rather may influence the accessibility of the DNA template by acting on a phosphorylation mechanism. Johnson and Hadden (1975) have reported that application of cGMP or cholinergic agents to intact lymphocytes stimulates phosphorylation of acidic nuclear proteins, while application of pgE_1, which elevates intracellular cAMP, inhibits this phosphorylation.

B Effects of Cyclic Nucleotides on Phosphoribosyl Pyrophosphate Synthetase

Phosphoribosyl pyrophosphate (PRPP) synthetase is the initial enzyme in the pathways for synthesis of both purine and pyrimidine nucleotides. Thus, it is a logical point at which to control nucleic acid metabolism. Chambers *et al.* (1974), and Hovi *et al.* (1975) have provided evidence that this enzyme is activated and/or induced during stimulation of lymphocytes with lectins. Hovi *et al.* (1975) reported a rapid transient rise in PRPP synthetase fifteen to thirty minutes after addition of the lectin. The time of stimulation of PRPP synthetase corresponds to the time at which a transient rise in the intracellular concentration of cGMP was observed by Hadden *et al.* (1972b). Chambers *et al.* (1974) found a slower increase in PRPP synthetase which could be largely blocked by an inhibitor of protein synthesis. The latter authors also reported that the activity of PRPP synthetase could be stimulated acutely with cGMP. Also evidence is provided that the effects of concanavalin A on the synthesis of purine precursors and on DNA synthesis can be antagonized by Bt_2-cAMP. Green and Martin (1974) demonstrated the direct stimulation of PRPP synthetase in the purified enzyme from hepatoma cells. Very low concentrations of cGMP (2×10^{-9} M) stimulate the purified enzyme. By itself cAMP has no effect on enzyme activity, but at a concentration of 10^{-6} M it can block the stimulatory effect of cGMP. These effects of the cyclic nucleotides do not require protein phosphorylation.

C Effects of cAMP on Ornithine Decarboxylase

The activity of ornithine decarboxylase, the initial enzyme in the pathway of polyamine biosynthesis, is high in rapidly proliferating tissues. Thus, it has been postulated that polyamines and ornithine decarboxylase participate in the regulation of mitosis. In a number of studies, for example, those by Theoharides and Cannelakis (1975), exogenous cAMP has been found to induce the synthesis of ornithine decarboxylase. Also, in synchronized tissue-cultured cells a rise in intracellelar cAMP precedes the induction of ornithine decarboxylase which occurs in late S phase (Russell and Stambrook, 1975).

VI IMMUNOLOGICAL PROCESSES

The immune response is commonly divided into afferent and efferent sections. The afferent section involves recognition of a foreign substance by a small number of lymphocytes which are sensitive to the antigens present on that substance, followed by amplification of the response through blast transformation and proliferation. The efferent portion involves interaction of members of this expanded pool of sensitized cells with the antigen to produce inflammation. The possible involvement of cyclic nucleotides in the afferent (proliferative) portion of the immune system has been discussed in the preceding paragraphs. Involvement of cyclic nucleotides in the efferent portion will be discussed below.

A Effect of cAMP on Antibody Production

A convenient method of assessing humoral immune response is to sensitize an animal to the surface antigens of sheep erythrocytes by injecting them, and then analysing the response by layering spleen cells from the sensitized animal onto normal sheep erythrocytes in agar. Following incubation of the layer with complement, lysis of red cells occurs in the vicinity of spleen cells which are secreting the antierythrocyte antibody. Melmon *et al.* (1974a) tested the effects of altering intracellular cAMP during the incubation of the spleen cells with sheep erythrocytes *in vitro*. They found that histamine, isoprenaline and pgE_2 substantially reduce the number of haemolytic plaques. Theophylline potentiates the effects of the adenylate cyclase stimulators. Cholera toxin produces a delayed increase in the cAMP content of the spleen cells, which is accompanied by inhibition of haemolytic plaque formation. Melmon *et al.* (1974b) were also able to show that most, if not all, of the antibody producing cells are adsorbed onto agarose–histamine beads, confirming that these cells bear histamine receptors.

B Effects of Cyclic Nucleotides on Lymphocyte-Mediated Cytotoxicity

If a tumour transplant or skin graft is performed using donor and recipient animals which are not histocompatible, the recipient animal produces thymus-derived (T) lymphocytes, which are specifically cytotoxic against cells bearing antigens of the donor. This cytotoxicity can be demonstrated *in vitro* by mixing donor cells which have been loaded with ^{51}Cr and lymphocytes from the sensitized recipient. Donor cells which are immunologically lysed release ^{51}Cr into the medium. Using this *in vitro* model of cellular immunity, Henney *et al.* (1972) found that elevation of cAMP in the 'attacking' lymphocytes inhibits the lymphocyte-mediated lysis of target cells. This inhibition can be produced by the adenylate cyclase stimulators histamine, isoprenaline, and PGE_1 or PGE_2, as well as by the phosphodiesterase inhibitor theophylline or by exogenous cAMP. Strom *et al.* (1973) extended this work by showing that the slow increase in cAMP produced by cholera toxin parallels the inhibition of cytotoxicity.

With cholera toxin, Strom *et al.* (1973) were able to show convincingly that the inhibitory effect is produced by elevated cAMP in the sensitized lymphocyte, and not in the target cell. These workers also showed that imidazole, which stimulates phosphodiesterase and thus decreases the concentration of cAMP, enhances cytotoxicity. In addition, elevation of cGMP with cholinergic agonists or with 8-bromo-cGMP stimulates cytotoxicity.

Another type of lymphocyte-mediated cytotoxicity involves reaction of non-thymic killer lymphocytes (K cells) with antibody-coated target cells. Garovoy *et al.* (1975) found that cyclic nucleotides had similar effects on this antibody-dependent cytotoxicity. Cholera toxin, prostaglandins or theophylline inhibit the cytotoxic reaction, while 8-bromo-cGMP slightly enhances the reaction.

Evaluation of the effects of cAMP on the cell-mediated immune response *in vivo* is much more difficult, since any effect on immunological function may be secondary to another cAMP-mediated process, such as secretion of adrenal cortical hormones. However, Warren *et al.* (1974) have provided substantial evidence that elevated cAMP does suppress the cellular immune response *in vivo*. They injected cholera toxin and found that, in parallel with an increase in the cAMP content of spleenic leucocytes, there is a reduction in the numbers of circulating lymphocytes and an inhibition of the delayed hypersensitivity skin reaction and granuloma formation around the insoluble antigen.

VII PHAGOCYTIC CELL FUNCTION

In 1955 a new group of cytoplasmic organelles—lysosomes—were described (de Duve *et al.*, 1955). Since their initial discovery, an important role in several processes has been assigned to lysosomes. Although intracellular digestion appears to be the main function of these organelles, they are also involved in autolysis and tissue injury. For a more detailed review of lysosomes and their diverse morphologies see the review by de Duve and Wattiaux (1966).

Cells engulf extracellular materials by the process called phagocytosis or endocytosis. The engulfed material, which forms a phagosome, is then exposed to the numerous acid hydrolases present within the lysosome by fusion of the two organelles. Digestion of the material normally occurs within the cell without exposing the cytoplasm, or exterior of the cell, to the lysosomal enzymes. However, under some circumstances, the lysosomal enzymes may enter the extracellular milieu. Since the acid hydrolases present within the lysosomes are very efficient catabolic enzymes, exposure of cells to these enzymes can lead to significant tissue injury.

Three theories have been presented to explain how the normally encapsulated lysosomal enzymes are released. One possible mechanism is that endocytosis disturbs the lysosomal membrane so much that fusion with the plasma membrane occurs causing direct enzyme extrusion to the outside of the cell. Another possibility is that there is incomplete closure of the plasma membrane after endocytosis and lysosomal fusion causing a leakage of some of the hydrolases. Finally, enzyme release could occur if foreign substances disrupt the lysosomal membrane. Enzyme release from the lysosome into the cytoplasm would then occur resulting in cell death. These anomalies may be responsible for the symptoms noted in rheumatoid arthritis, nephritis and vasculitis.

Reduction of lysosomal enzyme release could ameliorate the previously mentioned disorders. Catecholamines and adrenergic mechanisms have been implicated in the regulation of inflammation. cAMP behaves as a second messenger in many of the catecholamine-induced actions on cells. For example, the inhibition of the release of histamine and the slow-reacting substance of anaphylaxis (two mediators produced by IgE-sensitized cells and tissues in immediate hypersensitivity reactions) by catecholamines from monkey lung tissue is accompanied by increased levels of cAMP (Ishizaka *et al.*, 1970). In most cases, the inhibition of phosphodiesterases by methylxanthines potentiates the inhibitory effect of catecholamines. Also, anti-inflammatory actions of adrenaline have been shown in several models of acute and chronic inflammation.

The earliest work which revealed that cAMP is implicated in the release of lysosomal enzymes was reported by Weissmann *et al.* (1971b). In studying the lysosomal enzymes released from macrophages during the phagocytosis of zymosan, they showed that both PGE_1 and exogenously added cAMP produce a significant reduction in the release of glucuronidase or acid phosphatase from the macrophage. Potentiation of the effect was noted when the phosphodiesterase inhibitor theophylline was added simultaneously. These initial observations that compounds capable of increasing cAMP levels decrease enzyme extrusion have been followed by more extensive investigations.

Weissmann and co-workers (1971b) have described several effects that reflect the complicated regulatory mechanisms involved in the secretion of lysosomal enzymes. Using human leucocytes that had been converted with cytochalasin B to secretory-like cells (the lysosome fuses with the plasma membrane thus releasing the lysosomal enzymes to the outside of the cell), they found that pgE_1, histamine and isoprenaline, all of which raise cAMP levels, reduce the release of the acid hydrolases, such as glucuronidase (Zurier *et al.*, 1974). The effect of isoprenaline is blocked by propranolol. Cholera toxin also elevates cAMP levels in human polymorphonuclear cells with a concomitant reduction in lysosomal enzyme release. The α-adrenergic agonist phenylephrine is ineffective in inhibiting enzymic release. However, the cholinergic agonist carbamylcholine chloride produces an enhanced release of enzyme with an elevation of cGMP levels in the cell suspensions. This effect is blocked by atropine. Exogenously added cGMP also enhances release. Table 23.2 shows the percentage inhibition or enhancement of the glucuronidase release from human leucocytes, which have been treated with pharmacological agents.

The presence of a cAMP-dependent protein kinase in human polymorphonuclear leucocytes has been demonstrated (Tsung *et al.*, 1975). This enzyme is similar to other protein kinases with respect to its isoelectric point, K for cAMP and pH optimum. The presence of the protein kinase, along with other phospha-

Table 23.2 Pharmacological regulation by autonomic agonists of lysosomal enzyme release from human leucocytes

	enzyme release		
	β-glucuronidase[a]		lactate dehydrogenase[b]
compound (agonist ± antagonist)	% total	% change	% total
resting	1.1 ± 0.6	–	0.7 ± 0.4
stimulated (zymosan)	24.5 ± 1.7	–	1.1 ± 0.7
β-adrenergic			
isoprenaline	10.8 ± 1.0	−56.2	1.4 ± 0.7
isoprenaline + propranolol (10^{-6} M)	24.0 ± 1.9	0	1.7 ± 1.0
adrenaline (5×10^{-5} M)	14.7 ± 0.9	−40.0	1.7 ± 0.9
adrenaline + propranolol (10^{-6} M)	22.3 ± 1.1	0	1.8 ± 0.7
α-adrenergic			
phenylephrine	20.8 ± 1.4	−15.1	1.9 ± 1.1
phenylephrine + propranolol (10^{-6} M)	23.9 ± 1.3	0	2.4 ± 0.9
cholinergic			
carbachol	34.9 ± 2.6	+42.6	2.1 ± 0.7
carbachol + atropine (10^{-6} M)	23.9 ± 2.1	0	2.0 ± 1.1
cGMP	35.6 ± 3.0	+45.2	1.7 ± 1.0

Values = mean ± S.E.M. $N = 4$.
[a]β-glucuronidase is a lysosomal enzyme. Its release parallels the secretion of other lysosomal enzymes.
[b]Significant release of the cytoplasmic enzyme lactate dehydrogenase would indicate cytolysis. (Adapted from Zurier *et al.* (1974), with the permission of the publisher).

tases, suggests that the effects of cAMP may be mediated by the phosphorylation of an unknown substrate.

Ignarro and co-workers have also investigated lysosomal enzyme release from human neutrophils (Ignarro *et al.*, 1974a,b; Ignarro, 1974a). Their results indicate a very significant role for cGMP. During cell contact with both engulfable and non-engulfable immune reactants, agents which promote the accumulation of cAMP inhibit enzyme release, while compounds which promote the accumulation of cGMP enhance enzyme extrusion. The evidence from this work, however, suggests an even more important role of cGMP in that the levels of this cyclic nucleotide are raised upon contact of the cell with an immune reactant. Smith and Ignarro (1975) recently illustrated that the contact of human neutrophils with a soluble immune reactant or divalent cation ionophore (A-23187) causes firstly the influx of Ca^{2+} followed by increased intracellular levels of cGMP and then the secretion of lysosomal enzymes.

In summary, cAMP and cGMP are capable of regulating the extrusion of lysosomal enzymes from cells undergoing phagocytosis. The fluctuation in intracellular levels of these cyclic nucleotides regulates mechanisms which are important for the maintenance of inflammatory diseases, such as rheumatoid arthritis. More extensive research into the roles of Ca^{2+}, protein kinases and their phosphorylated substrates in these systems offers the possibility of understanding the mechanisms involved in certain acute and chronic diseases.

VIII RELEASE OF MEDIATORS OF IMMEDIATE HYPERSENSITIVITY

Before a discussion on the involvement of cyclic nucleotides in immediate hypersensitivity, a brief background of immunological mechanisms believed to occur in these reactions is needed. Mediators are released from sensitized cells during the immunologically induced state of immediate hypersensitivity. The cells implicated are either tissue-fixed mast cells or circulating basophils (Ishizaka and Ishizaka, 1971; Bennich *et al.*, 1971). These cells, depending on their source, contain storage granules of histamine and heparin. In addition, these cells are able to synthesize slow-reacting substance of anaphylaxis, eosinophil chemotactic factor of anaphylaxis, platelet activating factor, bradykinin and 5-hydroxytryptamine. These substances are actively released from the cells into the circulation where they have their physiological action. Initial exposure to an antigen stimulates synthesis of various antibody molecules with specific sites for binding that antigen. Sensitization of the cells is accomplished by the fixation of a particular antibody (IgE), to the cell membrane. The antibody has specific binding sites for some antigenic substance, and, upon a second exposure to the antigen, the cell-bound antibody–antigen complex initiates a sequence of events leading to the extrusion of mediators. Detailed investigations have been made on the binding of IgE to mast cells (Ishizaka *et al.*, 1973).

A variety of chemical signals which bind to specific membrane receptors influences most differentiated cells. A central problem in biochemistry is to explain in detail how such ligand–membrane interactions are transduced into intracellular alterations. More specifically, the central problem in immediate hypersentivity reactions is to understand the intracellular biochemical mechanisms which cause the release of pharmacological mediators initiated by an antigen–cell-fixed IgE antibody interaction.

The mechanisms of mediator release have been studied in whole lung fragments, human leucocytes (approximately 1% basophils) and isolated peritoneal mast cells (usually from the rat). The differences between purified cells and mixed-cell populations, such as tissues or leucocytes, are obvious, but the data at present indicate that the reaction mechanisms are rather similar.

Schild (1937) observed that adrenaline inhibits the antigen-induced release of histamine from sensitized guinea-pig lung. This report was relatively unnoticed until Lichtenstein and Margolis (1968) reported the inhibition of histamine release by catecholamines and methylxanthines from leucocytes incubated *in vitro*. These investigations were later extended to lung tissue (Assem and Schild, 1969). Using human lung tissue, sensitized with reaginic (IgE) serum, these workers were able to demonstrate that histamine release is inhibited by several catecholamines. Austen (1974) described the effects of adrenaline, isoprenaline and theophylline on release of the slow-reacting substance of anaphylaxis and histamine from sensitized monkey lung fragments. Synergism was observed when the β-agonist isoprenaline and theophylline were used together.

Table 23.3 Pharmacological regulation of histamine and slow-reacting substance of anaphylaxis (SRS-A) release from human lung tissue

	% change in cAMP	% change in histamine release	% change in SRS-A release
isoprenaline (10^{-6} M)	+1800	−80	N.D.
propranolol (5×10^{-5} M)	0	0	N.D.
isoprenaline (10^{-6} M) + propranolol (5×10^{-5} M)	+40	−40	N.D.
phenylephrine (10^{-7} M)	−30	+50	+20
phenylephrine (10^{-7} M) + propanolol (10^{-6} M)	−30	+140	+200
acetylcholine (10^{-8} M)	−15	+40	+40
carbachol (10^{-11} M)	−10	+30	+200
carbachol (10^{-10} M) + atropine (10^{-8} M)	N.D.	0	0
phenylephrine (10^{-8} M) + atropine (10^{-8} M)	N.D.	+50	+100
8-bromo-cGMP (10^{-5} M)	N.D.	+25	+140

N.D. = Not Determined.
(Adapted from Kaliner and Austen (1973) and Tauber *et al.* (1973), with the permission of the publishers).

Using human lung tissue, Kaliner *et al.* (1972) correlated changes in cAMP levels produced by β-agonists with inhibition of the release of histamine and the slow-reacting substance. Isoprenaline causes a twenty-fold increase in cAMP levels with a 75% decrease in histamine release. Propranolol, a β-antagonist, blocked this effect. In contrast to the effects noted with β-agonists (Tauber *et al.*, 1973), phenylephrine with propranolol causes a 20–30% decrease in cAMP levels with a 150–200% increase in mediator release. Acetylcholine and carbamylcholine show a marked enhancement (200%) of the release of histamine and the slow-reacting substance. Atropine blocks the cholinergic stimulation, but has no effect on the α-adrenergic enhancement. Although cGMP levels were not measured in these studies, cholinergic stimulants have been shown to raise cGMP levels in other tissues (George *et al.*, 1970). An analogue of cGMP, 8-bromo-cGMP, at 10^{-5} M also causes an enhancement of the immunological release of mediators. Therefore, compounds which raise cAMP levels inhibit mediator release, while compounds which elevate cGMP levels (or decrease cAMP) enhance mediator release. Table 23.3 summarizes some of these effects.

The sequence of biochemical events in the release of histamine from human lung tissue has been described by Kaliner and Austen (1973). Besides an energy-dependent step and at least two Ca^{2+}-dependent stages, a cAMP-sensitive step is present just prior to release. Increases in cAMP levels inhibit this step. Further delineation of the mechanisms involved are only possible using more purified cell populations as has been achieved using purified rat mast cells.

Kaliner and Austen (1947a), studying purified rat peritoneal mast cells, provided direct evidence that a bidirectional relationship exists between the concentration of cAMP and mediator release. Elevation of cAMP with PGE_1 and/or aminophylline inhibits reversed anaphylactic histamine release. This method of immunological inducement of histamine release shares the same biochemical mechanisms as actively or passively sensitized cells and results in a greater percentage mediator release. The most interesting aspect of this study is the observation that upon challenge, there is a rapid decrease (30–40% within thirty seconds) in intracellular cAMP levels which occurs with histamine release. The findings of Kaliner and Austen (1974) indicate that a depletion of cAMP may be a prerequisite for histamine release in the rat mast-cell system. Other workers using non-immunological methods of determining histamine release have noted similar findings.

Human leucocytes have been the area of interest for several researchers in the field of immediate hypersensitivity. The effects of catecholamines and methyl-anthines on inhibiting histamine release from these cells were first described by Lichtenstein and Margolis (1968). Basophils have been shown to be the only source of histamine in leucocytes (Pruzansky and Patterson, 1970), and, more recently, Lewis *et al.* (1975) have shown that basophilic leukaemia cells contain all the four mediators (histamine, the slow reacting substance and the eosinophil chemotactic factor of anaphylaxis and platelet activating factor) which have been described in human lung tissue and in nasal polyps (Kaliner *et al.*, 1973).

Evidence that cAMP is a physiologic inhibitor of histamine release from leucocytes is similar to that found in tissue fragments (Bourne *et al.*, 1972). β-Adrenergic agonists stimulate accumulation of cAMP by activating adenylate cyclase. This stimulation is coincident with an inhibition of histamine release, which is eliminated by propranolol. Methylxanthines potentiate the accumulation of cAMP and inhibitory effects on histamine release when used with either β-agonists or pgE_1. Bt_2-cAMP inhibits histamine release. Finally, cholera toxin produces a delayed increase in cAMP which correlates with inhibition of histamine release. Specific toxin antagonists eliminate both effects. Recently, Lichtenstein (1975) has suggested the importance of Ca^{2+} in the two-stage release of histamine from leucocytes. cAMP appears to be a critical determinant in the first Ca^{2+}-independent stage. In summary, the work by Lichtenstein and others has indicated a role for cAMP in regulating the release of histamine from sensitized leucocytes (containing 1% basophils—the cells responsible for histamine release). Although cGMP has not been studied in mast cells or leucocytes with regard to histamine release, further investigations may also involve this cyclic nucleotide in mediator release from cells.

IX CONTRACTION OF BRONCHIAL SMOOTH MUSCLE

A final aspect of the involvement of cyclic nucleotides in immediate hyper-sensitivity reactions is the role of cAMP and cGMP in smooth-muscle relaxation and contraction. β-Adrenergic agonists and/or phosphodiesterase inhibitors are drugs currently used in the treatment of asthma—a well documented example of acute (Type I) hypersensitivity (Coombs and Gell, 1968). Besides having ameliorating effects on the cells responsible for mediator release during anaphylaxis (that is inhibiting mediator release), agents which elevate cAMP also relax smooth muscle. This relaxation is needed to overcome the contraction induced by such mediators as histamine, the slow-releasing substance of anaphylaxis, 5-hydroxytryptamine and bradykinin. The effects of catecholamines on raising cAMP concentrations have been seen in smooth muscle preparations from blood vessels, uterus and intestine (Triner *et al.*, 1971). However, since in most instances, no correlation has been found between smooth muscle response and cyclic nucleotide levels, most of the evidence is indirect.

Recently, Murad and Kimura (1974) measured the cAMP and cGMP levels in smooth muscle (guinea-pig tracheal chains) preparations. Adrenaline and pgE_1 cause a dose-dependent increase in cAMP levels. Propranolol blocks the adrenaline stimulation while phenoxybenzamine, an α-adrenergic antagonist, is ineffective. Theophylline potentiates the effects of adrenaline. Histamine and carbamylcholine increase cGMP levels. These observations, coupled with the observation that exogenously added Bt_2-cAMP relaxes guinea-pig tracheal chains, have led to the conclusions that agents which increase cAMP levels relax tracheal smooth muscle and agents capable of stimulating contraction increase cGMP levels. Therefore, β-adrenergic agonists and phosphodiesterase inhibitors not only inhibit the continued release of the pharmacological mediators from mast cells (or basophils) which cause smooth muscle contraction, but also relax the tracheal smooth muscle to counteract the already released mediators. This dual effect of β-agonists and phosphodiesterase inhibitors has encouraged pharmaceutical companies to seek such compounds in order to alleviate acute immediate hypersensitivity diseases.

X CONCLUSIONS

In some cell types, such as fibroblasts and lymphocytes, cGMP appears to be a signal for proliferation, while cAMP appears to be a signal for greater specialization of cell function, and thus frequently acts as a brake on proliferation. In other cell types, such as regenerating liver and endocrine glands, high levels of cAMP appear to initiate proliferation. In the regulation of immunological function, the role of cGMP as a proliferative signal is crucial, especially for the afferent or amplifying limb of the immune system, and for the maintenance of normal levels of rapidly turning-over cells, such as granulocytes. A second major role of cGMP in the immune response appears to be as a signal for the secretion of immunological mediators. cAMP appears to oppose this action of cGMP. The contrast between the negative role of cAMP in immunological secretion and its positive role in the secretion of hormones is striking. The available data point to a parallel positive role for cGMP (and negative role for cAMP) in the growth of at least some malignant cells and also in the immune response which is postulated to combat malignancy (Prehn, 1974). It is hoped in the future to be able to manipulate these processes in the treatment of cancer. However the task is subtle and demanding. The various postulated suppressor functions (Gershon, 1974) add to the complexity of this undertaking, but may also provide the possibility for its success.

References

Abbas, A. K., Ault, K. A., Karnovsky, M. J. and Unanue, E. R. (1975) Non-random distribution of surface immunoglobulin on murine B-lymphocytes. *J. Immunol.*, **114**, 1197–1204

Abboud, F. M. and Eckstein, J. W. (1966) Active reflex vasodilation in man. *Fed. Proc.*, **251**, 1611–1617

Aberg, G. and Andersson, R. (1972) Studies on mechanical actions of mepivacaine (Carbocaine) and its optically active isomers on isolated smooth muscle: Role of Ca^{2+} and cyclic AMP. *Acta pharmac. tox.*, **31**, 321–336

Adachi, K., Yoshikawa, K., Halprin, K. M. and Levine, V. (1975) Prostaglandins and cyclic AMP in epidermis. *Br. J. Derm.*, **92**, 381–388

Ahlquist, R. P. (1948) A study of the adrenotropic receptors. *Am. J. Physiol.*, **153**, 586–600

Ahn, C. S., Athans, J. C. and Rosenberg, I. N. (1969) Stimulation of thyroid hormone secretion by dibutyryl-cyclic AMP. *Endocrinology*, **85**, 224–230

Ahn, C. S. and Rosenberg, I. N. (1970) Iodine metabolism in thyroid slices: Effects of TSH, dibutyryl-cyclic AMP, NaF and prostaglandin E_1. *Endocrinology*, **86**, 396–405

Ahren, K., Hjalmarson, A. and Isakson, O. (1971) Inotropic and metabolic effects of dibutyryl-cyclic adenosine 3′,5′-monophosphate in the perfused rat heart. *Acta physiol. scand.*, **82**, 79–90

Aiken, J. W. and Vane, J. R. (1972) Intrarenal prostaglandin release attenuates the renal vasoconstrictor activity of angiotensin. *J. Pharmac. exp. Ther.*, **184**, 678–687

Alberti, K. G. M. M., Christensen, N. J., Christensen, S. E., Hansen, A. P., Iversen, J., Lundback, K., Seyer-Hansenk, K. and Orskov, H. (1973) Inhibition of insulin secretion by somatostatin. *Lancet*, iv, 1299–1301

Albrecht, I., Hallbäck, M., Julius, S., Lundgren, Y., Stage, L., Weiss, L. and Folkow, B. (1975a) Arterial pressure, cardiac output and systemic resistance before and after pithing in normotensive and spontaneously hypertensive rats. *Acta physiol. scand.*, **94**, 378–385

Alford, R. H. (1970) Metal ion requirements for phytohemagglutinin induced transformation of human peripheral blood lymphocytes. *J. Immunol.*, **104**, 698–703

Allan, D. and Michell, R. H. (1975) Accumulation of 1,2-diacylglycerol in the plasma membrane may lead to echinocyte transformation of erythrocytes. *Nature, Lond.*, **258**, 348–349

Allen, D. O., Munshower, J., Morris, H. P. and Weber, G. (1971) Regulation of adenyl cyclase in hepatomas of different growth rates. *Cancer Res.*, **31**, 557–560

Allen, L. B., Eagle, N. C., Huffman, J. H., Shuman, D. A., Meyer, R. B., Jr. and Sidwell, R. W. (1974) Enhancement of interferon antiviral action in L-cells by cyclic nucleotides. *Proc. Soc. exp. Biol. Med.*, **146**, 580–584

Alonso, D. and Harris, J. B. (1965) Effect of xanthines and histamine on ion transport and respiration by frog gastric mucosa. *Am. J. Physiol.*, **208**, 18–23

Alonso, D., Rynes, R. and Harris, J. B. (1965) Effect of imidazoles on active transport by gastric mucosa and urinary bladder. *Am. J. Physiol.*, **208**, 1183–1190

Amer, M. S. (1972) Cyclic AMP and gastric secretion. *Am. J. dig. Dis.*, **17**, 945–957

Amer, M. S. (1973) Cyclic adenosine monophosphate and hypertension in rats. *Science, N.Y.*, **179**, 807–809

Amer, M. S. (1974) Cyclic guanosine 3′,5′-monophosphate and gallbladder contraction. *Gastroenterology*, **67**, 333–337

Amer, M. S. (1975) Cyclic nucleotides in disease: On the biochemical etiology of hypertension. *Life Sci.*, **17**, 1021–1038

Amer, M. S., Allen, W., Perhach, I. L., Ferguson, H. C. and McKinney, G. R. (1974a) Aberrations of cyclic nucleotide metabolism in the heart and vessels of hypertensive rats. *Proc. natn. Acad. Sci., U.S.A.*, **71**, 4930–4934

Amer, M. S. and Byrne, J. E. (1975) Interchange of adenylyl and guanylyl cyclases as an explanation for transformation of β to α-adrenergic responses in the rat atrium. *Nature, Lond.*, **256**, 421–424

Amer, M. S., Doba, N. and Reis, D. J. (1975) Changes in cyclic nucleotide metabolism in aorta and heart of neurogenically hypertensive rats: Possible trigger mechanisms of hypertension. *Proc. natn. Acad. Sci. USA*, **72**, 2135–2139

Amer, M. S., Gomoll, A. W., Perhach, J. L., Jr., Ferguson, H. C. and McKinney, G. R. (1974b) Aberrations of cyclic nucleotide metabolism in the hearts and vessels of hypertensive rats. *Proc. natn. Acad. Sci. U.S.A.*, **71**, 4930–4934

Amer, M. S. and Kreighbaum, W. E. (1975) Cyclic nucleotide phosphodiesterases: Properties, activators, inhibitors, structure–activity relationships and possible role in drug development. *J. pharm. Sci.*, **64**, 1–37

Amer, M. S. and McKinney, G. R. (1972) Studies with cholecystokinin *in vitro*. IV. Effects of cholecytokinin and related peptides on phosphodiesterase. *J. Pharmac. exp. Ther.*, **183**, 535–548

Amer, M. S. and McKinney, G. R. (1973) Possibilities for drug development based on the cyclic AMP system. *Life Sci.*, **13**, 753–767

Amer, M. S. and McKinney, G. R. (1974) Cyclic nucleotides and drug discovery. In *Annual Reports in Medicinal Chemistry*, Vol. 9 Ed. R. V. Heinzelman, Academic Press, New York, pp. 203–212

Amer, M. S. and McKinney, G. R. (1975) Cyclic nucleotides as mediators of drug action. In *Annual Report in Medical Chemistry*, **10**, Academic Press, New York, pp. 192–201

Amer, M. S., McKinney, G. R. and Akcasu, A. (1974c) Effect of glycyrrhetnic acid on the cyclic nucleotide system of the rat stomach. *Biochem. Pharmac.*, **23**, 3085–3092

Andén, N. E. (1972) Dopamine turnover in the corpus striatum and the limbic system after treatment with neuroleptic and anti-acetylcholine drugs. *J. Pharm. Pharmac.*, **24**, 905–906

Andén, N. E., Butcher, S. G., Corrodi, H., Fuxe, K. and Ungerstedt, U. (1970) Receptor activity and turnover of dopamine and noradrenaline after neuroleptics. *Europ. J. Pharmacol.*, **11**, 303–314

Andén, N. E., Carolsson, A. and Häggendal, J. (1969) Adrenergic mechanisms. *Ann. Rev. Pharmacol.*, **9**, 119–134

Andén, N. E., Dahlström, A., Fuxe, K., Larsson, K., Olson, L. and Ungerstedt, U. (1966) Ascending monoamine neurons to the telencephalon and diencephalon, *Acta physiol. scand.*, **67**, 313

Andén, N. E., Rubenson, A., Fuxe, K. and Hökfelt, T. (1967) Evidence for dopamine receptor stimulation by apomorphine. *J. Pharm. Pharmac.*, **19**, 627–629

Andén, N. E. and Stock, G. (1973) Effect of clozapine on the turnover of dopamine in the corpus striatum and in the limbic system. *J. Pharm. Pharmac.*, **25**, 346–348

Anderson, E. G., Haas, H. and Hosli, L. (1973a) Comparison of effects of noradrenaline and histamine with cyclic AMP on brain stem neurones. *Brain Res.*, **49**, 471–475

Anderson, W. B., Gallo, M. and Pastan, I. (1974) Adenylate cyclase activity in fibroblasts transformed by Kirsten and Maloney sarcoma virus. *J. biol. Chem.*, **249**, 7041–7048

Anderson, W. B., Johnson, G. S. and Pastan, I. (1973b) Transformation of chick embryo fibroblasts by wild-type and temperature-sensitive Rous sarcoma virus alters adenylate cyclase activity. *Proc. natn. Acad. Sci. USA*, **70**, 1055–1059

Anderson, W. B., Lovelace, E. and Pastan, I. (1973c) Adenylate cyclase is decreased in chick embryo fibroblasts transformed by wild-type and temperature-sensitive Schmidt–Ruppin sarcoma virus. *Biochem. biophys. Res. Commun.*, **52**, 1293–1299

Anderson, W. B., and Pastan, I. (1975) Altered adenylate cyclase activity: Its role in growth regulation and malignant transformation of fibroblasts. In *Advances in Cyclic Nucleotide Research*, Vol. 5, (Eds. G. I. Drummond, P. Greengard and G. A. Robison) Raven Press, New York, pp. 681–698

Anderson, W. B., Perlman, R. L. and Pastan, I. (1972) Effect of adenosine 3′,5′-monophosphate analogues on the activity of the cyclic adenosine 3′,5′-monophosphate receptor in *Escherichia coli. J. biol. Chem.*, **247**, 2717–2722

Anderson, W. B., Russell, T. R., Carchman, R. A. and Pastan, I. (1973d) Interrelationship between adenylate cyclase, adenosine 3′,5′-cyclic monophosphate phosphodiesterase activity, adenosine 3′,5′-cyclic monophosphate levels and growth of cells in culture. *Proc. natn. Acad. Sci. USA.*, **70**, 3802–3805

Andersson, R. G. G. (1972) Cyclic AMP and calcium ions in mechanical and metabolic responses of smooth muscles; influence of some hormones and drugs. *Acta. physiol. scand.*, suppl. **382**, 1–59

Andersson, R. (1973a) Cyclic AMP as a mediator of the relaxing action of papaverine, nitroglycerine, diazoxide and hydralazine in intestinal and vascular smooth muscle. *Acta pharmac. tox.*, **32**, 321–336

Andersson, R. (1973b) Role of cyclic AMP and Ca^{2+} in mechanical and metabolic events in isometrically contracting vascular smooth muscle. *Acta physiol. scand.*, **87**, 84–95

Andersson, R. and Mohme-Lundholm, E. (1970) Metabolic actions in intestinal smooth muscle associated with relaxation by adrenergic α and β-receptors. *Acta physiol. scand.*, **79**, 244–261

Andersson, R., Nilsson, K., Wikberg, J., Johansson, S., Mohme-Lundholm, E. and Lundholm, L. (1975) Cyclic nucleotides and the contraction of smooth muscle. In *Advances in Cyclic Nucleotide Research* Vol. 15, (Eds. G. I. Drummond, P. Greengard and G. A. Robison), Rowen Press, New York, pp. 491–518

Angel, A., Desai, K. S. and Halperin, M. L. (1971) Reduction in adipocyte ATP by lipolytic agents: Relation to intracellular free fatty acid accumulation. *J. Lipid Res.*, **12**, 203–211

Anisuzzaman, A. K. M., Lake, W. C. and Whistler, R. L. (1973) 4′-Thio-adenosine 3′,5′-cyclic phosphate and derivatives. Chemical synthesis and hydrolysis by phosphodiesterase. *Biochemistry*, **12**, 2041–2045

Anonymous (1974) Clinical observations on Bt_2-cAMP in treatment of psoriasis. *Chin. med. J.*, **4**, 201–204

Ansell, G. B., Dawson, R. M. C. and Hawthorne, J. N. (1973) *Form and Function of Phospholipids*. Elsevier Scientific, New York, p. 448.

Antonoff, R. S. and Ferguson, J. J., Jr. (1973) Photoaffinity labelling with cyclic nucleotides. *J. biol. Chem.*, **249**, 3319–3321

Aoki, K., Yamashita, K. and Hotta, K. (1975) Ability of Ca binding and release in subcellular membrane fraction of arterial muscle in normotensive and spontaneously hypertensive rats. *Proc. west. pharmac. Soc.*, **18**, 96–100

Aoki, K., Yamashita, K., Tomita, N., Tazumi, K., Kato, S., Yoshida, T., Shimazaki, H., Sato, I., Takikawa, K. and Hotta, K. (1974) Vascular smooth muscle contraction in hypertension. *Jap. Circul. J.*, **38**, 647

Aporti, F., Leon, A. and Toffano, G. (1975) Anti-phosphodiesterase activity of the coronary dilating agent hydrochloride carbocromene. *Pharmac. Res. Comm.*, **1**, 289–297

Appleman, M. M. and Terasaki, W. L. (1975) Regulation of cyclic nucleotide phosphodiesterase. In *Advances in Cyclic Nucleotide Research*, Vol. 5, (Eds. G. I. Drummond, P. Greengard and G. A. Robison), Raven Press, New York, pp. 153–162

Appleman, M. M., Thompson, W. J. and Russell, T. R. (1973) Cyclic nucleotide phosphodiesterases. In *Advances in Cyclic Nucleotide Research*, Vol. 3, (Eds P. Greengard and G. A. Robison), Raven Press, New York, pp. 65–98

Aprille, F. R., Lefkowitz, R. J. and Warshaw, J. B. (1974) [^{3}H]Norepinephrine binding and lipolysis by isolated fat cells. *Biochem. biophys. Acta*, **373**, 502–513

Arbuthnott, G. W., Attree, T. J., Eceleston, D., Loose, R. W. and Martin, M. J. (1974) Is adenylate cyclase the dopmaine receptor? *Med. Biol.*, **52**, 350–353

Argy, W. P., Handler, J. S. and Orloff, J. (1967) Ca^{2+} and Mg^{2+} effects on toad bladder responses to cyclic AMP, theophylline and ADH analogues. *Am. J. Physiol.*, **213**, 803–808

Ariëns, E. J., Van Rossum, J. M. and Simonis, A. M. (1957) Affinity, intrinsic activity and drug interactions. *Pharmac. Rev.*, **9**, 218–236

Armelin, H. A. (1973) Pituitary extracts and steroid hormones in the control of 3T3 cell growth. *Proc. natn. Acad. Sci.*, **701**, 2702–2706

Armstrong, K. J., Stoufer, I. E., van Inwegen, R. G., Thompson, W. I. and Robison, G. A. (1974) Effects of thyroid hormone deficiency on cyclic adenosine 3′:5′-monophosphate and control of lipolysis in fat cells. *J. biol. Chem.*, **249**, 4226–4231

Ashby, J. P. and Speake, R. N. (1975) Insulin and glucagon secretion from isolated islets of Langerhans. The effects of calcium ionophores. *Biochem. J.*, **150**, 89–96

Ashcroft, S. J. H., Randle, P. J. and Taljedal, I. B. (1972) Cyclic nucleotide phosphodiesterase activity in normal mouse pancreatic islets. *FEBS. Lett.*, **201**, 263–266

Ashman, D. F., Lipton, R. L., Melicow, M. M. and Price, T. D. (1963) Isolation of adenosine 3′,5′-monophosphate and guanosine 3′,5′-monophosphate from rat urine. *Biochem. Biophys. Res. Commun.*, **11**, 330–334

Aso, K., Deneau, D. G., Krulig, L., Wilkinson, D. I. and Farber, E. M. (1975a) Epidermal synthesis of prostaglandins and their effect on levels of cyclic adenosine, 3′,5′-monophosphate. *J. invest. Derm.*, **64**, 326–331

Aso, K., Orenberg, E. K. and Farber, E. M., (1975b) Reduced epidermal cyclic AMP accumulation following prostaglandin stimulation: Its possible role in the pathophysiology of psoriasis. *J. invest. Derm.*, **65**, 375–378

Aso, K., Orenberg, E. K., Rabinowitz, I. N. and Farber, E. M. (1974) The reduced levels of prostaglandins and the effect of prostaglandin stimulation on cyclic AMP accumulation in psoriatic epidermis. *J. invest. Derm.*, **62**, 545.

Assem, E. S. K. and Schild, H. (1969) Inhibition by sympathomimetic amines of histamine release induced by antigens in passively sensitized human lung. *Nature, Lond.*, **224**, 1028–1029

Atkinson, J. P., Greene, W. C., McNearney, T. A. and Parker, C. W. (1977) Studies on the stimulation of cyclic adenosine 3′,5′-monophosphate (cAMP) metabolism in human lymphocytes by latex polymers. *Exp. Cell Res.*, **99**, 395–407

Atkinson, J. P., Wedner, H. J. and Parker, C. W. (1975) Two novel stimuli of cyclic adenosine 3′,5′-monophosphate (cAMP) in human lymphocytes. *J. Immunol.*, **115**, 1023–1027

Atlas, D., Steer, M. L. and Levitzki, A. (1974) Stereospecific binding of propranolol and catecholamines to the β-Adrenergic Receptor. *Proc. natn. Acad. Sci. U.S.A.*, **71**, 4246–4248

Auerbach, R. (1974) Topical application of cAMP in psoriasis. *J. Am. med. Ass.*, **227**, 326–327

Aurbach, G. D. and Heath, D. A. (1974) Parathyroid hormone and calcitonin regulation of renal function. *Kidney Int.*, **6**, 331–345

Aurbach, G. D., Fedak, S. A., Woodward, C. J., Palmer, J. S., Hauser, D. and Troxler, F. (1974) β-Adrenergic receptor: Stereospecific interaction of iodinated β-blocking agent with high affinity sites. *Science, N.Y.*, **186**, 1223–1224

Austen, K. F., (1974) Reaction mechanisms in the release of mediators of immediate hypersensitivity from human lung. *Fed. Proc.*, **33**, 2256–2262

Bach, M. A. and Bach, J. F. (1973) Studies on thymus products. IV. The effects of cyclic nucleotides and prostaglandins on rosette-forming cells. Interaction with thymic factor. *Europ. J. Immunol.*, **31**, 778–783

Bannai, S. and Sheppard, J. R. (1974) Cyclic AMP, ATP and cell contact. *Nature, Lond,* **250**, 62–64

Bär, H-P. (1974a) Cyclic nucleotides and smooth muscle. In *Advances in Cyclic Nucleotide Research*, Vol. 4, (Eds. P. Greengard and G. A. Robison), Raven Press, New York, pp. 195–237

Bär, H.-P. (1974b) On the Kinetics and Temperature Dependence of Adrenaline–Adenylate Cyclase Interactions. *Molec. Pharmac.*, **10**, 597–604

Bär, H.-P. and Hechter, O. (1969a) Adenyl cyclase and hormone action. Effects of adrenocorticotropic hormone, glucagon, and epinephrine on the plasma membrane of rat fat cells. *Proc. natn. Acad. Sci., U.S.A.*, **63**, 350–356

Bär, H.-P. and Hecther, O. (1969b) Adenyl cyclase and hormone action. III. Calcium requirement for ACTH stimulation of adenyl cyclase. *Biochem. biophys. Res. Commun.*, **35**, 681–686

Bär, H.-P. and Hechter, O. (1969c) Adenyl cyclase assay in fat cell ghosts. *Analyt. Biochem.*, **29**, 476

Barber, E. D. and Bright, H. J. (1968) The rate of an allosteric process: Inhibition of homoserine dehydrogenase I from *E. coli* by threonine. *Proc. natn. Acad. Sci. U.S.A.*, **60**, 1363–1370

Barden, N. and Labrie, F. (1973) Receptor for thyrotropin-releasing hormone in plasma membranes of bovine anterior pituitary gland. *J. biol. Chem.*, **248**, 7601–7606

Basile, D. V., Wood, H. N. and Braun, A. C. (1973) Programming the cells for death under defined experimental conditions: Relevance to the tumor problem. *Proc. natn. Acad. Sci. USA*, **70**, 3055–3059

Bastomsky, C. H., Zakarya, M. and McKenzie, J. M. (1971) Thyroid hydrolysis of cyclic AMP as influenced by thyroid gland activity. *Biochim. biophys. Acta*, **230**, 286–295

Baudouin-Legros, M. and Meyer, P. (1973) Effects of angiotensin, catecholamines and cyclic AMP on calcium storage in aortic microsomes. *Br. J. Pharmac.*, **47**, 377–385

Baudry, M., Martres, M. P. and Schwartz, J. C. (1975) H_1 and H_2-Receptors in the histamine-induced accumulation of cyclic AMP in guinea-pig brain slices. *Nature Lond.*, **253**, 362–363

Bauer, R. J., Swiatek, K. R., Robins, R. K. and Simon, L. N. (1971) Adenosine 3′,5′-cyclic monophosphate derivatives II. Biological activity of some 8-substituted analogs. *Biochem. biophys. Res. Commun.*, **45**, 526–531

Baxter, J. D., Harris, A. W., Tomkins, G. M. and Cohn, M. (1971) Glucocorticoid receptors in lymphoma cells in culture: Relationship to glucocorticoid killing activity. *Science*, **171**, 189–191

Beavo, J. A., Bechtel, P. J. and Krebs, E. G. (1974) Activation of protein kinase by physiological concentrations of cyclic AMP. *Proc. natn. Acad. Sci. U.S.A.*, **71**, 3580–3583

Beavo, J. A., Rogers, N. L., Crofford, O. B., Hardman, J. G., Sutherland, E. W. and Newman, E. V. (1970) Effects of xanthine derivatives on lipolysis and on adenosine 3′,5′-monophosphate phosphodiesterase activity. *Molec. Pharmac.*, **6**, 597–603

Beck, L. (1961) Active reflex dilatation in the innervated perfused hind leg of the dog. *Am. J. Physiol.*, **201**, 123–128

Beck, L. (1965) Histamine as the potential mediator of active reflex vasodilation. *Fed. Proc.*, **24**, 1298–1310

Beck, N., Singh, H., Reed, S. W., Murdaugh, H. V. and Davis, B. B. (1974) Pathogenic role of cyclic AMP in the impairment of urinary concentrating ability in acute hypercalcemia. *J. Clin. Invest.*, **54**, 1049

Becker, E. L., Davis, A. T., Estensen, R. D. and Quie, P. G. (1972) Cytochalasin B:

IV. Inhibition and stimulation of chemotaxis of rabbit and human polymorphonuclear leukocytes. *J. Immunol.*, **108**, 396–402

Becker, E. L. and Henson, P. M. (1973) *In vitro* studies of immunologically induced secretion of mediators from cells and related phenomena, *Adv. Immunol.*, **17**, 93–193

Beer, B., Chasin, M., Glody, D. E., Vogel, J. R. and Horovitz, Z. P. (1972) Cyclic adenosine monophosphate phosphodiesterase in brain: Effect of anxiety. *Science, N.Y.*, **176**, 428–430

Belocopitow, E., Fernandez, M., Birnbaumer, L. and Torres, H. N. (1967) Comparative study of the different dependent forms of glycogen synthetase. *J. biol. Chem.*, **242**, 1227–1231

Benjamin, W. B. and Singer, I. (1975) Actions of insulin, epinephrine and dibutyryl-cAMP on fat cell protein phosphorylations. cAMP-dependent and independent mechanisms. *Biochemistry*, **14**, 3301–3309

Bennett, V., Mong, L. and Cuatrecasas, P. (1975a) Mechanism of activation of adenylate cyclase by *Vibrio cholerae* enterotoxin. *J. Membrane Biol.*, **24**, 107–129

Bennett, V., O'Keefe, E. and Cuatrecasas, P. (1975b) Mechanism of action of cholera toxin and the mobile receptor theory of hormone receptor–adenylate cyclase interactions. *Proc. natn. Acad. Sci. U.S.A.*, **72**, 33–37

Bennich, H., Gunnar, S. and Johansson, O. (1971) Structure and function of human immunoglobulin E. *Adv. Immunol.*, **13**, 1–55

Benoit, J. and Assenmacher, I. (1952) Influences de lésions hautes et basses de l'infundibulum sur la gonadostimulation chez le canard domestique. *C.r. hebd. Séanc. Acad. Sci.*, **235**, 1547–1549

Berridge, M. J. (1975a) The interaction of cyclic nucleotides and calcium in the control of cellular activity. In *Advances in Cyclic Nucleotide Research*, Vol. 6 (Eds. P. Greengard and G. A. Robison), Raven Press, New York, pp. 1–98.

Berridge, M. J. (1975b) The role of cyclic nucleotides and calcium in the control of secretion. *Proc. 6th Int. Congr. Pharmac., Helsinki*, **1**, 213–221

Berridge, M. J., Lindley, B. D. and Prince, W. T. (1975a) Role of calcium and cyclic AMP in controlling fly salivary gland secretion. In *Alfred Benzon Symposium VII. Secretory Mechanisms of Exocrine Glands.* (Ed. N. A. Thorn and O. H. Petersen), Munksgaard, Copenhagen, pp. 2101–2109

Berridge, M. J., Lindley, B. D. and Prince, W. T. (1975b) Membrane permeability changes during stimulation of isolated salivary glands of *Calliphora* by 5-hydroxytryptamine. *J. Physiol. Lond.*, **244**, 549–567

Berridge, M. J. and Prince, W. T. (1971) The electrical response of isolated salivary glands during stimulation with 5-hydroxytryptamine and cyclic AMP. *Phil. Trans. R. Soc. B.*, **262**, 111–120

Berridge, M. J. and Prince, W. T. (1972a) The role of cyclic AMP in the control of fluid secretion. In *Advances in Cyclic Nucleotide Research*, Vol. 1, (Eds. P. Greengard and G. A. Robison), Raven Press, New York, pp. 137–147

Berridge, M. J. and Prince, W. T. (1972b) Transepithelial potential changes during stimulation of isolated salivary glands with 5-hydroxytryptamine and cyclic AMP. *J. exp. Biol.*, **56**, 139–153

Bersimbaev, R. I., Argutinskaya, S. V. and Salganik, R. I. (1971) The stimulating action of gastrin pentapeptide and histamine on adenyl cyclase activity in rat stomach. *Experientia*, **27**, 1389–1390

Berti, F., Fumagalli, R., Folco, G. C., Omini, C. and Bernareggi, V. (1974) Role of cyclic 3′,5′-AMP on contraction and relaxation of perfused rat caudal artery. *Pharmac. Res. Commun.*, **61**, 519

Berti, F., Trabucchi, M., Bernareggi, V. and Fumagalli, R. (1972) The effects of prostaglandins on cyclic-AMP formation in cerebral cortex of different mammalian species. *Pharmac. Res. Commun.*, **4**, 253–259

Bevan, R. D., Purdy, R. E., Su, C. and Bevan, J. A. (1975) Evidence for an increase in adrenergic nerve function in blood vessels from experimental hypertensive rabbits. *Circulation Res.*, **37**, 503–508

Bieck, P. R., Oates, J. A., Robison, G. A. and Adkins, R. B. (1973) Cyclic AMP in regulation of gastric secretion in dogs and humans. *Am. J. Physiol.*, **224**, 158–164

Bilezikian, J. P. and Aurbach, G. D. (1973) A β-adrenergic receptor of the turkey erythrocyte. I. Binding of catecholamine and relationship to adenylate cyclase activity. *J. biol. Chem.*, **248**, 5577–5583

Bilezikian, J. C. and Aurbach, G. D. (1974) The effects of nucleotides on the expression of β-adrenergic adenylate cyclase activity in membranes from turkey erythrocytes. *J. biol. Chem.*, **249**, 157

Birnbaum, J. E., Sapp, T. M. and Moore, J. B., Jr. (1976) Effects of reserpine, EGF and cyclic nucleotide modulators on epidermal mitosis. *J. invest. Derm.*, **66**, 313–318

Birnbaumer, L. (1973) Hormone-sensitive adenylyl cyclases: Useful models for studying hormone receptor functions in cell-free systems. *Biochim. biophys. Acta*, **300**, 129–158

Birnbaumer, L., Nakahara, T. and Yang, P. C. (1974) Studies on receptor-mediated activation of adenylyl cyclase. II. Nucleotide and nucleoside regulation of the activities of the renal medullary adenylyl cyclase and their stimulation by neurohypophyseal hormones. *J. biol. Chem.*, **249**, 7857–7866

Birnbaumer, L. and Pohl, S. L. (1973) Relation of glucagon-specific binding sites to glucagon-dependent stimulation of adenylyl cyclase activity in plasma membranes of rat liver. *J. biol. Chem.*, **248**, 2056–2061

Birnbaumer, L., Pohl, S. L., Kraus, M. L. and Rodbell, M. (1970) The action of hormones on the adenyl cyclase system. *Adv. Biochem. Psychopharmac.*, **3**, 185–208

Birnbaumer, L., Pohl, S. L. and Rodbell, M. (1969) Adenyl cyclase in fat cells. I. Properties and the effects of adrenocorticotropin and fluoride. *J. biol. Chem.*, **244**, 2468–3476

Birnbaumer, L., Pohl, S. L. and Rodbell, M. (1971) The glucagon-sensitive adenyl cyclase system in plasma membranes of rat liver. II. Comparison between glucagon and fluoride-stimulated activities. *J. biol. Chem.*, **246**, 1857–1860

Birnbaumer, L., Pohl, S. L., Rodbell, M. and Sundby, F. (1972) The glucagon-sensitive adenylate cyclase system in plasma membranes of rat liver. VII. Hormonal stimulation: Reversibility and dependence on concentration of free hormone. *J. biol. Chem.*, **247**, 2038–2043

Birnbaumer, L. and Rodbell, M. (1969) Adenyl cyclase in fat cells. II. Hormone receptors. *J. biol. Chem.*, **244**, 3477–3482

Birnbaumer, L. and Yang, P.-C. (1974) Studies on receptor-mediated activation of adenylyl cyclases. III. Regulation by purine nucleotides of the activation of adenylyl cyclases from target organs for prostaglandins, luteinizing hormone, neurohypophyseal hormones and catecholamines. Tissue and hormone-dependent variations. *J. biol. Chem.*, **249**, 7867–7873

Birnbaumer, L., Yang, P.-C., Hunzicker-Dunn, M., Bockaert, J. and Duran, J. M. (1976) Adenylyl cyclase activities in ovarian tissues. I. Homogenization and conditions of assay in graafian follicles and corpora lutea of rabbits, rats, and pigs; Regulation by ATP and some comparative properties. *Endocrinology*, (in press)

Bishop, J. S. (1970) Inability of insulin to activate liver glycogen transferase D phosphatase in the diabetic pancreatectomized dog. *Biochim. biophys. Acta*, **208**, 208–218

Bitensky, M. W., Gorman, R. E. and Miller, W. H. (1971) Adenyl cyclase as a link between photon capture and changes in membrane permeability of frog photoreceptors. *Proc. natn. Acad. Sci. U.S.A.*, **68**, 561–562

Bitensky, M. W., Gorman, R. E. and Neufeld, A. H. (1972) Selective effects of insulin on hepatic epinephrine-responsive adenyl cyclase activity. *Endocrinology*, **90**, 1331–1335

Bitensky, M. W., Miki, N., Marcus, F. R. and Keirns, J. J. (1973) The role of cyclic nucleotides in visual excitation. *Life. Sci.*, **13**, 1451–1472

Bitensky, M. W., Wheeler, M. A., Mehta, H. and Miki, N. (1975) Cholera toxin activation of adenylate cyclase in cancer cell membrane fragments. *Proc. natn. Acad. Sci. U.S.A.*, **72**, 2572–2576

Bittner, R. R., Kraas, E., Beger, G. and Meves, M. (1972) Der Einfluß von Theophyllin auf die mit Gastrin stimulierte Magen-sekretion. *Z. Gastroent.*, **10**, 461–466

Bjorklaênd, A. and Steneir, U. (1972) Nerve growth factor: Stimulation of regenerative growth of central noradrenergic neurons. *Science N.Y.*, **175**, 1251–1253

Blat, C., Boix, N. and Harel, L. (1973) Inhibition by N^6-2′-*O*-dibutyryl 3′,5′-cyclic adenosine monophosphate of phosphate transport and metabolism in BHK 21C13 and BHK 21Py cells. *Cancer Res.*, **334**, 2104–2108

Blecher, M. (1971) Biological effects and catabolic metabolism of 3′,5′-cyclic nucleotides and derivatives in rat adipose tissue and liver. *Metabolism.*, **20**, 63–77

Blecher, M., Ro'Ane, J. T. and Flynn, P. D. (1970) Biochemical activities of tubercidin 3′,5′-cyclic monophosphate in rat epididymal adipose tissue. *Biochem. Pharmac.*, **20**, 249–251

Bliss, E. L., Ailion, J. and Zwanziger, J. (1968) Metabolism of norepinephrine, serotonin and dopamine in rat brain with stress. *J. Pharmac. exp. Ther.*, **1641**, 122–134

Bloch, A. (1974) Cytidine 3′,5′-monophosphate (cyclic CMP) I. Isolation from extracts of leukemia L-1210 cells. *Biochem. biophys. Res. Commun.*, **58**, 652–659

Bloch, A. (1975) Uridine 3′,5′-monophosphate (cyclic UMP) I. Isolation from rat liver extracts. *Biochem. biophys. Res. Commun.*, **64**, 210–218

Bloch, A., Dutschman, G. and Maue, R. (1974) Cytidine 3′,5′-monophosphate (cyclic (CMP) II. Initiation of leukemia L-1210 cell growth. *In vitro.*, **59**, 955–959

Bloch, A., Hromchak, R. and Henderson, E. S. (1975) Isolation of cytidine 3′,5′-monophosphate (cyclic CMP) From the urine of leukemic patients. *Proc. Am. Ass. Cancer Res.*, **16**, 191

Bloch, A. and Leonard, R. J. (1977) Formation of cytidine 3′,5′-monophosphate (cyclic CMP) in extracts of leukemia L-1210 cells. *Proc. Am. Ass. Cancer Res.*, (in press)

Block, A. (1975) Isolation of cytidine 3′,5′-monophosphate from mammalian tissues and body fluids and its effects on leukemia L-1210 cell growth in culture. In *Advances in Cyclic Nucleotide Research*, Vol. 5, (Eds. G. I. Drummond, P. Greengard and G. A. Robison), Raven Press, New York, pp. 331–338

Block, A. I., Feinberg, H., Herbyczynska-Cedro, K. and Vane, I. R. (1975) Anoxia-induced release of prostaglandins in rabbit isolated hearts. *Circulation Res.*, **36**, 34–42

Blonde, L., Wehmann, R. E. and Steiner, A. L. (1974) Plasma clearance rates and renal clearance of ^{3}H-labeled cyclic AMP and ^{3}H-labeled cyclic GMP in the dog. *J. Clin. Invest.*, **53**, 163

Bloom, F. E. (1974) To spritz or not to spritz: The doubtful value of aimless iontophoresis. *Life Sci.*, **14**, 1819–1834

Bloom, F. E. (1975) The role of cyclic nucleotides in central synaptic function. *Rev. Physiol. Biochem. Pharmac.*, **74**, 1–103

Bloom, F. E. and Giarman, N. J. (1968) Physiologic and pharmacologic considerations of biogenic amines in the nervous system. *Ann. Rev. Pharmac.*, **8**, 229–258

Bloom, F. E., Hoffer, B. J. and Siggins, G. R. (1971) Studies on norepinephrine containing afferents to Purkinje cells of rat cerebellum. I. Localization of the fibers and their synapses. *Brain Res., Osaka*, **25**, 501–521

Bloom, F. E., Siggins, G. R. and Hoffer, B. J. (1974a) Interpreting the failures to confirm the depression of cerebellar Purkinje cells by cyclic AMP. *Science, N.Y.*, **185**, 627–629

Bloom, F. E., Siggins, G. R., Hoffer, B. J., Segal, M. and Oliver, A. P. (1975) The role of cyclic nucleotides in the central synaptic actions of catecholamines. In Advances in Cyclic Nucleotide Research, Vol. 5, (Eds. G. Drummond, P. Greengard and G. A. Robinson), Raven Press, New York, pp. 603–618

Bloom, F. E., Wedner, H. J. and Parker, C. W. (1973) The use of antibodies to study cell structure and metabolism. *Pharm. Rev.*, **25**, 343–358

Bloom, S. R., Mortimer, C. H., Thorner, M. O., Besser, G. M., Hall, R., Gomez-Pan, A., Roy, W. M., Russel, R. C. G., Coy, D. H., Kastin, A. J. and Schally, A. V. (1974b) Inhibition of gastrin and gastric acid secretion by growth-hormone release-inhibiting hormone. *Lancet*, iv, 1106–1109

Blumberg, J. B., Taylor, R. E. and Sulser, F. (1975) Blockade by pimozide of a norepinephrine-sensitive adenylate cyclase in the limbic forebrain: Possible role of limbic noradrenergic mechanisms in the mode of action of antipsychotics. *J. Pharm. Pharmac.*, **27**, 127–128

Blumberg, J. B., Vetulani, J., Stawarz, R. J. and Sulser, F. (1976) The noradrenergic cyclic AMP-generating system in the limbic forebrain: Pharmacological characterization and possible role of limbic noradrenergic mechanisms in the mode of action of antipsychotics. *Europ. J. Pharmac.*, **37**, 357–366

Blume, A. J., Dalton, C. and Sheppard, H. (1973) Adenosine mediated elevation of cyclic 3′,5′-adenosine monophosphate concentrations in cultured mouse neuroblastoma cells. *Proc. natn. Acad. Sci. U.S.A.*, **70**, 3099–3102

Bockaert, J., Hunzicker-Dunn, M. and Birnbaumer, L. (1976) Hormone-stimulated desensitization of hormone-dependent adenylyl cyclase: Dual action of luteinizing hormone on pig graafian follicles. *J. biol. Chem.*, **251**, 2653–2663

Bockaert, J., Roy, C. and Jard, S. (1972) Oxytocin-sensitive adenylate cyclase in frog bladder epithelial cells. *J. biol. Chem.*, **247**, 7073–7081

Bockaert, J., Roy, C., Rajerison, R. and Jard, S. (1973) Specific binding of [^{3}H]lysine–vasopressin to pig kidney plasma membranes: Relationship of receptor occupancy to adenylate cyclase activation. *J. biol. Chem.*, **248**, 5922–5931

Boden, G., Sivitz, M. C., Owen, O. E., Essa-Koumar, N. and Landor, J. H. (1975) Somatostatin suppresses secretin and pancreatic exocrine secretion. *Science*, **190**, 163–165

Boeynaems, J. M., Van Sande, J., Pochet, R. and Dumont, J. E. (1974) The relation between adenylate cyclase activation and cAMP accumulation in the horse thyroid gland stimulated by thryotropin. *Molec. cell. Endocr.*, **1**, 135–155

Böhm, M., Bartel, S., Krause, E.-G., Wollenberger, A., Romaniuk, P., Porstmann, W., Parsi, R. A. and Witte, J. (1976) Cyclic AMP levels in coronary sinus and arterial blood of patients with ischemic heart disease before and after pacing of the heart. Deutsch. Gesundheitswesen, **31**, 1560–1566

Bohr, D. F. (1974) Reactivity of vascular smooth muscle from normal and hypertensive rats: Effect of several cations. *Fed. Proc.*, **33**, 127–132

Bolton, T. B. (1971) On the nature of the oscillations of the membrane potential slow wave produced by acetylcholine or carbachol in intestinal smooth muscle. *J. Physiol. Lond.*, **216**, 403–418

Borgeat, P., Labrie, F., Drouin, J., Belanger, A., Immer, H., Sestanj, K., Nelson, V., Gotz, M., Schally, A. V., Coy, D. H. and Coy, E. J. (1974) Inhibition of adenosine 3′,5′-monophosphate accumulation in anterior pituitary gland *in vitro* by growth-hormone release-inhibiting hormone. *Biochem. biophys. Res. Commun.*, **56**, 1025–1059

Borle, A. B. (1974) Cyclic AMP stimulation of calcium efflux from kidney, liver and heart mitochondria. *J. Membrane Biol.*, **16**, 221–236

Borman, L. S., Dumont, J. N. and Hsie, A. W. (1975) Relationship between cyclic AMP, microtubule organization, and mammalian cell shape. *Exp. Cell Res.*, **91**, 422–428

Boswell, K. H., Miller, J. O., Shuman, D. A., Sidwell, R. W., Simon, L. N. and Robins, R. K. (1973) Synthesis and biological activity of certain carbamoyl and alkoxycarbonyl derivatives of adenosine 3′,5′-cyclic phosphate. *J. med. Chem.*, **16**, 1075–1079

Bourne, H. R., Coffino, P., Hochman, J., Insel, P. A., Jones, P., Melmon, K. L. and

Tomkins, G. M. (1975a) Genetic and functional studies on the cyclic AMP-receptor in a cultured mammalian cell, *Proc. 6th. Int. Congr. Pharmac., Helsinki*, **1**, 223–232

Bourne, H. R., Coffino, P., Melmon, K. L., Tomkins, G. M. and Weinstein, Y. (1975b) Genetic analysis of cyclic AMP in a mammalian cell. In *Advances in Cyclic Nucleotide Research*, Vol. 5, (Eds. G. I. Drummond, P. Greengard, and G. A. Robison), Raven Press, New York, pp. 771–786

Bourne, H. R., Coffino, P. and Tomkins, G. M. (1975c) Selection of a variant lymphoma cell deficient in adenylate cyclase. *Science, N.Y.*, **187**, 750–752

Bourne, H. R., Lehrer, R. I., Cline, M. J. and Melmon, K. L. (1971) Cyclic 3′,5′-adenosine monophosphate in the human leukocyte: Synthesis, degradation, and effects on neutrophil candidacidal activity, *J. clin. Invest.*, **50**, 920–929

Bourne, H. R., Lichtenstein, L. M. and Melmon, K. L. (1972) Pharmacologic control of allergic histamine release *in vitro:* Evidence for an inhibitory role of 3′,5′-adenosine monophosphate. *J. Immunol.*, **108**, 695–705

Bourne, H. R. and Melmon, K. L. (1971) Adenyl cyclase in human leucocytes: Evidence for activation by separate β-adrenergic and prostaglandin receptors. *J. Pharmac. exp. Ther.*, **178**, 1–7

Bourne, H. R., Tomkins, G. M. and Dion, S. (1973) Regulation of phosphodiesterase synthesis: Requirement for cyclic adenosine monophosphate-dependent protein kinase. *Science, N.Y.*, **181**, 952–954

Bowen, V. and Lazarus, N. R. (1974) Insulin release from the perfused rat pancreas. Mode of action of tolbutamide. *Biochem. J.*, **142**, 385–389

Bowery, B. and Lewis, G. P. (1973) Inhibition of functional vasodilatation and prostaglandin formation in rabbit adipose tissue by indomethacin and aspirin. *Br. J. Pharmac.*, **47**, 305–314

Boyd, H., Louis, C. J. and Martin, T. J. (1974) Activity and hormone resonsiveness of adenyl cyclase during induction of tumors in rat liver with 3′-methyl-4-dimethyl amino-azobenzene. *Cancer Res.*, **34**, 1720–1725

Boyntoh, A. L., Whitfield, J. F., Isaacs, R. J. and Morton, H. J. (1974) Control of 3T3 cell proliferation by calcium. *In Vitro*, **10**, 12–17

Bradbury, E. M., Inglis, R. J., and Matthews, H. R. (1974) Control of cell division by very lysine rich histone (F_1) phosphorylation, *Nature, Lond.*, **247**, 257–261

Bradham, L. S. (1972) Comparison of the effects of Ca^{2+} and Mg^{2+} on the adenyl cyclase of beef brain. *Biochem. biophys. Acta*, **276**, 434–443

Bradham, L. S., Holt, D. A. and Sims, M. (1970) The effect of Ca^{2+} on the adenyl cyclase of calf brain. *Biochem. Biophys. Acta*, **201**, 250–260

Brandt, M., Gullis, R. J., Fischer, K., Buchen, C., Hamprecht, B., Moröder, L. and Wünsch, E. (1976) Enkephalin regulates the levels of cyclic nucleotides in neuroblastoma × glioma hybrid cells. *Nature, Lond.*, **262**, 311–313

Braun, T., Hechter, O. and Bär, H.-P. (1969) Lipolytic activity of ribo-nucleotide and deoxyribonucleotide-3′,5′-cyclic monophosphates in isolated rat fat cells. *Proc. Soc. exp. Biol. Med.*, **132**, 233–236

Braun, W., Lichtenstein, L. M. and Parker, C. W. (1974) *Cyclic AMP, Cell Growth and the Immune Response*, Springer-Verlag, New York

Braun-Falco, O. (1958) The histochemistry of psoriasis. *Ann. N.Y. Acad. Sci.*, **73**, 936–976

Braunwald, E., Ross, J. and Sonnenblick, E. H. (1967) *Mechanisms of Contraction of the Normal and Failing Heart*, J. and A. Churchill, London

Brazeau, P., Vale, W., Burgus, R., Ling, N., Butcher, M., Rivier, J. and Guillemin, R. (1973) Hypothalamic polypeptide that inhibits the secretion of immunoreactive pituitary growth hormone. *Science, N.Y.*, **179**, 77–79

Breckenridge, B. M., Burn, J. H. and Matschinsky, F. M. (1967) Theopylline epinephrine, and neostigmine facilitation of neuromuscular transmission. *Proc. natn. Acad. Sci. U.S.A.*, **57**, 1893–1897

Breckenridge, B. M. and Johnson, R. E. (1969) Cyclic 3′,5′-nucleotide phosphodiesterase in brain. *J. Histochem. Cytochem.*, **17**, 505–511

Brent, L. and Holborrow, J. (1974) *Progress in Immunology*, Vol. 2, North Holland Publishing Co., Amsterdam

Bressler, B. H., Phillis, J. W. and Kozachuk, W. (1975) Noradrenaline stimulation of a membrane pump in frog skeletal muscle. *Europ. J. Pharmac.*, **33**, 201–204

Bricker, L. A. and Levey, G. S. (1972) Autonomous cholesterol and fatty acid synthesis in hepatomas: Deletion of the adenosine 3′,5′-cyclic monophosphate control mechanism of normal liver. *Biochem. biophys. Res. Commun.*, **48**, 362–365

Brisson, G. R. and Malaisse, W. J. (1973) The stimulus–secretion coupling of glucose-induced insulin release. XI. Effects of theophylline and epinephrine on ^{45}Ca efflux from perifused islets. *Metabolism*., **22**, 455–465

Brisson, G. R., Malaisse-Lagae, F. and Malaisse, W. J. (1972) The stimulus–secretion coupling of glucose-induced insulin release. VII. A proposed site of action for adenosine 3′,5′-cyclic monophosphate. *J. clin. Invest.*, **52**, 232–241

Broadus, A. E., Hardman, J. G., Kaminsky, N. I., Ball, J. H., Sutherland, E. W. and Liddle, G. W. (1971) Extracellular cyclic nucleotides. *Ann. N.Y. Acad. Sci. U.S.A.*, **185**, 50–65

Broadus, A. E., Kaminsky, N. I., Northcutt, R. C., Hardman, J. G., Sutherland, E. W. and Liddle, G. W. (1970) Effects of glucagon on adenosine 3′,5′-monophosphate and guanosine 3′,5′-monophosphate in human plasma and urine. *J. clin. Invest.*, **49**, 2237–2245

Brody, M, J. (1966) Neurohormonal mediation of active reflex vasodilation. *Fed. Proc.*, **25**, 1583–1592

Brody, M. J. and Kadowitz, P. J. (1974) Prostaglandins as modulators of the autonomic nervous system. *Fed. Proc.*, **33**, 48–60

Brooker, G. (1973) Oscillation of cyclic adenosine monophosphate concentration during the myocardial contraction cycle. *Science, N.Y.*, **182**, 933–934

Brooker, G. (1975) Implications of cyclic nucleotide oscillations during the myocardial contraction cycle. In *Advances in Cyclic Nucleotide Research*, Vol. 5, (Eds. G. I. Drummond, P. Greengard and G. A. Robison), Raven Press, New York, pp. 435–452

Brooker, G. and Fichman, M. (1971) Chlorpropamide and tolbutamide inhibition of adenosine 3′,5′-cyclic monophosphate phosphodiesterase. *Biochem. biophys. Res. Commun.*, **42**, 824–828

Brooker, G., Thomas, L. J., Jr. and Appleman, M. M. (1968) The assay of adenosine 3′,5′-cyclic monophosphate in biological materials by enzymatic radioisotopic displacement. *Biochemistry*, **7**, 4177–4181

Brooks, J. C. and Siegel, F. L. (1973a) Purification of a Ca^{2+}-binding protein from beef adrenal medulla. *J. biol.Chem.*, **248**, 4, 189–4, 193

Brooks, J. C. and Siegel, F. L. (1973b) Calcium binding phosphoprotein: The principle acidic protein of mammalian sperm. *Biochem. biophys. Res. Commun.*, **55**, 710–716

Brostrom, C. O., Corbin, J. D., King, C. A. and Krebs, E. G. (1971a) Interaction of the subunits of cyclic AMP-dependent protein kinase of muscle. *Proc. natn. Acad. Sci. U.S.A.*, **68**, 2444–2447

Brostrom, C. O., Huang, Y. C., Breckenridge, B. McL. and Wolff, D. J. (1975) Identification of a calcium-binding protein as a calcium-dependent regulator of brain adenylate cyclase. *Proc. natn. Acad. Sci. U.S.A.*, **72**, 64–68

Brostrom, C. O., Hunkeler, F. L. and Krebs, E. G. (1971b) The regulation of skeletal muscle phosphorylase kinase by Ca^{2+}. *J. biol. Chem.*, **246**, 1961–1967

Brostrom, C. O. and Wolff, D. J. (1974) Calcium-dependent cyclic nucleotide phosphodiesterase from glial tumor cells. *Archs. Bioch. Biophys.*, **165**, 715–727

Brostrom, C. O. and Wolff, D. J. (1976) Calcium-dependent cyclic nucleotide phosphodiesterase from brain: Comparison of adenosine 3′,5′-monophosphate and guanosine 3′,5′-monophosphate as substrates. *Arch. Biochem. Biophys.*, **172**, 301–311

Brostrom, M. A., Kon, C., Olson, D. R. and Breckenridge, B. M. (1974) Adenosine 3′,5′-monophosphate in glial tumor cells treated with glucocorticoids. *Mol. Pharmac.*, **10**, 711–720

Brostrom, M. A., Reimann, E. M., Walsh, D. A. and Krebs, E. G. (1970) A cyclic AMP-stimulated protein kinase from cardiac muscle. *Adv. Enzyme Regul.*, **8**, 191–203

Brown, B. L., Elkins, R. P. and Albano, J. D. M. (1972) Saturation assay for cyclic AMP using endogenous binding protein. In *Advances in Cyclic Nucleotide Research*, Vol. 2, (Eds. P. Greengard, R. Paoletti and G. A. Robison), Raven Press, New York, pp. 25–40

Brown, E. M., Fedak, S. A., Woodward, C. J. and Aurbach, G. D. (1976) β-Receptor interactions: Direct comparison of receptor interaction and biological activity. *J. biol. Chem.*, **251**, 1239–1246

Brown, H. D., Chattopadhyay, S. K., Spjut, H. T., Spratt, J. S., Jr. and Pennington, S. N. (1969) Adenyl cyclase activity in dimethylamine biphenyl-induced breast carcinoma. *Biochim. biophys. Acta*, **192**, 372–375

Brown, H. D., Chattopadhyay, S. K., Morris, H. P. and Pennington, S. N. (1970) Adenyl cyclase activity in Morris hepatomas 7777, 779A, and 9618A. *Cancer Res.*, **30**, 123–126

Browning, E. T., Groppi, V. and Kon, C. (1974a) Papaverine. A potent inhibitor of respiration in C6 astrocytoma cells. *Molec. Pharmac.*, **10**, 175–181

Browning, E. T., Schwartz, J. P. and Breckenridge, B. M. (1974b) Norepinephrine-sensitive properties of C6 astrocytoma cells. *Molec. Pharmac.*, **10**, 162–174

Bruderlein, F., Humber, L. and Voith, K. (1975) Neuroleptic agents of the benzocyclo-heptapyridoisoquinoline series: I. The synthesis, stereochemical and structural requirements for activity of butaclamol and related compounds. *J. med. Chem.*, **18**, 185–188

Brunswick, D. J. and Cooperman, B. S. (1971) Photo-affinity labels for adenosine 3′,5′-cyclic monophosphate. *Proc. natn. Acad. Sci. U.S.A.*, **68**, 1801–1804

Brunswick, D. J. and Cooperman, B. S. (1973) Synthesis and characterization of photoaffinity labels for adenosine 3′,5′-cyclic monophosphate and adenosine 5′-monophosphate. *Biochemistry*, **12**, 4074–4078

Brush, J. S., Sutliff, L. S. and Sharma, R. K. (1974) Metabolic regulation and adenyl cyclase activity of adrenocortical carcinoma cultured cells. *Cancer Res.*, **34**, 1495–1502

Bryan, J. (1974) Biochemical properties of microtubules. *Fed. Proc.*, **33**, 152–157

Byrant, R. E. and Sutcliffe, M. C. (1974) The effect of 3′,5′-adenosine monophosphate on granulocyte adhesion. *J. clin. Invest.*, **54**, 1241–1244

Bublitz, C. (1973) Effects of lipids on cyclic-nucleotide phosphodiesterases. *Biochem. biophys. Res. Commun.*, **52**, 173–180

Buccino, R. A., Spann, J. F., Jr., Pool, P. E., Sonnenblick, E. B. and Braunwald, E. (1967) Influence of thyroid state on the intrinsic contractile properties and energy stores of the myocardium. *J. clin. Invest.*, **46**, 1669

Bülbring, E. and Tomita, T. (1969) Increase of membrane conductance by adrenaline in the smooth muscle of guinea-pig *Taenia coli. Proc. R. Soc., B.*, **172**, 89–102

Bülbring, E. and Tomita, T. (1970) Calcium and the action potential in smooth muscle. In *Calcium and Cellular Function*, (Ed. A. W. Cuthbert), Macmillan, London, pp. 249–260

Bullough, W. S. and Laurence, E. B. (1961) Stress and adrenaline in relation to the diurnal cycle of epidermal mitotic activity in adult male mice. *Proc. R. Soc. B.*, **154**, 540–556

Bunney, B. S. and Aghajanian, G. K. (1973) Electrophysiological effects of amphetamine in dopaminergic neurons. In *Frontiers in Catecholamine Research*, (Eds. E. Usdin and S. Snyder), Pergamon, New York, pp. 957–962

Bunney, B. S. and Aghajanian, G. K. (1974) A comparison of the effects of chlorpromazine, 7-hydroxychlorpromazine and chlorpromazine sulfoxide on the activity of central dopaminergic neurons. *Life Sci.*, **15**, 309–318

Bunney, B. S., Roth, R. H. and Aghajanian, G. K. (1975) Effects of molindone on central dopaminergic neuronal activity and metabolism: Similarity to other neuroleptics. *Psychopharmac. Comm.*, **4**, 349–358

Bunney, B. S., Walters, J. R., Roth, R. H. and Aghajanian, G. K. (1973) Dopaminergic neurons: Effect of antipsychotic drugs and amphetamine on single cell activity. *J. Pharmac. exp. Ther.*, **185**, 560–571

Bunney, W. E. and Davis, J. M. (1965) Norepinephrine in depressive reactions. *Archs. gen. Psychiat.*, **13**, 484–494

Buonassisi, V., Sato, G. and Cohen, A. I. (1962) Hormone-producing cultures of adrenal and pituitary origin. *Proc. natn. Acad. Sci. U.S.A.*, **48**, 1184–1190

Burgus, R., Dunn, T. E., Desiderio, D. and Guillemin, R. (1969) Structure moléculaire du facteur hypothalamique hypophysiotrope TRF d)origine ovine. Mise en évidence par spectrométrie de masse de la séquence PCA–his–Pro–NH_2. *C.r. Lebd. Séanc. Acad. Sci.*, **269**, 1870–1873

Bürk, R. R. (1968) Reduced adenyl cyclase activity in a polyoma virus-transformed cell line. *Nature Lond.*, **219**, 1272–1275

Burns, T. W., Langley, P. E. and Robison, G. A. (1975) Site of free fatty acid inhibition of lipolysis by human adipocytes. *Metabolism*, **24**, 265–276

Butcher, F. R. (1975) The role of calcium and cyclic nucleotides in α-amylase release from slices of rat parotid: Studies with the divalent cation ionophore A-23187. *Metabolism*, **24**, 409–418

Butcher, F. R., Becker, J. E. and Potter, V. R. (1971) Induction of tyrosine amino transferase by dibutyryl-cyclic AMP employing hepatoma cells in tissue culture. *Exp. Cell Res.*, **66**, 321–328

Butcher, F. R., Scott, D. F., Potter, V. R. and Morris, H. P. (1972) Endocrine control of cyclic adenosine 3′,5′-monophosphate levels in several Morris hepatomas. *Cancer Res.*, **32**, 2135–2140

Butcher, F. R., Thayer, M. and Goldman, J. A. (1977) Effect of adenosine 3′,5′-cyclic monophosphate derivatives on α-amylase release, protein kinase and cyclic nucleotide phosphodiesterase activity from rat parotid tissue. *Biochim. biophys. Acta*, (in press)

Butcher, R. W. (1970) Prostaglandins and cyclic AMP. *Adv. Biochem. Psychopharmac.*, **3**, 173–183

Butcher, R. W. and Baird, C. E. (1968) Effects of prostaglandins on cyclic AMP levels in fat and other tissues. *J. biol. Chem.*, **243**, 1713–1717

Butcher, R. W. and Baird, C. E. (1969) The regulation of cyclic AMP and lipolysis in adipose tissue by hormones and other agents. In *Drugs Affecting Lipid Metabolism*, (Eds. W. L. Holmes, L. A. Carlson and R. Paoletti), Plenum Press, New York, pp. 5–23

Butcher, R. W., Ho, R. J., Meng, H. C. and Sutherland, E. W. (1965) The measurement of adenosine 3′,5′-monophosphate in tissues and the role of the cyclic nucleotide in the lipolytic response of fat to epinephrine. *J. biol. Chem.*, **240**, 4515–4523

Butcher, R. W., Sneyd, J. G. T., Park, C. R. and Sutherland, E. J., Jr. (1966) Effect of insulin on adenosine 3′,5′-monophosphate in the rat epididymal fat pad. *J. biol. Chem.*, **241**, 1651–1653

Butcher, R. W. and Sutherland, E. W. (1962) Adenosine 3′,5′-phosphate in biological material. I. Purification and properties of cyclic 3′,5′-nucleotide phosphodiesterase and use of this enzyme to characterize adenosine 3′,5′-phosphate in human urine. *J. biol. Chem.*, **237**, 1244–1250

Byron, J. W. (1972) Evidence for a β-adrenergic receptor initiating DNA synthesis in hemopoietic stem cells. *Expl Cell Res.*, **71**, 228–232

Cailla, H. and Delaage, M. (1972) Succinyl derivatives of adenosine 3′,5′-cyclic monophosphate: Synthesis and purification. *Anal. Biochem.*, **48**, 62–72
Callingham, B. A. and Burgen, A. S. W. (1966) The uptake of isoprenaline by the perfused rat liver. *Mol. Pharmac.*, **3**, 37–42
Cannon, W. B. and Smith, P. E. (1922) Studies on the conditions of activity in endocrine glands. IX. Further evidence of nervous control of thyroid secretion. *Am. J. Physiol.*, **60**, 476–495
Capito, K. and Hedeskov, C. J. (1974) The effect of starvation on phosphodiesterase activity and the content of adenosine 3′,5′-cyclic monophosphate in isolated mouse pancreatic isles. *Biochem. J.*, **142**, 653–658
Carcham, R. A., Jaanus, S. D. and Rubin, R. P. (1971) The role of adrenocorticotropin and calcium in adenosine 3′,5′-phosphate production and steroid release from the isolated perfused cat adrenal gland. *Mol. Pharmac.*, **7**, 491–499
Carchman, R. A., Johnson, G. S. and Pastan, I. (1974) Studies on the levels of cyclic AMP in cells transformed by wild-type and temperature-sensitive Kirsten sarcoma virus. *Cell*, **1**, 59–64
Carenzi, A., Guidotti, A. and Costa, E. (1974) Adenyl cyclase in rat striatum and *N. accumbens*: Stimulation by dopamine receptor agonists. *Pharmacologist*, **16**, 287
Carr, R. H., Busse, W. W. and Reed, C. E. (1973) Failure of catecholamines to inhibit epidermal mitosis *in vitro*. *J. Allergy clin. Immunol.*, **51**, 255–262
Case, R. M. (1973) Calcium and gastrointestinal secretion. *Digestion*, **8**, 269–288
Casnellie, J. E. and Greengard, P. (1974) Guanosine 3′,5′-cyclic monophosphate-dependent phosphorylation of endogenous substrate proteins in membranes of mammalian smooth muscle. *Proc. natn. Acad. Sci., U.S.A.*, **71**, 1891–1895
Castor, C. W. (1974) Connective tissue activation. VI. The effects of cyclic nucleotides on human synovial cells *in vitro*. *J. Lab. clin. Med.*, **83**, 46–55
Castor, C. W. (1975a) Connective tissue activation. VII. Evidence supporting a role for prostaglandins and cyclic nucleotides. *J. Lab. clin. Med.*, **85**, 392–404
Castor, C. W. (1975b) Connective tissue activation. IX. Modification by pharmacological agents. *Arthritis Rheum.*, **18**, 451–460
Cehovic, G., Bayer, M. and Giao, N. (1976a) Synthesis of new cyclic nucleotides and their differential stimulatory effects on thyroid function in mice. *J. med. Chem.*, **19**, 899–903
Cehovic, G., Gabbai, A., Marcus, I. and Posternak, T. (1974) Derivatives of cyclo adenosine-3,5-phosphoric acid and their preparation. US Pat. No. 3,856,776
Cehovic, G., Giao, N. B. and Posternak, T. (1976b) The relative specific activity of new cyclic nucleotides on different endocrine functions. Abs Endocrine Soc.
Cehovic, G., Marcus, I., Gabbai, A. and Posternak, T. (1970) Etude de l'action de certains nouveaux analogues de l'AMP cyclique sur la liberation de croissance et de la prolactine *in vitro*. *C. R. Lebd. Séanc. acad. Sci.*, **271**, 1399–1401
Cehovic, G., Marcus, I., Vengadabady, S. and Posternak, T. (1968) Sur la préparation de l'acide iso-adénosine 3′,5′-phosphorique (iso-AMP cyclique) et sur certaines de ses propriétés biologiques. *C. R. Séanc. Phys. Hist. nat. Genève*, **3**, 135–139
Cehovic, G., Posternak, T. and Charollais, E. (1972) A study of the biological activity and resistance to phosphodiesterase of some derivatives and analogues of cyclic AMP. In *Advances Cyclic Nucleotide Research*, Vol. 1 (Eds. P. Greengard and G. A. Robinson), Raven Press, New York, pp. 521–540
Cerasi, E. (1975) Mechanisms of glucose stimulated insulin secretion in health and in diabetes. *Diabetologia*, **11**, 1–13
Chabardes, D., Imbert, M., Clique, A., Montegut, M. and Morel, F. (1975) PTH sensitive adenyl cyclase activity in different segments of the rabbit nephron. *Pflügers Arch.*, **354**, 229–239
Chader, G. J., Hertz, L. R. and Fletcher, T. (1974) Light-activation of phosphodiesterase activity in ROS. *Biochim. biophys. Acta*, **347**, 491–493

Chalazonitis, A. and Greene, L. A. (1974) Enhancement in excitability properties of mouse neuroblastoma cells cultured in the presence of dibutyryl-cyclic AMP. *Brain Res.*, **72**, 340–345

Chambers, D. A., Martin, D. W., Jr. and Weinstein, Y. (1974) The effect of cyclic nucleotides on purine biosynthesis and the induction of PRPP synthetase during lymphocyte activation. *Cell*, **3**, 373–380

Champion, S., Haye, B. and C. Jacquemin (1974) Cholinergic control by endogenous prostaglandins of cAMP accumulation under TSH stimulation in the thyroid. *FEBS Letters*, **46**, 289–292

Chang, K. J. and Cuatrecasas, P. (1974) ATP-Dependent inhibition of insulin stimulated glucose transport in fat cells. *J. biol. Chem.*, **249**, 3170–3180

Chang, K.-J., Marcus, N. A. and Cuatrecasas, P. (1974) Cyclic adenosine monophosphate-dependent phosphorylation of specific fat cell membrane proteins by an endogenous membrane-bound protein kinase. *J. biol. Chem.*, **249**, 6854–6865

Channing, C. P. and Kammerman, S. (1973) Characteristics of gonadotrophin receptors of porcine granulosa cells during follicle maturation. *Endocrinology*, **92**, 531–540

Charles, M. A., Lawecki, J., Pictet, R. and Grodsky, G. M. (1975) Insulin secretion. Interrelationships of glucose, cyclic adenosine 3′,5′-monophosphate, and calcium. *J. biol. Chem.*, **250**, 6134–6140

Charters, A. C., Chandler, J. G., Rosen, H. and Orloff, M. J. (1973) The role of cyclic AMP in gastric acid secretion. *Gastroenterology*, **64**, 708

Chasin, M., Harris, D. W., Phillips, M. B. and Hess, S. M. (1972) 1-Ethyl-4-(isopropylidenehydrazino-pyrazol-(3,4b)-pyridine-5-carboxylic acid, ethyl ester, hydrochloride (SQ-20009)—A potent new inhibitor of cyclic 3′,5′-nucleotide phosphodiesterases. *Biochem. Pharmac.*, **21**, 2443–2450

Chasin, M., Mamrak, I. and Samamego, S. G. (1974) Preparation and properties of a cell-free hormonally responsive adenylate cyclase from guinea-pig brain. *J. Neurochem.*, **22**, 1031–1038

Chasin, M., Mamrak, F., Samaniego, S. G. and Hess, S. M. (1973) Characteristics of the catecholamine and histamine-receptor sites mediating accumulation of cyclic adenosine3′,5′-monophosphate in guinea-pig brain. *J. Neurochem.*, **21**, 1415–1427

Chasin, M., Rivkin, I., Mamrak, F., Samaniego, G. and Hess, S. M. (1971) α- and β-Adrenergic receptors as mediators of accumulation of cyclic adenosine 3′,5′-monophosphate in specific areas of guinea-pig brain. *J. biol. Chem.*, **246**, 3037–3041

Chayoth, R., Epstein, S. M. and Field, J. B. (1973) Glucagon and prostaglandin E_1 stimulation of cyclic adenosine 3′,5′-monophosphate levels and adenylate cyclase activity in benign hyperplastic nodules and malignant hepatomas of ethionine-treated rats. *Cancer Res.*, **33**, 1970–1974

Chen, C. and Hirsch, J. G. (1972a) Restoration of antibody-forming capacity in cultures of nonadherent spleen cells by mercaptoethanol. *Science, N.Y.*, **176**, 60–61

Chen, C. and Hirsch, J. G. (1972b) The effects of mercaptoethanol and of peritoneal macrophages on the antibody-forming capacity of nonadherent mouse spleen cells *in vitro*. *J. exp. Med.*, **136**, 604–617

Cheung, W. Y. (1967) Properties of cyclic 3′.5′-nucleotide phosphodiesterase from rat brain. *Biochemistry*, **6**, 1079–1087

Cheung, W. Y. (1970) Cyclic 3′,5′-nucleotide phosphodiesterase: Demonstration of an activator. *Biochem. Biophys. Res. Commun.*, **38**, 533–538

Cheung, W. Y. (1971) Cyclic 3′,5′-nucleotide phosphodiesterase: Evidence for and properties of a protein activator. *J. biol. Chem.*, **246**, 2859–2869

Cheung, W. Y., Lin, Y. M., Liu, Y. P. and Smoake, J. J. (1975) Regulation of bovine brain cyclic 3′,5′-nucleotide phosphodiesterase by its protein activator. In *Cyclic Nucleotides in Diseases* (Ed. B. Weiss), University Park Press, Baltimore, pp. 321–350

Cheung, W. Y. and Williamson, J. R. (1965) Kinetics of cyclic adenosine monophosphate changes in rat heart following epinephrine. *Nature, Lond.*, **207**, 979–981

Chi, Y. and Francis, D. (1971) Cyclic AMP and calcium exchange in a cellular slime mould. *J. Cell Physiol.*, **77**, 169–174

Chiba, S., Kubota, K. and Hashimoto, K. (1972) Absence of chronotropic effects of dibutyryl cyclic adenosine 3′,5′-monophosphate on the dog S–A node. *Tokushima J. exp. Med.*, **107**, 103–105

Chidsey, C. A., Kaiser, G. A., Sonnenblick, E. H., Spain, J. F. and Braunwald, E. (1964) Cardiac norepinephrine stores in experimental heart failure in the dog. *J. clin. Invest.*, **43**, 2386–2394

Childers, S. R. and Siegel, F. L. (1975) Isolation and purification of a calcium-binding protein from electroplax of *Electrophorus electricus*. *Biochem. biophys. Acta*, **405**, 99–108

Chlapowski, F. J., Kelly, L. A. and Butcher, R. W. (1975) Cyclic nucleotides in cultured cells. In *Advances in Cyclic Nucleotide Research*, Volume 5 (Eds. G. I. Drummond, P. Greengard and G. A. Robison), Raven Press, New York, pp. 245–338

Chlouverakis, C. (1963) Parasympathomimetic agents and the metabolism of rat adipose tissue. *Metabolism*, **12**, 936–940

Cho-Chung, Y. S. (1974) *In vivo* inhibition of tumor growth by cyclic adenosine 3′,5′-monophosphate derivatives. *Cancer Res.*, **34**, 3492–3496

Cho-Chung, Y. and Berghoffer, B. (1974) The Role of cyclic AMP in neoplastic cell growth and regression II. Growth arrest and glucose-6-phosphate dehydrogenase isozyme shift by dibutryryl cyclic AMP. *Biochem. biophys. Res. Comm.*, **60**, 528–534

Cho-Chung, Y. and Clair, T. (1975) The role of cyclic AMP on neoplastic cell growth and regression. III. Altered cAMP-binding in DB-cAMP-unresponsive Walker 256 mammary carcinoma. *Biochem. biophys. Res. Commun.*, **64**, 768–772

Cho-chung, Y. S. and Gullino, P. M. (1974a) Effect of dibutyryl cyclic adenosine 3′,5′-monophosphate on *in vivo* growth of Walker 256 carcinoma: Isolation of responsive and unresponsive cell populations. *J. natn. Cancer Inst.*, **52**, 995–996

Cho-Chung, Y. S. and Guillino, P. M. (1974b) *In vivo* inhibition of growth of two hormone-dependent mammary tumors by dibutyryl cyclic AMP. *Science, N.Y.*, **183**, 87–88

Chrisman, T. D., Garbers, D. L., Parks, M. A. and Hardman, J. G. (1975) Characterization of particulate and soluble guanylate cyclases from rat lung. *J. biol. Chem.*, **250**, 374–381

Christ, E. J. and Nugteren, D. H. (1970) The biosynthesis and possible function of prostaglandins in adipose tissue. *Biochim. biophys. Acta*, **218**, 296–307

Christensen, L. F., Meyer, R. B., Jr., Miller, J. P., Simon, L N. and Robins, R. K. (1975) Synthesis and enzymic activity of 8-acyl and 8-alkyl derivatives of guanosine 3′,5′-cyclic phosphate. *Biochemistry*, **14**, 1490–1496

Christofferson, T., Morland, J., Osnes, J. and Elgjo, K. (1972) Hepatic adenyl cyclase: Alterations in hormone response during treatment with a chemical carcinogen. *Biochim. biophys. Acta*, **279**, 363–366

Chu, S.-H., Shiue, C.-Y. and Chu, M.-Y. (1974) Synthesis and biological activity of some 8-substituted seleno cyclic nucleotides and related compounds. *J. med. Chem.*, **17**, 406–409

Chu, S.-H., Shiue, C.-Y. and Chu, M.-Y. (1975) Synthesis and biological activity of some 8-substituted selenoguanosine cyclic 3′,5′-phosphates and related compounds. *J. med. Chem.*, **18**, 559–564

Chwang, A. K. and Sundaralingam, M. (1973) Molecular conformation of guanosine 3′,5′-cyclic monophosphate. *Nature (New biol.)*, **244**, 136–137

Claranello, R. D., Barchas, R. E., Byers, G. S., Stemmle, D. W. and Barchas, J. D. (1969) Enzymatic synthesis of adrenaline in brain. *Nature, Lond.*, **221**, 368–369

Clark, R. B., Gross, R., Su, Y. F. and Perkins, J. P. (1974) Regulation of adenosine 3′,5′-

monophosphate content in human astrocytoma cells by adenosine and the adenine nucleotides. *J. biol. Chem.*, **249**, 5296–5303

Clark, R. B., Morris, H. P. and Weber, G. (1973) Cyclic adenosine 3′,5′-monophosphate phosphodiesterase activity in normal, differentiating, regenerating, and neoplastic liver. *Cancer Res.*, **33**, 356–361

Clark, R. B. and Perkins, J. P. (1971) Regulation of adenosine 3′,5′-cyclic monophosphate concentration in cultured human astrocytoma cells by catecholamines and histamine. *Proc. natn. Acad. Sci., U.S.A.*, **68**, 2757–2760

Clark, R. B., Su, Y. F., Ortmann, R., Cubeddu, X., Johnson, G. L. and Perkins, J. P. (1975) Factors influencing the effect of hormones on the accumulation of cyclic AMP in cultured human astrocytoma cells. *Metabolism*, **24**, 343–358

Clement-Cormier, Y. C., Kebabian, J. W., Petzold, G. I. and Greengard, P. (1974) Dopamine-sensitive adenylate cyclase in mammalian brain: A possible site of action of antipsychotic drugs. *Proc. natn. Acad. Sci., U.S.A.*, **71**, 1113–1171

Clement-Cormier, Y. C., Parrish, R. G., Petzold, G. L., Kebabian, J. W. and Greengard, P. (1975) Characterization of a dopamine-sensitive adenylate cyclase in the rat caudate nucleus. *J. Neurochem.*, **25**, 143–149

Click, R. E., Benck, L. and Alter, B. J. (1972) Enhancement of antibody synthesis in vitro by mercaptoethanol. *Cell Immunol.*, **3**, 156–160

Cochrane, D. E. and Douglas, W. W. (1974) Calcium-induced extrusion of secretory granules (exocytosis) in mast cells exposed to 48/80 or the ionophores A23187 and X-537A. *Proc. natn. Acad. Sci., U.S.A.*, **71**, 408–412

Coffino, P., Bourne, H. R. and Tomkins, G. M. (1975a) Somatic genetic analysis of cyclic AMP action. Selection of unresponsive mutants. *J. Cell Physiol.*, **85**, 603–610

Coffino, P., Gray, J. W. and Tomkins, G. M. (1975b) Cyclic AMP, a nonessential regulator of the cell cycle. *Proc. natn. Acad. Sci., U.S.A.*, 72, 878–882

Cohen, P. (1974) The role of phosphorylase kinase in the nervous and hormonal control of glycogenolysis in muscle. *Biochem. Soc. Symp.*, **39**, 51–73

Cohen, S. and Taylor, J. M. (1972) Epidermal growth factor: Chemical and biological characterization. In *Epidermal Wound Healing* (Eds. H. I. Maibach and D. T. Rovee). Year Book Medical Publishers, Chicago, pp. 533–550

Cole, H. A. Perry, S. V. (1975) The phosphorylation of troponin I from cardiac muscle. *Biochem. J.*, **149**, 525–533

Coleman, R. (1973) Membrane-bound enzymes and membrane ultrastructure. *Biochim. biophys. Acta*, **300**, 1

Collard, J. G. and Temmink, J. H. M. (1976) Surface morphology and agglutinability with concanavalin A in normal and transformed murine fibroblasts. *J. Cell Biol.*, **68**, 101–112

Collins, J. H. (1974) Homology of myosin light chains and parvalbumins deduced from comparison of their amino sequences. *Biochem. biophys. Res. Commun.*, **58**, 301–308

Collins, J. H., Potter, J. D., Horn, M. J., Wilshire, G. and Jackson, N. (1973). The amino acid sequence of rabbit skelatel muscle troponin C: Gene replication and homology with calcium-binding proteins from carp and hake muscle. FEBS Letters, **36**, 268–272

Combret, Y. and Laudat, P. (1972) Adenyl cyclase activity in a plasma membrane fraction purified from "ghosts" of rat fat cells. *FEBS Letters*, **21**, 45–48

Connor, J. A., Prosser, C. L. and Weems, W. A. (1974) A study of pace-maker activity in intestinal smooth muscle. *J. Physiol., Lond.*, **240**, 671–701

Constantopoulos, A. and Najjar, V. A. (1973) The activation of adenylate cyclase. II The postulated presence of (*a*) adenylate cyclase in a phospho (inhibited) form; (*b*) a dephospho (activated) form with a cyclic adenylate-stimulated membrane protein kinase. *Biochem. biophys. Res. Commun.*, **53**, 794–799

Cooke, A. R., Chvasta, T. E. and Granner, D. K. (1974) Histamine, pentagastrin, methylxanthines and adenyl cyclase activity in acid secretion. *Proc. Soc. exp. Biol. Med.*, **147**, 674–678

Coombs, R. R. and Gell, P. G. (1968) Classification of allergic reactions responsible for clinical hypersensitivity and disease. In *Clinical Aspects of Immunology*, 2nd ed. (Eds. P. G. H. Gell and R. R. A. Coombs), Blackwell Scientific Publications, Oxford, pp. 575–596

Cooper, B., Partilla, J. S. and Gregerman, R. I. (1975) Adenylate cyclase of human fat cells. Expression of epinephrine-sensitive activation revealed by 5′-guanylylimidodiphosphate. *J. clin. Invest.*, **56**, 1350–1353

Cooper, R. H., Ashcroft, S. J. H. and Randle, P. J. (1973) Concentration of adenosine 3′,5′-cyclic monophosphate in mouse pancreatic islets measured by a protein-binding radioassay. *Biochem. J.*, **134**, 599–605

Cooperman, B. S. and Brunswick, D. J. (1973) On the photoaffinity labeling of rabbit muscle phosphofructokinase with O^2-(ethyl-2-diazomalonyl)adenosine 3′,5′-cyclic monophosphate. *Biochemistry*, **12**, 4079–4084

Corbin, J. D., Keely, S. L., Soderling, T. R. and Park, C. R. (1975) Hormonal regulation of adenosine 3′,5′-monophosphate-dependent protein kinase. In *Advances in Cyclic Nucleotide Research,* Vol. 5 (Eds. G. I. Drummond, P. Greengard and G. A. Robison), Raven Press, New York, pp. 265–279

Corbin, J. D., Reimann, E. M., Walsh, D. A. and Krebs, E. G. (1969) A cyclic AMP-stimulated protein kinase in adipose tissue. *Biochem. biophys. Res. Comm.*, **36**, 328–336

Corbin, J. D., Reimann, E. M., Walsh, D. A. and Krebs, E. G. (1970) Activation of adipose tissue lipase by skeletal muscle cyclic adenosine 3′,5′-monophosphate-stimulated protein kinase. *J. biol. Chem.*, **245**, 4849–4851

Cori, G. T. and Cori, C. F. (1952) Glucose-6-phosphatase of the liver in glucogen-storage disease. *J. biol. Chem.*, **199**, 661–667

Correze, C., Laudat, M. H., Laudat, P. and Nunez, J. (1974) Hormone-dependent lipolysis in fat cells from thyroidectomized rats. *Molec. cell. Endocrinol.*, **1**, 309–327

Costa, E., Carenzi, A., Guidotti, A. and Revuelta, A. (1973) Narcotic analgesics and the regulation of neuronal catecholamine stores. In *Frontiers in Catecholamine Research* (Eds. E. usdin and S. H. Snyder), Pergamon Press, New York, pp. 1003–1010

Costa, E. and Garattini, S. (Eds.) (1970) *International Symposium on Amphetamines and Related Compounds*. Raven Press, New York, p. 962

Costall, B. and Naylor, R. J. (1974) Mesolimbic involvement with behavioural effects indicating antipsychotic activity. *Europ. J. Pharmac.*, **27**, 46–58

Cotton, F. A., Gillen, R. B., Gohil, R. N., Hazen, E. E., Kirchner, C. R., Nagyvary, J., Rouse, J. P., Stanislowski, A. G., Stevens, J. D. and Tucker, P. W. (1975) Tumor-inhibiting properties of the neutral *p-O*-ethyl ester of adenosine 3′,5′-monophosphate in correlation with its crystal and molecular structure. *Proc. natn. Acad. Sci., U.S.A.*, **72**, 1335–1339

Coulson, R. and Bowman, R. H. (1974) Excretion and degradation of exogenous adenosine 3′,5′-monophosphate by isolated perfused rat kidney. *Life Sci.*, **14**, 545–556

Coulson, R., Bowman, R. H. and Roch-Ramel, F. (1974) The effects of nephrectomy and probenecid on *in vivo* clearance of anenosine 3′,5′-monophosphate from rat plasma. *Life Sci.*, **15**, 877–886

Cox, J. P. and Karnovsky, M. L. (1973) The depression of phagocytosis by exogenous cyclic nucleotides, prostaglandins, and theophylline. *J. Cell Biol.*, **59**, 480–490

Cramer, H., Johnson, D. G., Hanbauer, I., Silberstein, S. D. and Kopien, I. J. (1973) Accumulation of adenosine 3′,5′-monophosphate induced by catecholamines in the rat superior cervical ganglion *in vitro*. *Brain Res.*, **53**, 97–104

Cramer, H. and Lindl, T. (1974) Release of cyclic AMP from rat superior cervical ganglia after stimulation of synthesis *in vitro*. *Nature, Lond.*, **249**, 380–382

Cramer, H., Ng, L. K. Y. and Chase, T. N. (1972) Cyclic AMP: Probenecid-induced rise in human cerebrospinal fluid. *J. Neurochem.*, **19**, 1601–1602

Craig, S. W. and Cuatrecasas, P. (1975) Mobility of cholera toxin-receptors on rat lymphocyte membranes. *Proc. natn. Acad. Sci.*, U.S.A., **72**, 3844–3848

Crain, S. M. and Pollack, E. D. (1973) Restorative effects of cyclic AMP on complex bioelectric activities of cultures fetal rodent tissues after acute Ca^{2+} deprivation. *J. Neurobiol.*, **4**, 321–342

Criss, W. E., Murad, F. and Kimura, H. (1976) Properties of guanylate cyclase from rat kidney cortex and transplantable kidney tumors. *J. Cyclic Nucl. Res.*, **2**, 11–19

Cross, M. E. and Ord, M. G. (1971) Changes in histone phosphorylation and associated early metabolic events in pig lymphocyte cultures transformed by phytohemagglutinin or 6-*N*,2′-*O*-dibutyryl-adenosine 3′,5′-monophosphate. *Biochem. J.*, **124**, 241–248

Cuatrecasas, P. (1973) Cholera toxin–fat cell interaction and the mechanism of activation of the lipolytic response. *Biochemistry*, **12**, 3567–3577

Cuatrecasas, P., Hollenberg, M. D., Chang, K.-J. and Bennett, V. (1975a) Hormone receptor complexes and their modulation of membrane function. *Rec. Progr. Hormone Res.*, **31**, 37–84

Cuatrecasas, P. (1975b) Hormone receptors—their function in cell membranes and some problems related to methodology. In *Advances in Cyclic Nucleotide Research*, Vol. 5 (Eds. G. I. Drummond, P. Greengard and G. A. Robison), Raven Press, New York, pp. 79–104

Cuatrecasas, P., Jacobs, S. and Bennett, V. (1975) Activation of adenylate cyclase by phosphoramidate and phosphonate analogs of GTP: Possible role of covalent enzyme-substrate intermediates in the mechanism of hormonal activation. *Proc. natn. Acad. Sci. U.S.A.*, **72**, 1739–1743

Cuatrecasas, P., Tell, G. P. E., Sica, V., Parikh, I. and Chang, K.-J. (1974) Noradrenaline binding and the search for catecholamine receptors. *Nature, Lond.*, **247**, 92–97

Cubeddu, L. X., Barnes, E. M. and Weiner, N. (1974) Release of norepinephrine and dopamine-β-hydroxylase by nerve stimulation. *J. Pharmac. exp. Ther.*, **191**, 444–457

Cubeddu, L., Barnes, E. and Weiner, N. (1975) Release of norepinephrine and dopamine-β-hydroxylase by nerve stimulation. IV. An evaluation of a role for cyclic adenosine monophosphate. *J. Pharmac. exp. Ther.*, **193**, 105–127

Currie, B. L., Johansson, K. N., Greisbrokk, T., Folkers, K. and Bowers, C. Y. (1974) Identification and purification of factor A-GHRH from hypothalami which releases growth hormone. *Biochem. biophys. Res. Commun.*, **60**, 605–609

Curtis, D. R. (1964) Microelectrophoresis. In *Physical Techniques in Biological Research*, Vol. 5, Academic Press, New York

Cushman, S. W., Heindel, J. J. and Jeanrenaud, B. (1973) Cell-associated nonesterified fatty acid levels and their alteration during lipolysis in the isolated mouse adipose cell. *J. Lipid Res.*, **14**, 632–642

Czekalski, S., Loreau, N. Paillard, F., Ardaillou, R., Fillastre, J. P. and Mallet, E. (1974) Effect of bovine parathyroid hormone 1–34 fragment on renal production and excretion of adenosine 3′,5′-monophosphate in man. *Eur. J. Clin. Invest.*, **4**, 85

Dabrowski, M. P., Ryzewski, J., Dabrowska, B. K. and Ryzewska, A. G. (1974) Changes of *in vitro* phytohemagglutinin reactivity of rat lymphocytes influenced by cysteine. *Bull. Acad. pol. Sci., et II Sér, Sci. biol.*, **22**, 281–286

Dahlström, A. and Fuxe, K. (1964) Evidence for the existence of monoamine containing neurons in the central nervous system. *Acta physiol. scand.*, Suppl. 232

Dalton, C. and Hope, H. R. (1973) Inability of prostaglandin synthesis inhibitors to affect adipose tissue lipolysis. *Prostaglandins*, **4**, 641–651

Dalton, C. and Hope, W. C. (1974) Cyclic AMP regulation of prostaglandin biosynthesis in fat cells. *Prostaglandins*, **6**, 227–242

Daly, J. W. The role of cyclic nucleotides in the nervous system. In *Handbook of Psychopharmacology*. Vol. 5, (Eds. Iversen, L. L., Iversen, S. D., Snyder, S. H.), Plenum Press, New York, pp. 47–128

Daly, J. W. (1975) Cyclic adenosine 3′,5′-monophosphate: Role in the physiology and pharmacology of the central nervous system. *Biochem. Pharmac.*, **24**, 159–164

Daniel, V., Bourne, H. R. and Tomkins, G. M. (1973a) Altered metabolism and endogenous cyclic AMP in cultured cells deficient in cyclic AMP-binding proteins. *Nature (New Biol.)*, **244**, 167–169

Daniel, V., Litwack, G. and Tomkins, G. M. (1973b) Induction of cytolysis of cultured lymphoma cells by adenosine 3′,5′-cyclic monophosphate and isolation of resistant variants. *Proc. natn. Acad. Sci. U.S.A.*, **70**, 76–79

Daniel, D. E., Paton, D. M., Taylor, G. S. and Hodgson, B. J. (1970) Adrenergic receptors for catecholamine effects on tissue electrolytes. *Fed. Proc.*, **29**, 1410–1425

Danilo, P., Lebarhis, E. and Rosen, M. (1972) Effects of dibutyryl 3′,5′-cyclic AMP on electrophysiologic properties of canine Purkinje fibres. *5th Int. Congr. Pharmac., San Francisco*, p. 51

D'Armiento, M., Johnson, G. S. and Pastan, I. (1972) Regulation of adenosine 3′,5′-cyclic monophosphate phosphodiesterase activity in fibroblasts by intracellular concentrations of cyclic adenosine monophosphate. *Proc. natn. Acad. Sci. U.S.A.*, **69**, 459–462

D'Armiento, M., Johnson, G. S. and Pastan, I. (1973) Cyclic AMP and growth of fibroblasts. Effect of environmental pH. *Nature (New Biol.)*, **242**, 78–80

Das, I. (1974) Plasma adenosine 3′,5′-monophosphate in antiinsulin-treated rats. *Experientia*, **30**, 860

Davis, A. T., Estensen, R. and Quie, P. G. (1971) Cytochalasin B. III. Inhibition of human polymorphonuclear leukocyte phagocytosis. *Proc. Soc. exp. Biol. Med.*, **137**, 161–164

Davis, B. and Lazarus, N. R. (1975a) An *in vitro* system for elucidating mechanisms responsible for plasma membrane: β-Granule interaction. *Diabetologia*, **11**, 336–337

Davis, B. and Lazarus, N. R. (1975b) Regulation of 3′,5′-cyclic AMP-dependent protein kinase in the plasma membrane of cod (*Gadus callarius*) and mouse islets. *J. Membrane Biol.*, **20**, 301–318

Davoren, P. R. and Sutherland, E. W. (1963) Cellular location of adenyl cyclase in the pigeon erythrocyte. *J. biol. Chem.*, **238**, 3016–3023

Dean, P. M. (1974) Surface electrostatic-charge measurements on islet and zymogen granules: effects of calcium ions. *Diabetologia*, **10**, 427–430

Dean, P. M. and Matthews, E. K. (1970) Glucose-induced electrical activity in pancreatic islet cells. *J. Physiol. Lond.*, **210**, 255–264

Dean, P. M., Matthews, E. K. and Sakamoto, Y. (1975) Pancreatic islet cells: Effects of monosaccharides, glycolytic intermediates and metabolic inhibitors on membrane potential and electrical activity. *J. Physiol. Lond.*, **246**, 459–478

DeAsua, L. J., Clingan, D. and Rudland, P. S. (1975) Initiation of cell proliferation in cultured mouse fibroblasts by prostaglandin $F_{2\alpha}$. *Proc. natn. Acad. Sci. U.S.A.*, **72**, 2724–2728

DeAsua, L. J., Rozengurt, E. and Dulbeco, R. (1974) Kinetics of early changes in phosphate and uridine transport and cyclic AMP levels stimulated by serum in density-inhibited 3T3 cells. *Proc. natn. Acad. Sci. U.S.A.*, **71**, 96–98

DeAsua, L. J., Surian, E. S., Flawia, M. M. and Torres, H. N. (1973) Effects of insulin on the growth pattern and adenylate cyclase activity of BHK fibroblasts. *Proc. natn. Acad. Sci. U.S.A.*, **70**, 1388–1392

Deby, C. and Bacq, Z. M. (1962) Action de l'indométhacine, inhibiteur de la biosynthèse des prostaglandines, sur les lipides plasmatiques du rat. *C. r. Séanc. Soc. Biol.*, **166**, 750–753

de Duve, C., Pressman, B., Gianetto, R., Wattiaux, R. and Appelman, F. (1955) Tissue fractionation studies. *Biochem. J.*, **60**, 604–617

de Duve, C. and Wattiaux, R. (1966) Functions of lysosomes. *Ann. Rev. Physiol.*, **28**, 435–492 de Groot, L. J. and Stanbury, J. B. (1975) In *The Thyroid and Its Diseases*, J. Wiley and Sons, New York

Deguchi, T. and Axelrod, J. (1973) Supersensitivity and subsensitivity of β-adrenergic receptors in pineal gland regulated by catecholamine transmitter. *Proc. natn. Acad. Sci. U.S.A.*, **70**, 2411–2414

DeHaën, C. (1974) A new kinetic analysis of the effects of hormones and fluoride ion. *J. biol. Chem.*, **249**, 2756–2764

Dekker, A. and Field, J. B. (1970) Correlation of effects of thyrotropin, prostaglandins and ions on glucose oxidation, cyclic AMP and colloid droplet formation in dog thyroid slices. *Metabolism*, **19**, 453–464

Delapaz, R. L., Dickman, S. R. and Grosser, B. I. (1975) Effects of stress on rat brain adenosine cyclic 3′,5′-monophosphate *in vivo*. *Brain Res.*, **85**, 171–175

Delescluse, C., Colburn, N. H., Duell, E. A. and Voorhees, J. J. (1974) Cyclic AMP-elevating agents inhibit proliferation of keratinizing guinea-pig epidermal cells. *Differentiation*, **2**, 343–350

Delescluse, C., Fukuyama, K. and Epstein, W. L. (1976) Dibutyryl-cyclic AMP-induced differentiation of epidermal cells in tissue culture. *J. invest. Derm.*, **66**, 8–13

DeLorenzo, R. J. and Greengard, P. (1973) Activation by adenosine 3′,5′-monophosphate of a membrane-bound phosphoprotein phosphatase from toad bladder. *Proc. natn. Acad. Sci. U.S.A.*, **70**, 1831–1835

Demesy-Waeldele, F. and Stoclet, J. C. (1975) Papaverine, cyclic AMP and the dependence of the rat aorta on extracellular calcium. *Europ. J. Pharmac.*, **31**, 185–194

DeMeyts, P., Roth, J., Neville, D. M., Jr., Gavin, J. R. and Lesniak, M. A. (1973) Insulin interactions with its receptors: Experimental evidence for negative co-operativity. *Biochem. biophys. Res. Commun.*, **55**, 154–161

Dermer, G. D., Lue, J. and Neunstein, H. B. (1974) Comparison of surface material, cytoplasmic filaments, and intracellular junctions from untransformed and two mouse sarcoma virus-transformed cell line. *Cancer Res.*, **34**, 31–38

DeRobertis, E., Arnaiz, G. R. D. L., Alberici, M., Butcher, R. W. and Sutherland, E. W. (1967) Subcellular distribution of adenyl cyclase and cyclic phosphodiesterase in rat brain cortex. *J. biol. Chem.*, **242**, 3487–3496

Derry, D. J. and Howell, S. L. (1973) Rat anterior pituitary adenyl cyclase activity: GTP requirement of prostaglandin E_2 and E_1 and synthetic luteinizing hormone-releasing hormone activation. *Biochim. biophys. Acta*, **252**, 574–579

DeRubertis, F. R., Craven, P. A., Zenser, T. V. and Davis, B. B. (1976) Acetylcholine increases renal cortical content and urinary excretion of cyclic GMP. *Clin. Res.*, **24**, 398A

DeRubertis, F. R., Zenser, T. V., Adler, W. H. and Hudson, T. (1974) Role of cyclic adenosine 3′,5′-monophosphate in lymphocyte mitogenesis. *J. Immunol.*, **113**, 151–161

Desai, K. S., Li, K. C. and Angel, A. (1973) Bimodal effect of insulin on hormone-stimulated lipolysis: Relation to intracellular 3′,5′-cyclic adenylic acid and free fatty acids levels. *J. Lipid Res.*, **14**, 647–655

Devine, C. E., Somlyo, A. V. and Somlyo, A. P. (1972) Sarcoplasmic reticulum and excitation–contraction coupling in mammalian smooth muscles. *J. Cell Biol.*, **52**, 690–718

Devis, G., Somers, G. and Malaisse, W. J. (1975) Stimulation of insulin release by calcium. *Biochem. biophys. Res. Commun.*, **67**, 525–529

de Vellis, J. and Brooker, G. (1972) Effects of catecholamines on cultured glial cells: Correlation between cyclic AMP levels and lactic dehydrogenase induction. *Fed. Proc.*, **31**, 513

de Vellis, J., Inglish, D. and Brooker, G. (1974) Paradoxical effects of actinomycin D, acetoxycyclohexamide and methylisobutyl xanthine on cyclic AMP metabolism. *Fed. Proc.*, **33**, 507

Dhalla, N. S., Chernecki, W., Gandhi, S. S., McNamara, D. B. and Naimark, A. (1973) Cardiac and metabolic effects of dibutyryl-cyclic AMP in the intact dog heart and the isolated perfused rat heart. *Recent Advances in Studies of Cardiac Structure and Metabolism*, **34**, University Park Press, Baltimore, 233–250

Dhalla, N. S., Sulakhe, P. V., Khandelwal, R. L. and Olson, R. E. (1972) Adenylate cyclase activity in the perfused rat heart made to fail by substrate lack. *Cardiovasc. Res.*, **6**, 344–352

Diamantstein, T. and Ulmer, A. (1975) Stimulation by cyclic GMP of lymphocytes mediated by soluble factor released from adherent cells. *Nature, Lond.*, **256**, 418–419

Diamond, J. and Hartle, D. K. (1974) Cyclic nucleotide levels during carbachol-induced smooth muscle contractions. *Pharmacologist*, **16**, 273

Diamond, J. and Marshall, J. M. (1969) Smooth muscle relaxants: Dissocation between resting membrane potential and resting tension in rat myometrium. *J. Pharmac. exp. Ther.*, **168**, 13–20

Dianzani, F., Neri, P. and Zucca, M. (1972) Effect of dibutyryl-cyclic AMP on interferon production by cells treated with viral or nonviral inducers. *Proc. Soc. exp. Biol. Med.*, **140**, 1375–1378

Diaz-Buxo, J. A. and Knox, F. G. (1975) Effects of parathyroid hormone on renal function. *Proc. Mayo Clin.*, **50**, 537–541

Dietmann, K., Roesch, E., Schaumann, W. and Juhran, W. (1972) The effect of cyclic adenosine and guanosine 3′,5′-monophosphate on the normal and drug-increased coronary blood flow. *Biochem. Pharmac.*, **21**, 2193–2196

Dills, W. L., Jr., Beavo, J. A., Bechtel, P. J. and Krebs, E. G. (1975) Purification of rabbit skeletal muscle protein kinase regulatory subunit using cyclic adenosine-3′:5′-monphosphate affinity chromatography. *Biochem. biophys. Res. Commun.*, **62**, 70–77

Dills, W. L., Jr., Beavo, J. A., Bechtel, P. J., Myers, K. R., Sakai, L. J. and Krebs, E. G. (1976) Binding of adenosine 3′,5′-monophosphate-dependent protein kinase regulatory subunit to immobilized cyclic nucleotide derivatives. *Biochemistry*, **15**, 3724–3731

Dismukes, K. and Daly, W. (1974) Norepinephrine-sensitive systems generating adenosine 3′,5′-monophosphate: Increased responses in cerebral cortical slices from reserpine-treated rats. *Molec. Pharmac.*, **10**, 933–940

Dismukes, R. K. and Daly, J. W. (1975a) Accumulation of adenosine 3′,5′-monophosphate in rat brain slices: Effects of prostaglandins. *Life Sci.*, **17**, 199–210

Dismukes, R. K. and Daly, J. W. (1975b) Altered responsiveness of adenosine 3′,5′-monophosphate-generating systems in brain slices from adult rats after neonatal treatment with 6-hydroxydopamine. *Expl Neurol.*, **49**, 150–160

Dismukes, R. K., Ghosh, P., Creveling, C. R. and Daly, J. W. (1975) Altered responsiveness of adenosine, 3′,5′-monophosphate-generating systems in rat cortical slices after lesions of the medial forebrain bundle. *Exp Neurol.*, **49**, 725–735

Dismukes, R. K., Ghosh, P., Creveling, C. R. and Daly, J. W. (1976) Depletion of norepinephrine in guinea-pig brain. Lack of effect on responsiveness of norepinephrine-sensitive cortical cyclic AMP-generating systems. *Exp Neurol.*, **52**, 206–215

Dobson, J. G. and Mayer, S. E. (1973) Mechanisms of activation of cardiac glycogen phosphorylase in ischemia and anoxia. *Circulation Res.*, **33**, 412–420

Domschke, W., Classen, M. and Demling, L. (1972) Circadian rhythmicity of gastric secretion and cyclic 3′,5′-adenosine monophosphate contents of gastric mucosa in rats. *Scand. J. Gastroent.*, **7**, 39–41

Domschke, W., Domschke, S., Classen, M. and Demling, L. (1973) Histamine and cyclic 3′,5′-AMP in gastric acid secretion. *Nature, Lond.*, **24**, 454–455

Domschke, W., Domschke, S., Rösch, W., Classen, M. and Demling, L. (1974) Failure of pentagastrin to stimulate cyclic AMP accumulation in human gastric mucosa. *Scand. J. Gastroent.*, **9**, 467–471

Donnelly, T. E., Kuo, J. F., Reyes, P. L., Liu, Y. P. and Greengard, P. (1973) Protein kinase modulator from lobster tail muscle. Stimulatory and inhibitory effects of the modulator on the phosphorylation of substrate proteins by guanosine 3′,5′-monophosphate and adenosine 3′,5′-monophosphate-dependent protein kinases. *J. biol. Chem.*, **248**, 190–198

Doore, B. J., Bashor, M. M., Spitzer, N., Mawe, B. C. and Saier, H. (1975) Regulation of adenosine 3′,5′-monophosphate efflux from rat glioma cells in culture. *J. biol. Chem.*, **250**, 4371–4372

Dorrington, J. H. and Armstrong, D. T. (1975) Follicle-stimulating hormone stimulates estradiol-17β synthesis in cultured Sertoli cells. *Proc. natn. Scad. Sci. U.S.A.*, **72**, 2677–2681

Douglas, J. R., Johnson, E. M., Marshall, G. R., Heist, J., Hartman, B. K. and Needleman, P. (1975) Development and maintenance of renal hypertension in normal and guanethidine-sympathectomized rats. *Circulation Res. Suppl.*, **1**, 171–178

Douglas, W. W. (1968) Stimulus–secretion coupling. *Brit. J. Pharmac.*, **34**, 451–474

Douglas, W. W. (1975) Stimulus–secretion coupling in mast cells: Regulation of exocytosis by cellular and extracellular calcium. In *Calcium Transport in Contraction and Secretion* (Eds. E. Carafoli *et al.*), North Holland Publishing Co., Amsterdam, pp. 167–174

Dousa, T. P. (1972) Effect of renal medullary solutes on vasopressin-sensitive adenyl cyclase. *Am. J. Physiol.*, **222**, 21–24

Dousa, T. P. (1974) Interaction of lithium with vasopressin-sensitive cyclic AMP system of human medulla. *Endocrinology*, **95**, 1359–1366

Dousa, T. P. (1976) Drugs and other agents affecting renal adenylate cyclase-cyclic AMP system. In *Methods in Pharmacology IV: Renal Pharmacology* (Ed. M. Martinez-Moldonadeo), Plenum Press

Dousa, T. P. and Code, C. F. (1974) Effect of histamine and its methyl derivatives on cyclic AMP metabolism in gastric mucosa and its blockade by an H_2-receptor antagonist. *J. clin. Invest.*, **53**, 334–337

Dousa, T. P. and Hechter, O. (1970) The effect of NaCl and LiCl on vasopressin-sensitive adenyl cyclase. *Life Sci.*, **9**, 765

Dousa, T. P. and Valtin, H. (1976) Cellular actions of vasopressin in the mammalian kidney. *Kidney Int.*, **10**, 46–64

Dowd, F. and Schwartz, A. (1975) The presence of cyclic AMP-stimulated protein kinase substrates and evidence for endogenous protein kinase activity in various Na^+/K^+-ATPase preparations from brain, heart and kidney. *J. mol. Cell Cardiol.*, **7**, 483–497

Drizner, M., Neelon, F. A. and Lebowitz, H. E. (1973) Pseudopypoparathyroidism Type II: A possible defect in the reception of cyclic AMP signal. *New Engl. J. Med.*, **289**, 1056–1060

Drummond, G. I. (1967) Muscle metabolism. *Fortschritte Zool.*, **18**, 360–429

Drummond, G. I. and Duncan, L. (1970) Adenyl cyclase: In cardiac tissue. *J. biol. Chem.*, **245**, 976–983

Drummond, G. I., Duncan, L. and Hertzmann, E. (1966) Effect of epinephrine on phosphorylase b kinase in perfused rat hearts. *J. biol. Chem.*, **241**, 5899–5903

Drummond, G. I., Gilgan, M. W., Reiner, E. J. and Smith, M. (1963) Deocyribonucleoside-3′,5′-cyclic phosphates. Synthesis and acid-catalyzed and enzymic hydrolysis. *J. Am. chem. Soc.*, **86**, 1626–1630

Drummond, G. I. and Mah, Y. (1973) Metabolism and functions of cyclic AMP in nerve. In *Progress in Neurobiology* (Eds. G. A. Kerkut and J. W. Phillis), Pergamon Press, Oxford and New York, pp. 119–176

Drummond, G. I. and Powell, C. A. (1970) Analogues of adenosine 3′,5′-cyclic phosphate as activators of phosphorylase *b* kinase and as substrates for cylic 3′,5′-nucleotide phosphodiesterase. *Molec. Pharmac.*, **6**, 24–30

Drummond, G. I. and Severson, D. L. (1971) Biological actions of cyclic AMP analogs , *Ann. Rep. in Med. Chem.*, (Ed. G. V. Heinzelman), **1**, 215–226

Drummond, G. I. and Severson, D. L. (1974) Preparation and characterization of adenylate cyclase from heart and skeletal muscle. In *Methods of Enzymology*, Vol. 30 (Eds. J. G. Hardman and B. W. O'Malley), pp. 143–149

Drummond, G. I., Severson, D. L. and Duncan, L. (1971) Adenyl cyclase. Kinetic properties and nature of fluoride and hormone stimulation. *J. biol. Chem.*, **246**, 4166–4173

Dufau, M. L., Catt, K. L. and Tsuruhara, T. (1771) Gonadotrophin stimulation of testosterone production by rat testis *in vitro*. *Biochim. biophys. Acta.*, **252**, 574–579

Dufau, M. L., Watanabe, K. and Catt, K. J. (1973) Stimulation of cyclic AMP production by the rat testis during incubation with HCG *in vitro*. *Endocrinology*, **92**, 6–11

Dulbecco, R. and Elkington, J. (1975) Induction of growth in resting fibroblastic cell cultures by Ca^{2+}. *Proc. natn. Acad. Sci. U.S.A.*, **72**, 1584–1588

Dumont, J. E. (1971) The action of thyrotropin on thyroid metabolism. *Vitms. Horm.*, **29**, 287–412

Dumont, J. E., Willems, C., van Sande, J. and Neve., P. (1971) Regulation of the release of thyroid hormones: Role of cyclic AMP. *Ann. N.Y. Acad. Sci.*, **185**, 291–316

Dunham, E. W., Haddox, M. K. and Goldberg, N. D. (1974) Alteration of vein cyclic 3′,5′nucleotide concentrations during changes in contractility. *Proc. natn. Acad. Sci. U.S.A.*, **71**, 815–819

Dunnick, J. K. and Marinetti, G. V. (1971) Hormone action at the membrane level. III. Epinephrine interaction with the rat liver plasma membrane. *Biochim. biophys. Acta*, **249**, 122–134

DuPlooy, M., Michal, G., Weimann, G., Nelboeck, M. and Paoletti, R. (1971) Cyclophosphates I. Effect of various cyclophosphates on phosphorylase *b* kinase activation. *Biochim. biophys. Acta.*, **230**, 30–39

Durham, A. C. H. (1974) A unified theory of the control of actin and myosin in nonmuscle movements. *Cell*, **2**, 123–136

Eaglstein, W. H. and Weinstein, G. D. (1975) Prostaglandin and DNA synthesis in human skin: Possible relationship to ultraviolet light effects. *J. invest. Derm.*, **64**, 386–389

Ebadi, M. S. (1972) Firefly luminescence in assay of cyclic AMP. In *Advances in Cyclic Nucleotide Research*, Vol. 2 (Eds. P. Greengard, R. Paoletti and G. A. Robison), Raven Press, New York, pp. 89–110

Ebadi, M. S., Weiss, B. and Costa, E. (1971) Microassay of adenosine 3′,5′-monophosphate (cyclic AMP) in brain and other tissues by the luciferin–luciferase system. *J. Neurochem.*, **18**, 183–192

Ebashi, S. (1969) Ca ions as a basis of pharmacological action. *Proc. 4th Int. Congr. Pharmac*, 32–54

Ebashi, S. and Endo, M. (1968) Calcium ion and muscle contraction. *Progr. Biophys. molec. Biol.*, **18**, 123–183

Ebert, R., Hillebrandt, O. and Schwabe, U. (1974) Role of calcium and cyclic adenosine 3′,5′-monophosphate in the antilipolytic effect of tolbutamide and glibenclamide. *Naunyn-Schmiedebergs Arch. Pharmak.*, **286**, 181–194

Eccles, J. C. (1964) *The Physiology of Synapses.* Academic Press, New York

Eccles, J. C., Ito, M. and Szentagothai, J. (1967) *The Cerebellum as a Neuronal Machine.* Springer, Verlag, New York

Eccles, R. and Libet, B. (1961) Origin and blockase of the synaptic responses of curarized sympathetic ganglia. *J. Physiol., Lond.*, **157**, 484

Eckstein, F., Eimerl, S. and Schramm, M. (1976) Adenosine 3′,5′-cyclic phosphorothioate: An efficient inducer of amylase secretion in rat parotid slices. *FEBS letters*, **64**, 92–94

Eckstein, F., Simonson, L. P. and Bär, H.-P. (1974) Adenosine 3′,5,′-cyclic phosphorothioate: Synthesis and biological properties. *Biochemistry*, **70**, 3806–3810

Edelman, G. M., Yahara, I. and Wang, J. L. (1973) Receptor mobility and receptor cytoplasmic interactions in lymphocytes. *Proc. natn. Acad. Sci. U.S.A.*, **70**, 1442–1446

Egrie, J. C. and Siegel, F. L. (1975) Adrenal medullary cyclic nucleotide phosphodiesterase: Lack of activation by the calcium-dependent regulator. *Biochem. biophys. Res. Commun.*, **67**, 662–669

Eichhorn, J. H., Salzman, E. W. and Silen, W. (1974) Cyclic GMP response *in vivo* to cholinergic stimulation of gastric mucosa. *Nature, Lond.*, **248**, 238–239

Eisen, S. A., Lyle, L. R. and Parker, C. W. (1973) Antigen stimulation of [^{3}H] thymidine incorporation by subpopulations of human peripheral blood cells. *J. Immunol.*, **111**, 962–972

Eisen, S. A., Wedner, H. J. and Parker, C. W. (1972) Isolation of pure human peripheral blood T-lymphocytes using nylon-wool columns. *Immunol. Commun.*, **1**, 571–577

Elgjo, K. (1973) Epidermal chalone: Cell cycle specificity of two epidermal growth inhibitors. *Natn. Cancer Inst. Monogr.*, **38**, 71–76

Elliot, R. A. C. (1955) Tissue slice technique. In *Methods in Enzymology,* Vol. 1 (Eds. S. P. Colowick and N. O. Kaplan), Academic Press, New York, pp. 3–9

Ellis, E. and Hutchins, P. (1974) Cardiovascular responses to prostaglandin F_2 in spontaneously hypertensive rats. *Prostaglandins*, **7**, 345–353

Enero, M. A., Langer, S. Z., Rothlin, R. P. and Stefano, F. J. E. (1972) Role of the α-adrenoceptor in regulating noradrenaline overflow by nerve stimulation. *Brit. J. Pharmac.*, **44**, 672–688

England, P. (1975) Correlation between contraction and phosphorylation of the inhibitory subunit of troponin in perfused rat heart. *FEBS Letters*, **50**, 57–60

Entman, M. L. (1974) The role of cyclic AMP in the modulation of cardiac contractility. In *Advances Cyclic Nucleuotide Research,* Vol. 4, Raven Press, New York, pp. 163–193

Entman, M. L., Levey, G. S. and Epstein, S. E. (1969) Mechanism of action epinephrine and glucagon on the canine heart. *Circulation Res.*, **25**, 429–438

Epstein, S. E., Levey, G. S. and Skelton, C. L. (1971) Adenyl cyclase and cyclic AMP biochemical links in the regulation of myocardial contractility. *Circulation*, **43**, 437–450

Esber, H. J., Payne, I. J. and Bodgen, A. E. (1973) Variability of hormone concentrations and rations in commercial sera used for tissue culture. *J. natn. Cancer Inst.*, **50**, 559–562

Estensen, R., Hadden, J. W., Hadden, E. M., Touraine, F., Touraine, J. L., Haddox, M. K. and Goldberg, N. D. (1974) Phorbol myristate acetate: Effects of a tumor promoter on intracellular cyclic GMP in mouse fibroblasts and as a mitogen on human lymphocytes. In *The Cold Spring Harbor Symposium on the Regulation of Proliferation in Animal Cells* (Eds. B. Clarkson and R. Baserga), Cold Spring Harbor Laboratory, New York, pp. 627–534

Estensen, R. D., Hill, H. R., Quie, P. G., Hogan, N. and Goldberg, N. D. (1973) Cyclic GMP and cell movement. *Nature, Lond.*, **245**, 458–460

Eto, S., Wood, J. M., Hutchins, M. and Fleischer, N. (1974) Pituitary $^{45}Ca^{2+}$ uptake and release of ACTH, GH and TSH: Effect of verapamil. *Am. J. Physiol.*, **226**, 1315–1320

Exton, J. H., Lewis, S. B., Ho, R. J., Robison, G. A. and Park, C. R. (1971) The role of cyclic AMP in the interaction of glucagon and insulin in the control of liver metabolism. *Ann. N.Y. Acad. Sci.*, **185**, 85–558

Fabiato, A. and Fabiato, F. (1975) Relaxing and inotropic effects of cyclic AMP on skinned cardiac cells. *Nature, Lond.*, **253**, 556–558

Fain, J. N. (1973a) Biochemical aspects of drug and hormone action on adipose tissue. *Pharmac. Rev.*, **25**, 67–118

Fain, J. N. (1973b) Inhibition of adenosine cyclic 3′,5′-monophosphate accumulation in fat cells by adenosine, *N*^{6}-(phenylisopropyl)adenosine, and related compounds. *Molec. Pharmac.*, **9**, 595–604

Fain, J. N. (1975) Insulin as an activator of cyclic AMP accumulation in rat fat cells. *J. Cycl. Nucleot. Res.*, **1**, 359–366

Fain, J. N. and Butcher, F. R. (1976) Cyclic guanosine 3′,5′-monophosphate and the regulation of lipolysis in rat fat cells. *J. Cycl. Nucleotide Res.*, **2**, 71–78

Fain, J. N. and Czech, M. P. (1975) Glucocorticoid effects on lipid mobilization and adipose tissue metabolism. In *Handbook of Physiology*, Vol. VI, American Physiological Society, Washington, pp. 169–178

Fain, J. N., Dodd, A. and Novak, L. (1971) Enzyme regulation in gluconeogenesis and lipogenesis: Relationship of protein synthesis and cyclic AMP to lipolytic action of growth hormone and glucocorticoids. *Metabolism*, **20**, 109–118

Fain, J. N., Pointer, R. H. and Ward, W. F. (1972) Effects of adenosine nucleotides on adenylate cyclase, phosphodiesterase, cyclic adenosine monophosphate accumulation, and lipolysis in fat cells. *J. biol. Chem.*, **247**, 6866–6872

Fain, J. N., Psychoyos, S., Czernik, A. J., Frost, S. and Cash, W. D. (1973) Indomethacin, lipolysis and cyclic AMP accumulation in white fat cells. *Endocrinology*, **93**, 632–639

Fain, J. N. and Rosenberg, L. (1972) Antilipolytic action of insulin on fat cells. *Diabetes*, **21**, 414–425

Fain, J. N. and Saperstein, R. (1970) The involvement of RNA synthesis and cyclic AMP in the activation of fat cell lipolysis by growth hormone and glucocorticoids. In *Adipose Tissue: Regulation and Metabolic Functions* (Eds. B. Jeanrenaud and D. Hepp), Academic Press, New York, pp. 20–27

Fain, J. N. and Shepherd, R. E. (1975) Free fatty acids as feedback regulators of adenylate cyclase and cyclic 3′,5′-AMP accumulation in rat fat cells. *J. biol. Chem.*, **250**, 6586–6592

Fain, J. N. and Shepherd, R. E. (1976) Inhibition of adenosine 3′,5′-monophosphate accumulation in white fat cells by short chain fatty acids, lactate and β-hydroxybutyrate. *J. Lipid Res.*, **17**, 377–385

Fain, J. N., Tolbert, M. E. M., Pointer, R. H., Butcher, F. R. and Arnold, A. (1975) Cyclic nucleotides and gluconeogenesis by rat liver cells. *Metabolism*, **24**, 395–407

Fain, J. N. and Wieser, P. B. (1975) Effects of adenosine deaminase on cyclic adenosine monophosphate accumulation, lipolysis and glucose metabolism of fat cells. *J. biol. Chem.*, **250**, 1027–1934

Fakunding, J. L. and Means, A. R. (1975) Testicular protein kinase: Properties of two forms from cytosol. *Fed. Proc.*, **34**, 543

Fanger, M. W., Hart, D. A., Wells, J. V. and Nisonoff, A. (1970) Enhancement by reducing agents of the transformation of human and rabbit peripheral lymphocytes. *J. Immunol.*, **105**, 1043–1045

Farnebo, L. O. and Hamberger, B. (1971a) Drug-induced changes in the release of [^{3}H] noradrenaline from field-stimulated rat iris. *Br. J. Pharmac.*, **43**, 97–106

Farnebo, L. O. and Hamberger, B. (1971b) Drug-induced changes in the release of [^{3}H] monoamines from field-stimulated rat brain slices. *Acta physiol. scand. Suppl.*, **371**, 35–44

Fawcett, C. P., Beezley, A. E. and Wheaton, J, E. (1975) Chromatographic evidence for the existence of another species of luteinizing hormone-releasing factor (LRF). *Endocrinology*, **96**, 1311–1314

Ferrendelli. J. A., Chang, M. M. and Kinscherf, D. A. (1974) Elevation of cyclic GMP levels in central nervous system by excitatory and inhibitory amino acids. *J. Neurochem.*, **22**, 535–540

Ferrendelli, J. A., Kinscherf, D. A. and Chang, M.-M. (1975) Comparison of the effects of biogenic amines on cyclic GMP and cyclic AMP levels in mouse cerebellum *in vitro*. *Brain Res., Osaka*, **84**, 63–75

Ferrendelli, J. A., Kinscherf, D. A. and Kipnis, D. M. (1972) Effects of amphetamine, chlorpromazine and reserpine on cyclic GMP and cyclic AMP levels in mouse cerebellum. *Biochem. biophys. Res. Comm.*, **46**, 2114–2120

Ferrendelli, J. A., Steiner, A. L., McDougal, D. B. and Kipnis, D. M. (1070) The effect of oxotremorine and atropine on cGMP and cAMP levels in mouse cerebral cortex and cerebellum. *Biochem. biophys. Res. Commun.*, **41**, 1061–1067

Field, J. B. (1975) Thyroid-stimulating hormone and cyclic adenosine 3′,5′-monophosphate in the regulation of thyroid gland function. *Metabolism*, **24**, 381–393

Fikus, M., Kwast-Welfeld, J., Kazimierczuk, Z. and Shugar, D. (1974) Biochemical studies on some new analogues of adenosine-3′,5′-cyclic phosphate, including isoguanosine-3′,5′-cyclic phosphate. *Acta biochim. pol.*, **21**, 465–474

Finch, L. (1975) An increased reactivity in hypertensive rats unaffected by prolonged anti-hypertensive therapy. *Br. J. Pharmac.*, **54**, 437–443

Fisher, D. B. and Mueller, G. C. (1968) An early alteration in the phospholipid metabolism of lymphocytes by phytohemagglutinin. *Proc. natn. Acad. Sci. U.S.A.*, **60**, 1396–1402

Flaxman, B. A. and Harper, R. A. (1975) *In vitro* analysis of the control of keratinocyte proliferation in human epidermis by physiologic and pharmacologic agents. *J. invest. Derm.*, **65**, 52–59

Fleischer, N., Donald, R. A. and Butcher, R. W. (1969) Involvement of adenosine 3′,5′-monophosphate in release of ACTH. *Am. J. Physiol.*, **217**, 1286–1291

Folkers, K., Enzmann, F., Boler, J., Bowers, C. Y. and Schally, A. V. (1969) Discovery of modification of the synthetic tripeptide-sequence of the thyrotropin-releasing hormone having activity. *Biochem. biophys. Res. Commun.*, **37**, 123–126

Foreman, J. C., Hallett, M. B. and Mongar, J. L. (1975) ^{45}Calcium uptake in rat peritoneal mast cells. *Br. J. Pharmac.*, **55**, 283–284P

Foreman, J. C. and Mongar, J. L. (1975) Calcium and the control of histamine secretion from mast cells. In *Calcium Transport in Contraction and Secretion* (Eds. E. Carafoli *et al.*), North Holland Publishing Co., Amsterdam, pp. 175–184

Foreman, J. C., Mongar, J. L. and Gomperts, B. D. (1973) Calcium ionophores and movement of calcium ions following the physiological stimulus to a secretory process. *Nature, Lond.*, **245**, 249–251

Forn, J. and Krishna, G. (1971) Effect of norepinephrine, histamine, and other drugs on cyclic 3′,5′-AMP formation in brain slices of various animal species. *Pharmacology, Basel*, **5**, 193–204

Forn, J., Kruger, B. K. and Greengard, P. (1974) Adenosine 3′,5′-monophosphate content in rat caudate nucleus: Demonstration of dopaminergic and adrenergic receptors. *Science, N.Y.*, **186**, 1118–1120

Forn, J. and Valdecasas, F. G. (1971) Effects of lithium on brain adenyl cyclase activity. *Biochem. Pharmac.*, **20**, 2773–2779

Forscher, B. K. and Houck, J. C. (1973) Chalones: Concepts and current researches. *Natn. Cancer Inst. Monogr.* **38**

Forte, L. R., Nichols, G. A. and Anast, C. S. (1976) Renal adenylate cyclase and the inter-relationship between parathyroid hormone and vitamin D in the regulation of urinary phosphate and adenosine 3′,5′-monophosphate excretion. *J. Clin. Invest.*, **57**, 559–568

Foster, D. O. and Pardee, A. B. (1969) Transport of amino acids by confluent and nonconfluent 3T3 and polyoma virus-transformed 3T3 cells growing on glass cover slips. *J. biol. Chem.*, **244**, 2675–2681

Fox, R. H. and Hilton, S. M. (1958) Bradykinin formation in human skin as a factor in heat vasodilatation. *J. Physiol., Lond.*, **142**, 219–232

Frank, W. (1971) Cyclic 3′,5′-AMP and cell proliferation in cultures of embryonic rat cells. *Expl. Cell Res.*, **71**, 238—241

Franklin, T. J. and Foster, S. J. (1973) Leakage of cyclic AMP from human diploid fibroblasts in tissue culture. *Nature (New Biol.)*, **246**, 119–120

Franks, D. J., Perrin, L. S. and Malamud, D. (1974) Calcium ion: A modulator of parotid adenylate cyclase activity. *FEBS Letters*, **42**, 267–270

Frazer, A., Pandey, G., Mendels, J., Neeley, S., Kane, M. and Hess, M. E. (1974) The effect of tri-iodothyronine in combination with imipramine on [^{3}H]cyclic AMP production in slices of rat cerebral cortex. *Neuropharmacol.*, **13**, 1131–1140

Frazier, W. A., Boyd, E. F. and Bradshaw, R. A. (1974) Properties and specificity of binding sites for [^{125}I] nerve-growth factor in embryonic heart and brain. *J. biol. Chem.*, **249**, 5918–5923

Frazier, W. A., Ohlendorf, C. E., Boyd, L. F., Aloe, L., Johnson, E. M., Ferrendelli, J. A. and Bradshaw, R. A. (1973) Mechanism of action of nerve growth factor and cyclic AMP on neurite outgrowth in embryonic chick sensory ganglia: Demonstration of independent pathways of stimulation. *Proc. natn. Acad. Sci. U.S.A.*, **70**, 2448–2452

Fredholm, B. B. and Hedqvist, P. (1975) Indomethacin and the role of prostaglandins in adipose tissue. *Biochem. Pharmac.*, **24**, 61–66

Fredholm, B. B. and Rosell, S. (1970) Release of prostaglandin-like material from canine subcutaneous adipose tissue by nerve stimulation. *Acta physiol. scand.*, **79**, 18A

Free, C. A., Chasin, M. Paik, V. S. and Hess, S. M. (1971) Steroidogenic and lipolytic activities of 8-substituted derivatives of cyclic 3′,5′-adenosine monophosphate. *Biochemistry*, **10**, 3785–3789

Free, C. A., Chasin, M., Paik, V. S. and Hess, S. M. (1972) Structural requirements for steroidogenic and lipolytic activities of derivatives of cyclic 3′,5′-adenosine monophosphate. *Fed. Proc.*, **31**, 555

Freedman, R. and Hoffer, B. J. (1975) Phenothiazine antagonism of the noradrenergic inhibition of cerebellar Purkinje neurons. *J. Neurobiol.*, **6**, 277–288

Freedman, M. H., Raff, M. C. and Gomperts, B. (1975) Induction of increased calcium uptake in mouse T lymphocytes by concanavalin A and its modulation by cyclic nucleotides. *Nature Lond.*, **255**, 378–382

Freeman, J. J. and Sulser, F. (1972) Iprindole–amphetamine interactions in the rat: The role of aromatic hydroxylation of amphetamine in its mode of action. *J. Pharmac. exp. Ther.*, **183**, 307–315

French, S. W. and Palmer, D. S. (1973) Adrenergic supersensitivity during ethanol withdrawal in the rat. *Res. Commun. Chem. Pathol. Pharmac.*, **6**, 651–662

French, S. W., Palmer, D. S. and Narod, M. E. (1975a) Adrenergic subsensitivity of the cerebral cortex after reserpine treatment. *Fed. Proc.*, **34**, 297

French, S. W., Palmer, D. S. and Narod, M. E. (1975b) Effect of withdrawal from chronic ethanol ingestion on the cyclic AMP response of cerebral cortical slices using the agonists histamine, serotonin and other neurotransmitters. *Can. J. Physiol. Pharmac.*, **53**, 248–255

French, S. W., Reid, P. E., Palmer, D. S., Narod, M. E. and Ramey, C. W. (1974) Adrenergic subsensitivity of the rat brain during chronic ethanol ingestion. *Res. Comm. Chem. Pathol. Pharmacol.*, **9**, 575–578

Frieden, C. (1970) Kinetic aspects of regulation of metabolic processes. The histretic enzyme concept. *J. Biol. Chem.*, **245**, 5788–5799

Friedmann, N. (1972) Effects of glucogon and cyclic AMP on ion fluxes in the perfused liver. *Biochem. Biophys. Acta*, **274**, 214–225

Friedman, N., Somlyo, A. V. and Somlyo, A. P. (1971) Cyclic adenosine and guanosine monophosphate and glucagon: effect on liver membrane potentials. *Science*, **171**, 400–402

Frohlich, E. D. (1974) Hemodynamic concepts in hypertension. *Hosp. Practice*, **9**, 59–72

Froehlich, J. E. and Rachmeler, M. (1964) Inhibition of cell growth in the G_1 phase by adenosine 3′,5′-cyclic monophosphate. *J. Cell Biol.*, **60**, 249–257

Fujiwara, M., Kuchii, M. and Shibata, S. (1972) Differences of cardiac reactivity between spontaneously hypertensive and normotensive rats. *Eur. J. Pharmacol.*, **19**, 1–11

Fumagalli, R., Bernareggi, V., Berti, F. and Trabucchi, M. (1971) Cyclic AMP formation in human brain: An *in vitro* stimulation by neurotransmitters. *Life Sci.*, **10(I)**, 1111–1115

Furmanski, P. and Lubin, M. (1973) Cyclic AMP and the expression of differentiated properties *in vitro*. In *The Role of Cyclic Nucleotides in Carcinogenesis* (Eds. J. Schultz and H. G. Gratzer), Academic Press, New York, pp. 239–254

Furmanski, P., Silverman, D. J. and Lubin, M. (1971) Expression of differentiated functions in mouse neurolastoma mediated by dibutyryl-cyclic adenosine monophosphate. *Nature, Lond.*, **233**, 413–415

Furuta, Y., Yoshida, S., Inatome, T. and Tomomatsu, T. (1974) Calcium and sodium metabolism in experimental hypertensive rats. *Jap. Circul. J.*, 38, 1123–1126

Fuxe, K. and Hökfelt, T. (1969) Catecholamines in the hypothalamus and pituitary gland. In *Frontiers in Neuroendocrinology* (Eds. W. F. Ganong and L. Martini), Oxford University Press, New York, pp 47–69

Fuxe, K. and Ungerstedt, U. (1974) Action of caffeine and theophylline on supersensitive dopamine receptors. Considerable enhancement of receptor responses to treatment with DOPA and dopamine receptor agonists. *Med. Biol.*, **52**, 48–54

Gabbiani, G., Malaisse-Lagae, F., Blondel, B. and Orci, L. (1974) Actin in pancreatic islet cells. *Endocrinology*, **95**, 1630–1635

Gagliardino, J. J. and Martin, J. M. (1968) Stimulation of growth hormone secretion in monkeys by adrenaline, pitressin and adenosine 3′,5′-cyclic monophosphate. *Acta endocr., Copenh.*, **59**, 390–396

Gahwiler, B. H. (1976) Inhibitory action of noradrenaline and cyclic adenosine monophosphate in explants of rat cerebellum. *Nature, Lond.*, **259**, 483–484

Gale, R. P. and Zighelboim, J. (1974) Modulation of polymorphonuclear leukocyte-mediated antibody-dependent cellular cytotoxicity, *J. Immunol.*, **113**, 1793–1800

Gannon, M. C. and Nuttall, F. Q. (1975) Insulin stimulation of heart glycogen synthetase D phosphatase. *Diabetes*, **24** (Suppl. 2), 394

Garancis, J. C. (1968) Type II glycogenosis. Biochemical and electron micrographic study. *Am. J. Med.*, **44**, 289–330

Garbers, D. L. and Johnson, R. A. (1975) Metal and metal–ATP interactions with brain and cardiac adenylate cyclase. *J. biol. Chem.*, **250**, 8449–8456

Garelis, E. and Neff, N. H. (1974) Cyclic adenosine monophosphate: Selective increase in caudate nucleus after administration of L-DOPA. *Science, N.Y.*, **183**, 532–533

Garovoy, M. R., Strom, T. B., Kaliner, M. and Carpenter, C. B. (1975) Antibody-dependent lymphocyte-mediated cytotoxicity mechanism and modulation by cyclic nucleotides. *Cell Immunol.*, **20**, 197–204

Gaut, Z. N. and Huggins, C. G. (1966) Effect of epinephrine on the metabolism of inositol phosphatides in rat heart *in vivo*. *Nature, Lond.*, **212**, 612

George, W. J. and Busuttil, R. W. (1975) Relationship between myocardial cyclic nucleotide content and lysomal enzyme release in the anoxic heart following pretreatment with methylprednisolone. *Abstr. 6th Int. Congr. Pharmac., Helsinki*, p. 227

George, W. J., Polson, J. B., O'Toole, A. G. and Goldberg, N. D. (1970) Elevation of guanosine 3′.5′-cyclic phosphate in rat heart after perfusion with acetylcholine. *Proc. natn. Acad. Sci., U.S.A.*, **66**, 398–403

George, W. J., Wilkerson, R. D. and Kadowitz, P. J. (1973) Influence of acetylcholine on contractile force and cyclic nucleotide levels in the isolated perfused rat heart. *J. Pharmac. exp. Ther.*, **184**, 228–235

Gergely, J. (1964) *Biochemistry of Muscle Contraction.* Little Brown, Boston

Gerich, J. E., Lorenzi, M., Schneider, V., Karam, J. H., Rivier, J., Guillemin, R. and Forsham, P. H. (1974) Effects of somatostatin on plasma glucose and glucagon levels in human diabetes mellitus. *New Engl. J. Med.*, **291**, 544–547

Gerlach, A. and van Zwieten, P. A. (1969) Mechanical performance and calcium metabolism in rat isolated heart muscle after adrenalectomy. *Pfluegers Arch. Gs. Physiol.*, **311**, 96–108

Gershon, R. K. (1974) T-Cell control of antibody production. *Contemp. Top. Immunobiol.*, **3**, 1–40

Gertler, M. M., Saluste, E., Leetma, H. E. and Guthrie, R. G. (1971) The use of cyclic AMP in congestive heart failure and stroke. *Abstr. 1st Internat. Conf. Physiol. Pharmacol. cAMP, Milano*, p. 92

Gessa, G. L., Krishna, G., Forn, J., Tagliamonte, A. and Brodie, B. B. (1970) Behavioral and vegetative effects produced by dibutyryl-cyclic AMP injected into different areas of the brain. *Adv. Biochem. Psychopharmac.*, **3**, 371–381

Geyer, M. A. and Segal, D. S. (1973) Differential effects of reserpine and α-methyl-*p*-tyrosine on norepinephrine and dopamine-induced behavioral activity. *Psychopharmacologia*, **29**, 131–138

Gianutsos, G., Drawbaugh, R. B., Hynes, M. D. and Lal, H. (1974) Behavioral evidence for dopaminergic supersensitivity after chronic haloperidol. *Life Sci.*, **14**, 887–898

Giao, N.-B., Cehovic, G., Bayer, M., Gergely, H. and Posternak, T. (1974) Action comparée de trois dérivés thio-butyrylés de l'AMP cyclique sur la libération et la synthèse de la prolactine hypophysaire. *C.r. hebd. Séanc. Acad. Sci.*, **279**, 1705–1708

Gill, D. M. (1975) Involvement of nicotinamide adenine dinucleotide in the action of cholera toxin *in vitro*. *Proc. natn. Acad. Sci. U.S.A.*, **72**, 2064–2068

Gillen, R. C. and Nagyvary, J. (1976) Some biochemical properties of alkyl phosphotriesters of cyclic AMP. *Biochem. biophys. Res. Commun.*, **68**, 836–840

Gillespie, E. (1971) Colchicine binding in tissue slices: Decrease by calcium and biphasic effect of adenosine-3′,5′-monophosphate. *J. Cell Biol.*, **50**, 544–549

Gillespie, E., Valentine M. D. and Lichtenstein, L. M. (1974) Cyclic AMP metabolism in asthma: Studies with leukocytes and lymphocytes. *J. Allergy clin. Immunol.*, **53**, 27–33

Gilman, A. G. (1972) Protein binding assays for cyclic nucleotides. In *Advances in Cyclic Nucleotide Research* (Eds. P. Greengard, G. A. Robison and R. Paoletti), Vol. 2, Raven Press, New York, pp. 9–24

Gilman, A. G. (1974) Effect of 2-chloro-adenosine on cyclic AMP concentration in cultured cells. *Fed. Proc.*, **33**, 507

Gilman, A. G. and Minna, J. D. (1973) Expression of genes for metabolism of cyclic adenosine 3′,5′-monophosphate in somatic cells. I. Response to catecholamines in parental and hybrid cells. *J. biol. Chem.*, **248**, 6610–6617

Gilman, A. G. and Nirenberg, M. (1971a) Effect of catecholamine on the adenosine 3′,5′-cyclic monophosphate concentrations of clonal satellite cells of neurons. *Proc. natn. Acad. Sci. U.S.A.*, **68**, 2165–2168

Gilman, A. G. and Nirenberg, M. (1971b) Regulation of adenosine 3′,5′-cyclic monophosphate metabolism in cultured neuroblastoma cells. *Nature, Lond.*, **234**, 356–358

Gilman, A. G. and Rall, T. W. (1968) Factors influencing adenosine 3′,5′-phosphate accumulation in bovine thyroid slices. *J. biol. Chem.*, **243**, 5867–5871

Gilman, A. G. and Schrier, B. K. (1972) Adenosine cyclic 3′,5′-monophosphate in feta rat brain cell cultures. *Molec. Pharmac.*, **8**, 410–416

Girardier, L., Seydoux, J. and Clausen, T. (1968) Membrane potential of brown adipose tissue. A suggested mechanism for the regulation of thermogenesis. *J. gen. Physiol.*, **52**, 925–940

Glasgow, A., Polgar, P., Saporoschetz, I., Kim, H., Rutenberg, A. M. and Mannick, J. A. (1975) Phytohemagglutinin stimulation of human lymphocytes. *Clin. Immunol. Immunopath.*, **3**, 353–362

Glaviano, V. V. and Master, T. (1971) Inhibitory action of intracoronary prostaglandin E_1 on myocardial lipolysis. *Am. J. Physiol.*, **220**, 1187–1193

Glick, D., Katsumata, Y. and von Redlich, D. (1974) Quantitative histological distribution of adenosine-3′,5′-monophosphate in the rat stomach. Improved procedure for the luminescence assay. *J. Histochem. Cytochem.*, **22**, 395–400

Glinos, A. D. and Werrlein, R. J. (1972) Density-dependent regulation of growth in suspension cultures of L-929 cells. *J. Cell Physiol.*, **79**, 79–90

Gloor, P. (1955) Electrophysiological studies on the connections of the amygdaloid nucleus in the rat. *EEG Clin. Neurophysiol.*, **7**, 223–264

Gnegy, M. E., Costa, E. and Uzunov, P. (1976) Regulation of trans-synaptically elicited increase in 3′,5′-cyclic AMP by endogenous phosphodiesterase inhibitor. *Proc. Nat. Acad. Sci. U.S.A.*, **73**, 352–355

Godfraind, J. M. and Rumain, R. (1971) Cyclic adenosine monosphosphate and norepinephrine: Effect of Purkinje cells in rat cerebellar cortex. *Science, N.Y.*, **174**, 1257

Goggins, J. F., Johnson, G. S. and Pastan, I. (1972) The effects of dibutyryl-cyclic adenosine monophosphate on synthesis of sulfated acid mucopolysaccharides by transformed cells. *J. Biol. Chem.*, **247**, 5759–5764

Gohil, R. N., Gillen, R. G. and Nagyvary, J. (1974) Synthesis and properties of some cyclic AMP alkyl phosphotriesters. *Nucleic ACIDS Res.*, **12**, 1691–1701

Gold, H. K., Prindle, K. M., Levey, G. S. and Epstein, S. E. (1970) Effects of experimental heart failure on the capacity of glucagon to augment myocardial contractility and activate adenyl cyclase. *J. clin. Invest.*, **49**, 999–1006

Goldberg, A. L. and Singer, J. J. (1969) Evidence for a role of cyclic AMP in neuromuscular transmission. *Proc. natn. Acad. Sci. U.S.A.*, **64**, 134–141

Goldberg, N. D., Dietz, S. B. and O'Toole, A. G. (1969) Cyclic guanosine 3′,5′-monophosphate in mammalian tissues and urine. *J. biol. Chem.*, **244**, 4458–4466

Goldberg, N. D., Haddox, M. K., Dunham, E., Lopez, C. and Hadden, J. W. (1974) The Yin Yang hypothesis of biological control: opposing influences of cyclic GMP and cyclic AMP in the regulation of cell proliferation and other biological processes. In *Control of Proliferation in Animal Cells*, pp. 609–625

Goldberg, N. D., Haddox, M. K., Hartle, D. K. and Hadden, J. W. (1973a) The biological role of cyclic 3′,5′-guanosine monophosphate. *Proc. 5th Int. Cong. Pharmacol.*, **5**, 146–169

Goldberg, N. D., Haddox, M. K., Nicol, S. E., Glass, D. B., Sanford, C. H., Kuehl, F. A. and Estensen, R. (1975) Biologic regulation through opposing influences of cyclic GMP and cyclic AMP: The Yin Yang hypothesis.In *Advances in Cyclic Nucleotide Research*, Vol. 5 (Eds. G. I. Drummond P. Greengard, and G. A. Robison), Raven Press, New York, pp 307–330

Goldberg, N. D., O'Dea, R. F. and Haddox, M. K. (1973b) Cyclic GMP. In *Advances in Cyclic Nucleotide Research,* Vol. 3 (Eds. P. Greengard and G. A. Robison), Raven Press, New York, pp. 155–223

Goldfine, I. D., Gardner, J. D. and Neville, D. M. (1972a) Insulin action in isolated rat thymocytes. Binding of [^{125}I]insulin and stimulation of alpha-aminoisobutyric acid transport. *J. biol. Chem.*, **247**, 6919–6926

Goldfine, I. D., Perlman, R. and Roth, J. (1971) Inhibition of cyclic 3′,5′-AMP phosphodiesterase in islet cells and other tissues by tolbutamide. *Nature, Lond.*, **234**, 295–296

Goldfine, I. D., Roth, J. and Birnbaumer, L. (1972b) Glucagon receptors in β-cells. Binding of [^{125}I]-glucagon and activation of adenylate cyclase. *J. biol. Chem.*, **247**, 1211–1218

Goldstein, A. (1976) Opioid peptides (endorphins) in pituitary and brain. *Science*, **193**, 1081–1086

Goldstein, I. M., Brai, M., Osler, A. G. and Weissmann, G. (1973a) Lysosomal enzyme release from human leukocytes: Mediation by the alternate pathway of complement activation. *J. Immunol.*, **111**, 33–37

Goldstein, I., Hoffstein, S., Gallin, J. and Weissman, G. (1973b) Mechanisms of lysosomal enzyme release from human leucocytes: Microtubule assembly and membrane fusion induced by a component of complement. *Proc. natn. Acad. Sci. U.S.A.*, **70**, 2916–2920

Goldstein, M., Anagnoste, B. and Shirron, C. (1973c) The effect of trivastal, haloperidol and dibutyryl-cyclic AMP on [^{14}C]-dopamine synthesis in rat striatum. *J. Pharm. Pharmac.*, **25**, 348–351

Goldstein, M., Ebstein, B., Bronaugh, R. L. and Roberge, C. (1975) Stimulation of striatal tyrosine hydroxylase by cyclic AMP. In *Chemical Tools in Catecholamine Research II.* (Eds. O. Almgren, A. Carlsson and J. Engel), North-Holland Publishing Co., Amsterdam, pp. 257–264

Goldstein, R. E., Skelton, C. L., Levey, G. S., Glancy, D. L., Beiser, G. D. and Epstein, S. E. (1971) Effects of chronic heart failure on the capacity of glucagon to enhance contractility and adenyl cyclase activity of human papillary muscle. *Circulation*, **44**, 638–648

Goodman, D. B. P., Rasmussen, H., DiBella, F. and Guthrow, C. E. (1970) Cyclic adenosine 3′,5′-monophosphate stimulates phosphorylation of isolated neurotubule units. *Proc. natn. Acad. Sci., U.S.A.*, **67**, 652–659

Goren, E. N. and Rosen, O. M. (1971) The effect of nucleotides and a nondialysable factor on the hydrolysis of cyclic AMP by a cyclic nucleotide phosphodiesterase from beef heart. *Archs. Biochem. Biophys.*, **142**. 720–723

Goren, E. N. and Rosen, O. M. (1972) Purification and properties of a cyclic nucleotide phosphodiesterase from beef heart. *Archs. Biochem. Biophys.*, **153**, 384–397

Goridis, C., Massarelli, R., Sensenbrenner, M. and Mandel, P. (1974) Guanyl cyclase in chick embryo brain cell cultures: Evidence of neuronal localization. *J. Neurochem.*, **23**, 135–138

Goridis, C. and Virmaux, N. (1974) Light regulated guanosine 3′,5′-monophosphate phosphodiesterase of bovine retina. *Nature, Lond.*, **248**, 57–58

Goridis, C., Virmaux, N., Urban, P. F. and Mandel, P. (1973) Guanyl cyclase in a mammalian photoreceptor. *FEBS Letters*, **30**, 163–166

Gorman, R. R. (1975) Prostaglandin endoperoxides: Possible new regulators of cyclic nucleotide metabolism. *J. Cycl. Nucleot. Res.*, **1**, 1–9

Gorman, R. R., Hamberg, M. and Samuelsson, B. (1975) Inhibition of basal and hormone-stimulated adenylate cyclase in adipocyte ghosts by the prostaglandin endoperoxide prostaglandin H_2. *J. biol. Chem.*, **250**, 6460–6463

Gospodarowicz, D. (1974) Localization of a fibroblast growth factor and its effect alone and with hydrocortisone on 3T3 cell growth. *Nature, Lond.*, **249**, 123–127

Gospodarowicz, D. and Moran, J. S. (1974) Stimulation of division of sparse and confluent 3T3 cell populations by a fibroblast growth factor, dexamethasone, and insulin. *Proc. natn. Acad. Sci., U.S.A.*, **71**, 4584–4588

Granner, D. K. (1972) Protein kinase: Altered regulation in a hepatoma cell line deficient in adenosine 3′,5′-cyclic monophosphate-binding protein. *Biochem. biophys. Res. Commun.*, **46**, 1516–1522

Granner, D. K. (1974) Absence of high affinity adenosine 3′,5′-monophosphate binding sites from the cytosol of three hepatic-derived cell lines. *Archs. Biochem. Biophys.*, **165**, 359–368

Granner, D., Chase, L. R., Aurbach, G. D. and Tomkins, G. M. (1968) Tyrosine aminotransferase: Enzyme induction independent of adenosine 3′,5′-monophosphate. *Science, N.Y.*, **162**, 1018

Grant, G., Vale, W. and Guillemin, R. (1973a) Characteristics of the pituitary-binding sites for thyrotropin-releasing factor. *Endocrinology*, **92**, 1629–1633

Grant, G., Vale, W. and Rivier, J. (1973b) Pituitary-binding sites for [^{3}H] labelled luteinizing hormone releasing factor (LRF). *Biochem. biophys. Res. Commun.*, **50**, 771–778

Grantham, J. J. (1974) Action of antidiuretic hormone in the mammalian kidney. In *MTP International Review of Science; Kidney and Urinary Tract Physiology* (Ed. K. Thurau), University Park Press, Baltimore, **6**, 267–272

Graves, D. J., Hayakawa, T., Hovitz, R. A., Beckman, E. and Krebs, E. G. (1974) Studies on the subunit structure of trypsin-activated phosphorylase kinase. *Biochemistry*, **12**, 580–588

Greaves, M. F. and Bauminger, S. (1972) Activation of T and B lymphocytes by insoluble phytomitogens. *Nature, New Biol.*, **235**, 67–70

Greaves, M. W. and McDonald-Gibson, W. (1973) Effect of nonsteroid antiinflammatory and antipyretic drugs on prostaglandin biosynthesis by human skin. *J. invest. Derm.*, **61**, 127–129

Green, C. D. and Martin, D. W. (1974) A direct, stimulating effect of cyclic GMP on purified phosphoribosyl pyrophosphate synthetase and its antagonism by cyclic AMP. *Cell*, **2**, 241–245

Green. I. C., Howell, S. L., Montague, W. and Taylor, K. W. (1973) Regulation of insulin release from isolated islets of Langerhans of the rat in pregnancy. *Biochem. J.*, **134**, 481–487

Green, I. C. and Taylor, K. W. (1974) Insulin secretory response of isolated islets of Langerhans in pregnant rats: Effects of dietary restriction. *J. Endocr.*, **62**, 137–143

Green, J. D. and Harris, G. W. (1947) The neurovascular link between the neurohypophysis and adenohypophysis. *J. Endocr.*, **5**, 136–146

Green, R. D., Dale, M. M. and Haylett, D. C. (1972) Effect of adrenergic amines on the membrane potential of guinea-pig liver parenchymal cells in short term tissue culture. *Experientia*, **28**, 1073–1074

Greenberg, S. and Bohr, D. F. (1975) Venous smooth muscle in hypertension. Enhanced contractility of portal veins from spontaneously hypertensive rats. *Circulation Res.*, **36**, Suppl. I, 208–215

Greene, W. C. and Parker, C. W. (1975) A role for cytochalasin-sensitive proteins in the regulation of calcium transport in activated human lymphocytes. *Biochem. biophys. Res. Commun.*, **65**, 456–463

Greene, W. C., Parker, C. M. and Parker, C. W. (1975a) Effects of microtubular and microfilament reagents on amino acid transport in mitogen activated human lymphocytes. *Clin. Res.*, **23**, 410A

Greene, W. C., Parker, C. M. and Parker, C. W. (1976) Colchicine-sensitive structures and lymphocyte activation. *J. Immunol.* **117**, 1015–1022

Greene, W. C., Parker, C. M. and Parker, C. W. (1977) Opposing effects of mitogenic and nonmitogenic lectins on lymphocyte activation. Evidence that wheat germ agglutinin produces a negative signal. *J. biol. Chem.*, **251**, 4017–4025

Greengard, P. (1975) Presynaptic and postsynaptic roles of cyclic AMP and protein phosphorylation at catecholamingeric synapses. In *Chemical Tools in Catecholamine Research* II (Eds. O. Almgren, A. Carlsson and J. Engel.), North Holland Publishing Co., Amsterdam, pp. 249–256

Greengard, P., Hayaishi, O. and Colowick, S. P. (1969a) Enzymatic adenylation of pyrophosphate by 3′,5′-cyclic AMP-Reversal of the adenyl cyclase reaction. *Fed. Proc.*, **28**, 467

Greengard, P. and Kebabian, J. W. (1974) Role of cyclic AMP in synaptic transmission in the mammalian peripheral nervous system. *Fed. Proc.*, **33**, 1059–1068

Greengard, P. and Kuo, J. F. (1970) On the mechanism of action of cyclic AMP. In *Role of Cyclic AMP in Cell Function* (Eds. P. Greengard and E. Costa), Raven Press, New York, pp. 287–306

Greengard, P., Rudolph, S. A. and Sturtevant, J. M. (1969b) Enthalpy of hydrolysis of the 3′ bond of adenosine 3′,5′-monophosphates and guanosine 3′,5′-monophosphate. *J. biol. Chem.*, **244**, 4798–4800

Greenwood, F. C., Hunter, W. M. and Glover, J. S. (1963) The preparation of ^{131}I-labelled human growth hormone of high specific radioactivity. *Biochem. J.*, **89**, 114–123

Greer, M. A. and Solomon, D. H. (1974) Thyroid secretion. In *Handbook of Physiology*, Sect. VII, Vol. III. American Physiological Society, Washington, pp 135–146

Grenier, G., van Sande, J., Glick, D. and Dumont, J. E. (1974) Effect of ionophore A23187 on thyroid secretion. *FEBS Letters*, **49**, 96–99

Grey, N. J., Goldring, S. and Kipnis, D. M. (1970) The effect of fasting, diet and actinomycin D on insulin secretion in the rat. *J. clin. Invest.*, **49**, 881–889

Grill, V. and Cerasi, E. (1974) Stimulation by D-glucose of cyclic adenosine 3′,5′-monophosphate accumulation and insulin release in isolated pancreatic insulin from the isolated, perfused canine pancreas. *J. clin. Invest.*, **52**, 2102–2116

Grimm, W. and Marks, F. (1974) Effect of tumor-promoting phorbol ester on the normal and the isoproterenol-elevated level of adenosine 3′,5′-cyclic monophosphate in mouse epidermis *in vivo*. *Cancer Res.*, **34**, 3128–3134

Grossman, A. and Furchgott, R. G. (1964) The effects of frequency of stimulation and calcium concentration on ^{45}Ca exchange and contractility on the isolated guinea-pig auricle. *J. Pharmac. exp. Ther.*, **143**, 120–130

Guidotti, A., Mao, C. C. and Costa, E. (1973) Transsynaptic regulation of tyrosine hydroxylase in adrenal medulla: Possible role of cyclic nucleotides. In *Frontiers in Catecholamine Research* (Eds. E. Usdin and S. H. Snyder), Pergamon Press, New York, pp. 231–236

Gullis, R. J., Traber, J., Fischer, K., Buchen, C. and Hamprecht, B. (1975) Effects of cholinergic agents and sodium ions on the levels of guanosine and adenosine 3′,5′-cyclic monophosphates in neuroblastoma and neuroblastoma × glioma hybrid cells *FEBS Letters*, **59**, 74–79.

Gunaga, K. P., Kawano, A. and Menon, K. M. J. (1974) *In vivo* effect of estradiol benzoate on the accumulation of adenosine 3′,5′-cyclic monophosphate in the rat hypothalamus. *Neuroendocrinology*, **16**, 273–281

Gunaga, K. P. and Menon, K. M. J. (1973) Effect of catecholamines and ovarian hormones on cyclic AMP accumulation in rat hypothalamus. *Biochem. biophys. Res. Commun.*, **54**, 440–448

Guthrow, C. E., Rasmussen, H., Brunswick, D. L. and Cooperman, B. S. (1973) Specific photoaffinity labeling of the adenosine 3′,5′-cyclic monophosphate receptor in intact ghosts from human erythrocytes. *Proc. natn. Acad., Sci., U.S.A.*, **70**, 3344–3346

Guttler, R. B., Otis, C. L., Shan, J. W., Warren, D. W. and Nicoloff, J. T. (1975) The effect of thyroid hormone on adenylyl cyclase—a potential site for thyroid hormone action. In *Thyroid Hormone Metabolism*, pp. 201–211

Hadden, J. W., Coffey, R. G., Hadden, E. M., Lopez-Corrales, E. and Sunshine, G. H. (1975a) Effects of levamisole and imidazole on lymphocyte proliferation and cyclic nucleotide levels. *Cell. Immunol.*, **20**, 98–103

Hadden, J. W., Hadden, E. M., Haddox, M. K. and Goldberg, N. D. (1972) Guanosine 3′,5′-cyclic monophosphate: A possible intracellular mediator of mitogenic influence in lymphocytes. *Proc. natn. Acad. Sci. U.S.A.*, **69**, 3024–3027

Hadden, J. W., Hadden, E. M. and Goldberg, N. D. (1974) Cyclic GMP and cyclic AMP in lymphocyte metabolism and proliferation. In *Cyclic AMP, Cell Growth, and the Immune Response* (Eds. W. Braun, L. M. Lichtenstein and C. W. Parker), Springer-Verlag, New York, pp 237–246

Hadden, J. W., Johnson, E. M., Hadden, E. M., Coffey, R. G. and Johnson, L. D. (1975b) Cyclic GMP and lymphocyte activation. In *Immune Recognition* (Ed. A. S. Rosenthal), Academic Press, New York, pp. 359–389

Haddox, M. K., Newton, N. E., Hartle, D. K. and Goldberg, N. E. (1972) ATP(Mg^{2+}) induced inhibition of cyclic AMP reactivity with a skeletal muscle protein kinase. *Biochem. biophys. Res. Commun.*, **47**, 653–661

Häggendal, J. (1970) Some further aspects on the release of the adrenergic transmitter. In *New Aspects of Storage and Release Mechanisms of Catecholamines* (Eds. H. J. Schümann and G. Kroneberg), Springer Verlag, Berlin, pp. 100–108

Haggerty, D. F., Young, P. L., Popjak, G. and Carnes, W. H. (1973) Phenylalanine hydroxylase in cultured hepatocytes. I. Hormonal control of enzyme levels. *J. biol. Chem.*, **248**, 223–232

Haley, B. E. (1975) Photoaffinity of adenosine 3′,5′-cyclic monophosphate binding sites of human red cell membranes. *Biochemistry*, **14**, 3852–3857

Hall, R., Besser, G. M., Schally, A. V., Coy, D. H., Evered, D., Goldie, D. J., Kastin, A. J., McNeilly, A. S., Mortimer, C. H., Phenekos, C., Tunbridge, W. M. G. and Weightman, D. (1973) Action of growth hormone-release inhibitory hormone in healthy men and in acromegaly. *Lancet*, iii, 581–584

Halprin, K. M., Adachi, K., Yoshikawa, K., Levine, V., Mui, M. M. and Hsia, S. L. (1975). Cyclic AMP and psoriasis. *J. invest. Derml.*, **65**, 170–178

Ham, E. A., Zanetti, M. E., Goldberg, N. D. and Kuehl, F. A., Jr. (1974) Alterations in uterine cyclic GMP levels in the cycling rat. *Fed. Proc.*, **33**, 268

Hamberg, M. and Samuelsson, B. (1974) Prostaglandin endoperoxides. Novel transformations of arachidonic acid in human platelets. *Proc. natn. Acad. Sci. U.S.A.*, **71**, 3400–3404

Hamet, P., Kuchel, O. and Genest, J. (1973) Effect of upright posture and isoproterenol infusion on cyclic adenosine monophosphate excretion on control subjects and patients with labile hypertension. *J. clin. Endocr. Metab.*, **36**, 218–226

Hammers, R., Clarenbach, P., Lindl, T. and Cramer, H. (1977) Uptake and metabolism of cyclic AMP in rabbit choroid plexus *in vitro*. *Neuropharmacology*, **16**, 135–141

Hamilton, T. C. (1975) Influence of antihypertensive drug treatment on vascular reactivity in spontaneously hypertensive rats. *Br. J. Pharmac.*, **54**, 429–436

Hammarström, S., Hamberg, M., Samuelsson, B., Duell, E. A., Stawiski, M. and Voorhees, J. J. (1975) Increased concentrations of nonesterified arachidonic acid, 12L-hydroxy-5,8,10,14-eicosatetraenoic acid, prostaglandin E_2, and prostaglandin F_2 in epidermis of psoriasis. *Proc. natn. Acad. Sci. U.S.A.*, **72**, 5130–5134

Hamprecht, B. and Schultz, J. (1973) Influence of noradrenaline, prostaglandin E_1 and inhibitors of phosphodiesterase activity on levels of the cyclic adenosine 3′,5′-monophosphate in somatic cell hybrids. *Hoppe-Seyler's Z. physiol. Chem.*, **354**, 1633–1641

Hanoune, J., Lacombe, M. L. and Pecker, F. (1975) The epinephrine-sensitive adenylate cyclase of rat liver plasma membranes. *J. biol. Chem.*, **250**, 4559–4574

Harary, I., Hoover, F. and Farley, B. (1973) Catecholamine and dibutyryl- cyclic AMP effects on myosin adenosine triphosphatase in cultured rat heart cells. *Science, N.Y.*, **181**, 1061–1063

Hardman, J. G., Beavo, J. A., Gray, J. P., Chrisman, T. D., Patterson, W. D. and Sutherland, E. W. (1971) The formation and metabolism of cyclic GMP. *Ann. N.Y. Acad. Sci.*, **185**, 27–35

Hardman, J. G. and Sutherland, E. W. (1969) Guanyl cyclase, an enzyme catalysing the formation of guanosine 3′,5′-monophosphate from guanosine triphosphate. *J. biol. Chem.*, **244**, 6363–6370

Harper, J. E. and Brooker, G. (1975) Femtomole sensitive radioimmunoassay for cyclic AMP and cylic GMP after 2′*O*-acetylation by acetic anhydride in aqueous solution. *J. Cycl. Nucleot. Res.*, **1**, 207–218

Harper, R. A., Flaxman, B. A. and Chopra, D. P. (1974) Mitotic response of normal and psoriatic keratinocytes *in nitro* to compounds known to affect intracellular cyclic AMP. *J. invest. Derm.*, **62**, 384–387

Harris, D. N., Chasin, M., Phillips, M. B., Goldenberg, H., Samaniego, S. and Hess, S. M. (1973) Effect of cyclic nucleotides on activity of cyclic 3′,5′-monophosphate phosphodiesterase. *Biochem. Pharmac.*, **22**, 221–228

Harris, J. B. and Alonso, D. (1965) Stimulation of the gastric mucosa by adenosine 3′,5′-monophosphate. *Fed. Proc.*, **24**, 1368–1376

Harris, J. B., Nigon, K. and Alonso, D. (1969) Adenosine 3′,5′-monophosphate: Intracellular mediator for methylxanthine stimulation of gastric secretion. *Gastroenterology*, **57**, 377–384

Harris, J. E. (1976) β-Adrenergic receptor-mediated adenosine 3′,5′-monophosphate accumulation in the rat corpus striatum. *Molec. Pharmac.* **12**, 546–558

Harwood, J. P., Löw, H. and Rodbell, M. (1973) Stimulatory and inhibitory effects of guanyl nucleotides of fat cell adenylate cyclase. *J. biol. Chem.*, **248**, 6239–6245

Harwood, J. P. and Rodbell, M. (1973) Inhibition by fluoride ion of hormonal activation of fat cell adenylate cyclase. *J. biol. Chem.*, **248**, 4901–4904

Hashida, J., Terashina, H., Kin, K., Tei, T., Izeki, Y., Mikami, T., Ono, M., Uchiyama, T., Kita, T., Sato, M., Murakami, A., Kobayashi Y. and Kajiwara, N. (1974) Plasma catecholamines in spontaneously hypertensive rat assayed by double isotope method. *Jap. Heart J.*, **15**, 186

Haslam, R. J. (1973) Interactions of the pharmacological receptors of blood platelets with adenylate cyclase, *Ser. Haemat.*, **6**, 333–350

Haslam, R. J. (1975) Roles of cyclic nucleotides in platelet function. In *Biochemistry and Pharmacology of Blood Platelets* (Ed. J. Knight), Ciba Foundation Symposium, J. and A. Churchill, London, pp. 121–151

Haslam, R. J. and Goldstein, S. (1974) Adenosine 3′,5′-cyclic monophosphate in young and senescent human fibroblasts during growth and stationary phase *in vitro*: Effects of prostaglandin E_1 and of adrenaline. *Biochem. J.*, **144**, 153–263

Haslam, R. J. and McClenaghan, M. D. (1974) Effects of collagen and of aspirin on the concentration of guanosine 3′,5′-cyclic monophosphate in human blood platelets: Measurement by a prelabelling technique. *Biochem. J.*, **138**, 317–320

Haugaard, N. and Hess, M. E. (1965) Actions of autonomic drugs on phosphorylase activity and function of the heart. *Pharmac. Rev.*, **17**, 27–69

Hax, W. M. A., Van Venrooij, G. E. P. M. and Vossenberg, J. B. J. (1974) Cell communication: A cyclic AMP mediated phenomenon. *J. Membrane Biol.*, **19**, 253–266

Hayaishi, O., Greengard, P. and Colowick, S. P. (1971) On the equilibrium of the adenylate cyclase reaction. *J. biol. Chem.*, **246**, 5840–5843

Haylett, D. G. and Jenkinson, D. H. (1969) Effects of noradrenaline on the membrane potential and ionic permeability of parenchymal cells in the liver of the guinea pig. *Nature, Lond.*, **224**, 80–81

Hays, E. T., Dwyer, T. M., Horowitz, P. and Swift, J. F. (1974) Epinephrine action on sodium fluxes in frog striated muscle. *Am. J. Physiol.*, **227**, 1340–1347

Hedeskov, C. J. and Capito, K. (1975) The restoring effect of caffeine on the decreased sensitivity of the insulin secretory mechanism in mouse pancreatic islets during starvation. *Horm. Metab. Res.*, **7**, 1–5

Heidrick, M. L. and Ryan, W. L. (1971a) Metabolism of 3′,5′-cyclic AMP by strains of L cells. *Biochim. biophys. Acta*, **237**, 303–309

Heidrick, M. L. and Ryan, W. L. (1971b) Adenosine 3′,5′-cyclic monophosphate and contact inhibition. *Cancer Res.*, **81**, 1313–1315

Hellman, B. (1975) The significance of calcium for glucose stimulation of insulin release. *Endocrinology*, **97**, 392–398

Hellman, B., Idahl, L-A., Lernmark, A. and Taljedal, L-B. (1974) The pancreatic β-cell recognition of insulin secretagogues. XIII. Effects of sulphydryl reagents on cyclic AMP. *Biochim. biophys. Acta.*, **372**, 127–134

Hellman, B., Sehlin, J. and Taljedal, L-B. (1971a) Calcium uptake by pancreatic β-cells as measured with the aid of ^{45}Ca and mannitol-^{3}H. *Am. J. Physiol.*, **221**, 1795–1801

Hellman, B., Sehlin, J. and Taljedal, I. B. (1971b) The pancreatic β-cell recognition of insulin secretagogues. II. Site of action of tolbutamide. *Biochem. biophys. Res. Commun.*, **45**, 1384–1388

Henion, W. F., Sutherland, E. W. and Posternak T. H. (1967) Effects of derivatives of adenosine 3′,5′-phosphate on liver slices and intact animals. *Biochim. biophys. Acta*, **148**, 106–113

Henney, C. S., Bourne, H. R. and Lichtenstein, L. M. (1972) The role of cyclic 3′,5′-adenosine monophosphate in the specific cytolytic activity of lymphocytes. *J. Immunol.*, **108**, 1526–1534

Henquin, J.-C. and Lambert, A. E. (1974) Cationic environment and dynamics of insulin secretion. *Diabetes*, **23**, 933–942

Henry, P. D., Sobel, B. E., Kjekshus, J. K., Robison, A. and Stull, J. T. (1971) Cyclic-AMP in the perfused failing guinea-pig heart. *Proc. Soc. expl. Biol. Med.*, **137**, 768–771

Hepp, K. D. (1972) Adenylate cyclase and insulin action: Effect of insulin, non-suppressible insulin-like material, and diabetes on adenylate cyclase activity in mouse liver. *Europ. J. Biochem.*, **31**, 266–276

Hepp, K. D. and Renner, R. (1972) Insulin action on the adenyl cyclase system: Antagonism to activation by lipolytic hormones. *FEBS Letters*, **20**, 191–194

Herrera, E. (1975) Effects of sulfonylureas (tolbutamide, glipentide and glibenclamide) on *in vitro* glycerol metabolism in adipose tissue from rats. *Life Sci.*, **16**, 645–650

Hers, H. G. (1960) Amylo-1,6-glucosidase activity in tissues of children with glycogen storage disease. *Biochem. J.*, **76**, 69P

Hers, H. G. (1963) α-Glucosidase deficiency in generalized glycogen storage disease (Pompe's disease). *Biochem. J.*, **86**, 11–16

Hershko, A., Mamont, P., Shields, R. and Tomkins, G. M. (1971) Pleiotypic response. *Nature, New Biol.*, **232**, 206–211

Herting, D. C. and Steenbock, H. (1955) Vitamin D and gastric secretion. *J. Nutr.*, **57**, 469–482

Hess, H.-J. (1974) Biochemistry and structure activity studies with prazosin. In *Prazosin, Evaluation of a New Antihypertensive Agent* (Ed. D. W. K. Cotton), Excerpta Medica, pp. 3–15

Hess, M. E., Shanfeld, J. and Haugaard, N. (1962) The role of the autonomic nervous system in the regulation of heart phosphorylase in the open-chest rat. *J. Pharmac. exp. Ther.*, **133**, 191–196

Hickie, R. A., Walker, C. M. and Croll, G. A. (1974) Decreased basal cyclic adenosine 3′,5′-monophosphate levels in Morris hepatoma 5123 t.c.(h). *Biochem. biophys. Res. Commun.*, **59**, 167–173

Hidaka, H., Asano, T. and Shimamoto, T. (1975) Cyclic 3′,5′-AMP phosphodiesterase of rabbit aorta. *Biochim. biophys. Acta*, **377**, 103–116

Hidaka, T. and Kuriyama, H. (1969). Effects of catecholamines on the cholinergic neuromuscular transmission in fish red muscle. *J. Physiol., Lond.*, **201**, 61–71

Hilden, S. and Hokin, L. E. (1975) Active potassium transport coupled to active sodium transport in vesicles reconstituted from purified sodium and potassium ion-activated adenosine triphosphatase from the rectal gland of *Squalus acanthia*, *J. biol. Chem.*, **250**, 6290—6303

Hill, H. R., Estensen, R. D., Quie, P. G., Hogan, N. A. and Goldberg, N. D. (1975) Modulation of human neutrophil chemotactic responses by cyclic 3′,5′-guanosine monophosphate and cyclic 3′,5′-adenosine monophosphate. *Metabolism*, **24**, 447–456

Hirata, M. and Hayaishi, O. (1966) Enzymic formation of deoxyadenosine 3′,5′-phosphate. *Biochem. biophys Res. Commun.*, **24**, 360–364

Hirschhorn, R., Grossman, J. and Weissman, G. (1970) Effect of cyclic 3′,5′-adenosine monophosphate and theophylline on lymphocyte transformation. *Proc. Soc. exp. Biol. Med.*, **133**, 1361–1365

Hirschman, J. L. and Herfindal, E. T. (1971) Essential hypertension. *J. Am. Pharm. Ass.*, **11**, 555–561

Ho, H. C., Desai, R., Teo, T. S. and Wang, J. H. (1976) Catalytic and regulatory properties of two forms of bovine heart cyclic nucleotide phosphodiesterase. *Biochem. biophys, Acta.*, **429**, 461–474

Ho, H. C., Desai, R. and Wang, J. H. (1975a) Effect of Ca^{2+} on the stability of the protein activator of cyclic nucleotide phosphodiesterase. *FEBS Letters*, **50**, 374–377

Ho, R. J., Bomboy, J. D., Wasner, H. K. and Sutherland, E. W. (1975b) Preparation and characterization of a hormone antagonist from adipocytes. *Methods in Enzymology*, **39**, Academic Press, New York, pp. 431–438

Ho, R. J., Russell, T. R., Asakawa, T. and Hucks, M. W. (1975c) Inhibition of cyclic nucleotide phosphodiesterase activity by an endogenous factor. *J. Cycl. Nucleot. Res.*, **1**, 81–88

Ho, R. J. and Sutherland, E. W. (1971) Formation and release of a hormone antagonist by rat adipocytes. *J. biol. Chem.*, **246**, 6822–6827

Ho, R.-Y. and Sutherland, E. W. (1975) Action of feedback regulator on adenylate cyclase. *Proc. natn. Acad. Sci.*, U.S.A., **72**, 1773–1777

Hoefke, W., Kobinger, W. and Walland, A. (1975) Relationship between activity and structure in derivatives of clonidine. *Arzneimittel Forsch.*, **25**, 786–793

Hoffer, B. J., Freedman, R., Puro, D. and Woodward, D. J. (1975) Interaction of norepinephrine with cerebellar neuronal circuitry. *5th Ann. Mtg. Soc. Neurosci., New York*, 204

Hoffer, B. J., Freedman, R., Woodward, D. J., Daly, J. W. and Skolnick, P. (1976) Electrophysiological and biochemical interaction of fluphenazine with norepinephrine-sensitive cyclic adenosine 3′,5′-monophosphate generating systems in rat cerebella degranulated by X-irradiation. *Exp. Neurol.*, **51**, 653–667

Hoffer, B., Seiger, A., Freedman, R., Olson, L. and Taylor, D. (1977) Electrophysiology and cytology of hippocampal formation transplants in the anterior chamber of the eye. II. Cholinergic mechanisms. *Brain Res.*, **119**, 107–132

Hoffer, B. and Siggins, G. R. (1975) Electrophysiological techniques for the study of hormone action in the central nervous system. In *Methods in Enzymology*, Vol. 39 (Eds. J. Hadman and B. G. O'Malley), Academic Press, New York, pp. 429–442

Hoffer, B. J., Siggins, G. R. and Bloom, F. E. (1969) Prostaglandins E_1 and E_2 antagonize norepinephrine effects on cerebellar Purkinje cells: Microelectrophoretic study. *Science, N.Y.*, **166**, 1418–1420

Hoffer, B. J., Siggins, G. R., Oliver, A. P. and Bloom, F. E. (1971a) Cyclic AMP mediation of norepinephrine inhibition in rat cerebellar cortex: A unique class of synaptic responses. *Ann. N.Y. Acad. Sci.*, **185**, 531–549

Hoffer, B. J., Siggins, C. R., Oliver, A. P. and Bloom, F. E. (1973) Activation of the pathway from locus coeruleus to rat cerebellar Purkinje neurons: Pharmacological evidence of noradrenergic central inhibition. *J. Pharmac. exp. Ther.*, **184**, 553–569

Hoffer, B. J., Siggins, G. R., Woodward, D. J. and Bloom, F. E. (1971b) Spontaneous discharge of Purkinje neurons after destruction of catecholamine-containing afferents by 6-hydroxydopamine. *Brain Res., Osaka*, **30**, 425–430

Hofmann, F. and Sold, G. (1972) A protein kinase activity from rat cerebellum stimulated by guanosine-3′,5′-monophosphate. *Biochem. biophys. Res. Communic.*, **49**, 1100—1107

Hogberg, B. and Uvnäs, B. (1960) Further observations on the disruption of rat mesentery mast cells caused by compound 48/80, antigen—antibody reaction, lecithinase A and decylamine. *Acta. physiol. scand.*, **48**., 133—145

Hokin, L. E. (1969) Functional activity in glands and synaptic tissue and the turnover of phosphatidylinositol. *Ann. N.Y. Acad. Sci.*, **165**, 695

Hokin, M. R. and Hokin, L. E. (1953) Enzyme secretion and the incorporation of ^{32}P into phospholipids of pancreas slices. *J. biol. Chem.*, **203**, 967–977

Holian, O., Nyhus, L. M. and Bombeck, C. T. (1973) The effect of pentagastrin on adenyl cyclase. *Gastroenterology*, **64**, 746

Holla, S. W. J., Hollman, E. P. M. J., Mier, P. D., van der Staak, W. J. B. M., Urselmann, E. and Warndorff, J. A. (1972) Adenosine 3′,5′-cyclic monophosphate phosphodiesterase in skin. *Br. J. Derm.*, **86**, 147–149

Hollenberg, M. D. and Cuatrecasas, P. (1975) Insulin: interaction with membrane receptors and relationship to cyclic purine nucleotides and cell growth. *Fed. Proc.*, **34**, 1556–1563

Holley, R. W. (1975) Control of growth of mammalian cells in cell culture. *Nature, Lond.*, **258**, 487–490

Holley, R. W. and Kiernan, J. A. (1971) Studies on serum factors required by 3T3 and SV 3T3 cells. In *Ciba Foundation Symposium on Growth Control in Cell Cultures* (Eds. G. E. W. Wolstenholme and J. Knight), Churchill Livingstone, London, pp. 3–10

Holley, R. W. and Kiernan, J. A. (1974) Control of the initiation of DNA synthesis in 3T3 cells: Serum factors. *Proc. natn. Acad. Sci., U.S.A.*, **71**, 2908–2911

Holy, A. (1972) The use of modified nucleotide derivatives for the study of some nucleolytic enzymes. In *Mechanism and Control Properties of Phosphotransferases* (Eds. J. Hoffman and E. E. Bohm), Akademie-Verlag, Berlin, pp. 553–565

Honda, F. and Imamura, H. (1968) Inhibition of cyclic 3′,5′-nucleotide phosphodiesterase by phenothiazine and reserpine derivatives. *Biochim. Biophys. Acta.*, **161**, 267–269

Hong, C. I., Tritsch, G. L., Mittelman, A., Hebborn, P. and Chheda, G. B. (1975) Synthesis and antitumor activity of 5′-phosphates and cyclic 3′,5′-phosphates derived from biologically active nucleosides. *J. med. Chem.*, **18**, 465–473

Hopkins, H. A., Looney, W. B. and Kovacs, C. J. (1975) Increased urinary excretion of cyclic guanosine monophosphate in rats bearing Morris hepatoma 3924A. *Science, N.Y.*, **190**, 58–60

Hoppe, J. and Wagner, J. G. (1974) Synthesis and properties of N^6, C-8 and C-2 spin-labelled derivatives of adenosine cyclic 3′,5′-monophosphate. *Europ. J. Biochem.*, **48**, 519–525

Horn, A. S., Cuello, A. C. and Muller, R. J. (1974) Dopamine in the mesolimbic system of the rat brain: Endogenous levels and the effects of drugs of the uptake mechanism and stimulation of adenylate cyclase activity. *J. Neurochem.*, **22**, 265–270

Horn, A. S. and Phillipson, O. T. (1975) Noradrenaline-sensitive adenylate cyclase in rat limbic forebrain homogenates: Effects of agonists and antagonists. *Br. J. Pharmac.*, **55**, 299P—300P

Hornykievicz, O. (1966) Dopamine (3-hydroxytyramine) and brain function. *Pharman. Rev.*, **18**, 925–964

Hornykievicz, O. (1975) Parkinsonism induced by dopaminergic antagonists. *Adv. Neurol.*, **9**, 155–164

Horowitz, J. M., Horwitz, B. A. and Smith, R. E. (1971) Effect *in vivo* of norepinephrine on the membrane resistance of brown fat cells. *Experientia*, **27**, 1419

Horton, E. W. (1969) Hypothesis on the physiological roles of prostaglandins. *Physiol. Rev.*, **49**, -161

Horwitz, B. A. and Eaton, M. (1975) The effect of adrenergic agonists and cyclic AMP on the Na^+/K^+-ATPase activity of brown adipose tissue. *Europ. J. Pharmac.*, **34**, 241–245

Horwitz, B. A., Horowitz, J. M. and Smith, R. E. (1969) Norepinephrine-induced depolarization of brown fat cells. *Proc. natn. Acad. Sci., U.S.A.*, **64**, 113

Hoskins, D. D., Hall, M. L. and Munsterman, D. (1975) Induction of motility in immature bovine spermatozoa by cyclic AMP phosphodiesterase inhibitors and seminal plasma. *Biol. Reprod.*, **13**, 168–176

Hospkins, S. V. (1975) Reduction in isoprenaline-induced cyclic AMP formation in guinea-pig heart after exposure to isoprenaline or salbutamol. *Biochem. Pharmacol.*, **24**, 1237–1238

Hotz, J., Minner, H. and Ziegler, R. (1971a) Das Verhalten der Magensekretion des Menschen bei akuter ÄDTA-Hypocalcämie. *Verh. dt. Ges. inn. Med.*, **77**, 501–504

Hotz, J., Widmaier, F., Minner, H. and Ziegler, R. (1971b) Serum calcium and gastric secretion in chronic gastric fistula rat: Influence of parathyroid hormone and calcitonin. *Europ. J. clin. Invest.*, 1, 486–490

House, P. D. R., Poulis, P. and Weidemann, M. J. (1972) Isolation of a plasma membrane subfraction form rat liver containing an insulin-sensitive cyclic AMP phosphodiesterase. *Europ. J. Biochem.*, **24**, 429–437

Hovi, T., Allison, A. C. and Allsop, J. (1975) Rapid increase of phosphoribosyl pyrophosphate concentration after mitogenic stimulation of lymphocytes. *FEBS Letters*, **55**, 291–293

Hovi, T., Keski-Oja, J. and Vaheri, A. (1974) Growth control in chick embryo fibroblasts: No evidence for a specific role for cyclic nucleotides. *Cell*, **2**, 235–240

Howell, S. L., Green, I. C. and Montague, W. (1973) A possible role of adenylate cyclase in the long-term dietary regulation of insulin secretion from rat islets of Langerhans. *Biochem. J.*, **136**, 343–349

Howell, S. L. and Montague, W. (1973) Adenylate cyclase activity in isolated rat islets of Langerhans. *Biochim. biophys. Acta*, **320**, 44–52

Howell, S. L. and Montague, W. (1974) Regulation of guanylate cyclase in guinea-pig islets of Langerhans. *Biochem.* J., **142**, 379–384

Howell, S. L. and Montague, W. (1975) Regulation by nucleotides of ^{45}calcium uptake in homogenates of rat islets of Langerhans. *FEBS Letters*, **52**, 48–52

Howell, S. L., Montague, W. and Tyhurst, M. (1975) Calcium distribution in islets of Langerhans. *J. Cell Sci.*, **19**, 395–409

Hsie, A. W., Jones, C. and Puck, T. T. (1971) Further changes in differentiation state accompanying the conversion of Chinese hamster cells to fibroblastic form by dibutyryl adenosine cyclic 3′,5′-monophosphate and hormones. *Proc. natn. Acad. Sci. U.S.A.*, **68**, 1648–1652

Hsie, A. W. and Puck, T. T. (1971) Morphological transformation of Chinese hamster cells by dibutyryl adenosine-cyclic 3′,5′-monophosphate and testosterone. *Proc. natn. Acad. Sci. U.S.A.*, **68**, 358–361

Huang, K. P., Huang, F. L., Glinsmann, W. H. and Robinson, J. C. (1975) Regulation of glycogen synthetase activity by two kinases. *Biochem. biophys. Res. Commun.*, **65**, 1163–1169

Huang, L. C. and Huang, C. (1975) Rabbit skeletal muscle protein kinase: Conversion from cAMP-dependent to independent form by chemical perturbation. *Biochemistry*, **14**, 18–24

Huang, M. and Daly, J. W. (1972) Accumulation of cyclic adenosine monophosphate in incubated slices of brain tissue. I. Structure–activity relationships of agonists and antagonists of biogenic amines and of tricyclic tranquilizers and antidepressants. *J. med. Chem.*, **15**, 458–462

Huang, M. and Daly, J. W. (1974a) Interrelationships among levels of ATP, adenosine and cyclic AMP in incubated slices of guinea pig cerebral cortex: Effect of depolarizing agents, psychotropic drugs and metabolic inhibitors. *J. Neurochem.*, **23**, 393–404

Huang, M. and Daly, J. W. (1974b) Adenosine-elicited accumulation of cyclic AMP in brain slices: Potentiation by agents which inhibit uptake of adenosine. *Life. Sci.*, **14**, 489–503

Huang, M., Ho, A. K. S. and Daly, J. W. (1973a) Accumulation of adenosine cyclic 3′,5′-monophosphate in rat cerebral cortical slices: Stimulatory effect of α and β-adrenergic agents after treatment with 6-hydroxydopamine, 2,3,5-trihydroxyphenethylamine and dihydroxytryptamines. *Molec. Pharmac.*, **9**, 711–717

Huang, M., Gruenstein, E. and Daly, J. W. (1973b) Depolarizing-evoked accumulation of cyclic AMP in brain slices: Inhibition by exogenous adenosine deaminase. *Biochim. biophys. Acta*, **329**, 147–151

Huang, M., Shimizu, H. and Daly, J. W. (1971) Regulation of adenosine cyclic 3′,5′-phosphate formation in cerebral cortical slices: Interaction among norepinephrine, histamine, serotonin. *Molec. Pharmac.*, **7**, 155–162

Huang, T. S., Bylund, N. J., Stull, J. T. and Krebs, E. G. (1974) The amino acid sequences of the phosphorylated sites in troponin I from rabbit skeletal muscle. *FEBS Letters*, **42**, 249–252

Hubbard, W. C., Strecker, R. B. and Michelakis, A. M. (1974) Dissociation constant of the norepinephrine-receptor complex in normotensive and spontaneously hypertensive rats. *Pharmacologist*, **16**, 297

Hug, G. and Schubert, W. K. (1970) Type VI glycogenosis: Biochemical demonstration of liver phosphorylase deficiency. *Biochem. biophys. Res. Commun.*, **41**, 1178–1184

Hug, G., Schubert, W. K. and Chuck, G. (1969) Deficient activity of dephospho-phosphorylase kinase and accumulation of glycogen in liver. *J. clin. Invest.*, **48**, 704–715

Hug, G., Schubert, W. K. and Chuck, G. (1970) Loss of cyclic 3,5-AMP dependent kinase in skeletal muscle of a girl with deactivated phosphorylase and glycogenosis of liver and muscle. *Biochem. biophys. Res. Commun.*, **40**, 982–988

Hughes, R. G. and Kimball, A. P. (1972) Metabolic effects of cyclic 9-β- -arabino-furanosyladenine 3′,5′-monophosphate in L-1210 cells. *Cancer Res.*, **32**, 1791–1794

Hunt, W. A. and Majchrowicz, E. (1974) Alterations in the turnover of brain norepinephrine and dopamine in alcohol dependent rats. *J. Neurochem.*, **23**, 549–552

Hunzicker-Dunn, M. and Birnbaumer, L. (1976) Adenylyl cyclase activities in ovarian tissues. II. Regulation of responsiveness to LH, FSH, and PGE_1 in the rabbit. *Endocrinology* **99**, 185–197

Huston, R. B. and Krebs, E. G. (1968) Identification of phosphorylase kinase activating factor as a proteolytic enzyme. *Biochemistry*, **7**, 2116–2122

Huttunen, J. K., Steinberg, D. and Mayer, S. E. (1970) ATP-dependent and cyclic AMP-dependent activation of rat adipose tissue lipase by protein kinase from rabbit skeletal muscle. *Proc. natn. Acad. Sci. U.S.A.*, **67**, 290–295

Ignarro, L. J. (1974a) Nonphagocytic release of neutral protease and *O*-glucuronidase from human neutrophils. Regulation by autonomic neurohormones and cyclic nucleotides. *Arthrit. Rheum.*, **17**, 25–36

Ignarro, L. J. (1974b) Release of neutral protease and β-glucuronidase from human neutrophils in the presence of cartilage treated with various immunologic reactants. *J. Immunol.*, **113**, 298–308

Ignarro, L. J. (1974c) Regulation of lysosomal enzyme secretion: Role in inflammation. *Agents and Actions*, **4**, 241–258

Ignarrow, L. J. (1975) Regulation of lysosomal enzyme release by prostaglandins, autonomic neurohormones and cyclic nucleotides. In *Lysosomes in Biology and Pathology* (Eds. J. T. Dingle and R. T. Dean), Vol. 4, Elsevier, New York, pp. 481–523

Ignarro, L. J. (1976) Regulation of polymorphonuclear leucocyte, macrophage and platelet function. In *Immunopharmacology* (Eds, J. W. Hadden, F. Spreafico and S. Garattini) Plenum, New York (in press)

Ignarro, L. J. and Cech, S. Y. (1976) Bidirectional regulation of lysosomal enzyme secretion and phagocytosis in human neutrophils by guanosine 3′,5′-monophosphate and adenosine 3′,5′-monophosphate. *Proc. Soc. exp. Biol. Med.*, **151**, 448–452

Ignarro, L. J. and George, W. J. (1974a) Hormonal control of lysosomal enzyme release from human neutrophils: Elevation of cyclic nucleotide levels by autonomic neurohormones. *Proc. natn. Acad. Sci. U.S.A.*, **71**, 2027–2031

Ignarro, L. J., and George, W. J. (1974b) Mediation of immunologic discharge of lysosomal enzymes from human neutrophils by guanosine 3′,5′-monophosphate:

Requirement of calcium, and inhibition by adenosine 3′,5′-monophosphate. *J. exp. Med.*, **140**, 225–*238*

Ignarro, L. J. and George, W. J. (1975) Guanylate cyclase activity in purified human polymorphonuclear leukocytes. *Pharmacologist*, **17**, 270

Ignarro, L. J., Lint, T. F. and George, W. J. (1974a) Hormonal control of lysosomal enzyme release from human neutrophils. Effects of autonomic agents on enzyme release, phagocytosis, and cyclic nucleotide levels. *J. exp. Med.*, **139**, 1395—1414

Ignarro, L. J., Paddock, R. J. and George, W. J. (1974b) Hormonal control of neutrophil lysosomal enzyme release: Effect of epinephrine on adenosine 3′,5′-monophosphate. *Science, N.Y.*, **183**, 855–857

Illiano, G. and Cuatrecasas, P. (1971) Endogenous prostaglandins modulate lipolytic processes in adipose tissue. *Nature, New Biol.*, **234**, 72–74

Illiano, G. and Cuatrecasas, P. (1972) Modulation of adenylate cyclase activity in liver and fat cell membranes by insulin. *Science, N.Y.*, **175**, 906–908

Illiano, G., Tell, G. P. E., Siegel, M. I. and Cuatrecasas, P. (1973) Guanosine 3′.5′ cyclic monophosphate and the action of insulin and acetylcholine. *Proc. natn. Acad. Sci. U.S.A.*, **70**, 2443–2447

Illingworth, B. and Cori, G. T. (1952) Structure of glycogens and amylopectins. Normal and abnormal human glycogen. *J. biol. Chem.*, **199**, 653–660

Imbert, M., Chambardes, D., Mategut, M., Clique, A. and Morel, F. (1975) Vasopressin dependent adenylate cyclase in single segments of rabbit kidney tubule. *Pflügers Arch. Eur. J. Physiol.*, **357**, 173–186

Imura, H., Matsukura, S., Matsuyama, H., Setsuda, T. and Miyake, T. (1965) Adrenal steroidogenic effect of adenosine 3′,5′-monophosphate and its derivatives *in vivo*. *Endocrinology.*, **76**, 933–937

Ingbar, S. H. (1972) Autoregulation of the thyroid. *Mayo Clinic Proc.*, **47**, 814–823

Insel, P., Balakir, R. and Saktor, B. (1975a) The binding of cyclic AMP to renal brush border membranes. *J. Cyclic Nucl. Res.*, **1**, 107–122

Insel, P., Bourne, H. R., Coffino, P. and Tomkins, G. M. (1975b) Mutations affecting adenosine 3′,5-monophosphate-dependent protein kinase. *Fed. Proc.*, **34**, 263

Insel, P. A., Bourne, H. R., Coffino, P. and Tomkins, G. M. (1975c) Cyclic AMP-dependent protein kinase: pivotal role in regulation of enzyme induction and growth. *Science*, **190**, 896–898

Iriuchijima, J. (1973) Role of splanchnic nerves in spontaneously hypertensive rats. *Circulation J.*, **37**, 1251–1253

Isenberg, G. (1975) Is potassium conductance of cardiac Purkinje fibres controlled by [Ca^{2+}] ? *Nature*, Lond., **253**, 273–*274*

Ishikawa, E., Ishikawa, S., Davis, J. W. and Sutherland, E. W. (1969) Determination of guanosine 3′,5′-monophosphate in tissues and of guanyl cyclase in rat intestine. *J. biol. Chem.*, **244**, 6371–6376

Ishizaka, K. and Ishizaka, T. (1971) IgE and reaginic hypersensitivity. *Ann. N.Y. Acad. Sci.*, **190**, 443–456

Ishizaka, T., Ishizaka, K., Orange, R. and Austen, K. F. (1970) Pharmacologic inhibition of the antigen-induced release of histamine and slow reacting substance of anaphylaxis (SRS–A) from monkey lung tissues mediated by human IgE. *J. Immunol.*, **106**, 1267–1273

Ishizaka, T., Soto, C. S. and Ishizaka, K. (1973) Mechanisms of passive sensitization. Number of IgE molecules and their receptor sites on human basophil granulocytes. *J. Immunol.*, **111**, 500–511

Ito, Y., Kuriyama, H. and Tashiro, N. (1971) Effects of catecholamines on the neuromuscular junction of the somatic muscle of the earthworm *Pheretima communissima*. *J. exp. Biol.*, **54**, 167–186

Iversen, D. H. (1969) Chalones of the skin. In *Ciba Symposium on Homeostatic Regulators* (Eds. G. E. W. Wolstenholme and J. Knight), J. and A. Churchill, London, pp. 29–56

Iversen, L. L. (1975) Dopamine receptors in brain. *Science, N.Y.*, **188**, 1084–1089

Iversen, L. L., Horn, A. S. and Miller, R. J. (1975) Actions of dopaminergic agonists on cyclic AMP production in rat brain homogenates. *Adv. Neurol.*, **9**, 197–212

Iwatsubo, K. and Clouet, D. H. (1973) The effect of narcotic analgesic drugs on the levels and the rates of synthesis of cAMP in six areas of rat brain. *Fed. Proc.*, **32**, 536

Iwatsubo, K. and Clouet, D. H. (1975) Dopamine-sensitive adenylate cyclase of the caudate nucleus of rats treated with morphine or haloperidol. *Biochem. Pharmac.*, **24**, 1499–1503

Jaliashvili, T. A. and Chikvaidze, V. H. (1975) Fluorimetric ultramicro-determination of the dansyl derivatives of cyclic 3′,5′-AMP by thin layer chromatography. *Vop. med. Khim. Akad. med. Nauk SSSR*, **21**, 429–432

Jamison, R. L. (1974) Countercurrent systems. In *MTP International Review of Science: Kidney and Urinary Tract Physiology* (Ed. K. Thuran), University Park Press, Baltimore, **6**, 199–246

Jansons, V. K. and Burger, M. M. (1973) Isolation and characterization of agglutinin receptor sites. II. Isolation and partial purification of a surface membrane receptor for wheat germ agglutinin. *Biochim. biophys. Acta*, **291**, 127–135

Jard, S., Roy, C., Barth, T., Rajerison, R. and Bockaert, J. (1975) Antidiuretic hormone sensitive kidney adenylate cyclase. *Adv. Cycl. Nucl. Res.*, **5**, 31–52

Jarret, L., Reuter, M., McKeel, D. W. and Smith, R. M. (1971) Loss of adenyl cyclase hormone receptors during purification of fat cell plasma membranes. *Endocrinology*, **89**, 1186–1190

Jarett, L., Smith, R. M. and Crespin, S. R. (1974) Epinephrine binding to rat adipocytes and their subcellular fractions. *Endocrinology*, **94**, 719–725

Jastorff, B. (1975) *Regulation of Function and Growth of Eukariotic Cells by Intracellular Cyclic Nucleotides*, Plenum Press, New York, 1975

Jastorff, B. and Bär, H.-P. (1973) Effects of 5′-amido analogues of adenosine 3′,5′-monophosphate and adenosine 3′,5′-monophosphothioate on protein kinase, binding protein and phosphodiesterases. *Europ. J. Biochem.*, **37**, 497–504

Jastorff, B. and Freist, W. (1974) Synthesis and biological activities of cyclic AMP analogs modified in the 1,2, and 2′-positions. *Biorg. Chem.*, **3**, 103–113

Jawaharlal, K. and Berti, F. (1972) Effects of dibutyryl cyclic AMP and a new cyclic nucleotide on gastric acid secretion in the rat. *Pharmac. Res. Commun.*, **4**, 143–149

Jefferson, L. S., Exton, J. H., Butcher, R. W., Sutherland, E. W. and Park, C. R. (1968) Role of adenosine 3′,5′-monophosphate in the effects of insulin and anti-insulin serum on liver metabolism. *J. biol. Chem.*, **243**, 1031–1038

Jenkinson, D. H., Stamenovic, B. A. and Whitaker, B. D. L. (1968) The effect of noradrenaline on the end-plate potential in twitch fibres of the frog. *J. Physiol., Lond.*, **195**, 743–754

Jergil, B., Guilford, H. and Mosbach, K. (1973) Biospecific affinity chromatography of an adenosine 3′,5′-cyclic monophosphate-stimulated protein kinase (protamine kinase from trout testis) by using immobilized adenine nucleotides. *Biochem. J.*, **139**, 441–448

Johansson, B., Jonsson, O., Axelsson, J. and Wahlstrom, B. (1967) Electrical and mechanical characteristics of vascular smooth muscle response to norepinephrine and isoproterenol. *Circulation Res.*, **21**, 619–633

Johansson, K. N., Currie, B. L., Folkers, K. and Bowers, C. Y. (1974) Identification and purification of factor B-GHRH from hypothalami which releases growth hormone. *Biochem. biophys. Res. Commun.*, **60**, 610–615

Johansson, K. N., Greisbrokk, F., Currie, B. L., Hansen, J., Folkers, K. and Bowers, C. Y. (1975) Factor C-LHIH inhibits the luteinizing hormone from basal release and from synthetic LHRH and studies on purification of FSHRH. *Biochem. biophys. Res. Commun.*, **63**, 62–68

Johnson, E. M. and Hadden, J. W. (1975). Phosphorylation of lymphocyte nuclear acidic proteins: Regulation by cyclic nucleotides. *Science, N.Y.*, **187**, 1198–1200

Johnson, E. M., Maeno, H. and Greengard, P. (1971a) Phosphorylation of endogenous protein of rat brain by cyclic adenosine 3′,5′-monophosphate-dependent protein kinase. *J. biol. Chem.*, **246**, 7731–7739

Johnson, D. G., Thompson, W. J. and Williams, R. H. (1974) Regulation of adenyl cyclase from isolated pancreatic islets by prostaglandins and guanosine 5″-triphosphate. *Biochemistry*, **13**, 1920–1924

Johnson, G. S., Friedman, R. M. and Pastan, I. (1971b) Restoration of several morphological characteristics of normal fibroblasts in sarcoma cells treated with adenosine 3′,5′-cyclic monophosphate and its derivatives. *Proc. natn. Acad. Sci. U.S.A.*, **68**, 425–429

Johnson, G. S., Morgan, W. D. and Pastan, I. (1972) Regulation of cell motility by cyclic AMP. *Nature, Lond.*, **235**, 54–56

Johnson, G. S. and Pastan, I. (1972a) Role of 3′,5′-adenosine monophosphate in regulation of morphology and growth of transformed and normal fibroblasts. *J. natn. Cancer Inst.*, **48**, 1377–1383

Johnson, G. S. and Pastan, I. (1972b) Cyclic AMP increases adhesion of fibroblasts to the substratum. *Nature, New Biol.*, **236**, 247–249

Johnson, G. S. and Pastan, I. (1972c) N^6,O^2-Dibutyryl-adenosine 3′,5′-monophosphate induces pigment production in melanoma cells. *Nature, New Biol.*, **237**, 267–268

Johnson, L. D. and Abell, C. W. (1970) The effects of isoproterenol and cyclic adenosine 3′,5′-phosphate on phytohemagglutinin-stimulated DNA synthesis in lymphocytes obtained from patients with chronic lymphocytic leukemia. *Cancer Res.*, **30**, 2718–2723

Johnson, M., Jessup, R. and Ramwell, P. (1973) Ultraviolet light modification of the prostaglandin receptor. *Prostaglandins*, **4**, 593–605

Johnson, M. and Ramwell, P. W. (1973) Prostaglandin modification of membrane-bound enzyme activity: A possible mechanism of action? *Prostaglandins*, **3**, 703–719

Johnson, R. A. and Sutherland, E. W. (1973) Detergent-dispersed adenylate cyclase from rat brain: Effects of fluoride, cations, and chelators. -*J. biol. Chem.*, **248**, 5114–5121

Johsson, O., Lundgren, Y. and Wennergren, G. (1975) The distribution of sodium in aortic walls from spontaneously hypertensive and normotensive rats. *Acta physiol. scand.*, **93**, 548–552

Jones, D. J., Medina, M. A., Ross, D. H. and Stavinoha, W. B. (1974) Rate of inactivation of adenyl cyclase and phosphodiesterase: Determinants of brain cyclic AMP. *Life Sci.*, **14**, 1577–1585

Jones, D. R. and Phelan, E. L. (1974) Calcium and contraction of mesenteric arteries of hypertensive rats. *N.Z. Med. J.*, **80**, 119

Jones, G. H., Murthy, D. V. K., Tegg, D., Golling, R. and Moffatt, J. G. (1973) Analogs of adenosine 3′,5′-cycyclic phosphate. II. Synthesis and enzymatic activity of derivatives of 1,N^6-ethenoadenosine 3′,5′-cyclic phosphate. *Biochem. biophys. Res. Commun.*, **53**, 1338–1343

Jones, P. D. and Wakil, S. J. (1967) A requirement for phospholipids by the microsomal reduced diphosphopyridine nucleotide–cytochrome c reductase. *J. biol. Chem.*, **242**, 5267

Jonsson, C. E. and Ánggard, E. (1972) Biosynthesis and metabolism of prostaglandin E_2 in human skin. *Scand. J. clin. Lab. Invest.*, **29**, 289–296

Jordan, L., Lake, N. and Phillis, J. W. (1972) Mechanism of noradrenaline depression of cortical neurons: A species comparison. *Europ. J. Pharmac.*, **20**, 381–384

Jost, J-P. and Rickenberg, H. B. (1971) Biological activity of derivatives of cAMP. *Ann. Rev. Biochem.*, **40**, 741–774

Jungas, R. L. (1966) Role of cyclic-3′,5′-AMP in the response of adipose tissue to insulin. *Proc. natn. Acad. Sci., U.S.A.*, **56**, 757–763

Jurtshuk, P., Jr., Sekuzu, I. and Green, D. E. (1961) The interaction of the

D (-)β-hydroxybutyric apoenzyme with lecithin. *Biochem. biophys. Res. Commun.*, **6**, 76

Jutisz, M. and dela Llosa, M. P. (1970) Requirement of Ca and Mg ions for the *in vitro* release of follicle-stimulating hormone from rat pituitary glands and in its subsequent biosynthesis. *Endocrinology*, **86**, 761–768

Jutisz, M., Kerdelhue, B., Berault, A and de la Llosa, M. P. (1972) On the mechanism of action of the hypothalamic gonadotropin-releasing factors. In *Gonadotropins* (Eds. Saxena, Beling and Gandy). Wiley and Sons, New York, pp. 64–71

Jutisz, M. and Kraicer, J. (1975) Cellular regulation of adenohypophyseal function. In *Hypothalamic hormones. Structure, Synthesis and Biological Activity* (Eds. D. Gupta and W. Voelter), Verlag. Chemie, Weinheim, pp. 155–177

Kadowitz, P. J., Joiner, P. D. and Hyman, A. L. (1974) Influence of prostaglandins E_1 and $F_{2\alpha}$ on pulmonary vascular resistance in the sheep. *Proc. Soc. expt. Biol. Med.*, **145**, 1258–1261

Kagen, L. J. and Freedman, A. (1974) Studies on the effects of acetylcholine, epinephrine, dibutyryl-cyclic adenosine monophosphate, theophylline, and calcium on the synthesis of myoglobin in muscle cell cultures estimated by radioimmunoassay. *Expl. Cell Res.*, **88**, 135–142

Kakiuchi, K. S. and Marks, B, H. (1972) Adrenergic effects on pineal cell membrane potential. *Life Sci.*, **11**, 285–291

Kakiuchi, S. and Rall, T. W. (1968a) Studies on adenosine 3′,5′-phosphate in rabbit cerebral cortex. *Molec. Pharmac.*, **4**, 379–388

Kakiuchi, S. and Rall, T. W. (1968b) The influence of chemical agents on the accumulation of adenosine 3′,5′-phosphate in slices of rabbit cerebellum. *Molec.*, Pharmac., **4**, 367–378

Kakiuchi, S., Rall, T. W. and McIlwain, H. (1969) The effect of electrical stimulation upon the accumulation of adenosine 3′,5′-phosphate in isolated cerebral tissue. *J. Neurochem.*, **16**, 485–491

Kakiuchi, S. and Yamazaki, R. (1970a) Calcium-dependent phosphodiesterase activity and its activating factor from brain. *Biochem. biophys. Res. Commun.*, **41**, 1104–1110

Kakiuchi, S. and Yamazaki, R. (1970b) Stimulation of the activity of cyclic 3′,5′-nucleotide phosphodiesterase by calcium ion. *Proc. Jap. Acad.*, **46**, 387–392

Kakiuchi, S., Yamazaki, R. and Teshima, U. (1971) Cyclic 3′,5′-nucleotide phosphodiesterase, IV. Two enzymes with different properties from brain. *Biochem. biophys. Res. Commun.*, **42**, 968–974

Kakiuchi, S., Yamazaki, R. and Teshima, Y. (1972) Regulation of brain phosphodesterase activity. Ca^{2+} plus Mg^{2+}-dependent phosphodiesterase and its activating factor from rat brain. In *Advances in Cyclic Nucleotide Research*, Vol. 1, Raven Press, New York, 455–477

Kakiuchi, S., Yamazaki, R., Teshima, Y. and Uenishi, M. (1973) Regulation of nucleotide cyclic 3′,5′-monophosphate phosphodiesterase activity from rat brain by a modulator and Ca^{2+}. *Proc. natn. Acad. Sci. U.S.A.*, **70**, 3526–3530

Kakiuchi, S., Yamazaki, R., Teshima, Y. and Uenishi, K. (1975) Multiple cyclic nucleotide phosphodiesterase activities of rat tissues and occurrence of a calcium-plus-magnesium-ion-dependent phosphodiesterase and its protein activator. *Biochem. J.*, **146**, 109–120

Kaliner, M. and Austen, K. F. (1973) A sequence of biochemical events in the antigen-induced release of chemical mediators from sensitized human lung tissue. *J. exp. Med.*, **138**, 556–567

Kaliner, M. and Austen, K. F. (1974a) Cyclic AMP, ATP and reversed anaphylactic histamine release from rat mast cells. *J. Immunol.*, **112**, 664–674

Kaliner, M. and Austen, K. F. (1974b) Cyclic Nucleotides and Modulation of Effector Systems of Inflammation. *Biochem. Pharmac.*, **23**, 763–771

Kaliner, M., Orange, R. P. and Austen, K. F. (1972) Immunological release of histamine and slow reacting substance of anaphylaxis from human lung. IV. Enhancement by cholinergic and alpha adrenergic stimulation. *J. expt. Med.*, **136**, 556–567

Kaliner, M., Wasserman, S. and Austen, K. F. (1973) Immunologic release of chemical mediators from human nasal polyps. *New Engl. J. Med.*, **289**, 277–281

Kalisker, A., Rutledge, C. O. and Perkins, J. P. (1973) Effect of nerve degeneration by 6-hydroxy-dopamine on catecholamine-stimulated adenosine 3′,5′-monophosphate formation in rat cerebral cortex. *Molec. Pharmac.*, **9**, 619–629

Kaminskas, E. (1972) Serum-mediated stimulation of protein synthesis in Ehrlich ascites tumor cells. *J. biol. Chem.*, **247**, 5470–5476

Kaneko, T., Oka, H., Munemura, M., Saito, S. and Yanaihara, N. (1974a) Role of cyclic AMP in the mechanism of action of hypothalamic hypophysiotropic hormones. In *Psychoneuroendocrinology* (Ed. Hatotani), Karger, Basel, pp. 267–275

Kaneko, T., Oka, H., Munemura, M., Suzuki, S., Yasuda, H., Oda, T. and Yanaihara, N. (1974b) Stimulation of guanosine 3′,5′-cyclic monophosphate accumulation in rat anterior pituitary gland *in vitro* by synthetic somatostatin. *Biochem. biophys. Res. Commun.*, **61**, 53–57

Kaneko, T., Oka, H., Saito, S., Munemura, Musa, K., Oda, T., Yanaihara, N. and Yanaihara, C. (1973a) *In vitro* effects of synthetic somatotropin-release inhibiting factor on cyclic AMP level and GH release in rat anterior pituitary gland. *Endocrinology*, **20**, 535–538

Kaneko, T., Saito, S., Oka, H., Oda, T. and Yanaihara, N. (1973b) Effects of synthetic LH–RH and its analogs on rat anterior pituitary cyclic AMP and LH and FSH release. *Metabolism*, **22**, 77–80

Kaneko, T., Saito, S., Oka, H., Oda, T. and Yanaihara, N. (1973c) Effect of synthetic releasing hormones on cyclic AMP levels and hormone release from rat anterior pituitary tissue. In *Hypothalmic Hypophysiotropic Hormones* (Eds. Gaul and Rosemberg), Exerpta Medica, Amsterdam, pp. 198–203

Karl, R. C., Zawalich, W. S., Ferrendelli, J. A. and Matschinsky, F. M. (1975) The role of Ca^{2+} and cyclic adenosine 3′,5′-monophosphate in insulin release induced *in vitro* by the divalent cation ionophore A23187. *J. biol. Chem.*, **250**, 4575–4579

Karn, J., Johnson, E. M., Vidali, G. and Allfrey, V. G. (1974) Differential phosphorylation and turnover of nuclear acidic proteins during the cell cycle of synchronized HeLa cells. *J. biol. Chem.*, **249**, 667–677

Karppanen, H. O., Neuvonen, P. I., Bieck, P. R. and Westermann, E. (1974) Effect of histamine, pentagastrin and theophylline on the production of cyclic AMP in isolated gastric tissue of the guinea pig. *Naunyn-Schmiedebergs Arch. exp. Path. Pharmak.*, **284**, 15–23

Karppanen, H. O. and Westermann, E. (1973) Increased production of cyclic AMP in gastric tissue by stimulation of histamine (H_2)-receptors. *Naunyn-Schmiedebergs Arch. exp. Path. Pharmak.*, **279**, 83–87

Karobath, N. E. (1975) Tricyclic antidepressive drugs and dopamine-sensitive adenylate cyclase from rat brain striatum. *Europ. J. Pharmac.*, **30**, 159–163

Karobath, N. E. and Leitich, H. (1964) Antipsychotic drugs and dopamine- stimulated adenylate cyclase prepared from corpus striatum of rat brain. *Proc. Natn. Acad Sci. U.S.A.*, **71**, 2915–2918

Karsenti, E., Bornens, M. and Avrameas, S. (1975) Stimulation and inhibition of DNA synthesis in rat thymocytes: Action of concanavalin A and wheat germ agglutinin. *Europ. J. Immunol.*, **5**, 74–76

Kasbekar, D. K. (1973) Cyclic AMP-dependent restoration of gastric Histamin secretion following Ca^{2+} depletion. *Fed. Proc.*, **31**, 353

Katsch, G. and Kalk, H. (1925) Zum Ausbau der kinetischen Methode für die Untersuchung des Magenchemismus. *Klin. Wschr.*, **2**, 2190–2193

Katz, A. M. (1970) Contractile proteins of the heart. *Physiol. Rev.*, **50**, 63–158

Katz, A. M. and Repke, I. D. (1973) Calcium-membrane interaction in the myocardium: Effects of ouabain, epinephrine and 3′,5′ cyclic adenosine monophosphate. *Am. J. Cardiol.*, **31**, 293–201

Katz, A. M., Tada, M. and Kirchberger, M. A. (1975) Control of calcium transport in the myocardium by the cyclic AMP-protein kinase system. In *Advances in cyclic Nucleotide Research*, Vol. 5 (Eds. G. I. Drummond, P. Greengard and G. A. Robison), Raven Press, New York, pp. 453–472

Katz, A. M., Tada, M., Repke, D. I., Iorio, J. A. M. and Kirchberger, M. A. (1974) Adenylate cyclase: Its probable localization in sarcoplasmic reticulum as well as sarcolemma of the canine heart. *J. molec. cell. Cardiol.*, **61**, 73–78

Katz, B (1966) *Nerve, Muscle, and Synapse.* McGraw-Hill, New York

Kaufmann, F. C., Harkonen, M. H. A. and Johnson, E. (1972) Adenyl cyclase and phosphodiesterase activity in cerebral cortex of normal and undernourished neonatal rats. *Life Sci.* **11**, 613–621

Kaumann, A. J. and Birnbaumer, L. (1973) Adrenergic receptors in heart: Similarity of apparent affinities of beta-blockers for receptors mediating adenylyl cyclase activity and inotropic and chronotropic effects of catecholamines. *Acta physiol. Latinoam.*, **23**, 619–620

Kaumann, A. J. and Birnbaumer, L. (1974) Studies on receptor-mediated activation of adenylyl cyclases. IV. Characteristics of the adrenergic receptor coupled to myocardial adenylyl cyclase. Stereospecificity for ligands and determination of apparent affinity constants *J. biol. Chem.*, **249**, 7874–7885

Kazimierczuk, Z. and Shugar, D. (1973) Preparative photochemical synthesis of isoguanosine ribo and deoxyribonucleosides and nucleotides, and isoguanosine-3′,5′-cyclic phosphate, a new cAMP analogue. *Acta biochim. Pol.*, **20**, 395–402

Kebabian, J. W., Bloom, F. E., Steiner, A. L. and Greengard, P. (1975) Neurotransmitter-induced increases in cyclic nucleotides of postganglionic neurons in mammalian sympathetic ganglia: Immunocytochemical demonstration. *Science*, N.Y., **190**, 157–160

Kebabian, J. W. and Greengard, P. (1971) Dopamine-sensitive adenyl cyclase: Possible role in synaptic transmission. *Science, N.Y.*, **174**, 1346–2149

Kebabian, J. W., Petzold, G. L. and Greengard, P. (1972) Dopamine-sensitive adenylate cyclase in caudate nucleus of rat brain and its similarity to the 'dopamine receptor'. *Proc. natn. Acad. Sci. U.S.A.*, **69**, 2145–2149

Kebabian, J. W., Clement-Cormier, Y. C., Petzold, G. L. and Greengard, P. (1975b) Chemistry of dopamine receptors. *Adv. Neurol.*, **9**, 1—11

Keely, S. L., Corbin, J. D. and Park, C. R. (1975a) On the question of translocation of heart cyclic AMP-dependent protein kinase. *Proc. natn. Acad. Sci. U.S.A.*, **73**, 1501–1504

Keely, S. L., Corbin, J. D. and Park, C. R. (1975b) Regulation of cyclic AMP- dependent protein kinase. Regulation of the heart enzyme by epinephrine, glucagon, insulin and IBMX. *J. biol. Chem.*, **250**, 4832–4840

Keen, P. and McLean, W. G. (1972) Effects of dibutyryl-cyclic AMP on levels of dopamine-β-hydroxylase in isolated superior cervical ganglia. *Naunyn-Schmiedebergs. Arch. exp. Path. Pharmak.*, **275**, 465–469

Keirns, J. J., Freeman, J. and Bitensky, M. W. (1974) Cyclic adenosine monophosphate and clinical medicine. Carbohydrate and lipid metabolism. *Amer. J. med. Sci.*, **268**, 62–91

Kelly, L. A., Hall, M. S. and Butcher, R. (1974) Cyclic adenosine 3′,5′-monophosphate metabolism in normal and SV 40-transformed WI–38 cells. *J. biol. Chem.*, **249**, 5182–5187

Kerr, S. J., Brown, D. R., Sauk, J. J. and Sheppard, J. R. (1975) Hormonal and prostaglandin-induced cyclic AMP responses in cultured mouse melanoma cells. *J. Cell Biol.*, **67**, 207a

Khoo, J. C., Aguino, A. A. and Steinberg, D. (1974) The mechanism of activation of hormone-sensitive lipase in human adipose tissue. *J. clin. Invest.*, **53**, 1124–1131

Khoo, J. C. and Steinberg, D. (1974) Reversible protein kinase activation of hormone-sensitive lipase from chicken adipose tissue. *J. Lipid Res.*, **15**. 602–610

Khoo, J. C., Steinberg, D., Thompson, B. and Mayer, S. E. (1973) Hormonal regulation of adipocyte enzymes. *J. biol. Chem.*, **248**, 3823–3830

Khwaja, T. A., Boswell, K. H., Robins, R. K. and Miller, J. P. (1975) 8-Substituted derivatives of adenosine 3′,5′-cyclic phosphate require an unsubstituted 2′-hydroxyl group in the ribo configuration for biological activity. *Biochemistry*, **14**, 4238–4244

Kiessling, M., Lindl, T. and Cramer, H. (1975) Cyclic adenosine monophosphate in cerebrospinal fluid: Effects of theophylline, L-DOPA and a dopamine receptor stimulant in rats. *Arch. Psychiat. Nerv-kr.*, **220**, 325–333

Kim, J. K., Frohnert, P. P., Barnes, L. D., Hui, Y. S. F., Farrow, G. M. and Dousa, T. P. (1976a) Enzymes of cyclic AMP metabolism and action in renal cell carcinoma (RCC). *Clin. Res.*, **24**, 468A

Kim, J. K., Hui, Y. S. F., Fronhert, P. P., Barbes, L. D., Farrow, G. M. and Dousa, T. P. (1976b) Subcellular distribution of guanylate cyclase (GC), adenylate cyclase (AC), and cyclic 3′,5′-nucleotide phosphodiesterases in renal adenocarcinoma (RA) in man. *Clin. Res.*, **24**, (in press)

Kimberg, D. V. (1974) Cyclic nucleotides and their role in gastrointestinal secretion. *Gastroenterology*, **67**, 1023–1064

Kimura, H. and Murad, F. (1974) Evidence of two different forms of guanylate cyclase in rat heart. *J. biol. Chem.*, **249**, 6910–6916

King, C. A. and van Heyningen, W. E. (1973) Deactivation of cholera toxin by a sialidase-resistant monosialosylganglioside. *J. infect. Dis.*, **127**, 639–647

Kirchberger, M. A., Tada, M., Repke, D. I. and Katz, A. M. (1972) Cyclic adenosine 3′,5′-monophosphate-dependent protein kinase stimulation of calcium uptake by canine cardiac microsomes. *J. mol. cell. Cardiol.*, **4**, 673–680

Kirpekar, S. M. and Puig, M. (1971) Effect of flow-stop on noradrenaline release from normal spleens treated with cocaine, phentolamine or phenoxybenzamine. *Br. J. Pharmac.*, **43**, 359–369

Kish, V. M. and Kleinsmith, L. J. (1974) Nuclear protein kinases: Evidence for their heterogeneity, tissue specificity, substrate specificities and differential responses to adenosine 3′,5′-cyclic monophosphate. *J. biol. Chem.*, **249**, 750–760

Kishimoto, T., Miyake, T., Nishizawa, Y., Watanabe, T. and Yamamura, Y. (1975) Triggering mechanism of B lymphocytes. I. Effect of anti-immunoglobulin and enhancing soluble factor on differentiation and proliferation of B-cells. *J. Immunol.*, **115**, 1179–1184

Kissebah, A. H., Hope-Gill, H., Vydelingum, N., Tulloch, B. R., Clarke, P. V. and Frazer, T. R. (1975) Mode of insulin action. *Lancet*, i, 144

Klainer, L. M., Chi, Y. M., Friedberg, S. L., Rall, T. W. and Sutherland, E. W. (1962) Adenyl cyclase IV. The effects of neurohormones on the formation of adenosine 3′,5′-phosphate by preparations from brain and other tissues. *J. biol. Chem.*, **237**, 1239–1243

Klee, W. A., Sharma, S. K. and Nirenberg, M. (1975) Opiate receptors as regulators of adenylate cyclase. *Life Sci.*, **16**, 1869–1874

Klein, D. C., Berg, G. R. and Weller, J. (1970) Melatonin synthesis. Adenosine 3′,5′-monophosphate and norepine phrine stimulate *N*-acetyltransferase. *Science, N.Y.*, **168**, 979–980

Klein, D. C. and Parfitt, A. (1975) Ouabain blocks the adrenergic-cyclic AMP stimulation of pineal *N*-acetyltransferase activity. In *Cyclic Nucleotides in Disease* (Ed. B. Weiss), University Park Press, Baltimore, pp. 257–265

Klein, I., Fletcher, M. A. and Levey, G. S. (1973) Evidence for a dissociable glucagon binding site in a solubilized myocardial adenylate cyclase. *J. biol. Chem.*, **248**, 5552–5554

Klein, I. and Levy, G. S. (1971a) Activation of myocardial adenyl cyclase by histamine in guinea pig, cat, and human heart. *J. clin. Invest.*, **50**, 1012

Klein, I. and Levey, G. S. (1971b) Effect of prostaglandins on guinea-pig myocardial adenyl cyclase. *Metabolism*, **20**, 890

Klein, M. I. and Makman, M. H. (1971) Adenosine 3′,5′-monophosphate-dependent protein kinase of cultured mammalian cells. *Science, N.Y.*, **172**, 863–864

Kleitke, B., Sydow, H. and Wollenberger, A. (1976) Evidence of cyclic AMP-dependent protein kinase activity in isolated guinea-pig and rat mitochondria. *Acta biol. med. ger.*, K9–K17

Klenerova, V., Alabrecht, I. and Hynie, S. (1975) The activity of adenylate cyclase and phospho-diesterase in hearts and aortas of spontaneous hypertensive rats. *Pharmac. Res. Comm.*, **71**, 453–455

Klotz, U. and Stock, K. (1971) Evidence for a cyclic nucleotide phosphodiesterase with high specificity for cyclic uridine-3′,5′-monophosphate in rat adipose tissue. *Naunyn-Schmiedebergs Arch. exp. Path. Pharmak.*, **269**, 117–120

Klotz, U. and Stock, K. (1972) Influence of cyclic guanosine 3′,5′-monophosphate on the enzymatic hydrolysis of adenosine 3′,5′-monophosphate. *Naunyn-Schmiedebergs Arch. exp. Path. Pharmak.*, **274**, 54–62

Kneer, N. M., Bosch, A. L., Clark, M. G. and Lardy, H. A. (1974) Glucose inhibition of epinephrine stimulation of hepatic gluconeogenesis by blockade of the α-receptor function. *Proc. natn. Acad. Sci. U.S.A.*, **71**, 4523–4527

Knight, B. L. and Iliffe, J. (1973) The effect of glucose, insulin and noradrenaline on lipolysis, and on the concentrations of adenosine 3′,5′-monophosphate and adenosine 5′-triphosphate in adipose tissue. *Biochem. J.*, **132**, 77–82

Kobayashi, H. and Libet, B. (1968) Generation of slow postsynaptic potentials without increases in ionic conductance. *Proc. natn. Acad. Sci. U.S.A.*, **60**, 1304–1311

Kodama, T., Matsukado, Y. and Shimizu, H. (1973) The cyclic AMP systems of human brain. *Brain Res., Osaka*, **50**, 135–146

Kolena, J. and Channing, C. P. (1972) Stimulatory effects of LH, FSH and prostaglandins upon cyclic 3′,5′-AMP levels in porcine granulosa cells. *Endocrinology*, **90**, 1543–1550

Konijn, T. M. (1972) Cyclic AMP as a first messenger. In *Advances in Cyclic Nucleotide Research*, Vol. 1 (Eds. P. Greengard and G. A. Robison), Raven Press, New York, p. 17

Konijn, T. M. (1973) The chemotactic effect of cyclic nucleotides with substitutions in the base ring. *FEBS Letters*, **34**, 263–266

Konijn, T. M. and Jastorff, B. (1973) The chemotactic effect of 5′-amido analogues of adenosine cyclic 3′,5′-monophosphate in the cellular slime moulds. *Biochim. biophys. Acta.*, **304**, 774–780

Kono, T. (1970) Insulin effector systems of fat cells: Its destruction with trypsin and subsequent restoration. In *Adipose Tissue* (Eds. B. Jeanrenaud and D. Hepp), Academic Press, New York, pp. 108–111

Kono, T. and Barham, F. W. (1973) Effects of insulin on the levels of adenosine 3′,5-monophosphate and lipolysis in isolated rat epididymal fat cells. *J. biol. Chem.*, **248**, 7417–7426

Kono, T., Robinson, F. W. and Sarver, J. A. (1975) Insulin-sensitive phosphodiesterase: Its localization, hormonal stimulation, and oxidative stabilization. *J. biol. Chem.*, **250**, 7826–7835

Koopman, W. J., Orange, R. P. and Austen, K. F. (1970) Immunochemical and biologic properties of rat IgE. III. Modulation of the IgE-mediated release of slow-reacting substance of anaphylaxis by agents influencing the levels of cyclic 3′,5′-adenosine monophosphate. *J. Immunol.*, **105**, 1096–1102

Korenman, S. G., Bhalla, R. C., Sanborn, B. M. and Stevens, R. H. (1974) Protein kinase translocation as an early event in the hormonal control of uterine contraction. *Science, N.Y.*, **183**, 430–432

Koretz, S. H. and Marinetti, G. V. (1974) Binding of 1-norepinephrine to isolated rat fat cell plasma membranes. Evidence against covalent binding to catechol-*O*-methyl transferase. *Biochem. biophys. Res. Commun.*, **61**, 22–30

Koslow, S. H., Racagni, G. and Costa, E. (1974) Mass fragmentographic measurement of norepinephrine, dopamine, serotonin and acetylcholine in seven discrete nuclei of the rat tel-diencephalon. *Neuropharmacology*, **13**, 1123–1130

Kostopoulous, G. K., Limacher, J. J. and Phillis, J. W. (1975) Action of various adenine derivatives on cerebellar Purkinje cells. *Brain Res.*, **88**, 162–165

Kowal, J. (1970a) Adrenal cells in tissue culture. VII. Effect of inhibitors of protein synthesis on steroidogenesis and glycolysis. *Endocrinology*, **87**, 951–965

Kowal, J. (1970b) ACTH and the metabolism of adrenal cell cultures. *Recent Progr. Horm. Res.*, **26**, 623–676

Kowal, J. (1973) Adrenal cells in tissue culture. X. On the mechanism of the stimulation of steroidogenesis by cyclic cytidine monophosphate. *Endocrinology*, **93**, 461–468

Kowal, J. and Fiedler, R. (1969) Studies on adrenal cells in tissue culture II. Steroidogenic responses to nucleosides and nucleotides. *Endocrinology*, **84**, 1113–1117

Kowal, J. and Harano, Y. (1974) Adrenal cells in tissue culture: The effect of papaverine and amytal on steroidogenesis respiration and replication. *Archs. Biochem. Biophys.*, **163**, 466–475

Kowalski, K., Babiarz, D. and Burke, G. (1972b) Phagocytosis of latex beads by isolated thyroid cells: Effects of thyrotropin, prostaglandin E_1, and dibutyryl-cyclic AMP. *J. lab. clin. Med.*, **79**, 258–266

Kowalski, K., Babiarz, D., Sato, S. and Burke, G. (1972a) Stimulatory effects of induced phagocytosis on the function of isolated thyroid cells. *J. clin. Invest.*, **51**, 2800–2819

Kozireff, V. and De Visscher, M. (1974) Protein kinase activity of bovine thyroid microsomes. *Biochim. biophys. Acta*, **362**, 17–28

Krahl, M. E. (1972) Effects of insulin on synthesis of specific enzymes in various tissues. In *Insulin Action* (Ed. I. B. Fritz), Academic Press, New York, pp. 461–486

Kraicer, J. and Milligan, J. V. (1973) Effects of various secretagogues upon ^{42}K and ^{22}Na uptake during *in vitro* hormone release from the rat adenohypophysis. *J. Physiol., Lond.*, **232**, 221–237

Krakow, J. S. (1975) Cyclic adenosine monophosphate receptor. Effect of cyclic AMP analogues on DNA binding and proteolytic inactivation, *Biochim. Biophys. Acta.*, **383**, 345–350

Kram, R. Mamont, P. and Tomkins, G. M. (1973) Pleiotypic control by adenosine 3′,5′-cyclic monophosphate: A model for growth control in animal cells. *Proc. natn. Acad. Sci. U.S.A.*, **70**, 1432–1436

Kram, R. and Tomkins, G. M. (1973) Pleiotypic control by cyclic AMP: Interaction with cyclic GMP and possible role of microtubules. *Proc. natn. Acad. Sci. U.S.A.*, **70**, 1659–1663

Krasnow, S. and Grossman, M. I. (1949) Stimulation of gastric secretion in man by theophylline ethylenediamine. *Proc. Soc. exp. Biol. Med.*, **71**, 335–336

Krause, E.-G., Halle, W., Kallabis, E. and Wollenberger, A. (1970) Positive chronotropic response of cultured isolated rat heart cells to N^6-2-*O*-dibutyryl-3′,5′-adenosine monophosphate. *J. molec. cell. Cardiol.*, **1**, 1–10

Krause, E.-G., Halle, W. and Wollenberger, A. (1972) Effect of dibutyryl-cyclic GMP on cultured beating rat heart. In *Advances in Cyclic Nucleotide Research,* Vol. 1, Raven Press, New York, pp. 301–305

Krause, E.-G., Janiszewski, E., Bartel, S., Bogdanova, E. V., Karczewski, P. and Wollenberger, A. (1976) Cyclic AMP and cyclic GMP fluctuations during the cardiac cycle of the frog and the influence of sotalol and 6-hydroxydopamine. In *Progress in Pathophysiology* (Ed. J. Vascu), Internat. Publ. House of Czechoslovak Acad. Sci.

Krause, E.-G., Will, H., Pelouch, V. and Wollenberger, A. (1973) Cyclic AMP-dependent protein kinase activity in a cell membrane-enriched fraction of pig myocardium. *Acta Biol. Med. German.* **31**, K37–43

Krause, E.-G., Will, H., Schirpke, B. and Wollenberger, A. (1975) Cyclic AMP-enhanced protein phosphorylation and calcium binding in a cell membrane-enriched fraction from myocardium. In *Advances in Cyclic Nucleotide Research,* Vol. 5 (Eds. G. I. Drummond, P. Greengard and G. A. Robison), Raven Press, New York, pp. 473–490

Krause, E.-G. and Wollenberger, A. (1967) On the activation of phosphorylase *b* kinase in the acutely ischemic myocardium. *Acta Biol. Med. Germ.*, **19**, 381–393

Krebs, E. (1973) Protein kinases. *Curr. Top. Cell Regul.*, **5**, 99–133

Krebs, E. G. (1973) The mechanism of hormonal regulation by cyclic AMP. *Proc. 4th Intern. Congr. Endocrinol.*

Kreider, J. W., Rosenthal, M. and Lengle, N. (1973) Cyclic adenosine 3′,5′-monophosphate in the control of melanoma cell replication and differentiation. *J. natn. Cancer Inst.*, **50**, 555–558

Kreiner, P. W., Gold, C. J., Keirns, J. J., Brock, W. A. and Bitensky, M. W. (1973a) Hormonal control of melanocytes. MSH-sensitive adenyl cyclase in the Cloudman melanoma. *Yale J. Biol. Med.*, **46**, 583–591

Kreiner, P. W., Keirns, J. J. and Bitensky, M. W. (1973b) A temperature-sensitive change in the energy of activation of hormone-stimulated hepatic adenylyl cyclase. *Proc. natn. Acad. Sci. U.S.A.*, **70**, 1785–1789

Kretsinger, R. H. (1975) Hypothesis: Calcium modulated proteins contain EF hands. In *Calcium Transport in Contraction and Secretion* (Eds. E. Carafoli *et al.*), Elsevier Press, New York, pp. 469–478

Kretsinger, R. H. (1976) Calcium binding proteins. *Ann. Rev. Biochem.*, **45**, 239–266

Krishna, G., Forn, J., Voight, K., Paul, M. and Gessa, G. L. (1970a) Dynamic aspects of neurohormonal control of cyclic 3′,5′-AMP synthesis in brain. *Adv. Biochem. Psychopharmac.*, **3**, 155–172

Krishna, G., Harwood, J. P., Barber, A. J. and Jamieson, G. A. (1972) Requirement for guanosine triphosphate in the prostaglandin activation of adenylate cyclase of platelet membranes. *J. biol. Chem.*, **247**, 2253–2254

Krishna, G., Moskowitz, J., Dempsey, P. and Brodie, B. B. (1970) The effect of norepinephrine and insulin on brown fat cell membrane potentials. *Life Sci.*, **9**, 1353–1361

Krishnaraj, R. and Talwar, G. (1973) Role of cyclic AMP in mitogen-induced transformation of human peripheral Leukocytes *J. Immunol.*, **111**, 1010–1017

Kroeger, E. A. and Marshall, J. M. (1973) Beta-adrenergic effects on rat myometrium: Mechanisms of membrane hyperpolarization. *Am. J. Physiol.*, **225**, 1339–1345

Kroeger, E. A., Teo, T. S., Ho, H. C. and Wang, J. H. (1976) *Biochemistry of Smooth Muscle*, University Park Press, Baltimore

Krueger, B. K., Forn, J., Walters, J. R., Roth, R. H. and Greengard, P. (1976) Dopamine stimulation of adenosine 3′,5′-monophosphate formation in rat caudate nucleus: Effect of lesions of the nigro-neostriatal pathway. *Molec. Pharmac.*, **12**, 639–648

Kuba, K. (1970) Effects of catecholamines on the neuromuscular junction in the rat diaphragm. *J. Physiol., Lond.*, **211**, 551–570

Kuba, K. and Tomita, T. (1971) Noradrenaline action on nerve terminals in the rat diaphragm. *J. Physiol., Lond.*, **217**, 19–31

Kuczenski, R. (1975) Conformational adaptability of tyrosine hydroxylase in the regulation of striatal dopamine biosynthesis. *Adv. Biochem. Psychopharmac.*, **13**, 109–125

Kuehl, F. A. and Humes, J. L. (1972) Direct evidence for a prostaglandin receptor and its application to prostaglandin measurements. *Proc. natn. Acad. Sci. U.S.A.*, **69**, 480

Kuehl, F. A., Jr., Humes, J. L., Cirillo, V. S. and Ham, E. A. (1972) Cyclic AMP and prostaglandins in hormone action. In *Advances Cyclic Nucleotide Research*, Vol. 1 (Eds. P. Greengard and G. A. Robison), Raven Press, New York, pp. 493–502

Kukovetz, W. R. (1962) Kontraktilität und Phosphorylase-aktivität des Herzens bei ganglionärer Erregung nach adrenerger Blockade und unter Atropin. Naunyn-Schmiedebergs *Arch. exp. Path. Pharmak.*, **243**, 391–406

Kukovetz, W. R. and Pöch, G. (1970a) Inhibition of cyclic 3′, 5′-nucleotide-phosphodiesterase as a possible mode of action of papaverine and similarly acting drugs. *Naunyn-Schmiedebergs Arch. exp. Path. Pharmak.*, **267**, 189–194

Kukovetz, W. R. and Pöch, G. (1970b) Cardiostimulatory effects of cyclic 3′,5′-adenosine monophosphate and its acylated derivatives. *Naunyn-Schmiedebergs Arch. exp. Path., Pharmak.*, **266**, 236–290

Kukovetz, W. R. and Pöch, G. (1972) The positive inotropic action of cyclic AMP. In *Advances in Cyclic Nucleotide Research*, Vol. 1 (Eds. P. Greengard and G. A. Robison), Raven Press, New York, pp. 261–290

Kukovetz, W. R., Pöch, G. and Wurm, A. (1975) Quantitative relations between cyclic AMP and contraction as affected by stimulators of adenylate cyclase and inhibitors of phosphodiesterase. In *Advances in Cyclic Nucleotide Research*, Vol. 5 (Eds. G. I. Drummond P. Greengard, and R. A. Robison), Raven Press, New York, pp. 395–414

Kumar, R., Solomon, L. M., Schreckenberger, A. and Cobb, J. C. (1975) An evaluation of adenosine 3′,5′-cyclic monophosphate-dependent protein kinase activity in atopic dermatitis. *J. invest. Derm.*, **65**, 522–524

Kuo, J. F. (1974) Guanosine 3′,5′-monophosphate-dependent protein kinase in mammalian tissues. *Proc. natn. Acad. Sci. U.S.A.*, **71**, 4037–4041

Kuo, J. F. (1975a) Divergent actions of protein kinase modulator in regulating mammalian cyclic GMP-dependent and cyclic AMP-dependent protein kinases. *Metabolism*, **29**, 321–329

Kuo, J. F. (1975b) Changes in relative levels of guanosine-3′,5′-monophosphate-dependent and adenosine-3′,5′-monophosphate-dependent protein kinase in lung, heart, and brain of developing guinea-pigs. *Proc. natn. Acad. Sci. U.S.A.*, **72**, 2256–2259

Kuo, J. F. (1975c) Changes in activities of modulators of cyclic AMP-dependent and cyclic GMP-dependent protein kinases in pancreas and adipose tissue from alloxan induced diabetic rats. *Biochem. biophys. Res. Commun.*, **65**, 1214–1220

Kuo, J. F. and Davis, C. W. (1975) Decreased levels of cyclic GMP-dependent protein kinase in the heart of hypertensive rats and in the liver of diabetic mice. *Fed. Proc.*, **34**, 250

Kuo, J. F., Davis, C. E., Shoji, M. and Donnelly, T. E., Jr. (1966) Purification and general properties of guanosine 3′,5′-monophosphate-dependent protein kinase from guinea pig fetal lung. *J. biol. Chem.*, **251**, 1759–1766

Kuo, J. F. and Greengard, P. (1969) Cyclic nucleotide dependent protein kinases. Widespread occurrence of adenosine 3′,5′-monophosphate-dependent protein kinase in various tissues and phyla of the animal kingdom. *Proc. natn. Acad. Sci. U.S.A.*, **64**, 1349–1355

Kuo, J. F. and Greengard, P. (1970a) Stimulation of adenosine 3′,5′-monophosphate-dependent and guanosine 3′,5′-monophosphate-dependent protein kinases by some analogs of adenosine 3′,5′-monophosphate. *Biochem. biophys. Res. Commun.*, **40**, 1032–1038

Kuo, J. F. and Greengard, P. (1970b) Cyclic nucleotide-dependent protein kinases. VI. Isolation and partial purification of a protein kinase activated by guanosine 3′,5′-monophosphate. *J. biol. Chem.*, **245**, 2493–2498

Kuo, J. F. and Greengard, P. (1970c) Cyclic nucleotide-dependent protein kinase. VIII. An assay method for measurement of adenosine 3′,5′-monophosphate in various tissues and a study of agents influencing its level in adipose cells. *J. biol. Chem.*, **245**, 4067–4073

Kuo, J. F. and Greengard, P. (1974) Purification and characterization of cyclic GMP-dependent protein kinases. In *Methods in Enzymology*, Vol. 38 (Eds. J. G. Hardman and B. W. O'Malley), Academic Press, New York

Kuo, J. F., Kruger, B. K., Sanes, J. A. and Greengard, P. (1970) Cyclic nucleotide-dependent protein kinase. V. Preparation and properties of adenosine 3′,5′-monophosphate-dependent protein kinase from various bovine tissues. *Biochim. biophys. Acta*', **212**, 79–91

Kuo, J. F. and Kuo, W. N. (1973) Regulation by β-adrenergic receptor and muscarinic cholinergic receptor activation of intracellular cyclic AMP and cyclic GMP levels in rat lung slices. *Biochem. biophys. Res. Commun.*, **55**, 660–665

Kuo, J. F., Lee, T. P., Reyes, P. L., Walton, K. G., Donnelly, T. E. and Greengard, P. (1972) Cyclic nucleotide-dependent protein kinases. An assay method for the measurement of guanosine 3′,5′-monophosphate in various biological materials and a study of agents regulating its levels in heart and brain. *J. biol. Chem.*, **247**, 16–22

Kuo, J. F., Miyamoto, E. and Reyes, P. L. (1974a) Activation and dissociation of adenosine 3′,5′-monophosphate-dependent and guanosine 3′,5′-monophosphate-dependent protein kinases by various cyclic nucleotide analogs. *Biochem. Pharmac.*, **23**, 2011–2021

Kuo, T. H., Ou, C. T. and Tchen, T. T. (1975) The effect of calcium on the stimulation of corticosterone biosynthesis by dibutyryl-cAMP in cultures of ATOC cell line Y-1. *Biochem. biophys. Res. Commun.*, **65**, 190–195

Kuo, W. N., Hodgins, D. S. and Kuo, J. F. (1973) Adenylate cyclase in islets of Langerhans. Isolation of islets and regulation of adenylate cyclase activity by various hormones and agents. *J. biol. Chem.*, **248**, 2705–2711

Kuo, W-N., Hodgins, D. S. and Kuo, J. F. (1974b) Regulation by various hormones and agents of adenosine-3′,5′-monophosphate levels in islets of Langerhans of rats. *Biochem. Pharmacol.*, **23**, 1387–1391

Kurashina, Y., Takai, K., Suzuki-Hori, C., Okamoto, H. and Hayaishi, O. (1974) Adenylate cyclase from *Brevibacterium liquefaciens*. Reversibility and thermodynamic studies. *J. biol. Chem.*, **249**, 4824–4828

Kuroda, Y. and Kobayashi, K. (1975) Effects of adenosine and adenine nucleotides on the postsynaptic potential and on the formation of cyclic adenosine 3′,5′-monophosphate from radioactive adenosine triphosphate in guinea pig olfactory cortex slices. *Proc. Japan Acad.*, **51**, 495–500

Kurokawa, K. and Massry, S. G. (1973) Interaction between catecholamines and vasopressin on renal medullary cyclic AMP of rat. *Am. J. Physiol.*, **225**, 825–829

Kurtz, M. J., Polgar, P., Taylor, L. and Rutenburg, A. M. (1974) The role of adenosine 3′,5′-cyclic monophosphate in the division of WI-38 cells. *Biochem. J.*, **142**, 339–344

Kuschinsky, K. (1975) Dopamine-receptor sensitivity after repeated morphine administration to rats. *Life Sci.*, **17**, 43–48

Labrie, F., Baden, N., Poirier, G. and Lean, A. (1972) Binding of thyrotropin-releasing hormone to plasma membranes of bovine anterior pituitary gland. *Proc. natn. Acad. Sci. U.S.A.*, **69**, 283–287

Labrie, F., Gauthier, M., Pelletier, G., Borgeat, P., Lemay, A. and Gouge, J. J. (1973)

Role of microtubules in basal and stimulated release of growth hormone and prolactin in rat adenohypophysis *in vitro*. *Endocrinology*, **93**, 903–914

Labrie, F., Lemaire, S. and Courte, C. (1971a) Adenosine 3′,5′-monophosphate-dependent protein kinase from bovine anterior pituitary gland. I. Properties. *J. biol. Chem.*, **246**, 7239–7302

Labrie, F., Lemaire, S., Poirier, G., Pelletier, G. and Boucher, R. (1971b) Adenohypophyseal secretory granules. I. Their phosphorylation and association with protein kinase. *J. biol. Chem.*, **246**, 7239–7317

Lacombe, M. L. and Hanoune, J. (1974) Enhanced specificity of epinephrine binding by rat liver plasma membranes in the presence of EDTA. *Biochem. biophys. Res. Commun.*, **58**, 667–673

Lacy, P. E. and Malaisse, W. J. (1973) Microtubules and beta cell secretion. *Recent Prog. Horm. Res.*, **29**, 199–228

Lais, L. T. and Brody, M. J. (1975) Mechanism of vascular hyperresponsiveness in the spontaneously hypertensive rat. *Circulation Res.*, **35**, Suppl. I, 216–222

Lake, N. and Jordan, L. M. (1974) Failure to confirm cyclic AMP as second messenger for norepinephrine in rat cerebellum. *Science, N.Y.*, **183**, 663–664

Lake, N., Jordan, L. M. and Phillis, J. W. (1973) Evidence against cyclic adenosine 3′,5′-monophosphate (cAMP)-mediation of noradrenaline depression of cerebral cortical neurones. *Brain Res.*, **60**, 411–421

Lallemant, C., Seraydarian, K., Mommaerts, W. F. H. M. and Suh, M. (1975) A survey of the regulatory activity of some phosphorylated and dephosphorylated forms of troponin. *Archs. Biochem. Biophys.*, **164**, 367–371

Lamberts, S. W. J., Timmermans, H. A. T., Kramer-Blankestljn, M. and Birkenhager, J. C. (1975) The mechanism of the potentiating effect of glucocorticoids on catecholamine-induced lipolysis. *Metabolism*, **24**, 681–689

Lamy, F. and Dumont, J. E. (1974) Action of thyrotropin on phosphate incorporation into thyroid proteins *in vitro*. *Europ. J. Biochem.*, **45**, 171–179

Langan, T. A. (1971) Cyclic AMP and histone phosphorylation. *Ann. N.Y. Acad. Sci.*, **185**, 166–180

Langan, T. A. (1973) Protein kinases and protein kinase substrates. (Eds. P. Greengard and G. A. Robison), In *Advances in Cyclic Nucleotides Research*, Vol. 3, Raven Press, New York, pp. 99–154

Langer, C. A. (1973) Heart: Excitation–contraction coupling. *Ann. Rev. Physiol.*, **35**, 55–86

Langer, S. Z., Dubocovich, M. L. and Celuch, S. M. (1975) Prejunctional regulatory mechanisms for noradrenaline release elicited by nerve stimulation. In *Chemical Tools in Catecholamine Research* II. (Eds, O. Almgren, A. Carlsson, J. Engel), North-Holland Publishing Co., Amsterdam, pp. 183–191

Laragh, J. H. (1974) An approach to the classification of hypertensive states. *Hosp. Prac.*, **9**, 61–73

La Raia, P. J. and Morkin, E. (1974) Adenosine 3′,5′-monophosphate-dependent membrane phosphorylation. A possible mechanism for the control of microsomal calcium transport in heart muscle. *Circulation Res.*, **35**, 298–306

La Raia, P. J. and Reddy, W. J. (1969) Hormonal regulation of myocardial adenosine 3′,5′-monophosphate. *Biochim. biophys. Acta*, **177**, 189–195

La Raia, P. J. and Sonnenblick, E. H. (1971) Autonomic control of cardiac cAMP system. *Circulation Res.*, **28**, 377–384

Largis, E. E., Allen, D. O., Clark, J. and Ashmore, J. (1973) Isoproterenol and glucagon effects in perfused hearts from spontaneously hypertensive and normotensive rats. *Biochem. Pharmacol.*, **22**, 1735–1744

Larner, J. (1972) Insulin and glycogen synthase. *Diabetes*, **21**, 428–438

Larner, J., Huang, L. C., Brooker, G., Murad, F. and Miller, T. B. (1974) Inhibitor of protein kinase formed in insulin treated muscle. *Fed. Proc.*, **33**, 261

Larsson, L. B. (1964) Hereditary metabolic myopathy with paroxymal myoglobinuria due to abnormal glycolysis. *J. Neurol. Neurosurg. Psychiat.*, **27**, 361–380

Laugier, P., Posternak, T., Orusco, M., Cehovic, G. and Posternak, F. (1973) Essais de traitement du psoriasis par l'AMPc et un de ses dérivés. *Bull. Soc. fr. Derm. Syph.*, **80**, 632–636

Leclercq-Meyer, V., Marchand, J. and Malaisse, W. J. (1973) The effect of calcium and magnesium on glucagon secretion. *Endocrinology*, **93**, 1360–1370

Lee, T.-P., Busse, W. W., and Reed, C. E. (1974) Epidermal adenyl cyclase of human and mouse. *J. Allergy clin. Immunol.*, **53**, 283–287

Lee, T. P., Kuo, J. F. and Greengard, P. (1971) Regulation of myocardial cyclic AMP by isoproterenol, glucagon, and acetylchline. *Biochem. biophys. Res. Commun.*, **45**, 991–997

Lee, T. P., Kuo, J. F. and Greengard, P. (1972) Role of muscarinic cholinergic receptors in regulation of guanosine 3′,5′-cyclic monophosphate content in mammalian brain, heart muscle, and intestinal smooth muscle. *Proc. natn. Acad. Sci., U.S.A.*, **69**, 3287–3291

Lefebvre, P. J. and Luyckx, A. S. (1974) Effect of L8027, a new potent inhibitor of prostaglandin biosynthesis, on the metabolism and response to glucagon of rat adipose tissue. *Biochem. Pharmacol.*, **23**, 2119–2125

Lefkowitz, R. J. (1974) Stimulation of catecholamine-sensitive adenylate cyclase by 5′-guanylyl-imidodiphosphate. *J. biol. Chem.*, **249**, 6119–6124

Lefkowitz, R. J. (1975) Guanosine triphosphate binding sites in solubilized myocardium. *J. biol. Chem.*, **250**, 1006–1011

Lefkowitz, R. J. and Carol, M. G. (1975) Characteristics of 5′-guanylyl-imido-diphosphate-activated adenylate cyclase. *J. biol. Chem.*, **250**, 4418–4422

Lefkowitz, R. J. and Haber, E. (1971) A fraction of the ventricular myocardium that has the specificity of the cardiac beta-adrenergic receptor. *Proc. natn. Acad. Sci., U.S.A.*, **68**, 1773–1777

Lefkowitz, R. J., Haber, E. and O'Hara, D. (1972) Identification of the cardiac beta-adrenergic receptor protein: Solubilization and purification by affinity chromatography. *Proc. natn. Acad. Sci. U.S.A.*, **69**, 2828–2832

Lefkowitz, R. J. and Levey, G. S. (1972) Norepinephrine: Dissociation of β-receptor binding from adenylate cyclase activation in solubilized myocardium. *Life Sci.*, **11**, 821

Lefkowitz, R. J., Limbird, L. E., Mukherjee, C. and Caron, M. G. (1976) The beta-adrenergic receptor and adenylate cyclase. *Biochim. biophys. Acta*, **457**, 1–40

Lefkowitz, R. J., Mukherjee, C., Coverstone, M. and Caron, M. G. (1974b) Stereospecific [^{3}H](–)-alprenolol binding sites, β-adrenergic receptors and adenylate cyclase. *Biochem. biophys. Res. Commun.*, **60**, 703–709

Lefkowitz, R. J., O'Hara, D. and Warshaw, J. (1974a) Surface interaction of [^{3}H]norepinephrine with cultured chick embryo myocardial cells. *Biochem. biophys. Acta*, **332**, 317–328

Lefkowitz, R. J., Roth, J. and Pastan, I. (1970) Effects of calcium on ACTH stimulation of the adrenal; Separation of hormone binding from adenyl cyclase activation. *Nature, Lond.*, **228**, 864–866

Lefkowitz, R. J., Sharp, G. W. G. and Haber, E. (1973d) Specific binding of beta-adrenergic catecholamines to a subcellular fraction from cardiac muscle. *J. biol. Chem.*, **248**, 342–349

Legler, U.-F. and Sewing, K.-Fr. (1977) Effect of cyclic nucleotides on polysomal and ribosomal protein synthesis of the rat gastric mucosa. *Naunyn-Schmiedebergs Arch. Pharmak.*, (in press)

Leitner, J. W., Sussman, K. E., Vatter, A. E. and Schneider, F. H. (1975) Adenine nucleotides in the secretory granule fraction of rat islets. *Endocrinology*, **95**, 662–677

Lemaire, S., Pelletier, G. and Labrie, F. (1971) Adenosine 3′,5′-monophosphate-dependent protein kinase from bovine anterior pituitary gland. II. Subcellular distribution. *J. biol. Chem.*, **246**, 7303–7310

Lemay, A., Deschenes, M., Lemaire, S., Poirier, G., Poulin, L. and Labrie, F. (1974) Phosphorylation of adenohypophyseal plasma membranes and properties of associated protein kinase. *J. biol. Chem.*, **249**, 323–328

Lemay, A. and Labrie, F. (1972) Calcium-dependent stimulation of prolactin release in rat anterior pituitary *in vitro* by N^6-monobutyryl-adenosine 3′,5′-monophosphate. *FEBS Letters*, **20**, 7–10

Lepage, G. A. and Hersch, E. M. (1972) Cyclic nucleotide analogs as carcinostatic agents. *Biochem. biophys. Res. Comm.*, **46**, 1918–1922

Leray, F., Chambaut, A.-M., Perrenoud, M.-L. and Hanoune, J. (1973) Adenylate-cyclase activity of rat-liver plasma membranes: Hormonal stimulations and effect of adrenalectomy. *Europ. J. Biochem.*, **38**, 185–192

Lesko, L. and Marinetti, G. V. (1975) Hormone action at the membrane level. IV. Epinephrine binding to rat liver plasma membranes and rat epididymal fat cells. *Biochim. biophys. Acta*, **382**, 419–436

Levey, G. S. (1970) Solubilization of myocardial adenyl cyclase. *Biochem. biophys. Res. Commun.*, **38**, 86–92

Levey, G. S. (1971a) Solubilization of myocardial adenyl cyclase: Loss of hormone responsiveness and activation by phospholipids. *Ann. N.Y. Acad. Sci.*, **185**, 449–457

Levey, G. S. (1971b) Restoration of norepinephrine responsiveness of solubilized myocardial adenyl cyclase by phosphatidylinositol. *J. biol. Chem.*, **246**, 7405–7410

Levey, G. S. (1972) Phospholipids, adenylate cyclase and the heart. *J. molec. cell. Cardiol.*, **4**, 283–285

Levey, G. S. (1973) The role of phospholipids in hormone activation of adenylate cyclase. *Recent Prog. Horm. Res.*, **29**, 361

Levey, G. S. and Epstein, S. E. (1968) Activation of cardiac adenyl cyclase by thyroid hormone. *Biochem. biophys. Res. Commun.*, **33**, 990–995

Levey, G. S. and Epstein, S. E. (1969a) Activation of adenyl cyclase by glucagon in cat and human heart homogenates. *Circulation Res.*, **24**, 151

Levey, G. S. and Epstein, S. E. (1969b) Myocardial adenyl cyclase: Activation by thyroid hormones and evidence for two adenyl cyclase systems. *J. clin. Invest.*, **48**, 1669

Levey, G. S. and Epstein, S. E. (1969c) Myocardial adenyl cyclase: Activation by thyroid hormones and evidence for two adenyl cyclase systems. *J. clin. Invest.*, **48**, 1663–1669

Levey, G. S., Fletcher, M. A., Klein, I., Ruiz, E. and Schenk, A. (1974) Characterization of [^{125}HI]-glucagon binding in a solubilized preparation of cat myocardial adenylate cyclase. *J. biol. Chem.*, **249**, 2665–2673

Levey, G. S. and Klein, I. (1972) Solubilized myocardial adenyl cyclase: Restoration of histamine responsiveness by phosphatidylserine. *J. clin. Invest.*, **51**, 1578

Levey, G. S., Palmer, R. F., Lasseter, K. C. and McCarthy, J. (1971) Effect of tolbutamide on adenyl cyclase in rabbit and human heart and contractility of isolated rabbit atria. *J. clin. Endocrinol. Metab.*, **33**, 372–374

Levey, G. S., Schmidt, W. M. I. and Mintz, D. H. (1972) Activation of adenyl cyclase in a pancreatic islet cell adenoma by glucagon and tolbutamide. *Metabolism*, **21**, 93–98

Levey, J. V. and Killbrew, E. (1971) Inotropic effects of prostaglandin E_2 on isolated cardiac tissue. *Proc. Soc. exp. Biol. Med.*, **136**, 1227–1231

Levi-Montalcini, R. and Angeletti, P. U. (1968) Nerve growth factor. *Physiol. Rev.*, **48**, 534–569

Levine, R. A. and Washington, A. (1970) Glycogenolytic activity of cyclic 3′,5′-monophosphates in perfused rat liver. *Endocrinology*, **87**, 377–382

Levine, R. A. and Wilson, D. E. (1971) The role of cyclic AMP in gastric secretion. *Ann. N.Y. Acad. Sci.*, **185**, 363–375

Lewis, P. R. and Schute, C. D. (1967) The cholinergic limbic system: Projection to hippocampal formation, medial cortex, nuclei of the ascending cholinergic reticular system and the subfornical organ and supra-optic crest, *Brain*, **90**, 521–540

Lewis, R. A., Goetzl, E. T., Wasserman, S. I., Valone, F. H., Rubin, R. H. and Austen, K. F. (1975) The release of four mediators of immediate hypersensitivity from human leukemic basophils. *J. Immunol.*, **114**, 87–92

Li, H.-C. (1975) Protein phosphatase from canine heart: Evidence of four different fractions of the enzyme. *FEBS Letters*, **55**, 134–137

Liano, S., Lin, A. H. and Tymoczko, J. L. (1971) Adenyl cyclase of cell nuclei isolated from rat ventral prostrate. *Biochim. biophys. Acta*, **2301**, 535

Libet, B., Kobayashi, H. and Tanaka, T. (1975) Synaptic coupling in the production and storage of a neuronal memory trace. *Nature, Lond.*, **258**, 155–157

Lichtenstein, L. M. (1975) The mechanism of basophil histamine release induced by antigen and by the calcium ionophore A-23187. *J. Immunol.*, **114**, 1692–1699

Lichtenstein, L. M., Levy, D. A. and Ishizaka, K. (1970) *In vitro* reversed anaphalaxis: Characteristics of anti-IgE mediated histamine release. *Immunology*, **19**, 831–842

Lichtenstein, L. M. and Margolis, S. (1968) Histamine release: *In vitro* inhibition by catecholamines and methylxanthines. *Science, N.Y.,*, **163**, 902–903

Lima, A. O., Javierre, M. Q., Dias da Silva, W. and Camara, D. S. (1974) Immunological phagocytosis: Effect of drugs on phosphodiesterase activity. *Experientia*, **30**, 945–946

Limas, C. J., Ragan, D. and Freis, E. D. (1974) Effect of acute cardiac overload on intramyocardial cyclic 3′,5′-AMP: Relation to prostaglandin synthesis. *Proc. Soc. exp. Biol. Med.*, **147**, 103–105

Limbird, L., DeMeyts, P. and Lefkowitz, R. J. (1975) Beta-adrenergic receptors: Evidence for negative co-operativity. *Biochem. biophys. Res. Commun.*, **64**, 1160–1168

Limbird, L. E. and Lefkowitz, R. J. (1975) Myocardial guanylate cyclase: Properties of the enzyme and effects of cholinergic agonists *in vitro*. *Biochim. biophys. Acta*, **377**, 186–196

Lin, M. C., Salomon, Y., Rendell, M. and Rodbell, M. (1975a) The hepatic adenylate cyclase system. II. Substrate binding and utilization and the effects of magnesium ion and pH. *J. biol. Chem.*, **250**, 4246–4252

Lin, M. C., Wright, D. W., Hruby, V. J. and Rodbell, M. (1975b) Structure–function relationships in glucagon: Properties of highly purified Des-His^1-, monoiodo-, and [Des-Asn^{28}, Thr^{29}] (homoserine lactone^{27})-glucagon. *Biochemistry*, **14**, 1559–1563

Lin, S, Santi, D. V. and Spudich, J. A. (1974) Biochemical studies on the mode of action of cytochalasin B. *J. biol. Chem.*, **249**, 2268–2274

Lin, S. and Spudich, J. A. (1974a) Biochemical studies on the mode of action of cytochalasin B. *J. biol. Chem.*, **249**, 5778–5783

Lin, S. and Spudich, J. A. (1974b) Binding of cytochalasin B to a red cell membrane protein. *Biochem. biophys. Res. Commun.*, **61**, 1471–1476

Lin, Y. M., Liu, Y. P. and Cheung, W. Y. (1974d) Cyclic 3′,5′-nucleotide phosphodiesterase. Purification, characterization and active form of the protein activator from bovine brain. *J. biol. Chem.*, **249**, 4943–4954

Lin, Y. M., Liu, Y. P. and Cheung, W. Y. (1975) Cyclic 3′,5′-nucleotide phosphodiesterase: Ca^{2+}-dependent formation of bovine brain enzyme activator complex. *FEBS Letters*, **49**, 356–360

Lincoln, T. M. and Vaughan, G. L., (1975) The role of adenosine 3′,5′-monophosphate in the transformation of Cloudman mouse melanoma cells. *J. Cell Physiol.*, **86**, 543–552

Lindl, T. and Cramer, H. (1975) Evidence against dopamine as the mediator of the rise of cyclic AMP in the superior cervical ganglion of the rat. *Biochem. biophys. Res. Comm.*, **65**, 731–739

Lindl, T., Heinl-Sawaya, M. C. B. and Cramer, H. (1975) Compartmentation of an ATP substrate pool for histamine and adrenaline-sensitive adenylate cyclase in rat superior cervical ganglia. *Biochem. Pharmacol*, **24**, 947–950

Lindl, T., Teufel, E. and Cramer, H. (1976) 3′,5′-nucleotide phosphodiesterase in the superior cervical ganglion of the rat. Evidence for multiple forms and influence of the β-adrenergic receptor system in the electrophoretic pattern of the enzyme. *Hoppe-Seyler's Z. Physiol. Chem.*, *357*, 983–989

Ling, C. and Gerard, R. W. (1949) Normal membrane potential of frog sartorius fibres. *J. cell. comp. Physiol.*, **34**, 383–394

Lippmann, W., Pugsley, T. and Merker, J. (1975) Effect of butaclamol and its enantiomers upon striatal homovanillic acid and adenylcyclase of olfactory tubercle in rats. *Life Sci.*, **16**, 213–224

Lissitzky, S., Fayet, G., Giraud, A., Verrier, B. and Torresani, J. (1971) Thyrotrophin-induced aggregation and reorganization into follicles of isolated porcine thyroid cells. I. Mechanism of action of thyrotrophin and metabolic properties. *Europ. J. Biochem.*, **24**, 88–99

Liu, Y. P. and Cheung, W. Y. (1976) Cyclic 3′,5′-nucleotide phosphodiesterase. Ca^{2+} confers more conformation to the protein activator. *J. Biol. Chem.*, **251**, 4193–4198

Lloyd, J. V., Nishizawa, E. E. and Mustard, J. F. (1973) Effect of ADP-induced shape change on incorporation of [^{32}P] into platelet phosphatidic acid and mono, di and tri-phosphatidylinositol. *Br. J. Haematol.*, **25**, 77

Lo, H. K. and Levey, G. S. (1976) Glucagon-mediated stimulation of [^{32}P]-orthophosphate and [^{14}C]-serine incorporation into phosphatidylserine in cardiac muscle slices. *Endocrinology*, **98**, 251

Lochaya, S., Hamilton, J. C. and Mayer, J. (1963) Lipase and glycerokinase activities in the adipose tissue of obese hyperglycemic mice. *Nature, Lond.*, 182–183

Loeffler, L. J., Lovenberg, W. and Sjoerdsma, A. (1971) Effects of dibutyryl-3′,5′-cyclic adenosine monophosphate, phosphodiesterase inhibitors and prostaglandin E_1 on compound 48/80-induced histamine release from rat peritoneal mast cells *in vitro*. *Biochem. Pharmacol.*, **20**, 2287–2297

Londos, C. and Rodbell, M. (1975) Multiple inhibitory and activating effects of nucleotides and magnesium on adrenal adenylate cyclase. *J. biol. Chem.*, **250**, 3459–3465

Londos, C., Solomon, Y., Lin, M. C., Harwood, J. P., Schramm, M., Wolff, J. and Rodbell, M. (1974) 5′-guanylylimidodiphosphate, a potent activator of adenylate cyclase systems in ekaryotic cells. *Proc. natn. Acad. Sci., U.S.A.*, **71**, 3087–3090

Long, R. A., Szekeres, G. L., Khwaja, T. A., Sidwell, R. W., Simon, L. N. and Robins, R. K. (1972) Synthesis and antitumor and antiviral activities of 1-β-D-arabinofuranosylpyrimidine 3′,5′-cyclic phosphates. *J. med. Chem.*, **15**, 1215–1218

Loor, F., Forni, L. and Pernis, B. (1972) The dynamic state of the lymphocyte membrane. Factors affecting the distribution and turnover of surface immunoglobulins. *Europ. J. Immunol.*, **2**, 203–212

Loten, E. G. and Sneyd, J. G. T. (1970) An effect of insulin on adipose tissue adenosine 3′,5′-cyclic monophosphate phosphodiesterase. *Biochem. J.*, **120**, 187–193

Louis, L. H. (1969) Lipoatrophic diabetes: An improved procedure for the isolation and purification of a diabetogenic polypeptide from urine. *Metabolism*, **18**, 545–555

Louis, W. J., Doyle, A. E. and Anaverkar, S. (1973) Plasma norepinephrine levels in essential hypertension. *New Engl. J. Med.*, **288**, 599–601

Lovenberg, W., Bruckwick, E. and Hanbauer, I. (1975) Protein phosphorylation and regulation of catecholamine synthesis. In *Chemical Tools in Catecholamine Research* II. (Eds. O. Almgren, A. Carlsson and J. Engel,) North-Holland Publishing, Amsterdam, pp. 37–44

Luckasen, J. R., Sabad, A., Hauge, N. D., Goldberg, N. and Kersey, J. H. (1974a) Modulating influence of cAMP and cGMP on antigen (SRBC)-binding of murine lymphocytes. *Fed. Proc.*, **33**, 594

Luckasen, J. R., White, J. G. and Kersey, J. H. (1974b) Mitogenic properties of a calcium ionophore A23187. *Proc. natn. Acad. Sci. U.S.A.*, **71**, 5088–5090

Lugnier, C., Bertrand, Y. and Stoclet, J. C. (1972) Cyclic nucleotide phosphodiesterase inhibition and vascular smooth muscle relaxation. *Europ. J. Pharmac.*, **19**, 134–136

Lüllmann, H., Peters, T., Preuner, J. and Rüther, T. (1975) Influence of ouabain and dihydroouabain on the circular dichroism and cardiac plasmalemma microsomes. *Naunyn-Schmiedebergs Arch. exp. Path. Pharmak.*, **290**, 1–19

Lust, W. D. and Goldberg, N. D. (1970) Conditions affecting cyclic nucleotide levels in mouse brain. *Pharmacologist*, **12**, 290

Lyle, L. R., Liebhaber, H. R. and Parker, C. W. (1974) Concanavalin A induced changes in cyclic AMP in purified human lymphocytes. *Fed. Proc.*, **33**, 794

Lyle, L. R. and Parker, C. W. (1974) Cyclic AMP responses to concanavalin A in human lymphocytes. Evidence that the response involves specific carbohydrate receptors on the cell surface. *Biochemistry*, **13**, 5415–5420

Lynch, T. J., Tallant, A. and Cheung, W. Y. (1975) Separate genetic regulation of cyclic nucleotide phosphodiesterase and its protein activator in cultured mouse fibroblasts. *Biochem. biophys. Res. Commun.*, **63**, 967–970

Lynch, T. J., Tallant, E. A. and Cheung, W. Y. (1976) Ca^{2+}-dependent formation of brain adenylate cyclase–protein activator complex. *Biochem. biophys. Res. Commun.*, **68**, 616–625

Macchia, V., Tamburrini, O. and Pastan, I. (1970) Role of lecithin in the mechanism of TSH action. *Endocrinology*, **86**, 787

Macchia, V. and Varrone, S. (1971) Mechanism of TSH action. Studies with dibutyryl-cyclic AMP and dibutyryl-cyclic GMP. *FEBS Letters*, **13**, 342–344

Macchia, V. and Wolff, J. (1970) The effect of a purified preparation of lecithinase C on iodide transport. *FEBS Letters*, **10**, 219

MacLean, P. E. (1958) Contrasting functions of limbic neocortical systems of the brain and their relevance to psychophysiological aspects of medicine. *Am. J. Med.*, **26**, 611–626

Macleod, R. M. and Lehmeyer, J. E. (1970) Release of pituitary growth hormone by prostaglandins and dibutyryl-adenosine 3′,5′-monophosphate in the absence of protein synthesis. *Proc. natn. Acad. Sci. U.S.A.*, **67**, 1172–1179

Mackrell, D. J. and Sokal, J. E. (1969) Antagonism between the effects of insulin and glucagon on the isolated liver. *Diabetes*, **18**, 724–732

Madyastha, P. R., Barth, R. F. and Madyastha, K. R. (1975) Agglutination of leukemic and 2,4-dinitrophenyl-tagged normal human lumphocytes by wheat germ agglutinin. *J. natn. Cancer Inst.*, **54**, 597–600

Maeno, H. and Greengard, P. (1972) Phosphoprotein phosphatases from rat cerebral cortex. *J. biol. Chem.*, **247**, 3269–3277

Maeno, H., Johnson, E. M. and Greengard, P. (1971) Subcellular distribution of adenosine 3′,5′-monophosphate-dependent protein kinase in rat brain. *J. biol. Chem.*, **246**, 134–142

Maeno, H., Reyes, P. L., Veda, T., Rudolph, S. A. and Greengard, P. (1974)

Autophosphorylation of cyclic AMP-dependent protein kinase from bovine brain. *Archs. Biochem. Biophys.*, **164**, 551–559

Maeno, H., Ueda, T. and Greengard, P. (1975) Adenosine 3′,5′-monophosphate-dependent protein phosphatase activity in synaptic membrane fractions. *J. Cyc. Nucleto., Res.*, **1**, 37–48

Magaribuchi, M. and Kuriyama, H. (1972) Effects of noradrenaline and isoprenaline on the electrical and mechanical activities of guinea-pig depolarized *Taenia coli. Jap. J. Physiol.*, **22**, 253–270

Maguire, M. E., Goldmann, P. H. and Gilman, A. G. (1974) The reaction of [^{3}H] norepinephrine with particulate fractions of cells responsive to catecholamines. *Molec. Pharmac.*, **10**, 563–581

Mah, H. D. and Daly, J. W. (1975) Intracellular formation of analogs of cyclic AMP: Studies with brain slices labeled with radioactive derivatives of adenine and adenosine. *Biochim. biophys. Acta*, **404**, 49–56

Mah, H. D. and Daly, J. W. (1977) Adenosine-dependent formation of cyclic AMP in brain slices. *Pharmac. Res. Commun.* (in press)

Mahrle, G. and Orfanos, C. E. (1975) Ultrastructural localization and differentiation of membrane-bound ATP utilizing enzymes including adenyl cyclase in normal and psoriatic epidermis. *Br. J. Derm.*, **93**, 495–507

Maino, V. C., Green, N. M. and Crumpton, M. J. (1974) The role of calcium ions in initiating transformation of lymphocytes. *Nature, Lond.*, **251**, 324–327

Maino, V. C., Green, N. M. and Crumpton, M. J. (1975a) The role of divalent cations in the initiation of lumphocyte activation. In *Immune Recognition* (Ed. A. S. Rosenthal), Academic Press, New York, pp. 417–443

Maino, V. C., Hayman, M. J. and Crumpton, M. J. (1975b) Relationship between enhanced turnover of phosphatidylinositol and lumphocyte activation by mitogens. *Biochem. J.*, **146**, 246–252

Makman, M. H. (1971a) Conditions leading to enhanced response to glucagon, epinephrine, or prostaglandin by adenylate cyclase of normal and malignant cultured cells. *Proc. natn. Acad. Sci.*, U.S.A., **68**, 2127–2130

Makman, M. H. (1971b) Properties of adenylate cyclase of lymphoid cells. *Proc. natn. Acad. Sci., U.S.A.*, **68**, 885–889

Makman, M. H., Dvorkin, B. and Keehn, E. (1974) Hormonal regulation of cyclic AMP in aging and in virus-transformed human fibroblasts and comparison with other cultured cells. In *Cold Spring Harbor Symposium on the Regulation of Proliferation in Animal Cells* (Eds. B. Clarkson and R. Baserga), Academic Press, New York, pp. 649–663

Makman, M. H. and Klein, M. I. (1972) Expression of adenylate cyclase, catecholamine receptors and cyclic adenosine monophosphate-dependent protein kinase in synchronized cultures of Chang's liver cells. *Proc. natn. Acad. Sci., U.S.A.*, **69**, 456–458

Makman, M. H., Mishra, R. K. and Brown, H. (1975) Drug interactions with dopamine-stimulated adenylate cyclases of caudate nucleus and retina: Direct agonist effect of a piribedil metabolite. *Adv. Neurol.*, **9**, 213–222

Malaisse, W. J. (1973) Insulin Secretion: Multifactorial regulation for a single release process. *Diabetologia*, **9**, 167–173

Malaisse, W. J., Brisson, G. R. and Baird, L. E. (1973) Stimulus–secretion coupling of glucose-induced insulin release. X. Effect of glucose on ^{45}Ca efflux from perifused islets. *Am. J. Physiol.*, **224**, 389–394

Malaisse, W. J., Malaisse-Lagae, F. and Mayhew, D. (1967) A possible role for the adenylyclase system in insulin secretion. *J. clin. Invest.*, **46**, 1724–1734

Malawista, S. E., Gee, J. B. and Bensch, K. G. (1971) Cytochalasin B reversibly inhibits phagocytosis: Functional, metabolic, and ultrastructural effects in human blood leukocytes and rabbit alveolar macrophages. *Yale J. Biol. Med.*, **44**, 286–300

Malgieri, J. A., Shepherd, R. E. and Fain, J. N. (1975) Lack of feeback regulation of cyclic 3′,5′-AMP accumulation by free fatty acids in chicken fat cells. *J. biol. Chem.*, **250**, 6593–6598

Malkinson, A. M., Krueger, B. K., Rudolph, S. A., Casnellie, J. E., Haley, B. E. and Greengard, P. (1975) Widespread occurrence of a specific protein in vertebrate tissues and regulation by cyclic AMP of its endogenous phosphorylation and dephosphorylation. *Metabolism*, **24**, 331–341

Mandel, L. R., Kuehl, F. A. and Van Arman, C. G. (1975) The effect of prostaglandins on cyclic AMP in normal and psoriatic skin *in vitro. Fed. Proc.*, **34**, 764

Mandell, A. J. (1974) The role of adaptive regulation in the pathophysiology of psychiatric disease. *J. psychiat. Res.*, **11**, 173–179

Mandell, A. J. (1975) Neurobiological mechanisms of presynaptic metabolic adaptation and their organization: Implications for a pathophysiology of the affective disorders. *Adv. Biochem. Psychopharmac.*, **13**, 1–32

Manganiello, V. and Breslow, J. (1974) Effects of prostaglandin E_1 and isoproterenol on cyclic AMP content of human fibroblasts modified by time and cell density in subculture. *Biochim. Biophys. Acta*, **362**, 509–520

Manganiello, V., Breslow, J. and Vaughan, M. (1972) An effect of dexamethasone on the cyclic AMP content of human fibroblast stimulated by catecholamines and prostaglandin E_1. *J. clin. Invest.*, **51**, 60

Manganiello, V., Murad, F. and Vaughan, M. (1911) Effects of lipolytic and antilipolytic agents on cyclic 3′,5′-adenosine monophosphate in fat cells. *J. biol. Chem.*, **246**, 2195–2202

Manganiello, V. and Vaughan, M. (1972a) Prostaglandin E_1 effects on adenosine 3′,5′-cyclic monophosphate concentration and phosphodiesterase activity in fibroblasts. *Proc. natn. Acad. Sci., U.S.A.*, **69**, 269–273

Manganiello, V. and Vaughan, M. (1972b) An effect of dexamethasone on adenosine 3′,5′-monophosphate content and adenosine 3′,5′-monophosphate phosphodiesterase activity of cultured hepatoma cells. *J. clin. Invest.*, **51**, 2763–2767

Manganiello, V. and Vaughan, M. (1973) An effect of insulin on cyclic adenosine 3′,5′-monophosphate phosphodiesterase activity in fat cells. *J. biol. Chem.*, **248**, 7164–7170

Manganiello, V. C. and Vaughan, M. (1975) Inhibition of fat cell adenylate cyclase by fluoride. *J. biol. Chem.* (in press)

Mangla, J. C., Kim, Y. K. and Rubulis, A. K. (1974) Adenyl cyclase stimulation by aspirin in rat gastric mucosa. *Nature, Lond.*, **250**, 61–62

Manner, G. and Kuleba, M. (1974) Effect of dibutyryl-cyclic AMP on collagen and non-collagen protein synthesis in cultured human cells. *Connect. Tissue Res.*, **2**, 167–176

Mao, C. C., Guidotti, A. and Costa, E. (1974a) Inhibition by diazepam of the tremor and the increase of cerebellar cGMP content elicited by harmaline. *Brain Res.*, **83**, 526–529

Mao, C. C., Guidotti, A. and Costa, E. (1974b) The regulation of cyclic guanosine monophosphate in rat cerebellum: Possible involvement of putative amino acid neurotransmitters. *Brain Res.*, **79**, 510–514

Mao, C. C., Jacobson, E. D. and Shanbour, L. L. (1973) Mucosal cyclic AMP and secretion in the dog stomach. *Am. J. Physiol*, **225**, 893–896

Mao, C. C., Shanbour, L. L., Hodgins, D. S. and Jacobson, E. D. (1972) Adenosine 3′,5′-monophosphate (cyclic AMP) and secretion in the canine stomach. *Gastroenterology*, **63**, 427–438

Marcus, R. and Aurbach, G. D. (1971) Adenyl cyclase from renal cortex. *Biochim. biophys. Acta*, **242**, 410

Marks, F. and Grimm, W. (1972) Diurnal fluctuation and β-adrenergic elevation of cyclic AMP in mouse epidermis *in vivo. Nature, New Biol.*, **240**, 178–179

Marks, F. and Rebien, W. (1972) The second messenger system of mouse epidermis. I. Properties and β-adrenergic activation of adenylate cyclase *in vitro. Biochim. biophys. Acta*, **284**, 556–567

Martin, D. B. and Carter, J. R. (1970) Insulin-stimulated glucose uptake by subcellular particles from adipose tissue cells. *Science*, N.Y., **167**, 873–874

Martres, M. P., Baudry, M. and Schwartz, J. C. (1975) Subsensitivity of noradrenaline-stimulated cyclic AMP accumulation in brain slices of d-amphetamine treated mice. *Nature, Lond.*, **255**, 731–734

Mashiter, K., Mashiter, G. D., Hanger, R. L. and Field, J. B. (1973) Effect of cholera and *E. coli* enterotoxins on cyclic adenosine 3′,5′-monophosphate levels and intermediary metabolism in thyroid. *Endocrinology*, **92**, 541–552

Masui, H. and Garren, L. D. (1971) Inhibition of replication in functional mouse adrenal tumor cells by adrenocorticotropic hormone mediated by adenosine 3′,5′-cyclic monophosphate. *Proc. natn. Acad. Sci., U.S.A.*, **68**, 3206–3210

Matsuo, H., Baba, Y., Nair, R. M. G., Arimura, A. and Schally, A. V. (1971) Structure of the porcine LH and FSH-releasing hormone. I. The proposed amino acid sequence. *Biochem. biophys. Res. Commun.*, **43**, 1334–1339

Matsuzawa, H. and Nirenberg, M. (1975) Receptor-mediated shifts in cGMP levels in neuroblastoma cells. *Proc. natn. Acad. Sci., U.S.A.*, **72**, 3472–3476

Matthews, E. K. (1970) Electrical activity in islets cells and insulin secretion. *Acta. diabet. latinoam.*, **7**, 83–89

Matthews, E. K. (1975) Calcium, and stimulus-secretion coupling in pancreatic islet cells. In *Calcium Transport in Contraction and Secretion* (Eds. E. Carafoli *et al.*), North Holland Publishing Co., Amsterdam, pp. 203—210

Matthews, E. K. and Sakamoto, Y. (1975) Electrical characteristics of pancreatic islet cells. *J. Physiol.*, Lond., **246**, 421–437

Matussek, M. and Ladisich, W. (1969) Chronic administration of electroconvulsive shock and norepinephrine metabolism in the rat brain. III. Influence of acute and chronic electroshock upon drug-induced behavior. Psychopharmacologia, **15**, 305–309

Mawe, R., Doore, B., McCaman, M., Feucht, B. and Saier, M. (1974) Regulation of cyclic AMP excretion in bacteria and cultured animal cells. *J. Cell Biol.*, **63**, 211

May, C. D., Levine, B. B. and Weissmann, G. (1970) Effects of compounds which inhibit antigenic release of histamine and phagocytic release of lysosomal enzyme on glucose utilization by leukocytes in humans *Proc. Soc. exp. Biol. Med.*, **133**, 758–763

Mayer, S. E., Cotten, M. de V. and Moran, N. C. (1963) Dissociation of the augmentation of cardiac contractile force from the activation of myocardial phosphorylase by catecholamines. *J. Pharmac. exp. Ther.*, **139**, 275–282

Mayer, S. E., Namm, D. H. and Rice, L. (1970) Effect of glucagon on cyclic 3′,5′-AMP, phosphorylase activity and contractility of heart muscle of the rat. *Circulation Res.*, **26**, 225–233

Mayhew, D. A., Wright, P. H., and Ashmore, J. (1969) Regulation of insulin secretion. *Pharmacol. Rev.*, **21**, 183–212

McAfee, D. A. and Greengard, P. (1972) Adenosine 3′,5′-monophosphate: Electrophysiological evidence for a role in synaptic transmission. *Science, N.Y.*, **178**, 310–312

McAfee, D. A., Schorderet, M. and Greengard, P. (1971) Adenosine 3′,5′-monophosphate in nervous tissue: Increase associated with synaptic transmission. *Science., N.Y.*, **171**, 1156–1158

McDonald, T. F. and Sachs, H. G. (1975) Electrical activity in embryonic cell aggregates. *Pflügers Arch. ges Physiol.*, **354**, 165–176

McGiff, J. C. and Itskovitz, H. D. (1973) Prostaglandins and the kidney. *Circulation*, **33**, 479–488

McKinney, W. T. (1974) Animal models in psychiatry. *Biol. Med.*, **17**, 529–541

McNeill, J. H. and Muschek, L. D. (1972) Histamine effects on cardiac contractility, phosphorylase, and adenyl cyclase. *J. molec. cell Cardiol.*, **4**, 611

McNeill, J. H. and Verma, S. C. (1974) Stimulation of rat gastric adenylate cyclase by histamine and histamine analogues and blockade by burimamide. *B. J. Pharmac.*, **52**, 104–106

McNutt, N. S., Culp, L. A. and Black, P. H. (1973) Guanylate cyclase, cyclic GMP phosphodiesterase and cyclic GMP in cultured fibroblastic cells. *J. Cell Biol.*, **56**, 412–428

Medrano, E., Piras, R. and Mordoh, M. (1974) Effect of colchicine, vinblastine, and cytochalasin B on human lymphocyte transformation by phytohemagglutinin. *Expl. Cell Res.*, **86**, 295–300

Meech, R. W. (1974) The sensitivity of *Helix aspersa* neurons to injected calcium ion. *J. Physiol., Lond*, **237**, 259–277

Meinertz, T. H., Nawrath, H. and Scholz, H. (1973a) Influence of cyclization and acyl substitution on the inotropic effects of adenine nucleotides. *Naunyn-Schmiedebergs Arch. Pharmacol.*, **278**, 165–178

Meinertz, T., Nawrath, H. and Scholz, H. (1973b) Stimulatory effects of DB-cAMP and adrenaline on myocardial contraction and ^{45}Ca exchange. Experiments at reduced calcium concentration and low frequencies of stimulation. *Naunyn-Schmiedebergs Arch. Pharmacol.*, **279**, 327–338

Melander, A., Ericson, L. E. and Sundler, F. (1974) Sympathetic regulation of thyroid hormone secretion. *Life Sci.*, **14**, 237–246

Melmon, K. L., Bourne, H. R., Weinstein, Y., Shearer, G. M., Kram, J. and Bauminger, S. (1974a) Hemolytic plaque formation by leukocytes *in vitro*. Control by vasoactive hormones. *J. clin. Invest.*, **53**, 13–21

Melmon, K. L., Weinstein, Y., Shearer, G. M., Bourne, H. R. and Bauminger, S. (1974b) Separation of specific antibody-forming mouse cells by their adherence to insolubilized endogenous hormones. *J. clin. Invest.*, **53**, 22–30

Mendelsohn, J., Skinner, A. A. and Kornfeld, S. (1971) The rapid induction by phytohemagglutinin of increased α-aminoisobutyric acid uptake by lumphocytes. *J. clin. Invest.*, **50**, 818–826

Mertz, D. P. (1970) Nucleotidstoffwechsel und Magensäuresekretion. *Klin. Wschr.*, **48**, 831–838

Meyer, R. B., Jr. and Miller, J. P. (1974) Analogs of cyclic GMP: General methods of synthesis and the relationship of structure to enzymic activity. *Life Sci.*, **14**, 1019–1040

Meyer, R. B., Jr., Shuman, D. A. and Robins, R. K. (1973a) Synthesis of purine nucleoside 3′,5′-cyclic phosphoramidates. *Tetrahedron. Lett.*, **4**, 269–272

Meyer, R. B., Jr., Shuman, D. A., Robins, R. K., Bauer, R. J., Dimmitt, M. K. and Simon, L. N. (1972) Synthesis and biological activity of several 6-substituted 9-β-D-ribofuranosylpurine 3′,5′-cyclic phosphates. *Biochemistry*, **11**, 2704–2709

Meyer, R. B., Jr., Shuman, D. A., Robins, R. K., Miller, J. P. and Simon, L. N. (1973b) Synthesis and enzymic studies of 5-aminoimidazole and N^1 and N^6-substituted adenine ribonucleoside cyclic 3′,5′-phosphates prepared from adenosine cyclic 3′,5′-phosphate. *J. med. Chem.*, **16**, 1319–1323

Meyer, R. B., Jr., Uno, H., Robins, R. K., Simon, L. N. and Miller, J. P. (1975a) 2-Substituted derivatives of adenosine and inosine cyclic 3′,5′-phosphates. Synthesis, enzymic activity, and analysis of the structural requirements of the binding locale of the 2-substituent on bovine brain protein kinase. *Biochemistry*, **14**, 3315–3321

Meyer, R. B., Jr., Uno, H., Shuman, D. A., Robins, R. K., Simon, L. N. and Miller, J. P. (1975b) The synthesis of 2,6-disubstituted-9-β-ribofuranosylpurine cyclic 3′,5′-phosphates and the selectivity of cAMP and cGMP-specific enzymes to substituents in these positions. *J. Cycl. Nucleot. Res.*, **1**, 159–167

Meyer, S. E. and Krebs, E. G. (1970) Studies of the phosphorylation and activation of skeletal muscle phosphorylase and phosphorylase kinase *in vivo*. *J. biol. Chem.*, **245**, 3153–3160

Mian, A. M., Harris, R., Sidwell, R. W., Robins, R. K. and Khwaja, T. A. (1974) Synthesis and biological activity of 9-β-D-arabinofuranosyladenine cyclic 3′,5′-

phosphate and 9-β-D-arabinofuranosylguanine cyclic 3′,5′-phosphate. *J. med. Chem.*, **17**, 259–263

Michal, G., Mühlegger, K., Nelboeck, M., Thiessen, C. and Weimann, G. (1974) Cyclophosphates. VI. Cyclophosphates as substrates and effectors of phosphodiesterase. *Pharmac. Res. Commun.*, **6**,203–252

Michal, G., Nelbock, M. and Weimann, G. (1970) Cyclophosphates. III. Splitting of various cyclophosphates by phosphodiesterase from heart and adipose tissue. *Z. anal. Chem.*, **252**, 189–193

Michal, G., Weimann, G. and Nelboeck, M. (1973) *In vivo* metabolic and cardiovascular effects of new cyclophosphates. *Pharmac. Res. Comm.*, **5**, 87–99

Michell, R. H. (1975) Inositol phospholipids and cell surface receptor function. *Biochim. biophys. Acta*, **415**, 81–147

Mier, B. H. and Urselmann, E. (1970) The adenyl cyclase of skin. II. Adenyl cyclase levels in atopic dermatitis. *Br. J. Derm.*, **83**, 364–366

Miki, N., Baraban, J. M., Keirns, J. J., Boyce, J. J. and Bitensky, M. W. (1975) Purification and properties of the light-activated cyclic nucleotide phosphodiesterase of rod outer segments. *J. biol. Chem.*, **250**, 6320–6327

Miki, N., Keirns, J. J., Markus, F. R., Freeman, J. and Bitensky, M. W. (1973) Regulation of cyclic nucleotide concentrations in photoreceptors: An ATP-dependent stimulation of cyclic nucleotide phosphodiesterase by light. *Proc. natn. Acad. Sci.*, U.S.A., **70**, 3820–3824

Miki, N. and Yoshida, H. (1972) Purification and properties of cyclic AMP phosphodiesterase from rat brain. *Biochim. biophys. Acta*, **268**, 166–174

Miller, J. P., Beck, A. H., Simon, L. N. and Meyer, R. B. Jr. (1975a) Induction of hepatic tyrosine aminotransferase *in vivo* by derivatives of cyclic adenosine 3′,5′-monophosphate. *J. biol. Chem.*, **250**, 426–431

Miller, J. P., Boswell, K. H., Mian, A. M., Meyer, R. B., Jr., Robins, R. K. and Khwaja, T. A. (1976) 2′-Derivatives of guanosine and inosine cyclic 3′,5′-phosphates. Synthesis, enzymic activity, and the effect of 8-substituents. *Biochemistry*, **15**, 217–222

Miller, J. P., Boswell, K. H., Muneyama, K., Simon, L. N., Robins, R. K. and Shuman, D. A. (1973a) Synthesis and biochemical studies of various 8-substituted derivatives of guanosine 3′,5′-cyclic phosphate, inosine 3′,5′-cyclic phosphate, and xanthosine 3′,5′-cyclic phosphate. *Biochemistry*, **12**, 5310–5319

Miller, J. P., Boswell, K. H., Muneyama, K., Tolman, R. L., Scholten, M. B., Robins, R. K., Simon, L. N. and Shuman, D. A. (1973b) Activity of tubercidin, toycomycin, and sangivamycin-3′,5′-cyclic phosphates and related compounds with some enzymes of adenosine-3′,5′-cyclic phosphate metabolism. *Biochem. biophys. Res. Commun.*, **55**, 843–849

Miller, J. P., Boswell, K. H. and Robins, R. K. (1977a) 8-Alkylthio and 8-Arylthio-cAMP derivatives: Synthesis and enzymic activities. *Biochem. biophys. Res. Commun.* In press)

Miller, J. P., Boswell, K. H. and Robins, R. K. (1977b) Enzymic activities of 6,8-disubstituted derivatives of 9-β-D-ribofuranosylpurine cyclic 3′,5′-phosphate. *Biochemistry* (in press)

Miller, J. P., Christensen, L. F., Meyer, R. B., Jr., Kitano, S., Mizuno, Y, and Andrea, T. A. (1977c) Modification of the adenine ring of adenosine cyclic 3′,5′-phosphate. *J. Biol. Chem.* (in press)

Miller, J. P., Shuman, D. A., Scholten, M. B., Dimmitt, M. K., Stewart, C. M., Khwaja, T. A. Robins, R. K. and Simon, L. N. (1973c) Synthesis and biological activity of some 2′-derivatives of adenosine 3′,5′-cyclic phosphate. *Biochemistry*, **12**, 1010–1016

Miller, R. J., Horn, A. S. and Iversen, L. L. (1974a) The action of neuroleptic drugs on dopamine-stimulated adenosine cyclic 3′,5′-monophosphate production in rat neostriatum and limbic forebrain. *Molec. Pharmac.*, **10**, 759–766

Miller, R. J., Horn, A. S. and Iversen, L. L. (1974b) Stimulation of a dopamine-sensitive

adenylate cyclase in homogenates of rat striatum by a metabolite of piribedil (ET 495). *Naunyn-Schmiedebergs Arch. exp. Path. Pharmak.*, **282**, 213–216
Miller, R. J., Horn, A. S. and Iversen, L. L. (1975b) Effect of butaclamol on dopamine-sensitive adenylate cyclase in the rat striatum. *J. Pharm. Pharmac.*, **27**, 212–213
Miller, T. B., Exton, J. H. and Park, C. R. (1971a) A block in epinephrine-induced glycogenolysis in hearts from adrenalectomized rats. *J. biol. Chem.*, **246**, 3672–3678
Miller, T. B. and Larner, J. (1973) Mechanism of control of hepatic glycogenesis by insulin. *J. biol. Chem.*, **248**, 3483–3488
Miller, W. H. (1973) Cyclic nucleotides and photoreception. *Expl. Eye Res.*, **16**, 357–363
Miller, W. H., Gorman, R. E. and Bitensky, M. W. (1971b) Cyclic adenosine monophosphate: Function in photoreceptors. *Science, N.Y.*, **174**, 295–297
Miller, Z., Lovelace, E. and Pastan, M. G. I. (1975c) Cyclic guanosine monophosphate and cellular growth. *Science, N.Y.*, **190**, 1213–1215
Milligan, J. V. and Kraicer, J. (1971) ^{45}Ca uptake during the *in vitro* release of hormones from the rat adenohypophysis. *Endocrinology*, **89**, 766–773
Milligan, J. V. and Kraicer, J. (1974) Physical characteristics of the Ca compartments associated with *in vitro* ACTH release. *Endocrinology*, **94**, 435–443
Milligan, J. V., Kratter, J., Fawcett, C. P. and Illner, P. (1972) Purified growth hormone releasing factor increases ^{45}Ca uptake into pituitary cells. *Can. J. Physiol. Pharmacol.*, **50**, 613–617
Millis, A. J. T., Forrest, G. and Pious, D. A. (1972) Cyclic AMP in cultured human lymphoid cells: Relationship to mitosis. *Biochem. biophys. Res. Commun.*, **49**, 1645–1649
Millis, A. J. T., Forrest, G. A. and Pious, D. A. (1974) Cyclic AMP-dependent regulation of mitosis in human lymphoid cells. *Expl. Cell Res.*, **83**, 335–343
Minna, J. D. and Gilman, A. G. (1973) Expression of genes for metabolism of cyclic adenosine 3′,5′-monophosphate in somatic cells. II. Effects of prostaglandin E_1 and theophylline on parental and hybrid cells. *J. biol. Chem.*, **248**, 6618–6625
Minton, J. P., Wisenbaugh, T. and Matthews, R. H. (1974) Elevated cyclic AMP levels in human breast cancer tissue. *J. natn. Cancer Inst.*, **53**, 283–284
Mishra, R. K., Gardner, E. L., Katzman, R. and Makman, M. H. (1974) Enhancement of dopamine-stimulated adenylate cyclase activity in rat caudate after lesions in substantia nigra: Evidence for denervation supersensitivity. *Proc. natn. Acad. Sci., U.S.A.*, **71**, 3883–3887
Miyamoto, E. (1975) Protein kinases in myelin of rat brain: solubilization and characterization. *J. Neurochem.*, **24**, 503–512
Miyamoto, E., Kuo, J. F. and Greengard, P. (1969a) Adenosine 3′,5′-monophosphate-dependent protein kinase from brain. *Science, N.Y.*, **165**, 63–65
Miyamoto, E., Kuo, J. F. and Greengard, P. (1969b) Cyclic nucleotide-dependent protein kinases: I. Purification and properties of adenosine 3′,5′-monophosphate-dependent protein kinase from bovine brain. *J. biol. Chem.*, **244**, 6395–6402
Miyamoto, E., Petzold, G. L., Harris, J. S. and Greengard, P. (1971) Dissociation and concomitant activation of adenosine 3′,5′-monophosphate-dependent protein kinase by histone. *Biochem. biophys. Res. Commun.*, **44**, 305–312
Moellmann, G., Lerner, A. B. and Hendee, J. R. (1974) The mechanism of frog skin lightening by acetylcholine. *Gen. comp. Endocr.*, **23**, 45–51
Moens, W., Vokaer, A. and Kram, R. (1975) Cyclic AMP and cyclic GMP concentrations in serum and density-restricted fibroblast cultures. *Proc. natn. Acad. Sci., U.S.A.*, **72**, 1063–1067
Moir, A. J. G., Wilkinson, J. M. and Perry, S. V. (1974) The phosphorylation sites of troponin I from white skeletal muscle of the rabbit. *FEBS Letters*, **42**, 253–256
Mommaerts, W. F. H. M., Illingworth, B., Pearson, C. M., Guillory, R. J. and Seraydarian, K. (1959) A functional disorder of muscle associated with the absence of phosphorylase. *Proc. natn. Acad. Sci., U.S.A.*, **45**, 791–797

Montague, W. and Cook, J. R. (1971) The role of adenosine 3′,5′-cyclic monophosphate in the regulation of insulin release by isolated rat islets of Langerhans. *Biochem. J.*, **122**, 115–120

Montague, W. and Howell, S. L. (1975) Cyclic AMP and the physiology of the islets of Langerhans. In *Advances in Cyclic Nucleotides Research* (Eds. P. Greengard and G. A. Robison), Vol. 6, Raven Press, New York, pp. 201–243

Montague, W. and Howell, S. L. (1976) The mode of action of adenosine 3′,5′-cyclic phosphate in the regulation of insulin secretion. *Ciba Foundation Symposium*, **41**, 141–158

Montague, W., Howell, S. L. and Green, I. C. (1976) Insuline release and the microtubular system of the islets of Langerhans: Effects of insulin secretagogues on microtubule subunit pool size. *Horm. Metab. Res.*, **8**, 166–169

Moore, K. E. and Thornburg, J. E. (1975) Drug-induced dopaminergic supersensitivity. *Adv. Neurol.*, **9**, 93–104

Moore, L., Hurwitz, L. and Landon, E. J. (1974) Calcium-uptake activity of microsomes from normal and hypertensive rat aorta. *Pharmacologist*, **16**, 297

Morad, M. and Goldman, Y. (1973) Excitation-contraction coupling in heart muscle: Membrane control of development of tension. *Prog. Biophys. mol. Biol.*, **27**, 259–313

Morad, M. and Rolett, E. (1972) Relaxing effect of catecholamines on mammalian heart. *J. Physiol., Lond.*, **244**, 537–548

Morgane, P. J. and Stern, W. C. (1972) Relationship of sleep to neuroanatomical circuits, biochemistry and behavior. *Ann. N.Y. Acad. Sci.*, **193**, 95–111

Morgenroth, V. H., Hegistrand, L. R., Roth, R. H. and Greengard, P. (1975) Evidence for involvement of protein kinase in the activation by adenosine 3′,5′-monophosphate of brain tyrosine 3-monooxygenase. *J. biol. Chem.*, **250**, 1946–1948

Moskowitz, J. and Fain, J. N. (1970) Stimulation by growth hormone and dexamethasone of labeled cyclic ademosine 3′,5′-monophosphate accumulation by white fat cells. *J. biol. Chem.*, **245**, 1101–1107

Moyle, W. R. and Ramachandran, J. (1973) Effect of LH on steroidogenesis and cyclic AMP accumulation in rat Leydig cell preparations and mouse tumour Leydig cells. *Endocrinology*, **93**, 127–134

Muirhead, E. E., Germain, G. S., Leach, B. E., Brooks, B. and Stephenson, P. (1973) Renomedullary interstitial cells (RIC), prostaglandins (PG) and the antihypertensive function of the kidney. *Prostaglandins*, **3**, 581–594

Mukherjee, C., Caron, M. G., Coverstone, M. and Lefkowitz, R. J. (1975a) Identification of adenylate cyclase-coupled beta-adrenergic receptors in frog erythrocytes with (—)-[^{3}H]alprenolol. *J. biol. Chem.*, **250**, 4869–4876

Mukherjee, C., Caron, M. G. and Lefkowitz, R. J. (1975b) Catecholamine-induced subsensitivity of adenylate cyclase associated with loss of β-adrenergic receptor binding sites. *Proc. natn. Acad. Sci., U.S.A.*, **72**, 1945–1949

Muneyama, K., Bauer, R. J., Shuman, D. A., Robins, R. K. and Simon, L. N. (1971) Chemical synthesis and biological activity of 8-substituted adenosine 3′,5′-cyclic monophosphate derivatives. *Biochemistry*, **10**, 2390–2395

Muneyama, K., Shuman, D. A., Boswell, K. H., Robins, R. K., Simon, L. N. and Miller, J. P. (1974) Synthesis and biological activity of 8-haloadenosine 3′,5′-cyclic phosphates. *J. Carbohyd. Nucleos. Nucleot.*, **1**, 55–60

-Murad, F. (1973) Clinical studies and application of cyclic nucleotides. In *Advances in Cyclic Nucleotide Research*, Vol. 3 (Eds. P. Greengard and G. A. Robison), Raven Press, New York, p. 355

Murad, F., Chi, Y.-M., Rall, T. W. and Sutherland, E. W. (1962) Adenyl cyclase. III. The effect of catecholamines and choline esters on the formation of adenosine 3′,5′-phosphate by preparations from cardiac muscle and liver. *J. biol. Chem.*, **237**, 1233–1238

Murad, F. and Kimura, H. (1974) Cyclic nucleotide levels in incubations of guinea-pig trachea. *Biochim. biophys. Acta*, **343,** 275–286

Murad, F., Strauch, B. S. and Vaughan, M. (1969) The effect of gonadotropins on testicular adenyl cyclase. *Biochim. biophys. Acta*, **177**, 591–598

Murad, F. and Vaughan, M. (1969) Effect of glucagon on rat heart adenyl cyclase. *Biochem. Pharmacol*, **18**, 1053

Mustard, J. F. and Packham, M. A. (1970) Factors influencing platelet function: Adhesion, release and aggregation. *Pharmac. Rev.*, **22**, 97–187

Myers, R. D. (1974) *Handbook of Drug and Chemical Stimulation of the Brain.* Van Nostrand Rheinhold, New York

Nagaoka, A., Kikuchi, K. and Aramaki, Y. (1970) Participation of tissue electrolytes and water to the spontaneous hypertension in rats. *Jap. Circul. J.*, **34**, 489–497

Nagata, Y. and Burger, M. M. (1974) Wheat germ agglutinin. *J. biol. chem.*, **249**, 3116–3122

Nagyvary, J., Gohil, R. N., Kirchner, C. R. and Stevens, J. D. (1973) Studies on neutral esters of cyclic AMP. *Biochem biophys. Res. Comm.*, **55**, 1072–1077

Nahorski, S. R. (1975) Behavioral supersensitivity to apomorphine following cerebral dopaminergic denervation by 6-hydroxydopamine. *Psychopharmacologia*, **42**, 159–162

Nahorski, S. R., Rogers, K. J. and Smith, B. M. (1974) Histamine H_2-receptors and cyclic AMP in brain. *Life Sci.*, **15**, 1887–1894

Nair, K. G. (1966) Purification and properties of 3′,5′-cyclic nucleotide phosphodiesterase from dog heart. *Biochemistry*, **5**, 150–157

Nakahara, T. and Birnbaumer, L. (1974) Studies on receptor-mediated activation of adenylyl cyclases. V. Transient kinetics of the activation of beef renal medullary adenylyl cyclase by neurohypophyseal hormones. Estimation of apparent rate constants of the receptor hormone–interaction. *J. biol. Chem.*, **249**, 7886–7891

Nakajima, S., Hirschowitz, B. I. and Sachs, G. (1971) Studies on adenyl-cyclase in *Necturus* gastric mucosa. *Archs. Biochem. Biophys.*, **143**, 123–136

Nakajima, S., Shoemaker, R., Hirschowitz, B. I. and Sachs, G. (1970) Comparison of actions of aminophylline and pentagastrin on *Necturus* gastric mucosa. *Am. J. Physiol.*, **219**, 1259–1262

Namm, D. H., Mayer, S. E. and Maltbie, M. (1968) The role of potassium and calcium ions in the effect of epinephrine on cardiac cyclic 3′,5′-monophosphate, phosphorylase kinase and phosphorylase. *Mol. Pharmac.*, **4**, 522–530

Narumi, S. and Kanno, M. (1973) Effects of gastric acid stimulants and inhibitors on the activities of HCO_3^--stimulated, Mg^{2+}-dependent ATP-ase and carbonic anhydrase in rat gastric mucosa. *Biochim. biophys. Acta*, **311**, 80–89

Narumi, S. and Maki, Y. (1973) Possible role of cyclic AMP in gastric acid secretion in rat. *Biochim. biophys. Acta*, **311**, 90–97

Narumi, S. and Miyamoto, E. (1974) Activation and phosphorylation of carbonic anhydrase by adenosine 3′,5′-monophosphate-dependent protein kinases. *Biochim. biophys. Acta*, **350**, 215–224

Nath, J. Rebhun, L. I. (1973) Studies on the uptake and metabolism of adenosine 3′,5′-cyclic monophosphate and N^6,$O^{2'}$-dibutyryl 3′,5′-cyclic adenosine monophosphate in sea urchin eggs. *Expl. Cell Res.*, **82**, 73–78

Nathanson, J. and Greengard, P. (1976)) Cyclic Nucleotides and their role in nervous system function. *Physiol. Rev.* (in press)

Nauta, W. J. H. (1958) Hippocampal projections and related neural pathways to the midbrain in the cat. *Brain*, **81**, 319–340

Neer, E. J. (1973) Vasopressin-responsive, soluble adenylate cyclase from rat renal medulla. *J. biol. Chem.*, **248**, 3742–3744

Neer, E. J. (1974) The size of adenylate cyclase. *J. biol. Chem.*, **249**, 6527–6531

Nelboeck, M., Michal, G., Weimann, G., Paoletti, R. and Berti, F. (2973) Analogues of cyclic AMP and their physiological response. *Pure appl. Chem.* **35**, 411–437

Nesbitt, J., Russel, T. R., Miller, Z. and Pastan, I. (1975) Contact-inhibited revertant cell lines isolated from SV40-transformed cells. *Fed. Proc.*, **34**, 616

Neufeld, A. H. and Sears, M. L. (1975) Adenosine 3′,5′-monophosphate analogue increases the outflow facility of the primate eye. *Invest. Ophthalmol.*, **14**, 688–689

Neve, P., Ketelbant-Balasse, P., Willems, C. and Dumont, J. E. (1972) Effect of inhibitors of the micrctubules and microfilaments on dog thyroid slices *in vitro*. *Expl. Cell. Res.*, **74**, 227–244

Neville, D. M., Jr. (1968) Isolation of an organ specific protein antigen from cell-surface membrane of rat liver. *Biochim. biophys. Acta*, **154**, 540–552

Newcombe, D. S., Ciosek, C. P., Ishikawa, Y. and Fahey, J. V. (1975) Human synoviocytes: Activation and desentization by prostaglandins and l-epinephrine. *Proc. natn. Acad. Sci., U.S.A.*, **72**, 3124–3128

Nielsen, S. P. and Petersen, O. H. (1972) Transport of calcium in the perfused submandibular gland of the cat. *J. Physiol., Lond.*, **223**, 685–697

Nishi, S. and Koketsu, K. (1967) Origin of ganglionic inhibitory postsynaptic potentials. *Life Sci.*, **6**, 2049–2055

Noble, D. (1975) *The Initiation of the Heartbeat*, Clarendon Press, Oxford

Noonan, K. D. and Burger, M. M. (1973) Induction of 3T3-cell division at the monolayer stage. Early changes in the macromolecular process. *Expl. Cell Res.*, **80**, 405–414

Nybäck, H. V., Walters, J. R., Aghajanian, G. K. and Roth, R. H. (1975) Tricyclic antidepressants: Effects on the firing rate of brain noradrenergic neurons. *Europ. J. Pharmac.*, **32**, 302–312

Obata, K. (1974) Transmitter sensitivities of some nerve and muscle cells in culture. *Brain Res.* **73**, 71–88

Obata, K. and Yoshida, M. (1973) Caudate-evoked inhibition and actions of GABA and other substances on cat pallidal neurons. *Brain Res., Osaka*, **64** 455–459

O'Brien, J. R. and Strange, R. C. (1975) The release of cyclic AMP from the isolated perfused rat heart. *Biochem. J.*, **152**, 429–432

O'Dea, R. F., Haddox, M. K. and Goldberg, N. D. (1971) Interaction with phosphodiesterase of free and bound kinase complexed cyclic AMP. *J. biol. Chem.*, **246**, 6183–6190

Oey, J., Vogel, A. and Pollack, R. (1974) Intracellular cyclic AMP concentration responds specifically to growth regulation by serum. *Proc. natn. Acad. Sci., U.S.A.*, **71**, 694–698

Ohba, M., Sakamoto, Y. and Tomita, T. (1975) The slow wave in the circular muscle of the guinea-pig stomach. *J. Physiol., Lond.*, **253**, 505–516

Ohga, Y. and Daly, J. W. (1977) The accumulation of cyclic AMP and cyclic GMP in guinea-pig brain slices. Effect of calcium ions, norepinephrine and adenosine. *Biochim. Biophys. Acta* (in press)

Ohkura, H. and Hatton, N. (1975) Gastrin and histamine-activated adenyl cyclase in isolated rat parietal cell. In *Advances in Cyclic Nucleotide Research*, Vol. 5 (Eds. G. Drummond, P. Greengard and G. A. Robison), Raven Press, New York, pp. 819

Okada, Y. and Kuroda, Y. (1975) Inhibitory action of adenosine and adenine nucleotides on the postsynaptic potential of olfactory cortex slices of the guinea-pig. *Proc. Japan Acad.*, **51**, 491–494

Okazaki, T., Okazaki, A., Resiman, R. E. and Arbesman, C. E. (1975) Glycogenolysis and control of anaphylactic histamine release by cyclic adenosine monophosphate-related agents. *J. Allergy clin. Immunol.*, **56**, 253–261

O'Keefe, E. and Cuatrecasas, P. (1974) Cholera toxin mimics melanocyte stimulating hormone in inducing differentiation in melanoma cells. *Proc. natn. Acad. Sci., U.S.A.*, **71**, 2500–2504

Oliver, A. P. and Segal, M. (1974) Transmembrane changes in hippocampal neurons: Hyperpolarizing actions of norepinephrine, cyclic AMP and locus coeruleus. *Proc. Soc. Neurosci.*, **4**, 361

Oliver, J. M., Ukena, T. E. and Berlin, R. D. (1974) Effects of phagocytosis and colchicine on the distribution of binding sites on the cell surface. *Proc. natn. Acad. Sci., U.S.A.*, **71**, 394–398

Ono, M. and Hozumi, M. (1973) Effect of cytochalasin B on lymphocyte stimulation induced by concanavalin A or periodate. *Biochem. biophys. Res. Commun.*, **53**, 342–349

Opler, L. A. and Makman, M. H. (1972) Mediation by cyclic AMP of hormone-stimulated glycogenolysis in cultured rat astrocytoma cells. *Biochem. biophys. Res. Commun.*, **46**, 1140–1146

Orange, R. P., Austen, W. G. and Austen, K. F. (1971) Immunological release of histamine and slow-reacting substance of anaphylaxis from human lung. *J. exp. Med.*, **134**, 136–147

Orange, R. P., Kaliner, M. A., Laraia, P. J. and Austen, K. F. (1971) Immunological release of histamine and slow reacting substance of anaphylaxis from human lung. II. Influence of cellular levels of cyclic AMP. *Fed. Proc.*, **30**, 1725–1729

Orci, L., Gabbay, K. H. and Malaisse, W. J. (1972) Pancreatic beta-cell web: Its possible role in insulin secretion. *Science, N.Y.*, **175**, 1128–1130

Orenberg, E. K., Renson, J., Elliott, G. R., Barchas, J. D. and Kessler, S. (1975) Genetic determination of aggressive behavior and brain cyclic AMP. *Psychopharmac. Commun.*, **1**, 99–107, 1975

Orloff, J. and Handler, J. S. (1962) The similarity of effects of vasopressin, adenosine 3′,5′-phosphate (cyclic AMP) and theophylline on the toad bladder. *J. clin. Invest.*, **41**, 702–709

Ortiz, P. (1972) The inhibition of *E. coli* adenyl cyclase by ara-ATP. *Biochem. biophys. Res. Commun.*, **46**, 1728–1733

Osler, A. G., Lichtenstein, L. M. and Levy, D. A. (1968) *In vitro* studies of human reaginicallergy. *Adv. Immunol.*, **8**, 183–231

Osswald, H., and Jacobs, O. (1974) Cyclic guanosine 3′,5′-monophosphate induced diuresis in rats. *Naunyn-Schmiedeberg's Arch. Pharmacol.*, **284**, 207–214

Otten, J., Johnson, G. S. and Pastan, I. (1971) Cyclic AMP levels in fibroblasts: Relationship to growth rate and contact inhibition of growth. *Biochem. biophys. Res. Commun.*, **44**, 1192–1198

Otten, J., Johnson, G. S. and Pastan, I. (1972) Regulation of cell growth by cyclic adenosine 3′,5′-monophosphate. Effect of cell density and agents which alter cell growth on cyclic adenosine 3′,5′-monophosphate levels in fibroblasts. *J. biol. Chem.*, **247**, 7082–7087

Ottenjann, R., Nitzsche, R. and Rösch, W. (1971) Über den Einfluß von Theophyllin auf die durch Pentagastrin submaximal stimulierte gastrale Säuresekretion. *Klin. Wschr.*, **49**, 56–57

Øye, I. (1965) *The Mode of Action of Adrenaline in the Isolated Rat Heart*. Universitätsforlaget, Oslo

Øye, I. and Langslet, A. (1972) The role of cyclic AMP in the inotropic response to isoprenaline and glucagon. In *Advances in Cyclic Nucleotide Research*, Vol. 1, (Eds. P. Greengard and G. A. Robison), Raven Press, New York, pp. 291–300

Øye, I. and Sutherland, E. W. (1966) The effect of epinephrine and other agents on adenyl cyclase in the membrane of avian erythrocytes. *Biochim. biophys. Acta*, **127**, 347–354

Pace, C. S. and Price, S. (1974) Bioelectric effects of hexoses on pancreatic islet cells. *Endocrinology*, **94**, 142–147

Pačes, V. and Smrž, J. (1973) On the specificity of cyclic AMP action in *Escherichia coli*. *FEBS Letters*, **31**, 343–344

Pairault, J. and Laudat, M.-H. (1975) Selective identification of 'true' β-adrenergic receptors in the plasma membranes of rat adipocytes. *FEBS Letters*, **50**, 61–65

Palmer, G. C. (1972) Increased cyclic AMP response to norepinephrine in the rat brain following 6-hydroxydopamine. *Neuropharmacology*, **11**, 145–149

Palmer, G. C. and Burks, T. F. (1971) Central and peripheral adrenergic blocking actions of LSD and BOL. *Europ. J. Pharmacol.*, **16**, 113–116

Palmer, G. C. and Dail, W. G. (1975) Appearance of hormone-sensitive adenylate cyclase in the developing human heart. *Pediat. Res.*, **9**, 98–103

Palmer, G. C. and Manian, A. A. (1974a) A modification of the receptor component of adenylate cyclase in the rat brain by phenothiazine derivatives. *Neuropharmacology*, **13**, 851–866

Palmer, G. C. and Manian, A. A. (1974b) Effects of phenothiazines and phenothiazine metabolites on adenyl cyclase and the cyclic AMP response in the rat brain. In *The Phenothiazines and Structurally Related Drugs* (Eds. I. S. Forrest, C. J. Carr and E. Usdin), Raven Press, New York, pp. 749–767

Palmer, G. C., Robison, G. A., Manian, A. A. and Sulser, F. (1972) Modification by psychotropic drugs of the cyclic AMP response to norepinephrine in the rat brain *in vitro*. Psychopharmacologia, **23**, 201–211

Palmer, G. C., Robison, G. A. and Sulser, F. (1971) Modification by psychotropic drugs of the cyclic adenosine monophosphate response to norepinephrine in rat brain. *Biochem. Pharmac.*, **20**, 236–239

Palmer, G. C. and Scott, H. R. (1974) The cyclic AMP response to noradrenalin in young adult rat brain following post-natal injections of 6-hydroxydopamine. *Experientia*, **30**, 520–521

Palmer, G. C., Sulser, F. and Robison, G. A. (1973) Effects of neurohumoral and adrenergic agents on cyclic AMP levels in various areas of the rat brain *in vitro*. Neuropharmacology, **12**, 327–337

Panitz, N. (1974) Thesis, Technische Universität, Braunschweig, Germany

Panitz, N., Rieke, E., Morr, M., Wagner, K. G., Roesler, G. and Jastorff, B. (1975) The 3′-amido and 5′-amido analogues of adenosine 3′,5′-monophosphate: Interaction with cAMP-specific proteins. *Europ. J. Biochem.*, **55**, 415–422

Papeschi, R. A., Randrup, A. and Lal, S. (1974) Effect of ECT on dopaminergic mechanisms. *Psychopharmacologia*, **35**, 149–158

Pardee, A. B. (1974) A restriction point for control of normal animal cell proliferation. *Proc. natn. Acad. Sci., U.S.A.*, **71**, 1286–1290

Parfitt, A., Weller, J. L., Klein, D. C., Sakai, K. K. and Marks, B. H. (1975) Blockade by ouabain or elevated potassium ion concentration of the adrenergic and adenosine cyclic 3′,5′-monophosphate-induced stimulation of pineal serotonin *N*-acetyltransferase activity. *Molec. Pharmac.*, **11**, 241–245

Park, C. R., Lewis, S. B. and Exton, J. H. (1972) Relationship of some hepatic actions of insulin to the intracellular level of cyclic adenylate. *Diabetes*, **21**, Suppl. 2, 439–446

Parker, C. W. (1971) The nature of immunological responses and antigen-antibody interactions. In *Principles of Competitive Protein-Binding Assays* (Eds. W. D. Odell and W. H. Daughaday), J. B. Lippincott, Philadelphia, pp. 25–48

Parker, C. W. (1974a) Correlation between mitogenicity and stimulation of calcium uptake in human lymphocytes. *Biochem. biophys. Res. Commun.*, **61**, 1180–1186

Parker, C. W. (1974b) Complexities in the study of the role of cyclic nucleotides in lymphocyte responses to mitogens. In *Cyclic AMP, Cell Growth, and the Immune Response* (Eds. W. Braun, L. Lichtenstein and C. W. Parker), Springer-Verlag, New York, pp. 35–44

Parker, C. W. (1976) The role of cyclic AMP in immunological inflammation. *J. invest. Derm.*, **67**, 638–640

Parker, C. W., Dankner, R. E., Falkenhein, S. F. and Greene, W. C. (1976) Suggestive evidence for both stimulatory and inhibitory domains on human lymphocytes, as indicated by phospholipid turnover studies with wheat germ agglutinin and other lectins. *Immunol. Commun.*, **5**, 13–25

Parker, C. W. and Smith, J. W. (1973) Alterations in cyclic AMP metabolism in human bronchial asthma. I. Leukocyte responsiveness to β-adrenergic agents. *J. clin. Invest.*, **52**, 48—59

Parker, C. W., Snider, D. E. and Wedner, H. J. (1974b) The role of cyclic nucleotides in lymphocyte activation. In *Progress in Immunology*, Vol. 2 (Eds. L. Brent and J. Holborow), North-Holland Publishing Co., New York, pp. 85–94

Parker, C. W., Sullivan, T. J. and Wedner, H. J. (1974a) Cyclic AMP and the immune response. In *Advances in Cyclic Nucleotide Research*, Vol. 4 (Eds. P. Greengard and G. A. Robison), Raven Press, New York, pp. 1–79

Pastan, I., Anderson, W. B., Carchman, R. A., Willingham, M. C., Russell, T. R. and Johnson, G. S. (1974) Cyclic AMP and malignant transformation. In *Control of Proliferation in Animal Cells* (Eds. B. Clarkson and R. Baserga), Cold Spring Harbor Laboratory, New York, pp. 563–570

Pastan, I., Johnson, G. S. and Anderson, W. B. (1975) Role of cyclic nucleotides in growth control. *Ann. Rev. Biochem.*, **44**, 491–522

Pastan, I. and Katzen, R. (1967) Activation of adenyl cyclase in thyroid homogenates by thyroid stimulating hormone. *Biochem. biophys. Res. Comm.*, **29**, 792–798

Pastan, I. and Perlman, R. (1070) Cyclic adenosine monophosphate in bacteria. *Science, N.Y.*, **169**, 339–344

Pastan, I., Pricer, W. and Blanchette-Mackie, J. (1970) Studies on an ACTH-activated adenyl cyclase from a mouse adrenal tumor. *Metabolism*, **19**, 809–817

Pastan, I. and Wollman, S. H. (1967) Colloid droplets formation in dog thyroid *in vitro. J. Cell Biol.*, **35**, 262–266

Patten, R. L. (1970) The reciprocal regulation of lipoprotein lipase activity and hormone-sensitive lipase activity in rat adipocytes. *J. biol. Chem.*, **245**, 5557–5584

Paul, D., Henahan, M. and Walter, S. (1974) Changes in growth control and growth requirements associated with neoplastic transformation *in vitro. J. natn. Cancer Inst., U.S.A.*, **53**, 1499–1503

Pawelek, J., Sansone, M., Koch, N., Christie, G., Halaban, R., Hendee, J., Lerner, A. B. and Varga, J. M. (1975) Melanoma cells resistant to inhibition of growth by melanocyte stimulating hormone. *Proc. natn. Acad. Sci., U.S.A.*, **72**, 951–955

Pawelek, J., Wong, G., Sansone, M. and Morowitz, J. (1973) Molecular biology of pigment cells: Molecular controls in mammalian pigmentation. *Yale J. Biol. Med.*, **45**, 530–543

Pawlson, L. G., Lovell-Smith, C. J., Manganiello, V. C. and Vaughan, M. (1974) Effects of epinephrine, adrenocorticotrophic hormone, and theophylline on adenosine 3′,5′-monophosphate phosphodiesterase activity in fat cells. *Proc. natn. Acad. Sci., U.S.A.*, **71**, 1639–1642

Peake, G. T., Steiner, A. L. and Daughaday, W. H. (1972) Guanosine 3′,5′-cyclic monophosphate is a potent pituitary growth hormone secretagogue. *Endocrinology*, **90**, 212–216

Pechere, J. F., Demaille, J., Capony, J. P., Dutruge, E., Baron, G. and Pina, C. (1975) Muscular Parvalbumins. Some explorations into their possible biological significance. In *Calcium Transport in Contraction and Secretion* (Eds. E. Carafoli *et al.*), Elsevier, New York, pp. 459–467

Peery, C. V., Johnson, G. S. and Pastan, I. (1971) Adenyl cyclase in normal and transformed fibroblasts in tissue culture. *J. biol. Chem.*, **246**, 5785–5790

Penit, I., Jard, S. and Benda, P. (1974) Probenecid sensitive 3′,5′-cyclic AMP secretion by isoproterenol-stimulated glial cells in culture. *FEBS Letters*, **41**, 156–160

Perkins, J. P. (1973) Adenyl cyclase. In *Advances in Cyclic Nucleotide Research*, Vol. 3 (Eds. P. Greengard, R. Paoletti and G. A. Robison), Raven Press, New York, pp. 1–64

Perkins, J. P., MacIntyre, H. E., Riley, W. D. and Clark, R. B. (1971) Adenyl cyclase, phosphodiesterase, and cyclic AMP dependent protein kinase of malignant glial cells in culture. *Life Sci.*, **10**, 1069–1080

Perkins, J. P. and Moore, M. M. (1971) Adenyl cyclase of rat cerebral cortex: Activation by sodium fluoride and detergents. *J. biol. Chem.*, **246**, 62–68

Perkins, J. P. and Moore, M. M. (1973a) Characterization of the adrenergic receptors mediating a rise in cyclic 3′,5′-adenosine monophosphate in rat cerebral cortex. *J. Pharmac. exp. Therap.*, **185**, 371–378

Perkins, J. P. and Moore, M. M. (1973b) Regulation of the adenosine cyclic 3′,5′-monophosphate content of rat cerebral cortex: Ontogenetic development of the responsiveness to catecholamines and adenosine. *Molec. Pharmac.*, **9**, 774–782

Perrier, C. V. and Laster, L. (1969) Adenyl cyclase activity of guinea-pig gastric mucosa. *Clin. Res.*, **17**, 596

Perry, S. V. (1975) The contractile and regulatory proteins of the myocardium. In *Contraction and Relaxation of the Myocardium* (Ed. W. G. Nayler), Academic Press, London, pp. 29–77

Peterkofsky, B. and Prather, W. B. (1974) Increased collagen synthesis in Kirsten sarcoma virus-transformed Balb-3T3 cells grown in the presence of dibutyryl-cyclic AMP. *Cell*, **3**, 291–299

Peters, H. D., Dinnendahl, V. and Schönhöfer, P. S. (1975) Mode of action of antirheumatic drugs on the cyclic 3′,5′-AMP-regulated glycosaminoglycan secretion in fibroblasts. *Naunyn-Schmiedebergs Arch. exp. Path. Pharmak.*, **289**, 29–40

Peters, H. D., Karzel, K., Padberg, D., Schönhöfer, P. S. and Dinnendahl, V. (1974) Influence of prostaglandin E_1 on cyclic 3′,5′-AMP levels and glycosaminoglycan secretion of fibroblasts cultured *in vitro*. *Pol. J. Pharmac. Pharm.*, **26**, 41–49

Pfeiffer, D. R., Reed, P. W. and Lardy, H. A. (1974) Ultraviolet and fluorescent spectral properties of the divalent cation ionophore A-23187 and its metal ion complexes. *Biochemistry*, **13**, 4007–4014

Pfeuffer, T. and Helmreich, E. J. M. (1975) Activation of pigeon erythrocyte membrane adenylate cyclase by guanylnucleotide analogues and separation of a nucleotide binding protein. *J. biol. Chem.*, **250**, 867–876

Phillips, P. G., Furmanski, P. and Lubin, M. (1974) Cell surface interactions with concanavalin A. *Expl. Cell Res.*, **86**, 301–308

Phillis, J. W. and Kostopoulos, G. K. (1975) Adenosine as a putative transmitter in the cerebral cortex. Studies with potentiators and antagonists. *Life Sci.*, **17**, 1085–1094

Phillis, J. W., Kostopoulos, G. K. and Limacher, J. J. (1974) Depression of corticospinal cells by various purines and pyrimidines. *Can. J. Physiol. Pharmac.*, **52**, 1227–1229

Phillis, J. W., Kostopoulos, G. K. and Limacher, J. J. (1975) A potent depressant action of adenine derivatives on cerebral cortical neurons. *Europ. J. Pharmac.*, **30**, 125–129

Pierce, N. F., Greenough, W. B. and Carpenter, C. C. J. (1971) *Vibrio cholera* enterotoxin and its mode of action. *Bacteriol. Rev.*, **35**, 1–13

Pilkis, S. J., and Park, C. R. (1974) Mechanism of action of insulin. *Ann. Rev. Pharmacol.*, **14**, 365–388

Plagemann, P. G. and Erbe, J. (1974) The deoxyribonucleoside transport systems of cultured Novikoff rat hepatoma cells. *J. Cell Physiol.*, **83**, 337–344

Plunkett, W. and Cohen, S. S. (1975) Two approaches that increase the activity of analogs of adenine nucleosides in animal cells. *Cancer Res.*, **33**, 1547–1554

Pöch, G. and Kukovetz, W. R. (1972) Studies on the possible role of cyclic AMP in drug-induced coronary vasodilatation. In *Advances in Cyclic Nucleotide Research,* Vol. 1 (Eds. P. Greengard and G. A. Robison), Raven Press, New York, pp. 195–211

Pochet, R., Boeynaems, J. M. and Dumont, J. E. (1974) Stimulation by thyrotrophin of horse thyroid plasma membrane adenylate cyclase: Evidence of co-operativity. *Biochem. biophys. Res. Commun.*, **58**, 446—453

Pogo, B. G. T. (1972) DNA-dependent RNA polymerase activities in isolated lymphocyte nuclei. *J. Cell Biol.*, **53**, 635–641

Pohl, S. L., Birnbaumer, L. and Rodbell, M. (1969) Glucagon-sensitive adenyl cyclase in plasma membranes of hepatic parenchymal cells. *Science*, **164**, 566–567

Pohl, S. L., Birnbaumer, L. and Rodbell, M. (1971a) The glucagon-sensitive adenyl cyclase system in plasma membranes of rat liver. *J. biol. Chem.*, **246**, 1849–1856

Pohl, S. L., Krans, H. M. J., Kozyreff, V., Birnbaumer, L. and Rodbell, M. (1971b) The glucagon-sensitive adenyl cyclase system in plasma membranes of rat liver. VI. Evidence for a role of membrane lipids. *J. biol. Chem.*, **246**, 4447–4454

Poirier, G., Baden, N., Labrie, F., Borgeat, P. and DeLean, A. (1972) Partial purification and some properties of adenyl cyclase and receptor for TRH from anterior pituitary gland. *Proc. 4th Int. Congr. Endocrinol.*

Polgar, P., Vera, J. C., Kelley, P. R. and Rutenburg, A. M. (1973) Adenylate cyclase activity in normal and leukemic human leukocytes as determined by a radioimmunoassay for cyclic AMP. *Biochim. biophys. Acta*, **297**, 378–383

Pollack, R., Osborn, M. and Weber, K. (1975) Patterns of organization of actin and myosin in normal and transformed cultured cells. *Proc. natn. Acad. Sci., U.S.A.*, **72**, 994–998

Pomerantz, A. H., Rudolph, S. A., Haley, B. E. and Greengard, P. (1975) Photoaffinity labeling of a protein kinase from bovine brain with 8-azido-adenosine 3′,5′-monophosphate. *Biochemistry*, **14**, 3858–3862

Popa, G. T. and Fielding, J. (1930) A portal circulation from the pituitary to the hypothalamic region. *J. Anat.*, **65**, 88–91

Porohecky, L. S. (1974) Effect of ethanol on central and peripheral noradrenergic neurons. *J. Pharmac. exp. Ther.*, **189**, 380–391

Porter, K. D., Todaro, G. J. and Fonte, V. (1973) A scanning electron microscope study of surface features of vital and spontaneous transformants of mouse BALB/3T3 cells. *J. Cell Biol.*, **59**, 633–642

Posner, J. B., Stern, R. and Krebs, E. G. (1965) Effects of electrical stimulation and epinephrine on muscle phosphorylase, phosphorylase *b* kinase and cyclic AMP. *J. biol. Chem.*, **240**, 982–985

Posternak, T. and Cehovic, G. (1971) Derivatives and analogues of cyclic nucleotides. *Ann. N.Y. Acad. Sci.*, **185**, 42–49

Posternak, T., Marcus, I. and Cehovic, G. (1971) Préparation de nouveaux dérivés de l'AMPc (substitués en position C-8 et C-2) et étude de leur action sur la libération de l'hormone de croissance. *C. r. hebd. Séanc. Acad. Sci.*, **272**, 622–625

Posternak, T., Marcus, I., Gabbai, A. and Cehovic, G. (1969) Préparation et étude de quelques propriétes biologiques d'analogues de l'acide adenosine-3′,5′-phosphorique. *C. r. hebd. Séanc. Acad. Sci.*, **269**, 2409–2412

Posternak, F., Posternak, T., Orusco, M., Cehovic, G. and Laugier, P. (1976) Essais de traitement du psoriasis par injections intra-lesionnelles d'AMPc, de dibutyryl-8-thiol-AMPc, et de theophylline. *Bull. Soc. fr. Derm. Syph.* (in press)

Posternak, T., Sutherland, E. W. and Henion, W. F. (1962) Derivatives of cyclic 3′,5′-adenosine monophosphate. *Biochim. biophys. Acta.*, 558–560

Potter, L. T. (1967) Uptake of propranolol by isolated guinea-pig atria. *J. Pharmac. exp. Ther.*, **155**, 91–100

Powell, E. W. and Hines, G. (1974) The limbic sytem: An interface. *Behav. Biol.*, **12**, 149–164

Prasad, K. N. (1972) Morphological differentiation induced by prostaglandin in mouse neuroblastoma cells in culture. *Nature, New Biol.*, **236**, 49–52

Prasad, K. N. and Gilmer, K. N. (1974) Demonstration of dopamine-sensitive adenylate cyclase in malignant neuroblastoma cells and change in sensitivity of adenylate cyclase to catecholamines in "differentiated" cells. *Proc. natn. Acad. Sci., U.S.A.*, **71**, 2525–2529

Prasad, K. N., Gilmer, K. N., Sahu, S. K. and Becker, G. (1974) Demonstration of acetylcholine-sensititive adenyl cyclase in malignant neuroblastoma cells in culture. *Nature, Lond.*, **249**, 765–767

Prasad, K. N., Gilmer, K. N., Sahu, S. K. and Becker, G. (1975) Effect of neurotransmitters, guanosine triphosphate, and divalent ions on the regulation of adenylate cyclase activity in malignant and adenosine cyclic 3′,5′-monophosphate-induced "differentiated" neuroblastoma cells. *Cancer Res.*, **35**, 77–81

Prasad, K. N. and Kumar, S. (1973) Cyclic 3′, 5′-AMP phosphodiesterase activity during cyclic AMP-induced differentiation of neuroblastoma cells in culture. *Proc. Soc. exp. Biol. Med.*, **142**, 406–409

Prasad, K. N. and Mandal, B. (1973) Choline acetyltransferase levels in cyclic AMP and X-ray-induced morphologically differentiated neuroblastoma cells in culture. *Cytobiol.*, **8, 75**–80

Pratje, E. and Heilmeyer, L. M. G. (1972) Phosphorylation of rabbit muscle troponin and actin by a 3′,5′-cAMP-dependent protein kinase. *FEBS Letters*, **27**, 89–93

Pratt, R. M. and Martin, G. R. (1975) Epithelial cell death and cyclic AMP increase during palatal development. *Proc. natn. Acad. Sci., U.S.A.*, **72**, 874–877

Prehn, P. T. (1974) Immunological surveillance: Pro and con. *Clinical Immunobiology*, Vol. 2 (Eds. R. A. Good and F. Bach), pp. 191–203

Prince, W. T. and Berridge, M. J. (1973) The tole of calcium in the action of 5-hydroxytryptamine and cyclic AMP on salivary glands. *J. exp. Biol.*, **58**, 367–384

Prince, W. T., Berridge, M. J. and Rasmussen, H. (1972) Role of calcium and adenosine-3′,5′-cyclic monophosphate in controlling fly salivary gland secretion. *Proc. natn. Acad. Sci., U.S.A.*, **69**, 553–557

Prosser, C. L. (1974) Smooth muscle. *Ann. Rev. Physiol.*, **36**, 503–535

Pruzansky, J. J. and Patterson, R. (1970) Decrease in basophils after incubation with specific antigens of leukocytes from allergic donors. *Int. Archs. Allergy appl. Immunol.*, **38**, 522–526

Puchwein, G., Pfeuffer, T. and Helmreich, E. J. M. (1974) Uncoupling of catecholamine activation of pigeon erythrocyte membrane adenylate cyclase. *J. biol. Chem.*, **249**, 3232–3240

Puglisi, L., Berti, F. and Folco, G. C. (1972) Cyclic–GMP interaction with the parasympathetic system of isolated rat stomach. *Pharmac. Res. Commun.*, **4**, 227

Pull, I. and McIlwain, H. (1972a) Adenine derivatives as neurohumoral agents in the brain: The quantities liberated on excitation of superfused cerebral tissues. *Biochem. J.*, **130**, 975–981

Pull, I. and McIlwain, H. (1972b) Metabolism of[^{14}C]adenine and derivatives by cerebral tissues, superfused and electrically stimulated. *Biochem. J.*, **126**, 965–973

Puri, S. K., Cochin, J. and Volicer, L. (1975) Effect of morphine sulfate on adenylate cyclase and phosphodiesterase activities in rat corpus striatum. *Life Sci.*, **16**, 759–768

Purpura, D. P. and Shofer, R. J. (1972) Excitatory action of dibutyryl- cyclic adenosine monophosphate on immature cerebral cortex. *Brain Res.*, **38**, 179–181

Puszkin, E., Puszkin, S., Lo, L. W. and Tanenbaum, S. W. (1973) Binding of cytochalasin D to platelet and muscle myosin. *J. biol. Chem.*, **248**, 7754–7761

Rabinowitz, B., Kligerman, M. and Parmley, W. W. (1974) Plasma cyclic adenosine 3′,5′-monophosphate levels in acute myocardial infarction. *Am. J. Cardiol.*, **34**, 7–11

Rabinowitz, B., Parmley, W. W., Kligerman, M., Norman, J., Fujimura, S., Chiba, S. and Matloff, J. M. (1975) Myocardial and plasma levels of adenosine 3′,5′-cyclic phosphate: Studies in experimental myocardial ischemia. *Chest*, **68**, 69–74

Rabinowitz, M. and Zak, R. (1972) Biochemical and cellular changes in cardiac hypertrophy. *Ann. Rev. Med.*, **23**, 245–262

Racker, E. (1973) A new procedure for the reconstitution of biologically active phospholipid vesicles. *Biochem. biophys. Res. Commun.*, **55**, 224–230

Radzialowski, F. M. and Rosenberg, L. N. (1973) Effect of SC–19220, a prostaglandin inhibitor of the antilipolytic action of prostaglandin E_2, propranolol and insulin in the isolated rat adipocyte. *Life Sci.*, **12**, 337–343

Rajerison, R., Marchetti, J., Roy, C., Bockaert, J. and Jard, S. (1974) The vasopressin-sensitive adenylate cyclase of the rat kidney. Effect of adrenalectomy and corticosteroids on hormonal receptor-enzyme coupling. *J. Biol. Chem.*, **249**, 6390–6400

Rall, T. W. (1973) The metabolism and function of cyclic AMP in the central nervous system. In *Prostaglandins and Cyclic AMP* (Eds. R. H. Kahn and W. E. M. Lands), Academic Press, New York, pp. 57–72

Rall, T. W. and Gilman, A. G. (1970) The role of cyclic AMP in the nervous system. *Neurosci. Res. Bull.*, **8**, 221–323

Rall, T. W. and Sattin, A. (1970) Factors influencing the accumulation of cyclic AMP in brain tissue. *Adv. Biochem. Psychopharmac.*, **3**, 113–133

Rall, T. W. and Sutherland, E. W. (1958) Formation of cyclic adenine nucleotide by tissue particles. *J. biol. Chem.*, **232**, 1065

Rall, T. W. and Sutherland, E. W. (1962) Adenyl cyclase. II. The enzymatically catalysed formation of adenosine 3′,5′-phosphate and inorganic pyrophosphate from adenosine triphosphate. *J. biol. Chem.*, **237**, 1288–1232

Ramseyer, J., Kaslow, H. R. and Gill, G. N. (1974) Purification of the cAMP receptor protein by affinity chromatography. *Biochem. biophys. Res. Commun.*, **49**, 813–821

Rao, G. J. S., Del Monte, M. and Nadler, H. L. (1971) Adenyl cyclase activity in cultivated human skin fibroblasts. *Nature, New Biol.*, **232**, 253–255

Rapoport, B. and De Groot, L. J. (1972) Cyclic nucleotide dependent protein kinase in the thyroid. *Endocrinology*, **91**, 1259–1266

Rapoport, B., West, M. N. and Ingbar, S. H. (1975) Inhibitory effect of dietary iodine on the thyroid adenylate cyclase response to thyrotropin in the hypophysectomized rat. *J. clin. Invest.*, **56**, 516–519

Rappaport, S., Leterrier, J. F. and Nunez, J. (1971) Protéines phosphokinases du tissu thyroïdien, *Biochimie*, **53**, 721–726

Rasmussen, H. (1970) Cell communication, calcium ion, and cyclic adenosine monophosphate. *Science*, N.Y., **173**, 404–412

Rasmussen, H., Goodman, D. B. P. and Tenenhouse, A. (1972) The role of cyclic AMP and calcium in cell activation. *CRC Crit. Rev. Biochem.*, **1**, 95–148

Rasmussen, H., Jensen, P., Lake, W., Friedman, N. and Goodman, B. D. (1975) Cyclic nucleotides and cellular Ca^{2+} metabolism. In *Advances in Cyclic Nucleotide Research*, Vol. 5 (Eds. G. I. Drummond, P. Greengard and G. A. Robison), Raven Press, New York, pp. 375–394

Rasmussen, H. and Nagata, N. (1970) Hormones, cell calcium and cyclic AMP. In *Calcium and Cellular Function* (Ed. A. W. Cuthbert), Macmillan, London, pp. 198–213

Rasmussen, H. and Tenenhouse, A. (1968) Cyclic adenosine monophosphate, Ca and membranes. *Proc. natn. Acad. Sci.*, U.S.A., **59**, 1364–1370

Ratner, A. (1970) Stimulation of luteinizing hormone release *in vitro* by dibutyryl-cyclic AMP and theophylline. *Life Sci.*, **9**, 1221–1226

Ray, T. K. and Forte, J. G. (1974) Adenyl cyclase of oxyntic cells. Its association with different cellular membranes. *Biochim. biophys. Acta*, **363**, 320–339

Ray, T. K., Tomasi, V. and Marinette, G. V. (1970) Properties of adenyl cyclase in isolated plasma membranes of rat liver. *Biochim. biophys. Acta*, **211**, 20–30

Rebhun, L. I. and Villar-Palasi, C. (1973) Stimulation of purified muscle protein kinase by cyclic AMP and its butyrated derivatives. *Biochim. biophys. Acta*, **321**, 165–170

Reddy, Y. S., Ballard, D., Giri, N. Y. and Schwartz, A. (1973) Phosphorylation of cardiac native tropomyin and troponin: Inhibitory effect of actomyosin and possible presence of endogenous myofibrillar-located, cyclic AMP-dependent protein kinase. *J. molec. cell. Cardiol*, **5**, 461–471

Remington, J. A. and Klevecz, R. R. (1973) Hormone-treated CHO cells exit the cell cycle in the G_2 phase. *Biochim. biophys. Res. Commun.*, **50**, 140–146

Rendell, M., Salomon, Y., Lin, M. C., Rodbell, M. and Berman, M. (1975) The hepatic adenylate cyclase system. III. A mathematical model for the steady state kinetics of catalysis and nucleotide regulation. *J. biol. Chem.*, **250**, 4235–4260

Renner, R., Kemmler, W. and Hepp, K. D. (1974) Anatagonism of insulin and lipolytic hormones in the control of adenylate cyclase activity in fat cells. *Europ. J. Biochem.*, **49**, 129–141

Repine, J. E., White, J. G., Clawson, C. C. and Holmes, B. M. (1974) The influence of phorbol myristate acetate on oxygen consumption by polymorphonuclear leukocytes. *J. Lab. clin. Med.*, **83**, 911–920

Rethy, A., Tomasi, V., Trevisani, A. and Barnabei, O. (1972) The role of phophatidylserine in the hormonal control of adenylate cyclase of rat liver plasma membranes. *Biochim. biophys. Acta.*, **290**, 58

Rethy, A., Varci, L., Toth, F. D. and Boldogh, I. (1973) Abnormal distribution of adenylate cyclase. In *The Role of Cyclic Nucleotides in Carcinogenesis* (Eds. J. Schultz and H. G. Gratzer), Academic Press, New York, pp. 153–157

Reuter, H. (1967) The dependence of slow inward current in Purkinje fibres on the extracellular calcium concentration. *J. Physiol., Lond.*, **192**, 479–492

Reuter, H. (1973) Divalent cations as charge carriers in excitable membranes. *Prog. Biophys. molec. Biol.*, **26**, 3–43

Reuter, H. (1974a) Localization of beta adrenergic receptors, and effects of noradrenaline and cyclic nucleotides on action potentials, ionic currents and tension in mammalian cardiac muscle. *J. Physiol., Lond.*, **242**, 429–451

Reuter, H. (1974b) Exchange of calcium ions in the mammalian myocardium. *Circul Res.*, **34**, 599–605

Revankar, G. R., Huffman, J. H. Allen, L. B. Sidwell, R. W., Robins, R. K. and Tolman, R. L. (1975) Synthesis and antiviral activity of certain 5′-monophosphates of 9-D-arabinofuranosyladenine and 9-D-arabinofuranosylhypoxanthine. *J. med. Chem.*, **18**, 721–726

Richards, J. S. and Midgley, R. A. (1976) Protein hormone action: A key to understanding ovarian follicular and luteal development. Biol. Reprod. (in press)

Richelson, E. (1973) Stimulation of tyrosine hydroxylase activity in an adrenergic clone of mouse neuroblastoma by dibutyryl-cyclic AMP. *Nature, New Biol.*, **242**, 175–177

Rieke, E., Panitz, N., Eigel, A. and Wagner, K. G. (1975) On the detachment of the regulatory subunit of brain protein kinase from a cAMP-polyacrylamide gel. *Hoppe-Seyler's Z. physiol. Chem.*, **356**, 1177–1179

Rivkin, I., Rosenblatt, J. and Becker, E. L. (1975) The role of cyclic AMP in the chemotactic responsiveness and spontaneous motility of rabbit peritoneal neutrophils: The inhibition of neutrophil movement and the elevation of cyclic AMP levels by catecholamines, prostaglandins, theophylline and cholera toxin. *J. Immunol.*, **115**, 1126–1134

Rixon, R. H., Whitfield, J. F. and MacManus, J. P. (1970) Stimulation of mitotic activity in rat bone marrow and thymus by exogenous adenosine 3′,5′-monophosphate (cyclic AMP). *Expl. Cell Res.*, **63**, 110–116

Robinson, D. S. and Wing, D . D. (1970) Regulation of adipose tissue clearing factor lipase activity. In *Adipose Tissue* (Eds. B. Jeanrenaud and D. Hepp), Academic Press, New York, pp. 41–49

Robinson, S. E. and Sulser, F. (1975) Effect of metaclopramide on limbic noradrenergic and striatal and limbic dopaminergic mechanisms. *Pharmacologist*, **17**, 448

Robinson, S. E. and Sulser, F. (1976) The noradrenergic cyclic AMP generating system of the rat limbic forebrain and its stereospecificity for butaclamol. *J. Pharm. Pharmac.*, **28**, 645–646

Robison, G. A., Butcher, R. W. and Sutherland, E. W. (1967) Adenyl cyclase as an adrenergic receptor. *Ann. N.Y. Acad. Sci.*, **139**, 703

Robison, G. A., Butcher, R. W. and Sutherland, E. W. (1968) Cyclic AMP. *Ann. Rev. Biochem.*, **37**, 149–174

Robison, G. A., Butcher, R. W. and Sutherland, E. W. (1971) The catecholamines. In *Cyclic AMP* (Eds. G. A. Robison, R. W. Butcher and E. W. Sutherland), Academic Press, New York, pp. 146–231

Rodbard, D. (1974) Apparent positive co-operative effects in cyclic AMP and corticosterone production by isolated adrenal cells in response to ACTH analogues. *Endocrinology*, **9**, 1427–1437

Rodbell, M. (1965) Modulation of lipolysis in adipose tissue by fatty acid concentration in fat cells. *Ann. N.Y. Acad. Sci.*, **131**, 302–333

Rodbell, M. (1967) Metabolism of isolated fat cells. V. Preparation of 'ghosts' and their properties; adenyl cyclase and other enzymes. *J. biol. Chem.*, **242**, 5744–5750

Rodbell, M. (1075) On the mechanism of activation of fat cell adenylate cyclase by guanine nucleotides. *J. biol. Chem.*, **250**, 5826–5834

Rodbell, M., Birnbaumer, L. and Pohl, S. L. (1970) Adenyl cyclase in fat cells. III. Stimulation by secretin and the effects of trypsin on the receptors for lipolytic hormones. *J. biol. Chem.*, **245**, 718–722

Rodbell, M., Birnbaumer, L. and Pohl, S. L. (1971) Hormones, receptors, and adenyl cyclase activity in mammalian cells. In *The Role of Adenyl Cyclase and Cyclic 3′,5′-AMP in Biological Systems* (Eds. T. W. Rall, M. Rodbell and P. Condliffe), U.S. Govt. Printing Office, Washington, pp. 50–103

Rodbell, M., Birnbaumer, L., Pohl, S. L. and Krans, H. M. J. (1971c) The glucagon-sensitive adenyl cyclase system in plasma membranes of rat liver. V. An obligatory role of guanyl nucleotides in glucagon action. *J. biol. Chem.*, **246**, 1877–1882

Rodbell, M., Birnbaumer, L., Pohl, S. L. and Sundby, F. (1971b) The reaction of glucagon with its receptor: Evidence for discrete regions of activity and binding in the glucagon molecule. *Proc. natn. Acad. Sci., U.S.A.*, **68**, 909–913

Rodbell, M., Krans, H. M. J., Pohl, S. L. and Birnbaumer, L. (1971d) The glucagon-sensitive adenyl cyclase system in plasma membranes of rat liver. III. Binding of glucagon: Method of assay and specificity. *J. biol. Chem.*, **246**, 1861–1871

Rodbell, M., Krans, H. M. J., Pohl, S. L. and Birnbaumer, L. (1971e) The glucagon-sensitive adenyl cyclase system in plasma membranes of rat liver. IV. Effects of guanylnucleotides on binding of ^{125}I-glucagon. *J. biol. Chem.*, **246**, 1872–1876

Rodbell, M., Lin, M. C. and Salomon, Y. (1974) Evidence for interdependent action of glucagon and nucleotides on the hepatic adenylate cyclase system. *J. biol. Chem.*, **249**, 59–65

Rodbell, M., Lin, M. C., Salomon, Y., Londos, C., Harwood, J. P., Martin, B. R., Rendell, M. and Berman, M. (1975) Role of adenine and guanine nucleotides in the activity and response of adenylate cyclase systems to hormones: Evidence for multisite transition states. In *Advances in Cyclic Nucleotides Research* (Eds. G. T. Drummond, P. Greengard and G. A. Robison), Vol. 5, Raven Press, New York, pp. 3–29

Rodesch, F., Bogaert, C. and Dumont, J. E. (1974) Stimulation par l'hormone thyréotrope de la mobilisation du calcium thyroidien. *C. r. hebd. Séanc. Acad Sci.*, **278**, 931–934

Rodesch, F., Neve, P., Willems, C. and Dumont, J. E. (1969) Stimulation of thyroid metabolism by throtropin, cyclic 3′,5′-AMP, dibutyryl-cyclic 3′,5′-AMP and prostaglandin E_1. *Europ. J. Biochem.*, **8**, 26

R-O'Donnell, E. (1974) Stimulation and desensitization of macrophage adenylate cyclase by prostaglandins and catecholamines. *J. biol. Chem.*, **249**, 3615–3621

Rogers, M., Dismukes, K. and Daly, J. W. (1975) Histamine-elicited accumulations of cyclic adenosine 3′,5′-monophosphate in guinea-pig brain slices: Effect of H^1 and H_2-antagonists. *J. Neurochem.*, **25**, 531–534

Roisen, F. J., Murphy, R. A. and Braden, W. G. (1972) Dibutyryl-cyclic adenosine monophosphate stimulation of colcemid-inhibited axonal elongation. *Science, N.Y.*, **177**, 809–811

Roizen, M. F., Weise, V., Grobecker, H. and Kopin, I. J. (1975) Plasma catecholamines and dopamine-β-hydroxylase activity in spontaneously hypertensive rats. *Life Sci.*, **17**, 283–288

Roller, B. A., Hirai, K. and Defendi, V. (1974) Effect of cAMP on nucleoside metabolism. II. Cell cycle dependence of thymidine transport. *J. Cell Physiol.*, **84**, 333–342

Romero, P. J. and Whittam, R. (1971) The control by internal calcium of membrane permeability to sodium and potassium. *J. Physiol., Lond.*, **214**, 481–507

Roques, M., Tirard, A. and Torresani, L. (1973) Distribution subcellulaire de protéines kinases stimulables par l'AMP cyclique dans la thyroïde porc. *Biochimie*, **55**, 1421–1430

Rosen, O. M. and Erlichman, J. (1975) Reversible autophosphorylation of a cyclic AMP dependent protein kinase from bovine cardiac muscle. *J. biol. Chem.*, **250**, 7788–7794

Rosen, O. M. and Rosen, S. M. (1969) Properties of an adenyl cyclase partially purified from frog erthrocytes. *Archs. Biochem. Biophys*, **131**, 449–456

Rosen, O. M., Rubin, C. S. and Erlichman, J. (1973) Molecular characterisation of cyclic AMP dependent protein kinases derived from bovine heart and human erythrocytes. In *Protein Phosphorylation in Control Mechanisms*, Vol. 5 (Eds. F. Huijing and E. Y. C. Lee), pp. 67–79

Rosengurt, E. and Pardee, A. B. (1972) Opposite effects of dibutyryl-adenosine 3′,5′-cyclic monophosphate and serum on growth of Chinese hamster cells. *J. Cell Physiol.*, **80**, 273–279

Rosenthal, A. S. (1975) *Immune recognition* (Ed. A. S. Rosenthal), Academic Press, New York

Rosenthal, J. W. and Goldstein, S. (1975) The effect of insulin on basal and hormone-induced elevation of cyclic AMP content in cultured human fibroblasts. *J. Cell Physio.*, **85**, 235–242

Rosloff, B. N. and Davis, J. M. (1974) Effect of iprindole on norepinephrine turnover and transport, *Psychopharmacologia*, **40**, 53–64

Rosman, P. M., Agrawal, R., Goodman, A. D. and Steiner, A. L. (1974) Effects of angiotensin on cyclic GMP and cyclic AMP in human plasma. *Clin. Res.*, **22**, 712A

Roth, J. A. and Ivy, A. C. (1944) The effect of caffeine on gastric secretion in dog, cat and man. *Am. J. Physiol.*, **141**, 454–461

Rotrosen, J., Friedman, E. and Gershon, S. (1975) Striatal adenylate cyclase activity following reserpine and chronic chlorpromazine administration in rats. *Life Sci.*, **17**, 563–568

Rubalcalva, B. and Rodbell, M. (1973) The role of acidic phospholipids in glucagon action on rat liver adenylate cyclase. *J. biol. Chem.*, **248**, 3831

Rubin, C. S. and Rosen, O. M. (1975) Protein phosphorylation. *Ann. Rev. Biochem.*, **44**, 831–887

Rubin, B., O'Keefe, E. H., Waugh, M. H., Kotler, D. G., DeMaio, D. A. and Horovitz, Z. P. (1971) Activities *in vitro* of 8-substituted derivatives of adenosine-3′,5′-cyclic monophosphate on guinea-pig trachea and rat portal vein. *Proc. Soc. exp. Biol. Med.*, **137**, 1244–1248

Rubin, R. P. (1974) *Calcium and the Secretary Process.* Plenum Press, New York, pp. 125–149

Rubin, R. P., Carchman, R. A. and Joanus, S. D. (1972) Role of calcium and adenosine cyclic 3′,5′-phosphate in action of adrenocorticotropin. *Nature, New Biol.*, **240**, 150

Rubio, R., Bailey, C. and Villar-Palasi, C. (1975) Effects of cyclic AMP-dependent protein kinase on cardiac actomyosin. Increase in Ca^{++} sensitivity and possible phosphorylation of troponin I. *J. Cycl. Nucleot. Res.*, **1**, 143–150

Rudland, P. S., Gospodarowicz, D. and Seifert, W. (1974a) Activation of guanyl cyclase and intracellular cyclic GMP by fibroblast growth factor. *Nature, Lond.*, **250**, 741–742, 773–774

Rudland, P. S., Seeley, M. and Seifert, W. (1974b) Cyclic GMP and cyclic AMP in normal and transformed fibroblasts. *Nature, Lond.*, **251**, 417–419

Ruoff, H.-J. and Sewing, K.-Fr. (1973) Cyclic 3′,5′-adenosine monophosphate in the rat gastric mucosa after starvation, feeding and pentagastrin. *Scand. J. Gastroent.*, **8**, 241–243

Ruoff, H.-J. and Sewing, K.-Fr.: (1974) Rat gastric mucosal cAMP following cholinergic and histamine stimulation. *Europ. J. Pharmac.*, **28**, 338–343

Ruoff, H.-J. and Sewing, K.-Fr. (1975a) Adenylate cyclase and phosphodiesterase in rat gastric mucosa after starvation feeding and pentagastrin. *Naunyn-Schmiedebergs Arch. Pharmac.*, **288**, 147–153

Ruoff, H.-J. and Sewing, K.-Fr. (1975b) Influence of atropine, metiamide and vagotomy on cAMP of resting and stimulated gastric mucosa. *Europ. J. Pharmac.*, **32**, 227–232

Ruoff, H.-J. and Sewing, K.-Fr. (1976) Adenylate cyclase of the dog gastric mucosa: Stimulation by histamine and inhibition by metiamide. *Naunyn-Schmiedebergs Arch. Pharma.*, **294**, 207–208

Russell, D. H. and Stambrook, P. J. (1975) Cell cycle specific fluctuations in adenosine 3′,5′-cyclic monophosphate and polyamines of Chinese hamster cells. *Proc. natn. Acad. Sci.*, **72**, 1482–1486

Russell, T. and Pastan, I. (1973) Plasma membrane cyclic adenosine 3′,5′-monophosphate phosphodiesterase of cultured cells and its modification after trypsin treatment of intact cells. *J. biol. Chem.*, **248**, 5835–5840

Russell, T. R., Thompson, N. J., Schneider, F. W. and Appleman, M. M. (1972) 3′,5′-Cyclic adenosine monophosphate phosphodiesterase: Negative cooperativity. *Proc. natn. Acad. Sci., U.S.A.*, **69**, 1791–1795

Ryan, J. and Storm, D. R. (1974) Solubilization of glucagon and epinephrine-sensitive adenylate cyclase from rat liver plasma membranes. *Biochem. biophys. Res. Commun.*, **60**, 304–311

Ryan, W. L. and Durick, M. A. (1972) Adenosine 3′,5′-monophosphate and *N*6-2′-*O*-dibutyryl-adenosine 3′,5′-monophosphate transport in cells. *Science, N.Y.*, **177**, 1002–1003

Ryan, W. L. and Heidrick, M. L. (1974) Role of cyclic nucleotides in cancer. In *Advances in Cyclic Nucleotide Research,* Vol. 4 (Eds. P. Greengard and G. A. Robison), Raven Press, New York, pp. 81–116

Saavedra, T. M., Brownstein, M. and Axelrod, J. (1974) Serotonin distribution in the limbic system of the rat. *Brain Res.*, **79**, 437–441

Sachs, C., Pycock, C. and Jonsson, G. (1974) Altered development of central noradrenergic neurons during ontogeny by 6-hydroxydopamine. *Med. Biol.*, **52**, 55–65

Safran, E. M. and Galsky, A. G. (1974) The action of cyclic AMP on GA_3 controlled responses. VI. Characteristics of the promotion of light-inhibited seed germination in Phacelia tannacetifolia by GA_3 and cyclic 3′,-5′-adenosine monophosphate. *Pl. Cell Physiol.*, Tokyo, **15**, 527–532

Sahib, M. K., Jost, Y. C. and Jost, J. P. (1971) Role of cyclic adenosine 3′,5′-monophosphate in the induction of hepatic enzymes. III. Interaction of hydrocortisone and *N*6, *O*2-dibutyryl cyclic adenosine 3′,5′-monophosphate in the induction of tyrosine aminotransferase in cultured H-4-11-E hepatoma cells. *J. biol. Chem.*, **246**, 4539–4545

Sahyoun, N. and Cuatrecasas, P. (1975) Mechanism of activation of adenylate cyclase by cholera toxin. *Proc. natn. Acad. Sci., U.S.A.*, **72**, 3438–3442

Sakiai, K. and Marks, B. (1972) Adrenergic effects on pineal cell membrane potential. *Life Sci.*, **11**, 285–291

Sakai, T., Thompson, W. J., Lavis, V. B. and Williams, R. H. (1974) Cyclic nucleotide phosphodiesterase activities from isolated fat cells: Correlation of subcellular distribution with effects of nucleotides and insulin. *Archs. Biochem. Biophys.*, **162**, 331–339

Salmoiraghi. G. C. and Bloom, F. E. (1964) The pharmacology of individual neurons. *Science, N.Y.*, **144**, 493–497

Salmoiraghi, G. C. and Weight, F. (1967) Micromethods in neuropharmacology: An approach to the study of anesthetics. *Anesthesiology*, **28**, 54–64

Salomon, Y., Lin, M. C., Londos, C., Rendell, M. and Rodbell, M. (1975) The hepatic adenylate cyclase system. I. Evidence for transition states and structural requirements for guanine nucleotide activation. *J. biol. Chem.*, **250**, 4239–4245

Sams, D. J. and Montague, W. (1972) The role of adenosine 3′,5′-cyclic monophosphate in the regulation of insulin released by isolated rat islets of Langerhans. *Biochem. J.*, **122**, 115–120

Sams, D. J. and Montague, W. (1974) Possible involvement of adenosine 3′,5′-cyclic monophosphate in the mechanism of action of sulphonylureas on insulin secretion from islets of Langerhans. *Biochem. Soc. Trans.*, **2**, 411–412

Samuelsson, B., Granström, E., Green, K., Hamberg, M. and Hammarström, S. (1975) Prostaglandins. *Ann. Rev. Biochem.*, **44**, 669–695

Sandler, J. A., Gallin, J. I. and Vaughan, M. (1975) Effects of serotonin, carbamylcholine, and ascorbic acid on leukocyte cyclic GMP and chemotaxis, *J. Cell Biol.*, **67**, 480–484

Sanger, J. and Holtzer, H. (1972) Cytochalasin B: Effects on cell morphology, cell adhesion, and mucopolysaccharide synthesis. *Proc. natn. Acad. Sci., U.S.A.*, **69**, 253–257

Sartorelli, L., Galzigna, L., Rossi, C. R. and Gibson, D. M. (1969) Influence of lecithin on the activity of the GTP-dependent acyl-CoA synthetase. *Biochem. biophys. Res. Commun.*, **26**, 90

Satoyoshi, E. and Kowa, H. (1967) A myopathy due to glycolytic abnormality. *Archs. Neurol.*, **17**, 248–256

Sattin, A. (1975) Cyclic AMP accumulation in cerebral cortex tissue from inbred strains of mice. *Life Sci.*, **16**, 903–914

Sattin, A. and Rall, T. W. (1970) The effect of adenosine and adenine nucleotides on the cyclic adenosine 3′,5′-phosphate content of guinea-pig cerebral cortex slices. *Molec. Pharmac.*, **6**, 13–23

Sattin, A., Rall, T. W. and Zanella, J. (1975) Regulation of cyclic adenosine 3′,5′-monophosphate levels in guinea-pig cerebral cortex by interaction of α-adrenergic and adenosine receptor activity. *J. Pharmac. exp. Therap.*, **192**, 22–32

Sattler, J. and Wiegandt, H. (1975) Studies of the subunit structure of choleragen. *Europ. J. Biochem.*, **57**, 309–316

Schanberg, S. M., Stone, R. A., Kirshner, N., Gunnells, J. C. and Robinson, R. R. (1974) Plasma dopamine β-hydroxylase: A possible aid in the study and evaluation of hypertension. *Science, N.Y.*, **183**, 523–524

Schauder, P., Ebert, R. and Frerichs, H. (1975) 5′-Guanylylimidodiphosphate: A modulator of glucagon-induced insulin release from isolated rat pancreatic islets. *Biochem. Biophys. Res. Commun.*, **67**, 701–705

Scheit, K. H. (1974) Biological and spectroscopic properties of a fluorescent cyclic AMP analogue, 2-aminopurine nucleoside-3′,5′-cyclic phosphate. *J. Carboh. Nucleos. Nucleot.*, 385–399

Schild, H. (1937) Histamine release and anaphylactic shock in isolated lungs of guinea pigs. *Quart. J. exp. Physiol.*, **26**, 165–179

Schimmel, R. J. (1974) Responses of adipose tissue to sequential lipolytic stimuli. *Endocrinology*, **94**, 1372–1380

Schimmer, B. P. (1972) Adenylate cyclase activity in adrenocorticotropic hormone-sensitive and mutant adrenocortical tumor cell lines. *J. biol. Chem.*, 3134–3138

Schimmer, B. P., Ueda, K. and Sato, G. H. (1968) Site of action of adrenocorticotropic hormone (ACTH) in adrenal cell culture. *Biochem. biophys. Res. Commun.*, **32**, 806–810

Schlaeger, E. J. and Köhler, G. (1976) External cyclic AMP-dependent protein kinase activity in rat C-6 glioma cells. *Nature, Lond.*, **260**, 705—707

Schlender, K. K. and Reimann, E. M. (1975) Isolation of a glycogen synthetase I kinase that is independent of cAMP. *Proc. natn. Acad. Sci., U.S.A.*, **72**, 2197–2201

Schlender, K. K., Wei, S. H. and Villar-Palasi, C. (1969) UDP-glucose: Glycogen α-4-glucosyl-transferase I kinase activity of purified muscle protein kinase. *Biochim. biophys. Acta*, **191**, 272–278

Schmidt, M. J., Hopkins, J. T., Schmidt, D. E. and Robison, G. A. (1972) Cyclic AMP in brain areas: Effects of amphetamine and norepinephrine assessed through the use of microwave radiation as a means of tissue fixation. *Brain Res.*, **42**, 465–477

Schmidt, M. J., Palmer, E. C., Dettbarn, W.-D. and Robison, G. A. (1970) cAMP and adenyl cyclase in the developing rat. *Devl. Psychobiol.*, **3**, 53–67

Schmidt, M. J. and Robison, G. A. (1971) The effect of norepinephrine on cyclic AMP levels in discrete regions of the developing rabbit brain. *Life Sci.*, **10**, 459–464

Schmidt, M. J., Schmidt, D. E. and Robison, G. A. (1971) Cyclic adenosine monophosphate in brain areas: Microwave irradiation as a means of tissue fixation. *Science*, N.Y., **173**, 1142–1143

Schoenfeld, R. I. and Uretsky, N. J. (1972) Altered response to apomorphine in 6-hydroxydopamine treated rats. *Euro. J. Pharmac.*, **19**, 115–118

Schofield, J. G. (1967) Role of cyclic 3′,5′-adenosine monophosphate in the release of growth hormone *in vitro*. *Nature, Lond.*, **215**, 1382–1383

Schofield, J. G. (1971) Cytochalasin B and release of growth hormone. *Nature, New Biol.*, **234**, 215–216

Schölkens, B. A. and Steinbach, R. (1975) Increase of experimental hypertension following inhibition of prostaglandin biosynthesis. *Archs. int. Pharmacodyn. Thér.*, **214**, 328–334

Schönhöfer, P. S., Peters, H. D., Karzel, K., Dinnendahl, V. and Westhofen, P. (1974) Influence of antiphlogistic drugs on prostaglandin E_1 stimulated cyclic 3′,5′-AMP levels and glycosaminoglycan synthesis in fibroblast tissue cultures. *Pol. J. Pharmacol. Pharm.*, **26**, 51–60

Schönhöfer, P. S., Skidmore, I. F., Paul, M. I., Ditzion, B. R., Pauk, G. L. and Krishna, G. (1972) Effects of glucocorticoids on adenyl cyclase and phosphodiesterase activity in fat cell homogenates and the accumulation of cyclic AMP in intact fat cells. *Naunyn-Schmiedebergs Arch. Pharmac.*, **273**, 267–282

Schor, S. and Rozengurt, E. (1973) Enhancement by purine nucleosides and nucleotides of serum-induced DNA synthesis in quiescent 3T3 cells. *J. Cell Physiol.*, **81**, 339–346

Schorr, I., Rathnam, P., Saxena, B. B. and Ney, R. L. (1971) Multiple specific hormone receptors in the adenylate cyclase of an adrenocortical carcinoma. *J. biol. Chem.*, **246**, 5806–5811

Schramm, M., Feinstein, H., Naim, E., Land, M. and Lasser, M. (1972) Epinephrine binding to the catecholamine receptor and activation of the adenylate cyclase on erythrocyte membranes. *Proc. natn. Acad. Sci., U.S.A.*, **69**, 523–527

Schramm, M. and Naim, E. (1970) Adenyl cyclase of rat parotid gland. *J. biol. Chem.*, **245**, 3225–3231

Schramm, M. and Rodbell, M. (1975) A persistent active state of the adenylate cyclase system produced by the combined actions of isoproterenol and guanyl imidodiphosphate in frog erythrocyte membranes. *J. biol. Chem.*, **250**, 2232–2237

Schrier, B. K. and Shapiro, D. L. (1973) Effects of N^6-monobutyryl-cyclic AMP on glutamate decarboxylase activity in fetal rat brain cells and glial tumor cells in culture. *Expl. Cell Res.*, **80**, 459–462

Schultz, G., Böhme, E. and Munske, K. (1969) Guanyl cyclase. Determination of enzyme activity. *Life Sci.*, **8**, 1323–1332

Schultz, G. and Hardman, J. G. (1975) Regulation of cyclic GMP levels in the ductus deferens of the rat. In *Advances in Cyclic Nucleotide Research*, Vol. 5 (Eds. G. I. Drummond, P. Greengard and G. A. Robison), Raven Press, New York, pp. 339–351

Schultz, G., Hardman, J. G. and Hurwitz, L. (1975) Cyclic nucleotides and smooth muscle function. *Proc. Int. Congr. Pharmac., Helsinki*, Vol. 1; pp. 203–211

Schultz, G., Hardman, J. G., Schultz, K., Baird, C. E. and Sutherland, E. W. (1973) The importance of calcium ions for the regulation of guanosine 3′,5′-cyclic monophosphate levels. *Proc. natn. Acad. Sci., U.S.A.*, **70**, 3889–3893

Schultz, J. (1974a) Adenosine 3′,5′-monophosphate in guinea-pig cerebral cortical slices: Effect of benzodiazepines. *J. Neurochem.*, **22**, 685–690

Schultz, J. (1974b) Inhibition of 3′,5′-nucleotide phosphodiesterase in guinea-pig cerebral cortical slices. *Archs. Biochem. Biophys.*, **163**, 15–20

Schultz, J. (1975a) Cyclic adenosine 3′,5′-monophosphate in guinea-pig cerebral cortical slices: Possible regulation of phosphodiesterase activity by cyclic adenosine 3′,5′-monophosphate and calcium ions. *J. Neurochem.*, **24**, 495–501

Schultz, J. (1975b) Cyclic adenosine 3′,5′-monophosphate in guinea-pig cerebral cortical slices: Studies on the role of adenosine. *J. Neurochem.*, **24**, 1237–1242

Schultz, J. (1976) Psychoactive drug effects on a system which generates cyclic AMP in brain. *Nature*, **261**, 417–418

Schultz, J. and Daly, J. W. (1973a) Cyclic adenosine 3′,5′-monophosphate in guinea-pig cerebral cortical slices. I. Formation of cyclic adenosine 3′,5′-monophosphate from endogenous adenosine triphosphate and from radioactive adenosine triphosphate formed during a prior incubation with radioactive adenine. *J. biol. Chem.*, **248**, 843–852

Schultz, J. and Daly, J. W. (1973b) Cyclic adenosine 3′,5′-monophosphate in guinea-pig cerebral cortical slices. II. The role of phosphodiesterase activity in the regulation of levels of cyclic adenosine 3′,5′-monophosphate. *J. biol. Chem.*, **248**, 853–859

Schultz, J. and Daly, J. W. (1973c) Accumulation of cyclic adenosine 3′,5′-monophosphate in cerebral cortical slices from rat and mouse: Stimulatory effect of α and β-adrenergic agents and adenosine. *J. Neurochem.*, **21**, 1319–1326

Schultz, J. and Daly, J. W. (1973d) Adenosine 3′,5′-monophosphate in guinea-pig cerebral cortical slices: Effects of α and β-adrenergic agents, histamine, serotonin and adenosine. *J. Neurochem.*, **21**, 573–579

Schultz, J. and Daly, J. W. (1973c) Cyclic adenosine 3′,5′-monophosphate in guinea-pig cerebral cortical slices. III. Formation, degradation and reformation of cyclic adenosine 3′,5′-monophosphate during sequential stimulations by biogenic amines and adenosine. *J. biol. Chem.*, **248**, 860–866

Schultz, J. and Gratzner, H. G. (Eds.) (1973) *The Role of Cyclic Nucleotides in Carcinogenesis*, Academic Press, New York

Schultz, J. and Hamprecht, B. (1973) Adenosine 3′,5′-monophosphate in cultured neuroblastoma cells: Effect of adenosine, phosphodiesterase inhibitors, and benzodiazepines. *Naunyn-Schmiedebergs Arch. Pharmac.*, **278**, 215–225

Schultz, J. and Kleefeld, G. (1975) Stimulation of adenosine 3′,5′-monophosphate formation in guinea-pig cerebral cortical slices in a calcium free medium. *Naunyn-Schmiedebergs Arch. Pharmacol.*, **287**, 289–296

Schumm, D. E., Morris, H. P. and Webb, T. E. (1974) Early biochemical changes in phytohemagglutinin-stimulated peripheral blood lymphocytes from normal and tumor-bearing rats. *Europ. J. Cancer*, **10**, 107–113

Schwabe, U., Berndt, S. and Ebert, R. (1972) Activation and inhibition of lipolysis in isolated fat cells by various inhibitors of cyclic AMP phosphodiesterase. *Naunyn Schmiedebergs Arch. Pharmac.*, **273**, 62–74

Schwabe, U. and Daly, J. W. (1977) The role of calcium ions in accumulations of cyclic elicited by α and β-adrenergic agonists in rat brain slices. *J. Pharmac. exp. Ther.* (in press)

Schwabe, U. and Ebert, R. (1972) Different effects of lipolytic hormones and phosphodiesterase inhibitors on cyclic 3′,5′-AMP levels in isolated fat cells. *Naunyn Schmiedebergs Arch. exp. Path. Pharmacol.*, **274**, 287–298

Schwabe, U. and Ebert, R. (1974) Stimulation of cyclic adenosine 3′,5′-monophosphate accumulation and lipolysis in fat cells by adenosine deaminase. *Naunyn Schmiedebergs Arch. Pharmacol.*, **282**, 33–44

Schwabe, U., Ebert, R. and Erbler, H. C. (1973) Adenosine release from isolated fat cells and its significance for the effects of hormones on cyclic 3′,5′-AMP levels and lipolysis. *Naunyn Schmiedebergs Arch. Pharmac.*, **276**, 133–148

Schwabe, U., Ohga, Y. and Daly, J. W. (1976) The role of calcium in the regulation of cyclic nucleotide levels in brain slices of rat and guinea pig. *Biochim.* Biophys. *Acta* (in press)

Schwabe, U., Schöhöfer, P. S. and Ebert, R. (1974) Facilitation by adenosine of the action of insulin on the accumulation of cAMP, lipolysis and glucose oxidation in isolated fat cells. *Europ. J. Biochem.*, **46**, 537–545

Schwartz, A., Entman, M. L., Kaniike, K., Lane, L. K., van Winkle, W. B. and Bornet, E. P. (1977) The rate of calcium uptake into sarcoplasmic reticulum of cardiac muscle and skeletal muscle: Effect of cyclic AMP-dependent protein kinase and phosphorylase *b* kinase. *Biochim. biophys. Acta* (in press)

Schweizer, M. P. and Robins, R. K. (1973) NMR Studies on the conformation of nucleosides and 3′,5′-cyclic nucleotides. In *Conformation of Biological Molecules and Polymers, The Jerusalem Symposia on Quantum Chemistry and Biochemistry*. V. The Israel Academy of Sciences and Humanities Jerusalem

Scott, R. E. (1970) Effects of prostaglandins, epinephrine, and NaF on human leukocyte, platelet and liver adenyl cyclase. *Blood*, **35**, 514–516

Scott, T. W., Mills, S. C. and Freinkel, N. (1968) The mechanism of thyrotropin action in relation to lipid metabolism in thyroid tissue. *Biochem. J.*, **109**, 325

Scott, T. W., Freinkel, N., Klein, J. H. and Nitzan, N. (1970) Metabolism of phospholipids, neutral lipids and carbohydrates in dispersed thyroid cells. *Endocrinology*, **87**, 854–863

Secrist, J. A., Barrio, J. R., Leonard, N. J., Villar-Palasi, C. and Gilman, A. G. (1972) Fluorescent modification of adenosine 3′,5′-monophosphate: spectroscopic properties and activity in enzyme systems. *Science, N.Y.*, **177**, 279–280

Segal, D. S., Kuczenski, R. J. and Mandell, A. J. (1972) Strain differences in behavior and brain tyrosine hydroxylase activity. *Behav. Biol.*, **7**, 75–81

Segal, D. S., Kuczenski, R. and Mandell, A. J. (1974) Theoretical implications of drug-induced adaptive regulation for a biogenic amine hypothesis of affective disorder. *Biol. Psych.*, **9**, 147–159

Segal, M. and Bloom, F. E. (1974a) The action of norepinephrine in the rat hippocampus. II. Activation of the input pathway. *Brain Res.*, **73**, 99–114

Segal, M. and Bloom, F. E. (1974b) The action of norepinephrine in the rat hippocampus, I. Iontophoretic studies. *Brain Res.*, **72**, 79–97

Seidel, C. L., Schnarr, R. L. and Sparks, H. V. (1975) Coronary artery cyclic AMP content during adrenergic receptor stimulation. *Am. J. Physiol.*, **229**, 265–269

Seifert, W. and Paul, D. (1972) Levels of cyclic AMP in sparse and dense cultures of growing and quiescent 3T3 cells. *Nature, New Biol.*, **240**, 281–283

Seifert, W. and Rudland, P. S. (1974a) Possible involvement of cyclic GMP in growth control of cultured mouse cells. *Nature*, Lond., **248**, 138–140

Seifert, W. and Rudland, P. S. (1974b) Cyclic nucleotides and growth control in cultured mouse cells: Correlation of changes in intracellular 3′,5′-cGMP concentration with a specific phase of the cell cycle. *Proc. natn. Acad. Sci., U.S.A.*, **71**, 4920–4924

Seip, M. (1971) Generalized lipodystrophy. *Ergebn. inn. Med. Kinderheilk.*, **31**, 59–95

Senft, G., Schultz, G., Munske, K. and Hoffmann, M. (1968) Influence of insulin on cyclic-3,5-AMP phosphodiesterase activity in liver, skeletal muscle, adipose tissue and kidney. *Diabetologia*, **4**, 322–329

Severin, E. S., Kochetkov, S. N., Nesterova, M. V. and Gulyaev, N. N. (1974) Isolation of the regulatory subunit of pig-brain histone kinase by affinity chromatography on cyclic-AMP-containing adsorbent. *FEBS. Letters*, **49**, 61–64

Severin, E. S., Nesterova, M. V., Sashchemko, L. P., Rasumova, V. V., Tunitskaya, V. L., Kochetkov, S. N. and Gulyaev, N. N. (1975) Investigation of the adenosine 3′,5′-cyclic phosphate binding site of pig-brain histone kinase with the aid of some analogues of adenosine 3′,5′-cyclic phosphate. *Biochim. biophys. Acta*, **384**, 413–422

Severson, D. L., Drummond, G. I. and Sulakhe,P. V. (1972) Adenyl cyclase in skeletal muscle: Kinetic properties and hormonal stimulation. *J. biol. Chem.*, **247**, 2949

Sewing, K.-Fr., Ruoff, H.-J. and Ekerdt, R. (1976) Protein synthesis and the cyclic AMP system in the resting and stimulated gastric mucosa of rats. In *Stimulus–Secretion Coupling in the Gastrointestinal Tract* (Eds. R. M. Case and H. Goebell), MTP Press, Lancaster, pp. 193–195

Shahab, L., Haase, M., Schiller, U. and Wollenberger, A. (1969) Noradrenalinabgabe aus dem Hundeherzen nach vorübergehender Okklusion einer Coronararterie. *Acta biol. med. germ.*, **22**, 135–143

Shapiro, D. L. (1973) Morphological and biochemical alterations in fetal rat brain cells cultured in the presence of monobutyryl-cyclic AMP. *Nature*, Lond., **241**, 203–204

Sharma, S. K., Nirenberg, M. and Klee, W. A. (1975) Morphine receptors as regulators of adenylate cyclase activity. *Proc. natn. Acad. Sci.*, **72**, 590–594

Sharon, N. and Lis, H. (1972) Lectins: Cell-agglutinating and sugar-specific proteins. *Science, N.Y.*, **177**, 949–959

Sharp, G. W. G. (1975) Studies on the mechanism of insulin release. *Fed. Proc.*, **34**, 1537–1548

Shenkman, L., Imai, Y., Kataoka, K., Hollander, C. S., Wan, L., Tang, S. C. and Avruskin, T. (1974) Prostaglandins stimulate thyroid function in pregnant women. *Science, N.Y.*, **184**, 81–82

Sheppard, H. and Burghardt, C. R. (1974) The dopamine-sensitive adenylate cyclase of rat caudate nucleus. *Mol. Pharmac.*, **10**, 721–726

Sheppard, J. R. (1972) Difference in the cyclic adenosine 3′,5′-monophosphate levels in normal and transformed cells. *Nature, New Biol.*, **236**, 14–16

Sheppard, J. R. (1974) The role of cyclic AMP in the control of cell division. In *Cyclic AMP, Cell Growth, and the Immune Response* (Eds. L. Braun, L. M. Lichtenstein and C. W. Parker), Springer-Verlag, New York, pp. 290–301

Sheppard, J. R., Hudson, T. H. and Larson, J. R. (1975) Adenosine 3′,5′-monophosphate analogs promote a circular morphology of cultured schwannoma cells. *Science, N.Y.*, **187**, 179–181

Sheppard, J. R. and Prasad, K. N. (1973) Cyclic AMP levels and the morphological differentiation of mouse neuroblastoma cells. *Life Sci.*, **12**, 431–439

Sheppard, J. R. and Prescott, D. M. (1972) Cyclic AMP levels in synchronized mammalian cells. *Expl. Cell Res.*, **75**, 293–296

Sherwin, J. R. and Tong, W. (1975) Thyroidal autoregulation: Iodide-induced suppression of thyrotropin-stimulated cyclic AMP production and iodinating activity in thyroid cells. *Biochim. biophys. Acta*, **404**, 30–39

Shields, R. (1974) Control of cell growth by the cyclic nucleotide seesaw. *Nature, Lond.*, **252**, 11–12

Shill, J. P. and Neet, K. E. (1971) A slow transient kinetic process of yeast hexokinase. *Biochem. J.*, **123**, 283–285

Shimizu, H., Creveling, C. R. and Daly, J. W. (1970a) Effect of membrane depolarization and biogenic amines on the formation of cyclic AMP in incubated brain slices. *Adv. Biochem. Psychopharmac.*, **3**, 135–154

Shimizu, H., Creveling, C. R. and Daly, J. W. (1970b) Cyclic adenosine 3′,5′-monophosphate formation in brain slices: Stimulation by batrachotoxin, ouabain, veratridine and potassium ions. *Molec. Pharmacol.*, **6**, 184–188

Shimizu, H., Creveling, C. R. and Daly, J. W. (1970c) Stimulated formation of adenosine 3′,5′-cyclic phosphate in cerebral cortex: Synergism between electrical activity and biogenic amines. *Proc. natn. Acad. Sci., U.S.A.*, **65**, 1033–1040

Shimizu, H. and Daly, J. W. (1970) Formation of cyclic adenosine 3′,5′-monophosphate from adenosine in brain slices. *Biochim. biophys. Acta*, **222**, 465–473

Shimizu, H. and Daly, J. W. (1972a) Methods for the measurement of cyclic AMP in brain. In *Methods in Neurochemistry*, Vol. 2 (Ed. R. Fried), Marcel Dekker, New York, pp. 147–168

Shimizu, H. and Daly, J. W. (1972b) Effect of depolarizing agents on accumulation of cyclic adenosine 3′,5′-monophosphate in cerebral cortical slices. *Europ. J. Pharmac.*, **17**, 240–252

Shimizu, H., Daly, J. W. and Creveling, C. R. (1969) A radioisotopic method for measuring adenosine 3′,5′.cyclic monophosphate in incubated slices of brain. *J. Neurochem.*, **16**, 1609–1619

Shimizu, H., Tanaka, S., Suzuki, T. and Matsukado, Y. (1971) The response of human cerebrum adenyl cyclase to biogenic amines. *J. Neurochem.*, **18**, 1157–1161

Shoemaker, W. J., Balentine, L. T., Siggins, G. R., Hoffer, B. J., Henriksen, S. J. and Bloom, F. E. (1975) Characteristics of the release of cyclic adenosine 3′,5′-monophosphate from micropipets by microiontophoresis. *J. Cycl. Nucleot. Res.*, **1**, 97–106

Shuman, D. A., Miller, J. P., Scholten, M. B., Simon, L. N. and Robins, R. K. (1973) Synthesis and biological activity of some purine 5′-thio-5′-deoxynucleoside 3′,5′-cyclic phosphorothioates. *Biochemistry*, **12**, 2781–2786

Siddle, K. and Hales, C. N. (1974) The relationship between the concentration of adenosine 3′,5′-cyclic monophosphate and the anti-lipolytic action of insulin in isolated rat fat-cells. *Biochem. J.*, **142**, 97–103

Sidwell, R. W., Allen, L. B., Huffman, J. H., Khwaja, T. A., Tolman, R. L. and Robins, R. K. (1973a) Anti-DNA virus activity of the 5′-nucleotide and 3′,5′-cyclic nucleotide of 9-β-D-arabinofuranosyladenine. *Chemotherapy*, **19**, 325–340

Sidwell, R. W., Huffman, J. H., Allen, L. B., Meyer, R. B., Jr., Shuman, D A., Simon, L. N. and Robins, R. K. (1974a) *In vitro* antiviral activity of 6-substituted 9-β-D-ribofuranosylpurine 3′,5′-cyclic phosphates. *Antimicrob. Ag. Chemother.*, **5**, 652–657

Sidwell, R. W., Simon, L. N., Huffman, J. H., Allen, L. B., Long, R. A. and Robins, R. K. (1973b) DNA virus inhibitory activity of 1-β-D-arabinofuranosylcytosine-3′,5′-cyclic phosphate. *Nature, New Biol.*, **242**, 204–206

Sidwell, R. W., Simon, L. N., Witkowski, J. T. and Robins, R. K. (1974b) Antiviral activity of virazole: Review and structure-activity relationships. *Prog. Chemother.*, **2**, 889–903

Siggins, G. R., Battenberg, E. F., Hoffer, B. J., Bloom, F. E. and Steiner, A. L. (1973) Noradrenergic stimulation of cyclic adenosine monophosphate in rat Purkinje neurons: An immunocytochemical study. *Science, N.Y.*, **179**, 585–588

Siggins, G. R. and Henriksen, S. J. (1975) Analogs of cyclic adenosine monophosphate: Correlation of inhibition of Purkinje neurons with protein kinase activation. *Science, N.Y.*, **189**, 559–561

Siggins, G. R., Henriksen, S. J. and Landis, S. C. (1976a) Electrophysiology of Purkinje neurons in the weaver mouse: Iontophoresis of neurotransmitters and cyclic nucleotides, and stimulation of the nucleus locus coeruleus. *Brain Res.* (in press)

Siggins, G. R., Hoffer, B. J. and Bloom, F. E. (1969) Cyclic 3′,5′-adenosine monophosphate: Possible mediator for the response of cerebellar Purkinje cells to microelectrophoresis of norepinephrine. *Science*, **165**, 1018–1020

Siggins, G. R., Hoffer, B. J. and Bloom, F. E. (1971a) Prostaglandin–norepinephrine interactions in brain: Microelectrophoretic and histochemical correlates. *Ann. N.Y. Acad. Sci.*, **180**, 302–323

Siggins, G. R., Hoffer, B. J., Oliver, A. P. and Bloom, F. E. (1971b) Activation of a central noradrenergic projection to cerebellum. *Nature, Lond.*, **233**, 481–483

Siggins, G. R., Hoffer, B., Bloom, F. and Ungerstedt, U. (1976b) Cytochemical and electrophysiological studies of dopamine in the caudate nucleus. In *The Basal Ganglia* (Ed. M. Yahr). Raven Press, New York

Siggins, G. R., Hoffer, B. J. and Ungerstedt, U. (1974) Electrophysiological evidence for involvement of cyclic adenosine monophosphate in dopamine reponses of caudate neurons. *Life Sci.*, **15**, 779–792

Siggins, G. R., Oliver, A. P., Hoffer, B. J. and Bloom, F. E. (1876c) Cyclic adenosine monophosphate and norepinephrine: Effects on transmembrane properties of cerebellar Purkinje cells. *Science, N.Y.*, **171**, 192–194

Simantov, R. and Sachs, L. (1975) Temperature sensitivity of cyclic adenosine 3′,5′-monophosphate binding proteins and the regulation of growth and differentiation in neuroblastoma cells. *J. biol. Chem.*, **250**, 3236–3242

Simon, L. N., Shuman, D. A. and Robins, R. K. (1973) The chemistry and biological properties of nucleotides related to nucleoside 3′,5′-cyclic phosphates. In *Advances in Cyclic Nucleotide Research*, Vol. 3 (Eds. P. Greengard and G. A. Robison), Raven Press, New York, pp. 225–353

Singer, I. and Forrest, J. V. (1976) Drag-induced states of nephrogenic diabetes insipidus. *Kidney Int.*, **10**, 82–96

Singer, S. J. and Nicolson, G. L. (1972) The fluid mosaic model of the structure of cell membranes, *Science, N.Y.*, **175**, 720–731

Skala, J. P., Drummond, G. I. and Hahn, P. (1974) A protein kinase inhibitor in brown adipose tissue of developing rats. *Biochem. J.*, **138**, 195–199

Skelton, C. L., Levy, G. S. and Epstein, S. E. (1970) Positive inotropic effects of dibutyryl-adenosine 3′,5′-monophosphate. *Circulation Res.*, **26**, 35–43

Skolnick, P. and Daly, J. W. (1974a) Norepinephrine-elicited accumulation of adenosine 3′,5′-monophosphate in brain slices: Relationship to spontaneous behavioral activity and levels of brain tyrosine hydroxylase in several rat strains. *Science, N.Y.*, **184**, 175–177

Skolnick, P. and Daly, J. W. (1974b) The accumulation of adenosine 3′,5′-monophosphate in cerebral cortical slices of the quaking mouse, a neurologic mutant. *Brain Res.*, Amsterdam, **73**, 513–525

Skolnick, P. and Daly, J. W. (1975a) Functional compartments of adenine nucleotides serving as precursors of adenosine 3′,5′-monophosphate in mouse cerebral cortex. *J. Neurochem.*, **24**, 451–456

Skolnick, P. and Daly, J. W. (1975b) Stimulation of adenosine 3′,5′-monophosphate formation by α and β-adrenergic agonists in rat cerebral cortical slices: Effect of clonidine. *Molec. Pharmac.*, **11**, 545–551

Skolnick, P. and Daly, J. W. (1975c) Stimulation of adenosine 3′,5′-monophosphate formation in rat cerebral cortical slices by methoxamine: Interaction with an alpha-adrenergic receptor. *J. Pharmac. exp. Ther.*, **193**, 549–558

Skolnick, P., Daly, J. W., Freedman, R. and Hoffer, B. J. (1976) Interrelationship between catecholamine-stimulated formation of cyclic AMP in cerebellar slices and inhibitory effects on cerebellar Purkinje cells: Antagonism by neuroleptic compounds. *J. Pharmacol. exp. Therap.*, **197**, 280–292

Skolnick, P., Huang, M., Daly, J. and Hoffer, B. (1973) Accumulation of adenosine 3′,5′-monophosphate in incubated slices from discrete regions of squirrel monkey cerebral cortex: Effects of norepinephrine, serotonin and adenosine. *J. Neurochem.*, **21**, 237–240

Skolnick, P., Schultz, J. and Daly, J. W. (1975) Repetitive stimulation of cyclic adenosine 3′,5′-monophosphate formation by adrenergic agonists in incubated slices from rat cerebral cortex. *J. Neurochem.*, **24**, 451–456

Sloboda, R. D., Rudolph, S. A., Rosenbaum, J. L. and Greengard, P. (1975) Cyclic AMP-dependent endogenous phosphorylation of a microtubule-associated protein. *Proc. natn. Acad. Sci., U.S.A.*, **72**, 177–181

Smith, C. G. (1966) Regulation of cell metabolism: Role of cyclic AMP. Annual Reports in *Medicinal Chemistry.*, Vol. 1 (Ed. R. V. Heinzelman), Academic Press, New York, pp. 286–295

Smith, J. A. and Martin, L. (1973) Do cells cycle? *Proc. natn. Acad. Sci., U.S.A.*, **70**, 1263–1267

Smith, J. W., Steiner, A. L., Newberry, W. M. and Parker, C. W. (1969) The effect of dibutyryl-cAMP on human lymphocyte stimulation by phytohemagglutinin. *Fed. Proc.*, **28**, 566

Smith, J. W., Steiner, A. L., Newberry, W. M. and Parker, C. W. (1971a) Cyclic adenosine 3′-5′ monophosphate in human lymphocytes: Alterations after phytohemagglutinin stimulation. *J. clin. Invest.*, **50**, 432–441

Smith, J. W., Steiner, A. L. and Parker, C. W. (1971b) Human lymphocyte metabolism. Effects of cyclic and noncyclic nucleotides on stimulation by phytohemagglutinin. *J. clin. Invest.*, **50**, 442–448

Smith, R. J. and Ignarro, L. J. (1975) Bioregulation of lysosomal enzyme secretion from human neutrophils: Roles of guanosine 3′,5′-monophosphate and calcium in stimulus-secretion coupling. *Proc. natn. Acad. Sci., U.S.A.*, **72**, 108–112

Smoake, J. A., Song, S. Y. and Cheung, W. Y. (1974) Cyclic 3′,5′-nucleotide phosphodiesterase. Distribution and devclopmented changes of the enzyme and its protein activator in mammalian tissues and cells. *Biochim. biophys. Acta.*, **341**, 402–411

Snyder, D. S. (1975) Cutaneous effects of topical indomethacin, an inhibitor of prostaglandin synthesis, on UV-damaged skin. *J. invest. Derm.*, **64**, 322–325

Snyder, D. S. and Eaglstein, W. H. (1974) Intradermal anti-prostaglandin agents and sunburn. *J. invest. Derm.*, **62**, 47–50

Snyder, S., Greenberg, D. and Yamamura, H. T. (1974) Antischizophrenic drugs and brain cholinergic receptors. *Archs. gen. Psychiat.*, **31**, 58–61

So, L. and Goldstein, I. J. (1967) Protein–carbohydrate interaction. IV. Application of the quantitative precipitin method to polysaccharide–concanavalin A interaction. *J. biol. Chem.*, **242**, 1617–1622

Sobel, B. E. and Mayer, S. E. (1973) Cyclic adenosine monophosphate and cardiac contractility. *Circulation Res.*, **32**, 407–414

Soderling, T. R., Corbin, J. D. and Park, C. R. (1973) Regulation of cyclic AMP-dependent protein kinase. *J. biol. Chem.*, **248**, 1822–1829

Soifer, D. (1975) Enzymatic activity in tubulin preparations: Cyclic-AMP dependent protein kinase activity of brain microtubule protein. *J. Neurochem.*, **24**, 21–33

Soifer, D. and Hechter, O. (1971) Adenyl cyclase activity in rat liver nuclei. *Biochim. biophys. Acta*, **230**, 539

Sokol, W. N. and Beall, G. N. (1975) Leukocytic epinephrine receptors of normal and asthmatic individuals. *J. Allergy clin. Immunol.*, **55**, 310–324

Solomon, S. S. (1975) Effect of insulin and lipolytic hormones on cyclic AMP phosphodiesterase activity in normal and diabetic rat adipose tissue. *Endocrinology*, **96**, 1366–1373

Somlyo, A. P. and Somlyo, A. V. (1969) Pharmacology of excitation–contraction coupling in vascular smooth muscle and in avian slow muscle. *Fed. Proc.*, **28**, 1634–1642

Somlyo, A. P., Somlyo, A. V. and Smiessko, V. (1972) Cyclic AMP and vascular smooth muscle. In *Advances in Cyclic Nucleotide Research*, Vol. 1 (Eds. P. Greengard and G. A. Robison), Raven Press, New York, pp. 175–194

Somlyo, A. V., Haeusler, G. and Somlyo, A. P. (1970) Cyclic adenosine monophosphate: Potassium-dependent action on vascular smooth muscle membrane potential. *Science, N.Y.*, **169**, 490–491

Søndergaard, J., Wadskoy, S. A. and Kobaysi, T. (1975) Electron microscopic cytochemical demonstration of adenyl cyclase in psoriatic epidermis. *J. invest. Derm.*, **64**, 294

Spaulding, S. W. and Burrow, G. N. (1974) TSH regulation of cAMP-dependent protein kinase activity in the thyroid. *Biochem. biophys. Res. Comm.*, **59**, 386–391

Spiegel, A. M. and Aurbach, G. D. (1974) Binding of 5′-guanylyl-imidodiphosphate to turkey erythrocyte membranes and effects on beta-adrenergic-activated adenylate cyclase. *J. biol. Chem.*, **249**, 7630–7636

Spudich, J. A. (1972) Effects of cytochalasin B on actin filaments. *Cold Spring Harb. Symp. quant. Biol.*, **37**, 585–593

Sraer, J., Ardaillon, R., Loreau, N. and Sraer, J. D. (1974) Evidence for parathyroid hormone sensitive adenylate cyclase in rat glomeruli. *Mol. Cell Endocrinol.*, **1**, 285–294

Stämpfli, R. (1954) A new method for measuring membrane potentials with external electrodes. *Experientia*, **10**, 508–509

Starke, K. (1972a) α-Sympathomimetic inhibition of adrenergic and cholinergic transmission in the rabbit heart. *Naunyn-Schmiedebergs Arch. Pharmac.*, **274**, 18–45

Starke, K. (1972b) Influence of extracellular noradrenaline on the stimulation evoked secretion of noradrenaline from parasympathetic nerves: Evidence for a α-receptor mediated feedback inhibition of noradrenaline release. *Naunyn-Schmiedebergs Arch. Pharmac.*, **275**, 11–23

Starke, K. and Montel, H. (1973) α-Receptor mediated modulation of transmitter release from central noradrenergic neurons. *Naunyn-Schmiedebergs Arch. Pharmac.*, **279**, 53–60

Starkey, R. A., Cohen, S. and Orth, D. N. (1975) Epidermal growth factor: Identification of a new hormone in human urine. *Science, N.Y.*, **189**, 800–802

Stawarz, R. J., Hill, H., Robinson, S. E., Setler, P. Dingell, J. V. and Sulser, F. (1975) On the significance of the increase in homovanillic acid (HVA) caused by antipsychotic drugs in corpus striatum and limbic forebrain. *Psychopharmacologia*, **43**, 125–130

Stawiski, M. A., Powell, J. A., Lang, P. G., Schork, A., Duell, E. A. and Voorhees, J. J. (1975) Papaverine: Its effects on cyclic AMP *in vitro* and psoriasis *in vivo*. *J. invest. Derm.*, **64**, 124–127

Stawiski, M., Rusin, L., Schork, M. A., Burns, T., Duell, E. and Voorhees, J. (1976) Ro 20-1724 elevates epidermal cyclic AMP levels *in vitro* and improves psoriasis *in vivo*. *Clin. Res.*, **24**, 276A

Steer, M. L. and Levitzki, A. (1975) The control of adenylate cyclase by calcium in turkey erythrocyte ghosts. *J. biol. Chem.*, **250**, 2080–2084

Stegman, S. J., Fukuyama, K. and Epstein, W. L. (1976) Inhibition of the *in vivo* effects of concanavalin-A on mammalian epidermis by α-methyl-D-glycopyranoside. *J. invest. Derm.*, **66**, 17–21

Steinberg, D. and Huttunen, J. K. (1972) The role of cyclic AMP in activation of hormone-sensitive lipase of adipose tissue. In *Advances in Cyclic Nucleotide Research*, Vol. 1 (Eds. P. Greengard and G. A. Robison), Raven Press, New York, pp. 47–62

Steinberg, D. and Khoo, J. C. (1976) Reversible activation–deactivation of hormone-sensitive lipase and observations on its relation to lipoprotein lipase

Steinberg, D., Vaughan, M., Nestel, P. J. and Bergstrom, S. (1963) Effects of prostaglandin E opposing those of catecholamines on blood pressure and on triglyceride breakdown in adipose tissue. *Biochem. Pharmac.*, **12**, 764–766

Steiner, A. L., Kipnis, D. M., Utiger, R. and Parker, C. W. (1969) Radioimmunoassay for the measurement of adenosine 3′,5′-cyclic monophosphate. *Proc. natn. Acad. Sci., U.S.A.*, **64**, 367–373

Steiner, A. L., Parker, C. W. and Kipnis, D. M. (1972) Radioimmunoassay for cyclic nucleotides. I. Preparation of antibodies and iodinated cyclic nucleotides. *J. biol. Chem.*, **247**, 1106–1113

Steiner, A. L., Peake, G. T., Utiger, R. D., Karl, I. E. and Kipnis, D. M. (1970) Hypothalamic stimulation of growth hormone and thyrotropin release *in vitro* and pituitary 3′,5′-adenosine cyclic monophosphate. *Endocrinology*, **86**, 1354–1360

Stellwagen, R. H. (1972) Induction of tyrosine aminotransferase in HTC cells by N^6,O^{2}-dibutyryl-adenosine 3′,5′-monophosphate. *Biochem. biophys. Res. Commun.*, **47**, 1144–1150

Stellwagen, R. H. (1974) The effect of theophylline and certain other purine derivatives on tyrosine aminotransferase activity in hepatoma cells in culture. *Biochim. biophys. Acta*, **338**, 428–439

Stephenson, R. P. (1956) A modification of receptor theory. *Br. J. Pharmac.*, **11**, 379–393

Stevens, F. C., Walsh, M., Teo, T. S., Ho, H. C. and Wang, J. H. (1976) Comparison of calcium binding proteins. *J. biol. Chem.*, **351**, 4495–4500

Stitt, J. T. and Hardy, J. D. (1975) Microelectrophoresis of PGE_1 onto single units in the rabbit hypothalamus. *Am. J. Physiol.*, **222**, 240–245

Stock, K. and Prilop, M. (1974) Dissociation of catecholamine-induced formation of adenosine 3′,5′-monophosphate and release of glycerol in fat cells by prostaglandin E_1, E_2 and N^6-phenylisopropyladenosine. *Naunyn-Schmiedebergs Arch. Pharmacol.*, **282**, 15–31

Stoclet, J.-C., Pequet, M.-F. and Waeldele, G. (1971) On the mechanism of papaverine spasmolytic effect on normal and hypertensive rat aortic strips. *J. Pharmac. (Paris)*, **2**, 11–22

Stoff, J. S., Handles, J. S. and Orloff, J. (1972) The effect of aldosterone on the accumulation of adenosine 3′,5′ cyclic monophosphate in toad bladder epithelial cells in response to vasopressin and theophylline. *Proc. natn. Acad. Sci.*, U.S.A., **69**, 805–808

Stone, T. W., Taylor, D. A. and Bloom, F. E. (1975) Cyclic AMP and cyclic GMP may mediate opposite neuronal responses in the rat cerebral cortex. *Science, N.Y.*, **187**, 845–847

Strada, S. J. and Sulser, F. (1972) Effect of monoamine oxidase inhibitors on metabolism and *in vivo* release of [^{3}H]-norepinephrine from the hypothalamus. *Europ. J. Pharmac.*, **18**, 303–308

Strada, S. J., Uzunov, P. and Weiss, B. (1974) Ontogenetic development of a phosphodiesterase activator and the multiple forms of cyclic AMP phosphodiesterase of rat brain. *J. Neurochem.*, **23**, 1097–1103

Strange, R. C., Vetter, N., Rowe, M. J. and Oliver, M. F. (1974) Plasma cyclic AMP and total catecholamine during acute myocardial infarction in man. *Europ. J. clin. Invest.*, **4**, 115–121

Strom, T. B., Carpenter, C. B., Garovoy, M. R., Austen, K. F., Merrill, J. P. and

Kaliner, M. (1973) The modulating influence of cyclic nucleotides upon lymphocyte-mediated cytotoxicity. *J. exp. Med.*, **138**, 381–393

Strom, T. B., Deisseroth, A., Morganroth, J., Carpenter, C. B. and Merrill, J. P. (1972) Alteration of the cytotoxic action of sensitized lymphocytes by cholinergic agents and activators of adenylate cyclase. *Proc. natn. Acad. Sci., U.S.A.*, **69**, 2995–2999

Sturgill, T. W., Schrier, B. K. and Gilman, A. G. (1975) Stimulation of cyclic AMP accumulation by 2-chloroadenosine: Lack of incorporation of nucleoside into cyclic nucleotides. *J. Cycl. Nucleot. Res.*, **1**, 21–30

Sulakhe, P. V. and Dhalla, N. S. (1972) Adenyl cyclase activity in failing hearts of genetically myopathic hamsters. *Biochem. Med.*, **6**, 471–482

Sulakhe, P. V. and Drummond, G. I. (1974) Protein kinase-catalyzed phosphorylation of muscle sarcolemma. *Arch. Biochem. Biophys.*, **161**, 448–455

Sulakhe, P. V. and Phillis, J. W. (1975) The release of ^{3}H-adenosine and its derivatives from cat sensimotor cortex. *Life Sci.*, **17**, 551–556

Sulakhe, P. V., Sulakhe, S. J. and Leung, N. (1975) Properties of plasma membrane-associated and soluble guanylate cyclase of cardiac and skeletal muscle. *Circulation* **51/52** Supp. II: 246

Sulser, F. (1976) Tricyclic Antidepressants: Animals pharmacology (biochemical and metabolic aspects). In *Handbook of Psychopharmacology* (Eds. S. H. Snyder and L. L. Iversen), Plenum Press, New York

Sulser, F. and Sanders-Bush, E. (1971) Effect of drugs on amines in the CNS. *Ann. Rev. Pharmac.*, **11**, 209–230

Sun, I. Y., Shapiro, L. and Rosen, O. M. (1975) A specific cyclic guanosine 3′,5′-monophosphate-binding protein in *Caulobacter crescentus. J. biol. Chem.*, **250**, 6181–6184

Sundaralingam, M. and Abola, J. (1972) Sterochemistry of nucleic acids and their constituents. XXVII. The crystal structure of 5′-methylene-adenosine 3′,5′-cyclic monophosphonate monohydrate, a biologically active analog of the secondary hormonal messenger cyclic adenosine 3′,5′-monophosphate. Conformational "Rigidity" of the furanose ring in cyclic nucleotides. *J. Am. Chem. Soc.*, **94**, 5070–5076

Sung, C. P., Jenkins, B. C., Burns, L. R., Hackney, V., Spenney, J. G., Sachs, G. and Wiebelhaus, V. D. (1973) Adenyl and guanyl cyclase in rabbit gastric mucosa. *Am. J. Physiol.*, **225**, 1359–1363

Sung, C. P., Wiebelhaus, V D., Jenkins, B. C., -Adlercreutz, P., Hirschowitz, B. I. and Sachs, G. (1972) Heterogeneity of 3′,5′-phosphodiesterase of gastric mucosa. *Am. J. Physiol.*, **223**, 648–650

Sutherland, E. W. (1972) Studies on the mechanism of hormone action. *Science, N.Y.*, **177**, 401–408

Sutherland, E. W., Oye, I. and Butcher, R. W. (1965) The action of epinephrine and the role of the adenyl cyclase system in hormone action. *Recent Prog. Horm. Res.*, **21**, 623

Sutherland, E. W. and Rall, T. W. (1958) Fractionation and characterization of a cyclic adenine ribonucleotide formed by tissue particles. *J. biol. Chem.*, **232**, 1077–1091

Sutherland, E. W. and Rall, T. W. (1960) The relation of adenosine 3′,5′-phosphate and phosphorylase to the actions of catecholamines and other hormones. *Pharmac. Rev.*, **12**, 265–299

Sutherland, E. W., Rall, T. W. and Menon, T. (1962) Adenyl cyclase. I. Distribution, preparation and properties. *J. biol. Chem.*, **237**, 1220–1227

Sutherland, E. W. and Robison, G. A. (1966) The role of cyclic 3′,5′-AMP in responses to catecholamines and other hormones. *Pharmac. Rev.*, **18**, 145–161

Sutherland, E. W. and Robison, G. A. (1969) The role of cyclic AMP in the control of carbohydrate metabolism. *Diabetes*, **18**, 797–819

Sutherland, E. W., Robison, G. A. and Butcher, R. W. (1968) Some aspects of the biological role of adenosine 3′,5′-monophosphase (cyclic AMP). *Circulation*, **37**, 279–306

Sutherland, E. W., Robison, G. A. and Hardman, J. G. (1970) Some thoughts on the possible role of cyclic AMP in diabetes. *Nobel Symposium*, **13**, 137–154

Svennsson, T. H., Bunney, B. S. and Aghajanian, G. K. (1975) Inhibition of both noradrenergic and serotonergic neurons in brain by the α-adrenergic agonist clonidine. *Brain Res.*, **92**, 291–306

Sweet, C. S., Kadowitz, P. J. and Brody, M. J. (1971) A hypertensive response to the infusion of prostaglandin $F_{2\alpha}$ into the vertebral artery of the conscious dog. *Europ. J. Pharmac.*, **16**, 229–232

Swillens, S., van Cauter, E. and Dumont, J. E. (1974) Protein kinase and cyclic AMP: Significance of binding and activation constants. *Biochim. biophys. Acta*, **364**, 250–259

Swillens, S., van Sande, J., Pochet, R., Delbeke, D., Piccart, M., Paiva, M. and Dumont, J. E. (1976) Kinetics of cyclic AMP accumulation in dog thyroid slices. *Europ. J. Biochem.*, **62**, 87–93

Szabo, M. and Burke, G. (1972) Adenosine 3′,5′-cyclic phosphate phosphodiesterase from bovine thyroid. *Biochim. biophys. Acta*, **284**, 208–219

Szaduykis-Szadurski, L. and Berti, F. (1972) Smooth muscle relaxing activity of 8-bromo-guanosine-3′,5′monophosphate. *Pharmac. Res. Commun.*, **4**, 53–69

Szentivanyi, A. (1968) The beta adrenergic theory of the atopic abnormality in bronchial asthma. *J. Allergy*, **42**, 203–232

Tada, M., Kirchberger, M. A. and Katz, A. M. (1975) Phosphorylation of a 22,000-dalton component of the cardiac sarcoplasmic reticulum by adenosine 3′,5′-monophosphate-dependent protein kinase. *J. biol. Chem.*, **250**, 2640–2647

Tada, M., Kirchberger, M. A., Repke, D. I. and Katz, A. M. (1974) The stimulation of calcium transport in cardiac sarcoplasmic reticulum by adenosine 3′,5′-monophosphate-dependent protein kinase. *J. biol. Chem.*, **249**, 6174–6180

Takagi, K., Takayanagi, I. and Tomiyama, A. (1971) Action of dibutyryl-cyclic adenosine monophosphate on the intestinal smooth muscle. *Japan J. Pharmac.*, **21**, 271–273

Takai, K., Kurashina, Y., Suzuki-Hori, C., Okamoto, H. and Hayaishi, O. (1974) Adenylate cyclase from *Brevibacterium liquefaciens*. I. Purification, crystallization, and some properties. *J. biol. Chem.*, **249**, 1965–1972

Takamori, M., Ishii, N. and Mori, M. (1973) The role of cyclic 3′,5′-adenosine monophosphate in neuromuscular transmission. *Archs. Neurol.*, **29**, 420–424

Takayanage, I. and Takagi, K. (1973) The action of dibutyryl-cyclic GMP (N^6-2′-*O*-dibutyryl cyclic guanosine-3′,5′-monophosphate) on the ileum of guinea-pig. *Japan. J. Pharmac.*, **23**, 573–575

Takeuchi, T. and Manning, J. W. (1973) Hypothalamic mediation of sinus baroreceptor-evoked muscle cholinergic dilator response. *Am. J. Physiol.*, **224**, 1280–1287

Takeuchi, H., Yokoi, I., Mori, A. and Kohsaka, M. (1975) Effects of nucleic acid components and their relatives on the excitability of dopamine-sensitive giant neurones, identified in suboesophageal ganglia of the african giant snail (*Achatina fulica Ferussac*), *Gen. Pharmac.*, **6**, 77–85

Tan, K. B. and Sokol, F. (1974) Protein kinase stimulated by cyclic GMP in uninfected and simian virus 40-infected monkey kidney cells. *J. Virol.*, **13**, 234–236

Tannenbaum J., Tanenbaum, S. W., Lo, W. G., Godman, G. C. and Miranda, A. F. (1974) Binding and subcellular localization of tritiated cytochalasin D. *Expl. Cell Res.*, **91**, 47–56

Tarsy, D. and Baldessarini, R. L. (1974) Behavioral supersensitivity to apomorphine following chronic treatment with drugs which interfere with the synaptic function of catecholamines. *Neuropharmacology*, **13**, 927–940

Tarui, S., Okuno, G., Ikura, Y., Tanaka, T., Suda, M. and Nishikawa, M. (1965) Phosphofructokinase deficiency in skeletal muscle: A new type of glucogenosis. *Biochem. biophys. Res. Commun.*, **19**, 517–523

Tashijan, A. J. and Hoyt, R. F. (1972) Transient controls of organ-specific functions in pituitary cells in culture. In *Molecular Genetics and Developmental Biology* (Ed. M. Sussman), Prentice-Hall, Englewood Cliffs, pp. 353–387

Taube, H. D., Borowski, E., Endo, T. and Starke, K. (1976) Enkephalin: a potential modulator of noradrenaline release in rat brain. *Euro. J. Pharmacol.*, **38**, 377–440

Tauber, A., Kaliner, M., Stechschulte, D. and Austen, K. F. (1973) Immunologic release of histamine and slow-reacting substance of anaphylaxis from human lung. *J. Immunol.*, **111**, 27–32

Taunton, O. D., Roth, J. and Pastan, I. (1969) Studies on the adrenocorticotropic hormone-activated adenyl cyclase of a functional adrenal tumour. *J. biol. Chem.*, **244**, 247–253

Taylor, A. L., Davis, B. B., Pawlson, L. G., Josimovich, J. B. and Mintz, D. H. (1970) Factors influencing the urinary excretion of 3′,5′-adenosine monophosphate in humans. *J. clin. Endocr. Metab.*, **30**, 316–324

Taylor, R. B., Duffus, W. P., Raff, M. C. and de Petris, S. (1971) Redistribution and pinocytosis of lymphocyte surface immunoglobulin molecules induced by anti-immunoglobulin antibody. *Nature*, *New Biol.*, **233**, 225–229

Temple, R., Williams, J. A., Wilber, J. F. and Wolfe, J. (1972) Colchicine and hormone secretion. *Biochem. biophys. Res. Comm.*, **46**, 1454–1462

Teo, T. S. (1974) Purification and mechanism of action of bovine heart protein activator of cyclic nucleotide phosphodiesterase. Ph.D. Thesis. University of Manitoba

Teo, T. S. and Wang, J. H. (1973a) Mechanism of activation of a cyclic adenosine 3′,5′-monophosphate phosphodiesterase from bovine heart by calcium ions. Identification of the protein activator as a Ca^{2+} binding protein. *J. biol. Chem.*, **248**, 5950–5955

Teo, T. S., Wang, T. H. and Wang, J. H. (1973b) Purification and properties of the protein activator of bovine heart cyclic adenosine 3′,5′-monophosphate phosphodiesterase. *J. biol. Chem.*, **248**, 588–595

Teshima, Y. and Kakiuchi, S. (1974) Mechanism of stimulation of Ca^{2+} plus Mg^{2+}-dependent phosphodiesterase from rat cerebral cortex by the modulator protein and Ca^{2+}. *Biochem. biophys. Res. Commun.*, **56**, 489–495

Teshima, Y., Yamazaki, R. and Kakiuchi, S. (1974) Effect of ATP on the activity of nucleotide 3′,5′-monophosphate phosphodiesterase from brain. *J. Neurochem.*, **22**, 789–791

Theoharides, T. V. C. and Cannelakis, Z. N. (1975) Spermine inhibits induction of ornithine decarboxylase by cyclic AMP, but not by dexamethasone in rat hepatoma cells. *Nature, Lond.*, **255**, 733–734

Thoenen, H., Angeletti, P. U., Levi-Montalcini, R. and Kettler, R. (1971) Selective induction by nerve growth factor of tyrosine hydroxylase and dopamine-β-hydroxylase in the rat superior cervical ganglia. *Proc. natn. Acad. Sci., U.S.A.*, **68**, 1598–1602

Thoenen, H., Mueller, R. A. and Axelrod, J. (1969) Transsynaptic induction of adrenal tyrosine hydroxylase. *J. Pharmac. exp. Therap.*, **169**, 249–254

Thoenen, H. and Otten, U. (1975) Cyclic nucleotides and transsynaptic enzyme induction: Lack of correlation between initial cAMP increase, changes in cAMP/cGMP ratio and subsequent induction of tyrosine hydroxylase in the adrenal medulla. In *Chemical Tools in Catecholamine Research* (Eds. O. Almgren, A. Carlsson and J. Engel), North Holland Publishing Co., Amsterdam, pp. 275–282

Thomas, E. W., Murad, F., Looney, W. B. and Morris, H. P. (1973) Adenosine 3′,5′-monophosphate and guanosine 3′,5′-monophosphate. Concentrations in Morris hepatomas of different growth rates. *Biochim. biophys. Acta'*, **297**, 564–567

Thomas, R. C. (1972) Electrogenic sodium pump in nerve and muscle cells. *Physiol. Rev.*, **52**, 563–594

Thompson, M. J. and Jacobson, E. D. (1975) Regulation of rat gastric mucosal guanyl cyclase activity *Gastroenterology*, **68**, 998

Thompson, M. J., Williams, R. H. and Little, S. A. (1973) Activation of guanyl cyclase and adenyl cyclase by secretin. *Biochim. biophys. Acta.*, **302**, 329–337

Tobia, A. J., Adams, M. D., Miya, T. S. and Bousquet, W. F. (1970) Altered reflex vasodilatation in the hypertensive rat: Possible role of histamine. *J. Pharmac. exp. Ther.*, **175**, 619–626

Tobian, L. and Binton, J. T. (1952) Tissue cations and water in arterial hypertension. *Circulation*, **5**, 754–758

Todaro, G. J., Lazar, G. K. and Green, H. (1965) The initiation of cell division in a contact-inhibited mammalian cell line. *J. cell. comp. Physiol.*, **66**, 325–333

Tomasi, V., Koretz, S., Ray, T. K., Dunnick, J. and Marinetti, G. V. (1970) Hormone action at the membrane level. II. The binding of epinephrine and glucagon to the rat liver plasma membrane. *Biochim. biophys. Acta.*, **211**, 31–42

Tomasi, V., Rethy, A. and Trevisani, A. (1973) Soluble and membrane-bound adenylate cyclase activity in Yoshida ascites hepatoma. *Life Sci.*, **12**, 145–150

Tomita, T. (1970) Electrical properties of mammalian smooth muscle. In *Smooth Muscle* (Eds. E. Bülbring, A. F. Brading, A. W. Jones and T. Tomita), Edward Arnold, London, pp. 197–243

Tomita, T. and Watanabe, H. (1973). Factors controlling myogenic activity in smooth muscle. *Phil. Trans. R. Soc., B.*, **265**, 73–85

Traber, J., Fischer, K., Latzin, S. and Hamprecht, B. (1975) Morphine antagonises action of prostaglandin in neuroblastoma and neuroblastoma × glioma hybrid cells. *Nature, Lond.*, **253**, 120–122

Trabucchi, M., Longoni, R., Fresia, P. and Spano, P. F. (1975) Sulpiride: A study of the effects on dopamine receptors in rat neostriatum and limbic forebrain. *Life Sci.*, **17**, 1551–1556

Trautwein, W. (1963) Generation and conduction of impulses in the heart as affected by drugs. *Pharmac. Rev.*, **15**, 277–332

Triner, L., Nahas, G. G., Vulliemoz, Y., Overweg, N. I. A., Verosky, M., Habif, D. V. and Ngai, S. H. (1971) Cyclic AMP and smooth muscle function. *Ann. N.Y. Acad. Sci.*, **185**, 458–476

Triner, L., Vulliemoz, Y., Verosky, M., Habif, D. F. and Nahas, G. G. (1972) Adenyl cyclase–phosphodiesterase system in arterial smooth muscle. *Life Sci.*, **11**, 817–824

Triner, L., Vulliemoz, Y., Verosky, M. and Manger, W. M. (1975) Cyclic adenosine monophosphate and vascular reactivity in spontaneously hypertensive rats. *Biochem. Pharmac.*, **24**, 743–745

Troy, F. A., Vijay, I. K. and Kawakami, T. G. (1973) Cyclic 3′,5′-AMP-dependent and independent protein kinase levels in normal and feline sarcoma virus-transformed cells. *Biochem. biophys. Res. Commun.*, **52**, 150–158

Tsien, R. W., Giles, W. and Greengard, P. (1972) Cyclic AMP mediates the effect of adrenaline on cardiac Purkinje fibres. *Nature, New Biol.*, **240**, 181–183

Tsien, R. W. and Weingart, R. (1974) Cyclic AMP: Cell-to-cell movement and inotropic effect in ventricular muscle, studied by a cut-end method. *J. Physiol., Lond*, 95P–96P

Tsou, K. C., Yip, K. F. and Lo, K. W. (1974) 1,N^6-Etheno-2-aza-adenosine 3′,5′-monophosphate: A new fluorescent substrate for cycle nucleotide phosphodiesterase. *Analyt. Biochem.*, **60**, 163–169

Tsung, P. K., Sakamoto, T. and Weismann, G. (1975) Protein kinases and phosphatases from human polymorphonuclear leukocytes. *Biochem. J.*, **145**, 437–448

Tucci, J. R., Lin, T. and Kopp, L. (1973) Urinary cyclic 3′,5′-adenosine monophosphate levels in diabetes mellitus before and after treatment. *J. clin. Endocr. Metab.*, **37**, 832–835

Tufty, R. M. and Kretsinger, R. H. (1975) Tropin and paralbumin calcium binding regions predicted in myosin light chain and T_4 lysozyme. *Science, N.Y.*, **187**, 167–169

Tuganowski, W., Krause, M. and Korczak, K. (1973) The effect of dibutyryl-3′,5′-cyclic AMP on the cardiac pacemaker, arrested with reserpin and α-methyl-tyrosine. Naunyn-Schmiedebergs *Arch. Pharmac.*, **280**, 63–70

Ueda, T., Maeno, H. and Greengard, P. (1973) Regulation of endogenous phosphorylation of specific proteins in synaptic membrane fractions from rat brain by adenosine 3′,5′-monophosphate. *J. biol. Chem.*, **248**, 8295–8305

Unger, R. H. and Orci, L. (1975) The essential role of glucagon in the pathogenesis of diabetes mellitus. *Lancet*, *i*, 14–16

Uno, H., Meyer, R. B., Jr., Shuman, D. A., Robins, R. K., Simon, L. N. and Miller, J. P. (1976) The synthesis of some 1,8 and 2,8-disubstituted derivatives of adenosine cyclic 3′,5′-phosphate and their interaction with some enzymes of cyclic AMP metabolism. *J. med. Chem.*, **19**, 419–422

Uno, H., Meyer, R. B., Jr., Robins, R. K. and Miller, J. P. (1977) 2-Substituted derivatives of 1,N^6-etheno-cAMP. *J. Cycl. Nucleot. Res.* (in press)

Upton, G. V., Corbin, A., Mabry, C. C. and Hollingsworth, D. R. (1973) The etiology of lipoatrophic diabetes. In *Hypothalamic Hypophysiotropic Hormones: Clinical and Physiological Studies* (Eds. G. Carlos and E. Rosenberg), North Holland Publishing Co., Amsterdam, pp. 224–230

Usdin, E. and Snyder, S. H. (Eds.) (1973) Frontiers in catecholamine research. Pergamon Press, New York

Uzunov, P., Shein, H. M. and Weiss, B. (1974) Multiple forms of cyclic 3′,5′-AMP phosphodiesterase of rat cerebrum and cloned astrocytoma and neuroblastoma cells. *Neuropharmacology*, **13**, 377–391

Uzunov, P. and Weiss, B. (1971) Effects of phenothiazine tranquilizers on the cyclic 3′,5′-adenosine monophosphate system of rat brain. *Neuropharmacology*, **10**, 697–708

Uzunov, P. and Weiss, B. (1972a) Separation of multiple molecular forms of cyclic adenosine 3′,5′-monophosphate phosphodiesterase in rat cerebellum by polyacrylamide gel electrophoresis. *Biochim. biophys. Acta.*, **284**, 220–226

Uzunov, P. and Weiss, B. (1972b) Psychopharmacological agents and the cyclic AMP system of rat brain. In *Advances Cyclic Nucleotide Research*, Vol. 1, Raven Press, New York, pp. 435–453

Van den Berg, K. J. and Betel, I. (1971) Early increases of amino acid transport in stimulated lymphocytes. *Expl. Cell Res.*, **66**, 257–259

Van den Berg, K. J. and Betel, I. (1973) Increased transport of 2-aminoisobutyric acid in rat lymphocytes stimulated with concanavalin A. *Expl. Cell Res.*, **76**, 63–72

Van den Berg, K. J. and Betel, I. (1974) Correlation of early changes in amino acid transport and DNA synthesis in stimulated lymphocytes. *Cell. Immunol.*, **10**, 319–323

Van Inwegen, R. J., Strada, S. J. and Robison, G. A. (1975) Effects of prostaglandins and morphine on brain adenylate cyclase. *Life Sci.*, **16**, 1875–1876

van Rijn, H., Bevers, M. M., van Wijk, R. and Wicks, W. D. (1974) Regulation of phosphoenolpyruvate carboxykinase and tyrosine transaminase in hepatoma cell cultures. II. Comparative studies with H35, HTC, MH_1C_1 and RLC cells. *J. Cell Biol.*, **60**, 181–191

van Sande, J., Decoster, C. and Dumont, J. E. (1975) Control and role of cyclic 3′,5′-guanosine monophosphate in the thyroid. *Biochem. biophys. Res. Commun.*, **62**, 168–175

van Sande, J. and Dumont, J. E. (1975) Effects of thyrotropin, prostaglandin E_1 and iodide on cyclic 3′,5′-AMP concentrations in dog thyroid slices. *Biochim. biophys. Acta*, **313**, 320–328

van Sande, J. and Dumont, J. E. (1975) Determination of cyclic 3′,5′-AMP accumulation in resting and stimulated thyroid slices: Heterogenity of the ATP precursor pool. *Molec. Cell. Endocr.*, **2**, 289–301 van Sande, J., Grenier, G., Willems, C. and Dumont, J. E. (1975) Inhibition by iodide of the activation of the thyroid cyclic 3′,5′ AMP system. *Endocrinology*, **96**, 781–786

Van Scott, E. J. (1972) Tissue compartments of the skin lesion of psoriasis. *J. invest. Derm.*, **59**, 4–6

van Wijk, R., Wicks, W. D. and Clay, K. (1972) Effects of derivatives of cyclic 3′,5′-adenosine monophosphate on the growth, morphology and gene expression of hepatoma cells in culture. *Cancer Res.*, **32**, 1905–1911

Varga, J. M., Dipasquale, A., Pawelek, J., McGuire, J. S. and Lerner, A. B. (1974) Regulation of melanocyte-stimulating hormone action at the receptor level: Discontinuous binding of hormone to synchronized mouse melanoma cells during the cell cycle. *Proc. natn. Acad. Sci.*, U.S.A., **71**, 1590–1593

Vargiu, L. and Spano, P. F. (1971) Some central effects of a new derivative of cyclic 3′,5′-adenosine monophosphate. *Naunyn-Schmiedebergs Arch. Pharma.*, **269**, 410

Vatner, D. E. and Lefkowitz, R. J. (1973) [^{3}H]-Propranolol binding sites in myocardial membranes: Nonidentity with beta-adrenergic receptors. *Molec. Pharmac.*, **10**, 450–456

Vaughan, M. (1976) Effects of choleragen and fluoride on adenylate cyclase. In *Regulation of Function and Growth of Eukaryotic Cells by Intracellular Cyclic Nucleotides* (Eds. J. Dumont, R. W. Butcher and B. Brown), Plenum Press, London

Vaughan, M. and Steinberg, D. (1963) Effect of hormones on lipolysis and esterification of free fatty acids during incubation of adipose tissue *in vitro*. *J. Lipid Res.*, **4**, 193–199

Verrier, B., Fayet, G. and Lissitzky, S. (1974) Thyrotropin-binding properties of isolated thyroid cells and their purified plasma membranes. *Europ. J. Biochem.*, **42**, 355–365

Vetulani, J., Leith N. J. Stawarz, R. J. and Sulser, F. (1975) Effect of clonidine on the norepinephrine-sensitive cyclic AMP-generating system in slices of rat spinal cord, brain stem and limbic forebrain and on medial forebrain bundle stimulation. *Pharmacologist*, **17**, 196

Vetulani, J., Stawarz, R. J., Dingell, J. V. and Sulser, F. (1976a) A possible common mechanism of action of antidepressant treatments: Reduction in the sensitivity of the noradrenergic cyclic AMP generating system in the rat limbic forebrain. *Naunyn-Schmiedebergs Arch. Pharmacol.* (in press)

Vetulani, J., Stawarz, R. J. and Sulser, F. (1976b) Adaptive mechanisms of the noradrenergic cyclic AMP generating system in the limbic forebrain of the rat: Adaptation to persistent changes in the availability of norepinephrine. *J. Neurochem.*, **27**, 661–666

Vetulani, J. and Sulser, F. (1975) Action of various antidepressant treatments reduces reactivity of noradrenergic cyclic AMP-generating system in limbic forebrain. *Nature, Lond.*, **257**, 495–496

Villar-Pilasi, C., Larner, J. and Shen, L. C. (1971) Glycogen metabolism and the mechanism of action of cyclic AMP. *Ann. N.Y. Acad. Sci.*, **185**, 74–84

Viveros, O. H., Garleck, D. G. and Renkin, E. M. (1968) Sympathetic β-adrenergic vasodilatation in skeletal muscle of the dog. *Am. J. Physiol., Lond.*, **215**, 1218–1225

Volicer, L. and Hynie, S. (1971) Effect of catecholamines and angiotensin on cyclic AMP in rat aorta and tail artery. *Europ. J. Pharmac.*, **15**, 214–220

Volicer, L. and Hynie, S. (1971) Effect of catecholamines and angiotensin on cyclic AMP in rat aorta and tail artery. *Europ. J. Pharmac.*, **15**, 214–220

Von Hungen, K. and Roberts, S. (1973) Adenylate cyclase receptors for adrenergic neurotransmitters in rat cerebral cortex. *Europ. J. Biochem.*, **36**, 391–401

Von Hungen, K. and Roberts, S. (1974) Neurotransmitter-sensitive adenylate cyclase systems in the brain. In *Reviews of Neuroscience* (Eds. S. Ehrenpreis and I. J. Kopin), Vol. 1, Raven Press, New York, pp. 231–281

Von Hungen, K., Roberts, S. and Hill, D. F. (1975) Interactions between lysergic acid diethylamide and dopamine-sensitive adenylate cyclase systems in rat brain. *Brain Res.*, **94**, 57–66

von Loh, D. (1971) The effect of adrenergic drugs on spontaneously active vascular smooth muscle studied by long-term intracellular recording of membrane potential. *Angiologica*, **8**, 144–155

Von Voigtlander, P. F. (1974) Behavioral and biochemical investigation of dopamine supersensitivity induced by chronic neuroleptic treatment. *Fed. Proc.*, **33**, 578

Von Voigtlander, P. F., Boukma, S. J. and Johnson, G. A. (1973) Dopaminergic denervation supersensitivity and dopamine-stimulated adenyl cyclase activity. *Neuropharmacology*, **12**, 1081–1086

Von Voigtlander, P. F., Losey, E. G. and Triezenberg, H. J. (1975) Increased sensitivity to dopaminergic agents after chronic neuroleptic treatment. *J. Pharmac. exp. Ther.*, **193**, 88–99

Voorhees, J. J., Chambers, D. A., Duell, E. A., Marcelo, C. L. and Krueger, G. G. (1976) Molecular mechanisms in proliferative skin disease. *J. invest. Derm.*, **67**, 442–448

Voorhees, J. J., Duell, E. A., Bass, L. J., Powell, J. A. and Harrell, E. R. (1972a) The cyclic AMP system in normal and psoriatic epidermis. *J. invest. Derm.*, **59**, 144–120

Voorhees, J. J. and Duell, E. A. (1975) Imbalanced cyclic AMP–cyclic GMP levels in psoriasis. In *Advances in Cyclic Nucleotide Research*, Vol. 5 (Eds. G. I. Drummond, P. Greengard and G. A. Robison), Raven Press, New York, pp. 735–758

Voorhees, J. J., Duell, E. A., Bass, L. J., Powell, J. A. and Harrell, E. R. (1972b) Decreased cyclic AMP in the epidermis of lesions of psoriasis. *Archs. Derm.*, **105**, 695–701

Voorhees, J. J., Duell, E. A., Stawiski, M. and Harrell, E. R. (1974). Cyclic nucleotide metabolism in normal and proliferating epidermis. In *Advances in Cyclic Nucleotide Research*, Vol. 4 (eds. P. Greengard and G. A. Robison), Raven Press, New York, pp. 117–162

Voorhees, J., Kelsey, W., Stawiski, M., Smith, E., Duell, E., Haddox, M. and Goldberg, N. (1973) Increased cyclic GMP and decreased cyclic AMP levels in the rapidly proliferating epithelium of psoriasis. In *The Role of Cyclic Nucleotides in Carcinogenesis*, Vol. 6 (Eds. J. Schultz and D. G. Gratzner), Academic Press, New York, pp. 325–373

Voorhees, J. J., Marcelo, C. L. and Duell, E. A. (1975) Cyclic AMP, cyclic GMP, and glucocorticoids as potential metabolic regulators of epidermal proliferation and differentiation. *J. invest. Derm.*, **65**, 179–190

Vydelingum, N., Kissebah, A. H., Wynn, V., Simpson, A. (1975) The role of calcium in insulin action. V. Insulin, calcium, c-GMP and the regulation of protein synthesis in adipose tissue. *Diabetologia*, **11**, 382

Wagner, K., Roper, M. D., Leichtling, B. H., Wimalasena, J. and Wicks, W. D. (1975) Effects of 6 and 8-substituted analogs of adenosine 3′,5′-monophosphate on phosphoenolpyruvate carboxykinase and tyrosine aminotransferase in hepatoma cell cultures. *J. biol. Chem.*, **250**, 231–239

Wagner, R. C., Kreiner, P., Barrnett, R. J. and Bitensky, M. W. (1972) Biochemical characterization and cytochemical localization of a catecholamine-sensitive adenylate cyclase in isolated capillary endothelium. *Proc. natn. Acad. Sci., U.S.A.*, **69**, 3175–3179

Wagner, R., Rosenberg, M. and Estensen, R. (1971) Endocytosis in Chang liver cells. *J. Cell Biol.*, **50**, 804–817

Wahn, H. L., Lightbody, L. E., Tchen, T. T. and Taylor, J. D. (1975) Induction of neural differentiation in cultures of amphibian undetermined presumptive epidermis by cyclic AMP derivatives. *Science, N.Y.*, **188**, 366–369

Waisman, D., Stevens, F. C. and Wang, J. H. (1975) The distribution of the Ca^{2+}dependent protein activator of cyclic nucleotide phosphodiesterase in invertebrates. *Biochem. biophys. Res. Commun.*, **65**, 975–982

Walaas, O., Walaas, E. and Gronnerod, O. (1973) Hormonal regulation of cyclic AMP-dependent protein kinase of rat diaphragm by epinephrine and insulin. *Europ. J. Biochem.*, **40**, 465–477

Waldstein, S. S. (1966) Thyroid–catecholamine interrelations. *Ann. Rev. Med.*, **17**, 123–132

Walker, J. B. and Walker, J. P. (1973) Neurohumoral regulation of adenylate cyclase activity in rat striatum. *Brain Res.*, **54**, 386–390

Wallach, D. F. H. (1972) Medical and para-medical aspects of plasma membrane biology. In *The Plasma Membrane* (Ed. D. F. H. Wallach), Springer-Verlag, New York, pp. 108–144

Wallach, D. F. H. (1973) The role of the plasma membrane in disease processes. In *Biological Membranes*, Vol. 2 (Eds. D. Chapman and D. F. H. Wallach), Academic Press, London, pp. 253–294

Walsh, D. A. and Ashby, C. D. (1973) Protein kinases: Aspects of their regulation and diversity. *Recent Prog. Hormone Res.*, **29**, 329–359

Walsh, D. H., Ashby, C. D., Gonzales, C., Calkins, D., Fischer, E. H. and Krebs, E. G. (1971) Purification and characterization of a protein inhibitor of cyclic AMP-dependent protein kinase. *J. biol. Chem.*, **246**, 1977–1985

Walton, G. M. and Gill, G. N. (1973) Adenosine 3′,5′ monophosphate and protein kinase-dependent phosphorylation of ribosomal protein. *Biochemistry*, **12**, 2604–2611

Walton, K. G., De Lorenzo, R. J., Curran, P. F. and Greengard, P. (1975) Regulation of protein phosphorylation and sodium transport in toad bladder. *J. gen. Physiol.*, **65**, 153–177

Wang, J. H., Teo, T. S., Ho, H. C. and Stevens, F. C. (1975a) Bovine heart protein activator of cyclic nucleotide phosphodiesterase. In *Advances in Cyclic Nucleotide Research*, Vol. 5 (Eds. G. I. Drummond, P. Greengard and G. A. Robison), Raven Press, New York, pp. 179–194

Wang, J. H., Teo, T. S. and Wang, T. H. (1972) Hysteretic substrate activation of bovine heart cAMP phosphodiesterase. *Biochem. biophys. Res. Commun.*, **46**, 1306–1311

Wang, J. L., Gunther, G. R., Yahara, I., Cunningham, B. A. and Edelman, G. M. (1975b) Receptor–cytoplasmic interactions and lymphocyte activation. In *Immune Recognition* (Ed. A. S. Rosenthal), Academic Press, New York, pp. 473–489

Wang, J. L., McClain, D. A. and Edelman, G. M. (1975c) Modulation of lymphocyte mitogenesis. *Proc. natn. Acad. Sci., U.S.A.*, **72**, 1917–1921

Warbanow, W., Will-Shahab, L. and Wollenberger, A. (1975) Wiederherstellung der automatischen kontraktilen Tätigkeit K^+-gelähmter Herzmuskelzellen in Kultur durch Dibutyryl-3′,5′-adenosinmonophosphat. *Acta biol. med. Germ.*, **34**, 1553–1556

Ward, J. T., Adesola, A. O. and Welbourn, R. B. (1964) The parathyroids, calcium and gastric secretion in man and the dog. *Gut*, **5**, 173–183

Warren, K. S., Mahmoud, A. A. F., Boros, D. L., Rall, T. W., Mandel, M. A. and Carpenter, C. C. J. (1974) *In vivo* suppression by cholera toxin of cell-mediated and foreign body inflammatory response. *J. Immunol.*, **112**, 996–1007

Wasner, H. K. (1975) Regulation of protein kinase and phosphoprotein phosphatase by cyclic AMP and cyclic AMP antagonist. *FEBS Letters*, **57**, 60–63

Wasserman, S. I., Goetzl, E. L., Kaliner, M. and Austen, K. F. (1974) Modulation of the immunological release of the eosinophil chemotactic factor of anaphylaxis from human lung. *Immunology*, **26**, 677–684

Watanabe, A. M. and Besch, H. R. (1974) Cyclic adenosine monophosphate modulation of slow calciums in guinea-pig hearts. *Circulation Res.*, **35**, 316–324

Watenpaugh, K., Dow, J., Jensen, L. H. and Furberg, S. (1968) Crystal and molecular structure of adenosine 3′,5′-cyclic phosphate. *Science, N.Y.*, **159**, 206–207

Watson, J. (1975) The influence of intracellular levels of cyclic nucleotides on cell proliferation and the induction of antibody synthesis. *J. exp. Med.*, **141**, 97–111

Watson, J., Epstein, R. and Cohn, M. (1973) Cyclic nucleotides as intracellular mediators of the expression of antigen-sensitive cells. Structural similarities between the Ca^{2+}-dependent regulatory proteins of 3′,5′-cyclic nucleotide phosphodiesterase and actomyosine ATPase. *Nature, Lond.*, **246**, 405–409

Watterson, D. M., Harrelson, W. G., Keller, P. M., Sharief, F. and Vanaman, T. C. (1976) Structural similarities between the Ca^{2+}-dependent regulatory proteins of 3′,5′-

cyclic nucleotide phosphodiesterase and actinomycin ATPase. *J. biol. Chem.*, **251**, 4501–4513

Way, L. and Durbin, R. P. (1969) Inhibition of gastric acid secretion *in vitro* by prostaglandin E. *Nature, Lond.*, **221**, 874–875

Waymire, J. C., Weiner, N. and Prasad, K. N. (1972) Regulation of tyrosine hydroxylase activity in cultured mouse neuroblastoma cells: Elevation induced by analogs of adenosine 3′,5′-cyclic monophosphate. *Proc. natn. Acad. Sci., U.S.A.*, **69**, 2241–2245

Webb, D. R., Stites, D. P., Perlman, J., Austin, K. E. and Fudenberg, H. J. (1974) Cyclic AMP in the activation of human peripheral blood lymphocytes in immunologically deficient patients and in human lymphoid cell lines. In *Cyclic AMP, Cell Growth and the Immune Response* (Eds. W. Braun, L. Lichtenstein and C. W. Parker), Springer-Verlag, New York, pp. 55–76

Wedner, H. J., Hoffer, B. J., Battenberg, E., Steiner, A. L. and Bloom, F. E. (1972) A method for detecting intracellular cyclic adenosine monophosphate by immunofluorescence. *J. Histochem. Cytochem.*, **20**, 293–295

Wedner, H. J., Bloom, F. E. and Parker, C. W. (1975a) The role of cyclic nucleotides in lymphocyte activation. In *Immune Recognition* (Ed. A. S. Rosenthal), Academic Press, New York, pp. 337–357

Wedner, H. J., Dankner, R. and Parker, C. W. (1975b) Cyclic GMP and lectin induced lymphocyte activation. *J. Immunol.*, **115**, 1682–1687

Wedner, H. J. and Parker, C. W. (1975c) Protein phosphorylation in human peripheral lymphocytes—stimulation by phytohemagglutinin and N^6-monobutyryl-cyclic AMP. *Biochem. biophys. Res. Commun.*, **62**, 808–815

Wedner, H. J. and Parker, C. W. (1976) Lymphocyte activation. In *Progress in Allergy*, Vol. 20 (Eds. P. Kallós, B. H. Waksman, and A. de Weck), S. Karger, Basel, pp. 195–300

Weight, F. F. and Padjen, A. (1973a) Acetylcholine and slow synaptic inhibition in frog sympathetic ganglion cells. *Brain Res.*, **55**, 225–228

Weight, F. F., Petzold, G. and Greengard, P. (1974) Guanosine 3′,5′-monophosphate in sympathetic ganglia: Increase associated with synaptic transmission. *Science, N.Y.*, **186**, 942–944

Weight, F. F. and Votava, J. (1970) Slow synaptic excitation in sympathetic ganglion cells: Evidence for synaptic activation of potassium conductance. *Science, N.Y.*, **170**, 755–758

Weiner, N. (1970) Regulation of norepinephrine biosynthesis. *Ann. Rev. Pharmac.*, **10**, 273–290

Weinstein, G. D. and Frost, P. (1968) Abnormal cell proliferation in psoriasis. *J. invest. Derm.*, **50**, 254–259

Weinstein, Y., Chambers, D. A., Bourne, H. R. and Melmon, K. L. (1974) Cyclic GMP stimulates lymphocyte nucleic acid synthesis. *Nature, Lond.*, **251**, 352–357

Weinstein, Y., Segal, S. and Melmon, K. L. (1975) Specific mitogenic activity of 8-Br-guanosine 3′,5′-monophosphate (Br-Cyclic GMP) on B lymphocytes. *J. Immunol.*, **115**, 112–117

Weiss, B. (1975) Differential activation and inhibition of the multiple forms of cyclic nucleotide phosphodiesterase. In *Advances in Cyclic Nucleotide Research*, Vol. 5 (Eds. G. I. Drummond, P. Greengard and G. A. Robison), Raven Press, New York, pp. 195–211

Weiss, B. and Costa, E. (1967) Adenyl cyclase activity in rat pineal gland: Effects of chronic denervation and norepinephrine. *Science, N.Y.*, **156**, 1750–1752

Weiss, B. and Costa, E. (1968) Regional and subcellular distribution of adenyl cyclase and 3′,5′-cyclic nucleotide phosphodiesterase in brain and pineal gland. *Biochem. Pharmac.*, **17**, 2107–2116

Weiss, B., Fertel, R., Figlin, R. and Uzunov, P. (1974) Selective alteration of the activity of the multiple forms of adenosine 3′,5′-monophosphate phosphodiesterase of rat cerebrum. *Molec. Pharmac.*, **9**, 615–625

Weiss, B. and Greenberg, L. H. (1975) Cyclic AMP and brain function: Effect of psychopharmacologic agents on the cyclic AMP system. In *Cyclic Nucleotides in Disease* (Ed. B. Weiss), University Park Press, Baltimore, pp. 269–319

Weiss, B. and Strada, S. J. (1972) Neuroendocrine control of the cyclic AMP system of brain and pineal gland. In *Advances Cyclic Nucleotide Research*, Vol. 1 (Eds. P. Greengard and G. A. Robison), Raven Press, New York, pp. 357–374

Weissman, B. A., Daly, J. W. and Skolnick, P. (1975a) Diethylstilbestrol-elicited accumulation of cyclic AMP in incubated rat hypothalamus. *Endocrinology*, **97**, 1559–1566

Weissman, B. A. and Skolnick, P. (1975) Formation of adenosine 3′,5′-monophosphate in incubated rat hypothalamus by estrogenic compounds: Relationship to biologic potency and blockade by anti-estrogens. *Neuroendocrinology*, **18**, 27–35

Weissmann, G., Dukor, P. and Sessa, G. (1971a) Studies on lysosomes, mechanisms of enzyme release from endocytic cells and a model for latency *in vitro*. In *Immunopathology of Inflammation* (Eds. B. Forscher and J. C. Houck), Excerpta Medica, Amsterdam, pp. 107–117

Weissmann, G., Dukor, P. and Zurier, R. (1971b) Effect of cyclic AMP on release of lysosomal enzymes from phagocytes. *Nature, New Biol.*, **231**, 131–135

Weissmann, G., Goldstein, I., Hoffstein, S. and Tsung, P. K. (1975b) Reciprocal effects of cAMP and cGMP on microtubule-dependent release of lysosomal enzymes. *Ann. N.Y. Acad. Sci.*, **253**, 750–762

Weissmann, G., Zurier, R. B., Spieler, P. J. and Goldstein, I. M. (1971a) Mechanisms of lysosomal enzyme release from leukocytes exposed to immune complexes and other particles. *J. exp. Med.*, **134**, 149–165

Wells, J. N., Baird, C. E., Wu, V. J. and Hardman, J. G. (1975) Cyclic nucleotide phosphodiesterase activities of pig coronary arteries. *Biochim. biophys Acta*, **384**, 430–442

Werner, I., Peterson, G. R. and Shuster, L. (1971) Choline acetyltransferase and acetylcholinesterase in cultured brain cells from chick embryo. *J. Neurochem.*, **18**, 141–151

Wessels, N. K., Spooner, J. F., Ash, M. O., Bradley, M. A., Luduena, M. A., Taylor, E. L., Wrenn, J. T. and Yamada, K. M. (1971) Microfilaments in cellular and developmental processes. *Science, N.Y.*, **171**, 135–143

Westermann, E. and Stock, K. (1970) Inhibitors of lipolysis: Potency and mode of action of α-and β-adrenolytics, methoxamine derivatives, prostaglandin E_1, and phenylisopropyladenosine. In *Adipose Tissue Regulation and Metabolic Functions* (Eds. B. Jeanrenaud and D. Hepp), Academic Press, New York, London, pp. 47–54

Westermann, E., Stock, K. and Bieck, P. (1969) Phenylisopropyl-adenosin (PIA): Ein potenter Hemmstoff der Lipolyse *in vivo* und *in vitro*. *Medizin. Ernäh.*, **10**, 143–147

White, J. G., Goldberg, N. D., Estensen, R. D., Haddox, M. K. and Rao, G. H. R. (1973) Rapid increase in platelet cyclic 3′,5′-guanosine monophosphate (cGMP) levels in association with irreversible aggregation, degranulation, and secretion. *J. clin. Invest.*, **52**, 89a

White, L. E., Ignarro, L. J. and George, W. J. (1973) Stimulation of rat cardiac guanyl cyclase by acetylcholine. *Pharmacologist*, **15**, 157

Whitfield, J. F., MacManus, J. P., Boynton, A. L., Gillan, D. J. and Isaacs, R. J. (1974) Concanavalin A and the initiation of thymic lymphoblast DNA synthesis and the proliferation by a calcium-dependent increase in cyclic GMP level. *J. Cell Physiol.*, **84**, 445–458

Whitney, R. and Sutherland, R. (1972a) Requirement for calcium ions in lymphocyte transformation stimulated by phytohemagglutinin. *J. Cell Physiol.*, **80**, 329–338

Whitney, R. B. and Sutherland, R. M. (1972b) Enhanced uptake of calcium by transforming lymphocytes. *Cell. Immunol.*, **5**, 137–147

Whitney, R. B. and Sutherland, R. M. (1973) Kinetics of calcium transport in lymphocytes before and after stimulation by phytohemagglutinin. In *Proceedings of the Seventh Leucocyte Culture Conference* (Ed. F. Daguillard), Academic Press, New York, pp. 63–75

Wicks, W. D. (1974) Regulation of protein synthesis by cyclic AMP. In *Advances in Cyclic Nucleotide Research*, Vol. 4 (Eds. P. Greengard and G. A. Robison), Raven Press, New York, pp. 335–438

Wicks, W. D., Koontz, J. and Wagner, K. (1975) Possible participation of protein kinase in enzyme induction. *J. Cycl. Nucleot. Res.*, **1**, 49–58

Wicks, W. D., van Wijk, R., Clay, K. and Bearg, C. (1973) Regulation of growth rate, DNA synthesis and specific protein synthesis by derivatives of cyclic AMP in cultured hepatoma cells. *Miami Winter Symposia*, **6**, 103–126

Wickson, R. D., Boudreau, R. J. and Drummond, G. I. (1975) Activation of 3′,5′-cyclic adenosine monophosphate phosphodiesterase by calcium ion and a protein activator. *Biochemistry*, **14**, 669–675

Widström, A. and Cerasi, E. (1973) On the action of tolbutamide in normal man. Interaction of tolbutamide with glucagon, aminophylline and arginine in stimulating insulin release. *Acta Endocr.*, **72**, 532–544

Wieser, P. B. and Fain, J. N. (1975) Insulin, prostaglandin E_1, phenylisopropyladenosine and nicotinic acid as regulators of fat cell metabolism. *Endocrinology*, **96**, 1221–1225

Wilber, J., Peake, G. T., Mariz, I., Utiger, R. and Daughaday, W. H. (1968) Theophylline and epinephrine effects upon the secretion of growth hormone and thyrotropin. *Clin. Res.*, **16**, 277

Wilber, J. F., Peake, G. T. and Utiger, R. (1969) Thyrotropin release *in vitro*: Stimulation by cyclic 3′,5′-adenosine monophosphate. *Endocrinology*, **84**, 758–760

Wilchek, M., Salomon, Y., Lowe, M. and Selinger, Z. (1971) Conversion of protein kinase to a cyclic AMP independent form by affinity chromatography on N^6-caproyl 3′,5′-cyclic adenosine monophosphate-sepharose. *Biochem. biophys. Res. Commun.*, **45**, 1177–1184

Wildenthal, K. (1973) Maturation of responsiveness to cardioactive drugs. Differential effects of acetylcholine, norepinephrine, theophylline, tyramine, glucagon, and dibutyryl-cyclic AMP on atrial rate in hearts of fetal mice. *J. clin. Invest.*, **52**, 2250–2258

Wilkening, D. and Makman, M. H. (1975) 2-Chloroadenosine-dependent elevation of adenosine 3′,5′-cyclic monophosphate levels in rat caudate nucleus slices. *Brain Res.*, Amsterdam, **92**, 522–528

Will, H., Schirpke, B. and Wollenberger, A. (1976) Stimulation of Ca^{2+} uptake by cyclic AMP and protein kinase in sarcotubular-rich and sarcolemma-rich microsomal fractions from rabbit heart. *Acta biol. med. germ.*, **35**, 529–541

Willems, C., Rocmans, P. and Dumont, J. E. (1971) Calcium requirements in the action of thyrotropin on the thyroid. *FEBS Letters*, **14**, 323–325

Williams, B. J. and Pirch, J. H. (1974) Correlation between brain adenyl cyclase activity and spontaneous motor activity in rats after chronic reserpine treatment. *Brain Res.*, **68**, 227–234

Williams, J. A. and Chandler, D. (1975) Ca^{2+} and pancreatic amylase release. *Am. J. Physiol.*, **228**, 1729–1732

Williams, J. A. and Matthews, E. K. (1974a) Effects of ions and metabolic inhibitors on membrane potential of brown adipose tissue. *Am. J. Physiol.*, **227**, 981–986

Williams, J. A. and Matthews, E. K. (1974b) Membrane depolarization, cyclic AMP, and glycerol release by brown adipose tissue. *Am. J. Physiol.*, **227**, 987–992

Williamson, J. R. (1976) Epinephrine, cyclic AMP, calcium and myocardial contractility. In *Recent Advances in Studies on Cardiac Structure and Metabolism* (Eds. P.-E. Roy, et al.), University Park Press, Baltimore

Willingham, M. C., Carchman, R. A. and Pastan, I. (1973) A mutant of 3T3 cells with cyclic AMP metabolism sensitive to temperature change. *Proc. natn. Acad. Sci., U.S.A.*, **70**, 2906–2910

Willingham, M. C., Johnson, G. S. and Pastan, I. (1972) Control of DNA synthesis and mitosis in 3T3 cells by cyclic AMP. *Biochem. biophys. Res. Commun.*, **48**, 743–748

Willingham, M. C. and Pastan, I. (1975a) Cyclic AMP and cell morphology in cultured fibroblasts: Effects on cell shape, microfilaments and microtubule distribution, and orientation to substratum. *J. Cell Biol.*, **67**, 146–159

Willingham, M. C. and Pastan, I. (1975b) Cyclic AMP modulates microvillus formation and agglutinability in transformed and normal mouse fibroblasts. *Proc. natn. Acad. Sci.*, U.S.A., **72**, 1263–1267

Will-Shahab, L., Wollenberger, A. and Schulze, W. (1975) Modulation by thyroid hormone of catecholamine binding sites and of adenylate cyclase activity in heart muscle. *Proc. 9th FEBS Mtg, Budapest,* 1974, Vol. 7 (Ed. M. Willemann), North Holland Publishing Co., Amsterdam, pp. 107–127

Winand, R. J. and Kohn, L. D. (1975) Thyrotropin effects on thyroid cells in culture. Effects of trypsin on the thyrotropin receptor and on thyrotropin-mediated cyclic 3′,5′-AMP changes. *J. biol. Chem.*, **250**, 6534–6540

Wolfe, B. B., Zirrolli, J. A. and Molinoff, P. B. (1974) Binding of dl-[^{3}H]-epinephrine to proteins of rat ventricular muscle: Nonidentity with beta-adrenergic receptors. *Molec. Pharmac.*, **10**, 582–596

Wolff, D. J. and Brostrom, C. O. (1974) Calcium binding phosphoprotein from pig brain: Identification as a Ca^{2+} dependent regulator of rat brain nucleotide phosphodiesterase. *Archs. Biochem. Biophys.*, **163**, 349–358

Wolff, D. J. and Brostrom, C. O. (1976) Calcium-dependent cyclic nucleotide phosphodiesterase from brain: Identification of phospholipids as calcium-independent activators. *Archs. Biochem. Biophys.*, **173**, 720–731

Wolff, D. J. and Seigel, F. L. (1972) Purification of a calcium binding phosphoprotein from pig brain. *J. biol. Chem.*, **247**, 4180–4185

Wolff, J., Berens, S. C. and Jones, A. B. (1970) Inhibition of thyrotropin-stimulated adenyl cyclase activity of beef thyroid membranes by low concentration of lithium ions. *Biochem. biophys. Res. Commun.*, **39**, 77–82

Wolff, J. and Cook, G. H. (1973) Activation of thyroid membrane adenylate cyclase by purine nucleotides. *J. biol. Chem.*, **248**, 350–355

Wolff, J. and Cook, G. H. (1975) Choleragen stimulates steroidogenesis and adenylate cyclase in cells lacking functional hormone receptors. *Biochim. biophys. Acta*, **413**, 283–290

Wolff, J. and Jones, A. B. (1971) The purification of bovine thyroid plasma membranes and the properties of membrane-bound adenyl cyclase. *J. biol. Chem.*, **246**, 3939–3947

Wollenberger, A. (1964) Rhythmic and arrhythmic contractile activity of single myocardial cells cultured *in vitro. Circulation Res.*, **14/15** supp., 184–201

Wollenberger, A. (1975) The role of cyclic AMP in the adrenergic control of myocardium. In *Contraction and Relaxation of the Heart* (Ed. W. G. Nayler), Academic Press, London, pp. 113–190

Wollenberger, A., Babski, E. B., Krause, E.-G., Genz, S., Blohm, D. and Bogdanova, E. V. (1973) Cyclic changes in levels of cyclic AMP and cyclic GMP in frog myocardium during the cardiac cycle. *Biochem. biophys. Res. Commun.*, **55**, 446–452

Wollenberger, A. and Kleitke, B. (1973) Ribonucleic acid and protein synthesis in rat heart mitochondria isolated after aortic constriction, strenuous physical exercise, total myocardial ischemia, and theophylline treatment. In *Myocardial Metabolism: Recent Advances in Studies on Cardiac Structure and Metabolism*, Vol. 3 (Ed. N. S. Dhalla), University Park Press, Baltimore, pp. 535–550

Wollenberger, A., Krause, E.-G. and Heier, G. (1969) Stimulation of 3′,5′-cyclic AMP formation in dog myocardium following arrest of blood flow. *Biochem. biophys. Res. Commun.*, **36**, 664–670

Wollenberger, A. and Schulze, W. (1976) Cytochemical studies on sarcolemma: Na^+, K^+-

adenosine triphosphatase and adenylate cyclase. *Recent Adv. Stud. Cardiac Structure and Metabolism*, **9**, 101–115

Wollenberger, A., Schulze, W. and Krause, E.-G. (1973) Cytochemical examination of the effect of histamine on adenylate cyclase in guinea pig heart tissue. *J. mol. cell. Cardiol.*, **5**, 427–431

Wollenberger, A., Will-Shahab, L., Krause, E.-G., Genz, S., Warbanow, W. and Nitschkoff, S. (1973) Effect of acute ischemia on myocardial cyclic AMP, phosphorylase *a* and lactate levels in various forms of cardiac hypertrophy. Correlation with cardiac norepinephrine stores. *Recent Advances in Studies on Cardiac Structure and Metabolism,* Vol. 3 (Ed. N. S. Dhalla), University Park Press, Baltimore, pp. 535–550

Wollheim, C. B., Blondel, B., Trueheart, P. A., Renold, A. E. and Sharp, G. W. G. (1975) Calcium-induced insulin release in monolayer cultures of the endocrine pancreas. *J. biol. Chem.*, **250**, 1354–1360

Wollin, A., Code, C. F. and Dousa, T. P. (1976) Interaction of prostaglandins and histamine with enzymes of cyclic AMP metabolism from guinea-pig gastric mucosa. *J. Clin. Invest.*, **57**, 1548–1553

Wong, G., Pawelek, J., Sansone, M. and Morowitz, J. (1974) Response of mouse melanoma cells to melanocyte stimulating hormone. *Nature, Lond.*, **248**, 351–354

Wood, H. N. and Braun, A. C. (1973) 8-Bromoadenosine 3′,5′-cyclic monophosphate as a promoter of cell division in excised tobacco pith parenchyma tissue. *Proc. natn. Acad. Sci., U.S.A.*, **70**, 447–450

Woodin, A. M. and Wieneke, A. A. (1963) The accumulation of calcium by the polymorphonuclear leucocyte treated with staphylococcal leucocidin and its significance in the extrusion of protein. *Biochem. J.*, **87**, 487–495

Woodward, D. J., Hoffer, B. J. and Altman, J. (1974) Physiological and pharmacological properties of Purkinje cells in rat cerebellum degranulated by postnatal X-irradiation. *J. Neurobiol.*, **5**, 283–304

Wray, H. L., Gray, R. R. and Olsson, R. A. (1973) Cyclic adenosine 3′,5′-monophosphate-stimulated protein kinase and a substrate associated with cardiac sarcoplasmic reticulum. *J. biol. Chem.*, **248**, 1496–1498

Wray, V. P. and Walborg, E. F., Jr. (1971) Isolation of tumor cell surface binding sites for concanavalin A and wheat germ agglutinin. *Cancer Res.*, **31**, 2072–2079

Wulff, V. J. (1971) The effect of cyclic AMP on *Limulus* lateral eye retinular cells. *Vision Res.*, **11**, 1493–1495

Wulff, V. J. (1973) The effect of cyclic AMP and aminophylline on *Limulus* lateral eye retinular cells. *Vision Res.*, **13**, 2335–2344

Yamashita, K. and Field J. B. (1972a) Cyclic AMP-stimulated protein kinase prepared from bovine thyroid glands. *Metabolism*, **21**, 150–158

Yamashita, K. and Field, J. B. (1972b) Elevation of cyclic guanosine 3′,5′-monophosphate levels in dog thyroid slices caused by acetylcholine and sodium fluoride. *J. biol. Chem.*, **247**, 7062–7066

Yamashita, K. and Field, J. B. (1973) The role of phospholipids in TSH stimulation of adenylate cyclase in thyroid plasma membranes. *Biochim. biophys. Acta*, **304**, 686

Yamasaki, Y., Fujiwara, M. and Toda, N. (1974) Effects of intracellularly applied cyclic 3′,5′-adenosine monophosphate and dibutyryl cyclic 3′,5′-adenosine monophosphate on the electrical activity of sino-atrial nodal cells of the rabbit. *J. Pharmacol. exp. Therap.*, **190**, 15–20

Yarbrough, G. G. (1975) Supersensitivity of caudate neurons after repeated administration of haloperidol. *Europ. J. Pharmac.*, **31**, 367–369

Yoshikawa, K., Adachi, K., Halprin, K. M. and Levine, V. (1975a) On the lack of response to catecholamine stimulation by the adenyl cyclase system in psoriatic lesions. *Br. J. Derm.*, **92**, 619–624

Yoshikawa, K., Adachi, K., Halprin, K. M. and Levine, V. (1975b). Is the cyclic AMP in psoriatic epidermis low? *Br. J. Dermatol.*, **93**, 253–258

Yoshinaga, M., Yoshinaga, A. and Waksman, B. H. (1972) Regulation of lymphocyte responses in vitro: Potentiation and inhibition of rat lymphocyte responses to antigen and mitogen by cytochalasin B. *Proc. natn. Acad. Sci.*, U.S.A., **69**, 3251–3255

Young, N. M. (1974) The effects of maleylation on the properties of concanavalin A. *Biochim. biophys. Acta*, **336**, 46–52

Zanella, J., Jr. and Rall, T. W. (1973) Evaluation of electrical pulses and elevated levels of potassium ions as stimulants of adenosine 3′,5′-monophosphate (cyclic AMP) accumulation in guinea-pig brain. *J. Pharmac. exp. Ther.*, **186**, 241–251

Zawalich, W. S., Karl, R. C., Ferrendelli, J. A. and Matschinsky, F. M. (1975) Factors governing glucose induced elevation of cyclic 3′,5′ AMP levels in pancreatic islets. *Diabetologia*, **11**, 231–235

Zeilig, C. E., Johnson, R. A., Friedman, D. L. and Sutherland, E. W. (1972) Cyclic AMP concentrations in synchronized HeLa cells. *J. Cell Biol.*, **55**, 296a

Ziboh, V. A. (1975) Prostaglandins and their biological significance in the skin. *Int. J. Derm.*, **14**, 485–493

Zierler, K. L. (1972) Insulin, ions, and membrane potentials. *Handbook of Physiology, Endocrinology, Endocrine Pancreas*, American Physiological Society, Washington, pp. 347–368

Zigmond, S. H. and Hirsch, J. G. (1972a) Cytochalasin B: Inhibition of D-2-deoxyglucose transport into leukocytes and fibroblasts. *Science N.Y.*, **176**, 1432–1434

Zigmond, S. H. and Hirsch, J. G. (1972b) Effects of cytochalasin B on polymorphonuclear leucocyte locomotion, phagocytosis and glycolysis. *Expl. Cell Res.*, **73**, 383–393

Zinman, B. and Hollenberg, C. H. (1974) Effect of insulin and lipolytic agents on rat adipocyte low K_m cyclic adenosine 3′,5′-monophosphate phosphodiesterase. *J. biol. Chem.*, **249**, 2182–2187

Zivkovic, B., Guidotti, A. and Costa, E. (1974) Effects of neuroleptics on striatal tyrosine hydroxylase: Changes in affinity for the pteridine cofactor. *Molec. Pharmac.*, **10**, 727–735

Zivkovic, B., Guidotti, A. and Costa, E. (1975) Stereospecificity of dopamine receptors involved in the regulation of the kinetic state of tyrosine hydroxylase in striatum and nucleus accumbens. *J. Pharm. Pharmac.*, **27**, 359—360

Zor, U., Kaneko, T., Lowe, I. P., Bloom, G. and Field, J. B. (1969a) Effects of TSH and prostaglandins on thyroid adenyl cyclase activation and cyclic adenosine 3′,5′-monophosphate. *J. biol. Chem.*, **244**, 5189–5195

Zor, U., Kaneko, T., Schneider, H. P. G., McCann, S. M. and Field, J. B. (1970) Further studies of stimulation of anterior pituitary cyclic 3′,5′-monophosphate formation by hypothalamic extract and prostaglandins. *J. biol. Chem.*, **245**, 2883–2888

Zor, U., Kaneko, T., Schneider, H. P. G., McCann, S. M., Lowe, I. P., Bloom, G., Borland, B. and Field, J. B. (1969b) Stimulation of anterior pituitary adenyl cyclase activity and adenosine 3′,5′-cyclic phosphate by hypothalamic extract and prostaglandin E_1. *Proc. natn. Acad. Sci., U.S.A.*, **63**, 918–925

Zurier, R. B., Hoffstein, S. and Weissmann, G. (1973a) Cytochalasin B: Effect on lysosomal enzyme release from human leukocytes. *Proc. natn. Acad. Sci., U.S.A.*, **70**, 844–848

Zurier, R., Hoffstein, S. and Weissmann, G. (1973b) Mechanisms of lysosomal enzyme release from human leukocytes. I. Effect of cyclic nucleotides and colchicine. *J. Cell Biol.*, **58**, 27–41

Zurier, R., Weissmann, G., Hoffstein, S., Kammerman, S. and Tai, H. H. (1974) Mechanisms of lysosomal enzyme release from human leukocytes. *J. clin. Invest.*, **53**, 297–309

Index